PATENT LAW IN GLOBAL PERSPECTIVE

PATENT LAW IN GLOBAL PERSPECTIVE

Edited by Ruth L. Okediji

Margo A. Bagley

OXFORD
UNIVERSITY PRESS

OXFORD
UNIVERSITY PRESS

Oxford University Press is a department of the University of Oxford. It furthers the University's objective of excellence in research, scholarship, and education by publishing worldwide.

Oxford New York
Auckland Cape Town Dar es Salaam Hong Kong Karachi Kuala Lumpur Madrid Melbourne Mexico City Nairobi New Delhi Shanghai Taipei Toronto

With offices in
Argentina Austria Brazil Chile Czech Republic France Greece Guatemala Hungary Italy Japan Poland Portugal Singapore South Korea Switzerland Thailand Turkey Ukraine Vietnam

Published in the United States of America by
Oxford University Press
198 Madison Avenue, New York, NY 10016

Library of Congress Cataloging-in-Publication Data

Patent law in global perspective / Edited by Ruth L. Okediji, Margo A. Bagley.
pages cm
Includes bibliographical references and index.
ISBN 978-0-19-933427-8 ((hardback) : alk. paper)
1. Patent laws and legislation. 2. Patents (International law) I. Okediji, Ruth L., editor of compilation. II. Bagley, Margo A., editor of compilation.
K1505.P3747 2014
346.04'8—dc23

2013035975

---

9 8 7 6 5 4 3 2 1

Printed in the United States of America on acid-free paper

Note to Readers
This publication is designed to provide accurate and authoritative information in regard to the subject matter covered. It is based upon sources believed to be accurate and reliable and is intended to be current as of the time it was written. It is sold with the understanding that the publisher is not engaged in rendering legal, accounting, or other professional services. If legal advice or other expert assistance is required, the services of a competent professional person should be sought. Also, to confirm that the information has not been affected or changed by recent developments, traditional legal research techniques should be used, including checking primary sources where appropriate.

(Based on the Declaration of Principles jointly adopted by a Committee of the American Bar Association and a Committee of Publishers and Associations.)

To Tade O. Okediji with love; I thank God
upon every remembrance of you
—R.L.O.

To Samuel S. Bagley and Samuel S. Bagley II;
Two "good and perfect gifts" from God
—M.A.B.

CONTENTS

Acknowledgments xi
Preface xiii
Introduction xv
Table of Abstracts xix

1. Public Welfare and the International Patent System 1
Ruth L. Okediji

PART ONE: *Global Patent Law and the Political Economy of Harmonization*

2. Intellectual Property Lawmaking, Global Governance, and Emerging Economies 53
Rochelle C. Dreyfuss

3. US Executive Branch Patent Policy, Global and Domestic 85
Arti K. Rai

4. Transnational Legal Ordering and Access to Medicines 97
Gregory Shaffer and Susan K. Sell

5. The Limits of Substantive Patent Law Harmonization 127
Graham Dutfield

PART TWO: *Global Approaches to Subject Matter Standards and Patentability*

6. Patent Barbarians at the Gate: The Who, What, When, Where, Why and How of US Patent Subject Matter Eligibility Disputes 149
Margo A. Bagley

7. Patent Law's Problem Children: Software and Biotechnology in Transatlantic Context 187
Dan L. Burk

8. Patenting Plants: A Comparative Synthesis 213
Mark D. Janis

9. Enablement and Written Description 243
Matthew Fisher

PART THREE: *Patents, Institutions, and Innovation Pathways*

10. Indigenous Developmental Networks and the Non-developmental State: Making Intellectual Property Work for Indigenous People Without Patents 287
Peter Drahos

11. Observing the Patent System in Social and Political Perspective: A Case Study of Europe 321
Shobita Parthasarathy and Alexis Walker

12. Toward a Theory of Regulatory Exclusivities 345
John R. Thomas

PART FOUR: *Exceptions and Limits to Patent Protection*

13. A False Sense of Security Offered by Zero-Price Liability Rules? Research Exceptions in the United States, Europe, and Japan in an Open Innovation Context 379
Esther van Zimmeren and Geertrui van Overwalle

14. Exhaustion and Patent Rights 419
Christopher Heath

15. A New Approach to the Compulsory License Conundrum 485
Cynthia M. Ho

16. Balancing "Incentive to Innovate" and "Protection of Competition": An African Perspective on Intellectual Property Rights and Competition Law 515
Mor Bakhoum

PART FIVE: *TRIPS Compliance, Patent Enforcement, and Patent Remedies*

17. Patentability Criteria as TRIPS Flexibilities: The Examples of China and India 541
Daniel J. Gervais

18. Proof of Progress: The Role of the Inventive Step/Non-obviousness Standard in the Indian Patent Office 571
Feroz Ali Khader and Srividhya Ragavan

19. Pharmaceutical Patent Enforcement: A Development Perspective 603
Shamnad Basheer, Jay Sanklecha and Prakruthi Gowda

20. A Research Agenda for the Comparative Law and Economics of Patent Remedies 637
Thomas F. Cotter

21. The Rule of Patent Law (RPL) as Established by the TRIPS Agreement and Its Role of Promoting Trade Rather than Invention 673
Nuno Pires de Carvalho

Index 699

ACKNOWLEDGMENTS

The road to a finished monograph is filled with lessons in patience, humility, and sacrifice. The cost is shared by many along the way whose faithful efforts ensure that the end goal is as intellectually rewarding as it is successful. We are immensely grateful to the wonderful group of scholars who contributed to this volume. Their willingness to join us in addressing various patent law topics from a variety of theoretical perspectives and to offer original contributions has made this book far richer and of greater scholarly consequence than was foreshadowed in our modest proposal.

Like any long journey, despite a sensible timeline, careful preparation, and diligent oversight of the editorial process, there were surprises along the way and opportunities to take shortcuts around the vision we outlined for this intellectual enterprise. We were encouraged to stay the course by a dedicated team of former editors at Oxford, starting with Lori Wood whose gentle prodding first gave us the idea for this work, Kevin Pendergast who remained contagiously indefatigable about the importance of the subject and our approach to it, and finally Matt Gallaway, who ensured our smooth transition to a new editorial team and staff at OUP. Our thanks to Jennifer Gong, Alden Domizio, and others at Oxford for an effective production process and assistance with the finishing touches to the manuscript.

Tucker Chambers and Bethany Mihalik, University of Minnesota Law School Class of 2014, Allyson Schmitt Hicks and Alicia Smith, University of Minnesota Law School Class of 2013, and Joseph Schaffner, University of Virginia School of Law Class of 2013, devoted many hours of excellent research and editing support; Tomas Felcman, Associate Director of the University of Minnesota IP and Development Program, ensured compliance with various substantive and technical requirements, including overall editorial management in finalizing the manuscript; and Bria Goldman supervised and coordinated the many component parts of the submission process.

Finally, our deepest thanks go to our husbands and families who endured with us throughout the entire process including above all, to the Father of all good gifts without whose grace and mercy neither our lives nor this book would have been possible.

Ruth L. Okediji

Margo A. Bagley

PREFACE

Six centuries after making its debut in Venice, Italy, the patent system remains as critically important to national and international economic interests as it is contentious. As patent laws became widespread throughout the eighteenth and nineteenth centuries, so did a greater appreciation and concern for the domestic welfare objectives and broader social effects of patent grants. These concerns were reflected in the earliest national patent laws in Europe, which, to varying degrees, expressed an instrumentalist view of patents as a means to improve the lives and welfare of citizens. The general conditions of patentability characteristic of these pioneer patent regimes—a new invention (novelty), usefulness (utility), time limits on protection (duration), and in many cases, domestic working requirements—are facially consistent with the strong overarching concern that patents should secure technological and economic advantages that inure to the benefit of the patent-granting jurisdiction. The incorporation of several of these standards into modern patent laws, and in to the international patent system, in part reflects the resilience of the public interest as a significant consideration in the design and application of patent laws.

There is little controversy over the general account of patent law's primary goal of promoting public welfare and the importance of patentability criteria and patent policy in reinforcing that objective. There also is little doubt that the grant and exercise of patent rights continues to produce negative externalities both domestically and in the global context. Patent protection has consequences for the innovation system per se, such as in influencing norms that govern scientific research and practice, in raising the costs of access to basic scientific data, and in shaping the pace and extent of private R & D investments. Patent protection also influences the design of regulatory processes for safeguarding public health and safety, access to medicines, healthcare, and other public goods. Further, patent protection increasingly implicates controversial issues in science such as climate change, the food supply, and genetic engineering.

Against this dynamic backdrop and normative complexity of national patent laws, global efforts to harmonize patent protection across borders have become a mainstay of the multilateral economic system, further compounding the nature and extent of trade-offs associated with global patent cooperation. The conclusion of the Agreement on Trade Related Aspects of Intellectual Property Rights (TRIPS) in 1994 ostensibly moved the international community toward greater convergence in patent law standards and, with this, offered hope that the international protection of patents would facilitate access to markets, engender greater cooperation among firms, and produce a stable legal environment with respect to trade in patented goods. Instead, the combined effects of domestic policy trade-offs, competing social welfare considerations, and enduring differences in policy tools to address the negative effects of patent laws *nationally* has produced unexpected, and at times unintended, consequences. These effects are visible both in countries still struggling with basic economic development goals and in those that assumed that enforceable patent standards could replace or effectively delimit the exercise of national policy discretion in response to welfare considerations associated with access to patented products.

As this book demonstrates, very little is settled in patent law and policy, even in those countries with a long history of patent protection. The scholars who have contributed to this volume tackle, among other things, enduring and new questions that revolve around *how* to identify the national public interest with respect to the role of the patent system in innovation and access to new inventions; *what* constitutes a legitimate exercise of sovereign discretion over the public welfare effects of patent protection in light of competing public international law regimes; *which* national agencies and institutions are best suited to assess and interpret the implications of patent enforcement under local conditions; and *when* national or regional implementation of international rules should logically produce divergent practices, particularly with regard to morally sensitive inventions.

The contours of domestic patent laws have always been subject to intensely negotiated trade-offs and compromises among and between industries, institutions, and agencies advancing competing (and, at times, complementary) national interests or priorities. When viewed in a global perspective, and in light of mandatory international patent rules, the wisdom of many of these trade-offs and their benefit—whether to inventors or the public—is far more contestable. As this book also suggests, the consolidation of patent norms in international instruments means that experiments with national patent policy take place and are immediately scrutinized in international public spaces, heightening the political stakes over decisions about what constitutes legitimate incorporation of international rules in domestic law, and what tools can be used to adjust those global rules to fit national circumstances.

This book reveals a global motif of patent law and policy that is as challenging and complex as it is unpredictable. We hope that readers will find, as we did, the arguments and lessons embodied in this collection rich, insightful, thought-provoking, and instructive.

INTRODUCTION

Since the first Patent Act in Venice in 1474, through the Agreement on Trade Related Aspects of Intellectual Property Rights (TRIPS Agreement) in 1994, the patent system has been used by countries to accomplish a variety of national objectives, not the least of which has been as a means to secure competitive gains in regional and global trade. The intensified twentieth century efforts to secure export markets by enticing as many countries as possible to participate in a harmonized normative patent law framework have revealed significant limitations in the institutional capacity and political willingness of countries to fully cede the architecture of domestic patent law to a singular trade ethos. At the same time, patent law in many countries faces new and increasingly complex issues that are, in theory, governed by the overarching global patent *acquis*. Some of these new challenges involve moral and ethical considerations, such as patents over life forms and other biotechnology inventions that can only be meaningfully considered in reference to specific sociocultural values. Others implicate policy choices regarding the cost to governments of purchasing patented items to address public and social welfare needs, particularly in developing and least-developed countries. Moreover, enduring questions about the role of patents in innovation and how best to facilitate a robust climate for scientific research and development remain critical to the growth of even the most developed economies. Patent agencies, policymakers, and courts have to confront these questions while carefully maneuvering policy options at the interface of national and international patent protection. In our view, addressing these issues effectively requires analysis of, and some deference to, the domestic environment in which patent law operates.

This book analyzes and explores the scope and effects of various efforts by states and other international actors to reconstruct or, at times resist, the implicit hierarchy of international patent norms within the context of national patent law and policy. For almost every topic and subject addressed, comparative and theoretical approaches reflect the dynamic and complex navigation between the imperatives of the patent regime and the strong public welfare considerations that permeate any understanding of the role of patents in modern society. Authors were not compelled to write on certain themes or topics; nonetheless the contributions invariably fell in clusters that addressed major issues at the frontiers of patent law and policy.

The book is organized around five broad themes: ***Part I****—Global Patent Law and the Political Economy of Harmonization*, ***Part II****—Global Approaches to Subject Matter Standards and Patentability*, ***Part III****—Patents, Institutions, and Innovation Pathways*, ***Part IV****—Exceptions and Limitations to Patent Protection*, and ***Part V****—TRIPS Compliance, Patent Enforcement, and Patent Remedies*. The introductory chapter explores gaps in the international patent system that arguably have fueled global concerns about the welfare effects of patent law harmonization. It draws lessons from the failure of patent harmonization efforts to address public welfare objectives and outlines core elements of a "canon" that could inform a progressive framework for global patent policy while accounting for differences in economic development, national competitive strategies, institutional agility, and rule-of-law tenets.

In Part I, Professors Dreyfuss, Dutfield, Shaffer, and Sell explore various aspects of international patent harmonization, describing problems endemic to the harmonization process while also delineating new methods and opportunities for developing and least-developed countries to contribute effectively in shaping and influencing global patent law norms. Professor Rai adds to this new landscape an assessment of tactical differences in US approaches to global patent law standards, and discusses which of the leading emerging economies is more likely to influence US positions in international patent policy.

In Part II, Professors Bagley, Burk, and Janis examine the chronically complex issue of patent subject matter eligibility and comparative approaches thereto among the major patent-granting jurisdictions, focusing particularly on software, biotechnology patents, plant patent protection, and morally sensitive inventions. Nowhere are the normative divergences of patent law more striking, nor the competing national policy priorities more explicitly stated than in this collection of chapters that address some of the thorniest issues in the national application of patent law standards. Professor Fisher then offers a rich and compelling historical and comparative analysis of the enablement and written description standards, highlighting their policy objectives—communicating an invention to the interested public and providing the quid pro quo for patent exclusivity—and differences in the theory and practice of these disclosure standards in major patent-granting jurisdictions.

Part III of the book explores various innovation pathways that exist both in tension with and complementary to the patent system, offering possibilities for imagining multiple forms of policy levers to encourage a diversity of inventive activity and governance models. Professors Drahos and Thomas address institutional and regulatory approaches to promoting and protecting innovation outside the patent system, such as regulatory exclusivities and approaches to recognize traditional forms of innovation in communities for whom patent protection is implausibly delinked from social and cultural values. These two chapters uniquely highlight the virtues and problems of non-patent regimes with broader implications for innovation policy and competition over cultural knowledge assets. Professor Parthasarathy and Ms. Walker write on the historical and sociological factors that have shaped institutional and agency responses to controversial innovations in the European context, demonstrating the practical effects of civic engagement with the patent system in the design and implementation of patent law and policy. Their research and interviews with officials and other stakeholders reveal specific social and political factors that helped facilitate incorporation of concerns expressed by the public in the administration and governance of the European Patent Office.

Part IV of the book deals with limits and exceptions to patents, addressing the incoherence of national laws on this topic and negative consequences for nascent transnational innovation models. The chapters discuss the prospects of an improved and more coherent construction of various patent law limitations, and demonstrate how vital such limits are for promoting welfare, protecting markets, and encouraging new forms of innovation. Professors van Zimmeren and van Overwalle provide a comparative analysis of the research exception, emphasizing the importance of greater convergence among leading jurisdictions to foster transborder collaboration and provide greater security for downstream researchers. Dr. Heath offers an in-depth analysis of the intellectual foundations of the exhaustion doctrine, tracing its historical evolution and providing an analysis of comparative legal sources (some not previously translated into English) and judicial decisions dealing with questions involving repair and/or reproduction of patented items, contractual limitations, and the scope of the doctrine. Professor Ho offers a fresh, social science–based approach to the controversial question of compulsory licensing, identifying and distinguishing distinct policy strands that could better inform cases where explicit public interest objectives are at stake. In a similar vein, Dr. Bakhoum focuses on the prospective role of competition law as a policy instrument to balance competing interests in free markets and innovation, particularly in the context of access to medicines in sub-Saharan Africa. He explores the legal opportunities to engage in such a balancing act in the context of the *GlaxoSmithKline* case brought before the South African Competition Commission, and suggests approaches that could serve to strengthen the role of non-patent agencies in facilitating access to public goods.

Finally in Part V, scholars examine domestic compliance with the TRIPS Agreement, patent enforcement, and patent remedies. These chapters reveal the diversity of approaches to TRIPS standards, and a degree of legal innovation by emerging countries such as China and India in the national implementation of TRIPS requirements. Professor Gervais explores how undefined terms in the TRIPS Agreement provide important space for national policy experimentation in implementing the fundamental patentability requirements of inventive step and industrial applicability. In their chapter, Mr. Khader and Professor Ragavan argue that the non-obviousness standard as implemented by the Indian Patent Office is a heightened standard compared to that of the United States, analyzing grant opinions to illustrate how this standard operates in practice in India. Professor Basheer and Messrs. Sanklecha and Gowda offer an analysis of pharmaceutical patent enforcement in India. Specifically, they examine procedural aspects of patent litigation and argue for the elimination of the temporary injunction phase as an important step in ensuring that the public is not denied access to pharmaceutical products when the integrity of the patent is in question. Their insightful proposals for change underscore the importance of effective enforcement procedures both to the patent holder and the public, and the unusual convergence of interests that should make their development-friendly proposals politically salient in other developing countries. Next, Professor Cotter sets forth a methodological approach to the study of patent remedies, challenging scholars and policymakers to consider the insights offered by comparative law and economics to better understand the nature and implications of divergent national approaches to two major patent remedies: permanent injunctions and damages. And finally, Mr. de Carvalho argues that despite the strong rules established by the TRIPS Agreement, there still remains little correlation between the goals of the Agreement and the rules established therein. Assessing TRIPS compliance by both developed and developing countries, he concludes that the promotion of innovation is far less central to the TRIPS regime than the elimination of trade barriers for pharmaceutical inventions.

The chapters in this book collectively demonstrate that patent law is best understood and, ultimately, defended, in a national context. Nevertheless, patent law is undeniably moored in a global framework that exerts pressure along a continuum of policy choices available to countries as they grapple with numerous issues in the design of their national laws. Overall, the book offers insights from comparative, historical, and institutional analyses that reveal the intricate past, present, and future of patent law in its multifaceted role as an agent of innovation, a facilitator of private investment in scientific research, and, ultimately, an unrivaled instrument of contemporary international economic relations.

TABLE OF ABSTRACTS

TITLE AND AUTHOR(S)	ABSTRACT
INTRODUCTION	
Chapter 1 **Public Welfare and the International Patent System** *Ruth L. Okediji*	International patent harmonization initiatives have become a mainstay of contemporary international economic relations, as has resistance to those efforts. Since the conclusion of the TRIPS Agreement, resistance to patent harmonization has evolved into normatively complex and coordinated challenges to the patent system at national and international levels. These challenges derive from a number of "gaps" between the goals of the patent system and the social costs of patent protection, particularly in the international context. For example, none of the prevailing arguments for patent harmonization meaningfully address the role of patents in ensuring the supply of public goods, the national welfare implications of international patent protection, or how globalized patent norms fit within a broader international public law framework. Historically, a variety of national policies and less invasive patent harmonization regimes downgraded the effects of these gaps. Combined trade and intellectual property obligations under the TRIPS Agreement and its progeny have, however, significantly constrained policy options, while also unevenly increasing the costs of participation in a liberalized global economy. These harmonization

TITLE AND AUTHOR(S)	ABSTRACT
	gaps reveal that the current global patent system lacks the institutional and policy capacity to manage the national welfare consequences of contemporary patent harmonization processes, but strips countries of the right to do so without fear of trade reprisals. This chapter explores the implications and consequences of patent harmonization gaps and outlines elements of a core set of remedial principles—a global patent law canon. Future efforts at international patent cooperation must substantially accommodate tenets of this canon or face continual crises of legitimacy that engender new, dynamic forms of resistance to global patent norms.
PART I—GLOBAL PATENT LAW AND THE POLITICAL ECONOMY OF HARMONIZATION	
Chapter 2 **Intellectual Property Lawmaking, Global Governance, and Emerging Economies** *Rochelle C. Dreyfuss*	This chapter discusses the role that emerging economies could play in rendering intellectual property law and lawmaking more responsive to changing conditions. At present, neither the North nor the South is likely to challenge the accommodations made in the WTO TRIPS Agreement. In the North, the politics of change are complex; the South largely lacks expertise. But emerging economies have the political will to improve access to the world's intellectual output on behalf of their poorest citizens. At the same time, they have growing creative sectors and thick legal and political cultures capable of striking new and imaginative balances between proprietary and access interests. Because the goals of these economies are best served by partnering with least-developed countries and members of civil society interested in intellectual property issues, these nations also have an

TITLE AND AUTHOR(S)	ABSTRACT
	incentive to improve other types of access norms—norms of participation, fairness, and transparency in international lawmaking. These nations are, in other words, in a unique position to contribute to the reforms that are the cornerstone of the global administrative law agenda.
Chapter 3 **US Executive Branch Patent Policy, Global and Domestic** *Arti K. Rai*	This chapter assesses the relationship between domestic patent policy positions taken by the US executive branch and its international positions. As the chapter details, in recent years one important area of apparent divergence between domestic and international positions has been biopharmaceutical patent policy. The chapter analyzes dynamics within the executive branch that lead to this divergence and concludes that it probably represents a relatively stable equilibrium. Outside the biopharmaceutical arena, however, China's aggressive, highly government-centered patent and innovation policies may create challenges, both in terms of chilling agency efforts to advocate for proactive approaches to domestic innovation and in ongoing negotiations over the Trans-Pacific Partnership Agreement.
Chapter 4 **Transnational Legal Ordering and Access to Medicines** *Gregory Shaffer and Susan K. Sell*	The politics of pharmaceutical patent protection increasingly involve the interaction of different transnational legal orders (TLOs), and, in particular, a TLO focused on trade and intellectual property protection, and a TLO focused on the human right to health. These transnational legal orders involve recursive interaction between legal norm-making and implementation at the international and national levels. This chapter shows how rival transnational legal orders frame a country's approach to intellectual property protection. Understanding

TITLE AND AUTHOR(S)	ABSTRACT
	the operation and interaction of these TLOs provides an important framework for countries to develop strategies to navigate among the multiple pressures placed upon them.
Chapter 5 **The Limits of Substantive Patent Law Harmonization** *Graham Dutfield*	Patent law harmonization at the level of principles, rules, and institutional structures is a highly ambitious goal. Its practical intent is to facilitate secure patent coverage on a global scale and for the longest period possible, and to accelerate and cheapen the process of attaining it. Economically the stakes are very high, not just for businesses but also for nations. Generally speaking, harmonization based on high protection standards benefits leading innovator countries and large businesses already enjoying strong market power assisted by their ownership of sizeable intellectual property portfolios. Follower nations and businesses seeking to enhance their innovation levels prefer, understandably, to maintain a greater freedom to copy than such harmonization would allow. This chapter reviews recent harmonization efforts at the international level and explains their lack of success despite the apparent economic and political dominance of advocates and the relative weakness of opponents.
PART II—GLOBAL APPROACHES TO SUBJECT MATTER STANDARDS AND PATENTABILITY	
Chapter 6 **Patent Barbarians at the Gate: The Who, What, When, Where, Why and How of US Patent Subject Matter Disputes** *Margo A. Bagley*	Patent subject matter determinations continue to present a moving target in the United States, primarily for biotechnology, software, business methods, and diagnostic methods, with court decisions, patent office policies, and legislative initiatives sometimes

TITLE AND AUTHOR(S)	ABSTRACT
	expanding, and sometimes contracting, eligibility boundaries. Moreover, decisions in the United States often have repercussions for patenting abroad. This chapter explores changes in the who (players), what (disputes), when (timing), where (venues), why (theories), and how (mechanisms) of patent eligibility challenges fueling these developments now and for the foreseeable future.
Chapter 7 **Patent Law's Problem Children: Software and Biotechnology in Transatlantic Context** *Dan L. Burk*	The law of patentable subject matter has been in flux now for nearly three decades. On both sides of the Atlantic, courts, administrative agencies, and legislators have struggled with the application of patent law to biotechnology and computer software, the "problem children" of modern patent law. Although this ongoing struggle has not yet produced definitive solutions to the quandaries posed by these technologies, it may have at last defined the problem. Both biological and digital code are increasingly treated under similar standards, suggesting the convergence on a consistent doctrine for information technologies. This chapter traces the major developments in software and biotechnology patenting in Europe and North America, demonstrating the parallel approaches taken by jurisdictions on both sides of the Atlantic, as well as the common issues posed by both technologies.
Chapter 8 **Patenting Plants: A Comparative Synthesis** *Mark D. Janis*	This chapter surveys the state of the law regarding utility patent protection for plants. It considers the practices of the major patenting jurisdictions (the United States, Europe, China, Japan, and South Korea), along with others where agriculture is important to the domestic economy (Canada, Australia, and India, for example).

TITLE AND AUTHOR(S)	ABSTRACT
	It assesses whether the lack of multilateral harmonization in plant protection is providing opportunities for innovation in national legal rules governing utility patents for plants. It examines approaches to the threshold question of utility patent eligibility for plants and plant-related inventions and then considers other substantive patent law issues that are emerging, or are likely to emerge, in jurisdictions that permit some level of utility patent protection for plants. It finds abundant evidence of diversity in approaches to the question of patent protection for plants, but concludes that it is doubtful whether this diversity exemplifies robust and progressive innovation in legal thinking.
Chapter 9 **Enablement and Written Description** *Matthew Fisher*	This chapter considers the birth and development of the patent specification as the chosen vessel for communicating an invention to the public, and the interpretation of the standards to which is it now held. It examines the genesis of, and recent developments in, the English and US patent jurisprudence and legislative initiatives concerning enablement, sufficiency, and written description, and considers how well the theoretical and policy justifications underlying the various disclosure requirements mesh with the law in practice.
PART III—PATENTS, INSTITUTIONS, AND INNOVATION PATHWAYS	
Chapter 10 **Indigenous Developmental Networks and the Non-developmental State: Making Intellectual Property Work for Indigenous People Without Patents** *Peter Drahos*	Indigenous people face problems of economic underdevelopment, including in wealthy states such as Australia, Canada, New Zealand, and the United States. The knowledge assets of indigenous people form a potential source of income, but this potential has largely not been realized. State intellectual property systems struggle

TITLE AND AUTHOR(S)	ABSTRACT
	to accommodate the nature of indigenous peoples' innovation. Intellectual property rules relocate indigenous knowledge assets from a cosmological context of place and duties to an open public domain. Faced by extractive state property orders that are non-developmental for them, the chapter argues that indigenous people have to form developmental networks to manage their assets. Drawing on fieldwork from Australia, the chapter outlines how these networks might operate. Secrecy and trademarks form the core of an adaptive strategy. More important than particular systems, however, is that an indigenous developmental network builds capacity by enrolling scientists into its network. The chapter argues that by focusing on the creation of trust within the network, an indigenous group lessens the chances of opportunistic behavior with its assets and creates the conditions for a sharing of its knowledge assets that increase its chances of developmental gains.
Chapter 11 **Observing the Patent System in Social and Political Perspective: A Case Study of Europe** *Shobita Parthasarathy and Alexis Walker*	Most legal scholars treat the major patent systems across the world as substantially similar. These systems reward inventors with intellectual property rights for a limited period of time, bring scientific and legal knowledge to bear on the assessment of patent applications, and have substantially similar approaches to patentability. But this is not the full story. Using the European patent system as a case study, this chapter argues that when patent systems are understood through a social and political lens, they exhibit many important differences. We demonstrate that the European patent system's distinct history and social and political context have shaped its policies and

TITLE AND AUTHOR(S)	ABSTRACT
	practices, but also its responses to the growing activist challenges that have plagued patent systems over the past three decades. Understanding these differences, and the reasons for them, enhances our understanding of both the patent system's role in the governance of innovation and our ability to predict its evolution in the face of controversy over its responsibility to the public interest.
Chapter 12 **Toward a Theory of Regulatory Exclusivities** *John R. Thomas*	The patent system has traditionally supported pharmaceutical innovation through the grant of exclusive rights. Yet another proprietary interest, the so-called regulatory exclusivity, has played an increasingly prominent role in motivating the research, development, and distribution of medical treatments. Surprisingly, academic literature has yet to provide a sustained account of the relationship between these two sorts of intellectual property rights. This chapter makes a modest effort to fill this policy gap by asserting that regulatory exclusivities should be framed as *sui generis* rights that complement, rather than supplement, the patent system. In particular, the incentive of a regulatory exclusivity is most appropriate where the patent framework is incapable of promoting activity that lies in the interest of public health. In contrast, the case for a regulatory exclusivity is weakest when it merely promotes the sorts of innovation the patent system was designed to protect, or, perhaps more importantly, secures protection for potentially patentable innovation that falls short of core patentability requirements.

TITLE AND AUTHOR(S)	ABSTRACT
PART IV—EXCEPTIONS AND LIMITS TO PATENT PROTECTION	
Chapter 13 **A False Sense of Security Offered by Zero-Price Liability Rules? Research Exceptions in the US, Europe, and Japan in an Open Innovation Context** *Esther van Zimmeren & Geertrui Van Overwalle*	It is widely accepted that research exceptions are one of the legitimate exceptions to the exclusivity granted by patent rights. In principle, a robust research exception can fuel follow-on innovation, ward off "patent thicket" problems, and mitigate refusals to license. Yet, a comparative analysis of the research exception in the US, Europe, and Japan shows that, as far as a general research exception does exist, its boundaries often are unclear and confusing. This chapter revisits the role and the impact of the research exception in an "open innovation" setting. The core research question the chapter seeks to address is how open and collaborative innovation challenges the concept, rationale, and scope of the research exception, and whether current national research exceptions can effectively deal with regional and global, large-scale collaborations. The chapter applies a two-step approach. First, it describes and analyzes the content and scope of the current research exceptions in the US, Europe, and Japan. Next, it assesses and evaluates the current regimes using the property rights/liability rules dichotomy as an analytical framework, enabling us to detect deficiencies and suggest routes for improvement.
Chapter 14 **Exhaustion and Patent Rights** *Christopher Heath*	The concept of exhaustion of patent rights refers to the free circulation of patented goods, normally after their first marketing. The scope of such freedom depends on the doctrinal foundation of the exhaustion concept either as an implied license or an absolute limit on the patent right as such.

TITLE AND AUTHOR(S)	ABSTRACT
	The issue of exhaustion is of significant commercial interest in relation to the repair and/or reproduction of patented articles, to contractual limitations, and to the issue of whether exhaustion should be domestic or international in reach. This chapter provides an in-depth exploration of these issues in their historical context, and provides an analysis of international treaties, recent decisions from several jurisdictions, and comparative legal sources.
Chapter 15 **A New Approach to the Compulsory License Conundrum** *Cynthia M. Ho*	This chapter addresses the controversial topic of compulsory licenses by considering the international legal framework in conjunction with insights from social science research. The chapter suggests that there are different perspectives of patent policies which serve as "schemas" through which the same facts or law are viewed differently. This chapter explains and illustrates these schema to provide an explanation for some seeming misrepresentations even by scholars. This approach provides a fresh examination of a contentious topic and may also help provide insights into other areas where patent law and policy are in tension with public health or other social policy objectives.
Chapter 16 **Balancing "Incentive to Innovate" and "Protection of Competition": An African Perspective on Intellectual Property Rights and Competition Law** *Mor Bakhoum*	This chapter discusses the interface between IP and competition law from a development perspective, with a particular emphasis on the circumstances and needs of countries in Sub-Saharan Africa. It analyzes the "essential facility" doctrine raised in the *GlaxoSmithKline* case in South Africa, which was settled by the South African Competition Commission against the back-drop of EU case law. The chapter reimagines *GlaxoSmithKline* and provides insights on

TITLE AND AUTHOR(S)	ABSTRACT
	the legal challenges and economic implications the case would have raised had South African authorities followed the approach set forth by the European Court of Justice in the *Magill* case. As evidenced by the "innovation surplus" requirement set out in *Magill*, EU law emphasizes the need to promote innovation as a core consideration in the application of the essential facility doctrine. However, in countries that are net importers of technology and where local innovation is limited or nonexistent, the innovation surplus requirement is difficult to meet. At the same time, access to technology is often more relevant than local innovation for many developing countries, particularly in relation to the pressing needs these countries face in the areas of public health and food security.
PART V—TRIPS COMPLIANCE, PATENT ENFORCEMENT, AND PATENT REMEDIES	
Chapter 17 **Patentability Criteria as TRIPS Flexibilities: The Examples of China and India** *Daniel J. Gervais*	This chapter situates the "semantic" flexibilities following from the use of undefined terms in the TRIPS Agreement in the broader context of other forms of policy space available to WTO Members in their effort to implement TRIPS and calibrate such implementation to maximize innovation outcomes, while trying to minimize welfare and transaction costs. It considers two specific undefined terms, non-obviousness/inventive step and utility/industrial applicability, and how those definitions may be implemented in national law to embody domestic policies specific to a country's priorities. Such national design flexibility in applying these international standards may produce significantly different

TITLE AND AUTHOR(S)	ABSTRACT
	outcomes and better align patent law to the innovation goals with which it is often associated without compromising interests in making public goods readily accessible. Definitional flexibility, as seen in India and China, allows for policy experimentation in supporting research in new fields of scientific inquiry, providing adequate levels of incentives for those industries that depend more heavily on patents and, ultimately, in assuring that the patent system produces net gains that are meaningful in the specific context of each country.
Chapter 18 **Proof of Progress: The Role of the Inventive Step/Non-obviousness Standard in the Indian Patent Office** *Feroz Ali Khader & Srividhya Ragavan*	Though much has been written about the recent developments in Indian patent law, scholarship has not looked beyond the famed §3(d) and compulsory licensing. This chapter introduces a new protagonist in Indian patent law: the new standard of non-obviousness. The authors argue that the heightened standard operates as a more objective tool to measure the inherently subjective concept of non-obviousness. By requiring patent applicants to demonstrate technical advancement, the law mandates them to show technical superiority over the prior art, which the authors refer to as the "proof of progress" requirement. The new standard is explained and compared with the non-obviousness standard in the United States. Uniquely focusing on the grant opinions of the patent office involving pharmaceutical patent applications, the authors show how the new standard operates in practice.

TITLE AND AUTHOR(S)	ABSTRACT
Chapter 19 **Pharmaceutical Patent Enforcement: A Development Perspective** *Shamnad Basheer, Jay Sanklecha & Prakruthi Gowda*	In the post-TRIPS era, India is facing an exponential increase in the number of pharmaceutical patent litigations brought before her courts. However, given institutional constraints and the relative unfamiliarity of courts with complex patent disputes, there is a grave threat that erroneous injunctions will issue, harming generic manufacturers sued by drug originators and, more importantly, patients who rely on low-cost medications. This chapter advocates policy proposals to minimize the possibility of such erroneous injunctions. The authors caution against the adoption of the low *prima facie* threshold for the grant of a temporary injunction, as advocated in the UK. They recommend a stricter standard that calls for an assessment of the relative merits of each party's case before the grant of an interim injunction. However, as it is nearly impossible to make such an assessment without resorting to a mini-trial in complex cases, the authors argue that in such cases, the temporary injunction phase should be done away with and courts should move directly to the trial stage. In order to guard against excessive delays that effectively gut patent value, they also propose specific measures to expedite trials.
Chapter 20 **A Research Agenda for the Comparative Law and Economics of Patent Remedies** *Thomas F. Cotter*	Over the past two decades, a small but growing number of law-and-economics scholars have begun to apply the standard tools of economic analysis to a field that had long evaded scrutiny by the law-and-economics community, namely the field of comparative law. Until recently, however, comparative law and economics

TITLE AND AUTHOR(S)	ABSTRACT
	scholars have devoted relatively little attention to the law of intellectual property—a gap that is in some ways surprising, given the professedly instrumental character of (much of) IP law and policy. This chapter presents a modest effort at remedying this state of affairs by presenting a proposed research agenda for a comparative law and economics analysis of one specific set of issues within the law of intellectual property: the law of patent remedies. The chapter lays out a proposed protocol for the use of economic analysis to better understand, evaluate, and critique the law of patent remedies as it exists both in the United States and abroad. It focuses on the two principal remedies for patent infringement, permanent injunctions and monetary damages. In particular, the chapter outlines the similarities and differences among various nations' approaches to injunctions and damages for patent infringement; suggests some possible explanations for, and consequences of, the perceived differences; and proposes some areas which deserve further examination.
Chapter 21 **The Rule of Patent Law (RPL) as Established by the TRIPS Agreement and its Role in Promoting Trade Rather Than Invention** *Nuno Pires de Carvalho*	The standards of patent protection (here designated as RPL, or the Rule of Patent Law), as established by Section 5 of Part II of the TRIPS Agreement, have been set by Article 7 as a tool to contribute to the promotion of technological innovation. How successful has Article 7 been in leading WTO Members to achieve these objectives? Not very. Actually, there is no direct correlation between the RPL and levels of innovation.

TITLE AND AUTHOR(S)	ABSTRACT
	High RPL levels are important to promote innovation but alone they cannot lead to it. The success of RPL in promoting invention and innovation depends on external factors. After all, Article 7 corresponds to the desires of a number of WTO Members, but it is not the main objective of the Agreement. The truth is that the main goal of RPL, as well as of the TRIPS Agreement in general, is to prohibit discriminatory treatment against pharmaceutical inventions, and thereby eliminate non-tariff barriers to international trade that resulted from lack or insufficiency of protection. And in this regard, available data indicate that the RPL has been very successful.

1 PUBLIC WELFARE AND THE INTERNATIONAL PATENT SYSTEM

Ruth L. Okediji*

Introduction

Like other categories of intellectual property, the cross-border protection of patents intensified as part of a broader tapestry of trade relations and international commerce among sovereign states. Historically, the complex interplay between general economic policies and the exclusive privileges granted to inventors through patents obscured an important distinction in the means adopted by countries to promote their external foreign trade interests and those utilized directly to improve the welfare of citizens at home.[1] For most countries, patents fell firmly in the latter category, despite collateral benefits for trade.[2]

---

* William L. Prosser Professor of Law, University of Minnesota Law School.

1. *See, e.g.*, James Harrison, Encouraging Innovation in the Eighteenth and Nineteenth Centuries: The Society of Arts and Patents, 1754–1904 (2006) (providing a rich account of the role of the Society of Arts and Patents in innovation and technical progress in the United Kingdom to address national needs, including its influence on the evolution of UK patent laws as a means of encouraging public access to innovation and technical knowledge).

2. *Id.* at 2 (describing England's concern about dependence on imports of saltpeter, an essential ingredient of gunpowder); *id.* at 14–15 (discussing the proposals by William Shipley that led to the establishment of the Society of Arts and Patents. The Society was a nongovernmental organization that established a scheme to distribute rewards for qualifying inventions. The purpose was to encourage "liberal Sciences," the "polite Arts" and "every useful Manufactury." In addition to meeting national innovation needs, Shipley described the scheme as a means to "render Great Britain a school of instruction as it is already the centre of traffic to the greatest part of the known world"); *see also* Christine MacLeod, Inventing the Industrial Revolution 21–24 (1988) (discussing the use of patents as a source of government revenue with regard to imports).

As patent laws became widespread and protectionist trade practices fell into disrepute, explicit consideration of the welfare effects of patent protection also intensified as part of broader efforts to facilitate national economic growth and competitiveness.[3] Patents for inventions had survived the abolition of monopolies[4] and the dismantling of mercantilist policies in major European countries, but usually were subject to explicit stipulations that patent grants should not undermine competition in domestic markets, other state interests, or the general public welfare.[5] Nevertheless, the first efforts at multilateral patent harmonization, which started amid debates about the justifications for patent protection, did not harness global patent norms to the welfare interests that distinguished domestic patent policies. Indeed, by the conclusion of the Agreement on Trade-Related Aspects of Intellectual Property Rights (TRIPS Agreement)[6] in 1994, the international patent system had undergone a shift from a benign tolerance of domestic public interest tenets to an explicit restraint of such interests at the national level.[7] Recent setbacks in efforts to strengthen intellectual property harmonization[8] and, in developing and least-developed countries, continued skepticism over the beneficial effect of patents, are in part attributable to these policy deficits in the international framework.

Substantive patent harmonization efforts are unlikely to abate in the near future, and neither will conflict over the appropriate scope and terms of such harmonization.[9] The resulting gridlock, already evident in the World Intellectual

---

3. *See generally* Harrison, *supra* note 1; MacLeod, *supra* note 2.

4. The leading example is Great Britain's Statute of Monopolies, 1624, 21 Jac. 1, c. 3, § 6.

5. *Id.* (protecting letters patent and grants of privileges "so long as also they be not contrary to the law, mischievous to the state by raising prices of commodities at home, or hurt of trade, or generally inconvenient"); *see also* MacLeod, *supra* note 2, ch. 10.

6. Agreement on Trade-Related Aspects of Intellectual Property Rights, Apr. 15, 1994, 33 I.L.M. 1197 (1994) [hereinafter TRIPS Agreement or TRIPS].

7. The TRIPS Agreement incorporated a three-step test that constrains the scope of permissible exceptions to patent rights. Under Article 30, "Members may provide limited exceptions to the exclusive rights conferred by a patent, provided that such exceptions do not unreasonably conflict with a normal exploitation of the patent and do not unreasonably prejudice the legitimate interests of the patent owner, taking account of the legitimate interests of third parties."

8. *See, e.g.*, Stop Online Piracy Act, H.R. 3261, 112th Cong. (2011) [hereinafter SOPA]; PROTECT IP Act (Preventing Real Online Threats to Economic Creativity and Theft of Intellectual Property Act), S. 968, 112th Cong. (2011) [hereinafter PIPA]; Anti-Counterfeiting Free Trade Agreement, Dec. 3, 2010, 50 I.L.M 243 (2011), *available at* http://www.mofa.go.jp/policy/economy/i_property/pdfs/acta1105_en.pdf.

9. *See* Jerome H. Reichman & Rochelle Cooper Dreyfuss, *Harmonization Without Consensus: Critical Reflections on Drafting a Substantive Patent Law Treaty*, 57 Duke L.J. 85 (2007); *see also* Graham Dutfield, *The Limits of Substantive Patent Law Harmonization*, *in* Patent Law in Global Perspective ch. 5 (Ruth L. Okediji & Margo Bagley eds., 2013).

Property Organization (WIPO) where the work program of the Standing Committee on Patents (SCP) was at an impasse for years,[10] will prove costly for all countries. It is important, then, to explore prospects for a more credible approach to patent harmonization, or at least one that attends to the possibility of meaningful attainment of the public welfare goals of disparately situated countries.

In Section I of this chapter, I outline various points of tension in the way the public interest has been addressed in the domestic and international context. In matters of national patent policy, there is typically no question that the primary focus is on what policy initiatives best advance the national interest, with consideration of alternative or supplementary policies to spur innovation.[11] Further, efforts to address the social costs associated with the patent system often coincide with periodic policy reviews motivated by industry concerns or shifts in national economic trends and priorities.[12] No comparable analysis undergirds the international patent system.[13] Instead, the focus tends to be on what consequences one

---

10. Kaitlin Mara, *Standing Committee on the Law of Patents to Reconvene After Two Year Hiatus,* Intell. Prop. Watch (June 19, 2008, 11:24 AM), http://www.ip-watch.org/2008/06/19/standing-committee-on-the-law-of-patents-to-reconvene-after-two-year-hiatus; Rachel Marusuk Hermann, *WIPO Patent Committee Moves Quickly Through Agenda; Heavy Lifting to Come,* Intell. Prop. Watch (Feb. 26, 2013, 3:01 AM), http://www.ip-watch.org/2013/02/26/wipo-patent-committee-moves-quickly-through-agenda-heavy-lifting-to-come; Rachel Marusuk Hermann, *WIPO Patent Law Committee Cinches Agreement on Future Work,* Intell. Prop. Watch (Mar. 1, 2013, 5:51 PM), http://www.ip-watch.org/2013/03/01/wipo-patent-law-committee-cinches-agreement-on-future-work; *see 2006 Patent Cooperation Treaty Conference: Transcript of Proceedings,* 32 Wm. Mitchell L. Rev. 1603, 1645–46 (2006).

11. A series of important studies by the National Academies (NAS) explore this question. *See, e.g.,* A Patent System for the 21st Century (Stephen A. Merrill, Richard C. Levin & Mark B. Myers eds., 2004); Comm. on Prospering in the Global Econ. of the 21st Century: An Agenda for American Science and Tech., Nat'l Acad. of Sciences, Nat'l Acad. of Eng'g, Inst. of Med., Rising Above the Gathering Storm: Energizing and Employing America for a Brighter Economic Future (2007); Comm. on Comparative Innovation Policy, 21st Century Innovation Systems for Japan and the United States: Lessons from a Decade of Change (Sadao Nagaoka et al. eds., 2009); *see also Europe 2020 Flagship Initiative Innovation Union,* COM (2010) 546 final (June 10, 2010).

12. *See, e.g., Europe 2020 Flagship Initiative Innovation Union, supra* note 11; Adam B. Jaffe, *The U.S. Patent System in Transition: Policy Innovation and the Innovation Process* (Nat'l Bureau of Econ. Research, Working Paper No. 7280, 1999), *available at* http://www.nber.org/papers/w7280.pdf.

13. *But see* U.K. Comm'n on Intellectual Prop. Rights, Integrating Intellectual Property Rights and Development Policy (2002), *available at* http://www.cipr.org.uk/papers/pdfs/final_report/CIPRfullfinal.pdf. The Commission was established by the UK government "to look at how intellectual property rights might work better for poor people and developing countries." *See Home,* Commission on Intell. Prop. Rights, http://www.cipr.org.uk/home.html (last visited Sept. 7, 2013).

country's policy choices have on the competitive advantage of another country (or group of countries). Even more challenging, there are few agreed criteria to inform evaluation, much less a determination, of what constitutes the global public interest in relation to patent harmonization. Is the "global public interest" the sum of national public interests set in a cooperative framework? Or is it similar to classic theories of free trade, a globalized *national* interest that could result from unilateral adherence to international patent doctrines regardless of how other countries behave?

How the global public interest is constructed matters precisely because the results of patent harmonization processes are often disconnected from the local realities of the domestic users who pay the social and economic price for stronger patent standards. This is especially true for countries with no local constituencies to support the patent system, and where political environments allow little to no room for reciprocal bargains among industries affected by trade liberalization, as is the case in the developed countries.[14] Too little effort has been made in patent harmonization processes to differentiate between the *potential benefits* of harmonized patent norms and the *actual properties* of the international patent system. Even when supported by stylized empirical models, traditional justifications for stronger patent harmonization based on benefits from trade liberalization, foreign direct investment flows, or dissemination of technical information have not overcome global public disenchantment with the patent system. In fact, to the extent patent harmonization is linked to globalization processes, these justifications may only serve to deepen the malaise. A new approach to future harmonization efforts is therefore necessary.

In Section II, I review the principal justifications underlying patent harmonization under the Paris Convention for the Protection of Industrial Property[15] and the structural evolution of harmonization trends through the conclusion of the TRIPS Agreement. I briefly examine the consequences of these justifications for national policy entrepreneurship and the development-related costs they entailed. In particular, I underscore the effect of the most-favored-nation (MFN) and national treatment (NT) clauses in a patent harmonization framework.

14. J. Michael Finger, *The Doha Agenda and Development: A View from the Uruguay Round* 11-15 (Asian Dev. Bank Econ. & Research Dep't, Working Paper No. 21, 2002), *available at* http://www.adb.org/sites/default/files/pub/2002/wp021.pdf (noting that in the United States, "[o]vercoming resistance from potential losers [from trade negotiations] has been, in practice, part power politics, e.g., where negotiating authority must be specifically granted, using export industries to win more Congressional votes than the opposition could rally. It has been, in other part, compensation. Adjustment assistance is the straightforward example, the familiar political coin of public works has also been used").

15. Paris Convention for the Protection of Industrial Property, Mar. 20, 1883, as revised at Stockholm, July 14, 1967, 21 U.S.T. 1583, 828 U.N.T.S. 305 [hereinafter, Paris Convention].

These core non-discrimination principles, laudable as they seemed in theory, served in practice to incubate pervasive problems of modern patent harmonization initiatives. I note recent Supreme Court decisions in India and Kenya[16] that creatively limit the public welfare costs of foreign patent protection. I suggest that these decisions represent new forms of resistance to the uneven effects of obligations arising from patent harmonization.

In Section III, I suggest an approach to future harmonization processes that could mediate the gaps between national patent policy interests and trade-focused patent harmonization. I identify three core principles that should guide harmonization efforts at regional or multilateral levels, and propose that these minimum elements constitute a global patent law canon. In my view, such a canon will become increasingly indispensable for securing more predictable and socially responsive national patent law policies, while also providing a construct for the global public welfare that could alleviate the scale and scope of resistance to the patent system.

I. Patents and Public Welfare in the Domestic and International Context

A. Weaknesses of the Patent Harmonization Process

The classic assertion that patent grants are justified principally in relation to the benefits that new inventions contribute to the public welfare is no longer uncontroversial. Since the conclusion of the TRIPS Agreement[17] there has been a renewed and more forceful emphasis on patent law's relationship to the public interest. Simply put, the monopoly privileges characteristic of patents *cost* something to the public. The exclusivity that is so central to the incentive structure of the patent system typically means that the prices of patented goods are above

---

16. *See infra* Section II.

17. There is an extensive literature on the impact of TRIPS on developing countries, on access to public goods, on innovation, and on the domestic welfare interests of developed countries as well. For some leading works, see International Public Goods and Transfer of Technology Under a Globalized Intellectual Property Regime (Keith Maskus & Jerome H. Reichman eds., 2005); Keith E. Maskus, Intellectual Property Rights in the Global Economy (2000); J.H. Reichman, *Universal Minimum Standards of Intellectual Property Protection Under the TRIPS Component of the WTO Agreement, in* Intellectual Property and International Trade: The TRIPS Agreement 21 (C.M. Correa & A. Yusuf eds., 1998); *see also* Keith E. Maskus, Private Rights and Public Problems: The Global Economics of Intellectual Property in the 21st Century (2012); Maureen A. O'Rourke, *Toward a Doctrine of Fair Use in Patent Law*, 100 Colum. L. Rev. 1177 (2000).

marginal cost,[18] limiting the population that can afford access to them. There are also indirect costs to society overall.[19] Exclusivity raises market entry costs and impedes downstream research activities, and can limit markets for substitute goods.[20] Moreover, the patent system requires costly administration.[21]

Like other forms of monopolies, the negative effects and problems of the patent system are well-known.[22] Accordingly, for many developed countries, patent "policy" often amounts to no more than a careful calibration of trade-offs necessary to ensure that the benefits of strong incentives to invent outweigh the systemic costs of the patent system. There are a variety of tools for achieving and maintaining the necessary balance between the means and ends of patent protection. Within patent law itself, important doctrines police the boundaries of patentability, including subject matter eligibility and other qualifying criteria,[23] to ensure that ideas and knowledge that constitute the building blocks of basic science are not improperly privatized or enclosed.[24] Statutory and judicially created limitations and exceptions to patent rights also provide important avenues of access to protected inventions, facilitate continued research efforts in the field

---

18. William M. Landes & Richard A. Posner, The Economic Structure of Intellectual Property Law 22 (2003); Richard A. Posner, *Intellectual Property: The Law and Economics Approach*, 19 J. Econ. Persp. 2, 57, 58 (2005).

19. To quote Posner, "[p]atents are a source of great social costs, and only occasionally of commensurate benefits." Richard Posner, *Pharmaceutical Patents—Posner*, The Becker-Posner Blog (Dec. 12, 2004, 9:50 PM), http://www.becker-posner-blog.com/2004/12/pharmaceutical-patents--posner.html.

20. *See, e.g.*, Amy C. Waltz, *Closing the Deal: Making the Right Congressional Decision About Patent Settlement Agreements*, 5 Ind. Health L. Rev. 155, 173 (2008) (discussing barriers to market entry that generic drug manufacturers face); Wendy Thai, *Toward Facilitating Access to Patent Research Tools*, 6 Minn. J.L. Sci. & Tech. 373, 385 (2004) (discussing upstream patents impeding downstream research); *see also* Christopher Heath, *Exhaustion and Patent Rights, in* Patent Law in Global Perspective ch. 14 (Ruth L. Okediji & Margo Bagley eds., 2013).

21. *See, e.g.*, USPTO, Performance and Accountability Report: Fiscal Year 2012, at 107 (2012), *available at* http://www.uspto.gov/about/stratplan/ar/USPTOFY2012PAR.pdf (noting that the USPTO Patent Program cost $2,079,357,000 in 2012).

22. *See, e.g.*, A Patent System for the 21st Century, *supra* note 11; *see also* James Bessen & Michael Meurer, Patent Failure: How Judges, Bureaucrats, and Lawyers Put Innovation at Risk (2008); Mark A. Lemley, *Fixing the Patent Office* (SIEPR Discussion Paper No. 11-014, 2012).

23. See TRIPS, *supra* note 6, art. 27(1) for the minimum standards of patentability required for all WTO members.

24. Mayo Collaborative Servs. v. Prometheus Labs., Inc., 132 S. Ct. 1289, 1293 (2012) ("[M]onopolization of those tools through the grant of a patent might tend to impede innovation more than it would tend to promote it.").

of the patented technology,[25] and incentivize the development of substitute and complementary products.[26] Other legal regimes, such as antitrust, constrain overreaching in the exercise of patent rights, secure competitive markets, and enhance consumer welfare.[27] In short, an array of policy levers operates in tandem, and sometimes in tension, with the patent system in order to ensure that it delivers net gains to offset its tax on the public.

If securing and advancing the domestic public interest requires a consistent search for the optimal arrangement of institutions, rules, and tools to induce the desired welfare outcomes from patent protection, no less can be true of the international patent system. Efforts at patent harmonization have tended to proceed on the assumption that the trade-offs reflected in the patent system of one country have (or should have) equal salience in the social context and public welfare calculus of others.[28] In focusing principally on transplanting domestic patent norms and standards from the developed countries to the multilateral level, prevailing approaches to patent harmonization have long avoided at least two difficult, but critically important, considerations.

First, patent harmonization efforts have rarely considered whether alternative policies, or a likely combination of macroeconomic policies, may better serve to induce innovation in countries with less experience with the patent system. The question of what precisely the global patent system is meant to accomplish *in its own terms* beyond strategic trade gains, and what appropriate interface should exist between national and international patent norms, remains largely ambiguous, amid growing concern that even national patent systems have increasingly dubious effects on innovation.[29] Moreover, despite the strengthening of enforceable patent norms in the TRIPS Agreement,[30] there remain important differences of approach

---

25. Rebecca S. Eisenberg, *Patents and the Progress of Science: Exclusive Rights and Experimental Use*, 56 U. Chi. L. Rev. 1017 (1989); *see also* Panel Report, *Canada—Patent Protection of Pharmaceutical Products*, 75–76, WT/DS114/R (Mar. 17, 2000) [hereinafter *Canada—Pharmaceuticals*] (noting the importance of ongoing research to society and that the experimental use exception operates to prevent a chilling effect on such research).

26. Heath, *supra* note 20 (exploring the patent exhaustion doctrine with regard to, inter alia, commercial interests in relation to the repair and/or reproduction of patented articles).

27. *See* U.S. Dep't of Justice & Fed. Trade Comm'n, Antitrust Guidelines for the Licensing of Intellectual Property 2 (1995), *available at* http://www.justice.gov/atr/public/guidelines/0558.pdf.

28. *See* Dongwook Chun, *Patent Law Harmonization in the Age of Globalization: The Necessity and Strategy for a Pragmatic Outcome*, 93 J. Pat. & Trademark Off. Soc'y 127, 129 (2011) (noting that developed countries "argue that only by adopting strong intellectual property protection can developing countries hope to attract foreign direct investment essential to their economic development").

29. *See* sources cited *supra* note 11; *see also* Bessen & Meurer, *supra* note 22, at 16.

30. *See* TRIPS Agreement, *supra* note 6, pt. II, § 5.

in the developed countries on fundamental issues of patent law, including subject matter scope,[31] exceptions to patent rights,[32] and various other elements of patent policy.[33] The powerful defense of the patent system by the developed countries in various fora,[34] their staunch resistance to any effort to reshape its current configuration,[35] and ongoing political and capital investment by patent-intensive industries in ever stronger patent harmonization,[36] belies the significance of these differences in national patent systems. Moreover, the intense policy debates, resourceful proposals, and solutions, which—though complex and politically charged—are nevertheless integral to an informed national patent policy, are usually absent from negotiations in multilateral fora.[37] The results of these negotiations become the source of,

---

31. Margo A. Bagley, *Patent Barbarians at the Gate: The Who, What, When, Where, Why and How of US Patent Subject Matter Disputes, in* PATENT LAW IN GLOBAL PERSPECTIVE ch. 6 (Ruth L. Okediji & Margo Bagley eds., 2013); Dan L. Burk, *Patent Law's Problem Children: Software and Biotechnology in Transatlantic Context, in* PATENT LAW IN GLOBAL PERSPECTIVE ch. 7 (Ruth L. Okediji & Margo Bagley eds., 2013).

32. Esther van Zimmeren & Geertrui Van Overwalle, *A False Sense of Security Offered by Zero-Price Liability Rules? Research Exceptions in the US, Europe, and Japan in an Open Innovation Context, in* PATENT LAW IN GLOBAL PERSPECTIVE ch. 13 (Ruth L. Okediji & Margo Bagley eds., 2013).

33. Anthony D. Sabatelli & J.C. Rasser, *Impediments to Global Patent Law Harmonization*, 22 N. KY. L. REV. 579, 587–88 (1995) (discussing important areas of international nonuniformity in patent law from the United States' standpoint).

34. *See, e.g.*, David Kappos, Director, USPTO, Keynote Address at the Center for American Progress (Nov. 20, 2012), *available at* http://www.uspto.gov/news/speeches/2012/kappos_CAP.jsp.

35. *See* Reichman & Dreyfuss, *supra* note 9.

36. *See* Press Release, Coalition for 21st Century Patent Reform Seeks to Protect American Innovation and Competitiveness (Mar. 21, 2007), *available at* http://www.patentsmatter.com/press/release_032107.php (supporting harmonization of the U.S. patent system with those of major trading partners); Testimony of Franklin J. Vargo, Vice President for International Economic Affairs, National Association of Manufacturers (Dec. 9, 2009), *available at* http://www.nam.org/~/media/B97951F9FFF548F9B9B27DD757076188/NAM_Testimony_on_Global_IPR___12_9_09pdf.pdf (supporting an aggressive IPR agenda with international trading partners).

37. There are many studies available on a variety of issues in the global patent system. In particular, the establishment at WIPO of the Office of the Chief Economist has been significantly helpful in providing high quality economic analysis of the patent system. *See, e.g.*, Interview with Carsten Fink, Chief Economist, WIPO (Dec. 2010), *available at* http://www.wipo.int/wipo_magazine/en/2010/06/article_0001.html. However, such studies rarely are reflected in the normative debates that feed directly into the harmonization processes, or in the vast empire of technical assistance projects undertaken in developing countries. As suggested in Section III of this chapter, some additional institutional innovation in WIPO could go a long way in reshaping the nature and effectiveness of the debate over the proper function and role of patent systems, including how to better construct a global patent order that more meaningfully responds to human and national needs.

and in some cases the exact text of, national patent legislation in developing and least-developed countries.

At a minimum, these differences among leading Organization for Economic Co-operation and Development (OECD) countries, and the various institutional (and normative) "fixes" required to adjust their domestic patent policies incrementally in response to new dynamics,[38] reveal just how fragile are the presumptions on which the global patent system is premised, even as it exacts significant costs on the current and future welfare interests of more vulnerable societies.[39] Of course, part of the challenge is the scale of investment required to transform the status quo. Real change to the patent system, no matter how reasonable, rational, or required for the welfare of innovators and users, entails high political stakes.[40] Even powerful governments such as in the United States struggle to make serious policy changes,[41] settling instead for valuable, but limited, adjustments.[42] Mere hints of patent reform in almost any jurisdiction are sure to attract pressure, lobbying, and threats of reprisals from multiple stakeholders on all sides.[43] Any meaningful adjustment to national patent laws has become a risky proposition of international import.[44] And rarely can such efforts offer sufficient

38. *See* sources cited *supra* note 11.

39. *See generally* Reichman & Dreyfuss, *supra* note 9.

40. Camilla A. Hrdy, *Dissenting State Patent Regimes*, 3 IP THEORY 78, 78 (2013), *available at* http://www.repository.law.indiana.edu/ipt/vol3/iss2/2 ("Congress will not be able to [solve patent law's fundamental problems] in the near future due to institutional incompetence and lobbying by powerful national interest groups who are heavily invested in the current system and resist reforms that might reduce their profits, or increase their competitors' [profits], even though reforms might make the system work better for everyone in the long run."); Brooks Boliek, *Patent Bill Hits Resistance*, POLITICO (Feb. 3, 2011, 4:34 AM), www.politico.com/news/stories/0211/48734.html ("The legislation splits the high tech community.").

41. Timothy B. Lee, *Mostly Pointless Patent Reform Bill Goes to Obama for Signature*, ARSTECHNICA (Sept. 8, 2011, 4:48 PM), http://arstechnica.com/tech-policy%20/2011/09/mostly-pointless-patent-reform-bill-goes-to-obama-for-signature ("Congress has been wrestling with patent reform legislation since 2005. When we last took an in-depth look at the issue in 2008, we concluded that the legislation then under consideration was not the kind of serious patent reform the nation needs. This year's legislation is weaker still.").

42. *See, e.g.*, *GAO Report Finds Patent Trolls Are Not the Problem with Litigation*, INVENTIVE STEP: A BLOG DEDICATED TO A DISCUSSION OF PATENT LAW ISSUES & STRATEGIES (Aug. 26, 2013, 11:10 AM), http://inventivestep.net/2013/08/26/gao-report-finds-patent-trolls-are-not-the-problem-with-litigation ("The AIA was passed as a bellweather of tremendous change and improvement in the patent system. The AIA, however, does little to address the real issues surrounding the patent system....").

43. Patent reform efforts in Brazil offer the latest example. *See* Ana Mano, *Brazilian Government Inches Closer to Reform of Patent Law and Related Legislation*, FINANCIAL TIMES, Mar. 21, 2013 (describing diverse interests and disagreements over key aspects of the reform).

44. Ryan O'Quinn & Sanya Sukduang, *Drug Patents Under Fire in Brazil*, PHARMEXECBLOG (June 4, 2013), http://blog.pharmexec.com/2013/06/04/drug-patents-under-fire-in-brazil

short-term returns with which to assuage diverse interest group concerns about the long-term wisdom of fundamental reforms.

Second, patent harmonization efforts have ignored the moral power of other international regimes that affect how global patent norms are applied in particular domestic conditions.[45] Until the TRIPS Agreement, the international patent system had never been purposefully linked to other international legal regimes,[46] despite a common emphasis in national patent laws on ensuring appropriate policy limits on the scope of patent rights, and notwithstanding the anti-discriminatory impulse that largely fueled the Paris Convention.[47] The classic view of the Paris Convention as a "free-standing" treaty[48] suggests a vision of intellectual property as a law unto itself, insulated from a broader public international law complex.[49] In reality, however, non-patent norms can and do affect the nature and scope of international patent obligations, just as is the case with domestic patent policy.[50] The failure to situate patent harmonization norms within a broader global social framework meant that important questions about what kind of institutions are needed to ensure net gains from the translation of global norms into unique domestic conditions were usually ignored. This left the task of assuring the consistency of patent norms with domestic public welfare exclusively to national institutions and processes, even though the very import of harmonization is to curtail the scope of national prerogatives.

("ANVISA faced opposition to its prior consent policies both inside and outside Brazil. Brazil has appeared on the annual 'Special 301' watch list issued by the Office of the United States Trade Representative each year since 2000, which identifies countries with IP policies that concern the U.S. Government.").

45. Peter K. Yu, *Currents and Crosscurrents in the International Intellectual Property Regime*, 38 Loy. L.A. L. Rev. 323, 375–408 (2004) (discussing five major crosscurrents' impact on the international intellectual property regime).

46. In addition to being "nested" within the trade regime, the TRIPS Agreement makes explicit reference to the role of competition law. *See* TRIPS Agreement, *supra* note 6, art. 40.

47. *See infra* Section II.

48. *Canada—Pharmaceuticals, supra* note 25, at 81.

49. Laurence R. Helfer, *Toward a Human Rights Framework for Intellectual Property*, 40 U.C. Davis L. Rev. 971, 980 (2007) ("Since its inception in the late nineteenth century, the development of intellectual property protection rules occurred in a uni-modal international regime confined to intellectual property-specific diplomatic conferences and conventions.").

50. *See* Michael A. Carrier, *Unraveling the Patent-Antitrust Paradox*, 150 U. Pa. L. Rev. 761 (2002); *Competition and Patents*, WIPO, http://www.wipo.int/patent-law/en/developments/competition.html (last visited Aug. 22, 2013); *Public Health and Patents*, WIPO, http://www.wipo.int/patent-law/en/developments/publichealth.html (last visited Aug. 22, 2013).

B. The Role of Domestic Actors and Institutions

In developed countries, the involvement of numerous public and private actors has been a staple feature of the development of domestic patent policy.[51] Beyond contributions to legislative initiatives, domestic actors play an indispensable role in ensuring that the welfare calculus embedded in the principle that patents should serve society prevails over the short-term and often myopic interests that otherwise dominate domestic political processes. Local institutions act as mediating agents at the interface of global obligations and national priorities; they perform important roles as "shock absorbers," assessing and ensuring that international obligations can be sustained within the cultural and legal fabric of the local economy. As public choice scholars have demonstrated, the institutional context in which global patent norms are implemented can sometimes be of greater importance to securing the interests of society at large than the nature of the patent norms themselves.[52] This institutional context is precisely what is missing in the international framework and at the national level in many developing or least-developed countries.

A meaningfully accessible range of institutions with carefully defined mandates has proven important in the public welfare experiments of mature patent systems. When decisions by patent offices run afoul of constitutional prerogatives, courts offer an alternative venue to correct these welfare slippages and restore appropriate balance between societal interests and patent rights.[53] Disputes over patentable subject matter illustrate this corrective function of institutional actors, especially the judiciary.[54] Moreover, when judicial decisions upend the preferred political or tactical calculus embodied in legislative compromises, the legislative branch or administrative agencies may choose to intervene. A few notable examples in the United States include the Hatch Waxman Act[55] as a response to *Roche v. Bolar*[56]

---

51. *See* Arti K. Rai, *US Executive Branch Patent Policy, Global and Domestic, in* PATENT LAW IN GLOBAL PERSPECTIVE ch. 3 (Ruth L. Okediji & Margo Bagley eds., 2013).

52. *See, e.g.*, Daniel K. Tarullo, *Norms and Institutions in Global Competition Policy*, 94 AM. J. INT'L L. 478, 479 (2000) ("[T]he meaning, as well as the effectiveness, of a proposed competition-policy arrangement will depend substantially on the institutional context in which it is grounded.").

53. 35 U.S.C. § 141 (2012).

54. *See, e.g.*, Ass'n for Molecular Pathology v. Myriad Genetics, Inc., 133 S. Ct. 2107 (2013); Diamond v. Chakrabarty, 447 U.S. 303 (1980); Harvard College v. Canada (Comm'r of Patents), [2002] 4 S.C.R. 45, 2002 SCC 76 (Can.).

55. Drug Price Competition and Patent Term Restoration Act of 1984, Pub. L. No. 98-417, 1984 Stat. 1538 (codified as amended in scattered sections of 21 and 35 U.S.C.).

56. Roche Prods., Inc. v. Bolar Pharm. Co., Inc., 733 F.2d 858 (Fed. Cir. 1984).

and the CREATE Act[57] as a response to *Oddzon v. Just Toys*.[58] Similarly, in 2007, a revision to the European Patent Convention in response to *Lenzing AG's European Patent (UK)*[59] took effect to address a procedural concern raised in that case.[60] A mix of institutions, each responsible for specific policy objectives, provides important checks and balances that, overall, keep the patent system socially meaningful and effective.[61] In short, institutional diversity, competency, and agility play a significant role in policing the balance among patent incentives, international patent obligations, and the domestic public good.

The importance of the institutional context in which patent policy is implemented is already evident in some developing countries. For example, in the face of great public outcry over access to medicines to address South Africa's HIV pandemic, the government sought legal reprieve and domestic policy space from the South African Competition Commission[62] as a forum for addressing economic barriers to access to antiretrovirals.[63] It did not attack the validity of the relevant patents before the South African Patent Office (SAPTO) or question patent norms more generally. In India, the Supreme Court denied patent protection for the Novartis cancer drug Glivic on the grounds that it failed to satisfy tailor-made eligibility requirements of the Indian Patent Act.[64] The decision

57. Cooperative Research and Technology Enhancement (CREATE) Act, Pub. L. No. 108-453 (enacted Dec. 10, 2004) (codified as amended in 35 U.S.C. § 103(c)).

58. Oddzon Prods., Inc. v. Just Toys Inc, 122 F.3d 1396 (Fed. Cir. 1997).

59. [1997] R.P.C. 245.

60. *See* Convention on the Grant of European Patents (European Patent Convention) art. 112(a), Oct. 5, 1973, 1065 U.N.T.S. 199 (as amended Nov. 29, 2000) [hereinafter EPC 2000].

61. *See* Laurence R. Helfer & Karen J. Alter, *The Influence of the Andean Intellectual Property Regime on Access to Medicines in Latin America*, *in* BALANCING WEALTH AND HEALTH: THE BATTLE OVER INTELLECTUAL PROPERTY AND ACCESS TO MEDICINES IN LATIN AMERICA (Rochelle Dreyfuss & César Rodríguez-Garavito eds., forthcoming 2014) ("Our research reveals how the Andean Tribunal, domestic administrative agencies, and generic drug companies contributed to building an IP rule of law in the Andean Community that balanced IP rights against public health goals.").

62. William W. Fisher & Cyrill P. Rigamonti, *The South Africa AIDS Controversy: A Case Study in Patent Law and Policy*, HARV. L. SCH. 18 (Feb. 10, 2005), http://cyber.law.harvard.edu/people/tfisher/South%20Africa.pdf.

63. *Group Challenges Cost of Aids Drugs in South Africa:* Makhathini v. GlaxoSmithKline S. Afr., 10 ANDREWS AIDS LITIG. REP. 4 (Mar. 2003) ("In a complaint filed with the South Africa Competition Commission, the AIDS Healthcare Foundation says GSK's refusal to license key AIDS drugs such as AZT, lamivudine and Abacavir to a generic manufacturer deprives thousands of South Africans the treatment they need and results in as many AIDS deaths there daily.").

64. *See* Novartis AG v. Union of India, Nos. 2706-2716 (India 2013), *available at* http://supremecourtofindia.nic.in/outtoday/patent.pdf.

may reflect a higher bar for inventive activity in India as some commentators suggest,[65] although a stronger inventive step requirement certainly would be a pro-innovation measure for all national patent systems.[66] But the more important point about the Novartis case is that it reflects India's domestic welfare calculus. Undoubtedly, this is a calculus that encompasses India's generic pharmaceutical industry and its role in the supply of affordable medicines to markets in the developing world. Both the South African and Indian cases illustrate the significance of domestic institutions for filling gaps created by global patent law, while still complying with global obligations.[67]

Substantive patent harmonization processes have long ignored questions of institutional mediators, to their peril. In another recent national battlefront over patent protection with important international implications, the Kenyan High Court overturned an anti-counterfeiting statute[68] supported by the United States and the pharmaceutical sector.[69] The conflation of counterfeit drugs with generic drugs has been a particularly shrewd tactic to stymie the rise of generic competition in the supply of essential medicines.[70] According to the Court, the statute

---

65. *See* Feroz Ali Khader & Srividhya Ragavan, *Proof of Progress: The Role of the Inventive Step/Non-obviousness Standard in the Indian Patent Office*, *in* PATENT LAW IN GLOBAL PERSPECTIVE ch. 18 (Ruth L. Okediji & Margo Bagley eds., 2013) (arguing that India has adopted a stronger obviousness standard).

66. *See, e.g.*, Sadao Nagaoka, *Reform of Patent System in Japan and Challenges*, *in* 21ST CENTURY INNOVATION SYSTEMS FOR JAPAN AND THE UNITED STATES: LESSONS FROM A DECADE OF CHANGE 153, 166 (Sadao Nagaoka et al. eds., 2009) (advocating higher patentability standards, particularly the inventive step standard, in the context of dealing with patent thickets and stating that a higher standard also shifts "patenting activities to pioneering inventions").

67. *See* Shamnad Basheer, Jay Sanklecha & Prakruthi Gowda, *Pharmaceutical Patent Enforcement: A Development Perspective*, *in* PATENT LAW IN GLOBAL PERSPECTIVE ch. 19 (Ruth L. Okediji & Margo Bagley eds., 2013) (proposing changes to the Indian trial process to ensure a balance between the goals of access to medicines and TRIPS compliance in light of India's specific institutional context and constraints).

68. Anti-Counterfeit Act, No. 13 (2008), KENYA GAZETTE SUPPLEMENT NO. 97, 261.

69. *World Anti-counterfeit Day*, U.S. EMBASSY BLOG: NAIROBI (June 12, 2012), http://blogs.usembassy.gov/nairobi/2012/06/12/world-anti-counterfeit-day; *see also* Suleiman Mbatiah, *Kenya: Pharmaceutical Companies Pushing Anti-counterfeit Law*, INTER PRESS SERVICE NEWS AGENCY (June 14, 2010), http://www.ipsnews.net/2010/06/kenya-pharmaceutical-companies-pushing-anti-counterfeit-law (stating that multinational pharmaceutical companies pushed for the legislation).

70. Jonathan Liberman, *Combating Counterfeit Medicines and Illicit Trade in Tobacco Products: Minefields in Global Health Governance*, 40 J.L. MED. & ETHICS 326, 330 (2012) (describing complaints by developing countries that WHO "was involved in activities in which the term 'counterfeit' was employed in ways that confused issues of quality, safety and efficacy . . . with matters of intellectual property protection"); *see also* Council for Trade-Related Aspects of Intellectual Property Rights, *Communication from the United States: Securing*

did not effectively distinguish between counterfeit drugs and generic medicines, and thus could undermine the fundamental human right to health guaranteed by the Kenyan Constitution.[71] This thinly disguised effort to ensconce patent interests in alternative "national" policy domains produced an unexpected legal bequest for public welfare activists. It forced a determination of what interests should prevail where conflict is unavoidably implicated by a competing set of policy options—in this instance access to medicines and patent protection. The decision of the Kenyan High Court is especially significant because it acknowledged a formal hierarchy in the relationship between public health and patent law regimes, a precedent patent owners certainly would have preferred to avoid.

The political authority of the Kenyan High Court decision makes it even more difficult to question, and is thus of more profound tactical significance to critics of patent harmonization. It is a decision by a *national* institution, applying *national* constitutional principles to a *national* law ostensibly unconnected with the global patent system. It illustrates the potential countervailing power of domestic institutions, which are primarily concerned about national welfare outcomes when dealing with the demands of international harmonization processes.[72]

International organizations can fulfill a similar role in addressing national welfare gaps created by patent harmonization[73] particularly when national

---

Supply Chains Against Counterfeit Goods, IP/C/W/570 (May 31, 2012). India's response to the U.S. submission squarely confronts the claims raised by such imprecise conflations of intellectual property infringement and counterfeit drugs. *See India's Statement on TRIPS*, IP-WATCH.ORG (June 2012), http://www.ip-watch.org/weblog/wp-content/uploads/2012/06/Indian-Statement-TRIPS-June-2012.pdf ("The issues of IP enforcement are distinct from issues of quality and safety. The TRIPS Agreement itself makes no connection between 'counterfeit' and issues of quality and safety.... Further, IP enforcement is not an appropriate or an effective framework to deal with issues of quality and safety of products. In fact making a link between IP enforcement and quality and safety is wholly inappropriate because it marginalizes the role of other agencies that are actually mandated with the responsibility of ensuring standards, quality and safety.... We would therefore like to stress that the issues pertaining to the products of compromised, safety and efficacy do not fall within the purview of the TRIPS Agreement and therefore are beyond the mandate provided to the TRIPS Council. Enforcement measures cannot guarantee products of quality but excessive and unreasonable enforcement measures can surely undermine access to affordable medicines.").

71. Patricia Asero Ochieng v. Attorney General, (2009) Pet. No. 409, ¶¶ 75–78 (H.C.K.) (Kenya).

72. *Id.* ¶ 66 (Kenya) ("The state's obligation with regard to the right to health therefore encompasses not only the positive duty to ensure that its citizens have access to health care services and medication but must also encompass the negative duty not to do anything that would in any way affect access to such health care services and essential medicines. Any legislation that would render the cost of essential drugs unaffordable to citizens would thus be in violation of the state's obligations under the Constitution.").

73. *See* Ruth L. Okediji, *The Role of WIPO in Access to Medicines*, *in* BALANCING WEALTH AND HEALTH: THE BATTLE OVER INTELLECTUAL PROPERTY AND ACCESS TO MEDICINES IN LATIN AMERICA (Rochelle Dreyfuss & César Rodríguez-Garavito eds., forthcoming 2014).

institutions are weak or nonexistent. What kind of international organization should serve in such a capacity is, however, another question. Intellectual property subject matter, and how each is regulated, is uniquely tied to the structure of the domestic economy, reflecting choices consistent with overarching policies that support markets, producers, and consumers in their domestic productive activities. When a regulatory agency or court is called upon to provide oversight or judge the behavior of a patent, copyright, or trademark owner, the legitimacy and efficacy of such private actions are viewed primarily through the lens of the domestic objectives that inform the proprietary interests at issue, and how the exercise of those interests align with other policies that define fundamental national economic values. For most developed countries, patent laws, like other intellectual property laws, were forged in the fire of cross-cutting domestic needs and priorities.

There is a question whether an international organization created to help reduce barriers to trade is the right (or best) kind of organization to assist countries to undertake the required economic and development analysis that should precede the implementation of globalized intellectual property norms.[74] Professor Finger has suggested that this is a function for which development institutions are more suited.[75] As he observes:

> It is possible that the "comparative advantage" of multilateral negotiations to support correct policy choices on trade restrictions does not carry over to the construction of behind-the-border regulations [like intellectual property rights] and institutions that provide the basic business environment.... As an instrument to help the international community make correct decisions about the reduction of trade barriers, multilateral negotiations did not need a capacity for economic analysis.... Furthermore, implementation requires no investment. While the politics of reaching agreement might be difficult, implementing the lower tariff rates requires no more than an official document instructing customs agents accordingly.
>
> Development institutions were created to take on different issues with more difficult economics. To create or replace a transport system or an education system is different from reaching an agreement to mutually reduce import restrictions in two key ways: it requires real investment, and it brings one into the realm where some alternatives have higher

74. Finger, *supra* note 14, at 9–10.

75. *Id.* at 9.

> rates of return than others. These matters require a different set of tools, one that includes project design and cost-benefit analysis.... [W]ith a development institution legal obligations are country-specific and project-specific. As such development institutions are more suited to the one-off problems and trial-error rhythm of what is needed to build behind-the-border regulations and institutions for countries at different levels of development than is WTO's generic approach to legal obligation. With negotiations, legal obligation comes before project design and cost-benefit analysis; with development institutions it comes after. When the objective was reduction of import restrictions, this difference did not matter. When the objective is to set up behind-the-border regulations and institutions, it does matter.[76]

Prior to the TRIPS Agreement, WIPO could have assumed a meaningful role in addressing welfare gaps engendered by patent harmonization. Until recently, however, WIPO was far removed from the policy debates surrounding the welfare costs of the patent system.[77] Those debates, which focused on the development impact of the international patent system, started in the United Nations in the 1970s shortly after the independence of many developing and least-developed countries.[78] In that context, WIPO was forced to formally acknowledge the link between patent harmonization and economic development when Brazil introduced a resolution before the General Assembly to address the concerns of developing countries about the international patent system.[79] Thereafter, WIPO embarked on technical assistance and capacity-building projects aimed largely at compelling allegiance to the existing normative balance in the Paris Convention—a balance that, although better than what obtains now in the TRIPS Agreement, was tailored to the extant welfare interests of the developed countries.

---

76. *Id.* at 9–10.

77. Ahmed Abdel Latif, *Change and Continuity in the International Intellectual Property System: A Turbulent Decade in Perspective*, 3 WIPO J. 36 (2011); Okediji, *supra* note 73.

78. Ruth L. Okediji, *The International Relations of Intellectual Property: Narratives of Developing Country Participation in the Global Intellectual Property System*, 7 SING. J. INT'L & COMP. L. 315 (2003).

79. *See* Draft Resolution, U.N. Doc. A/C.2/L.565 (Nov. 8, 1961). As adopted, the Resolution was different from the Draft. *See* G.A. Res. 1713 (XVI), 16 U.N. GAOR, Supp. (No. 17) at 20, U.N. Doc. A/5100 (1962). The Resolution called for a study of the effects of patents on developing countries, covering topics such as compulsory licenses, technology transfer, and working requirements. *See* United Nations, The Role of Patents in the Transfer of Technology to Developing Countries, U.N. Doc. E/3861/Rev.1 (1964).

The failure of WIPO to address concerns of significance to developing and least-developed countries, combined with its unrelenting advocacy of increasingly stronger intellectual property norms, even in the poorest developing countries, lasted through the conclusion of the TRIPS Agreement. The backlash to the TRIPS Agreement, and the dissatisfaction of developing and least-developed countries with WIPO's response to the Doha Declaration on Public Health,[80] eventually led to the establishment of the WIPO Development Agenda, whose principal mandate is to integrate development considerations into its massive technical assistance operations.[81] In sum, the adverse reaction to the TRIPS Agreement extended beyond the WTO to disrupt, at least on paper, the long-standing ethos that undergirded WIPO's uncompromised position on intellectual property harmonization. Although the jury is still out on the impact of the WIPO Development Agenda,[82] there is no question that it is at least reflective of the deep-seated resentment to forms of intellectual property harmonization that yield no room for public welfare interests to counteract the social costs of intellectual property protection in domestic settings.

II. Public Welfare from Paris to Punta del Este

A. Rationalizing the Justifications for Patent Rights on a Global Stage

So how exactly did the patent harmonization processes manage to elide a functional construct of the global public interest? Despite the adoption of patent laws in many countries at about the same time during the mid-nineteenth century, no spontaneous convergence of patent norms occurred. From its genesis, international patent relations reflected specific preferences related to underlying justifications for patent protection around which similar-minded countries coalesced.[83] Similarity in patent law principles did not, however, mean consistency or agreement on all aspects of patent protection and administration. No uniform approach

80. *See* Inaamul Haque, *Doha Development Agenda: Recapturing the Momentum of Multilateralism and Developing Countries*, 17 Am. U. Int'l L. Rev. 1097, 1121 (2002) (providing examples of developing country dissatisfaction).

81. *See Development Agenda for WIPO*, WIPO, http://www.wipo.int/ip-development/en/agenda (last visited Aug. 30, 2012).

82. Peter K. Yu, *A Tale of Two Development Agendas*, 35 Ohio N.U. L. Rev. 465, 566 (2009) (discussing concerns with the WIPO Development Agenda).

83. Stephen P. Ladas, The International Protection of Industrial Property 13 (1930); Heinrich Kronstein & Irene Till, *A Reevaluation of the International Patent Convention*, 12 Law & Contemp. Probs. 765 (1947).

to patent protection existed, and national laws did not protect the same inventions, offer the same kind of rights, or impose the same limits and conditions on protection.[84] In particular, protection for the rights of foreigners was significantly unsatisfactory. Foreigners either were precluded from acquiring patent rights or, where allowed, were subject to numerous obligations that, more often than not, led to lesser levels of protection than enjoyed by citizens.[85] In particular, reciprocity was the basis for extending protection to foreigners; protection was offered only if the inventor's home country also extended protection to citizens of the patent-granting country. The rationale for this discriminatory approach was trade protectionism—recognizing protection for foreign patents led to restrictions that prevented free riding of foreign know-how and inventions by local industries, a tactic that had specifically been encouraged under mercantilist regimes.[86] Very early on, then, a commitment to non-discrimination principles served as a shortcut to integrate free-trade archetypes into the patent system, and would be an important anchor for the trade-related arguments that justified the TRIPS Agreement over a century later.[87]

The specter of patent protection extending equally to foreigners under a new kind of international agreement galvanized two critical aspects of the Paris Convention. First, it focused attention on the necessity for a clear articulation of the basis for patent protection. Second, it established patent protection as part of the toolkit used to engender peace and cooperation among nations.[88] In the discussions leading up to the conclusion of the Paris Convention, two competing justifications for patent protection were debated. The first, championed by Switzerland, Germany, and Holland, viewed patent rights strictly as an instrument

---

84. LADAS, *supra* note 83, at 23–30 (discussing divergences in national patent laws before the Paris Convention).

85. *Id.* at 15.

86. *E.g.*, Hilton v. Guyot, 159 U.S. 113, 202–34 (1895) (establishing rules of reciprocity in the U.S. judicial system).

87. J.H. Reichman, *Universal Minimum Standards of Intellectual Property Protection Under the TRIPS Component of the WTO Agreement*, 29 INT'L LAW. 345, 347–51 (1995) (discussing application of MFN and NT provisions in TRIPS).

88. Charles B. Rangel, Essay, *Moving Forward: A New, Bipartisan Trade Policy That Reflects American Values*, 45 HARV. J. ON LEGIS. 377, 378–79 (2008) ("Congress recognized in the Reciprocal Trade Agreements Act of 1934 and in subsequent legislation that the elimination of trade barriers and unfair trade practices could contribute to economic prosperity in the United States and around the world, and that international economic interdependence could help to promote international peace. As President Franklin D. Roosevelt's Secretary of State once said, the expansion of international trade 'dovetailed with peace; high tariffs, trade barriers, and unfair economic competition, with war.' "). For a critique of this position, see KATHERINE BARBIERI, THE LIBERAL ILLUSION: DOES TRADE PROMOTE PEACE? (2002) (testing hypotheses relating to the relation of trade to conflict).

of government regulation,[89] and the second, which ultimately prevailed, was the American view of patents as private property.[90] These justifications were not just negotiating ploys but represented a constitutional basis for the nature and extent of harmonization. Viewing patents as *private* property attached great significance to the humanity of the inventor and the rights granted over the invention. The protection of private property rights needed to reflect the general ethos of prevailing principles of international comity, which required acknowledgment of the rights of individuals protected by the state.[91] Some efforts to justify strong protection of intellectual property on human rights grounds have roots in this idea of patents as private property.[92] In this regard, non-discrimination principles served to advance a free-trade premise in the nascent push for multilateral patent harmonization under the Paris Convention. Statements issued by various pro-patent governments in the discussions leading up to the Paris Convention were rife with linkages between patent protection as a symbol of mature economies and the importance of consistency among the domestic laws of nations to further commercial and technical progress.[93] Here also was early evidence of the idea that one nation's technological bounty, in this case the United States, could be experienced only by countries brave enough to adopt similar patent legislation.[94]

As commentators note, the push for strong international patent protection was successful in large part because other countries viewed the industrial success of the United States as a function of its patent system.[95] No reference was made to the supplementary macroeconomic policies that made the patent system

89. Kronstein & Till, *supra* note 83; James Thuo Gathii, *The Structural Power of Strong Pharmaceutical Patent Protection in U.S. Foreign Policy*, 7 J. GENDER RACE & JUST. 267, 274–5 (2003) (describing competing views of patents before and at the Paris Convention).

90. Kronstein & Till, *supra* note 83, at 775.

91. Hilton v. Guyot, 159 U.S. 113, 164 (1895) (defining the doctrine of international comity as "the recognition which one nation allows within its territory to the legislative, executive, or judicial acts of another nation, having due regard both to international duty and convenience, and to the rights of its own citizens, or of other persons who are under the protection of its laws").

92. *See, e.g.*, Laurence R. Helfer, *The New Innovation Frontier? Intellectual Property and the European Court of Human Rights*, 49 HARV. INT'L L.J. 1, 2 (2008) (explaining that the right of property brings intellectual property under the purview of an international human rights court).

93. *See* CTR. FOR INT'L LEGAL STUDIES, 3 WORLD INTELLECTUAL PROPERTY RIGHTS AND REMEDIES § 60:2 (2013); M. Bruce Harper, *Trips Article 27.2: An Argument for Caution*, 21 WM. & MARY ENVTL. L. & POL'Y REV. 381, 389 (1997).

94. Kronstein & Till, *supra* note 83.

95. *Id.*

a plausible poster child for U.S. technological growth.[96] The prevailing culture of infringement of literary and artistic works, which made access to knowledge easily available and cheap to diffuse in the United States, undoubtedly enhanced the domestic capacity to absorb technical knowledge.[97] To the extent patents played a central role in late nineteenth century technological acceleration in the United States, it was in combination with key policy variables that made patents in discrete circumstances extremely valuable in establishing national industries of great strength.[98] Important inventions such as the Bunsen burner and X-ray machines, which were influential in the growth of key industries, were not patented but instead made freely available to the public,[99] thus allowing easy market entry and reducing the costs of follow-on innovation. A similar pattern occurred in England, where not just major inventions, but also small, technologically significant inventions were made freely available to the public.[100] Throughout the history of today's leading technology exporters, the presence of a public domain that, in effect, functioned as an innovation commons, from which later generations of patented technologies were enabled, is a shared characteristic.[101] Pioneering technologies supported by government subsidies or dedicated

---

96. *See* FRANK DOBBIN, FORGING INDUSTRIAL POLICY: THE UNITED STATES, BRITAIN, AND FRANCE IN THE RAILWAY AGE 3 (1994) (discussing free market policy in the United States); *Economic Growth and the Early Industrial Revolution*, U.S. HISTORY: PRE-COLUMBIAN TO THE NEW MILLENNIUM, http://www.ushistory.org/us/22a.asp (last visited Aug. 24, 2013) (attributing nineteenth century economic growth in part to the factory system, wage labor, improved transportation system for raw materials, Bank of the United Sates establishment, state charters providing special benefits for private corporations, immigrants, and the Erie Canal).

97. *See* R. Anthony Reese, *Innocent Infringement in U.S. Copyright Law: A History*, 30 COLUM. J.L. & ARTS 133, 135–44 (2007) (discussing the limited scope of early copyright protection).

98. *See Economic Growth and the Early Industrial Revolution*, *supra* note 96.

99. *See An Aside on the Bunsen Burner*, CHEMTEAM: A TUTORIAL FOR HIGH SCHOOL CHEMISTRY, http://www.chemteam.info/Electrons/Bunsen-Burner.html (last visited Aug. 24, 2013); Joachim Pietzsch, *About the Nobel Prize in Physics 1901: Perspectives*, NOBELPRIZE.ORG, http://www.nobelprize.org/nobel_prizes/physics/laureates/1901/perspectives.html (last visited Aug. 24, 2013).

100. HARRISON, *supra* note 1, at xv–xxiii (discussing the importance and examples of secondary developments that, though seemingly trivial, "often make the difference between modest and greater success.... They may also avoid serious hold-ups in the development of major innovations."); *id.* at xxii.

101. DAVID MOWERY, RICHARD NELSON, BHAVEN SAMPAT & ARVIDS ZIEDONIS, IVORY TOWER AND INDUSTRIAL INNOVATION (2004); NAT'L SCI. FOUND. & NAT'L CTR. FOR SCI. & ENG'G STATISTICS, SURVEY OF FEDERAL FUNDS FOR RESEARCH AND DEVELOPMENT tbl.1 (2012), *available at* http://www.nsf.gov/statistics/nsf13326/pdf/tab1.pdf ("Federal obligations for research and development, by character of work, and for R&D plant: FYs 1951–2012."); *Policy: Investing in Research, Development and Innovation*, GOV.UK, https://www.gov.uk/government/policies/investing-in-research-development-and-innovation/supporting-pages/science-and-research-funding (last updated Mar. 5, 2013) (explaining the importance of UK public investment in science and research); Sadao Nagaoka & Kenneth

to the public that improved infrastructure/transportation, developed efficient manufacturing techniques for traditional industries, and addressed public health crises[102] established an environment in which patents could be meaningfully related to national growth and competitiveness. The national technological success of the United States that weighed so heavily in influencing the theoretical approach to patent harmonization was not only a product of a patent-saturated innovation culture. Instead, it was one in which the supply of *public goods* by *private inventors*, combined with other strategic government policies, including discrimination against foreign intellectual property owners that lowered the cost of access to knowledge, enhanced national technological capacity and allowed patents to be successfully leveraged as an instrument of trade policy.

B. Non-discrimination as a Basis of Patent Harmonization

The environment for patent harmonization in the latter part of the nineteenth century should also be understood in the broader context of the development of public international law. As technology and commerce brought greater contact between new states emerging in Europe and elsewhere, there was a need for rules of conduct, or a "law of nations," to regulate sovereign interaction by defining states' obligations toward one another.[103] Although the first expression of non-discrimination—the most-favored nation (MFN) provision—dates back to the twelfth century,[104] its use as a basis for widespread commercial interactions greatly expanded with the expansion of patents in Europe and the decline of mercantilism.[105] Its most liberal form, the unconditional MFN, reached its height in Europe at the same time of the discussions leading up to the conclusion of the Paris Convention in 1883.[106]

Flamm, *Introduction: The Chrysanthemum Meets the Eagle: The Co-evolution of Innovation Policies in Japan and the United States*, *in* 21ST CENTURY INNOVATION SYSTEMS FOR JAPAN AND THE UNITED STATES: LESSONS FROM A DECADE OF CHANGE 3–9 (Sadao Nagaoka et al. eds., 2009) (describing policies in the United States and Japan that facilitated commercialization of research efforts, including government-funded research). In the United States, the scope of the research commons changed dramatically with the adoption of the Bayh-Dole Act. *See* Bayh-Dole Act, Pub. L. No. 96-517, 94 Stat. 3015 (codified in 35 U.S.C. §§ 200–212).

102. *See Edward Jenner*, HISTORY LEARNING SITE, http://www.historylearningsite.co.uk/edward_jenner.htm (last visited Aug. 24, 2013) (stating that Jenner's smallpox vaccine was not patented).

103. *See* Douglas J. Sylvester, *International Law as Sword or Shield? Early American Foreign Policy and the Law of Nations*, 32 N.Y.U. J. INT'L L. & POL. 1 (1999).

104. *See* SUBCOMM. ON INTL. TRADE, S. COMM. ON FIN., 93D CONG., COMPILATION OF 1973 STUDIES PREPARED BY THE EXECUTIVE BRANCH (Comm. Print 1974).

105. *Id.*

106. *See* Justus D. Doenecke & Michael R. Adamson, *Most-Favored Nation Principle*, ENCYCLOPEDIA OF THE NEW AMERICAN NATION, http://www.americanforeignrelations.

The second rule of non-discrimination in international trade, the national treatment obligation, also emerged in the late nineteenth century as a pillar of international commercial relations.[107] The principle requires treatment of imported or foreign goods or persons on terms no less favorable than afforded to domestically produced goods and citizens. Although there was no customary international law obligation to extend non-discriminatory treatment to nations,[108] the concept of non-discrimination as a principle applicable to private property and the rights of individuals was widely embraced. Accordingly, in countries where foreigners could apply for patents, they were typically subject to the same conditions and obligations as nationals[109] and so in this sense they enjoyed a measure of formal equality. The MFN principle was not included in the Paris Convention, but the national treatment principle ensured that the advantages of domestic patent laws applied equally to foreign patentees.[110]

The ostensibly equalizing force of non-discrimination principles in the Paris Convention did not preclude countries from orienting patent policies to favor domestic industry or advance national interests. Under the Paris Convention, the primacy of national patent law was never in question.[111] And as trade scholars acknowledge, commitments to free trade have never completely quelled the mercantilist impulse.[112] Numerous and significant escape routes, exceptions, safeguards, and limitations to the free-trade principle exist in theory, and are greatly manipulated in practice, allowing countries to strategically advance national perceived interests even when these are inconsistent with the free-trade imperative. The use of intellectual property rights to discriminate against foreign-owned

---

com/E-N/Most-Favored-Nation-Principle-Mfn-treatment-in-practice-1887-1933-conditional-and-unconditional-mfn-within-protectionism-and-export-expansionism.html (last visited Aug. 24, 2013).

107. Thomas Cottier & Matthias Oesch, *Direct and Indirect Discrimination in WTO Law and EU Law* (Swiss Nat'l Centre of Competence in Research, Working Paper No. 2011/16, 2011).

108. George Schwarzenberger, *Equality and Discrimination in International Economic Law (I)*, 25 Y.B. World Aff. 163, 164–65 (1971).

109. Ladas, *supra* note 83, at 33–34.

110. Mattel, Inc. v. MCA Records, Inc., 296 F.3d 894, 908 (9th Cir. 2002) ("The Paris Convention does not provide substantive rights but ensures 'national treatment.'" (quoting 5 McCarthy on Trademarks and Unfair Competition § 29:25 (4th ed. 2013))).

111. Ladas, *supra* note 83, at 217 ("[W]ith the exception of the special advantages and rights accorded to foreign patentees by articles 4, 4b, 5, nd 5b, of the Convention, *the municipal law of each country controls in all questions in the matter of patents*.") (emphasis added).

112. *See* Anup Shah, *Criticisms of Current Forms of Free Trade*, Global Issues (Mar. 31, 2006), http://www.globalissues.org/article/40/criticisms-of-current-forms-of-free-trade ("We are constantly told that we live in a world of global free trade capitalism, and yet we see that while free markets are preached (in Adam Smith's name), mercantilism still appears to be practiced.").

technologies is particularly well known. In the 1980s, as rapid acceleration of Japanese technological capacity created significant competitive pressure on U.S. industries, an array of trade-policy mechanisms was deployed to protect domestic firms.[113] For example, the computer industry, with government support, brought a GATT antidumping case against Japanese supercomputer sales, which kept Japanese firms out of the U.S. market for over seven years.[114] In addition to such cases, retaliatory tariffs, quotas, and patent infringement claims lodged at the International Trade Commission (ITC) were deployed to protect the U.S. domestic market.[115] Indeed, the European Community in 1989 successfully challenged Section 337 of the U.S. Tariff Act of 1930,[116] claiming that it violated the national treatment principle by providing less favorable treatment to imported products alleged to infringe U.S. patents, than that accorded to like products of U.S. origin, in violation of the GATT Article III national treatment obligation.[117] The Uruguay Round Implementation Act responded to this ruling and made changes to align ITC procedures with GATT rules,[118] but the GATT-consistency of these changes is yet to be determined.

Discrimination in favor of citizens is arguably a core function and justification for the Westphalian state,[119] so that some level of discrimination against foreigners should be tolerated in any construct of free trade as the GATT system had to confront.[120] Yet, the non-discrimination ideal provides a fundamental mooring of the multilateral trade system, and is "[p]erhaps the most important 'basic' principle" of the TRIPS Agreement.[121] But non-discrimination in trade entails costs of

---

113. Nagaoka & Flamm, *supra* note 101, at 6.

114. *Id.* at 9.

115. *Id.* at 6–9 (describing the rise of Japanese technological prowess, U.S. policy responses, and the use of trade policy and claims of patent infringement before the ITC and federal courts to protect U.S. domestic industry from foreign competition).

116. Panel Report, *United States – Section 337 of the Tariff Act of 1930*, L/6439 (Nov. 7, 1989), GATT B.I.S.D. (36th Supp.) at 345 (1989).

117. Peter D. Trooboff & Kenneth W. Abbott, *International Decision: GATT Dispute Settlement Panel*, 84 AM. J. INT'L L. 274, 275 (1990).

118. *See* David A. Gantz, *A Post-Uruguay Round Introduction to International Trade Law in the United States*, 12 ARIZ. J. INT'L & COMP. L. 1, 108 (1995).

119. *See* Robert A. Schapiro, *In the Twilight of the Nation-State: Subnational Constitutions in the New World Order*, 39 RUTGERS L.J. 801, 804–06 (2008) (laying out the premises of a Westphalian state).

120. John W. Head, *Making International Trade Less Foreign: A "Nutshell" for Nonspecialists on the Changing Rules Governing International Trade*, J. KAN. B. ASS'N, Dec. 1992, at 42, 44 (discussing the erosion of the GATT's non-discrimination principle).

121. Reichman, *supra* note 87, at 347–51 (discussing application of MFN and NT provisions in TRIPS).

a fundamentally different political and economic nature than non-discrimination in intellectual property harmonization.[122] Certainly, in the nineteenth-century setting of European internecine inter-state rivalries and conquests, protection for foreigners on equal terms required different domestic policy considerations. On the political front, countries could selectively avoid conferring any benefit to citizens of countries with which they were in conflict, or did not otherwise have any diplomatic relations. For this reason, reciprocity as a basis for cross-border patent relations worked remarkably well to reinforce relations between friendly trading partners and political allies.[123] This was evidenced in the numerous treaties of Friendship, Commerce, and Navigation (FCN) in which industrial property protection sometimes featured.[124]

In addition to maintaining the status quo of patent law in the inventor's and patent-granting country, a reciprocal basis for cross-border patent relations also encouraged countries to play favorites, a tendency that was viewed as a possible contributor to international tensions.[125] The variability of these FCN treaties, and the fact that they imposed no obligations on state parties to upgrade their national patent laws, reflected the inadequate, patchwork state of affairs for international intellectual property protection generally, and highlighted the need for new, multilateral arrangements.[126] In this regard, international patent harmonization was a way to induce changes to improve and strengthen the national laws of the parties.

---

122. Finger, *supra* note 14, at 10–16 (describing differences in the nature of GATT/WTO legal obligations and legal obligations of development institutions, and arguing that the difference entails different tools and analysis that matter fundamentally when the objective is to set up "behind-the-border regulations and institutions."); *id.* at 10.

123. John F. Coyle, *The Treaty of Friendship, Commerce and Navigation in the Modern Era*, 51 Colum. J. Transnat'l L. 302, 303 (2013) (describing uses of FCN treaties, along with their decline, and noting that "[t]he bilateral treaty of friendship, commerce and navigation (FCN treaty) was once a staple of international diplomacy"); Richard L. Doernberg, *Legislative Override of Income Tax Treaties: The Branch Profits Tax and Congressional Abrogation of Authority*, 42 Tax Law. 173, 202 (1989) (stating that FCN treaties "may be useful for improvement of general diplomatic and political relations").

124. *See, e.g.*, Treaty of Commerce and Navigation, U.S.–Japan, art. XVI, Nov. 22, 1894, 1 U.S.T. 352 (stating that citizens of the contracting parties "shall enjoy in the territories of the other the same protection as native citizens or subjects in regard to patents, trade-marks and designs").

125. For example, the invitation to the Vienna Conference stated: "Moreover a continuance of hitherto antagonistic views and measures would scarcely conduce to the preservation of general harmony; and if, for example, Patent protection were maintained in one country, so as to attract thereby skilled operatives from another, then the danger of disturbance of the International industrial balance might readily be apprehended." Kronstein & Till, *supra* note 83, at 768.

126. Ladas, *supra* note 83, at 15.

The economic tradeoffs of non-discrimination were (and still are) far more difficult to assess. The benefit to a country of granting patent protection to foreign imports is largely dependent on a number of national conditions and factors difficult to measure with any degree of consistency or certainty.[127] In the ideal situation where patents truly serve a teaching function,[128] disseminate useful technical information, and facilitate further research activity, patent protection of citizens and foreigners alike could contribute to building the fundamental capabilities so critical to long-term economic growth.[129] In such an ideal state, the requirements for patentability, carefully designed and working in coordination with national macroeconomic policies, should facilitate welfare gains equally from the protection of domestic and foreign goods. Thus, even under the radically restructured terms of patent harmonization under the TRIPS Agreement, domestic patent policy may still be used to build capacity and indirectly favor national inventors as the United States did in the pre-TRIPS era.[130] This requires effective exercise of normative flexibilities, including local working requirements and effective oversight of domestic patent laws in a way that assures that the patent is "good," that valuable technical information is effectively disclosed, and that local industries and research institutions have the capacity to absorb and use such knowledge.[131]

But even with balancing maneuvers, the domestic effect of patent protection for foreign imports has to be acknowledged. If the issue was just competitive fairness, that is one thing; it is quite another to use patents to limit or foreclose opportunities for imitation, hinder access to knowledge for economic growth, and offer no technological benefits domestically. The Paris Convention did not address how countries could respond to new competitive pressures created by affording patents to domestic and foreign inventors on the same terms. This was perhaps not as much of a challenge because countries remained largely free

---

127. Edith Tilton Penrose, The Economics of the International Patent System 116–17, 226 (1951) (arguing that the costs of granting patents on foreign imports outweigh benefits).

128. A number of initiatives at WIPO are directed at addressing this problem. *See, e.g.*, *Standing Committee on the Law of Patents*, WIPO, http://www.wipo.int/patent-law/en/scp.htm (last visited Aug. 24, 2013) (detailing sessions, discussions, and studies on patent issues); *see also* Sean B. Seymore, *The Teaching Function of Patents*, 85 Notre Dame L. Rev. 621 (2010) (suggesting ways to improve the role of patents as a source of technical information to promote the diffusion of knowledge).

129. Dani Rodrik, *The Past, Present and Future of Economic Growth* (Global Citizen Found., Working Paper 1, 2013).

130. *See, e.g.*, 35 U.S.C.A. § 104 (2006) (repealed 2011) (allowing evidence only of inventive activity in the United States in priority contests); Thomas L. Irving & Stacy D. Lewis, *Proving a Date of Invention and Infringement After GATT/TRIPS*, 22 AIPLA Q.J. 309 (1994).

131. Maskus, *supra* note 17.

to join the treaty and there were no enforcement mechanisms. Further, the Paris Convention recognized important limitations to patent rights, such as allowing countries to impose local working requirements,[132] compulsory licensing,[133] and exclusions of subject matter protection, all of which were either drastically constrained or completely eliminated under the TRIPS Agreement.

In sum, the Paris Convention retained important deference to national policy fiats and incorporated provisions that could address the unique social costs of foreign patent protection. This was important for the stability and credibility of the Paris Union and patent policy in general. But additionally, domestic patent policy in leading developed countries addressed national needs and interests in less transparent ways, in some cases using trade institutions and trade policy to block competing imports on specious grounds of patent infringement.[134] This institutional and policy flexibility greatly lowered the risks and costs of patent harmonization for countries such as the United States, making the unevenness of harmonization far less obvious as a problem of patent norms. The ability to advance national interests with an insistence on strong patent norms internationally, while defending national markets against competitive imports through domestic trade policy, has made it possible for several developed countries to secure national welfare goals without sacrificing harmonization strategies. But as recent history has shown, this approach will not always be tenable even for developed countries.[135] Put starkly, protection of foreign patent applicants under the TRIPS Agreement effectuated something no trade or patent policy alone would have allowed—extraction of high rents, limitation of competition, imposition of costs to the public, and foreclosure of opportunities to domestic innovators more likely to address domestic industry needs and areas vital to growth. Under these conditions, non-discrimination principles, at least in the short run, establish the basis for combining the worst of trade policy (dislocation of local industries and exacerbation of inequalities in society)[136] and the worst of

---

132. Paris Convention, *supra* note 15, art. 5(A)(2).

133. Paris Convention, *Id.* art. 5.

134. Nagaoka & Flamm, *supra* note 101, at 6.

135. *See, e.g.*, In the Matter of Certain Electronic Devices, Including Wireless Communication Devices, Portable Music and Data Processing Devices, and Tablet Computers, USITC Inv. No. 337-TA-794 (June 2013). *But see* Letter from Michael B.G. Froman, U.S. Trade Rep., to Irving A. Williamson, Int'l Trade Comm'n Chairman (Aug. 3, 2013), *available at* http://www.ustr.gov/sites/default/files/08032013 Letter_1.PDF.

136. Diana Tussie & Carlos Aggio, *Economic and Social Impacts of Trade Liberalization*, *in* Coping with Trade Reforms: A Developing-Country Perspective on the WTO Industrial Tariff Negotiations 89 (2006) (describing short-term "adjustment outcomes" of liberalized trade as including "a reduction in employment and output, the loss

patent policy (high entry costs, high costs of products, limited competition, minimal knowledge diffusion, and limited national policy space).

C. The Consolidation of Patent Harmonization and Trade Policy: The TRIPS Agreement

Reciprocity and discrimination as a basis for international patent protection formally ceased with the conclusion of the Paris Convention in 1883. In theory, this was a welcome development; it provided stability and a general framework for patent protection that replaced the patchwork of treaties with a set of goals and objectives specific to the use of patents to advance commerce and industry. The Paris Convention was not an effort to aggregate the common provisions in national laws, of which there were not very many.[137] Instead, it purposefully introduced norms based on the vindicated private property rationale, while limiting those based on competing views of patents as instruments of state policy. It was not that public policy did not matter under the Paris Convention; indeed, provisions were directed at accommodating states' concerns in this regard.[138] But overall, the structure and norms imposed by the Paris Convention were focused on a system of cross-border cooperation and not the development of a tool to advance *national* public policy objectives.[139] Instead, national patent policy governed national interests, except in those key areas required to ensure the cross-border protection of patent applicants.[140]

This instrumentalist approach to the design of the Paris Convention was consistent with the primacy of national patent policy and the calculated decision to relegate public welfare decisions to the domestic framework. If national patent policy reflected an optimal balance between protection and limitations to patent rights, assimilating foreigners into that scheme should, in theory, cause no more harm than protection for domestic inventions. The important point, then, was ensuring a balance between the scope of harmonization and the limits on national policy freedoms. Indeed, this was the strategic focus of the Paris Convention. The greater the scope of harmonization, the less room would be available for national action to favor local industries. The tension between the scope of harmonization

---

of industry- and firm-specific human capital, and macroeconomic instability arising from balance-of-payments difficulties or reductions in government revenue").

137. LADAS, *supra* note 83.

138. *See, e.g.*, Paris Convention, *supra* note 15, art. 5.

139. LADAS, *supra* note 83.

140. *See supra* note 123.

and the room for national policy initiatives remains one of the most contentious issues in international patent law under the TRIPS Agreement.[141]

When the changed competitive stakes of the knowledge economy demanded a radical upgrade of the Paris Convention, the comparative advantage of the knowledge-based economies of OECD countries was championed as a basis for a more disciplined approach to intellectual property protection.[142] The patent provisions under TRIPS were significantly more robust than those under the Paris Convention.[143] Significantly, with a transparent motive of securing the knowledge-intensive industries of the United States and Europe, the new, mandatory patent provisions were strategically designed to secure import markets in the burgeoning economies of the Global South, in particular India and Brazil. This carefully calibrated use of global patent harmonization marked a powerful turning point in international patent relations. The equality norms that had separately informed both the GATT regime and the Paris Convention merged to create a formidable constraint on the structural and policy options of developing and least-developed countries to pursue both long- and short-term growth strategies.[144] The TRIPS provisions not only heightened the social cost of international patent protection, they justified elimination of competition in sectors such as the pharmaceutical and biotechnology industries that are key to addressing public health needs.[145]

Discriminatory approaches to intellectual property regulation are omnipresent in the international framework. In particular, a structural form of reciprocity as a basis for cross-border intellectual property relations has re-emerged as a vital part of contemporary international economic relations in the form of bilateral and regional free-trade or investment treaties.[146] These ubiquitous arrangements allow

---

141. *See WIPO Conference on Intellectual Property and Public Policies Wraps Up*, WIPO (July 14, 2009), http://www.wipo.int/pressroom/en/articles/2009/article_0023.html.

142. The extensive literature on the TRIPS Agreement chronicles the scholarly analysis of the trade and intellectual property linkage. *See, e.g.*, Frederick M. Abbott, Commentary, *The International Intellectual Property Order Enters the 21st Century*, 29 VAND. J. TRANSNAT'L L. 471, 477 (1996); Laurinda L. Hicks & James R. Holbein, *Convergence of National Intellectual Property Norms in International Trading Agreements*, 12 AM. U. J. INT'L L. & POL'Y 769, 784 (1997); Anne D. Waters, *Trade, Intellectual Property, and the Development of Central and Eastern Europe: Filling the GATT Gap*, 26 VAND. J. TRANSNAT'L L. 927 (1993).

143. See TRIPS Agreement, *supra* note 6, pt. II, § 2.

144. Rodrik, *supra* note 129.

145. See TRIPS Agreement, *supra* note 6, pt. II, § 2.

146. *See* Susan Sell, *Intellectual Property and Public Policy in Historical Perspective: Contestation and Settlement*, 38 LOY. L.A. L. REV. 267, 319 (2004) ("This historical overview suggests that, in many respects, everything old is new again."); Peter Drahos, *Securing the Future of Intellectual Property: Intellectual Property Owners and Their Nodally Coordinated Enforcement Pyramid*, 36 CASE W. RES. J. INT'L L. 53, 57 (2004) ("Bilaterally, however, the U.S. has been

the developed countries to cherry-pick select developing and least-developed countries for preferential trade terms, which typically include stronger intellectual property norms than those imposed by the TRIPS Agreement.[147]

III. Toward a Global Welfare Canon

A. The Shared View of the Importance of Patent Protection

The list of dire consequences for human welfare, economic growth, and development attributed to intellectual property harmonization under the TRIPS Agreement, particularly its disproportionate impact on developing countries, has been extensive. With few exceptions, these observations have also been generalized across sectors. But as development economists and trade scholars observe, and even as the TRIPS bargain itself suggested, the costs of stronger legal protection for knowledge goods could in the short term be mitigated by other growth strategies. In theory, trade-offs that negatively impact one industry may be offset by gains in another.[148] Dynamic shifts in the production frontier, changes in technology and technical leakage, and improvements in policies that promote structural changes,[149] among other things, may yet produce growth benefits for developing and least-developed countries despite burdens imposed by patent harmonization.[150] Indeed, many developing countries experienced higher levels of growth in the last decade despite costs associated with the TRIPS Agreement.[151] But growth gains, unless significant enough to offset deep poverty and improve

---

busily negotiating free trade agreements (FTAs) with countries that it sees as being important regional models.").

147. *See Sell*, *supra* note 146, at 319 ("Desperate for foreign investment, many countries have signed onto foreign investment agreements that require them to offer much higher standards of protection than are incorporated in TRIPS.").

148. Finger, *supra* note 14, at 14 (capturing this dynamic aptly in the phrase "All Reciprocity is Local"); *id.* at 14–15 (discussing ways in which domestic sectors have been compensated from losses associated with GATT concessions, challenges of domestic politics, and establishment of domestic bargains. He notes that the intellectual property/market access exchange in the Uruguay Round "brings with it a more challenging domestic politics than do more traditional trade agreements. It demands a broader domestic payoff from winners to losers than trade negotiations have supported in the past. It also requires that the benefiting domestic constituency be created by the exchange, something trade negotiations have never done").

149. *See* U.N. Conference on Trade & Dev. (UNCTAD), Economic Development in Africa Report 2012: Structural Transformation and Sustainable Development in Africa (2012) [hereinafter, UNCTAD 2012].

150. Daron Acemoglu & James Robinson, Why Nations Fail: The Origins of Power, Prosperity, and Poverty (2012).

151. Rodrik, *supra* note 129.

the lives of ordinary citizens in the long term, will not obviate the long-term impact of the intellectual property system, or its short-term costs on development aspirations. Long-term economic growth—requiring sustained investments in public goods such as education, infrastructure, and health—and short-term growth associated with industrialization, both require technology inputs of different types. There simply is no growth path that will not require some access to the technological knowledge that the intellectual property system protects. To quote Keith Maskus, "[i]n the 21st century, knowledge truly is the basis on which human life will improve."[152]

Undoubtedly, the TRIPS Agreement entails significant costs for technology-importing countries not only in terms of the financial resources required to implement TRIPS provisions,[153] but also in the form of increased rent payments to technology-exporting countries.[154] The human cost of the TRIPS Agreement is also significant. Not only has price-related inadequate access to medicines cost human lives and much suffering, but overly strong patent rights also impose systemic developmental costs or negative externalities. The peculiar mix of trade and intellectual property interests under the WTO fundamentally affects the nature and effectiveness of economic-growth strategies that nations can pursue, with serious implications for future generations. For example, the kind of imitation-adaptation-improvement strategy successfully used by Japan to build its technological capacity,[155] or subsidies of basic technological inputs and policy

---

152. MASKUS, *supra* note 17, at 1.

153. U.N. CONFERENCE ON TRADE & DEV. (UNCTAD), THE TRIPS AGREEMENT AND DEVELOPING COUNTRIES 36, U.N. Sales No. 96.II.D.10 (1996) (providing estimates of costs of TRIPS implementation in some developing countries). For just three countries (Chile, Egypt, and Bangladesh) the estimated fixed costs of compliance came to slightly over $1.7 million, with recurring costs of roughly 3 million. *Id.* Upfront costs for TRIPS compliance for the poorest countries were estimated at $1.5–$2 million upfront, exclusive of annual recurring costs. *Id.* Other studies suggested that these costs would be higher. *See, e.g.*, THE INT'L BANK FOR RECONSTRUCTION & DEV. & THE WORLD BANK, GLOBAL ECONOMIC PROSPECTS AND THE DEVELOPING COUNTRIES 136 (2002), *available at* http://siteresources.worldbank.org/INTGEP2002/Resources/gep2002complete.pdf (citing J. Michael Finger & P. Schuler, Implementation of Uruguay Round Commitments: The Development Challenge (1999) (manuscript) (1999)); *see also* Finger, *supra* note 14. As Professor Finger states, "[t]hrough TRIPS developing countries took on as legal obligation a cost of $60 billion per year, but there is no legal obligation in the agreement on any Member to provide anything in exchange." Finger, *supra* note 14, at 8, 13.

154. THE INT'L BANK FOR RECONSTRUCTION & DEV. & THE WORLD BANK, *supra* note 153, at 136 (discussing changes in rent transfers as a result of TRIPS). According to the study, "[i]f TRIPS were fully implemented, rent transfers to major technology creating countries—particularly the United States, Germany, and France— . . . would amount to more than $20 billion." *Id.* at xvii.

155. Nagaoka & Flamm, *supra* note 101 (noting a cycle of "American product innovation, followed by Japanese imitation, adaptation, and improvement" throughout the 1960s and early 1970s).

support such as the United States pursued and continues to support,[156] or protection of domestic markets and restrictions on imports as China implemented,[157] are far less possible, and arguably illegal, under the WTO arrangements.[158]

There are important economic studies indicating that under certain conditions there could be positive welfare benefits from effective and strong intellectual property protection in developing countries.[159] Empirical and survey evidence from firms across countries has been useful in supporting these highly stylized findings.[160] However, the structural significance of heightened intellectual property protection across all subject matter categories has been largely ignored in studies exploring the effects of intellectual property protection as part of trade liberalization. It is the *combined* effect of harmonized patent, copyright, and trademark regimes that is particularly pernicious for the dynamic welfare concerns of developing and least-developed countries.

Prior to the TRIPS Agreement, heightened harmonization efforts did not proceed simultaneously across all intellectual property fields—copyright *and* industrial property—as occurred under the TRIPS Agreement. The freedom to maneuver between varying levels of global harmonization and protection, especially with regards to patents and copyrights, facilitated national development strategies guided by very specific priorities. Countries such as the United States could deny copyright protection to foreign works as a means of investing in human capital and building institutions as part of a long-term growth strategy,[161] while adopting strong patent

---

156. *Id.* at 4–5, 7–9 (describing changes in U.S. innovation policy in the 1980s through various government subsidies and programs, many of which are still in place today).

157. Helene Cooper & Mark Landler, *U.S. Shifts Focus to Press China for Market Access*, N.Y. Times, Jan. 19, 2011, at A1.

158. Maskus, *supra* note 17, at 7 (noting that enhanced global intellectual property protection under the TRIPS Agreement "diminishes the ability of follower enterprises anywhere, but especially in emerging-market and developing economies, to copy and reverse engineer those elements").

159. *See, e.g.*, Comm'n on Intellectual Prop. Rights, Integrating Intellectual Property Rights and Development Policy 7 (2002) ("Whether IPRs are a good or bad thing, the developed world has come to an accommodation with them over a long period. Even if their disadvantages sometimes outweigh their advantages, by and large the developed world has the national economic strength and established legal mechanisms to overcome the problems so caused. Insofar as their benefits outweigh their disadvantages, the developed world has the wealth and infrastructure to take advantage of the opportunities provided. It is likely that neither of these holds true for developing and least developed countries."). *See generally* Maskus, *supra* note 17.

160. Keith E. Maskus & Mohan Penubarti, *How Trade-Related are Intellectual Property Rights?*, 39 J. Int'l Econ. 227 (1995) (discussing evidence about differing levels of patent protection influencing trade flows).

161. *See* Barbara A. Ringer, *The Role of the United States in International Copyright—Past, Present, and Future*, 56 Geo L.J. 1050, 1051 (1968).

rights as part of their broader industrial policy. Patent rights could be used successfully to pursue growth from structural transformations that gave rise to new, productive industries such as in the case of biotechnology,[162] while copyright laws contributed to the macroeconomic environment necessary for dynamic growth patterns to emerge.[163] Similarly, Brazil deliberately focused on the development of its patent policies, rather than copyright, as a focus for its industrialization and economic development strategy.[164] The experiences of most OECD countries show that access to education, which is facilitated by robust limitations on copyright in addition to educational subsidies, played a significant role in building human and institutional capacity for long-term growth.[165] There clearly is a strong link between investments in education and knowledge dissemination and the successful utilization of the patent system to enhance technical knowledge and absorption, adaptation, or imitation of protected technologies.[166]

Against this backdrop, the demand for more flexible patent regimes, access to medicines to address public health, and other patent-specific proposals, simply are insufficient to address the development goals and aspirations that engender resentment of global patent harmonization. Paradoxically, these demands for flexibility reinforce a narrative of intellectual property rights as a vital source of economic development, such that the inability to access knowledge goods

---

162. *See, e.g.*, Esteban Burrone, *Patents at the Core: The Biotech Business*, WIPO, http://www.wipo.int/sme/en/documents/patents_biotech.htm (last visited Aug. 30, 2013) ("The business model of biotech firms often relies heavily on intellectual property rights, in particular patents, as they are often the most crucial asset they own in a sector that is extremely research-intensive and with low imitation costs. Investors in biotech companies are generally well aware of the centrality of patents and the survival of such companies may very well depend on their ability to convince investors that they have a solid IP strategy and that risks are reduced to a minimum.").

163. *See* World Intellectual Prop. Org., Copyright + Creativity = Jobs and Economic Growth (2012), *available at* http://www.ip-watch.org/weblog/wp-content/uploads/2012/02/WIPO-Copyright-Economic-Contribution-Analysis-2012-FINAL-230-2.pdf (analyzing WIPO studies on the economic contribution of the copyright industries).

164. I am thankful to Kenneth Nobrega of the Brazilian Ministry of Foreign Affairs for help on this point. *See* Brazilian Mission Communication in Preparation for the Stockholm Conference, Del.Bras.Gen./Conf./249/1967 (Apr. 18, 1967).

165. *See* Org. for Econ. Co-operation and Dev., Education at a Glance 2013: OECD Indicators 126 (2013), *available at* http://www.oecd.org/edu/eag2013%20(eng)—FINAL%2020%20June%202013.pdf ("Higher educational achievement benefits both individuals and society, not only financially, but in the well-being with which it is also associated. For individuals, having a higher education improves chances for employment and reduces the risk of unemployment."); Susan Isiko Štrba, International Copyright Law and Access to Education in Developing Countries (2012).

166. *See* UNCTAD 2012, *supra* note 149, at 81–82 (discussing the importance of "leapfrogging" technological stages in pursuit of development goals in Africa).

leaves little room for sustainable growth. Such a position implicitly recognizes the importance of protecting intellectual property rights to induce the commercialization of new technologies. At first blush, this may seem to place developing and least-developed countries in a contradictory space in the international ecosystem—if access to knowledge goods is critically important for economic development, so must be the system of rights that encourage production of the goods in the first place. But it is not really the protection of rights as such that is at issue, but instead the terms on which such protection is extended globally that matters most for national development policies.

The patent system has historically proven valuable and important to innovation, and its importance to the growth strategy of most industrialized countries remains high despite strong concerns that the domestic system of major countries needs significant changes.[167] Notwithstanding these challenges domestically, the patent system's political entrenchment in the free-market enterprise of major economies makes it unlikely that the emphasis on a strong international patent system will cease. Accordingly, it is important to consider how to respond to the spirited resistance to further patent harmonization.

B. Minding the Public Welfare Gaps on a Global Stage

Reducing the costs—both political and economic—of patent harmonization efforts is beneficial to all countries. In light of this, global patent harmonization should accommodate at least three principles: (1) address the provision of global public goods, (2) improve coherence and coordination with complementary international regimes, and (3) ensure that patent protection for imports delivers local welfare benefits. Further, future patent harmonization efforts must show reasonable deference to domestic compliance methods and technical assistance. Moreover, rather than filtering resources through technical assistance or "capacity building" programs that have historically proven ineffective, aid programs should instead be directed at facilitating the efforts of institutional mediators to enable innovation in developing and least-developed countries. Where such institutions do not exist or are weak, technical assistance should help establish and strengthen them. These three principles and additional two policy proposals constitute a core set of tenets and ideals—a global patent law canon of sorts—on which future patent harmonization efforts should proceed.

---

167. Bessen & Meurer, *supra* note 22; Lemley, *supra* note 22.

1. Addressing the Provision of Global Public Goods in the Global Patent Framework

One of the most significant "gaps" in the global patent law framework is its failure to recognize the effect of stronger harmonized rules on the provision of public goods such as education, environmental protection, and public health.[168] There has been no more effective stage for the theatre of concern and resistance to the TRIPS Agreement than the global outcry over access to medicines.[169] The mandatory extension of patent rights to pharmaceutical products, including the protection of clinical test data,[170] and arguably greater restrictions on mechanisms such as compulsory licensing and local working requirements,[171] predictably created a system in which greater rent payments would flow to technology-exporting countries.[172] This much was to be expected. In addition, however, governments were further constrained in the scope of explicitly available limitations that could be placed on patent rights than under the Paris Convention.[173] Finally, the TRIPS Agreement transferred the costs of enforcement of foreign patent rights to domestic constituencies and institutions by requiring countries to make their laws publicly available and notify them to the TRIPS Council,[174] making oversight of compliance easier to accomplish and enforce, at least in theory. Combined, these normative and administrative changes to the international patent system introduced uncertainty about what *domestic* mechanisms can legitimately be pursued to supply access to medicines, education, and other public goods. In short, the four corners of the TRIPS Agreement established a paramount right to police and constrain the policies of other countries in managing the supply of public goods, and also served as a barrier to access to those goods.

---

168. Keith E. Maskus & Jerome H. Reichman, *The Globalization of Private Knowledge Goods and the Privatization of Global Public Goods*, *in* INTERNATIONAL PUBLIC GOODS AND TRANSFER OF TECHNOLOGY UNDER A GLOBALIZED INTELLECTUAL PROPERTY REGIME 3 (2005).

169. Okediji, *supra* note 73 (describing access to medicines as "the poster child of all that is wrong with IP harmonization under the [TRIPS] Agreement").

170. TRIPS Agreement, *supra* note 6, art. 39(3).

171. Paul Champ & Amir Attaran, *Patent Rights and Local Working Under the WTO TRIPS Agreement: An Analysis of the U.S.–Brazil Patent Dispute*, 27 YALE J. INT'L L. 365 (2002).

172. KEITH MASKUS, PRIVATE RIGHTS AND PUBLIC PROBLEMS: THE GLOBAL ECONOMICS OF INTELLECTUAL PROPERTY IN THE 21ST CENTURY 34 (2012) (noting that the implication of TRIPS "is that the innovative enterprises in the wealthier countries gain a greater ability to control distribution and pricing of their goods across a wider spectrum of countries, bringing potentially large profits to those nations").

173. TRIPS Agreement, *supra* note 6, art. 30.

174. *Id.* art. 63(1)–(2).

The widespread resistance to the TRIPS Agreement over public health concerns demonstrates the importance of domestic policy options being explicitly integrated into patent harmonization toolkits. The Doha Declaration on the TRIPS Agreement and Public Health,[175] a direct result of the popular mobilization against TRIPS, specifies clearly what governments can do with regard to addressing public health needs. The Declaration states, for example, that "the TRIPS Agreement does not and should not prevent members from taking measures to protect public health. Accordingly, . . . the Agreement can and should be interpreted and implemented in a manner supportive of WTO members' right to protect public health and, in particular, to promote access to medicines for all."[176] Although industry interests have had some success in pushing back on the practical utility of the Doha Declaration,[177] the text nonetheless provides a powerful legal basis for national policies directed at supplying public goods. Future harmonization efforts ignore this at their peril.

At least one WTO dispute panel has recognized that the TRIPS Agreement, while unprecedented in scope, still does not represent the full range of interests implicated by its provisions. In *Canada—Patent Protection for Pharmaceuticals*,[178] the panel acknowledged that exceptions to patent rights recognized in Article 30 "amount[] to a recognition that . . . patent rights . . . would need certain adjustments."[179] Subsequently, the Doha Declaration confirmed the existence of such policy room to "include" rights of governments to read "*each* provision of the TRIPS Agreement . . . in the light of the object and purpose of the Agreement as expressed, in particular, in its objectives and principles."[180] Specifically, "each member has the right to grant compulsory licences and the freedom to determine the grounds upon which such licences are granted; . . . the right to determine what constitutes a national emergency or other circumstances of extreme urgency, it being understood that public health crises, including those relating to HIV/AIDS, tuberculosis, malaria and other epidemics, can represent a national emergency or other circumstances of extreme urgency and . . . each member is free

---

175. World Trade Organization, Declaration on the TRIPS Agreement and Public Health, WT/MIN(01)/DEC/W/2 (2001) [hereinafter Doha Declaration].

176. *Id.* ¶ 4.

177. *Cf.* Alan O. Sykes, TRIPS, *Pharmaceuticals, Developing Countries, and the Doha "Solution,"* 3 CHI. J. INT'L L. 47, 54–55 (2002) (discussing developing nations' issues with the Doha Declaration).

178. *Canada—Pharmaceuticals*, *supra* note 25.

179. *Id.* at 154, ¶ 7.26.

180. *See* Doha Declaration, *supra* note 175, ¶ 5(a) (emphasis added).

to establish its own regime for exhaustion without challenge, subject to the MFN and national treatment provisions...."[181]

The failure to explicitly address public goods concerns in the TRIPS Agreement created a gap now filled partly by the Doha Declaration. As an international instrument in its own right, the Declaration is not an interpretive guide for Article 30 of the TRIPS Agreement, although it should have salience in any analysis of Article 30 by future panels.[182] Rather, read in conjunction with the panel's reasoning in *Canada—Patent Protection for Pharmaceuticals*, the Declaration arguably speaks to the overall balance between the TRIPS Agreement and national policies. Both the panel and the Declaration recognized that there are "legitimate interests" beyond the rights of patent owners that must be considered to accomplish the social benefits attributable to patent protection.[183] Arguably then, the failure in the TRIPS Agreement to address the supply of public goods could also be filled by policy measures that fall within the scope of the underlying reasons that produced the Declaration. The significant ongoing efforts by international organizations, the emergence of public-private partnerships (PPPs), and other institutional efforts to combat public health concerns, confirm that singular reliance on markets to address the provision of global public goods is misplaced.[184] National governments are an indispensable component of the global framework to supply public goods and they must have the policy

---

181. *Id.* ¶ 5(b)–(d).

182. *Canada—Pharmaceuticals*, *supra* note 25, at 150, ¶ 7.14 ("The Panel noted that, in the framework of the TRIPS Agreement, which incorporates certain provisions of the major pre-existing international instruments on intellectual property, the context to which the Panel may have recourse for purposes of interpretation of specific TRIPS provisions, in this case Articles 27 and 28, is not restricted to the text, Preamble and Annexes of the TRIPS Agreement itself, but also includes the provisions of the international instruments on intellectual property incorporated into the TRIPS Agreement, as well as any agreement between the parties relating to these agreements within the meaning of Article 31(2) of the Vienna Convention on the Law of Treaties."); *see also id.* at 150, ¶ 7.15 ("As a consequence of *the extended context that has to be taken into account when interpreting provisions of the TRIPS Agreement*, the Panel, in considering the negotiating history of the TRIPS Agreement, concluded that *interpretation may go beyond the negotiating history of the TRIPS Agreement proper* and also inquire into that of the incorporated international instruments on intellectual property.") (emphasis added).

183. *Id.* at ¶¶ 7.69–7.73 (rejecting the "EC argument based solely on the patent owner's legal rights pursuant to Article 28.1, without reference to any more particular normative claims of interest"); *id.* ¶ 7.73; *see also id.* at 76 (noting explicitly that Article 30 encompasses the interests of society at large).

184. Maskus & Reichman, *supra* note 168, at 8; *see also* Inge Kaul, Pedro Conceição, Katell Le Goulven & Ronald U. Mendoza, *How to Improve the Provision of Global Public Goods*, *in* Providing Global Public Goods 185 (Inge Kaul, Pedro Conceição, Katell Le Goulven & Ronald U. Mendoza eds., 2003).

room to do so. Any global patent agreements that preclude or unduly constrain this will undoubtedly face continuous resistance and disruption.

2. Coherence with Complementary International Regimes

Intellectual property harmonization has negatively affected the implementation of the normative obligations of public health and environmental law regimes. Examples include the unrelenting effort of big tobacco companies to thwart the key elements of the first international public health treaty, the WHO Framework Convention for Tobacco Control (FCTC),[185] on grounds that, among other things, plain packaging requirements violate the intellectual property rights recognized by the Paris Convention and the TRIPS Agreement;[186] lawsuits by Gerber in Guatemala protesting the country's efforts to comply with the 1981 WHO Code on the Marketing of Breast-Milk Substitutes;[187] or concerns by developing country governments over intellectual property "prospecting" of flu samples to develop vaccines under the auspices of the 2005 WHO International Health Regulations.[188] These examples demonstrate how significantly intellectual

---

185. WHO Framework Convention on Tobacco Control, at v, May 21, 2003–June 29, 2004, WHA 56.1, 2302 U.N.T.S. 166, 42 I.L.M. 518, *available at* http://www.who.int/tobacco/framework/WHO_FCTC_english.pdf.

186. *See* Generic Packaging, Supplementary Submission to Senate Comm., Submission to the Industry Inquiry from W.D. & H.O. Wills (Aust.) Ltd., The Tobacco Growing and Manufacturing Industries 32 (Jan. 1994), *available at* http://legacy.library.ucsf.edu/tid/kpk33a99/pdf (stating that the proposed packaging regulations "amount to a severe infringement of internationally-registered intellectual property rights"); *see also* Jacqueline Maley, *Tobacco Giant Won't Quit Fight on Plain Packaging*, Sunday Morning Herald, Jan. 20, 2013, http://www.smh.com.au/national/health/tobacco-giant-wont-quit-fight-on-plain-packaging-20130119-2czy6.html#ixzz2dPnXMy1M; Nathalie Thomas, *Tobacco Industry to Fight "Standardised" Cigarette Packs in Europe,* Telegraph, May 2, 2013 (citing industry claims that plain packaging will flood the market with black market cigarettes); James Reilly, *Tobacco Industry's Campaign Against Plain Packaging Is Based on Fear for Its Profits*, Irish Times, July 26, 2013 (noting that the industry's "insistence now is that its branded packs represent intellectual property that must be protected" and observing that "[i]n my view it would be a strange society that would allow a situation where intellectual property rights supersede the rights and expectations of our children to health and wellbeing").

187. *See* World Health Organization, International Code of Marketing of Breast Milk Substitutes (1981), *available at* http://www.who.int/nutrition/publications/code_english.pdf.

188. World Health Org., International Health Regulations 1 (2005); David Fidler, *Influenza Virus Samples, International Law, and Global Health Diplomacy*, 14 Emerging Infectious Diseases 88–94 (2008) ("Developing countries provided information and virus samples to the WHO-operated system; pharmaceutical companies in industrialized countries then obtained free access to such samples, exploited them, and patented the resulting products, which the developing countries could not afford."); *see also WHO-Linked Centre Lays Patent Claim Related to Bird Flu Virus*, Third World Network (Aug. 9, 2008), http://www.twnside.org.sg/title2/health.info/2008/twnhealthinfo20080804.htm.

property norms have been isolated from the broader public international law framework, and the long-standing insularity of WIPO, the principal institution within which most intellectual property harmonization processes unfold.[189]

The debate about supply of public goods and the use of intellectual property rights to forestall efforts to more easily make such goods available has been heavily laced with human rights theory and advocacy. WIPO has maintained a polite distance from human rights arguments that could have any import on the scale of its core intellectual property harmonization mandate.[190] And although the Organization has recently directed attention to global issues such as public health and climate change,[191] there remains little evidence of any significant reorientation of patent harmonization goals in response to the demands that these regimes have made on the global intellectual property system at large. To the contrary, developed countries have insisted on insulating the intellectual property system from the normative demands of other public international law regimes that might hint at, much less suggest, anything contrary to absolute protection of rights. For example, the climate change negotiations under the United Nations Framework Convention on Climate Change (UNFCCC) have been stymied by an impasse over the role of intellectual property in access to and use of climate change technologies.[192] Similarly, as noted earlier, intellectual property rights have been used to erect costly obstacles to the robust normative implementation of international public health law norms.[193]

---

189. With the exception of the TRIPS Agreement, the most significant harmonization processes in intellectual property have occurred within WIPO.

190. *See* Convention Establishing the World Intellectual Property Organization art. 3, July 14, 1967, 2186 U.N.T.S. 121 [hereinafter WIPO Convention].

191. Abdel Latif, *supra* note 77; *see also Global Challenges and Intellectual Property: Unlocking the Potential of Innovation and IP for an Equitable World*, WIPO, www.wipo.int/globalchallenges/en (last visited Sept. 2, 2013) ("WIPO's work in the area of Global Challenges aims to emphasize the positive relationship between innovation and intellectual property (IP), and how IP can best be used for economic and social development. More specifically, WIPO harnesses expertise in the crucial global policy debates related to IP, focusing on three global challenges: Climate change; Public health; Food security.").

192. Ahmed Abdel Latif, Keith Maskus, Ruth Okediji, Jerome Reichman & Pedro Roffe, *Overcoming the Impasse on Intellectual Property and Climate Change at the UNFCCC: A Way Forward* (ICTSD Programme on Innovation, Technology and Intellectual Property, Policy Brief No. 11, 2011) (noting that intellectual property rights have been a "particularly divisive" issue in the UNFCCC process.); *see also* Joshua D. Sarnoff, *The Patent System and Climate Change*, 16 VA. J.L. & TECH. 301, 306 (2011) ("Concerns over the patent system and climate change have already caused serious political tensions.").

193. *See, e.g.*, David P. Fidler, *International Law and Global Public Health*, 48 U. KAN. L. REV. 1, 29 (1999) ("Public health experts have raised concerns that the TRIPS- heightened protection of pharmaceutical patents will adversely affect access to patented drugs in developing countries by raising prices.").

From its beginning in the Paris Convention, patent harmonization has evolved in parallel with normative frameworks in public international law, even human rights and environmental law regimes, which logically should influence and shape the interpretation of national obligations under, and compliance with, harmonized patent norms.[194] Shortly after the conclusion of the Uruguay Round, developing and least-developed countries sought to leverage human rights and environmental law obligations as a countervailing force to the TRIPS Agreement. This occurred primarily at the institutional level where resistance to the Agreement was organized around a tactic of forum shifting[195] to unorthodox fora for intellectual property discussions, such as the Convention on Biological Diversity (CBD),[196] the United Nations Conference on Trade and Development (UNCTAD), and the WHO. While the specialized competencies of these institutions nominally involve intellectual property issues, the welfare impact of patent harmonization on their core missions has, paradoxically, provided policy space for these organizations to exert important and practical influence in countries dealing with the challenges of patent harmonization. UNCTAD and the WHO have been particularly active in producing data, studies, and policy instruments directed at both the development concerns associated with TRIPS obligations[197] and the welfare goals intrinsically linked to a balanced patent system.[198] Further, the WHO has been actively engaged in the creation of

---

194. Laurence R. Helfer, *Human Rights and Intellectual Property: Conflict or Coexistence?*, 5 MINN. INTELL. PROP. REV. 47, 47 (2003) ("Human rights and intellectual property, two bodies of law that were once strangers, are now becoming increasingly intimate bedfellows. For decades the two subjects developed in virtual isolation from each other. But in the last few years, international standard setting activities have begun to map previously uncharted intersections between intellectual property law on the one hand and human rights law on the other.").

195. Laurence R. Helfer, *Regime Shifting: The TRIPs Agreement and New Dynamics of International Intellectual Property Lawmaking*, 29 YALE J. INT'L L. 1 (2004).

196. *See* Council for Trade-Related Aspects of Intellectual Property Rights, *The Relationship Between the TRIPS Agreement and the Convention on Biological Diversity: Summary of Issues Raised and Points Made*, IP/C/W/368 (Feb. 8, 2006).

197. UNCTAD 2012, *supra* note 149, at 87 (describing the role of the international policy agenda in Africa's growth as including the "design the international trade regime and intellectual property rights regime in a way that facilitates the sustainable development process."). *See generally id.* ch. 3.

198. *See, e.g.*, Catherine Saez, *UNCTAD Handbook: IP and the CBD Protocol on Genetic Resources*, INTELL. PROP. WATCH (Apr. 30, 2013, 11:55 AM), http://www.ip-watch.org/2013/04/30/unctad-handbook-ip-and-the-cbd-protocol-on-genetic-resources (discussing a forthcoming handbook by UNCTAD on the interface between the Nagoya Protocol (adopted under the Convention on Biological Diversity), which establishes rules to prevent misappropriation of genetic resources, and the international intellectual property system); Catherine Saez, *UNCTAD: IPRs in Health, Research, Cosmetics, Meet Access & Benefit Sharing*, INTELL. PROP. WATCH (Apr. 29, 2012, 5:32 PM) (discussing UNCTAD's Ad Hoc Working Expert Group on the Development Dimensions of Intellectual Property and the World Health

new agencies and mechanisms to address supply-side constraints of developing and least-developed countries regarding innovation to address tropical diseases, the creation of vaccines, and access to medicines.[199] Recently, WIPO has also been more active in the global debate over the role of the intellectual property system in the provision of public goods and in addressing global challenges such as climate change.[200] It purposefully, if carefully, works in partnership with other international organizations to address and analyze specific topics, promote dialogue, and suggest policy initiatives.[201] The Organization has undergone important institutional changes to mainstream these concerns into its organizational activities.[202]

It is becoming clearer that patent norms must cohere with other normative obligations of WTO member states.[203] Patent harmonization processes will not

---

Organization's Pandemic Influenza Preparedness (PIP) Framework). "The PIP Framework brings together Member States, industry, other stakeholders and WHO to implement a global approach to pandemic influenza preparedness and response. Its key goals include: to improve and strengthen the sharing of influenza viruses with human pandemic potential; and to increase the access of developing countries to vaccines and other pandemic related supplies." *Pandemic Influence Preparedness (PIP) Framework*, WORLD HEALTH ORGANIZATION (2011), http://www.who.int/influenza/pip/en. The PIP Framework was adopted in 2011 by the World Health Assembly. *Id.*

199. A leading example is the Medicines Patent Pool, an international nonprofit organization, which was spun off from the WHO affiliate UNITAID. The mission of the Pool is to "bring down the prices of HIV medicines and facilitate development of better-adapted HIV medicines, such as simplified 'fixed-dose combinations' and special formulations for children, by creating a pool of relevant patents for licensing to generic manufacturers and product development partnerships." *See Mission: Increasing Access to Affordable, Appropriate Medicines*, MEDICINES PATENT POOL, http://www.medicinespatentpool.org/who-we-are2/mission (last visited Sept. 2, 2013).

200. Abdel Latif, *supra* note 77.

201. For a recent, highly publicized example, see WORLD HEALTH ORG. (WHO), WORLD INTELLECTUAL PROP. ORG. (WIPO) & WORLD TRADE ORG. (WTO), PROMOTING ACCESS TO MEDICAL TECHNOLOGIES AND INNOVATION: INTERSECTIONS BETWEEN PUBLIC HEALTH, INTELLECTUAL PROPERTY, AND TRADE (2012), *available at* http://www.wto.org/english/res_e/booksp_e/pamtiwhowipowtoweb13_e.pdf; *see also Access to Medicines: Patent Information and Freedom to Operate: A Joint Technical Symposium by WHO, WIPO and WTO*, WIPO NEWS AND EVENTS (Feb. 18, 2011), http://www.wipo.int/meetings/en/2011/who_wipo_wto_ip_med_ge_11/program.html.

202. *See Global Challenges and Intellectual Property: Unlocking the Potential of Innovation and IP for an Equitable World*, WIPO, www.wipo.int/globalchallenges/en (last visited Sept. 2, 2013).

203. For example, the Medicines Patent Pool has negotiated several important licenses to ensure low-cost access to HIV drugs. *See* Kelly Burke, *Medicines Patent Pool, Roche Strike HIV-Related Medicine Pricing Agreement*, IP-WATCH (Aug. 5, 2013), http://www.ip-watch.org/2013/08/05/medicines-patent-pool-roche-strike-hiv-related-medicine-pricing-agreement (reporting an agreement with Roche to discount its drug, valganciclovir, at up to 90 percent cheaper for 138 countries than existing prices. The drug is an oral treatment for a viral infection that can cause blindness in people with HIV/AIDS). Roche agreed to provide valganciclovir at a discounted

attract the possible complementary force of other legal regimes when the norms advocated serve not only to forestall the delivery of global public goods, but also to undermine the normative standards upon which those regimes have been credibly built. Moreover, any international goodwill produced by efforts to address the provision of public goods outside the patent harmonization process will inure not to the benefit of the patent system but to the importance of resisting its effects.

Evidence of cooperation by international organizations and effective assistance to address challenges related to patent harmonization are important and welcome on the global stage. Such institutional arrangements, soft-law mechanisms centered on multilayered cooperation among states, international organizations, and the private sector, will likely be permanent features of the international intellectual property landscape for the foreseeable future.[204] Although these developments may have little immediate or short-term effects on the scope of the underlying patent norms that are the key focus of resistance to patent harmonization efforts, they can and already have shown signs of a gentle, slower, but insistent alteration of the international environment.

3. Harmonization and Innovation: Patents for Patent (not Trade) Policy

Despite its limitations, the Paris Convention left important policy space for countries to mediate the impact of international patent protection domestically.[205] This included flexibility to design national patent systems consistent with national priorities and economic growth strategies. By severely curtailing the options for national patent policy, the TRIPS Agreement has had the effect of reorienting national patent policy solely to accomplish foreign trade goals. The benefits of stronger global patent protection for developed countries are clear.

---

price "in an effort to scale-up the screening, diagnosis, and treatment of HIV-related CMV" in 138 developing countries. *Id.* According to the article, the Roche agreement is, unlike others negotiated by the Medicines Pool, "a pricing agreement that aims to make the treatment more readily available and affordable, creating an environment that would support generic competition...." *Id.*; *see also* Brittany Ngo, *Medicines Patent Pool Signs New Licence Agreements*, IP-WATCH (June 26, 2013), http://www.ip-watch.org/2013/06/26/medicines-patent-pool-signs-new-licence-agreements; William New, *Medicines Patent Pool Signs Deal with Indian Generics Producer*, IP-WATCH (Oct. 11, 2011), http://www.ip-watch.org/2011/10/11/medicines-patent-pool-signs-deal-with-indian-generics-producer.

204. *See, e.g.*, *Pandemic Influence Preparedness (PIP) Framework*, *supra* note 198; *Mission: Increasing Access to Affordable, Appropriate Medicines*, *supra* note 199.

205. GHC BODENHAUSEN, GUIDE TO THE APPLICATION OF THE PARIS CONVENTION FOR THE PROTECTION OF INDUSTRIAL PROPERTY 9, 70, 116 (1968) (discussing compulsory licenses and other limits to industrial property rights to address national interests); JOHN WALKER BAXTER, JOHN P. SINNOT & WILLIAM COTREAU, WORLD PATENT LAW AND

According to one report, the TRIPS Agreement alone was worth more to the United States than the possible gains from the entire Round.[206] But these are *trade* benefits. The protection-disclosure calculus that is so central to domestic patent policy in technology-exporting countries has to bear some relevance to the patent or innovation environment in the protecting country. In other words, granting patent protection for foreign patents should yield some technological *and* social benefits to the protecting country. The disconnection between the costs—both real and perceived—of foreign patent protection and the gains locally is a large part of the surprising resiliency behind the pushback to the TRIPS Agreement.

The policies needed to promote innovation and those necessary to facilitate trade may overlap, but are distinct. Patent harmonization purely for export trade with real costs to the public interest in the importing countries simply will not be sustainable in a global environment in which the "public" interest requirement for the protection of patent rights is a globally coordinated construct.[207] Where patent harmonization only secures markets in developing and least-developed countries, global patent rules, along with trade policies, impose a global division of labor that reinforces existing disparities in economic development prospects. These countries open their markets to technology-intensive products, while limiting their ability to engage in imitation, adaptation, or adoption of meaningful constraints on the exercise of patent rights in their markets.

How might harmonization for the sake of patent (and not just trade) policy occur? At a minimum, it is crucial that the patents being protected globally reflect real innovative advance. Dysfunctional patent systems in major technology-exporting countries are an important part of the resistance to harmonized patent norms. Patentability standards now incorporated in Article 28(1) of the TRIPS Agreement should be credibly applied by the patent offices of the developed countries. Each patentability criterion serves as a vector for discrete aspects of societal welfare. For example, appropriate boundaries of subject-matter

---

PRACTICE § 8 (2007) (discussing the local working requirements of section 5A of the Paris Union after TRIPS and laying out approaches used by some countries).

206. Finger, *supra* note 14 (stating that "the net gains from changed patent obligations with the gains from Uruguay Round liberalization of tariffs on industrial goods by *all* WTO Members shows that TRIPS-patents are worth *13 times* more to the US than is the Uruguay Round tariff package on industrial goods. On the other side of the ledger, for the three developing countries for which . . . estimates are available, TRIPS-patents bring increased claims *against* them several times larger than what they will gain from Uruguay Round tariff liberalization on industrial goods").

207. David Held & Anthony McGrew, *Political Globalization: Trends and Choices, in* PROVIDING GLOBAL PUBLIC GOODS 185, 187 (Inge Kaul, Pedro Conceição, Katell Le Goulven & Ronald U. Mendoza eds., 2003) ("[P]olitical communities confront a growing number of policy issues that concern them all. . . . Political decisions and actions in one part of the world can quickly have worldwide ramifications . . . . Associated with this "stretching" of

eligibility ensure that there is a public domain from which other innovators can draw basic ideas, facts, and discoveries; the non-obviousness/inventive step is generally viewed as an important standard to ensure benefits to society in the form of a new product or service to improve life and well-being.[208] Well-functioning patent systems and offices can reinforce the legitimacy of domestic patent policies and thus provide an important underpinning for normative harmonization. In short, the innovation benefits of patents in technology-exporting countries serve as an important social metric for the global public. The more skeptical of the system the domestic public in the originating jurisdiction is, the less acceptable patent harmonization will be to the global public and certainly, the domestic public in the patent-receiving jurisdiction.

In addition to connecting patent protection with local needs, the administration of the patent system itself must also reflect the socioeconomic realities of the protecting country. Where there is no constituency benefitting from patent protection, it is much easier for resistance to patent harmonization to be sustained.[209] Further, it should be possible to harmonize standards that arguably improve the efficacy of the patent system for society. For example, there could be efforts to consider harmonization on issues regarding patent quality, or to explore ways to leverage patent information disclosed in one jurisdiction to trigger appropriate scrutiny in another, rather than insist on strict adherence to the principle of independence of patents.[210] The protracted efforts of developing countries in WIPO to advance an international instrument on genetic resources with an obligation that patent applicants disclose the origin of genetic material used in the invention is an example of how new kinds of information disclosure in patent documents could simultaneously serve the legitimate interests of innovation, human rights, and development.[211] Further, developing and least-developed countries should be able to leverage important information about the degree of innovation, patent quality,

---

politics is a . . . deepening of global processes such that 'action at a distance' permeates the social conditions and cognitive worlds of specific places or policy communities.").

208. *See* Rebecca Greendyke, *No Patent for You!: How* KSR v. Teleflex*'s Nonobviousness Test Conflicts with the Scientific Method and Removes the Incentive to Innovate*, 35 U. Dayton L. Rev. 413, 431 (2010) ("The [non-obviousness] requirement exists to prevent inventors from obtaining patent protection for combination inventions that remove important building blocks from the public domain."); Benjamin N. Roin, *Unpatentable Drugs and the Standards of Patentability*, 87 Tex. L. Rev. 503, 532 (2009) (discussing the purpose of the non-obviousness requirement).

209. Finger, *supra* note 14, at 15 ("For intellectual property rights to take root, a constituency to support the reforms that the agreement demands must be built.").

210. Paris Convention, *supra* note 15, art. 4*bis*.

211. *See* World Intellectual Prop. Org., WIPO Technical study on Patent Disclosure Requirements Related to Genetic Resources and Traditional

or other concerns about a patent in assessing the relevance of such information for their own domestic environment. The point simply is that patent harmonization could be extended to matters that have important significance to the goals and objectives of patent law, thus making harmonization processes more responsive and relevant to concerns about the patent system and its benefit to society.

C. Policy Proposals

1. Domestic Public Interest and Modes of Compliance

The Paris Convention did not impose specific enforcement prerequisites beyond those required under customary international law that countries are to implement treaties in good faith and in light of their purpose and objective.[212] This rule remains salient under the TRIPS Agreement, subject to the review of an international dispute settlement process.[213] Article 1(1) of TRIPS states clearly that "Members shall be free to determine the appropriate *method* of implementing the provisions of this Agreement within their own legal system and practice."[214] This ostensible deference to national policy could be seen as a veiled constraint on the exercise of sovereign prerogative in the assessment of appropriate domestic trade-offs. Normative harmonization embodies within its structure an implicit commitment to seek convergence in both the recognition of the standard and in *how* the standard is implemented. Put differently, when normative harmonization is obtained by legal fiat, an expectation that national implementation shall be guided by the experiences (or expectations) of patent-mature economies is firmly associated with the pedigree of the formal text. This expectation undoubtedly corresponds with the degree of political and economic investment by the developed economies in the Uruguay Round negotiations.

KNOWLEDGE: STUDY NO. 3 (2004), *available at* http://www.wipo.int/export/sites/www/freepublications/en/tk/786/wipo_pub_786.pdf.

212. This rule of customary international law was codified in the Vienna Convention on the Law of Treaties. Vienna Convention on the Law of Treaties, May 23, 1979, 1155 U.N.T.S. 331; *see also* Michael A. Santoro, *Human Rights and Human Needs: Diverse Moral Principles Justifying Third World Access to Affordable HIV/AIDS Drugs*, 31 N.C. J. INT'L L. & COM. REG. 923, 925 n.8 (2006) ("There were no specific enforcement mechanisms or dispute resolution procedures under the Paris Convention.").

213. Tania Voon, *Flexibilities in WTO Law to Support Tobacco Control Regulation*, 39 AM. J.L. & MED. 199, 215 (2013) (stating that WTO dispute resolution bodies must apply customary rules of interpretation of public international law, including Articles 31 and 32 of the VCLT, and citing Appellate Body Report, *United States—Countervailing Duties on Certain Corrosion-Resistant Carbon Steel Flat Products from Germany*, ¶ 61, WT/DS213/AB/R (Nov. 28, 2002)).

214. TRIPS Agreement, *supra* note 6, art. 1(1) (emphasis added).

One WTO dispute panel has suggested that the trade-offs countries made to achieve consensus on the TRIPS Agreement are reflected in the "four corners" of the text alone and no more.[215] But later panels have expanded on this initial approach and recognized the need and possibility for additional adjustments to reflect specific national applications of the bargain between patent owners and society.[216] It is true that the TRIPS Agreement is a critical component of a political bargain exchanging stronger intellectual property protection for the developed countries and greater access for agricultural goods for the developing and least-developed countries.[217] But this grand bargain is of a fundamentally different nature than the trade-offs that are nested within and across each of the constituent agreements that constitute the Final Act of the Uruguay Round.[218] For the TRIPS Agreement, managing the net effect of its provisions to foster innovation and meet specific public welfare concerns was deferred to national implementation methods, as stated in Article 1 of the Agreement. Notwithstanding, the TRIPS Agreement created an expectation that the sheer unprecedented scope of enforceable legal obligations would be sufficient to impose or require specific modalities of compliance closer to those of the developed countries. And indeed, early cases interpreting the TRIPS Agreement illustrated the difficulty in determining when national political or economic interests could legitimately define how to implement TRIPS obligations.

The seminal dispute between the United States and India[219] over the latter's mode of implementing the "mailbox" and "pipeline" obligations in TRIPS[220] is illustrative. The WTO Appellate Body determined that India's purported implementation of its TRIPS obligations required a "sound legal basis" in Indian law.[221]

---

215. Panel Report, *India—Patent Protection for Pharmaceutical and Agricultural Chemical Products*, WT/DS50 (Sept. 5, 1997), at 17–18 (stating in the context of the doctrine of reasonable expectations that "[t]he legitimate expectations of the parties to a treaty are reflected in the language of the treaty").

216. *Canada—Pharmaceuticals*, *supra* note 25, at 154, ¶ 7.26 ("In the Panel's view, Article 30's very existence amounts to a recognition that the definition of patent rights contained in Article 28 would need certain adjustments.").

217. *See* Susan K. Sell, *Trade Issues and HIV/AIDS*, 17 Emory Int'l L. Rev. 933, 944 (2003) ("A more benign interpretation of the adoption of the TRIPS Agreement suggests a 'win-win' bargain in which developing countries acquiesced to the inclusion of the TRIPS Agreement in exchange for promises of expanded market access for their textiles and agricultural goods.").

218. *See* Daniel Gervais, The TRIPS Agreement: Drafting History and Analysis (4th ed. 2012) (reviewing the evolution of each article in TRIPS and discussing compromises made).

219. Appellate Body Report, *India—Patent Protection for Pharmaceutical and Agricultural Chemical Products*, WT/DS50/AB/R (Dec. 19, 1997) [hereinafter *India—Patent Protection*], *available at* http://www.wto.org/english/tratop_e/dispu_e/tripab.pdf.

220. *See* TRIPS Agreement, *supra* note 6, arts. 70(8)(a), 70(9).

221. *India—Patent Protection*, *supra* note 219, ¶ 58.

The "administrative instructions" India offered as its method of compliance had no clear legal provenance particularly in the teeth of mandatory provisions in India's Patent Act denying pharmaceutical patent protection.[222] As a matter of public international law, the Appellate Body made clear that questions of compliance with international obligations are subject to evaluation by the appropriately designated body.[223] But in stating correctly that no WTO member country can be the sole arbiter of whether its laws are compliant with international obligations,[224] the Appellate Body conflated substance and *method* of implementation, arguably bringing the latter for the first time within the reach of global normative harmonization.

2. Improving Domestic Institutions to Promote Innovation

If patent policy is used for innovation and not just trade, harmonization efforts could encourage the development of local institutions to support innovation. As suggested in Section I above, domestic institutions are at the forefront of defining compliance with or resistance to global patent norms. Often, the lines between the two are indistinguishable, particularly where national public interest arguably is best served by defending local industries, in clear violation of the free-trade ethos that dominates contemporary harmonization efforts. Recently, the Obama administration vetoed an import ban issued by the ITC on certain Apple products after finding that the products infringed on certain Samsung standard essential patents (SEPs) patents.[225] The executive branch defended its rare exercise of veto power, inter alia, on the grounds that the patent holders were gaining undue advantage in the U.S. market and cited "potential harm to consumers and competitive conditions in the U.S. economy."[226] Although this particular calculus smacks of protectionism or, worse, discrimination, it arguably is consistent with the long-standing impulse to prioritize domestic inventors

222. *Id.* ¶¶. 66–69.

223. *Id.* (holding that Article 1.1 of TRIPs allows member nations to "determine the appropriate method of implementing the provisions of this Agreement" in the context of their domestic legal systems); J.H. Reichman, *The TRIPs Agreement Comes of Age: Conflict or Cooperation with Developing Countries?*, 32 CASE W. RES. J. INT'L L. 441, 446 (2000) (explaining that "the Appellate Body also manifested considerable deference to local law").

224. *India–Patent Protection*, *supra* note 219, ¶ 58.

225. Letter from Michael B.G. Froman, *supra* note 135. Injunctive relief for SEPs is controversial, and complicates the characterization of the President's action as protectionist.

226. Brent Kendall & Ian Sherr, *Obama Administration Vetoes Ban on Sale of Some Apple iPhones, iPads*, WALL ST. J., Aug. 4, 2013, http://online.wsj.com/article/SB10001424127887324136204578646192008412934.html.

and/or domestic industry. Although commentators have pointed to the tension between such administrative actions and the United States' leadership in patent harmonization, such actions are not facially inconsistent with TRIPS, and are not entirely incongruent with the historic role of patent protection. Although a court decision may ultimately facilitate a more even-handed outcome, the point remains that a diversity of institutional routes to protect patent rights also offers a diversity of means to recalibrate national interests and international obligations.

Decisions regarding what patent-related actions are necessary to promote or protect the public interest are most defensible when competency is spread across competent institutions. In their absence, the social cost and risks of implementation of global patent norms are materially amplified. The situation is particularly acute in developing and least-developed countries where wholesale imitation of the patent laws of the developed countries often constitutes the chief form of compliance with the TRIPS Agreement.[227] Such copycat methods are the result of a complex set of factors that include direct political pressure from the developed countries, the shrewd and pervasive norm-setting activities of WIPO working in tandem with national and regional patent offices,[228] and the limited institutional capacity of many least-developed countries. These calculated efforts to embed homogeneity in the domestic patent laws of developing and least-developed countries take place in the absence of strong domestic institutions or local constituencies to support norm transplantation. These conditions heighten the political pressure to resist patent harmonization, and fuel concerns that globalized norms wreak havoc on societal interests. In the absence of domestic institutions with credibility to translate global norms into local realities, domestic compliance with harmonized norms will always be debated, contested, and resisted.

To the extent the primary interest in normative harmonization is stability[229] and a degree of predictability in the application of shared patent norms, the primary basis for harmonization should not be facile normative convergence, but rather consideration of what principal objectives the patent system must accomplish to justify compelling trade-offs at the national level. The international patent system requires a welfare balance *of its own* in relation to other international regimes. At the same time, it must be sufficiently flexible to

227. Carolyn Deere, The Implementation Game: The TRIPS Agreement and the Global Politics of Intellectual Property Reform in Developing Countries 37–38 (2008); *see also* WIPO Model Law for Developing Countries or Inventions, Volume 1: Patents (1979), *available at* http://keionline.org/sites/default/files/Model_law_patents_vol1_1979.pdf.

228. *See generally* Deere, *supra* note 227.

229. Ladas, *supra* note 83, at 15.

facilitate independent analysis by countries and cannot impose its own welfare calculus on sovereign states, whether developed or developing. Arguably, this is why harmonization efforts in all intellectual property regimes are characterized by a focus on enumerated minimum rights, the interpretation and application of which are left to national institutions.[230] In the same regard, determinations about appropriate limitations and exceptions to patent rights are made principally at the domestic level, subject to Article 30 of the TRIPS Agreement. But limitations and exceptions, no matter how well-designed, cannot in themselves shoulder the burden of advancing public welfare through the patent system or of curtailing its most harmful effects. The real challenge is that resistance to patent norms is linked to welfare considerations that are themselves dynamic. A shifting calculus of national interests is inevitable as technological frontiers evolve and produce new technologies, new industries, and new actors. To the extent that the domestic welfare calculus is dynamic and requires accommodation in the global patent system, specific instances of unpredictable resistance to the effects of global patent law norms may be a fixture of the global system for the foreseeable future.

The international patent system cannot credibly impose its welfare calculus on participating sovereign countries, but it can and should help inform how select trade-offs can enhance dynamic welfare gains by ensuring that the patent system is responsive to and constrained by other policy considerations relevant to specific national contexts. This suggests that *national* patent law and policy must have appropriate "freedom to operate" consistent with local development needs and goals. In the alternative, domestic patent policy will continuously face pressure to address national needs and interests in less-transparent ways. The absence of a minimum set of standards that mediate the costs of patent protection and assure that patent grants reflect effective applications of inbuilt balancing mechanisms is a significant cause of pervasive resistance to patent harmonization. In essence, just like the implementation of domestic patent norms, the strength of the social and cultural context in which institutional competence over patent law is exercised is the best vector for a more authentic effort to correlate the implementation of harmonized norms with concerns related to the public interest.[231]

---

230. A similar structure exists in international copyright agreements. *See* Ruth L. Okediji, *The International Copyright System: Limitations, Exceptions and Public Interest Considerations for Developing Countries* 30-31 (ITCSD Issue Paper No. 15, 2006).

231. Shobita Parthasarathy & Alexis Walker, *Observing the Patent System in Social and Political Perspective: A Case Study of Europe, in* PATENT LAW IN GLOBAL PERSPECTIVE ch. 11 (Ruth L. Okediji & Margo Bagley eds., 2013).

Conclusion

An inevitable consequence of living in a globalized political ecosystem is the ease with which "publics" are galvanized around common causes. Continual post-TRIPS resistance to the scope of pharmaceutical patent protection, and unremitting outrage over patent-related barriers to access to essential drugs are cogent examples of the consequences of such indifference, as was the rapid demise of imperial intellectual property efforts such as ACTA and SOPA. Patent harmonization efforts must adopt a canon that can credibly respond to the reality and power of coordinated efforts to resist the patent system and to help manage the interplay between general economic policies and the exclusive privileges granted to inventors.

The international patent framework is insufficiently equipped with mechanisms for addressing patent-related regulatory failures; its capacity to influence development-related patent considerations is still debatable; and it is unclear whether it possesses the necessary agility to adjust its rules, institutions, and processes to deal meaningfully with global challenges engendered by the implementation and exercise of patent rights. Yet, patent harmonization continues to proceed in regional, bilateral, and multilateral fora. Resistance to patent harmonization also continues to manifest in unpredictable and seemingly unrelenting forms. Challenges to patent harmonization arise from national and international constituencies joined by a common experience of the role of the patent system in the under-supply of public goods, or a concern over the long-term effects of patent protection on public welfare.

The theories that justify national patent law cannot easily be extrapolated to the international patent system. Yet, harmonization has a tendency to penalize the practical exercise of state power to serve its citizens differently from the terms of a treaty. The traditional justifications for patent protection have become less compelling in the countries that most favor strong patent harmonization, and far less credible in those that oppose it. The very impulses that required recognition of unfair advantages for countries that did not offer adequate patent protection have also engendered an approach to the global public interest organized around resistance to the negative effects of patent rights. This nascent construct of the global public interest is the sum of perceived and actual considerations to which the patent system must be accountable in terms of what positive gains it delivers to the public, what costs are exacted in return, and what the processes are by which those global norms are created and extended. Future patent harmonization cannot effectively proceed without a deliberate engagement of the various elements of a proposed global canon consisting of public goods issues, coherence with other international law regimes, and ensurance of a connection between what patents accomplish in specific national contexts and what they impose in different institutional and policy settings. To successfully proceed in the future, patent harmonization simply cannot afford to ignore the gaps in the current system.

GLOBAL PATENT LAW AND THE POLITICAL ECONOMY OF HARMONIZATION

2 INTELLECTUAL PROPERTY LAWMAKING, GLOBAL GOVERNANCE, AND EMERGING ECONOMIES

Rochelle C. Dreyfuss*

Introduction

It is no secret that the Agreement on Trade-Related Aspects of Intellectual Property Rights (TRIPS Agreement) is fraying around the edges.[1] Developing countries feel shortchanged. Although the Agreement's patent, copyright, and trade secrecy requirements raise the cost of, and impede access to, the information products generated in the North, TRIPS fails to offer any protection to the traditional knowledge, genetic resources, and folklore that constitute much of the intellectual wealth of the South. Yet to materialize are the Agreement's vaulted promises: technical assistance in writing TRIPS-compatible legislation that is adapted to developing countries' needs, and foreign investments that can help these countries prosper.

Developed countries are, in some ways, equally stymied. The technological landscape is changing in ways that necessitate revision of the intellectual property system. In some cases, higher levels of protection are required (the ease of digital distribution is, for example, challenging territorially based enforcement mechanisms). In other situations, reductions in existing obligations may be desirable (for instance, the

---

* Pauline Newman Professor of Law, New York University School of Law. This chapter is a revised and updated version of a paper written for the conference on Globalization and Legal Governance convened jointly by NYU Law School's Global Administrative Law Project and Tsinghua Law School in Beijing, May 22–23, 2009. The author wishes to thank Darryl Tam for his research assistance and the Filomen D'Agostino and Max E. Greenberg Research Fund of NYU for financial support.

1. Agreement on Trade-Related Aspects of Intellectual Property Rights, Apr. 15, 1994, 33 I.L.M. 1197 (1994) [hereinafter TRIPS Agreement].

same digital technologies facilitate forms of open innovation that are less dependent on intellectual property protection). However, the politics of change are complex, and the duty to conform to TRIPS makes it even harder to respond effectively to new domestic realities. TRIPS was largely written at the behest of the owners of standard forms of intellectual property (patents, copyrights, trademarks);[2] it creates rights for producers, but says little about the rights of users.[3]

Although TRIPS could arguably be altered to deal with the difficulties that the World Trade Organization (WTO) members are experiencing, lawmaking within the WTO is at an impasse. In other parts of the WTO, adjudication by the Dispute Settlement Body (DSB)[4] is taking up the slack. But successive panels (and in some instances, the Appellate Body) have interpreted the TRIPS Agreement so rigidly, member states have been left with insufficient room to maneuver, especially with regard to improving access.[5] For problems that require a coordinated solution, many members of the WTO are beginning to give up on the trade framework. In Larry Helfer's words, they are shifting to other regimes,[6] including regional and bilateral arrangements;[7] instruments administered by UN-related entities, such as the World Health Organization (WHO) and particularly the World Intellectual Property Organization (WIPO)[8] human rights

---

2. *See, e.g.*, SUSAN K. SELL, PRIVATE POWER, PUBLIC LAW: THE GLOBALIZATION OF INTELLECTUAL PROPERTY RIGHTS (2003).

3. *See, e.g.*, Henning Grosse Ruse-Khan & Annette Kur, *Enough Is Enough—The Notion of Binding Ceilings*, *in* INTELLECTUAL PROPERTY RIGHTS IN A FAIR WORLD TRADE SYSTEM (Annette Kur & Marianne Levin eds., 2011).

4. Understanding on Rules and Procedures Governing the Settlement of Disputes, Apr. 15, 1994, Marrakesh Agreement Establishing the World Trade Organization, Annex 2, 33 I.L.M. 1226 (1994) [hereinafter DSU].

5. *See, e.g.*, Graeme B. Dinwoodie & Rochelle C. Dreyfuss, *Diversifying Without Discriminating: Complying with the Mandates of the TRIPS Agreement*, 13 MICH. TELECOMM. & TECH. L. REV. 445 (2007) [hereinafter Dinwoodie & Dreyfuss, *Diversifying*]; Graeme B. Dinwoodie & Rochelle C. Dreyfuss, *International Intellectual Property Law and the Public Domain of Science*, 7 J. INT'L ECON. L. 431 (2004) [hereinafter Dinwoodie & Dreyfuss, *Public Domain*]; Graeme B. Dinwoodie & Rochelle C. Dreyfuss, *TRIPS and the Dynamics of International Property Lawmaking*, 36 CASE W. RES. J. INT'L L. 95 (2004) [hereinafter Dinwoodie & Dreyfuss, *Dynamics*].

6. Laurence R. Helfer, *Regime Shifting: The TRIPS Agreement and New Dynamics of International Intellectual Property Lawmaking*, 29 YALE J. INT'L L. 1, 20–22 (2004).

7. Timothy P. Trainer, *Intellectual Property Enforcement: A Reality Gap (Insufficient Assistance, Ineffective Implementation)?*, 8 J. MARSHALL REV. INTELL. PROP. L. 47 (2008); Office of the United States Trade Representative, Trans-Pacific Partnership (TPP): 13th Round, *available at* http://www.ustr.gov/tpp (lasted visited Sept. 5, 2013) (listing, also, all US free trade agreements).

8. *See generally* Graeme B. Dinwoodie & Rochelle C. Dreyfuss, *Designing a Global Intellectual Property System Responsive to Change: The WTO, WIPO, and Beyond*, 46 HOUS. L. REV. 1187 (2009).

agreements;[9] negotiations over the Convention on Biological Diversity;[10] the Anti-Counterfeiting Trade Act, and more recently, the Trans Pacific Partnership Agreement.[11] The result is cacophony: thickets of rights,[12] conflicting demands,[13] disputes that perpetually cycle,[14] and new uncertainties.[15]

In this environment, emerging economies such as India, Brazil, and China may well hold the key to the future.[16] On the one hand, significant numbers of residents in these countries suffer from the same access problems experienced by less-developed economies. On the other hand, these nations have growing creative sectors that are beginning to enjoy the benefits of strong intellectual property protection. Because countries in this "emerging Middle" internalize the problems of both the North and the South, they are forced to acquire the political will needed to accommodate the demands of all kinds of intellectual property producers and

---

9. Laurence R. Helfer, *The New Innovation Frontier? Intellectual Property and the European Court of Human Rights*, 49 HARV. INT'L L.J. 1 (2008).

10. Convention on Biological Diversity (CBD) art. 15(1), June 5, 1992, 31 I.L.M. 818 (1992).

11. Anti-Counterfeiting Trade Agreement, Dec. 3, 2010, 50 I.L.M 243 (2011), *available at* http://www.mofa.go.jp/policy/economy/i_property/pdfs/acta1105_en.pdf; Peter K. Yu, *Six Secret (and Now Open) Fears of ACTA*, 64 SMU L. REV. 975 (2011).

12. *See, e.g.*, Sabrina Safrin, *Hyperownership in a Time of Biotechnological Promise: The International Conflict to Control the Building Blocks of Life*, 98 AM. J. INT'L L. 641, 646–58 (2004) (describing overlap between rights protected by TRIPS and the CBD).

13. *See, e.g.*, Panel Report, *European Communities—Protection of Trademarks and Geographical Indications for Agricultural Products and Foodstuffs*, WT/DS174/R (Mar. 15, 2005) [hereinafter *EC-GI*].

14. *See, e.g.*, Case C-482/09, Budějovický Budvar v. Anheuser-Busch, Inc., 2011 E.C.R. I-8701; Case T-225/06, Budějovický Budvar v. OHIM—Anheuser-Busch, 2008 E.C.R. II-3555; Anheuser-Busch Inc. v. Portugal, App. No. 73049/01, 2007-I Eur. Ct. H.R. 39 (2007); Case C-245/02, Anheuser-Busch Inc. v. Budějovický Budvar, 2004 ECR I-10989 Anheuser-Busch v. Budějovický Budvar [1984] F.S.R. 413 (Ch. App.) (U.K.).

15. *See* Decision by the Arbitrators, *European Communities—Regime for the Importation, Sale and Distribution of Bananas—Recourse to Arbitration by the European Communities Under Article 22.6 of the DSU*, WT/DS27/ARB/ECU (Mar. 24, 2000) [hereinafter *EC—Bananas*] (suggesting that it was not within the jurisdiction of WTO arbitrators to assess whether measures authorized under the WTO Agreements might result in noncompliance with obligations under WIPO Conventions).

16. *See generally* Daniel Benoliel & Bruno Salama, *Towards an Intellectual Property Bargaining Theory: The Post-WTO Era*, 32 U. PA. J. INT'L L. 265 (2010) (discussing the East Asian "tigers"); Jerome H. Reichman, *Intellectual Property in the Twenty-First Century: Will the Developing Countries Lead or Follow?*, 46 HOUS. L. REV. 1115 (2009); Gregory Shaffer, Michelle Ratton Sanchez & Barbara Rosenberg, *The Trials of Winning at the WTO: What Lies behind Brazil's Success*, 41 CORNELL INT'L L.J. 383 (2008); Peter K. Yu, *Access to Medicines, BRICS Alliances, and Collective Action*, 34 AM. J.L. & MED. 345 (2008).

users—that is, to strike the types of delicate balances that are currently eluding policymakers elsewhere. Equally important, these countries have a thick legal and political culture and can ably defend their domestic legislation in international circles. As emerging economies move into a leadership position in establishing new practices and advancing their pro-access views, they are sure to challenge the preeminent role of the North in setting world norms for intellectual property protection.

The engagement of emerging economies in international intellectual property lawmaking could produce another key benefit. As power has shifted away from states to international organizations, such as the World Bank, the International Monetary Fund, and the WTO, and even to nongovernment actors, such as the Basel Committee on Banking Supervision, the legitimacy of lawmaking at the global level has become a target of increasing concern. Many question the legitimacy of norms imposed by nondemocratic, unaccountable entities;[17] others, as noted previously, are worried about the effects of multiple overlapping centers of norm development.[18] Among these "global discontents," emerging nations are in a unique position. Because they represent the demands of intellectual property consumers as much as they do intellectual property creators, their goals are best served by partnering with groups currently underrepresented in the international arena, including least-developed countries; nongovernmental organizations involved in areas such as healthcare, where access to protected works is key; and other members of civil society interested in challenging the current primacy of the holders of standard intellectual property rights. Thus, while furthering their own interests in access to knowledge products, emerging economies may also work to improve another type of access—the ability to participate in a fair and transparent process of international lawmaking. Furthermore, so long as they take consistent positions at all negotiations over intellectual property instruments, emerging economies could contribute to the harmonization and integration of international norms, and thereby reduce the problems associated with regulatory overlap. These nations could, in other words, contribute in an important way to the development of governance reforms, such as those that are the cornerstone of the global administrative law agenda.[19]

This chapter begins, in Section I, by describing the problems currently plaguing intellectual property law for countries at both extremes of the development spectrum. Section II then examines the solutions both domestic and international, that are emerging from the Middle. Section III considers how the

---

17. *See, e.g.*, Robert O. Keohane, *Global Governance and Democratic Accountability*, *in* TAMING GLOBALIZATION: FRONTIERS OF GOVERNANCE 130 (2003); Michael S. Barr & Geoffrey P. Miller, *Global Administrative Law: The View from Basel*, 10 EUR. J. INT'L L. 15 (2006).

18. *See supra* notes 6–8.

19. *See* Benedict Kingsbury, Nico Krisch & Richard B. Stewart, *The Emergence of Global Administrative Law*, 68 LAW & CONTEMP. PROBS, Summer 2005, at 15.

participation of these countries in international negotiations over intellectual property can influence global lawmaking norms.

I. The Problem

Post the Uruguay Round, the intellectual property system has been experiencing serious difficulties. The TRIPS Agreement sets out minimum standards of protection and enforcement obligations for rights long-protected by multilateral instruments currently administered by WIPO, including patents and trademarks, which remain the subjects of the Paris Convention,[20] and copyrights, which are covered by the Berne Convention.[21] TRIPS also codifies commitments to protect certain intellectual properties that previously received less strong international recognition, including geographical indications, sound recordings, industrial designs, trade secrets, and integrated circuits.[22] And because it is a part of the WTO framework, TRIPS introduces a compliance mechanism into the international intellectual property realm: it makes the Agreement subject to the new, judicialized dispute settlement provisions of the WTO.[23] To accommodate the interests of countries new to these obligations, the Agreement staggers the dates for full compliance in accordance with levels of development.[24] It also permits limited exceptions and provides a framework for compulsory patent licensing.[25] In addition, the Agreement promises technical assistance[26] and, for the

---

20. Paris Convention for the Protection of Industrial Property, Mar. 20, 1883, as revised at Stockholm, July 14, 1967, 21 U.S.T. 1583, 828 U.N.T.S. 305 [hereinafter Paris Convention]; TRIPS Agreement, *supra* note 1, arts. 16–21, 27–33.

21. Berne Convention for the Protection of Literary and Artistic Works, Sept. 9, 1886, as revised at Paris on July 24, 1971, and amended on Sept. 29, 1979, S. Treaty Doc. No. 99-27, 1161 U.N.T.S. 3 [hereinafter Berne Convention]; TRIPS Agreement, *supra* note 1, arts. 9–13.

22. TRIPS Agreement, *supra* note 1, arts. 22–24, 14, 25–26, 39, 35–38. The enforcement obligations are found in arts. 41–61.

23. *Id.* art. 64.

24. *Id.* art. 65. Under the original timetable, the transition periods would be over. However, they have been extended for least-developed countries, Decision of the Council for TRIPS of 27 June 2002, *Extension of the Transition Period under Article 66.1 of the TRIPS Agreement for Least-Developed Country Members for Certain Obligations with Respect to Pharmaceutical Products*, IP/C/25 (July 1, 2002); Decision of the Council for TRIPS of 29 November 2005, *Extension of the Transition Period under Article 66.1 for Least-Developed Country Members*, IP/C/40 (Nov. 30, 2005); and Decision of the Council for TRIPS of 11 June 2013, Extension of the Transition Period under Article 66.1 of the TRIPS Agreement for Least-Developed Country Members, IP/C/64 (Jun. 12, 2013).

25. TRIPS Agreement, *supra* note 1, arts. 13, 17, 26.2, 30, 31, 41–61.

26. *Id.* art. 67.

least developed, technology transfer.[27] Nevertheless, almost immediately after the Agreement went into force, troubles emerged on both sides of the development divide.

A. Developing Countries

No sooner was the ink dry on the TRIPS Agreement than it became clear that developing countries had struck a bad deal, for the trade-off to which they had agreed—access to foreign markets in exchange for raising intellectual property levels—turned out to be something of a losing proposition. The profits available on commodities do not offset the supracompetitive prices charged for protected knowledge products. As a result, the fruits of contemporary innovation efforts are beyond the reach of most of the population of these states.[28] Without strong creative industries of their own and absent any recognition for the information wealth of the South (genetic resources, folklore, traditional knowledge), many nations have found it difficult to persuade their citizens to enact implementing legislation in the time periods set out in the TRIPS Agreement. Furthermore, many of these countries had expected that TRIPS would represent a ceiling to demands by the North—that TRIPS would, in effect, replace the unilateral trade sanctions previously used by the United States to induce countries to strongly protect intellectual property within their borders.[29] In reality, however, these demands did not cease when TRIPS went into force. Instead, developed countries have imposed "TRIPS-plus" requirements through unilateral threats of trade sanctions and through bilateral negotiations, for instance, over free trade agreements (FTAs), European Partnership Agreements (EPAs), and bilateral investment treaties (BITs).[30]

---

27. *Id.* art. 66.2.

28. *See, e.g.*, Jerome. H. Reichman & Rochelle Cooper Dreyfuss, *Harmonization without Consensus: Critical Reflections on Drafting a Substantive Patent Law Treaty*, 57 DUKE L.J. 85 (2007).

29. *See* DANIEL GERVAIS, THE TRIPS AGREEMENT: DRAFTING HISTORY AND ANALYSIS 12-2 (3d ed. 2008); *see also* ANDREAS LOWENFELD, INTERNATIONAL ECONOMIC LAW, § 8.8 (2d ed. 2008); JOHN CROOME, RESHAPING THE WORLD TRADE SYSTEM: A HISTORY OF THE URUGUAY ROUND 113 (1998); *cf.* Joel P. Trachtman, *The Constitutions of the WTO*, 17 EUR. J. INT'L L. 623, 633 (2006) (noting that the United States agreed to give up unilateralism under Section 301 of its Trade Act in exchange for a stronger dispute settlement system).

30. *See, e.g.*, Henning Grosse Ruse-Khan, *Protecting Intellectual Property under BITs, FTAs, and TRIPS: Conflicting Regimes or Mutual Coherence*, *in* EVOLUTION IN INVESTMENT TREATY LAW AND ARBITRATION (Kate Miles & Chester Brown eds., 2011); Peter Drahos, *BITS and BIPS: Bilateralism in Intellectual Property*, 4 J. WORLD INTELL. PROP. 791 (2001). As Drahos notes, the United States also raises the stakes through Section 301 of its Trade Act, 19 U.S.C. § 2411. *See* Drahos, *supra*, at 791; *see also* AXEL BORRMANN, FRIEDRICH-EBERT-STIFTUNG ET AL., THE IMPACT OF ACP/EU ECONOMIC PARTNERSHIP AGREEMENTS ON ECOWAS

These problems are, by now, well recognized and the focus of considerable international consternation.[31] For these purposes, the real question is causation: Why has it proved so difficult for countries to take advantage of the accommodations TRIPS provided, step to the technological frontier, and enjoy the benefits of strong intellectual property protection? Why have countries so easily succumbed to obligations exceeding TRIPS minimum standards? Part of the answer is surely the absence of promised foreign investments as well as technical and legal assistance. More poignantly, however, it appears that the drafters of TRIPS seriously underestimated the obstacles inherent in the development process. As economists know, information is sticky—it is not easily transmitted or learned. Without sufficient technological infrastructure and absorptive capacity, the promises of TRIPS cannot be easily realized.

The essential medicines problem provides a dramatic illustration. TRIPS negotiators actually considered the possibility that patents would pose an access problem. Article 31 of the Agreement therefore provides members with the power to license third parties in situations where the patent holder abuses its position or otherwise fails to adequately supply the market. The provision, however, includes a limitation: Article 31(f) mandates that "any such use shall be authorized predominantly for the supply of the domestic market of the Member authorizing such use." For states that can manufacture their own pharmaceuticals, this provision works as envisioned.[32] But many members lack the capacity to manufacture drugs at the level of purity required for medical use and cannot, therefore, consistent with this obligation cope with the health problems of their populations.

In the Doha Round, that problem was alleviated in the short run by waiving Article 31(f) for the production of medicines, subject to a series of requirements that protects right holders from the spread of cheap pharmaceuticals into the North. In the long run, the TRIPS Agreement is to be amended to include a similar set of conditions.[33] Indeed, the Ministerial Conference used the occasion of granting this dispensation to stress the WTO's commitment to development and to allow members to engage in interest-balancing.[34] Nonetheless, the WTO missed the underlying

---

Countries: An Empirical Analysis of the Trade and Budget Effects: Final Report (2004).

31. *See, e.g.*, Amy Kapczynski, *The Access to Knowledge Mobilization and the New Politics of Intellectual Property*, 117 Yale L.J. 804 (2008); Reichman & Dreyfuss, *supra* note 28.

32. Brazil and Thailand have both issued (or threatened to issue) compulsory licenses to reduce the cost of healthcare. *See* Robert C. Bird & Daniel R. Cahoy, *The Emerging BRIC Economies: Lessons from Intellectual Property Negotiation and Enforcement*, 5 Nw. J. Tech. & Intell. Prop. 400 (2007).

33. TRIPS Agreement, *supra* note 1, art. 31bis.

34. World Trade Organization, Ministerial Declaration of 14 November 2001, WT/MIN(01)/DEC/1, 41 I.L.M. 746 (2002) [hereinafter Doha Declaration], *available at*

lesson: without a sufficiently strong technological infrastructure, developing countries will continue to have trouble capturing the benefits of TRIPS. They cannot become net producers of intellectual property until they establish creative industries that operate at the intellectual frontier. And they cannot reach that point without training and education. As Maggie Chon has argued, educational material must be accessible,[35] and as Jerry Reichman has suggested, training opportunities are equally critical.[36] To learn from the technological assistance received and become creative in its own right, a country needs to engage in "fair following." The essential medicines crisis produced by Article 31(f) highlighted the infrastructure problem, but by only providing for access, it failed to solve the long-term problem of generating absorptive capacity.

To be sure, TRIPS appears to include the flexibilities members need to make education and training opportunities available. For copyrighted materials, TRIPS incorporates the Berne Convention, which includes a fair amount of room to maneuver: it leaves the concepts of a "work of authorship," "originality," and the idea/expression dichotomy to national legislation. It also includes an Appendix that empowers member states to create local translations of works when copyright holders do not voluntarily meet their needs.[37] TRIPS also carries over and expands on an exceptions clause found in the Berne Convention. Thus, it permits members to enact exemptions in "special cases," if they do not "conflict with normal exploitation" or "unreasonably prejudice legitimate interests."[38] On the patent side, the height of the inventive step is not specified,[39] the terms "making, using, offering for sale, selling or importing"[40] are not defined,

---

http://www.wto.org/english/thewto_e/minist_e/min01_e/mindecl_e.htm; World Trade Organization, Declaration on the TRIPS Agreement and Public Health, WT/MIN(01)/DEC/2 (2001), *available at* http://www.wto.org/english/thewto_e/minist_e/min01_e/mindecl_trips_e.htm. *See generally* Frederick M. Abbott & Jerome H. Reichman, *The Doha Round's Public Health Legacy: Strategies for the Production and Diffusion of Patented Medicines under the Amended TRIPS Provisions*, 10 J. Int'l Econ. L. 921 (2007).

35. *See* Margaret Chon, *Intellectual Property "From Below": Copyright and Capability for Education*, 40 U.C. Davis L. Rev. 803 (2007) [hereinafter Chon, *From Below*]; Margaret Chon, *Intellectual Property and the Development Divide*, 27 Cardozo L. Rev. 2821 (2006).

36. Reichman, *supra* note 16.

37. TRIPS Agreement, *supra* note 1, art. 9.1; Berne Convention, *supra* note 21 (Special Provisions Regarding Developing Countries).

38. TRIPS Agreement, *supra* note 1, art. 13. Berne Convention, *supra* note 21, art. 9(2), reads slightly differently and covers only the reproduction right. *See generally* P. Bernt Hugenholtz & Ruth L. Okediji, *Conceiving an International Instrument on Limitations and Exceptions to Copyright: Final Report* (2008), *available at* http://www.ivir.nl/publicaties/hugenholtz/final-report2008.pdf.

39. TRIPS Agreement, *supra* note 1, art. 27.1.

40. *Id.* art. 28.

and whole classes of inventions can be excluded from patenting requirements.[41] There are also the aforementioned compulsory licensing opportunities and an "exceptions" provision similar to the one available for copyright.[42] Astute lawyers should be able to utilize these flexibilities to design domestic intellectual property systems that allow the populace to hone their creative and technological skills, and perhaps even work on incremental improvements that meet other local needs.[43]

The rub, however, is the need for astute lawyering. It is not only information necessary to establish creative industries that is sticky; so too is the knowledge needed to foster a legal community capable of utilizing the Agreement's flexibilities effectively. For example, both the Berne Appendix and the accommodation on medicines worked out in the Doha Round are underutilized.[44] The problem appears to be that these provisions interpose significant hoops, including the need to issue precise declarations, notifications, and regulations, along with a commitment to continued monitoring. Without lawyers to oversee implementation, these restrictions can quickly turn into insurmountable obstacles. Furthermore, neither the Appendix nor the Doha waiver/amendment was drafted in a manner conducive to use. Both, for example, give the right holder the power to end the conditions under which the third party is permitted to operate. It is difficult to imagine that anyone would be willing to translate works or gear up to supply drugs if the right holder can simply march in as soon as it realizes there is a viable market, and obliterate the investment. With a stronger legal culture, developing countries would likely have insisted on accommodations that avoid these problems.[45]

Developing countries face an equivalent problem with respect to their domestic legislation. The best strategy for creating educational and training opportunities is far from evident; good legal (and economic) skills are needed to figure out the

---

41. *Id.* arts. 27.2–27.3.

42. The patents provision permits members to enact "limited" exemptions so long as they "do not unreasonably conflict with a normal exploitation," and "do not unreasonably prejudice the legitimate interest of the patent owner, taking account of the legitimate interests of third parties." TRIPS Agreement, *supra* note 1, art. 30. There are similar exceptions provisions for trademarks (art. 17) and industrial designs (art. 26.2).

43. Rochelle C. Dreyfuss, *Fostering Dynamic Innovation, Development and Trade: Intellectual Property as a Case Study in Global Administrative Law*, 2009 Acta Juridica 237.

44. *See, e.g.*, Jillian C. Cohen-Kohler, *The Morally Uncomfortable Global Drug Gap*, 82 Clinical Pharmacology & Therapeutics 610 (2007).

45. Girish R. Kumar, *South–South Cooperation Strategy in Dealing with the Identified Common Challenges in Intellectual Property* (Feb. 19, 2009), *available at* http://ssrn.com/abstract=1346304.

appropriate rules.[46] Furthermore, any departure from the intellectual property regime prevalent in the North could attract a challenge in the WTO. It takes a sophisticated and well-funded legal team to bargain for a waiver or defend national laws in the TRIPS Council, before a dispute resolution panel, or at the Appellate Body. Making matters even worse, to date, few disputes reaching the DSB have involved developing countries and none have involved least-developed nations. Accordingly, there has been little opportunity to give these countries guidance on the sorts of moves they can safely regard as compatible with international obligations.

A developing nation that can "lawyer up" would be at a real advantage not only in regard to drafting and defending legislation, but in that it could also be proactive: it could insist on renegotiating the Agreement to deal with its development problems. It could work to revamp the exceptions provisions and add standards to protect the users (and not just the producers) of intellectual property. It could even bring its own challenges in the DSB and require developed countries to live up to their express obligations to engage in technology transfer. Indeed, Susy Frankel suggests that WTO members might even consider challenging unilateral attempts to impose TRIPS-plus obligations as nonviolation complaints.[47]

B. Developed Countries

Admittedly, TRIPS has had much more positive effects in the North, where many creative sectors are benefiting from the added revenues available on world markets. But even here, the dynamics of information production are changing in ways that make revision of both national laws and the TRIPS Agreement desirable. The soaring cost of medicine, which is making patient access in the North almost as difficult as it is in the South, has led experts to question the operation of the patent system. In the life sciences, there is concern that patenting has moved too far upstream. Patents over genes, protein sequences, diagnostic correlations, and other fundamental knowledge allow patentees to dominate broad technological opportunities.[48] The thickets of rights produced are difficult and costly to

---

46. *See* Reichman & Dreyfuss, *supra* note 28 (noting that it is possible to devise regimes that promote local training by either raising or lowering the height of the inventive step).

47. Susy Frankel, *Challenging TRIPS-Plus FTAs: The Potential Utility of Non-violation Complaints*, 12 J. Int'l Econ. L. 1023 (2009).

48. *See, e.g.*, Nat'l Research Council of the Nat'l Academies, Reaping the Benefits of Genomic and Proteomic Research: Intellectual Property Rights, Innovation, and Public Health (2006); Sec'y's Advisory Comm. on Genetics, Health and Soc'y, *Report on Gene Patents and Licensing Practices and Their Impact on Patient Access to Genetic Tests, available at* http://oba.od.nih.gov/oba/sacghs/reports/SACGHS_patents_report_2010.pdf. Compare these materials to Justice Breyer's dissent from the dismissal of certiorari in *Laboratory Corp. of America Holdings v. Metabolite Laboratories, Inc.*, 548 U.S.

negotiate. Furthermore, through such measures as the United States' Bayh-Dole Act (which has attracted imitators),[49] universities have become players in the patent enterprise.[50] Their involvement in the commercialization of basic science has led to the patenting of information that would have formerly gone into the public domain and has, in the United States, prompted courts to restrict defenses to infringement that formerly eased the need for costly negotiations.[51]

Outside the life sciences, there are also many concerns about overprotection. In the information technology industry, the patent-to-product ratio is extremely high, meaning that products often involve many features that are separately patented and that are meant to inter-operate with other products that are also subject to multiple patent rights. With many patents required to discover and make novel products, transaction costs rise, as does the probability of holdouts. Compounding the problem, many of the players in this sector find that they must acquire patents for defensive purposes, leading to a situation where patents can cost more to acquire and negotiate than they generate in new revenue.[52]

In the copyright sector, there are also many who question the need for strong protection. As the cost of reproducing works has declined through digitization, the need to provide incentives to intermediaries, such as publishers and recording companies, has diminished.[53] The success of Linux and other open-source and open-data initiatives, along with the work of Eric von Hippel and Kathy Strandburg on user-innovation, has demonstrated that there are

---

124 (2006) and the recent decision in Ass'n for Molecular Pathology v. Myriad Genetics, Inc., 133 S. Ct. 2107 (2013).

49. *See, e.g.*, Pallava Bagla, *India Allows Government Scientists to Own Companies*, 323 SCI. 1278 (2009); Kazuhiro Asakawa, *Transition in Japan's Biotechnology Sector: Institutional–Organizational Co-evolution*, *in* STRUGGLES FOR SURVIVAL: INSTITUTIONAL AND ORGANIZATIONAL CHANGES IN JAPAN'S HIGH-TECH INDUSTRIES (Kazuhiro Asakawa ed., 2005), *available at* http://www.springerlink.com/content/l614956178355143/fulltext.pdf.

50. *See also* Arti K. Rai & Rebecca S. Eisenberg, *Bayh–Dole Reform and the Progress of Biomedicine*, 66 L. & CONTEMP. PROBS, Winter/Spring 2003, at 289; Katherine J. Strandburg, *Curiosity-Driven Research and University Technology Transfer*, 16 ADVANCES IN THE STUDY OF ENTREPRENEURSHIP, INNOVATION, & ECON. GROWTH 97 (2005) (noting changing pressures in academia).

51. *See* Madey v. Duke Univ., 307 F.3d 1351, 1362 (Fed. Cir. 2002); *see also* Embrex, Inc. v. Serv. Eng'g Corp., 216 F.3d 1343 (Fed. Cir. 2000). *See generally* Rochelle Dreyfuss, *Protecting the Public Domain of Science: Has the Time for an Experimental Use Defense Arrived?*, 46 ARIZ. L. REV. 457 (2004).

52. JAMES BESSEN & MICHAEL J. MEURER, PATENT FAILURE: HOW JUDGES, BUREAUCRATS, AND LAWYERS PUT INNOVATORS AT RISK 138–44 (2008). Patent trolls may make the situation even worse. *See* Jeremiah Chan & Matthew Fawcett, *Footsteps of the Patent Troll*, 10 INTELL. PROP. L. BULL. 1 (2005).

53. *See* Paul Ganley, *The Internet, Creativity and Copyright Incentives*, 10 J. INTELL. PROP. RTS. 188,193 (2005) (noting how the Internet is causing traditional intermediary incentives to become outdated).

significant creative activities that do not rely directly on standard intellectual property rights.[54] Indeed, it is becoming clear that many of the new types of creativity can be hampered by strong rights regimes. For example, appropriation art and mash-ups are possible only if artists can use the work of others (and, in fact, those who work in those areas utilize Creative Commons licensing to reduce the salience of intellectual property protection).[55] There are also many new players in this sector, including Internet service providers, such as Earthlink, which act as conduits to the Web; information service providers, such as YouTube and Flickr, which permit users to generate and share their own creative works; and search engines and other information aggregators, such as Google and eBay. Because these entities make novel uses of intellectual property and cannot efficiently monitor the works they transmit, they would all be severely constrained under a strict application of traditional intellectual property law.[56]

To be sure, the obstacles that the South faces in attempting to reduce the levels of intellectual property protection do not exist in the North, where there is a substantial legal and political culture ready to identify new strategies, devise new laws, and make the case for them. But in the North, there is also a forceful counter-narrative. University involvement in patenting is said to promote technology transfer.[57] For industries, such as pharmaceuticals, where patents protect products that are costly to invent but easily copied, intellectual property rights remain a necessity. Digitization cuts both ways. It reduces the costs of publication, but it also gives rise to new information enterprises (database compilation, for example)—and thus to new demands for protection.[58] Furthermore, the ease of distribution means that there are entire industries (journalism is an example) that are at risk of going out of business unless new types of intellectual property

---

54. *See* Eric von Hippel, Democratizing Innovation 79–91 (2005) (discussing why users freely reveal their innovations even though standard intellectual property rights are available); Katherine J. Strandburg, *Users as Innovators: Implications for Patent Doctrine*, 79 U. Colo. L. Rev. 467 (2008).

55. *See* Lawrence Lessig, Remix 226–27 (2008); Madhavi Sunder, *IP*^{3}, 59 Stan. L. Rev. 257, 303–09 (2006).

56. Graeme B. Dinwoodie, Rochelle C. Dreyfuss & Annette Kur, *The Law Applicable to Secondary Liability in Intellectual Property Cases*, 42 N.Y.U. J. Int'l L. & Pol. 201 (2009); Christopher M. Newman, *Infringement as Nuisance*, 59 Cath. U. L. Rev 61 (2009).

57. David C. Mowery et al., Ivory Tower and Industrial Innovation: University–Industry Technology Transfer Before and After the Bayh-Dole Act in the United States 33–34 (2004).

58. Council and Parliament Directive 96/9 of the European Parliament and of the Council on the Legal Protection of Databases, 1996 O.J. (L 77) 20 (EC).

rights are created, the coverage of existing rights is broadened, or enforcement mechanisms are strengthened.[59]

With equally strong arguments on both sides of the debate over intellectual property rights, the result is a political stalemate. For example, passage of patent reform in the United States was stalled for many years. The pharmaceutical industry is concerned that any attempt to deal with the patent-to-product ratio problem will undermine their rights; the research tool industry has obstructed proposals for a research exemption that would ease the degree to which patents on fundamental discoveries can chill downstream research. Similarly, any attempt to lower the infringement risks facing Internet intermediaries runs headlong into copyright holders' concerns about rampant piracy.[60]

TRIPS complicates the matter. Negotiations over change dissolve into arguments over what TRIPS permits. On the patent side, compromises are difficult to reach because TRIPS includes a commitment to technological neutrality.[61] As the DSB has elevated that requirement to a structural feature of the Agreement, it is not immediately evident how to create one law for sectors such as pharmaceuticals, where strong protection is needed, and a different one for information technology, where protection is getting in the way of progress.[62] On the copyright side, that structural impediment is absent. But, as Graeme Dinwoodie and I demonstrated, the combination of local and international review makes negotiated compromises vulnerable to unraveling, always in the direction of raising the level of intellectual property protection.[63] Knowing the DSB's potential for upsetting domestic compromises, the parties are less likely to enter into them in the future.

---

59. Digitization has also put pressure on the TRIPS Agreement's core commitment to territoriality. The ubiquity of infringement, and the possibility of moving the locus of infringement to "information havens" or suing in "information hells" has also put new pressures on domestic courts, and has led to demands for new kinds of rights, such as robust doctrines of contributory infringement. *See* Rochelle C. Dreyfuss, *Resolving Patent Disputes in a Global Economy*, *in* PATENT LAW AND THEORY: A HANDBOOK OF CONTEMPORARY RESEARCH 610 (Toshiko Takenaka ed., 2008).

60. *See, e.g.*, *Obama Administration Asked for "Balance" in Intellectual Property Appointments*, 77 PAT., TRADEMARK & COPYRIGHT J. (BNA) 631 (Apr. 10, 2009) [hereinafter *Obama Administration*] (describing a letter from a coalition including librarian associations, the consumer electronics industry, the Electronic Frontier Foundation (EFF), Knowledge Ecology International (KEI), and the Wikimedia Foundation asking President Obama to take account of "the diversity of stakeholders affected by IP policy" in making his appointments to positions with influence over intellectual property matters).

61. TRIPS Agreement, *supra* note 1, art. 27.1.

62. Panel Report, *Canada—Patent Protection of Pharmaceutical Products*, WT/DS114/R (Mar. 17, 2000) [hereinafter *Canada—Pharmaceuticals*]. *But see* Dinwoodie & Dreyfuss, Diversifying, *supra* note 5 (suggesting alternative ways to conceptualize remedial measures).

63. Dinwoodie & Dreyfuss, *supra* note 5. For example, in the Sony Bono Copyright Act, those interested in strong copyright protection won a twenty-year extension of the copyright term. In exchange, they gave up control over the rebroadcasting of music in small drinking and

II. The Role of Emerging Nations in Developing Intellectual Property Norms

The situation for emerging economies is quite different from that of either the North or the South. Nations such as Argentina, Brazil, China, and India have reached a point in their development where they are beginning to see the benefits of intellectual property protection.[64] At the same time, all of these countries have populations with profound access needs. Thus, they are forced to compromise. In order to earn the supracompetitive returns available on world markets for cutting-edge products, they must identify areas where intellectual property rights will spur domestic industries to become innovative.[65] But the presence of a substantial population living in poverty means that these countries must simultaneously protect access to essential information products. Accordingly, these countries necessarily have the will to enact the sort of compromises that are elusive in the North. As important, many of these countries have or are developing a strong legal infrastructure[66] and a few have even attained considerable market power.[67] Thus, they also have the skills and the political heft needed to navigate the TRIPS Agreement—to write legislation, negotiate waivers, and defend their

---

eating establishments. A challenge to term extension went to the Supreme Court. *See* Eldred v. Ashcroft, 537 U.S. 186 (2003). However, the exception for rebroadcasts did not receive the same deference in the DSB and was held to be inconsistent with TRIPS obligations. *See* Panel Report, *United States—Section 110(5) of US Copyright Act*, WT/DS160/R (June 15, 2000) [hereinafter *US—110(5)*].

64. *See, e.g.*, Francesco LaForgia, Fabio Montobbio & Luigi Orsenigo, *IPRs and Technological Development in Pharmaceuticals, Who Is Patenting What in Brazil after TRIPS?*, *in* THE DEVELOPMENT AGENDA: GLOBAL INTELLECTUAL PROPERTY AND DEVELOPING COUNTRIES 293 (Neil Weinstock Netanel ed., 2008); WORLD INTELLECTUAL PROP. ORG. (WIPO), WORLD PATENT REPORT: A STATISTICAL REVIEW 25 (2008) (demonstrating that China is rapidly becoming a prolific filer of utility model patents); WIPO, Provisional Committee on Proposals Related to a WIPO Development Agenda, Fourth Session Draft Report 15, PCDA/4/3/Prov.2 (June 11–15, 2007), *available at* http://www.wipo.int/edocs/mdocs/mdocs/en/pcda_4/pcda_4_3_prov_2.doc [hereinafter PCDA] (tracing the growth of pharmaceutical patenting in China and other countries). *See generally* ALEXANDER STACK, INTERNATIONAL PATENT LAW: COOPERATION, HARMONIZATION AND INSTITUTIONAL ANALYSIS OF WIPO AND THE WTO (2010).

65. In theory, intellectual property rights available in other markets could equally well spur innovation. In practice, however, local incentives appear to remain important.

66. *See, e.g.*, Brandt Goldstein, *Notes From China: Legal Education Playing Catch-Up, in a Hurry*, WALL ST. J.L. BLOG (June 5, 2008), http://blogs.wsj.com/law/2008/06/05/notes-from-china-legal-education-playing-catch-up-in-a-hurry; Shaffer et al., *supra* note 16 (documenting the rise in legal expertise for dealing with trade matters in Brazil).

67. Benoliel & Salama, *supra* note 16, particularly stress market power in modeling the power of emerging nations to exert influence on the way in which the TRIPS Agreement is implemented.

actions. In short, these countries have the interest, power, and capacity to exert a strong influence on the future development of intellectual property norms.

A. Domestic Legislation

On the question of tailoring law to domestic needs, India provides a case in point. As Lionel Bently demonstrated in connection with the debate over translation rights, India has a long history of negotiating its way through a morass of domestic politics, colonial rule, and international obligations.[68] Its recent Patent Act amendments utilized those talents to good effect. Immediately prior to TRIPS, India recognized patents on pharmaceutical processes but not on pharmaceutical products.[69] As a result, it developed a strong industry adept at finding manufacturing methods that reduce the cost of generic drugs and increase supply of both Indian and foreign markets. The industry also served as a training ground for budding Indian scientists.

The TRIPS patent obligations—specifically, the obligations to maintain technological neutrality and to protect products as well as processes—threatened this arrangement.[70] However, India recovered nicely. In the Patent Act, drafted to comply with TRIPS, it provided that "new form[s] of a known substance which do[] not result in the enhancement of the known efficacy," as well as "new use[s] for a known substance" that do not significantly improve efficacy, are not patentable.[71] The statute, Section 3(d) of the Indian Patent Act, assures the continued availability of generic drugs because it limits medicines to a single twenty-year patent term, even when new applications or formulations are developed.[72] This constraint not only sustains the generic drug sector, it also gives the industry the freedom to find new uses for old medicines (including, arguably, cures for neglected diseases) and to find treatments that are more easily administered.[73] Thus, the statute permits India to service the broader community of

68. Lionel Bently, *Copyright, Translations, and Relations between Britain and India in the Nineteenth and Early Twentieth Centuries*, 82 Chi.-Kent L. Rev. 1181 (2007).

69. Shamned Basheer, *"Policy Style" Reasoning at the Indian Patent Office*, 3 Intell. Prop. Q. 309 (2005).

70. TRIPS Agreement, *supra* note 1, art. 27.1.

71. The Patents Act, §3 (d), No. 39 of 1970, (Universal 2005) (India). An explanation states that "salts, … metabolites, … combinations and other derivatives of known substance shall be considered to be the same substance, unless they differ significantly in properties with regard to efficacy," *Id.*

72. *See* Shamned Basheer & Prashant Reddy, *The "Efficacy" of Indian Patent Law: Ironing Out the Creases in Section 3(d)*, 5 Scripted 232 (2008).

73. Of course, the new product will not be patentable, but that may not matter to Indian pharmaceutical companies, which do not rely on the same "blockbuster" models as Northern

needy patients. In an action by Novartis concerning the protectability of the drug Gleevac, India has already defended this statute before its local courts.[74] Although the Indian tribunals did not reach the question whether the provision is compatible with TRIPS, India has a good argument that the Agreement leaves it free to define "inventive step" as it sees fit. Whether matters will get to that stage is, however, an open issue: so far, India has managed to sidestep a challenge—in part, by provoking a public protest that has restrained Novartis from having the issue brought before the DSB.[75]

India has been equally active in protecting its cultural heritage, which includes information of considerable current value. In some instances, India has decided to capture the benefits for itself. For example, it has instituted strong protection for geographical indications, with the expectation that it will be able to label goods with these signals of local, traditional production and thereby raise their prices on world markets.[76] In other situations, India has decided that the better strategy is to put the information into the public domain. In order to prevent "biopiracy"—the incorporation of local knowledge into foreigners' patents—it has developed a digital library of traditional knowledge. The information is available in five languages (English, German, French, Japanese, and Spanish), and is classified in a manner that permits patent examiners around the world to consider the information prior art, thus making it ineligible for patent protection.[77]

Of course, a single set of domestic laws does not amount to a change in international law. However, India is acting in parallel with—or in some instances, setting an example for—other emerging economies. Not all these nations take the same approach or legal position that India does on every issue. For example, Brazil has solved its access problems by the judicious use of threats to award compulsory licenses.[78] The Andean states have approached the biopiracy situation

companies and have lower operating costs. *See* David W. Opderbeck, *Patents, Essential Medicines, and the Innovation Game*, 58 VAND. L. REV. 501 (2005).

74. For a description of the case, see Allessandra Acuri & Rosa Castro, *How Innovative Is Innovative Enough? Reflections on the Interpretation of Article 27 TRIPS from* Novartis v. Union of India (Soc'y of Int'l Econ. Law, Working Paper No. 52/08, 2008), *available at* http://papers.ssrn.com/sol3/papers.cfm?abstract_id=1159821.

75. *See, e.g.*, Letter from Berne Declaration Group to Novartis (Apr. 26, 2007), *available at* http://www.evb.ch/en/p25011413.html (showing signatures by over seventy-five organizations, including many NGOs involved in medical issues, and individuals from around the world).

76. Suresh C. Srivastava, *Geographical Indications and Legal Framework in India*, 38 ECON. & POL. WKLY. 4022 (2003).

77. Traditional Knowledge Digital Library, *available at* http://www.tkdl.res.in/tkdl/langdefault/common/home.asp?GL=Eng (last visited Sept. 5, 2013).

78. *See, e.g.*, Benoliel & Salama, *supra* note 16.

by using bilateral negotiations to develop information-sharing programs—that is, to insure that the proceeds from any intellectual property rights developed from local knowledge are shared with the local communities.[79] On the other hand, the Andean countries are similarly opposed to what they call "second use patents."[80] Furthermore, India's views also have champions within developed countries. India is joined by the EC in its interest in promoting local trade with geographical indications.[81] And because the patenting of new forms and new uses of old drugs facilitates "evergreening" (perpetuation of patent protection through successive applications on minor variations), there are also groups in the North intent on ending the practice.[82] Clearly, new norms can develop as particular approaches are imitated and shared. As the next sections suggest, these practices also achieve international recognition as they are defended in international courts and put on the agendas of international organizations.

B. International Adjudication

In theory, there are several international tribunals with adjudicatory authority that extends to intellectual property cases. In practice, however, the WTO's dispute settlement procedure is the main locus of dispute resolution. The European Court of Human Rights has had a few intellectual property cases, but none directly implicating the substantive questions that are plaguing the intellectual property community.[83] The International Court of Justice is also theoretically available for disputes involving the Paris and Berne Conventions, but it has never heard a case.[84] In contrast, the DSB has been heavily involved in determining the capacity of WTO members to respond to new situations and to adjust their laws to local needs.[85] Its ability

---

79. *See* Rafael T. Boza, *Protecting Andean Traditional Knowledge and Biodiversity Perspectives under The U.S.-Peru Trade Promotion Agreement*, 16 CURRENTS: INT'L TRADE L.J. 76 (2008).

80. CARLOS CORREA, INTEGRATING PUBLIC HEALTH CONCERNS INTO PATENT LEGISLATION IN DEVELOPING COUNTRIES (2000).

81. Sachin Chaturvedi, *India, the European Union and Geographical Indications: Convergence of Interests and Challenges Ahead*, 4 S. ASIA ECON. J. 99 (2003).

82. *See, e.g.*, Thomas A. Faunce & Joel Lexchin, *"Linkage" Pharmaceutical Evergreening in Canada and Australia*, 4 AUSTL. & N.Z. HEALTH POL'Y 8 (2007).

83. LAURENCE R. HELFER & GRAEME W. AUSTIN, HUMAN RIGHTS AND INTELLECTUAL PROPERTY: MAPPING THE GLOBAL INTERFACE (2011); Helfer, *supra* note 9.

84. Paris Convention, *supra* note 20, art. 28(1); Berne Convention, *supra* note 21, art. 33(1).

85. *See* Appellate Body Report, *India—Patent Protection for Pharmaceutical and Agricultural Chemical Products*, WT/DS50/AB/R (Dec. 19, 1997) [hereinafter *India—Pharmaceuticals*]; Appellate Body Report, *Canada—Term of Patent Protection*, WT/DS170/AB/R (Sept. 18, 2000); *Canada—Pharmaceuticals*, *supra* note 62; *US—110(5)*, *supra* note 63; Appellate

to articulate international norms is somewhat constrained in that adjudicators can hear only disputes arising under the WTO agreements.[86] However, that restriction is somewhat mitigated by the Vienna Convention's references to subsequent practices and relevant rules of international law.[87]

More limiting is the way in which the cases have been presented. Thus, although less-developed countries may well have viable complaints (for example, regarding the failure of developed countries to comply with their obligations to transfer technology, to provide technical assistance, or to regard TRIPS as setting a limit on demands for intellectual property protection),[88] no action by a developing country has ever become the subject of a formal adjudication. Nor have these countries often been named as respondents.[89] The result is that adjudicators have mainly been called upon to elucidate the meaning of TRIPS in cases that feature developed economies on both sides of the dispute. Because, as noted earlier, international representation of developed countries has been largely dominated by high protectionist interests, none of the litigants in these disputes has had a strong incentive to argue for interpretations that maximize TRIPS flexibilities.

Especially telling are the cases that involve the three "exceptions" provisions for copyright (art. 13), patents (art. 30), and trademarks (art. 17). These measures most heavily implicate members' ability to balance user and producer interests, yet the three cases—*US—110(5)* (involving certain establishments'

Body Report, *United States—Section 211 Omnibus Appropriations Act of 1998*, WT/DS176/AB/R (Jan. 2, 2002); Panel Report, *European Communities—Protection of Trademarks and Geographic Indication for Agricultural Products and Foodstuffs*, WT/DS174/R (Mar. 15, 2005) [hereinafter *EC—GI*]; Panel Report, *China—Measures Affecting the Protection and Enforcement of Intellectual Property Rights*, WT/DS362/R (Jan. 26, 2009) [hereinafter *China—Enforcement*]. Technically, there was one other dispute as both the United States and the European Communities brought an *India—Pharmaceuticals* challenge. There are also two disputes that involved intellectual property tangentially in that retaliation was sought in the intellectual property sector. *See EC—Bananas*, *supra* note 15 (involving Ecuador); Decision by the Arbitrator, *United States—Measures Affecting the Cross-Border Supply of Gambling and Betting Services*, WT/DS285/ARB (Dec. 21, 2007) (involving Antigua). Finally, one case raised national treatment claims under TRIPS, but it was somewhat peripheral to the principal dispute and barely addressed by the DSB. *See* Report of the Panel, *Indonesia—Certain Measures Affecting the Automobile Industry*, WT/DS54/R (July 2, 1998) (adopted Jul. 23, 1998).

86. DSU, *supra* note 4, art. 1.

87. Vienna Convention on the Law of Treaties, art. 31(3)(b)–(c), May 23, 1969, 1155 U.N.T.S. 331, 8 I.L.M. 679; GRAEME B. DINWOODIE & ROCHELLE C. DREYFUSS, A NEOFEDERALIST VISION OF TRIPS: BUILDING A RESILIENT INTERNATIONAL INTELLECTUAL PROPERTY SYSTEM (2012); Dinwoodie & Dreyfuss, *supra* note 8.

88. *See* Frankel, *supra* note 47.

89. To be sure, developing countries have been involved in earlier stages of the dispute resolution process. Their cases have settled, often in ways favorable to development. However, these settlements could be viewed as waivers and not as establishing new norms of behavior.

unauthorized rebroadcasts of copyrighted works); *Canada—Pharmaceuticals* (involving the ability of generic producers to stockpile and test patented pharmaceuticals) and *EC—GI* (on the relationship between trademarks and geographical indications)—all involved developed countries on both sides and yielded decisions that heavily circumscribe members' abilities to safeguard access interests. *EC—GI,* which involved the trademark exceptions provision (which differs in several respects from the patent and copyright provisions), was less limiting than *Canada—Pharmaceuticals* and *US—110(5),* but all three panels ignored the domestic rationales for the challenged legislation,[90] they considered the various parts of the tests cumulatively (which meant that the interests of third parties were not reached),[91] and they largely refused to interpret terms such as "normal," "legitimate," "prejudice," and "unreasonable" normatively.[92] In the patents case, the exceptions provision was also subjected to a separate analysis on whether it was technologically neutral—a move that makes it hard for countries to simultaneously fulfill the obligation to create only exceptions that are "limited."[93]

There have, however, been two (technically, three) cases involving emerging economies, and they were resolved quite differently. In two cases involving India, the United States and the European Communities challenged the efficacy of provisions enacted to protect pharmaceuticals during TRIPS' transition period; in a dispute against China, the United States raised the question of compliance with TRIPS' enforcement obligations.[94] Although the cases were not uniform wins for the respondent, each resolution makes clear just what a difference the participation of an emerging economy can make. Thus, although *India—Pharmaceuticals* ultimately held that India had not properly safeguarded the interests of pharmaceutical patent holders, India won on two profoundly important points. First, the complainants had argued that during the negotiations, India had agreed to protect

---

90. For example, the copyright exception test, like the Berne Convention, uses the term "special." That could have been used to examine the justification for the measure; instead, it was taken to mean "clearly defined." *See US—110(5), supra* note 63, ¶¶ 6.107–6.110. On options for examining the justifications, see Graeme B. Dinwoodie, *The Development and Incorporation of International Norms in the Formation of Copyright Law,* 62 Ohio St. L.J. 733, 751 n.73 (2001); Dinwoodie & Dreyfuss, *Dynamics, supra* note 5.

91. *See Canada—Pharmaceuticals, supra* note 62, ¶ 7.21.

92. *See, e.g.,* Jane C. Ginsburg, *Toward Supranational Copyright Law? The WTO Panel Decision and the "Three Step Test" for Copyright Exemptions,* 187 Revue Internationale du Droit d'Auteur 17 (2001). To be sure, the *Canada—Pharmaceuticals* panel considered the practices of other states to determine the patent holder's legitimate interests, but it was not persuaded that they demonstrated a consensus position. *See Canada—Pharmaceuticals, supra* note 62, ¶¶ 7.78–7.82.

93. *Canada—Pharmaceuticals, supra* note 62, ¶ 6.69.

94. *India—Pharmaceuticals, supra* note 85; *China—Enforcement, supra* note 85.

patentability in a specific way. The Appellate Body, however, firmly rejected the notion that expectations that were not codified into the final agreement could be enforced through a violation complaint.[95] Second, the Appellate Body emphasized that where there is no controlling international norm, member states are owed substantial deference in the way they choose to implement their obligations.[96]

Similarly, China did not win the entire *China—Enforcement* case. Rather, the panel found that China was not adequately protecting the censored elements in copyrighted works and that it needed to do more to prevent counterfeit goods from entering the channels of commerce. Nonetheless, the panel carefully preserved China's national prerogatives. It gave it extensive leeway to determine how to dispose of infringing goods and where to set the threshold for criminal enforcement. It emphasized that the wording of the enforcement provisions requires members to give judicial authorities the *power* to order particular remedies, but does not require specific results.[97] The panel also held that TRIPS remedies are not exclusive.[98] Although it found that "the Agreement contains substantive obligations that are not simply matters of national discretion,"[99] it stressed that TRIPS is a *minimum* standards regime and that it gives members freedom to determine the most appropriate method of implementing their obligations.[100]

Especially with respect to the question of whether China had set its threshold for criminal enforcement too high, the difference between this case and *US—110(5)* is striking. In *US—110(5)*, the panel engaged in fairly speculative determinations of the harm caused by the unauthorized rebroadcasts.[101] In considering the impact of the challenged practice on the right holders' future markets, it refused to take account of how the establishments that had been allowed to rebroadcast music for free would react if they were required to pay for authorization.[102] In contrast, the *China—Enforcement* panel measured the impact of

95. *India—Pharmaceuticals, supra* note 85, ¶ 45.

96. *Id.* ¶¶ 46, 59.

97. *China—Enforcement, supra* note 85, ¶¶ 7.236, 7.331.

98. *Id.* ¶ 7.240.

99. *Id.* ¶ 7.507.

100. *Id.* ¶¶ 7.513 (citing art. 1.1), 7.602.

101. A measure of the panel's willingness to speculate can be seen by comparing the estimate of the EC copyright holders' loss ($53.65 million per year) in the decision with the findings at arbitration (€1,219,900 per year, at a time when the US dollar and the Euro were close to parity). Award of the Arbitrators, *United States—Section 110(5) of the US Copyright Act—Recourse to Arbitration under Article 25 of the DSU*, WT/DS160/ARB25/1 (Nov. 9, 2001).

102. *Id.* ¶¶ 4.9–4.15.

China's enforcement obligations by examining the profit to infringers, and not the impact on right holders. Careful attention was paid to the specifics of how China's local markets operate.[103] Speculative harm was excluded: the panel held the United States to strict evidentiary standards and even rejected assertions concerning the difficulty of finding information in China.[104]

Given the wording of the TRIPS Agreement, the DSB is likely to be far more deferential to member states in cases involving enforcement in general, and criminal liability in particular.[105] Nonetheless, the ability of China—aided in certain respects by the participation of three other emerging economies (Brazil, Argentina, and China Taiwan)[106] —to protect the concepts of minimum standards, deference, and TRIPS flexibilities is highly significant. Should India be put to the defense of Section 3(d), equally important principles might be elucidated. First, because India's main defense is likely to be that it has the right to determine the meaning of "inventive step," a decision in its favor would establish the proposition that the many undefined terms in the TRIPS Agreement are, indeed, left open for domestic interpretation.[107] Moreover, because the Indian provision—although facially neutral—is clearly intended to have its principal impact on pharmaceuticals,[108] a decision for India would also provide guidance on the meaning of the prohibition on "discrimination as to... the field of technology." As we saw, the *Canada—Pharmaceutical* panel's elevation of this provision to structural importance makes it difficult for countries to utilize the patents exception provision and to reconcile the competing interests of the various sectors of the patent industries.[109] The full participation of emerging countries in the adjudicatory process should, in sum, lead to the articulation of a balanced set of norms that will benefit both developed and developing countries.[110]

---

103. *China—Enforcement, supra* note 85, ¶ 7.577.

104. *Id.* ¶¶ 7.626–7.632.

105. For example, TRIPS, *supra* note 1, arts. 1.1 and 41.5, make clear that members' duties regarding implementation are determined contextually. *See generally* Rochelle C. Dreyfuss & Andreas F. Lowenfeld, *Two Achievements of the Uruguay Round: Putting TRIPS and Dispute Settlement Together*, 37 VA. J. INT'L L. 275 (1997).

106. *See, e.g., China—Enforcement, supra* note 85, ¶¶ 7.203, 7.205–.206, 7.211.

107. *See supra* text accompanying notes 37–40.

108. Section 3(d) speaks only of incremental innovation. However, it is accompanied by an official Explanation that mentions "metabolites," a term unique to the pharmaceutical field.

109. *See supra* text accompanying notes 61–62.

110. These decisions—which emphasize national autonomy and deference—support the notion, suggested by Susy Frankel, *supra* note 47, that unilaterally imposed TRIPS-plus obligations (in the form of FTAs and BITS) run afoul of the structural core of the Agreement. *Cf.* Panel Report, *United States—Sections 301–310 of the Trade Act of 1974*, WT/DS152/R (Dec. 22, 1999) (limiting US use of § 301 actions).

C. International Lawmaking

Although the previous section suggests that emerging economies could establish new international intellectual property norms through international adjudication, that procedure is expensive and fraught with uncertainties. Most of the cases involving TRIPS issues have been decided only at the panel level. Accordingly, the Appellate Body's views remain largely unknown. Besides, there is no general principle of stare decisis on which domestic lawmakers can rely. International lawmaking is thus a core source of norm-development.

Recent deliberations of the WTO and WIPO demonstrate how significant participation by increasingly sophisticated emerging nations can be. India, for example, tried hard to limit the inclusion of intellectual property into WTO negotiations during the Uruguay Round.[111] Although it was unsuccessful at that time, its recent accomplishments at the WTO have been substantial. India was active in negotiating the Doha Declaration,[112] and the international initiatives that it has instigated since that time are well-documented and closely followed.[113] It is also trying to simplify the post-Doha waiver/amendment procedures, which—as we saw—are complicating the ability of developing countries to import generic pharmaceuticals.[114] At the same time, it is raising before the TRIPS Council other questions concerning generic drug importation.[115] Furthermore, India is working on protecting its own intellectual contributions. It is pursuing its interest in increasing protection for geographical indications and in protecting traditional knowledge from biopiracy (or, alternatively, in promoting benefit sharing).[116]

---

111. GERVAIS, *supra* note 29, at 16 (§ 1.18); JAYASHREE WATAL, INTELLECTUAL PROPERTY RIGHTS IN THE WTO AND DEVELOPING COUNTRIES (2001).

112. *See, e.g.*, World Trade Organization (WTO), Preparations for the Fourth Session of The Ministerial Conference, Communication from India, WT/GC/W/459 (Nov. 6, 2001); T.N. Srinivasan, *India at the Doha Round* (Sept. 2003), *available at* http:/www.econ.yale.edu/~srinivas/India%20in%20the%20Doha%20Round.pdf.

113. *See, e.g.*, Amy Kapczynski, *Harmonization and Its Discontents: A Case Study of TRIPS Implementation in India's Pharmaceutical Sector*, 97 CAL. L. REV. 1571, 1625–26 (2009); Janice M. Mueller, *The Tiger Awakens: The Tumultuous Transformation of India's Patent System and the Rise of Indian Pharmaceutical Innovation*, 68 U. PITT. L. REV. 491 (2007).

114. *See supra* text accompanying notes 33, 46.

115. *See, e.g.*, EU—Seizure of Generic Drugs in Transit–Request for Consultations, WT/DS408/1 (May 19, 2010); *EU, India Drop Generics Dispute to Focus on FTA Talks*, FDANEWS (Jan. 24, 2011), *available at* http://fdanews.com/newsletter/article?issueId=14404&articleId=133690.

116. *See* Kasturi Das, *International Protection of India's Geographical Indications with Special Reference to "Darjeeling" Tea*, 9 J. WORLD INTELL. PROP. 459 (2006); GERVAIS, *supra* note 29, at 43, 389 (§§ 1.46, 1.50); GERVAIS, *supra* note 29, at 71–74 (§§ 1.91–1.94); GERVAIS, *supra* note 29, at 76–80 (§§ 1.97–1.100) (proposing an amendment to the TRIPS Agreement on biological resources and traditional knowledge).

It is also at the forefront on the protection of biological materials under the Convention on Biological Diversity and has made various proposals for resolving conflicts between TRIPS and the CBD.[117] India is, of course, not alone in these initiatives. On the generic drug issues, it is joined by Brazil;[118] on biological diversity, its allies include Brazil, China, Ecuador, and South Africa.[119] For geographical indications, it is aligned with the EC (and opposed by Argentina).[120]

For emerging nations, WIPO is an even more attractive venue. It has a more open structure than the WTO, and in recent times has been more active than the WTO in considering new norm-building initiatives. Although its stated mission is to "promote intellectual property throughout the world," one of the reasons that negotiations over intellectual property shifted to the WTO was because WIPO was perceived to be more inclined toward balance: developing countries not only have an equal vote, they cannot be bought off with side-payments in the form of access to world markets for its commodities. WIPO also has long-standing expertise in intellectual property matters. Most important, since the shift in focus to the WTO, WIPO has had to work hard to maintain its relevance; receptivity to the needs of those badly represented in the WTO appears to be one of its new niches.[121] Because there are both formal and informal ties between WIPO and the WTO, developments in WIPO can both create freestanding intellectual property norms and exert considerable influence on the interpretation of the TRIPS Agreement.[122]

---

117. *Trips Council Discusses Biodiversity, Health In Informal Mode,* Int'l Centre for Trade and Sustainable Dev. (Mar. 10, 2007), http://ictsd.net/i/ip/39087. For the proposals, see WTO, Article 27.3b, Traditional Knowledge, Biodiversity, WTO, *available at* http://www.wto.org/english/tratop_e/TRIPs_e/art27_3b_e.htm.

118. *Statement by Brazil at TRIPS Council: Public Health Dimension of TRIPS Agreement,* Knowledge Ecology Int'l (Mar. 4, 2009), http://www.keionline.org/blogs/2009/03/04/brazilian-intervention-at-trips-council.

119. *See supra* note 117. *See generally* Daniel Gervais, *(Re)implementing TRIPS to Foster Innovation,* 12 J. World Int'l Prop. 348 (2009).

120. *See supra* note 116; *see also* Press Release, Europa, WTO Talks: EU Steps Up Bid for Better Protection of Regional Quality Products (Aug. 28, 2003), *available at* http://europa.eu/rapid/pressReleasesAction.do?reference=IP/03/1178; WTO, Proposed Draft TRIPS Council Decision on the Establishment of a Multilateral System of Notification and Registration of Geographical Indications for Wines and Spirits, TN/IP/W/10/Rev.2 (July 24, 2008), *available at* http://docsonline.wto.org/imrd/directdoc.asp?DDFDocuments/t/tn/ip/W10R2.doc.

121. *See* Dinwoodie & Dreyfuss, *supra* note 87; Dinwoodie & Dreyfuss, *supra* note 8. WIPO's Working Group on Traditional Knowledge, Genetic Resources and Traditional Cultural Expressions/Folklore is an example of an effort to develop new niche specialties. *See Traditional Knowledge,* WIPO, http://www.wipo.int/tk/en (last visited Oct. 11, 2013).

122. *See* TRIPS Agreement, *supra* note 1, arts. 63, 68; Agreement Between the World Intellectual Property Organization and the World Trade Organization, 35 I.L.M. 754

Three developments illustrate the power that emerging economies have in this environment. First, the TRIPS exceptions provisions, which received such a constrained interpretation by the DSB, are a particular focus. WIPO has made several studies of national limitations and exceptions (L&Es) to copyright protection and has recently conducted a similar survey on patents.[123] Supplemented by an analogous effort by Chile on behalf of APEC,[124] these materials demonstrate the many ways in which emerging nations have managed to improve access to protected works without substantially sacrificing the interests of intellectual property holders. The compilations thus provide templates for similarly situated economies, as well as for developing countries that lack the legal resources to devise L&Es for themselves. The compilations are also valuable to those groups in the North that are interested in promoting access interests. As important, WIPO's authoritative collection of L&Es will help future TRIPS adjudicators understand the need for balance and the kinds of approaches necessary to keep domestic laws responsive to changes in the creative and technological environment. The *US—110(5)* and *Canada—Pharmaceuticals* panels evinced interest in state practices;[125] as noted earlier, the Vienna Convention also deems state practices to be relevant interpretive data.

Second, WIPO issues more formalized reports on how its own instruments should be interpreted. The significance of this type of "soft lawmaking" is controversial: although these reports are the product of considerable consultation and expert deliberation, they are not subjected to a vote by the membership, and in some instances, they arguably go beyond simple interpretation and push the law past the norms set by the WIPO conventions. The DSB has, accordingly, been reluctant to consider these reports, even when they deal with measures that have been incorporated by reference into the TRIPS Agreement.[126] But that could

(Dec. 22, 1995), *available at* http://www.wipo.int/treaties/en/agreement/pdf/trtdocs_wo030.pdf; Dinwoodie & Dreyfuss, *supra* note 8; Ruth Okediji, *WIPO–WTO Relations and the Future of Global Intellectual Property Norms*, 39 NETH. Y.B. INT'L L. 69 (2008).

123. Standing Committee on Copyright and Related Rights, WIPO Study on Limitations and Exceptions of Copyright and Related Rights in the Digital Environment, SCCR/9/7 (Apr. 5, 2003) (prepared by Sam Ricketson); Standing Committee on Patents, WIPO Study on Exclusions from Patentable Subject Matter and Exceptions and Limitations to the Rights, SCP/15/3 (Oct. 2010) (prepared by Lionel Bently et al.); Hugenholtz & Okediji, *supra* note 38.

124. Standing Committee on Copyright and Related Rights, Draft Report, 16th Session, March 10–12, 2008, WIPO Doc. SCCR/16/3 Prov. (Sept. 5, 2008), *available at* http://www.wipo.int/edocs/mdocs/copyright/en/sccr_16/sccr_16_3_prov.doc.

125. *See e.g., US—110(5), supra* note 63, ¶ 6.55; *Canada—Pharmaceuticals, supra* note 62, ¶ 7.82.

126. *See, e.g., China—Enforcement, supra* note 85, ¶¶ 7.567, 7.585.

easily change. Emerging nations participate vigorously in the promulgation of these reports and clearly signal the points that are contentious. These signals could be used by policymakers in other countries, by nations seeking to avoid unilateral sanctioning, and by TRIPS adjudicators to differentiate between the provisions that provide useful interpretation and those that overreach the agreed frontier.

An example is the 1999 Report on the scope of the obligation to protect well-known trademarks.[127] One part of this report tackled an issue that many domestic courts have confronted: the definition of a well-known mark; another sets out a rule requiring member states to protect marks from dilution. Consensus was achieved on the definition, but Argentina (among other emerging states) refused to join in the dilution recommendation, on the theory that it increased the level of trademark protection well beyond the likelihood-of-confusion requirement set out in the Paris Convention and the TRIPS Agreement.[128] Although it would be possible for policymakers at both national and international levels to ignore the report entirely, it has been suggested that a better course would be to follow the unanimous recommendations, but to use Argentina's substantial reservations as a reason to reject reliance on the dilution provisions.[129]

Third, and potentially most far-reaching, Argentina and Brazil have engaged WIPO in an ambitious "Development Agenda."[130] Launched in the aftermath of the Doha Declaration, this initiative (as its name suggests) is primarily directed at fostering development: integrating the intellectual property goals of WIPO with the Millennium Development Goals of the United Nations more generally, eliminating the "digital divide" and the "knowledge gap" among UN members, and promoting more effective technology transfer and engagement in "development oriented" technical assistance.[131] Led initially by a Provisional Committee on Proposals Related to a WIPO Development Agenda (PCDA), the initiative's

---

127. Joint Recommendation Concerning Provisions on the Protection of Well-Known Marks, WIPO Doc. Publication 833(E) (Sept. 20–29, 1999), *available at* http://www.wipo.int/about-ip/en/development_iplaw/pub833.htm.

128. *See* Joint Resolution Concerning Provisions on the Protection of Well-Known Marks, ¶ 8, WIPO Doc. A/34/13 (Aug. 4, 1999), *available at* http://www.wipo.int/meetings/en/doc_details.jsp?doc_id=1101. In contrast, the United States used the TRIPS Agreement as an excuse to ratchet up the level of trademark protection to include dilution. *See* Graeme B. Dinwoodie et al., International Intellectual Property Law and Policy 211 (2d ed. 2008).

129. Dinwoodie & Dreyfuss, *supra* note 8.

130. Proposal by Argentina and Brazil for the Establishment of a Development Agenda for WIPO, WIPO Doc. WO/GA/31/11 (Aug. 27, 2004), *available at* http://www.wipo.int/edocs/mdocs/govbody/en/wo_ga_31/wo_ga_31_11.doc.

131. *Id.*

proponents have now attained full status as the Committee on Development and Intellectual Property (CDIP), with regular meetings on a comprehensive set of recommendations.[132]

As the Development Agenda has matured, its ambitions have evolved. They now deal with many of the issues at the overlap between the interests of emerging and developing countries. These include proposals on protecting traditional knowledge and genetic resources, on drafting competition law, and on building technological and institutional infrastructure.[133] Notably, the initiatives also encompass many of the access problems that are equally vexing lawmakers in the North. For example, the CDIP was instructed to "[c]onsider the preservation of the public domain within WIPO's normative processes... [, to] deepen the analysis of the implications and benefits of a rich and accessible public domain,"[134] and to review the flexibilities available under international intellectual property law.[135] To these ends, the CDIP has proposed studies of how the trademark, copyright, and patent systems protect the public domain and access interests; it is commissioning work on the interface between intellectual property and competition law, and developing proposals to revise licensing practices in order to improve access to protected works.[136] The CDIP also commissioned economists to study the intellectual property system is examining proposals on the preservation of library materials, exclusions to patentable subject matter, and improving usage and dissemination of patented information.[137]

These efforts are ongoing;[138] it remains to be seen whether they can be accomplished in a freestanding agreement and whether they will require amendments to TRIPS. Either way, the initiative that Argentina, Brazil, India, and other emerging economies have shown will have a significant impact on the development of the

---

132. *See* WIPO, Conferences, Meetings and Seminars, *available at* http://www.wipo.int/meetings/en/details.jsp?meeting_id=17382 (last visited Dec. 1, 2012).

133. *See, e.g.*, PCDA, *supra* note 64, at Annex 1, Clusters A and B.

134. *Id.* ¶ 17 (Cluster B); *see also id.* ¶ 20.

135. *Id.* ¶¶ 17, 22.

136. Committee on Development and Intellectual Property (CDIP), WIPO Doc. CDIP/3/3 (Mar. 18, 2009), *available at* http://www.wipo.int/edocs/mdocs/mdocs/en/cdip_3/cdip_3_3.doc.

137. CDIP, WIPO Doc. CDIP/3/5 (Mar. 23, 2009), *available at* http://www.wipo.int/edocs/mdocs/mdocs/en/cdip_3/cdip_3_5.doc.

138. *See, e.g.*, CDIP, Progress Reports, WIPO Doc. CDIP/8/2 (Nov. 13–18, 2011), *available at* http://www.wipo.int/edocs/mdocs/mdocs/en/cdip_6/cdip_6_2.pdf; *Projects for Implementation of Development Agenda Recommendations*, WIPO, http://www.wipo.int/ip-development/en/agenda/projects.html (last visited Sept. 5, 2013).

world's norms regarding access to creative materials and the opportunities to participate in intellectual pursuits.

III. The Role of Emerging Nations in Developing Global Administrative Law Norms

Emerging—both as an economy and onto the world stage—is not an easy task, and most countries do not accomplish it alone. For one, they derive substantial strength from building coalitions with like-minded nations. As we saw, Brazil and Argentina jointly proposed the Development Agenda for WIPO; India has partnered with Argentina in opposing protection for second-use patents; the "Chindia" construct suggests how often China's interests align with those of Southeast Asia.[139] Emerging economies also frequently partner with less-developed countries, and occasionally, as in India's alignment with the EC on geographical indications, with developed nations as well.[140]

Civil society has also played a key role in emerging countries' success, both in terms of tailoring domestic legislation to local needs and in furthering intellectual property agendas on the world stage. For example, Greg Shaffer and his coauthors have demonstrated how Brazil's ascendancy has depended strongly on the combined efforts of a professionalized political corps, well-functioning trade associations, an academy interested in fostering legal education, and NGOs, such as Doctors Without Borders and Oxfam, with specific technical expertise on intellectual property matters.[141] Duncan Matthews has likewise documented how civil society helps provide emerging economies with the information and technical knowledge they need to follow the broad array of issues with which they are concerned.[142]

Significantly, many of the groups involved in what Amy Kapczynski calls "A2K mobilization"[143] are also involved in efforts at the far ends of the development spectrum. They are active in helping developing countries deal with their international obligations. In that role, they shore up the natural alliances between

---

139. *See, e.g.*, Peter Engardio, Chindia: How China and India Are Revolutionizing Global Business (2006).

140. *See, e.g.*, Yu, *supra* note 16.

141. Shaffer et al., *supra* note 16.

142. Duncan Matthews, *The Role of International NGOs in the Intellectual Property Policy-Making and Norm-Setting Activities of Multilateral Institutions*, 82 Chi.-Kent L. Rev. 1369, 1374 (2007); *see also* Dinwoodie & Dreyfuss, *supra* note 87.

143. Kapczynski, *supra* note 31. A2K stands for access to knowledge. *Id.*; *see also* Margaret Chon, *A Rough Guide to Global Intellectual Property Pluralism*, *in* Working within the Boundaries of Intellectual Property Protection (Rochelle C. Dreyfuss, Harry First & Diane Zimmerman eds., 2010).

the Middle and the South.[144] And because some of the leading figures come out of the free software movement, the farmers' rights movement, the open genomics movement, the Creative Commons, the organizations of librarians, and the like, these organizations also create "back door" alliances between emerging countries and the low-protectionist factions that are having difficulty making headway in the political arena of the North.[145]

But as Duncan Matthews has shown, protectionist interests have fought back. The United States and the European Union use their entrenched positions to devise negotiation strategies that mute the voices of those involved in protecting access interests.[146] One example is the move from WIPO to the WTO, which diluted the power of developing countries to resist an international instrument requiring high levels of intellectual property protection. The WTO's inhospitality to civil society participation is another. Even more important is the United States' continued use of Special 301 actions, including priority watch lists to threaten sanctions if standards are not raised to US-approved levels. And the proliferation of FTAs, BITs, and EPAs is an especially pernicious development. These deals are often struck with developing countries at a time in their development when they cannot resist Northern demands. The agreements assure intellectual property protection in markets that are so small, it is hard to believe that improving trading opportunities is their main objective. Rather, their intent seems to be to ratchet up apparent global norms on the level of protection. They increase the number of states with an (ostensibly) shared view on the meaning of TRIPS and, depending on how the most-favoured-nation provision in the TRIPS Agreement is interpreted, they may even indirectly raise world standards.[147] In addition, it has been suggested that Special 301 actions, FTAs, BITs, and EPAs are aimed at splitting coalitions—at using side agreements to buy off states that would benefit from advocating lower levels of protection and preventing them from participating in international lawmaking protective of access interests.[148]

Faced with this strategy, emerging nations have no choice but to fight for more than just intellectual property norms that suit their needs: they must also protect

---

144. *See, e.g.*, Yu, *supra* note 16.

145. *Cf. Obama Administration*, *supra* note 60 (describing the organizations working to improve balanced lawmaking in the United States). An example is the paper cited in *infra* note 148, on the effects of bilateralism. It was written by an academic who was commissioned by Oxfam as part of its Cut the Cost of Medicines Campaign.

146. Duncan Matthews, Globalising Intellectual Property Rights: The TRIPs Agreement (2003).

147. TRIPS Agreement, *supra* note 1, art. 4; Frankel, *supra* note 47; Dinwoodie & Dreyfuss, *supra* note 87, at 157.

148. Peter Drahos, *Bilateralism in Intellectual Property*, Oxfam (2001), http://www.anu.edu.au/fellows/pdrahos/reports/pdfs/biltateralism_ip.pdf.

their partners' ability to participate in international negotiations. They must, in other words, advocate for precisely the same transparency, participation, and accountability principles that are at the core of the global administrative law program. For example, the PCDA's initial agenda proposed a new mandate: "WIPO shall conduct informal, open and balanced consultations, as appropriate, prior to any new norm-setting activities, through a member-driven process, promoting the participation of experts from Member States, particularly developing countries and LDCs."[149] That agenda also contained a cluster of governance goals that emphasized cooperation by "relevant international organizations."[150] These included measures that ensure "the wide participation of civil society,"[151] steps to "improve WIPO's role in finding partners to fund and execute projects for IP-related assistance in a transparent and member-driven process,"[152] as well as commitments to operate "in a manner open and transparent to all Members" and to provide advance notice and agendas for all meetings.[153] The CDIP is beginning to implement these ideas in the work programs of WIPO's various Standing Committees.[154]

Although the emphasis has thus far related to procedural protections that *facilitate* participation, rampant resort to bilateral negotiations suggests that there may also be a problem making sure that other states (and their allies) are *disposed* to act in ways that further access interests. Maggie Chon, relying on work by Amartya Sen and Martha Nussbaum, has suggested that the global intellectual property system also needs a set of meta-norms designed to foster human capabilities (e.g., health, nourishment, literacy, and basic mathematical and scientific training).[155] Although her arguments are made on fairness and distributive justice grounds, such norms could also operate strategically. If these meta-norms were adopted, easy targets could no longer be picked off with bilateral agreements. They would have, in effect, pre-committed to safeguard the access interests of their weakest citizens. Likewise, northern protectionists would have less room to maneuver. In addition to appealing to broader human rights arguments, limits

---

149. Provisional Committee on Proposals, *supra* note 64, Annex I, Cluster B, ¶ 21.

150. *Id.* ¶ 39 (Cluster E).

151. *Id.* ¶ 42.

152. *Id.* ¶ 43.

153. *Id.* ¶ 44.

154. CDIP, *supra* note 137.

155. Chon, *From Below*, *supra* note 35 (citing Martha C. Nussbaum, *Human Capabilities, Female Human Beings*, *in* WOMEN, CULTURE, AND DEVELOPMENT: A STUDY OF HUMAN CAPABILITIES 61 (Martha C. Nussbaum & Jonathan Glover eds., 1995); AMARTYA SEN, DEVELOPMENT AS FREEDOM (1999)).

could also be created from within intellectual property law itself. Indeed, several academics have made such proposals.[156] Spearheaded by Annette Kur and Marianne Levin, a Nordic group of professors has put forward specific suggestions for amending the TRIPS Agreement to impose ceilings on the level of protection that a WTO member may enact.[157]

Emerging nations could produce yet another procedural benefit. Although the appeal to multiple international organizations is often conceptualized as "regime shifting," it may be more productive to think of international intellectual property lawmaking organically, as a network of nodes, related through assorted links. Thus, whereas the WTO and WIPO take center stage in generating international intellectual property law, the World Health Organization has a strong interest in issues regarding access to medicines,[158] UNCTAD and UNIDO are involved in development issues,[159] UNESCO is engaged in improving technology transfer,[160] UNEP is encouraging environmental technologies,[161] and the OECD is working on patent pooling as well as other patent law issues.[162]

---

156. *See, e.g.*, Graeme B. Dinwoodie, *The International Intellectual Property Law System: New Actors, New Institutions, New Sources*, 98 PROC. AM. SOC'Y INT'L L. 213, 219 (2004); Rochelle C. Dreyfuss, *TRIPS—Round II: Should Users Strike Back?*, 71 U. CHI. L. REV. 21 (2004); *see also* DINWOODIE & DREYFUSS, *supra* note 87 (proposing the recognition of an international intellectual property *acquis*).

157. *Proposed Amendments to TRIPS*, *in* INTELLECTUAL PROPERTY RIGHTS IN A FAIR WORLD TRADE SYSTEM, *supra* note 3, at 455.

158. *See, e.g.*, WORLD HEALTH ORG. (WHO), COMM'N ON INTELLECTUAL PROP. RIGHTS, INNOVATION & PUB. HEALTH (CIPIH), PUBLIC HEALTH, INNOVATION AND INTELLECTUAL PROPERTY RIGHTS (2006), *available at* http://www.who.int/intellectualproperty/report/en/index.html.

159. Ruth Okediji, *The International Copyright System: Limitations, Exceptions and Public Interest Considerations for Developing Countries* 13–14 (2006), *available at* http://unctad.org/en/Docs/iteipc200610_en.pdf; Ruth Okediji, *UNCTAD-ICSTD Dialogue on Moving the Pro-Development IP Agenda Forward: Preserving Public Goods in Health, Education and Learning* (2004), *available at* http://www.iprsonline.org/unctadictsd/bellagio/docs/Okideiji_Bellagio4.pdf; Dwijen Rangnekar, *UNCTAD/ICTSD Capacity Building Project, Geographical Indications: A Review of Proposals at the TRIPS Council* (2002), *available at* http://www.iprsonline.org/unctadictsd/docs/GI%20paper.pdf.

160. *See, e.g.*, Lucie Guibault, *The Nature and Scope of Limitations and Exceptions to Copyright and Neighbouring Rights with Regard to General Interest Missions for the Transmission of Knowledge: Prospects for Their Adaptation to the Digital Environment*, UNESCO E-COPYRIGHT BULL. (Oct.–Dec. 2003), *available at* http://portal.unesco.org/culture/en/files/17316/10874797751l_guibault_en.pdf/l_guibault_en.pdf.

161. *See* Kumar, *supra* note 45; *Environmental Governance,* UN ENV'T. PROGRAMME, http://www.unep.org/environmentalgovernance (last visited Oct. 11, 2013).

162. Birgit Verbeure, Esther van Zimmeren, Gert Matthijs & Geertrui Van Overwalle, *Patent Pools and Diagnostic Tests*, 24 TRENDS IN BIOTECH. 115 (2006); OECD, PATENTS AND

Emerging nations—and importantly, the civil society organizations on which they rely—act as the connective tissue in that they participate in the many fora in which their interests are at stake.[163] Although they may press different interests in different places (depending on the receptivity of the organization), the discussions foster cross-fertilization of ideas, which in turn promotes consensus on both intellectual property principles and meta-norms of substantive fairness and distributive justice. In that way, the increasing salience of emerging nations on the world stage could lead to a reduction in the cacophony produced by regulatory overlap.

In fact, attempts to create more stable international rules through cooperation by intergovernmental organizations are already underway. The Development Agenda calls for WIPO to "intensify its cooperation on IP related issues with UN agencies... and other relevant international organizations... in order to strengthen coordination for maximum efficiency in undertaking development programs."[164] The CDIP has already enlisted UNESCO and UNCTAD in its efforts to develop a coordinated program among the divisions within WIPO, soliciting international input as it goes about furthering its agenda.[165]

To some extent, emerging economies can also improve cooperation on their own. As I have suggested elsewhere, one way to deal with overlapping interests is by developing concepts of institutional hierarchy, authority, and deference.[166] Because these nations are enmeshed in deliberations within several organizations simultaneously, they can use their involvement to further these principles. For example, Argentina and India, which are both robust participants in WIPO and the WTO, also have representatives on the WHO's Commission on Intellectual Property Rights.[167] Presumably, these positions can be used to enlist WHO's expertise over medicine to advance the same access goals that these nations are trying to achieve in the WTO and at WIPO. At the same time, they can also influence the WTO and WIPO to accept WHO's authority on such matters as making determinations about when nations are enduring health emergencies that require access to specific essential medicines.

INNOVATION: TRENDS AND POLICY CHALLENGES (2004), *available at* http://www.oecd.org/science/sci-tech/24508541.pdf.

163. *Cf.* Matthews, *supra* note 142.

164. Provisional Committee on Proposals, *supra* note 64, Annex I, Cluster E, ¶ 40.

165. CDIP, *supra* note 137.

166. Dreyfuss, *supra* note 43; DINWOODIE & DREYFUSS, *supra* note 87.

167. *See supra* note 158.

Conclusion

With their large class of impoverished citizens and their increasing industrial muscle, emerging economies internalize all of the conflicts within the international intellectual property system. As they find solutions to their domestic problems and forge international alliances with similarly situated economies, they are positioned to take a leadership role in developing a sounder intellectual property system—a system that recognizes the richness of the creative environment and the changing needs of both producers and consumers of knowledge products. Such a system would acknowledge not only cutting-edge science, but also traditional knowledge. It would support incentive-based knowledge production systems, but it would also make space for curiosity- and creativity-driven pursuits. Most important, it would be attentive to distributive justice and human development. As individual nations fully emerge as intellectual powerhouses, it would not be surprising if they move to the "other side" on various intellectual property issues and begin to champion the same views developed countries pursue today.[168] Hopefully, however, new countries will move into their roles and take up the challenge of keeping intellectual property law responsive to the needs of both producers and consumers of knowledge products.

As important, the same forces that are pushing the likes of India, Brazil, China, and Argentina to find better accommodations among creative groups, users, and producers also propels them to foster better access to international lawmaking. The governance principles that they create may be as important a legacy as the intellectual property norms that they are currently advocating.

168. *See, e.g.*, Benoliel & Salama, *supra* note 16 (showing that the domestic calculus—and hence the overall bargaining power—of emerging countries changes as their creative sectors attain dominant positions in the economy).

3 US EXECUTIVE BRANCH PATENT POLICY, GLOBAL AND DOMESTIC

Arti K. Rai*

Introduction

Over the last few decades, the US executive branch has been a central player in numerous struggles over the content of global patent law. Although the venue for these struggles has frequently shifted, the executive branch has consistently favored strong patent rights. In this chapter, I discuss how US global positioning compares to positions taken by executive branch agencies that focus on domestic policy. I also discuss how the rise of China may affect not only US global and domestic patent policy but also larger questions of "innovation policy."

I. A Brief History of US Efforts

Starting in the 1980s, the US and like-minded countries engineered a transfer of patent law discussions from the World Intellectual Property Organization (WIPO) to a venue, the World Trade Organization (WTO), that they viewed as more friendly to strong patent rights. At the WTO, these countries succeeded in promulgating a suite of minimum intellectual property standards in the 1994 Agreement on Trade-Related Aspects of Intellectual Property Rights (TRIPS Agreement).[1]

---

* Elvin R. Latty Professor, Duke Law School. From 2009 to 2010, I served as the Administrator for Policy and External Affairs at the US Patent and Trademark Office (USPTO). This contribution relies only on publicly available information, however. I thank Larry Helfer and Ruth Okediji for helpful comments. Richard Lee provided exceptional research assistance.

1. *See* SUSAN K. SELL, PRIVATE POWER, PUBLIC LAW: THE GLOBALIZATION OF INTELLECTUAL PROPERTY RIGHTS (2003); Jerome H. Reichman, *Universal*

Subsequently, when Geneva-based institutions not only proved resistant to further increases in global patent protection,[2] but actually emerged as fora for pushing back on the TRIPS Agreement,[3] the United States shifted its efforts to bilateral and plurilateral agreements. The United States currently has "TRIPS-plus" intellectual property provisions in bilateral treaties negotiated with numerous developing countries, including Bahrain, Cambodia, Chile, Colombia, Costa Rica, Jordan, Nicaragua, Oman, Panama, Peru, and Vietnam.[4]

On the plurilateral patent front, the United States has also been active, if less immediately successful. Although some scholars had argued that the Anti-Counterfeiting Trade Agreement (ACTA) could inhibit access to generic drugs by encouraging parties to adopt border measures to interdict drugs suspected of patent or trademark infringement,[5] the treaty's border measures ultimately excluded patents.[6] Activity now centers on the Trans-Pacific Partnership Agreement (TPP), a plurilateral agreement being negotiated among the United States, Australia, New Zealand, Chile, Brunei, Singapore, Peru, Vietnam, Malaysia, and (most recently) Japan.

The patent provisions of these bilateral and plurilateral agreements focus on pharmaceuticals.[7] The provisions of the draft TPP agreement appear representative. Although an official draft of the TPP has not been made public, leaked drafts include a host of requirements related to pharmaceuticals.[8] Perhaps the most significant is a requirement that signatories allow so-called "secondary" patents. These are

---

Minimum Standards of Intellectual Property Protection under the TRIPS Component of the WTO Agreement, 29 INT'L LAWYER 345 (1995).

2. Rochelle Dreyfuss & Jerome H. Reichman, *Harmonization without Consensus: Critical Reflection on Drafting a Substantive Patent Law Treaty*, 57 DUKE L.J. 86 (2007).

3. Laurence R. Helfer, *Regime Shifting: The TRIPS Agreement and the New Dynamics of International Intellectual Property Making*, 29 YALE J. INT'L L. 1 (2004).

4. Susan K. Sell, *TRIPS Was Never Enough: Vertical Forum Shifting, FTAs, ACTA, and TPP*, 18 J. INTELL. PROP. L. 447 (2011).

5. Cynthia Ho, *Global Access to Medicine: The Influence of Competing Patent Perspectives*, 34 FORDHAM INT'L L.J. 1 (2011); Henning Grosse Ruse-Khan, *A Trade Agreement Creating Barriers to International Trade? ACTA Border Measures and Goods in Transit*, 26 AM. U. INT'L L. REV. 645 (2011).

6. In any event, the decision by the European Parliament to reject ACTA renders these concerns less pressing. *See* Press Release, European Parliament, European Parliament Rejects ACTA (July 4, 2012), *available at* http://www.europarl.europa.eu/news/en/pressroom/content/20120703IPR48247/html/European-Parliament-rejects-ACTA.

7. *See, e.g.*, Dominican Republic–Central America–United States Free Trade Agreement (CAFTA-DR), arts. 15.9–15.10, Aug. 5, 2004; United States–Oman Free Trade Agreement, arts. 15.8–15.9, Jan. 19, 2006.

8. *See* Trans-Pacific Partnership (TPP), Intellectual Property Chapter, Draft—Feb. 10, 2011, *available at* http://keionline.org/sites/default/files/tpp-10feb2011-us-text-ipr-chapter.pdf

patents that cover not the drug's primary active chemical compound but, instead, modified forms of the compound (e.g., salts, isomers), new uses of the compound, or new methods of using the compound.[9] According to one leaked draft, the TPP's secondary patent provision states that patents on new forms, uses, and methods "may satisfy the criteria for patentability, even if such invention does not result in the enhancement of the known efficacy of that product."[10] This provision appears to be a direct response to Article 3(d) of the Indian Patent Act, which prohibits granting patents on "the mere discovery of a new form of a known substance which does not result in the enhancement of the known efficacy of that substance. . . ."[11]

Other significant TPP provisions include a mandate of data exclusivity to the originator firm that supplies the clinical data for the national regulatory authority's approval of a drug. The draft agreement calls for five years of exclusivity from the date of marketing approval of a new "pharmaceutical product"[12] and three years from the date of marketing approval based on new "clinical information."[13] The TPP draft also calls for legal linkage between patents and approval by a regulatory authority of generic marketing. Specifically, prior to securing approval, the generic firm must show that patents relevant to the drug have expired, are invalid, or are not infringed.[14] This legal linkage has the potential to delay generic marketing, as does a TPP provision calling for patent term extension based on regulatory delays.[15] Finally, TPP signatories are supposed to agree to limits on pre-grant opposition proceedings and to border measure provisions to interdict goods suspected of patent infringement, including in-transit goods.[16]

[hereinafter TPP1]; Trans-Pacific Partnership (TPP), Intellectual Property Chapter (Selected Provisions), Draft—Sept. 2011, *available at* http://www.citizenstrade.org/ctc/wp-content/uploads/2011/10/TransPacificIP1.pdf [hereinafter TPP2].

9. *See generally* Michael Burdon & Kristie Sloper, *The Art of Using Secondary Patents to Improve Protection*, 3 Int'l J. Med. Marketing 226 (2003); European Commission, Pharmaceutical Sector Inquiry: Final Report 184-92 (2009), *available at* http://ec.europa.eu/competition/sectors/pharmaceuticals/inquiry/staff_working_paper_part1.pdf.

10. TPP1, *supra* note 8, art. 8.

11. The Patents (Amendment) Act, No. 15 of 2005, § 3(d) (Universal 2005).

12. TPP2, *supra* note 8, art. 9.2(a).

13. *Id.* art. 9.2(c).

14. *Id.* art. 9.5.

15. *Id.* art. 8.6(b).

16. TPP1, *supra* note 8, arts. 8.7, 14.4; *see also* Medecins Sans Frontieres, *Trading Away Health: How the U.S.'s Intellectual Property Demands for the Trans-Pacific Partnership Agreement Threaten Access to Medicines*, Aug. 2012, *available at* http://aids2012.msf.org/wp-content/uploads/2012/07/TPP-Issue-Brief-IAC-July2012.pdf.

II. Comparing International and Domestic Positions

To a significant extent, the pharmaceutical patent positions being advocated by the United States internationally overlap with current US law. As discussed further below, US law allows at least certain types of secondary patents. United States' law also confers five- and three-year data exclusivity terms for originator applications, establishes a linkage between patenting and approval of generics by the Food and Drug Administration (FDA), contains patent term extensions for delays in both patent processing at the US Patent and Trademark Office (USPTO) and regulatory approval delays at the FDA, and prohibits pre-grant opposition.

The extent to which the US-style approach to pharmaceutical patents is appropriate for poorer countries is the subject of tremendous dispute. Presumably, however, US executive branch advocacy for patent rights abroad *stronger* than those for which it advocates domestically would be even more controversial. Thus it is worth examining whether the US international posture is in fact more "pro-patent" than its domestic posture.

One candidate for such inconsistency is the US non-obviousness requirement. As compared with its application in others areas of technology, judicial application of the non-obviousness requirement in pharmaceuticals has been relatively robust.[17] Notably, the executive branch has consistently supported a strong non-obviousness requirement. In fact, in 2007, the executive branch was influential in persuading the US Supreme Court to issue an opinion, *KSR v. Teleflex*,[18] making it clear to lower courts that non-obviousness was an important gatekeeper requirement for all patent applications.[19]

The executive branch's domestic support for non-obviousness as a muscular requirement is arguably in tension with its international position that secondary patents on new formulations "may satisfy the criteria for patentability, even if such invention does not result in the enhancement of the known efficacy of that product."[20] Although US non-obviousness law does not deny secondary formulation patents, it typically subjects them to a requirement that the formulation have unexpected properties.[21] Moreover, during the Obama administration, the executive branch brought a Supreme Court case arguing that US antitrust law, US patent law, and the US "Hatch–Waxman" regime for patent linkage and generic

---

17. Arti Rai, *Building a Better Innovation System: Combining Facially Neutral Patent Standards with Regulation of End Product Therapeutics*, 45 HOUS. L. REV. 1037 (2008).

18. 550 U.S. 398 (2007).

19. Arti Rai, *Who's Afraid of the Federal Circuit?*, 121 YALE L.J. ONL. 135 (2011).

20. TPP1, *supra* note 8, art. 8.

21. Rai, *supra* note 17.

entry are all intended to encourage validity challenges by generics to secondary pharmaceutical patents that might be non-obvious or otherwise invalid.[22] Over the strong objections of the branded pharmaceutical industry, the Obama administration also successfully argued in favor of allowing generic entry in cases where the generic states it will not market its product for uses indicated in secondary method-of-use patents.[23]

Ultimately, in part because of executive branch advocacy of robust generic entry through Hatch–Waxman, the actual time to generic entry has remained stable over the past decade despite increases in secondary patenting.[24] In contrast, no suggestion that signatories are permitted to constrain secondary patenting through robust generic challenge appears in the language of the leaked TPP drafts. Thus, there is arguably some tension in the US executive branch's domestic and international positions.

A realist who assumes that states pursue only their immediate material interest (with more or less success depending on their military and economic clout)[25] would presumably find little reason for congruence between domestic and international positions. Even so, delving into the institutional dynamics that lead to divergent domestic and international positions highlights the agendas, and relative power, of different agencies.

Consider first the domestic context. Although certain agencies and White House offices may be quite supportive of strong domestic patent rights, they compete for influence on questions of domestic patent policy with agencies typically more wary of patent rights. The latter group includes the Antitrust Division of the Department of Justice (DOJ), the Federal Trade Commission (FTC), and the Department of Health and Human Services (HHS).[26] On the specific question of pharmaceutical patent policy, DOJ Antitrust, FTC, and HHS have all committed themselves quite publicly to facilitating domestic generic entry so as to achieve low-cost access for US patients.[27] Although the commitment of the

---

22. Brief for the Petitioner, FTC v. Actavis, Inc., 133 S. Ct. 2223 (2013) (No. 12-416), *available at* http://www.justice.gov/atr/cases/f291700/291720.pdf.

23. Arti Rai, *Use Patents, Carve-Outs, and Incentives—A New Battle in the Drug Patent Wars*, 367 N. Engl. J. Med. 491 (2012).

24. Scott C. Hemphill & Bhaven N. Sampat, *Evergreening, Patent Challenges, and Effective Market Life in Pharmaceuticals*, 31 J. Health Econ. 327 (2012).

25. Anne-Marie Slaughter, *International Relations, Principal Theories*, *in* Max Planck Encyclopedia of Public International Law (R. Wolfrum ed., 2011).

26. Rai, *supra* note 23.

27. *See, e.g.*, Fed. Trade Comm'n (FTC) & Dep't of Justice (DOJ), Improving Health Care: A Dose of Competition (2004), *available at* http://www.ftc.gov/reports/healthcare/040723healthcarerpt.pdf.

DOJ Antitrust and the FTC is largely ideological, agencies such as HHS, as well as powerful White House offices such as the Office of Management and Budget, have a direct budgetary interest in costs incurred by public insurance programs.

By contrast, although an executive branch constituency for low-cost global access does exist, its strength is limited. Domestically oriented agencies obviously do not have a strong financial stake in global access. To the extent they support such a goal in principle, they are likely to manifest such support through laudable but modest programs that emphasize voluntary incentives for patent holders. During the first Obama administration, for example, the USPTO established a program under which entities that demonstrate successful deployment of technology for humanitarian use can receive accelerated patent processing.[28]

International humanitarian programs, such as the President's Emergency Plan for AIDS Relief (PEPFAR), which provides antiretroviral (ARV) drugs to 5.1 million of the 8 million individuals currently receiving ARVs in low- and middle-income countries, do have a direct financial stake in low-cost global access. Congress scrutinizes the activities of PEPFAR (and the Department of State office that manages PEPFAR) to determine whether it is procuring ARV drugs at a low per-patient cost.[29] To fulfill this goal, PEPFAR relies heavily on generic drug producers such as India. Indeed, according to a recent Government Accountability Office (GAO) report, from 2005 to 2011, PEPFAR's reliance on generic ARVs increased from 15 percent by volume to 98 percent.[30] This increase saved PEPFAR approximately $934 million, allowing the program to expand coverage rapidly and to include World Health Organization–recommended products, such as fixed-dose combination products that do not have an equivalent branded formulation.[31]

Particularly in times of national economic stress, however, the power of these constituencies is not comparable to that of economically oriented offices such as the Office of the United States Trade Representative (USTR). Although USTR is a small office with a budget of less than $100 million, its career staff is highly skilled, it is committed to strong global IP rights, and it has close connections

---

28. Quentin Palfrey & Hillary Chen, *Rewarding Inventors Who Work to Solve Global Challenges*, WHITEHOUSE.GOV (Feb. 8, 2012 10:22 AM), http://www.whitehouse.gov/blog/2012/02/08/rewarding-inventors-who-work-solve-global-challenges (discussing the launch of the Patents for Humanity program).

29. U.S. GOV'T ACCOUNTABILITY OFFICE, GAO-13-345, PRESIDENT'S EMERGENCY PLAN FOR AIDS RELIEF (2013), *available at* http://www.gao.gov/products/GAO-13-345.

30. *Id.*

31. *Id.*

with US corporate leadership. The USTR thus exercises significant influence across changes in political leadership.

From a realist perspective, the ability of other countries to challenge the United States is also limited. Although certain large emerging economies such as India might invoke the manner in which US domestic law diverges from TPP as rhetorical support for their own efforts to develop alternatives to TPP-type regimes,[32] mere compelling rhetoric is unlikely to diminish US influence, particularly in the bilateral and plurilateral context.

Even perspectives that emphasize the role of multilateral institutions in mediating longer-term, iterated bargaining between states might deem the US approach a relatively stable equilibrium. On this view, deviations between the positions the United States is advocating internationally and domestically are sufficiently narrow that they may fail to impress even international institutions that have a "pro-access" agenda. Moreover, from the standpoint of delivering a clear public message, both these institutions and civil society groups may prefer to emphasize not narrow technical differences between US global and domestic approaches but instead what they see as direct links between the strong patenting and lack of access.

Going forward, the situation may change if the US healthcare spending dilemma becomes sufficiently untenable that more dramatic moves are made with respect to biopharmaceutical patents. But for the moment the United States is likely to be able to manage any tensions in its global and domestic positions.

I turn next to the potentially disruptive role of China. Because China's development strategy has generally been "pro-patent," and because it has specifically refrained from playing a leadership role in global controversies over access to medicines, China's rise probably will not disrupt the equilibrium that the United States has achieved on biopharmaceutical patents. Nonetheless, as discussed below, China is likely to play a role in numerous domestic deliberations outside the biopharmaceutical arena. On the global stage, China's relationship to the American-led TPP negotiations will be an issue worth watching.

III. The Rise of China

For several reasons, China's rise is qualitatively different from what has come before, at least since the rise of the Japanese economy in the 1970s and 1980s. Most obviously, unlike other nations, China represents a serious threat to US global economic dominance. Additionally, in contrast to large emerging

---

32. Amy Kapczynski, *Harmonization and Its Discontents: A Case Study of TRIPS Implementation in India's Pharmaceutical Sector*, 97 Cal. L. Rev. 1571 (2009).

economies such as India and Brazil, which have generally favored weaker patent protection, China has become substantially more aggressive than the United States in encouraging its firms to seek patents.

That said, the Chinese approach toward patents has not been unilaterally "pro-patent." To the contrary, in the context of standard setting and patents, China has sometimes taken positions that could be seen as "anti-patent," perhaps on the theory that standard setting represents a context where aggressive patent assertions have the potential to create particularly acute "holdup" problems for follow-on innovation.[33] Relatedly, China appears to view patents as an integral piece of a broader innovation policy. Indeed, since 2006, China has been implementing a sweeping "innovation strategy" in which patents represent one lever. By contrast, the US executive branch has historically resisted discussions of innovation policy, lest such discussions be seen as veering too close to an endorsement of industrial policy.

How the unique case of China's patent and innovation policy affects US positions domestically and internationally is a story that is only beginning to unfold. After discussing China's patent and innovation policy, I offer some preliminary thoughts.

IV. China's Patent and Innovation Policy

In 2011, the Chinese State Intellectual Property Office (SIPO) promulgated a National Patent Development Strategy in which it set forth a goal of two million patent applications by 2015. By way of comparison, the USPTO had 25 percent of that figure (540,000 applications) in 2012. To be sure, the two million figure misses important distinctions between invention patents that have been examined for inventiveness and appropriate scope by an administrative agency (so-called invention patents) and utility models available without examination. The figure also fails to distinguish between domestic invention patents and invention patents filed by firms based in other countries.

However, even if we focus only on domestic invention patents, China's trajectory is notable. Since 2000, most of the growth in China's domestic patent applications has come from domestic invention patents.[34] As a consequence, although foreign and domestic inventors filed approximately similar numbers of invention patents until 2005, by 2008 the domestic inventor percentage of applications

33. Mark Lemley & Carl Shapiro, *Patent Holdup and Royalty Stacking*, 85 Tex. L. Rev. 1991 (2007).

34. Albert Hu & Gary Jefferson, *A Great Wall of Patents: What Is Behind China's Recent Patent Explosion?*, 90 J. Dev. Econ. 57 (2009).

filed had increased to 67.1 percent. Additionally, by 2008, the absolute number of domestic invention patent grants was equal to foreign invention patent grants. By 2011, SIPO was ahead of the USPTO with respect to domestic invention patent applications filings and grants in the respective offices. SIPO had 415,829 domestic invention patent applications, as compared with 247,750 for the USPTO, and 112,347 domestic invention grants, as compared with 108,626 for the USPTO.[35]

Another traditional proxy for patents that are likely to have commercial value is patents filed at WIPO via the Patent Cooperation Treaty (PCT). China's numbers here are also rising rapidly. In 2011, China occupied the fourth spot, behind the United States, Japan, and Germany.[36]

In part, these increasing numbers reflect a deliberate strategy by the Chinese government to encourage patenting. According to the US International Trade Commission (USITC), provincial governments offer payments to Chinese companies to support the filing of patents in China and overseas.[37] Chinese companies may also qualify for tax relief, research awards, and government contracts through patenting.[38] In some cases, China is giving specific funds to small and medium-sized firms that file abroad. For example, in October 2009, the Chinese Ministry of Finance announced a 100 million RMB fund to help small and medium enterprises (SMEs), as well as public research institutions, file abroad.[39]

That said, the firms most responsible for high levels of Chinese patenting both domestically and abroad are large information and communications technology (ICT) firms.[40] Indeed, in 2011, two Chinese telecommunications firms, ZTE Corp. and Huawei Technologies, had the largest and third-largest number of PCT applications respectively. For these firms, the Chinese government may be interested not so much in innovation per se as in the need to assemble defensive patent portfolios in order to compete in the global ICT marketplace.

---

35. Peter Yu, *Five Oft-Repeated Questions About China's Recent Rise as a Patent Power*, 2013 CARDOZO L. REV. DE NOVO 78 (2013).

36. WORLD INTELLECTUAL PROP. ORG. (WIPO), WORLD INTELLECTUAL PROPERTY INDICATORS 63 (2012), *available at* http://www.wipo.int/export/sites/www/freepublications/en/intproperty/941/wipo_pub_941_2012.pdf.

37. U.S. INT'L TRADE COMM'N (USITC), Pub. No. 4199, China: IP Infringement, Indigenous Innovation Policies and Framework for Measuring the Effects on the U.S. Economy (2010), *available at* http://www.usitc.gov/publications/332/pub4199.pdf.

38. *Id.*

39. Tu Lei, *Ministry of Finance to Support Overseas Patents*, ALIBABA.COM (Oct. 13, 2009, 8:02 AM), http://news.alibaba.com/article/detail/business-in-china/100182805-1-mof-support-overseas-patents.html.

40. Markus Eberhardt, Christian Helmers & Zhihong Yu, *Is the Dragon Learning to Fly? An Analysis of the Chinese Patent Explosion*, (Univ. of Oxford Ctr. for the Study of African Econs. Working Paper Series, No. 2011-15, 2011).

The Chinese government has not, however, been uniformly "pro-patent." To the contrary, in 2009, the government proposed regulations that would have required patented technology relevant to national standards to be subject to compulsory licensing either royalty-free or at below-market rates.[41] Although the government then acceded to commentary by foreign firms strongly opposing these proposed regulations, it still manages standard setting centrally.[42]

China's patent strategy is just one part of an overall fifteen-year innovation strategy. The strategy, summarized in a 2006 document entitled the "National Medium and Long Term Plan for Science and Technology Development," not only specifies domestic invention patent goals (further refined in the 2011 patent plan noted earlier) but more broadly calls for China to become an "innovation-oriented society" by 2020.[43] The approach suggested by the plan includes: an emphasis on "zuzhu chuangxin"—a somewhat ambiguous concept often translated as "indigenous innovation,"[44] prioritizing certain areas of frontier technology, the development of an "integrated national system of institutions supportive of R&D", increased funding for R&D, and assimilation of foreign technology.

V. The US Reaction

Thus far, the US executive branch has reacted most strongly, at least at the rhetorical level, to the Chinese "indigenous innovation" strategy. It has particularly objected to the government procurement preferences for Chinese firms, which form part of the strategy. These objections may form part of the basis for US resistance to Chinese participation in the TPP negotiations.

The executive branch's reaction to the notable growth in Chinese patent applications and patents has emphasized quality problems, particularly in utility applications that are not examined. For example, even the White House Office of the Intellectual Property Enforcement Coordinator, which traditionally advocates for strong intellectual property protection internationally, has publicly expressed concern about the quality of "unexamined utility model patents."[45] Presumably

41. USITC, *supra* note 37.

42. *Id.* U.S. Int'l Trade Comm'n (USITC), China: Effects of IP Infringement and Indigenous Innovation Policies on the U.S. Economy (2011) *available at* http://www.usitc.gov/publications/332/pub4226.pdf.

43. Cong Cao, Richard P. Suttmeier & Denis F. Simon, *China's 15-Year Science and Technology Plan*, 59 Physics Today 38–43 (2006).

44. Yu, *supra* note 35.

45. Mark MacCarthy, *China's Utility Model System: A Perfect Storm for Patent Trolls*, Digital Discourse (Sept. 20, 2012), http://www.siia.net/blog/index.php/2012/09/chinas-utility-model-patent-system-a-perfect-storm-for-patent-trolls (last visited Sept. 5, 2013).

these concerns are also being expressed in private bilateral meetings, and will become more acute as China's patent litigation docket continues to increase significantly and we see more cases of Chinese firms using patents arguably of dubious quality to sue US firms.

Agreement across the executive branch about poor Chinese patent quality could bolster efforts to improve patent quality domestically. Alternatively, the executive branch might feel compelled to compete with China on patent applications and grants. The domestic effect in the United States of Chinese efforts in the areas of standard setting and innovation policy could also be significant. In standard setting, the executive branch has, since 1998, had a formal policy of deferring to private sector efforts. Specifically, Office of Management and Budget (OMB) Circular A-119, entitled "Federal Participation in the Development and Use of Voluntary Consensus Standards and in Conformity Assessment Activities," directs agencies that need to adopt standards generally to defer to decisions made by private sector standard-setting bodies.[46] This deferential policy has, however, been criticized for being outdated, particularly in light of increasing concerns about holdup when private standard-setting bodies have unclear policies on patents. Even so, the executive branch has not amended Circular A-119.

A recent statement by OMB, USTR, and the White House Office of Science and Technology Policy (OSTP) does indicate that, in certain areas of "national priority," such as health information technology and promoting clean energy, federal agencies might be more proactive with respect to standards.[47] Overall, however, the document reaffirms a commitment to private-sector leadership in the area. A desire on the part of certain executive branch offices (and like-minded congressional leaders) to present a strong counter-example to China may constrain domestic action on standards. Similarly, the need to draw a stark contrast with China may represent at least a rhetorical obstacle to the adoption of highly proactive executive branch innovation policies.

Ultimately, concerns about drawing contrasts with China are probably less important than purely domestic factors such as pro-market ideology, political gridlock, or dramatic fiscal constraints. Even so, of any country that could affect US patent and innovation policy in the domestic context, China is the likeliest candidate.

---

46. Office of Mgmt. & Budget, Exec. Office of the President, OMB Circular No. A-119, Federal Participation in the Development and Use of Voluntary Consensus Standards and in Conformity Assessment Activities, 63 Fed. Reg. 8546 (Feb. 19, 1998), *available at* http://www.whitehouse.gov/omb/circulars_a119_a119fr.

47. Memorandum from the Office of Mgmt. & Budget, U.S. Trade Rep. & Office of Science & Tech. Pol'y (Jan. 17, 2012), *available at* http://www.whitehouse.gov/sites/default/files/omb/memoranda/2012/m-12-08.pdf.

On the global patent stage, one key question to watch will be China's relationship to the TPP negotiations. With respect to the possibility of Chinese entry, US executive branch views will be influenced in part by wariness regarding the impact of Chinese patent and innovation policy on the interests of US firms. By the same token, any aggressive move by the United States to exclude China may have unintended consequences in terms of causing China to more aggressively pursue patent and innovation strategies that explicitly favor Chinese firms.

4 TRANSNATIONAL LEGAL ORDERING AND ACCESS TO MEDICINES

Gregory Shaffer and Susan K. Sell*

Introduction

One can no longer speak meaningfully about national law without speaking of law in its transnational context. Transnational legal ordering now spans public regulatory law, private business law, and human rights law. These transnationally shaped legal domains interact. For example, at the international level, one cannot speak meaningfully about the international law and politics of trade and intellectual property (IP) law under the World Trade Organization (WTO) without considering other transnational legal norms, such as economic and social rights. The issues that these legal domains cover intersect across levels of social organization, from national legal systems to international regimes.

This chapter proceeds in five sections. Section I examines the complications of normatively assessing intellectual property protection in light of trade-offs involving multiple goals, including knowledge creation, trade governance, economic development, and public health. Section II discusses the development and interaction of different transnational legal orders (TLOs), by which we refer to legal orders that go all the way down from international law to national law and local legal practice, and that interact recursively in both bottom-up

* Gregory Shaffer is Melvin C. Steen Professor, University of Minnesota Law School, and Affiliated Professor of Political Science, University of Minnesota; Susan Sell is Professor of Political Science and International Affairs at George Washington University. We wish to thank participants at conferences in Bologna, Italy, and at FLACSO in Buenos Aires, for their comments. We also wish to thank Julia Norsetter for her valuable research assistance.

and top-down ways.[1] Different transnational legal orders reflect diverse, and potentially conflicting, values and priorities that affect national law and practice. The ensuing sections address the tensions between a TLO that prioritizes trade and intellectual property protection, and one that foregrounds human rights values, including public health. Section III presents the rise of the TLO for intellectual property, and the struggles that continue to shape it. Section IV examines the rise of an overlapping and rival TLO based on the human right to health. This emergent TLO directly challenges some of the central tenets of the trade and intellectual property TLO. Suddenly states are pushing back and forth over what they had appeared to have "settled" multilaterally under the Agreement on Trade-Related Aspects of Intellectual Property Rights (TRIPS Agreement). Competing values that once traveled along separate but parallel tracks are now intersecting and directly clashing, forcing hard policy choices. Section V concludes with strategies that developing countries can adopt, and have adopted, to address the challenges of implementing these two transnational legal orders to advance their development goals, including access to medicines.

I. Normatively Evaluating IP Protection

From a normative perspective, there is no singular optimal level of intellectual property protection either globally or nationally. Without transnational legal ordering, the level that a nation chooses will vary based on the values and priorities that it seeks to advance. In part, these choices will reflect its level of development and its comparative advantage in innovation and imitation. Different forms of intellectual property protection are important for the creation and diffusion of knowledge for economic development. Yet the proper level of intellectual property protection is an extremely difficult and contested empirical question.[2] Too little protection can undercut incentives for creative and innovative activity, and thus result in a reduction of social welfare. Too much protection can obstruct innovation and raise costs for consumers. Thus too much protection may reduce both the production of knowledge and the ability of the public to use it. Intellectual property protection may jeopardize the pursuit of other social goals as well, such as access to medicines. Policymakers must navigate this tension

---

1. For two broader projects on transnational legal ordering and transnational legal orders respectively, see TRANSNATIONAL LEGAL ORDERS (Terence Halliday & Gregory Shaffer eds., forthcoming 2013); Gregory Shaffer, *Transnational Legal Ordering and State Change*, *in* TRANSNATIONAL LEGAL ORDERING AND STATE CHANGE (Gregory Shaffer ed., 2013) [hereinafter Shaffer, *Transnational*].

2. *See, e.g.*, KEITH MASKUS, INTELLECTUAL PROPERTY RIGHTS IN THE GLOBAL ECONOMY (2000).

and balance the different policy goals that reflect divergent values and priorities. In short, determining the appropriate scope and level of intellectual property protection raises both empirical issues and value choices.

The debates over pharmaceutical patent protection and the WTO TRIPS Agreement raise at least five potentially competing normative priorities with profound distributive implications. These are: monopoly incentives to induce innovation; free trade; economic development, public health; and the right to life and human dignity.[3] We discuss each in turn.

Knowledge has some public goods attributes. Once knowledge enters the public domain it is no longer excludable, and one's consumption of it does not diminish its availability. And yet, among the main challenges that intellectual property policy confronts is how to generate this knowledge, both efficiently and equitably. States have granted temporary monopoly privileges such as patents in order to provide incentives to innovate. The offering of such legal protection removes knowledge from the public domain for a defined time period.[4]

A legal order of liberalized trade also provides benefits. Although it is true that liberalized trade produces both winners and losers and may give rise to distributional conflicts, one can make a normative case that trade liberalization results in a wider variety of products being made available at lower prices. Liberalized trade rewards efficient production and exchange. A liberal trade legal order also helps to provide predictability to encourage long-term investment and increased productivity. It further helps countries to avoid the kind of tit-for-tat protectionist measures that gave rise to the drastic reduction in global trade during the 1930s that had devastating economic and political consequences. A stable and predictable legal order of liberalized trade can help countries avoid the temptation of engaging in trade discrimination during financial crises.[5]

A legal order can also prioritize economic development goals in which poorer countries may "free ride" on the innovations of others, benefitting from imitating their products and making them available for their citizens at reduced

---

3. For an exploration of this issue in the context of WTO dispute settlement, see Gregory Shaffer, *Recognizing Public Goods in WTO Dispute Settlement: Who Participates? The Case of TRIPS and Pharmaceutical Patent Protection*, 7 J. INT'L ECON. L. 459 (2004).

4. Knowledge can be subject to some excludability, as through trade secrets, so that it is not a pure public good. *See, e.g.*, Joseph E. Stiglitz, *Knowledge as a Global Public Good*, *in* GLOBAL PUBLIC GOODS: INTERNATIONAL COOPERATION IN THE 21ST CENTURY 306–25 (Inge Kaul, Isabelle Grunberg & Marc Stern eds., 1999) [hereinafter GLOBAL PUBLIC GOODS] (labeling knowledge an "impure public good").

5. *See, e.g.*, Nancy Birdsall & Robert Lawrence, *Deep Integration and Trade Agreements: Good for Developing Countries*, *in* GLOBAL PUBLIC GOODS, *supra* note 4, at 128, 133.

prices. Imitation and reverse engineering are core mechanisms of this strategy.[6] Through this process, these countries can develop new industries that eventually can become competitive innovators, as happened with Japan and Korea, and now appears to be occurring with China. At a later time, these countries may switch their policies toward granting greater intellectual property protection to further innovation.

Public health has been viewed as a public good because all individuals benefit from the global curtailment and eradication of diseases and they do not diminish that good when they benefit from it.[7] Incentives for innovation, such as patents, can contribute to the development of drugs and treatments to combat disease. Although ideally everyone could benefit equally from effective advances in medicine, in practice these goods have not been equally available to everyone, in large part because of their cost.

Finally from a human rights perspective, the availability of disease-eradicating medicines can be viewed as an integral component of the human right to life and dignity. The normative priority of the human right to life and dignity can compete or conflict with other important values. In particular, sharp tensions between human rights and property rights produce political, social, and economic controversies over pharmaceutical patent protection, as well as issues regarding the funding of medicines compared to other policy initiatives in a world of scarcity. "Rights" language, whether applied to human rights or property rights, implies a sense of entitlement in which the right trumps other policy goals in terms of hierarchy.

The recognition and enforcement of patent rights under TRIPS and other intellectual property treaties can generate incentives for the production of knowledge and new drugs for public health and the protection of human life. But the HIV/AIDS pandemic underscores how the protection of pharmaceutical patent rights may also undermine the benefits of liberalized trade by restricting consumer access through monopoly pricing power and staving off generic drug competition. These policies can reduce public health options for containing diseases, and raise human rights concerns insofar as lack of access leads to unnecessary and preventable deaths.

Societies inevitably must make choices among conflicting normative priorities.[8] The key institutional question is: Who decides the appropriate balancing

---

6. Susan K. Sell, Power and Ideas: North–South Politics of Intellectual Property and Antitrust (1998).

7. *See, e.g.*, Todd Sandler & Daniel G. Arce, *A Conceptual Framework for Understanding Global and Transnational Goods for Health*, 23 Fiscal Stud. 195 (2002).

8. *See* Gregory Shaffer, *International Law and Global Public Goods in a Legal Pluralist World*, 31 Eur. J. Int'l L. 669 (2012).

among policy goals?[9] Choices among liberalized trade, patent protection, economic development, public health, and human rights reflect different values, priorities, perspectives, and generate considerable uncertainty. Policymakers must make institutional choices despite these trade-offs. The rise of different transnational legal orders in different substantive domains help frame approaches to these normative and institutional questions.

II. The Rise of Transnational Legal Orders

Social orders increasingly are legalized transnationally, reflecting processes of economic and cultural globalization.[10] As a result, issues that formerly were viewed as national in scope have been transformed to have transnational (and sometimes global) dimensions.[11] Whether the social order concerns protections for intellectual property rights or public health, removal of barriers to trade or observance of human rights, the ordering of responses to these issues inside the state and across national frontiers increasingly involves legal norms that transcend national boundaries. We can define a transnational legal order (TLO) as a *collection of formalized legal norms and associated organizations and actors that authoritatively order behavior that spans national jurisdictions.*[12]

The concept of the creation of TLOs addresses the social and political construction of problems and their resolution. The construction of a problem is not a "natural" one, but involves actors with particular perceptions, including of their own interests, which advance particular imaginings of solutions to the problem to create a new order. They increasingly do so by working to create and institutionalize a legal order that is transnational in scope. Different transnational legal orders emerge that vary in terms of geographic scope and legal scope. TLOs can vary geographically, ranging from bilateral to regional, plurilateral, multilateral, and global agreements. Such range is illustrated by law applying to the European Union, to North America under the North American Free Trade Agreement (NAFTA), to parts of South America under the Andean Pact, to countries spanning the Pacific under the proposed Trans-Pacific Partnership Agreement,

---

9. Gregory Shaffer, *Comparative Institutional Analysis and a New Legal Realism*, 2013 Wisc. L. Rev. 607 (2013).

10. *See* Shaffer, *Transnational*, *supra* note 1.

11. Shaffer, *supra* note 8.

12. "Associated organizations and actors" is construed broadly to include any organization or social formation, including networks, and actors may refer both to collective actors and to individuals whose activities and careers cross national boundaries. *See* Transnational Legal Orders, *supra* note 1; Shaffer, *Transnational*, *supra* note 1.

and to countries spanning the globe under the WTO and the United Nations (UN). They also vary by substantive legal focus, such as over trade, investment, finance, environmental protection, intellectually property, public health, and human rights. These TLOs, of varying legal and geographic scope operate simultaneously, sometimes incognizant of each other, sometimes in competition with each other, and sometimes in antagonistic interaction with each other. They can thus be seen as varying along two dimensions: (1) the degree of settlement of the transnational legal norms, and (2) the extent of issue alignment among existing TLOs.[13]

The politics of international trade and intellectual property law in relation to pharmaceutical patents increasingly involves the interaction of such transnational legal orders. In particular, it involves the interaction of TLOs focused on trade and intellectual property protection, and TLOs focused on human rights. These transnational legal orders involve recursive interaction between transnational legal norm-making and national implementation. They may feature conflicts among TLOs not only transnationally but also within nation states.

The issue of access to medicines, in particular, has been characterized by a high degree of contestation and a significant lack of alignment among international institutions over the last decade. The result has been significant tension among TLOs that previously were distinct and segregated, those of intellectual property, trade, and the human right to health. As Laurence Helfer shows, one can see the interaction of these TLOs in terms of three time periods. In the first period, the TLOs were separate and did not interact, and their norms were relatively open-ended and unsettled transnationally. In the second period, an intellectual property–trade TLO (IP TLO) became predominant and its norms appeared to be relatively settled transnationally. In the third period, the IP TLO on pharmaceutical patents became unsettled in light of developments in a parallel TLO on the economic and social right to healthcare.[14]

Before the mid-1990s, the IP and economic and social rights TLOs were entirely distinct and relatively unsettled. Under the IP TLO, there was considerable debate within the World Intellectual Property Organization (WIPO) as to what form of special and differential treatment should be applied to developing countries, and there were few substantive requirements for patents other than application of non-discrimination norms. Similarly, under the UN International Covenant on Economic, Social and Cultural Rights (ICESCR), the scope of an

---

13. Transnational Legal Orders, *supra* note 1.

14. *See* Laurence Helfer, *Pharmaceutical Patents and the Human Right to Health: The Contested Evolution of the Transnational Legal Order on Access to Medicines*, *in* Transnational Legal Orders, *supra* note 1; *see also* Laurence R. Helfer & Graeme W. Austin, Human Rights and Intellectual Property: Mapping the Global Interface (2011).

economic and social right to health was contested among states and civil society actors. Yet, although there was contestation within these TLOs, there was no interaction between them. States were thus largely unconstrained in adapting the intellectual property and right to health legal norms that they desired, including regarding access to medicines.

From the mid-1990s until around 2000, in light of the incorporation of intellectual property into the WTO package of agreements, and the negotiation of new bilateral and plurilateral agreements containing TRIPS-plus provisions,[15] countries around the world adopted new intellectual property laws, including for the patenting of pharmaceutical products. There appeared to be settlement as to the applicable legal norms in this domain, which many observers and civil society actors decried. As a result, countries were considerably more constrained in providing inexpensive drugs to patients pursuant to a right to health.

Since around 2000, a new transnational legal order for the economic and social right of access to medicines has become institutionalized. It has done so as a result of a number of factors, and in particular the resistance of states, nonstate actors, and international human rights and health organizations to the IP TLO. The salience of the AIDS epidemic, where improved access to medicines could save millions of lives, helped to precipitate the emergence of this rival TLO. Powerful business interests supported by their home states nonetheless continued to push for greater IP protection, including through bilateral and plurilateral agreements, leading to considerable normative dissensus and contestation, as well as competing constraints for developing countries from these rival transnational legal orders.

III. The Construction of the TLO for Intellectual Property

Law is often viewed as producing social order out of conflict. But in studying law, we also need to assess how law reflects power. Law, in this sense, is Janus-faced, involving both the creation of normative order and the reflection of power. Legal realist scholars have long noted this aspect of law, but it has often been elided in legal scholarship.[16] As the political scientist Michael Barnett writes:

> Power and legitimacy [such as legal legitimacy] . . . are not conflicting concepts but rather are complementary ones. The powerful, too, want their

---

15. "TRIPS-plus" refers to requirements that exceed the minimum standards mandated by the TRIPS Agreement.

16. *See* Hanoch Dagan, *The Realist Conception of Law*, 57 U. Toronto L.J. 607 (2007); Victoria Nourse & Gregory Shaffer, *Varieties of New Legal Realism: Can a New World Order Prompt a New Legal Theory?*, 95 Cornell L. Rev. 62 (2009).

> actions to be viewed as legitimate, if only to maintain their power and further their interests. Even the powerful, in this view, cannot act in an expedient and narrowly self-interested manner and must observe international society's underlying rules and norms.[17]

The United States and the European Union use a variety of means to shape intellectual property rules transnationally. They use the carrot of granting access to their valuable markets, and the stick of threatening to withdraw such access, in order to pressure countries to raise intellectual property standards. They have the material resources to engage in negotiations in multiple fora to advance their interests, and are the protagonists in creating institutions that shape and constrain what can be discussed within them. They have the knowledge resources to shape understandings, including through the provision of technical assistance and capacity building, which frame issues in particular ways. For example, representatives of pharmaceutical trade associations work with US and EU officials to draft "model" laws and to teach as "faculty" in workshops organized by WIPO on intellectual property law and its enforcement.[18]

The very concept of "intellectual property" is a social and political construction. Alternative organizing concepts could be used and institutionalized through law, such as "access to knowledge," or "knowledge for development," but the concept of "intellectual property" has been globalized. The concept is, for example, at odds with many traditional and indigenous approaches to knowledge and ideas, and countries have varied over time in the extent of intellectual property protection available in light of their stage of development. The original signatories to the Berne and Paris Conventions, respectively, for copyright and patent protection were primarily developed countries.[19] Over time, countries have adopted intellectual property laws that mirrored those of the former colonizer, or tailored intellectual property laws to reflect perceptions of domestic interests.[20]

---

17. Michael N. Barnett, *Bringing in the New World Order: Liberalism, Legitimacy, and the United Nations*, 49 World Pol. 526, 544 (1997).

18. *See* Carolyn Deere, The Implementation Game: The TRIPS Agreement and the Global Politics of Intellectual Property in Developing Countries (2009).

19. Brazil, Ecuador, El Salvador, Guatemala, for example, were the only developing country signatories to the 1883 Paris Convention for the Protection of Industrial Property. *Id.* at 37.

20. *See* Ruth L. Okediji, *The International Relations of Intellectual Property: Narratives of Developing Country Participation in the Global Intellectual Property System*, 7 Sing. J. Int'l & Comp. L. 316, 326–37 (2011); Deere, *supra* note 18, at 37–38 (contrasting Latin American

For example, a number of developing countries would not issue patents on pharmaceutical products and agricultural chemicals.[21] In fact, before 1970, the United States was among a small minority of countries that offered patents on pharmaceutical products.[22]

Much has been written about the strategies of the powerful that led to the creation of the TRIPS Agreement.[23] In a strategic move, the United States integrated intellectual property into trade policies during the 1980s, using the threat of withdrawing market access to constrain the ability of developing countries to resist demands for greater intellectual property protection. The United States used, and continues to use, its Special 301 procedure under which countries are placed on watch lists of varying priority regarding their intellectual property protections.[24] From 1985 to 1995, at least eighteen countries reformed national laws to strengthen patent protection due to other measures taken by the United States and European Union.[25] Working closely with private trade associations, these strategies eventually facilitated the inclusion of the TRIPS Agreement as part of the WTO's single package of agreements.[26] A country could not benefit from WTO market access opportunities unless it also agreed to be bound by the TRIPS Agreement's IP requirements.

---

adaptation of intellectual property rules for national purposes with African countries copying the laws of the colonial metropole).

21. By the 1970s, India's patent laws allowed patents on the methods and processes related to medicines, but not on the medicines themselves. Brazil did not permit patents on pharmaceutical processes and products from 1971 to 1996. DEERE, *supra* note 18, at 40.

22. *Id.*

23. *See, e.g.*, PETER DRAHOS & JOHN BRAITHWAITE, INFORMATION FEUDALISM: WHO OWNS THE KNOWLEDGE ECONOMY? (2003) [hereinafter DRAHOS & BRAITHWAITE, INFORMATION FEUDALISM]; CHRISTOPHER MAY & SUSAN SELL, INTELLECTUAL PROPERTY RIGHTS: A CRITICAL HISTORY (Lynne Rienner ed., 2005); SUSAN SELL, PRIVATE POWER, PUBLIC LAW: THE GLOBALIZATION OF INTELLECTUAL PROPERTY RIGHTS (2003).

24. OFFICE OF THE U.S. TRADE REP., EXEC. OFFICE OF THE PRESIDENT, 2012 SPECIAL 301 REPORT (2013), *available at* www.ustr.gov/sites/default/files/2012 Special 301 Report_0.pdf.

25. A number of US and EU domestic laws allowed for the coercion of developing countries. The Generalized System of Preferences (GSP) Renewal Act (1984), for example, authorized the US president to withdraw tariff concessions from developing countries with weak intellectual property protection. At the time, 140 countries received US GSP benefits. DRAHOS & BRAITHWAITE, INFORMATION FEUDALISM, *supra* note 23, at 88.

26. Agreement on Trade-Related Aspects of Intellectual Property Rights, Apr. 15, 1994, Marrakesh Agreement Establishing the World Trade Organization, Annex 1C, 33 I.L.M. 81 (1994) [hereinafter TRIPS Agreement].

A. The TRIPS Agreement and Its Flexibilities

The TRIPS Agreement set forth unprecedented strong mandatory, minimum standards of intellectual property rules, and applied them to countries regardless of level of development, although with different deadlines for implementation, which have been extended for least-developed countries.[27] The Agreement obliges WTO members to implement these minimum standards for most categories of intellectual property. For patents, it requires protection for all fields of technology for a minimum of twenty years.[28] It also requires states to ensure that private intellectual property rights holders can take "effective action" against intellectual property infringement, and help enforce intellectual property rights through border measures and criminal law.[29] The liberal trade economist Jagdish Bhagwati thus writes that the TRIPS Agreement "positions the WTO primarily as a collector of intellectual property–related rents on behalf of multinational corporations."[30] The financial burden of the agreement's implementation and enforcement is considerable. World Bank economists have estimated that the agreement has cost less-developed countries $60 billion a year.[31]

Paradoxically, although the TRIPS Agreement imposes significant obligations on states to protect the holders of intellectual property rights, commentators now commonly refer to the flexibilities in its provisions to counter US and EU demands for ever greater intellectual property requirements. Although the most stringent at the time it was signed, the TRIPS Agreement provides for interpretive options, such as what constitutes novelty and an inventive step for

---

27. Deadlines for implementation are varied and have been revised; developing countries were granted a five-year transition period until January 2000. Least-developed countries (LDCs) were initially expected to implement TRIPS by 2006, a deadline that was later extended to 2013 after a request for extension. *See* DEERE, *supra* note 18, at 67–69 tbl.3.1. Negotiations for a further extension for the LDCs are currently underway. William New, IP-WATCH (May 24, 2013), www.ip-watch.org/2013/05/24/ldc-fight-for extension-of-trips-transition-continues ("At WTO, LDC Fight for Extension of TRIPS Transition Continues").

28. The Paris Convention is silent on the question of patent duration. DEERE, *supra* note 18, at 66.

29. TRIPS Agreement, *supra* note 26, art. 41.

30. Jagdish Bhagwati, *What It Will Take to Get Developing Countries into a New Round of Multilateral Trade Negotiations*, INST. FOR AGRIC. & TRADE POL. (Sept. 23, 2002), http://www.iatp.org/files/What_It_Will_Take_to_Get_Developing_Countries_.htm.

31. *See* J. Michael Finger, *Introduction and Overview*, *in* POOR PEOPLE'S KNOWLEDGE: PROMOTING INTELLECTUAL PROPERTY IN DEVELOPING COUNTRIES 4 (J. Michael Finger & Philip Schuler eds., 2004). *See also* WORLD BANK, GLOBAL ECONOMIC PROSPECTS AND THE DEVELOPING COUNTRIES 136 (2002) ("Bangladesh anticipated one-time costs of administrative TRIPS compliance (drafting legislation) amounting to $250,000, and over $1.1 million in annual costs for judicial work, equipment, and enforcement efforts.").

purposes of granting a patent. Among the most important of the exceptions for patents for developing countries are: the right of parallel importation, the right to grant compulsory licenses, and "Bolar" exceptions for generic drug companies to prepare a drug under patent for marketing authorization once the patent expires.[32]

The TRIPS Agreement restricts the use of compulsory licensing in certain respects, and in certain circumstances, for predominant use for only the domestic market.[33] The explosion of the AIDS crisis and international protests against the effect of pharmaceutical patents on public health placed pressure on the WTO to facilitate the use of compulsory licenses and other TRIPS "flexibilities." In 2001, WTO members negotiated the Doha Declaration on Public Health[34] and they adopted a waiver in August 2003 that enables any member country to import pharmaceutical products under a compulsory license (although the conditions of the waiver are still contested for being too stringent).[35] This decision could

---

32. TRIPS Agreement, *supra* note 26, arts. 6, 8.1, 8.2, 30, 31; CARLOS CORREA, INTEGRATING PUBLIC HEALTH CONCERNS INTO PATENT LEGISLATION IN DEVELOPING COUNTRIES 70–77, 93–100 (2000), *available at* http://apps.who.int/medicinedocs/pdf/h2963e/h2963e.pdf.

33. Article 6 of the 2001 Doha Declaration underlined the fact that TRIPS Article 31(f) might limit the ability of countries with insufficient manufacturing capabilities to import cheaper generics, as the products manufactured under compulsory licenses must be "predominantly" for the domestic market. World Trade Organization (WTO), Declaration on the TRIPS Agreement and Public Health, WT/MIN(01)/DEC/2 (2001) [hereinafter Doha Declaration], *available at* www.who.int/medicines/areas/policy/tripshealth.pdf; *see also* Carlos M. Correa, *Public Health and Patent Legislation in Developing Countries*, 3 TUL. J. TECH. & INTELL. PROP. 1, 45–47 (2001).

34. Doha Declaration, *supra* note 33.

35. This waiver is subject to the terms of the "30 August Decision." *See* DEERE, *supra* note 18, at 82. Additionally, on December 6, 2005, the General Council adopted a decision to amend TRIPS in an attempt to solve problems posed by Article 31(f); however, the amendment will not become part of the Agreement until two-thirds of WTO members ratify it. *See* Press Release, WTO, Members OK Amendment to Make Health Flexibility Permanent (Dec. 6, 2005), *available at* http://www.wto.org/english/news_e/pres05_e/pr426_e. The original deadline was December 2007, which was later extended to December 2009, and then to December 2011. *See* WTO, Amendment of the TRIPS Agreement, Decision of 6 December 2005, WT/L/641 (2005), *available at* http://www.wto.org/english/tratop_e/trips_e/wtl641_e.htm; WTO, Amendment of the TRIPS Agreement—Extension of the Period for the Acceptance by Members of the Protocol Amending the TRIPS Agreement, Decision of 18 December 2007, WT/L/711 (2007), *available at* http://www.wto.org/english/tratop_e/trips_e/wt-l-711_e.pdf; WTO, Amendment of the TRIPS Agreement—Second Extension of the Period for the Acceptance by Members of the Protocol Amending the TRIPS Agreement, Decision of 17 December 2009, WT/L/785 (2009), http://www.wto.org/english/tratop_e/trips_e/wt-l-785_e.pdf. As of December 2011, forty-one countries had ratified the amendment. *See WTO Members Accepting Amendment of the TRIPS Agreement*, WTO (Sept. 10, 2013), http://www.wto.org/english/tratop_e/trips_e/amendment_e.htm. "Countries wanting to import under the 'paragraph 6' system

assist those developing countries that lack the domestic manufacturing capacity to produce pharmaceuticals themselves under a compulsory license. A number of developing countries have since issued compulsory licenses, mostly related to public health or government use.[36] Perhaps even more important, some larger developing countries have used the threat of compulsory licenses to gain leverage in negotiation with patent holders to lower prices.[37]

The "Bolar" exception (named after the US *Roche v. Bolar* case) permits generic drug producers to make use of patented materials during the patent term for the purpose of acquiring marketing approval and obtaining registration of a generic version of a branded drug. This exception helps to expedite bringing generic versions to market upon expiration of the patent.[38] Because it facilitates a more rapid entry of lower-cost generic drugs, the Bolar exception can lower prices for both government health agencies and individual consumers. A WTO panel in the *Canada—Patent Protection of Pharmaceutical Products* case interpreted Article 30 of the TRIPS Agreement to permit such exceptions.[39] Developing countries, however, so far have not widely adopted Bolar-type exceptions, particularly

---

to have to notify the WTO in two ways. They have to announce once that they intend to make use of the system, and then they have to supply information each time they use it." *Notifications by Importing WTO Members,* WTO, http://www.wto.org/english/tratop_e/trips_e/public_health_notif_import_e.htm (last visited Sept. 5, 2013) (listing Rwanda as a state that has imported pursuant to paragraph 6).

36. These countries include Brazil, Ghana, Guinea, Indonesia, Malaysia (three times), Mozambique, Rwanda, Swaziland, Taiwan, Thailand (four times), Zambia, and Zimbabwe. DEERE, *supra* note 18, at 83. Importantly, developed countries have also referenced government use as a basis to interfere with patent rights for public health purposes. *See* Jerome H. Reichman & Catherine Hasenzahl, *Non-voluntary Licensing of Patented Inventions: Historical Perspective, Legal Framework Under TRIPS, and an Overview of the Practice in Canada and the USA,* (ICSTD/UNCTAD Project on IPRs and Sustainable Dev., Issue Paper 5, 2002), *available at* ictsd.net/downloads/2008/06/cs_reichman_hasenzahl.pdf.

37. *See* Frederick M. Abbott & Jerome H. Reichman, *The Doha Round's Public Health Legacy: Strategies for the Production and Diffusion of Patented Medicines under the Amended TRIPS Provisions,* 10 J. INT'L ECON. L. 921, 951 (2007) ("[T]he Brazilian government has used the threat of compulsory licenses to pressure foreign multinational patent holders to significantly lower the prices charged for [antiretroviral treatments]....").

38. *See generally* DEERE, *supra* note 18, at 81; Jerome H. Reichman, *Intellectual Property in the Twenty-First Century: Will the Developing Countries Lead or Follow?*, 46 HOUS. L. REV. 1115, 1138 (2009) (describing a Chinese Bolar exception "which permits generic producers to reverse-engineer patented medicines to conduct clinical trials prior to the expiration of the patent").

39. *See* Report of the Panel, *Canada—Patent Protection of Pharmaceutical Products* WT/DS114/R (Mar. 17, 2000), *available at* http://www.wto.org/english/tratop_e/dispu_e/7428d.pdf; *see also* DEERE, *supra* note 18, at 81.

because many lack a generic industry. A 2006 survey of 106 developing countries showed that less than 10 had adopted explicit Bolar-type provisions.[40]

In contrast, a much greater number of developing countries have permitted parallel imports of pharmaceutical products. Pharmaceutical companies provide opportunities for parallel imports whenever they practice price discrimination across diverse markets. A patent holder may charge "x" for a drug in one territorial market, and "x+100" for that same drug in another territorial market. Parallel importing provides a country with the opportunity to import a drug from the territorial market that offers the "x" price, rather than the "x+100" price, thereby containing costs. Such practices create an incentive for developing countries to import patented products from lower-priced markets. A 2006 sample of fifty-four developing countries showed that thirty-three permitted parallel imports. The countries varied, however, as to whether they recognized "exhaustion" of intellectual property rights upon first sale in the world or only in a region.[41] TRIPS Article 6 permits countries to choose their preferred exhaustion system. Under a national or domestic exhaustion regime, when an intellectual property rights owner sells a good in a foreign market, she retains her right to prevent that genuine good from entering the domestic market where the intellectual property right has not been "exhausted." In other words, she has a right to prevent parallel importation. Generally, international exhaustion regimes, in contrast to national ones, should expand access to medicines at a lower price, because they facilitate parallel importation and deny the patent holder the right to block such imports because a first sale of the good anywhere in the world exhausts the domestic intellectual property right.[42]

B. The Proliferation of Bilateral and Plurilateral Agreements after TRIPS

For leading industrialized countries such as the United States, the TRIPS Agreement represented only a foundation on which further intellectual property protection could be built. They considered the TRIPS standards to be a floor, constituting absolute minimum standards, whereas many developing countries

---

40. Deere, *supra* note 18, at 79–81. For example, "Argentina's patent law states that any third party may use a product or process protected by patent prior to its expiration 'to obtain the information required for the approval of a product or process by the competent authority' so that it may be marketed following the patent expiration." *Id.*

41. *Id.* at 75.

42. Carsten Fink, *Comment: Competition Law as a Means of Containing Intellectual Property Rights*, *in* International Public Goods and Transfer of Technology under a Globalized Intellectual Property Regime 771 (Keith Maskus & Jerome Reichman eds., 2005).

considered them to be a ceiling above which they would not have to go. At the conclusion of the TRIPS negotiations, a leading US advocate proclaimed that "we got 95% of what we wanted."[43]

In subsequent developments, the 5 percent that TRIPS proponents failed to achieve has prominently featured in US and EU bilateral and plurilateral agreements. These agreements have eliminated many of the flexibilities provided in the TRIPS Agreement.[44] For example, many of them prohibit the practice of parallel importation of cheaper patented drugs and restrict the conditions under which countries may issue compulsory licenses. These provisions are examples of the exact features that US- and OECD-based pharmaceutical firms sought, but failed to obtain, in the TRIPS negotiations. These new agreements are invariably TRIPS-plus agreements, because they expand requirements from the base that the TRIPS Agreement set, and TRIPS-minus agreements, in that they eliminate many TRIPS flexibilities.[45] As a result, these bilateral, regional, and plurilateral intellectual property agreements reduce access to lower-cost products and may jeopardize generic competition.

With respect to essential medicines, TRIPS-minus provisions *limit* a number of pertinent flexibilities under the TRIPS Agreement, such as parallel importation, and compulsory licenses. To restrict parallel importation, TRIPS-minus provisions may provide the patent owner with an exclusive right to prohibit parallel importing contractually, thus curtailing a country's ability to access more-affordable patented drugs,[46] as under the US FTAs with Australia, Morocco, and Singapore.[47] The granting of compulsory licenses may be limited to narrowly specified conditions, considerably more restrictive than under the TRIPS Agreement.[48] The US FTAs with

---

43. SUSAN K. SELL, PRIVATE POWER, PUBLIC LAW: THE GLOBALIZATION OF INTELLECTUAL PROPERTY RIGHTS 48 (2003).

44. Susan K. Sell, *TRIPS Was Never Enough: Vertical Form Shifting, FTAS, ACTA, AND TPP*, 18 J. INTELL. PROP. L. 447, 448 (2011). Deere discusses a number of other tools used to push developing countries to implement stronger intellectual property protection, including bilateral investment treaties, Special 301 threats, diplomatic threats, and industry pressure, such as direct lobbying by the pharmaceutical sector in developing countries. *See* DEERE, *supra* note 18, at 154–64. The WTO website indicates that, "[a]s of 15 November 2011, some 505 RTAs (counting goods and services notifications separately) have been notified to the GATT/WTO." *Regional Trade Agreements*, WTO, http://www.wto.org/english/tratop_e/region_e/region_e.htm (last visited Sept. 5, 2013).

45. *See, e.g.*, Sell, *supra* note 44, at 448.

46. *Id.* at 453–54.

47. *See* DEERE, *supra* note 18, at 338–40 (listing the US FTAs with Singapore, Morocco, and Australia as containing provisions where "patent holders may limit parallel imports of pharmaceutical products through licensing contracts").

48. Sell discusses the detrimental effect of the interplay between TRIPS-plus provisions. For example, the ability of a state to issue compulsory licenses is affected by data exclusivity provisions and drug registration/patent linkage provisions. As a result, "prospective licensees are

Australia, Jordan, Singapore and Vietnam, for example, contain provisions that limit compulsory licenses to situations such as national emergencies and public noncommercial use, and as an antitrust remedy.[49]

The TRIPS-plus provisions include expanded data-exclusivity requirements, patent linkage for marketing approvals, and lengthier patent terms.[50] Some FTAs have included more stringent data-exclusivity requirements designed to force generic manufacturers to generate their own test data, as opposed to being able to rely on the findings of the brand name companies.[51] The aim is to further delay generic competition and thus prolong the payment of higher prices to US and EU right holders. The United States has FTAs with provisions that expand test data protection for pharmaceutical products with at least thirteen developing countries, including Vietnam, Jordan, Chile, Singapore, Morocco, Australia, Bahrain, and the countries of the Dominican Republic–Central American Free Trade Agreement (DR–CAFTA): that is, Costa Rica, El Salvador, Guatemala, Honduras, Nicaragua, and the Dominican Republic.[52] These data-exclusivity provisions significantly extend TRIPS Article 39.3, which requires WTO members to protect undisclosed test data against unfair competition.[53]

Many FTAs also require that the registration of medicines by health authorities be linked to patent protection. As a consequence, governmental authorities cannot register a drug that remains under patent protection. The treaties require regulatory authorities to notify the patent holder if a generic competitor attempts to register a generic version of a drug. This enrolls regulatory authorities into the patent enforcement process, mimicking the US Hatch–Waxman regime. Agreements that contain provisions prohibiting marketing approval of a generic drug during the patent term, unless authorized by the patent owner, include the US FTAs with Singapore, Chile, Morocco, Australia, Bahrain, and the countries of the DR–CAFTA.[54]

---

unlikely to replicate test data, and government cannot normally wait until a new set of test data has been developed." Sell, *supra* note 44, at 454.

49. *See* Deere, *supra* note 18, at 338–40.

50. *See* Carlos M. Correa, *Implications of Bilateral Free Trade Agreements on Access to Medicines*, 84 Bull. World Health Org. 399, 401 (2006); *see also* Sell, *supra* note 44, at 452–55.

51. For example TRIPS-plus data exclusivity requirements are included in US FTAs with Morocco, Bahrain, Singapore, Australia, and DR–CAFTA. *See* Carsten Fink & Patrick Reichenmiller, *Tightening TRIPS: Intellectual Property Provisions of U.S. Free Trade Agreements*, *in* Trade, Doha, and Development: A Window into the Issues 291 (Richard Newfarmer ed., 2006).

52. *See* Deere, *supra* note 18, at 338–40; Sell, *supra* note 44, at 453.

53. *See* Sell, *supra* note 44, at 453.

54. *See* Deere, *supra* note 18, at 338–40. TRIPS-plus US FTAs include: United States–Chile Free Trade Agreement, U.S.-Chile, Jan. 1, 2004, *available at* www.ustr.gov/trade-agreements/free-trade-agreements/chile-fta/final-text; United States–Singapore Free Trade Agreement,

In combination, many US FTAs prohibit generic producers from challenging a patent until after it has been granted. United States' proposals for the Trans-Pacific Partnership negotiations would expressly forbid pre-grant opposition.[55] Pre-grant opposition provisions in national laws permit generic producers to contest the patent application as it is filed. Without pre-grant opposition, generic producers wishing to challenge the validity of a drug patent have to wait until the patent has been granted, and may have to engage in a costly and time-consuming judicial process. These provisions thus effectively block a generic company from obtaining marketing authorization from health authorities during the period that a patent is contested, again creating delays for generic companies and stalling competition.[56]

In addition, a number of FTAs expand the patent term to make up for delays in the application process. They thus may significantly extend the twenty-year period of patent protection provided under the TRIPS Agreement. The United States has FTAs with Vietnam, Jordan, Singapore, Chile, Morocco, Australia, Bahrain, and the DR–CAFTA, which all contain provisions that provide patent term extensions for delays caused by regulatory approval processes.[57]

The bilateral TRIPS-plus agreements are complemented by a unique "certification" process pursuant to which in order for the agreement to take effect, the US executive must certify to the US Congress that the foreign country has modified its laws in compliance with the agreement.[58] This certification process,

---

U.S.-Singapore, Jan. 1, 2004, *available at* http://www.ustr.gov/trade-agreements/free-trade-agreements/singapore-fta/final-text; United States–Morocco Free Trade Agreement, U.S.-Morocco, Jun. 15, 2004, *available at* http://www.ustr.gov/trade-agreements/free-trade-agreements/morocco-fta/final-text; United States–Australia Free Trade Agreement, U.S.-Australia, Jan. 1, 2005, *available at* http://www.ustr.gov/trade-agreements/free-trade-agreements/australian-fta/final-text; United States–Bahrain Free Trade Agreement, U.S.-Bahrain, Jan. 11, 2006, *available at* http://www.ustr.gov/trade-agreements/free-trade-agreements/bahrain-fta/final-text; United States–Central America–Dominican Republic Free Trade Agreement, Aug. 5, 2004, (CAFTA-DR), *available at* http://www.ustr.gov/trade-agreements/free-trade-agreements/cafta-dr-dominican-republic-central-america-fta/final-text.

55. Trans-Pacific Partnership (TPP), Intellectual Property Chapter, Draft—February 10, 2011, *available at* http://keionline.org/sites/default/files/tpp-10feb2011-us-text-ipr-chapter.pdf. For discussion, see Sean Flynn, Brook Baker, Margot Kaminski & Jimmy Koo, *The U.S. Proposal for an Intellectual Property Chapter in the Trans-Pacific Partnership Agreement*, 28 Am. U. Int'l L. Rev. 106 (2012).

56. *See* Deere, *supra* note 18, at 338.

57. *See id.* at 338–40; *see also* Correa, *supra* note 50, at 401.

58. For example, the US Congress passed an implementation bill in the United States–Central America–Dominican Republic Free Trade Agreement that requires the president to certify that the signatories have complied with the FTA provisions before its entry into force. Dominican Republic–Central America–United States Free Trade Agreement Implementation Act § 101, Pub. L. No. 109–53, 119 Stat. 462 (2005); *see also Bush Secures CAFTA Vote in Last Hours with Renewed Textile Pledge*, Inside U.S. Trade (July 29, 2005), http://wtonewsstand.com/Inside-US-Trade/Inside-U.S.-Trade-07/29/2005/menu-id-445.html.

in practice, requires the foreign country to show its draft legislation to the US executive and receive its approval. The United States typically demands changes in such legislation, triggering a new, one-sided negotiation over that country's implementing legislation. The aim of the US executive is, once more, to reduce any flexible interpretations of the agreement's requirements. In contrast, and asymmetrically, under US law, the bilateral agreement will have no effect if it conflicts with any existing or subsequent US law.[59]

These TRIPS-plus agreements are part of a US strategy to divide developing countries. Those countries that have already increased intellectual property protections under an FTA with the United States have nothing to lose from their being multilateralized (and perhaps much to gain as other developing countries and their constituencies will then have to pay increased royalties and license fees to US companies as well). Over time, as developing country holdouts become isolated, a tipping point could be reached through which the more stringent legal requirements become multilateralized.[60]

The United States so far has been unsuccessful in negotiating enhanced patent protections in multilateral fora since the TRIPS Agreement. Thus the United States has pursued not only bilateral FTAs, but also plurilateral agreements. In particular, it initially failed in its efforts to negotiate a Substantive Patent Law Treaty (SPLT) under the auspices of WIPO.[61] The United States then led efforts with Japan to strengthen intellectual property enforcement through a new Anti-Counterfeiting Trade Agreement (ACTA), which the USTR calls "the highest-standard plurilateral agreement ever achieved concerning the enforcement of intellectual property rights."[62] Signed by eight countries on October 1, 2011,[63] non-signatories may be

---

59. Frederick Abbott, *Intellectual Property Provisions of Bilateral and Regional Trade Agreements in Light of U.S. Federal Law* 3–5 (UNCTAD-ICTSD Project on IPRs and Sustainable Dev., Issue Paper No. 12, 2006), *available at* http://papers.ssrn.com/so13/papers.cfm?abstract_id=1912621.

60. Peter Drahos, *BITS and BIPS: Bilateralism in Intellectual Property*, 4 J. World Intell. Prop. 791 (2001).

61. *See, e.g.*, Sell, *supra* note 44, at 455 ("In intellectual property norm setting, developing countries successfully rebuffed the US and European efforts to pursue a TRIPS-plus Substantive Patent Law Treaty (SPLT) at WIPO."). After painstaking negotiations to resume formal discussion, in late 2010 the Standing Committee on the Law of Patents agreed to begin work on multilateral agreements on exceptions and limitations and patent quality. William New & Kaitlin Mara, *WIPO Returns to Substantive Patent Law Talks After 5 Years with Balance*, IP-Watch (Oct. 16, 2010, 12:58 PM), http://www.ip-watch.org/2010/10/16/wipo-returns-to-substantive-patent-law-talks-after-5-years-with-balance; Kimberlee Weatherall, *Three Lessons from ACTA and Its Political Aftermath*, 35 Suffolk Transnat'l L. Rev. 575 (2012).

62. Anti-Counterfeiting Trade Agreement (ACTA), Dec. 3, 2010, 50 I.L.M 243 (2011), *available at* http://www.ustr.gov/acta (last visited Sept. 5, 2013).

63. The signatories were the United States, Australia, Canada, Korea, Japan, New Zealand, Morocco, and Singapore. *Id.*

affected because the United States undoubtedly will press for countries to adopt its provisions as a condition for new FTAs.[64] Indeed, the Office of the United States Trade Representative (USTR) maintains that "it looks forward to partnering with developing countries through ACTA, and cooperating with ACTA partners to provide technical assistance to developing countries."[65]

Many commentators have noted the unbalanced nature of ACTA, finding that it favors right holders.[66] Unlike the WTO Doha negotiating text[67] and the WIPO Development Agenda,[68] ACTA includes no development dimension. It lacks language concerning limitations and exceptions for fair use of copyrights, and early drafts of ACTA would have endorsed in-transit seizures of generic drugs.[69] It creates a new ACTA Committee to oversee intellectual property enforcement under its terms. The United States now views ACTA as the new (raised) floor for intellectual property enforcement.

ACTA constitutes a "club" agreement in which only industrialized countries interested in strong intellectual property protection were invited to participate.[70]

---

64. Sell, *supra* note 44, at 456 ("Inevitably, TRIPS-plus ACTA provisions will reappear in bilateral and regional trade agreements going forward in an effort to raise global standards of protection.").

65. Office of the United States Trade Rep., Exec. Office of the President, Trade Facts: Anti-Counterfeiting Trade Agreement (Aug. 4, 2008), *available at* http://www.ustr.gov/sites/default/files/uploads/factsheets/2008/asset_upload_file760_15084.pdf.

66. *See*, e.g., Sell, *supra* note 44, at 457; Sean Flynn, *Learning from ACTA: Toward a Positive Agenda for TPP*, INFOJUSTICE.ORG (Mar. 3, 2012), http://www.infojustice.org/archives/8650; Margot E. Kaminski, *An Overview and the Evolution of the Anti-Counterfeiting Trade Agreement*, 21 ALB. L.J. SCI. & TECH. 385, 394–95 (2011); J. Janewa Oseitutu, *Value Divergence in Global Intellectual Property Law*, 87 IND. L.J. 1639, 1667–69 (2012); Andrew Rens, *Enforcement Theater: The Enforcement Agenda and the Institutionalization of Enforcement Theater in the Anti-Counterfeiting Trade Agreement*, 35 SUFFOLK TRANSNAT'L L. REV. 553, 556 (2012).

67. World Trade Organization (WTO), Ministerial Declaration of 18 December 2005, WT/MIN (05)/DEC (2005), *available at* www.wto.org/english/thewto_e/minist_e/min05_e/final_text_e.htm.

68. World Intellectual Prop. Org. (WIPO), *Proposal by Argentina and Brazil for the Establishment of a Development Agenda for WIPO*, WO/GA/31/11, Aug. 27, 2004 [hereinafter WIPO, Proposal] *available at* http://www.wipo.int/edocs/mdocs/govbody/en/wo_ga_31/wo_ga_31_11.pdf; WIPO, *Report of the Provisional Committee on Proposals Related to a WIPO Development Agenda*, A/43/13 Rev., Sept. 17, 2007, *available at* www.wipo.int/edocs/mdocs/govbody/en/a_43/a_43_13_rev.doc.

69. Kaminski, *supra* note 66, at 394–95 ("The only time ACTA mentions limitations or exceptions is in its discussion of technological circumvention measures, stating that 'appropriate' limitations or exceptions may be maintained or adopted by parties in providing for remedies for the circumvention of technological measures."); Weatherall, *supra* note 61, at 577.

70. Daniel Gervais, *International Decision*, 103 AM. J. INT'L L. 549, 555 (2009); Peter K. Yu, *ACTA and Its Complex Politics*, 3 WIPO J. 1 (2011)[hereinafter Yu, *ACTA*]; Peter K. Yu, *TRIPS*

Argentina, Brazil, India, and China were thus excluded. The draft texts were also kept from the public so that there was less opportunity to shape debates. The final agreement, moreover, was not even being presented to the US Congress for ratification, but rather went into effect as a so-called sole "executive agreement" on the grounds that it is a trade agreement, and not an intellectual property agreement.[71] However, it is a trade agreement in name only, and not in substance.[72] Had it been negotiated as an intellectual property agreement under WIPO's auspices, the treaty would have had to be presented to the Senate, requiring a two-thirds supra-majority approval under the Treaty Presentment Clause of the US Constitution. The future of ACTA, nonetheless, remains uncertain after the European Parliament overwhelmingly voted against it. Hundreds of thousands of Europeans took to the streets to protest ACTA in February 2012, and the European Union has suspended ratification.[73]

The United States is now actively negotiating to join another plurilateral agreement, the Trans-Pacific Partnership (TPP) Agreement. Leaked negotiating texts indicate that the United States aims to introduce more TRIPS-plus provisions.[74] In these negotiations, the United States now looks to ACTA and the 2011 US–Korea FTA as providing the new "gold standard" for intellectual property protection and enforcement.[75] In sum, industrialized countries continue to work with the pharmaceutical and entertainment industries to build an enhanced intellectual property transnational legal order.

---

and Its Achilles' Heel, 18 J. Intell. Prop. L. 479, 513 (2011) [hereinafter Yu, *TRIPS*]. On club goods, see Richard Cornes & Todd Sandler, The Theory of Externalities, Public Goods, and Club Goods (1996).

71. Jack Goldsmith & Lawrence Lessig, *Anti-Counterfeiting Agreement Raises Constitutional Concerns*, Wash. Post, Mar. 26, 2010, http://www.washingtonpost.com/wp-dyn/content/article/2010/03/25/AR2010032502403.html (discussing the Obama administration's assertion that ACTA will be adopted as a sole executive agreement).

72. Sell, *supra* note 44, at 456 (noting ACTA is an intellectual property treaty, a fact counterintuitive in light of its name).

73. Sean Flynn, *supra* note 66; Susan K. Sell, *Revenge of the "Nerds": Collective Action Against Intellectual Property Maximalism in the Global Information Age*, 15 Int'l Studies Rev. 67 (2013).

74. For a detailed analysis, see Flynn et al., *supra* note 55.

75. *U.S. TPP IPR Proposal Offers Middle Ground Between "May 10" Korea FTA*, Inside U.S. Trade (Oct. 27, 2011), http://insidetrade.com/Inside-US-Trade/Inside-U.S.-Trade-10/28/2011/us-tpp-ipr-proposal-offers-middle-ground-between-may-10-korea-fta/menu-id-710.html.

IV. The Emergence of a Countervailing TLO for Economic and Social Rights

Implementing the TRIPS Agreement has been more of a challenge than the United States and Europe initially contemplated.[76] It has been so, in large part, because a countervailing process of transnational legal ordering has emerged with a different normative frame, one with a human rights focus.[77] Developing countries, non-state actors, and UN-based organizations have advanced this frame at the international and national levels both to counter the push for ever-stronger intellectual property protections and to spur recognition and application of a right to health in public policy more generally. As the right to health becomes institutionalized nationally and transnationally, it too can be viewed as a TLO. It interacts with and rivals the TLO for intellectual property protection when the two overlap in addressing an issue such as access to life-saving medicines.

At the international level, developing countries have found support from the World Health Organization (WHO) Assembly, the UN Committee on Economic, Social and Cultural Rights, the UN High Commissioner on Human Rights (UNHCHR), the UN Special Rapporteur on the Right to Health, and the Joint UN Programme on HIV/AIDS. The WHO Constitution establishes that "the enjoyment of the highest attainable standard of health is one of the fundamental rights of every human being."[78] The right to health also is incorporated in multiple human rights instruments, including the 1948 Universal Declaration on Human Rights, the International Covenant on Economic Social and Cultural Rights, and the United Nations Human Rights Council.[79] The UNHCHR identified access to essential medicines as "a vital component of fulfilling the right to health."[80] The UN Millennium Development Goals (MDG) include access to essential medicines; the MDG Gap Task Force Report 2012 advised that "developing countries should carefully assess possible adverse impacts on access to medicines when adopting TRIPS-Plus provisions."[81]

---

76. *See, e.g.*, Yu, *TRIPS, supra* note 70 (discussing the particular difficulty TRIPS enforcement provisions have posed).

77. *See* Lisa Forman, *"Rights" and Wrongs: What Utility for the Right to Health in Reforming Trade Rules on Medicines?*, 10 Health & Hum. Rts. 37 (2008); Helfer, *supra* note 14.

78. World Health Org., World Intellectual Property Org. & World Trade Org., Promoting Access to Medical Technologies and Innovation: Intersections between Public Health, Intellectual Property and Trade 40 (2013) [hereinafter Promoting Access] (quoting the Constitution of the WHO).

79. *Id.* at 40–41.

80. *Id.* at 42.

81. *Id.* at 43.

Such rights advocacy helped to build alternative normative framings and pressure for the amendment and interpretation of the TRIPS Agreement in light of social welfare concerns, as reflected in the negotiation of the Doha Declaration on TRIPS and Public Health in 2001. Egypt, Brazil, and Argentina, supported in parallel by powerful campaigns by nongovernmental organizations (NGOs), proposed and succeeded in establishing the WIPO Development Agenda.[82] The WIPO Development Agenda has been viewed as "a call for restraint on the part of developed countries in their crusade for ever-stronger IP protection."[83] After six years of debate, WIPO adopted a series of recommendations related to the Development Agenda in 2007, and temporarily suspended negotiations for a new Substantive Patent Law Treaty that industrialized countries had promoted.[84] Even though critics contend that WIPO remains part of a transnational legal culture focused on intellectual property protection as defined predominantly by industrialized country legal systems, WIPO's Development Agenda reflects a move to a more development-friendly approach to intellectual property, endorsing access to technology and benefit-sharing.[85]

Institutionalizing the human right to health has gained momentum at the regional and national levels as well. Laurence Helfer and Karen Alter have analyzed the Andean Community's successful regional pro-access to medicines decisions in the Andean Tribunal.[86] They demonstrate that Andean governments *required* the adoption of TRIPS flexibilities in implementation. The Andean Tribunal has been able to combat TRIPS-plus provisions by enforcing bans on pipeline patents and second-use patents, and by placing limits on data exclusivity.[87]

---

82. DEERE, *supra* note 18, at 128; WIPO, Proposal *supra* note 68.

83. DEERE, *supra* note 18, at 128. On the various ways to frame an understanding of the WIPO Development Agenda, see THE DEVELOPMENT AGENDA: GLOBAL INTELLECTUAL PROPERTY AND DEVELOPING COUNTRIES (Neil Netanel ed., 2009) [hereinafter DEVELOPMENT AGENDA].

84. DEERE, *supra* note 18, at 132. However, as of this writing, negotiations on a Substantive Patent Law Treaty have been renewed by WIPO. New & Mara, *supra* note 61.

85. *See, e.g.*, Sell, *supra* note 44, at 450; Peter K. Yu, *A Tale of Two Development Agendas*, 35 OHIO N.U. L. REV. 456, 515–22 (2009) (describing the Development Agenda's evolution into an "agenda on pro development reform of the international intellectual Property system"). For more cautious assessments, see Ruth Okediji, *History Lessons for the WIPO Development Agenda*, *in* DEVELOPMENT AGENDA, *supra* note 83, at 137–161; Keith Maskus, The *WIPO Development Agenda: A Cautionary Note*, *in* DEVELOPMENT AGENDA, *supra* note 83, at 163–174.

86. Laurence Helfer & Karen Alter, *The Influence of the Andean Intellectual Property Regime on Access to Medicines in Latin America*, *in* BALANCING WEALTH AND HEALTH: GLOBAL ADMINISTRATIVE LAW AND THE BATTLE OVER INTELLECTUAL PROPERTY AND ACCESS TO MEDICINES IN LATIN AMERICA (Rochelle Dreyfuss & César Rodríguez-Garavito eds., forthcoming 2014).

87. *Id.*

Most important, the development of a rival economic and social rights frame is reflected at the national level in the growth of new transnational constitutional law movements.[88] Although there is a small body of literature on the role of constitutionalism in international law,[89] it is this bottom-up transnational constitutional movement that is the key development, representing a sweep of judicialization and rights adjudication around the world.[90] This movement diffuses constitutional law norms across national jurisdictions that can be used to interpret international law commitments, such as those under the TRIPS Agreement and the ICESCR. As a joint report of the WHO, WIPO and the WTO notes "[b]y 2009, 135 countries had incorporated aspects of the right to health in their national constitutions."[91]

Brazil, India, and South Africa have enjoyed some notable success in implementing provisions facilitating access to medicines and the use of TRIPS flexibilities. They all successfully have warded off legal challenges from originator drug firms and their government sponsors.[92] India, for example, defended itself against the Swiss pharmaceutical firm Novartis's challenge to India's right to implement flexibilities to prevent evergreening.[93] In April 2013 the Indian Supreme Court upheld the Indian Patent Office's decision to deny Novartis a patent on

---

88. *See, e.g.*, VICKI JACKSON, CONSTITUTIONAL ENGAGEMENT IN A TRANSNATIONAL ERA (2009); Tom Ginsburg, *Locking in Democracy: Constitutions, Commitment and International Law*, 38 N.Y.U. J. INT'L L. & POL. 707 (2006).

89. JEFFREY L. DUNOFF & JOEL P. TRACHTMAN, RULING THE WORLD?: CONSTITUTIONALISM, INTERNATIONAL LAW, AND GLOBAL GOVERNANCE (2009); JAN KLABBERS ET AL., THE CONSTITUTIONALIZATION OF INTERNATIONAL LAW (2009).

90. *See* Ran Hirschl, *The Judicialization of Politics*, *in* THE OXFORD HANDBOOK OF LAW AND POLITICS 126–27 (Keith E. Whittington, R. Daniel Kelemen & Gregory A. Caldeira eds., 2008); Tom Ginsburg & Mila Veerstteg, *The Global Spread of Constitutional Review: An Empirical Assessment*, J.L. ECONS. & ORG. (forthcoming 2014). As Ginsburg and Veersteeg write, "[b]y our account, some 38% of all constitutional systems had constitutional review in 1951; by 2011, 72% of the world's constitutions gave courts the power to supervise implementation of the constitution and set aside legislation for incompatibility.... Arguably, this trend is one of the most important phenomena in late 20th and early 21st century government." *Id.* For a political science analysis of diffusion from below, see Ann Towns, *Norms and Social Hierarchies: Understanding International Policy Diffusion "From Below,"* 66 INT'L ORG. 179 (2012).

91. PROMOTING ACCESS, *supra* note 78, at 40.

92. *See, e.g.*, HELFER & AUSTIN, *supra* note 14; Smita Narula, *The Rights-Based Approach to Intellectual Property and Access to Medicines: Parameters and Pitfalls* (N.Y.U. Pub. L. and Legal Theory Working Papers, Paper No. 299, 2011), *available at* http://lsr.nellco.org/nyu_plltwp/299.

93. *See, e.g.*, Sandeep Rathod, *Evergreening: A Status Check in Selected Countries*, 7 J. GENERIC MEDS. 227 (2010)

its drug Gleevec.[94] India and Brazil also lodged complaints at the WTO against the European Union and the Netherlands, in May 2010, for seizing generic drugs in transit. In its complaint, India invoked relevant TRIPS provisions as well as the Doha Ministerial Declaration on TRIPS and Public Health and Article 12(1) of the ICESCR, "which recognizes the right of all persons to the enjoyment of the highest attainable standard of physical and mental health."[95] India and Brazil obtained a favorable settlement as a result.[96]

Smaller countries such as Ecuador also have been champions of promoting access to health. For example, a coalition of local generic drug producers, social, environmental, and indigenous organizations enjoyed the strong support of Ecuador's President Correa, who rejected a proposed US–Ecuador FTA. In doing so, they framed intellectual property rights as "an urgent health issue."[97]

Around the world, national courts have become bolder in overseeing human rights protections, including economic and social rights, such as the right to health. National courts in Argentina, Colombia, Costa Rica, Ecuador, El Salvador, Peru, South Africa, Kenya, and Venezuela have recognized HIV/AIDS patients' rights to access to medicines.[98] As Laurence Helfer points out, "a 2006 study identified seventy-one cases from twelve countries invoking a right of access to medicines, with

94. After a seven-year legal battle, the Indian Supreme Court upheld a decision of the Indian Patent Office to refuse to grant a patent on the Swiss pharmaceutical company Novartis's Gleevec. *See* Novartis AG v. Union of India, Nos. 2706–16 (S.C. Apr. 1, 2013) (India), *available at* http://judis.nic.in/supremecourt/imgs1.aspx?filename=40212.nic.in/outtoday/patent.pdf. Under Indian patent law, Section 3(d) requires that patents only be granted on new and innovative compounds that have demonstrated increased therapeutic efficacy. Patents (Amendment) Act, No. 15 of 2005, § 3(d) (Universal 2005). This provision aims to curtail the pharmaceutical firms' practice of evergreening. *See* Ellen't Hoen, *A Victory for Global Public Health in the Indian Supreme Court*, J. PUB. HEALTH POL'Y 1, 1–5 (2013).

95. Lisa Forman, *An Elementary Consideration of Humanity? Linking Trade-Related Intellectual Property Rights to the Human Right to Health in International Law*, 14 J. WORLD INTELL. PROP. 168 (2011) (quoting Indian complaint).

96. Whether the settlement goes far enough is still a matter of debate. *See* Brook K. Baker, *Settlement of India/EU WTO Dispute re Seizures of In-Transit Medicines: Why the Proposed EU Border Regulation Isn't Good Enough*, (Am. Univ. Washington Coll. Law, PIJIP Research Paper No. 2012-02, 2012), *available at* http://digitalcommons.wcl.american.edu/research/24.

97. Antoni Verger & Barbara van Paassen, *Human Development* vis-à-vis *Free Trade: Understanding Developing Countries' Positions in Trade Negotiations on Education and Intellectual Property Rights*, 2012 REV. OF INT'L POL. ECON. 22.

98. Helfer, *supra* note 14; Patricia Asero Ochieng, (2009) Pet. No. 409 (H.C.K.) (Kenya), *available at* http://www.aidslawproject.org/wp-content/uploads/2012/04/Judgment-Petition-No-409-of-2009-Anti-counterfeit-case.pdf. The Kenyan High Court struck down a US-sponsored anti-counterfeiting law. The Court cited the right to health and determined that the law would have curbed access to generic medicines.

a success rate of 83 percent."[99] This national move in developing countries of incorporating a human rights frame can also contribute, and in a number of countries already has contributed, to interpreting and shaping the definition of intellectual property rights at the national level.

V. Strategies for Developing Countries

The development of a rival human rights transnational legal order has created pressure to shape patent law in developing countries for social ends. Toward this end, developing countries can adopt pragmatic strategies to shape and tailor intellectual property legal norms and practices. This chapter concludes by raising five complementary strategies: two concerning substantive matters and three institutional ones.[100]

A. Substantive Strategies

A number of developing countries and their stakeholders have adopted two complementary strategies to shape intellectual property norms. On the one hand, they have participated in the development of a rival TLO that can affect the interpretation and application of intellectual property law, that of economic and social rights. On the other hand, they may apply long-established US and EU exceptions to intellectual property protections in new ways to address broader social goals, strategically building from the foreign to meet local needs.

1. Developing Alternative Normative Frames for the Interpretation of International Intellectual Property Norms

As John Braithwaite and Peter Drahos contend in their book *Global Business Regulation*, less powerful stakeholders in global debates fare better if they invest resources in developing principle-based normative frames. These frames can then be combined with technical expertise to apply them. Otherwise, such expertise will tend to work within normative frames created by powerful actors, such as the United States and European Union, and diffused through technical assistance

99. *Id.*

100. These strategies complement those proposed by others for the implementation of intellectual property regimes within developing countries, such as Professors Carlos Correa, Rochelle Dreyfuss, César Rodríguez-Garavito, and Amy Kapczynski. *See, e.g.*, Frederick Abbott, *The TRIPS-Legality of Measures Taken to Address Public Health Crises: Responding to USTR–State–Industry Positions That Undermine the WTO*, *in* The Political Economy of International Trade Law (Daniel Kennedy & James Southwick eds., 2002); Carlos Correa, Intellectual Property Rights, the WTO and Developing Countries: The TRIPS Agreement and Policy Options (2000).

programs, whether through WIPO, WTO, bilateral, or private sector programs. The United States, for example, arguably has the strongest protections of free speech in the world under the First Amendment of the US Constitution, which it does not compromise in applying policies in overlapping policy domains. Developing countries can do the same using the normative frames of human rights, consumer protection, access to medicines, and access to knowledge in applying intellectual property law. As is already evident in the right-to-health context, these frames can be transnationalized through the support of allied civil society groups and international institutions, furthering the development of a rival TLO that can be used for local ends.

2. Citing US and EU precedents while Transforming Them for National Purposes

As the French child psychologist Jean Piaget writes, "to comprehend is to invent."[101] Developing countries are required to implement international intellectual property law that they have signed and ratified, which sets forth particular concepts, standards, and rules. They can do so, however, in light of their development needs by inventing while implementing, by adapting while adopting. The US and EU legal systems for intellectual property include exceptions that reflect US and EU social goals. Developing countries can strategically cite these exceptions conceptually, but apply them in new ways, thereby building from the foreign in a manner that advances the local. In this way, developing countries can help to foil US and EU pressures by noting that they are applying exceptions recognized in US and EU law itself, and applied by their courts,[102] even if US and EU courts in practice would apply the exceptions differently. For example, Indian courts have limited the granting of injunctive relief to curtail access to generic medicines, creatively citing US jurisprudence for support. In a 2008 case, *Roche v. Cipla*, the Indian court determined that it would apply the normal Indian standards for granting a preliminary injunction, which are more limited than the standards demanded by intellectual property right holders.[103] Interestingly, the Indian court creatively cited a US case, *eBay v. MercExchange*,[104] and a non-precedential US Federal Circuit decision that noted public health concerns, to affirm the denial of an injunction in a case involving a drug-eluting

---

101. *See* JEAN PIAGET, OÙ VA L'ÉDUCATION? COMPRENDRE, C'EST INVENTER (1973). Roberto Unger appears to have adapted this concept slightly while attributing it to Piaget. *See* ROBERTO MANGABEIRA UNGER, DEMOCRACY REALIZED 96 (2000) ("To imitate, wrote Piaget, is to invent.").

102. Reichman, *supra* note 38.

103. F. Hoffmann-La Roche Ltd. v. Cipla Ltd., (2008) 37 P.T.C. 71 (Delhi H.C.) (India).

104. 547 U.S. 388 (2006).

stent. The Indian court concluded that harm to third parties would be irreparable and denied a preliminary injunction, even though it appears that US courts would have regularly granted it under US case law.[105]

Additional opportunities may exist for using US precedents and practices in the area of antitrust/competition law that have sought to balance protecting innovation while trying to constrain healthcare costs. The US Federal Trade Commission (FTC) has fined brand-name companies that pursue strategies of product launches that "'serve no purpose other than to undermine the ability of a generic to compete.'"[106] The FTC is more likely to pursue a case in which the plaintiff can demonstrate: "(1) reduced consumer choice; (2) additional conduct aimed at preventing generic substitution, and (3) little or no demonstrable medical benefit from substitution."[107] Such strategies could support efforts to prevent evergreening and other market manipulations to stave off generic competition. While hardly a panacea, adopting and implementing strong competition laws to reduce intellelctual property-related abuses of monopoly positions is a strategy that has been advocated and implemented in a number of countries, such as South Africa.[108]

B. Institutional Strategies

Institutional strategies complement substantive ones. Developing countries have implemented a number of institutional strategies to advance their positions and thwart pressure on them. They have invested in developing local technical expertise that can build links regionally and transnationally.

1. Developing Local Expertise

Developing countries need to develop not only normative frames for addressing the trade–human rights–intellectual property nexus, they also need technical expertise to apply it. In this way, they become entrepreneurs and not simply

---

105. Amy Kapczynski, *Harmonization and Its Discontents: A Case Study of TRIPS Implementation in India's Pharmaceutical Sector*, 97 Calif. L. Rev. 1571, 1607 n.212 (2009) (citing F. Hoffmann-La Roche Ltd. v. Cipla Ltd., (2008) 37 P.T.C. 71 (Delhi H.C.) (India).

106. Seth Silber & Kara Kuritz, *Product Switching in the Pharmaceutical Industry: Ripe for Antitrust Scrutiny?*, 7 J. Generic Medicines 119, 129 (2010) (quoting FTC Chairman Jon Leibowitz).

107. *Id.* at 127.

108. *See* Carsten Fink, *Comment I: Competition Law as a Means of Containing Intellectual Property Rights*, *in* International Public Goods and Transfer of Technology Under a Globalized Intellectual Property Regime 772 (Keith E. Maskus & Jerome H. Reichman eds., 2005); Heinz Klug, *Access to Medicines and the Transformation of the South African State*, *in* Transnational Legal Ordering and State Change (Gregory Shaffer ed., 2013).

adapters of US and EU legal norms that are institutionalized through international law and technical assistance implementation programs. This local intellectual property expertise will need to be broad-based, embedded within government institutions, and include representatives of generic producers of pharmaceuticals, academics, and civil society.[109] In this way, developing countries can better institutionalize alternative frames for addressing the trade–human rights–intellectual property nexus at the national level.

Governments need to create strong institutions with professional expertise. Patent offices, for example, must make critical determinations in approving patents, as must courts in reviewing challenges to them. Peter Drahos has written extensively about the development of expertise in developing country patent offices in order for them to make independent decisions and not simply follow analyses made in the United States and European Patent Office (EPO).[110] The Indian legal system, as well as a number of Latin American systems, for example, provides for important substantive and procedural innovations,[111] but these provisions will only have meaning if institutions apply them in practice. Institutions are most likely to implement them in practice if they are pressed to do so by legal advocates working with the generic pharmaceutical sector and civil society, supported by academics.[112]

2. Pooling Resources

One way that expertise can be built and diffused is by pooling resources through regional, transnational, and international alliances specializing in trade-related

---

109. Brazil has built considerable expertise, for example, in WTO law. *See* Gregory Shaffer, *The Trials of Winning at the WTO: What Lies Behind Brazil's Success*, 41 CORNELL INT'L L.J. 383 (2008); *see also* BALANCING WEALTH AND HEALTH: THE BATTLE OVER INTELLECTUAL PROPERTY AND ACCESS TO MEDICINES IN LATIN AMERICA (Rochelle Dreyfuss & César Rodríguez-Garavito eds., forthcoming 2014) (focusing on this need and examples within the Latin American context); GREGORY SHAFFER & JAMES NEDUMPARA, INDIA AND THE WTO (2013).

110. *See, e.g.*, Peter Drahos, *"Trust Me": Patent Offices in Developing Countries*, 34 AM. J.L. & MED. 151 (2008).

111. *See, e.g.*, Paola Bergallo & Agustina Ramón Michel, *The Recursivity of Global Lawmaking in the Fight for an Argentine Policy on Drug Patents*, *in* BALANCING WEALTH AND HEALTH, *supra* note 109; Kapczynski, *supra* note 105 (addressing Indian innovations in (1) the substantive evaluation of patents in terms of what constitutes novelty, (2) the substantive evaluation of patents in terms of what constitutes an inventive step, (3) the inclusion of broad standing rights to challenge a patent before it is issued through an administrative process, (4) the limits on the grant of injunctive relief in favor of right holders, and (5) the enhancement of rights to challenge patents under competition law).

112. BALANCING WEALTH AND HEALTH, *supra* note 109.

intellectual property issues.[113] The Andean Pact provides an example of how such expertise can be institutionalized.[114] Regional centers can provide a forum for the sharing of experiences and the identification of best practices. These centers can work with academics, or be tied to academic institutions. UNCTAD's development-oriented initiatives can serve as a node in this network.[115] However, regional institutions can be captured if they are not developed in a bottom-up manner, as Carolyn Deere demonstrates in her study of the French West African organization for intellectual property, OAPI (Organisation Africaine de la Propriété Intellectuelle).[116] Thus, broad-based initiatives that include health activists and generic pharmaceutical producers are, once more, central.

3. Coordinating with Allies in Industrialized Countries

Finally, developing countries will need to continue to coordinate with private parties and government authorities in the United States and Europe to undercut industry pressure in the formation of US and EU negotiating positions and strategies. International negotiations involve a two-level game in which national constituencies compete in the formation of national positions, and those national positions are then advanced in international negotiations.[117] Developing countries can work with US and European political allies to alter the US and European domestic political calculus. If developing countries cannot neutralize the clout of large pharmaceutical firms in the formation of US and European positions, then developing countries will face the full brunt of US and European pressure in the negotiation and enforcement of pharmaceutical patent rights.

In a world of asymmetric power, developing countries enhance the prospects of their success if other US and European constituencies offset the pharmaceutical industry's pressure on US and European trade authorities to aggressively advance industry interests. Domestic and international nongovernmental advocates, such as ACT UP, Doctors Without Borders, and Oxfam, have, directly

---

113. *See also* Peter Drahos, *When the Weak Bargain with the Strong: Negotiations in the World Trade Organization*, 8 INT'L NEGOTIATION 79 (2003).

114. Laurence R. Helfer et al., *Islands of Effective International Adjudication: Constructing an Intellectual Property Rule of Law in the Andean Community*, 103 AM. J. INT'L L. 1 (2009).

115. DEERE, *supra* note 18, at 137–40.

116. *Id.*

117. *See, e.g.*, DOUBLE-EDGED DIPLOMACY: INTERNATIONAL BARGAINING AND DOMESTIC POLITICS (Peter B. Evans, Harold K. Jacobson & Robert D. Putnam eds., 1993); Robert Putnam, *Diplomacy and Domestic Politics: The Logic of Two-Level Games*, 42 INT'L ORG. 427 (1988). For an application to the US and EU contexts regarding trade, see GREGORY SHAFFER, DEFENDING INTERESTS: PUBLIC–PRIVATE PARTNERSHIPS IN WTO LITIGATION (2003).

and indirectly, been allies. They raise fundamental moral issues that are picked up in the media and that help to hold US and EU political leaders accountable. They also harness the public's self-interest over the cost of prescription drugs and public officials' struggles to finance healthcare commitments within the United States and Europe themselves, potentially "destabilizing the consensus that US business elites had built around TRIPS."[118]

This strategy has worked in a number of cases. The United States backed off from challenging South Africa's and Brazil's patent laws, primarily in the context of US domestic political pressures. AIDS activists gathered at Vice President Gore's presidential campaign stops holding placards for the nightly news and chanting "Gore's greed kills" to press the administration to change its policy on South Africa.[119] The Bush administration withdrew the United States' claim against Brazil's compulsory licensing provisions under Brazil's patent law following similar protests.[120] United States Trade Representative Robert Zoellick again deflected US pharmaceutical industry pressures in agreeing to the Declaration on the TRIPS Agreement and Public Health at Doha.[121] United States' negotiators were similarly pressed by a Democratic Congress to alter its negotiating positions in the FTAs with Columbia and Panama, ultimately concluded in 2011, to retain more flexibility for providing access to medicines.[122] In short, when TRIPS issues become politicized domestically within the United States and Europe,

---

118. John Braithwaite & Peter Drahos, Global Business Regulation 576 (2001).

119. *See* Doug Ireland, *AIDS Drugs for Africa*, Nation, Oct. 4, 1999, at 5; Steven Meyers, *South Africa and U.S. End Dispute over Drugs*, N.Y. Times, Sept. 18, 1999, at A8. Vice-President Gore was co-chairman of the US–South Africa Bi-National Commission on pharmaceutical issues.

120. *See, e.g.*, *U.S., Brazil End WTO Case on Patents, Split on Bilateral Process*, 19 Inside U.S. Trade 1, 2 (2001). Doctors Without Borders and Oxfam launched campaigns in the United States and Europe against US policies. *See, e.g.*, *Drug Companies vs. Brazil: The Threat to Public Health*, Oxfam (May 1, 2001), http://policy-practice.oxfam.org.uk/publications/drug-companies-vs-brazil-the-threat-to-public-health-114469.

121. *See also* Gary Yerkey & Daniel Pruzin, *Agreement on TRIPS/Public Health Reached at WTO Ministerial in Doha*, 18 Int'l Trade Rep. 1817 (2001).

122. *U.S. TPP Environment Proposal Follows "May 10," but May Have Different Effects*, 29 Inside U.S. Trade 1, 9 (2011), ("The May 10 deal was negotiated between House Democrats and the Bush administration and strengthened labor rights and environmental protections in trade deals with Colombia, Panama and Peru, while also offering these countries additional flexibilities in the area of intellectual property rights in order to promote access to medicines."). James Love at Knowledge Ecology International in the United States, for example, has been relentless in the struggle for access to medicines and knowledge both domestically and transnationally. *See, e.g.*, James Love, Director, Knowledge Ecology International, *available at* http://keionline.org/jamie (including a biography and selected publications of James Love).

developing countries retain greater leeway to develop intellectual property policies to fit their own needs.

Conclusion

The construction of transnational legal orders provides normative ordering that addresses issues perceived to be problems that are transnational in scope. It thereby constrains and facilitates domestic law and practice. Developing countries can benefit from transnational legal ordering, but such ordering also can reflect the particular interests of powerful countries and constituencies within them. As regards the transnational legal ordering of trade and intellectual property, constituencies within developing countries increasingly demand that their countries retain the flexibility to implement intellectual property protection in a manner that respects human rights and development concerns.

Paradoxically, these countries profit from, and are also constrained by, a rival legal order that is also transnational in scope—that of the economic and social right to health. On the one hand, a number of developing countries have adopted strategies to deploy this rival TLO to advance their goals. Countries such as India and Brazil, for example, have attempted to adapt patent law in innovative ways, in part to meet the demands of constituencies within them to further a human right to healthcare. On the other hand, such constituencies have pressed the government to provide health services on human rights grounds that the government otherwise would have foregone, whether because of budgetary priorities or otherwise. Today, national law is made in a transnational context, and national policies are being adapted accordingly. To understand and participate effectively in legal processes, one increasingly needs to think transnationally.

5 THE LIMITS OF SUBSTANTIVE PATENT LAW HARMONIZATION

Graham Dutfield*

Introduction

National patent laws throughout the world are more abundant and similar than they have ever been. Gone are the days when countries could "just say no" to patents, as Holland and Switzerland did for much of the late nineteenth century.[1] Neither can countries exclude fields of technology, for example, by excluding patents on chemicals and medicines.[2] This is what many European nations did until the late twentieth century, including Germany and Switzerland, who were among the most advanced in the world in these fields.[3]

Despite this general trend toward uniformity, there are limits to how far countries are prepared to go, at least in the form of treaty-making. So far, two serious efforts to achieve substantive patent law harmonization by treaty at the World Intellectual Property Organization (WIPO) have failed.[4] In consequence, countries are not required to have identical

---

* Professor of International Governance, University of Leeds School of Law.

1. ERIC SCHIFF, INDUSTRIALIZATION WITHOUT NATIONAL PATENTS: THE NETHERLANDS, 1869–1912, SWITZERLAND, 1850–1907 (1971).

2. *See* Agreement on Trade-Related Aspects of Intellectual Property Rights, art. 27, ¶ 1, Apr. 15, 1994, Marrakesh Agreement Establishing the World Trade Organization, Annex 1C, 33 I.L.M. 81 (1994) [hereinafter TRIPS Agreement].

3. For example, France allowed pharmaceuticals to be patented only in 1960, Ireland in 1964, Germany in 1968, Japan in 1976, Switzerland in 1977, Italy and Sweden both in 1978, and Spain in 1992. *See* Graham Dutfield & Uma Suthersanen, *Harmonisation or Differentiation in Intellectual Property Protection? The Lessons of History*, 23 PROMETHEUS 131, 135 (2005).

4. Adam J. Sedia, *Storming the Last Bastion: The Patent Reform Act of 2007 and Its Assault on the Superior First-to-Invent Rule*, 18 DEPAUL J. ART, TECH. & INTELL. PROP. L. 79, 113 (2007) ("A good illustration of this point is WIPO's 'deep harmonization' plan, which foundered in 1996, was resurrected in 2000,

patent laws or policies. Nor do they have to treat patent applications or grant patents in the same ways. For example, countries may continue to vary in terms of the extent of limitations and exceptions their patent laws provide. A broad patent on a particular invention may be granted and upheld in one jurisdiction whereas a patent on the same invention elsewhere may be granted on the basis of narrower and fewer claims.[5] It may even later be invalidated by a court.[6] Alternatively, it may be rejected at the application stage for a range of possible reasons, such as claiming unpatentable subject matter or for being immoral.[7] Why is it that despite so much convergence and globalization, substantive patent law harmonization by treaty has stalled and currently appears unlikely to succeed? One reason, and the main focus of this chapter, is that countries have divergent interests that are hard to reconcile through multilateral negotiations. Additionally, issue-linkage bargaining is difficult to achieve in a forum dedicated solely to intellectual property.[8] Unlike at the World Trade Organization (WTO), demanders are simply unable to offer anything tempting to opponent countries that would induce enough of them to break ranks and shift sides. A third reason is that demander nations themselves have been unable to harmonize their own laws.[9] The European and US patent systems continue to vary

---

and stalled again in 2003 because some member nations attempted to use harmonization to *weaken* international patent protection by failing to adhere to some provisions of TRIPS.") (emphasis added).

5. The various Apple iPhone patents are a good example of this phenomenon. *See* Stephen McJohn, *Top Tens in 2012: Patent, Trademark, Copyright and Trade Secret Cases*, 11 Nw. J. Tech. & Intell. Prop. 173, ¶ 28 (2013).

6. *Id.* ¶¶ 28–29 (describing different outcomes in the United States and the United Kingdom).

7. For example, in 1988, the first-ever animal patent was granted, on the oncomouse. The patent was extremely broad, claiming "a transgenic nonhuman mammalian animal" having an activated oncogene sequence. U.S. Patent No. 4,736,866 (filed June 22, 1984) (describing and claiming the Harvard oncomouse). Product and process claims were allowed. In Canada, the product claims of the equivalent patent were rejected by the Supreme Court, allowing only claims to the methods. *Harvard College v. Canada (Commissioner of Patents)*, [2002] 4 S.C.R. 45. In Europe, the scope of the patent was reduced in stages from nonhuman mammal to rodent to mouse on the basis of morality and *ordre public*. The Harvard/Onco-mouse [1989] OJ EPO 451; The Harvard/Onco-mouse [1990] OJ EPO 490 (TBA); The Harvard/Onco-mouse [1991] EPOR 525; Harvard/Transgenic Animals, T315/03, [2005] EPOR 31; *President & Fellows of Harvard Coll. v. British Union for the Abolition of Vivisection*, [2005] OJ EPO 246, 247.

8. Michael P. Ryan, Knowledge Diplomacy: Global Competition and the Politics of Intellectual Property 92 (1998).

9. Ruth L. Okediji, *TRIPS Dispute Settlement and the Sources of (International) Copyright Law*, 49 J. Copyright Soc'y U.S.A. 585 (2001); Donald P. Harris, *TRIPS and Treaties of Adhesion Part II: Back to the Past or a Small Step Forward?*, 2007 Mich. St. L. Rev. 185, 212.

in certain important respects.[10] Last but definitely not least, those behind the harmonization drive have found other ways to pursue their goals, and these have been quite effective.

I. What Is Substantive Patent Law Harmonization?

Much has been written about harmonization of patent law, though not all of it has been favorable.[11] As a word, harmonization is hard to dislike, or at least to admit to disliking. Who is against harmony? And yet in the patent law context, anti-harmonization appears to be a perfectly reasonable and pragmatic stance to take for most, if not all, developing countries. But even for the developed countries, there is a deep unwillingness in some quarters to go much further, especially if this is to be done by a legally binding multilateral instrument.

So what *is* patent law harmonization? How can we know, before the event, whether there is too little harmonization, too much, or just enough? At what level of patent "strength" should the harmonization regime be calibrated if all contracting parties are to benefit more or less optimally? To what extent should rule and interpretative flexibility be eliminated if the purpose is to benefit all parties? Apart from the first, I would argue that these are imponderable questions. If this is true, then polite decline is an entirely rational response to any invitation to negotiate.

Harmonization at the level of principles, rules, and institutional structures is a highly ambitious goal. Its practical intent is twofold: (1) facilitating secure patent coverage on a global scale and for the longest period possible; and (2) accelerating and cheapening the process of attaining it. Seen this way, the stakes are self-evidently very high. Unsurprisingly, advocacy from some stakeholders has been very determined. These are obviously those who stand to benefit the most. It is hardly controversial to say that the primary beneficiaries of the standardization of patent law and practice are the transnational corporations, who are best placed to generate vast numbers of patent families covering the same inventions throughout the

10. Okediji, *supra* note 9 at 633. ("Of the TRIP[S] disputes so far, one-third have been settled informally, and most of these were between two developed countries.").

11. John H. Barton, *Issues Posed by a World Patent System*, *in* INTERNATIONAL PUBLIC GOODS AND TRANSFER OF TECHNOLOGY UNDER A GLOBALIZED INTELLECTUAL PROPERTY REGIME 617 (Jerome Reichman & Keith E. Maskus eds., 2005); PETER DRAHOS, THE GLOBAL GOVERNANCE OF KNOWLEDGE: PATENT OFFICES AND THEIR CLIENTS (2010); Graham Dutfield, *Is the World Ready for Substantive Patent Law Harmonization? A Lesson from History*, *in* DEATH OF PATENTS 228 (Peter Drahos ed., 2005); Morten Walløe Tvedt, *One Worldwide Patent System: What's in It for Developing Countries?*, 31 THIRD WORLD Q. 277 (2010).

world.[12] This is as intended.[13] Presumably those national economies where the headquarters of such firms reside stand to benefit too, hence the supportiveness of certain governments.

It is well known that patent rights are territorial. Thus, for example, although the European Patent Office (EPO) grants "European patents," these are really bundles of national patents such that an EPO-granted patent may be upheld in one national court but found invalid in another.[14] Similarly, Patent Cooperation Treaty (PCT) patent applications do not lead to the grant of "world patents."

Taken to its extreme, full harmonization would entail the following:

- Textually identical legislation.
- Identical interpretation of claims by granting offices made possible by sharing of search and examination reports.
- National courts that reliably come to the same decisions on any given patent concerning validity, scope, infringement, limitations, and defenses.

The absolute highest stage of patent law harmonization would be a unitary global system that operates: (1) to the greatest possible extent as if the whole world were a single patent law jurisdiction; and (2) on the basis of the principles, rules, and practices of the United States, Europe, or Japan (or a mixture of the three). It follows that total global patent law harmonization would mean that a patent granted somewhere would be in force everywhere else too.

II. The World Intellectual Property Organization and Harmonization

Substantive patent law harmonization is not a new idea. In 1966, the United States President's Commission on the Patent System declared "the establishment of a

12. Dongwook Chun, *Patent Law Harmonization in the Age of Globalization: The Necessity and Strategy for a Pragmatic Outcome*, 93 J. PAT. & TRADEMARK OFF. SOC'Y 127, 134 (2011).

13. SUSAN K. SELL, PRIVATE POWER, PUBLIC LAW: THE GLOBALIZATION OF INTELLECTUAL PROPERTY RIGHTS 46 (2003).

14. One exception is Francophone Africa, where the Organisation Africaine de la Propriété Intellectuelle (OAPI) provides unitary patents valid in all member countries. Gerald J. Mossinghoff & Vivian S. Kuo, *World Patent System Circa 20XX, A.D.*, 80 J. PAT. & TRADEMARK OFF. SOC'Y 523, 538 (1998); Christopher J. Harnett & Amanda F. Wieker, *The EU Unitary Patent and Unified Patent Court: Simplicity and Standardization, Challenge, and Opportunity*, INTELL. PROP. & TECH. L.J., Apr. 2013, at 15 (discussing the European Union's envisioned unitary patent regime).

universal patent" as being highly desirable.[15] It is unlikely to be coincidental that the United States was then at its peak in terms of global economic power and technological dominance. Europe had only just emerged fully from post–World War II reconstruction, and Japan had not yet become a rival in terms of industrial production and innovation. It was, of course, the height of the Cold War, and the Soviet Union was an advanced nation in terms of space technology. But the rivalry was far more political and military than it was commercial. The Eastern Bloc's impact on world trade was tiny, producing few manufactured goods that the rest of the world wanted to buy. How self-evidently beneficial for the United States, then, if its companies—which together held a large portion of the world's effective patents—could achieve inexpensive and legally certain global patent coverage in line with the domestic rules to which it had become so well accustomed?

By the mid-1990s, at the level of statutory law, a certain amount of harmonization had been achieved by way of the WTO Agreement on Trade-Related Aspects of Intellectual Property Rights (TRIPS Agreement).[16] As with the earlier US proposal for a harmonized global patent system, the idea for introducing substantive intellectual property norms in the General Agreement on Tariffs and Trade (GATT) came from the United States, which aggressively promoted negotiations on intellectual property in the GATT Uruguay Round launched in 1986.[17] Again, this is not coincidental. Although the United States remained dominant, its technological supremacy was by then threatened by the emergence of Japan and the growing strength of the so-called "Asian Tigers" that had become major producers and exporters of relatively complex manufactured goods such as automobiles, ships, and television sets, to name just a few. Holding countries to high minimum intellectual property standards is important to technological leaders at any time, but especially when their dominance is faced with challenges

---

15. James E. Rogan, Under Sec'y of Commerce for Intellectual Prop. and Dir. of US Patent and Trademark Office, The Global Recognition of Patent Rights: An Agenda for the 21st Century, Address Before the WIPO Conference on the International Patent System (Mar. 26, 2002), *available at* http://www.usembassy.it/file2002_03/alia/a2032608.htm.

16. In terms of multilateral instruments, the international law of patents consists mainly of the following: Paris Convention for the Protection of Industrial Property, the Patent Cooperation Treaty (PCT), the Patent Law Treaty (PLT), and the TRIPS Agreement. *See* Paris Convention for the Protection of Industrial Property, Mar. 20, 1883 (as revised at Stockholm, July 14, 1967), 21 U.S.T. 1583; 828 U.N.T.S. 305; Patent Cooperation Treaty, June 19, 1970, 28 U.S.T. 7645; 1160 U.N.T.S. 231; TRIPS Agreement, *supra* note 2.

17. Sell, *supra* note 13; Peter Drahos, *Global Property Rights in Information: The Story of TRIPS at the GATT*, 13 Prometheus 6 (1995); Ruth L. Okediji, *Public Welfare and the Role of the WTO: Reconsidering the TRIPS Agreement*, 17 Emory Int'l L. Rev. 819, 842–44 (2003).

from countries becoming successful at importing, absorbing, and adapting foreign technologies, thanks in part to permissive (that is, weak) patent and trade secrecy laws.

However, although GATT was useful in achieving commitments to minimum standards of protection and enforcement, it was not the right forum to seek substantive harmonization, which is a far more technical matter. Consequently, the United States managed to achieve a division of labor in which GATT provided a forum to discuss minimum intellectual property standards including patents, while WIPO was the place to discuss patent law harmonization.

From 1985, WIPO held negotiations on patent harmonization, including administrative formalities and substantive law elements. A Diplomatic Conference took place in 1991 at which WIPO presented a draft Patent Law Treaty.[18] The Conference resulted in deadlock, for two main reasons. The first was the determination of the US Congress to stick to first-to-invent in response to lobbying by independent inventor groups.[19] The second was that industry groups in some European countries objected to having an international grace period for novelty.[20] WIPO was requested to lower its ambitions, and in consequence the Patent Law Treaty adopted in 2000 dealt only with procedural matters.[21]

Nonetheless, as the century came to an end, with the completion of the Uruguay Round and the establishment of the WTO to administer its resultant agreements, interest in patent law harmonization enjoyed a revival. Japan itself became a technological leader, but an insecure one, being anxious about the other emerging economies in Asia. Predictably, it too became a harmonization enthusiast. In 1997, a commission set up to advise the Japan Patent Office recommended that the country take the lead in promoting a global patent system.[22] Needless to say, such a global system would have to be consistent with its own high levels of protection. This is somewhat ironic given that, just a few decades earlier, the

18. Mark S. Cohen, *Japanese Patent Law and the WIPO Patent Law Harmonization Treaty: A Comparative Analysis*, 4 FORDHAM INTELL. PROP. MEDIA & ENT. L.J. 847, 852 (1994).

19. Christina Bohannan & Herbert Hovenkamp, *IP and Antitrust: Reformation and Harm*, 51 B.C. L. Rev. 905, 971 (2010); *see also* Joshua Lerner, *Small Businesses, Innovation, and Public Policy*, *in* ARE SMALL FIRMS IMPORTANT? THEIR ROLE AND IMPACT 201, 211–12 (Zoltan J. Acs ed., 1999).

20. Heinz Bardehle, *A New Approach to Worldwide Harmonization of Patent Law*, 81 J. PAT. & TRADEMARK OFF. SOC'Y 303, 304–05 (1999).

21. Donald P. Harris, *TRIPS and Treaties of Adhesion Part II: Back to the Past or a Small Step Forward?*, 2007 MICH. ST. L. REV. 185, 212 (2007).

22. *See* COMM'N ON INTELLECTUAL PROP. RIGHTS, TOWARDS THE ERA OF INTELLECTUAL CREATION: CHALLENGES FOR BREAKTHROUGH. THE REPORT OF THE COMMISSION ON INTELLECTUAL PROPERTY RIGHTS IN THE TWENTY-FIRST CENTURY (1997).

government's technology licensing policy was quite aggressive and foreign companies frequently felt shortchanged by the Japanese patent system.[23]

Having seen TRIPS negotiated and being administered elsewhere, and then receiving so much attention, WIPO viewed the harmonization issue as an opportunity to recover its influence. Thus, in response to interest from certain corporate associations, professional and expert groups such as the *Association Internationale pour le Protection de la Propriété Industrielle* (AIPPI), and developed countries governments, WIPO accepted a mandate to revive the Substantive Patent Law Treaty (SPLT) process. In 2001, the Organization's Standing Committee on the Law of Patents began to debate the text of a draft treaty, along with accompanying patent office regulations and practice guidelines.[24] WIPO's ambitions about this were reflected in statements suggesting as a future possibility "the establishment of basic principles regulating an ideal global patent system, according to which a patent granted in a civil procedure would have effect in different countries, and it would co-exist with existing national patents system."[25]

WIPO's Standing Committee took the view that the least controversial issues should be dealt with first, whereas such matters as first-to-file versus first-to-invent and the grace period, both of which were stumbling blocks due to differences between the United States and Europe, could be negotiated later.[26] However, it turned out that there was little of importance in the first draft presented in May 2001 that was *not* controversial. The reasons for this and why the SPLT process came to a halt are explained in the next section after a more general discussion on the most common grounds for opposing harmonization.

III. A Clash of Interests? History, Development, Politics

There are three general arguments commonly made against harmonization, each of which seems to be intuitively plausible. I call these (1) the "levels-of-development," (2) the "differentiation," and (3) the "nobody-can-be-sure-what's-best-for-*any*-country" arguments. Most commonly, they are

23. *See, e.g.,* Chalmers Johnson, Japan: Who Governs? The Rise of the Developmental State 74–75 (1995).

24. *See generally* Toshiko Takenaka, *The Best Patent Practice or Mere Compromise? A Review of the Current Draft of the Substantive Patent Law Treaty and a Proposal for a "First-To-Invent" Exception for Domestic Applicants*, 11 Tex. Intell. Prop. L.J. 259 (2003).

25. Shozo Uemura, *WIPO Update: Patent Law Harmonization and the Grace Period*, 6 CASRIP Publication Series: Rethinking Int'l Intell. Prop. 263, 268 (2000).

26. Takenaka, *supra* note 24, at 271–72.

promoted on the basis of economic analysis and the selective deployment of historical examples.[27]

A. Anti-Harmonization: The "Levels-of-Development" Argument

According to this argument, the "strength" of patent protection should ideally be commensurate with a country's level of development. It needs to be mentioned in passing that "strong" and "weak" are frequently used in relation to patent law, but they tend not to be explained. By "strong" I refer to the extent of enforceability of the rights; the absence of exceptions to patentability by subject matter or technological field; and the absence of limitations such as compulsory licensing, Bolar exemptions, and legal parallel importation. Generally speaking, the jurisdictions with the "strongest" patent rights, which will inevitably provide the basic model for the SPLT, are the United States, Europe, and Japan, though it is worth mentioning that in some developing nations, unlike in the United States and several European countries, patent infringement may be treated as a crime. Recent economic studies[28] suggest that the realization of substantive patent law harmonization along such lines could potentially harm the economic development interests of some, if not many, countries.[29] Empirical research by Kim on the experience of South Korea, a highly successful rapid developer, led to the conclusion that "strong IPR protection will hinder rather than facilitate technology transfer and indigenous learning in the early stage of industrialization when learning takes place through reverse engineering and duplicative imitation of mature foreign products."[30] He also concluded that it is "only after countries have accumulated sufficient indigenous capabilities with extensive science and technology infrastructure to undertake creative imitation [in the later stage that] IPR protection becomes an important element in technology transfer and industrial activities."[31]

---

27. For historical evidence, see Dutfield & Suthersanen, *supra* note 3. *See also* CHRISTOPHER MAY & SUSAN K. SELL, INTELLECTUAL PROPERTY RIGHTS: A CRITICAL HISTORY (2006).

28. Carlos M. Correa & Sisule F. Musungu, *The WIPO Patent Agenda: The Risks for Developing Countries* (Trade-Related Agenda, Dev. & Equity, Working Paper No. 12, 2002); Jerome H. Reichman & Rochelle Cooper Dreyfuss, *Harmonization without Consensus: Critical Reflections on Drafting a Substantive Patent Law Treaty*, 57 DUKE L.J. 85 (2007).

29. In this discussion I put to one side social welfare–related issues, such as the effects of patents on prices of medicines and agricultural inputs and products. These are hardly insignificant but I am dealing primary with development in terms of economic, industrial, and technological improvement.

30. Linsu Kim, *Technology Transfer and Intellectual Property Rights: The Experience of Korea*, UNCTAD-ICTSD, vii (June 2003), *available at* http://www.iprsonline.org/resources/docs/Kim%20-%20ToT%20and%20IPRs%20-%20Blue%202.pdf.

31. *Id. See generally* Ferris K. Nesheiwat, *The Adoption of Intellectual Property Standards beyond TRIPS—Is It a Misguided Legal and Economic Obsession by Developing Countries?*, 32

Essentially, harmonization based on the regimes existing in leader countries will mostly benefit medium-sized and large firms based in these nations rather than those in follower countries. *Relatively close* follower economies with the capacity to absorb and adapt foreign technologies and imitate complex products and technologies—but that are still some distance from the technology frontier—are likely to be held back the most. At the risk of sounding rhetorical, wealthy countries' attempts to impose patent harmonization could be construed as trying to "kick away the ladder" that they used to climb out of poverty and into developed-nation country status.[32] If so, the most technologically advanced developing countries are the ones most likely to oppose substantive patent law harmonization. A recent article surveying the recent economic literature does not prove the validity of such a perspective, but neither does it do much to allay the concern. Indeed, the authors state ominously that "the current move to harmonize the IPR system internationally, such as the TRIPS [A]greement, may have unexpected but grave consequences on developing countries."[33] What we might expect regarding the stance of the relatively close followers is more or less what we have seen in reality. Leader countries such as the United States and Japan have been generally positive about substantive patent law harmonization, while countries such as Brazil and India have been skeptical.[34] More underdeveloped developing countries tend to be unenthusiastic too. As with Brazil and India, they are concerned about social welfare and biodiversity-related implications, but unlike them, priorities driven by industrial considerations or issues tend to be less important given that their absorptive capacities, and thus their catchup prospects, at least in the short run, are lower.

It is one thing to say that relatively advanced developing countries would prefer weak patent regimes in order to advance their capacities to innovate through imitation-derived technological learning, and then strengthen them later when they are more innovative. It is quite another thing to assume that such a policy

---

Loy. L.A. Int'l & Comp. L. Rev. 361 (2010); Llewellyn Joseph Gibbons, *Do as I Say (Not as I Did): Putative Intellectual Property Lessons for Emerging Economies from the Not So Long Past of the Developed Nations*, 64 SMU L. Rev. 923 (2011).

32. Ha-Joon Chang, Kicking Away the Ladder: Development Strategy in Historical Perspective (2002). The expression was coined by the nineteenth-century German economist Friedrich List.

33. Hiroyuki Odagiri, Akira Goto, Atsushi Sunami & Richard R. Nelson, *Introduction, in* Intellectual Property Rights, Development, and Catch Up: An International Comparative Study 1–2 (Hiroyuki Odagiri, Akira Goto, Atsushi Sunami & Richard R. Nelson eds., 2010).

34. Graham Dutfield, *Knowledge Diplomacy and the New Intellectual Property Fundamentalism*, *in* Interpreting and Implementing the TRIPS Agreement: Is it Fair? 31 (Justin Malbon & Charles Lawson eds., 2008); Dutfield & Suthersanen, *supra* note 3.

works just because it rings true and many governments have favored it.[35] For one thing, there are many potentially relevant factors other than just the patent system. For another, the links between patent systems and economic development involve hugely complicated factors that defy straightforward analysis.[36] As the authors of a collection of national studies on intellectual property cautiously conclude:

> IPR does affect catch-up but the effects are multi-dimensional and it is misleading to say simply that IPR helped catch-up or that it hindered catch-up. The effects vary significantly across industries, with pharmaceuticals clearly being the industry most affected by patents whereas, in other industries, it sometimes appears that patents hardly matter. At any rate, IPR is only one of the factors that influence the speed of catch-up and other policies, such as education policy to create a sufficient pool of human resources and coordination among related policies (including science and technology policy, industrial policy, and trade and FDI policy), are equally (or even more) important. Accordingly, TRIPS may affect some industries, most likely pharmaceuticals, in some countries but otherwise may have little effect.[37]

All things considered, this argument is persuasive and is probably the most influential anti-harmonization perspective.

B. The "Differentiation" Argument

The late economist Sanjaya Lall produced a study finding ample evidence that "the need for IPRs varies with the level of development."[38] Based in part on the

35. In the cases of Argentina and Thailand, the weak-IP-for-development strategy appears to have had little effect. *See* Patarapong Intarakumnerd & Peera Charoenporn, *The Roles of IPR Regime on Thailand's Technological Catching Up*, *in* Intellectual Property Rights, Development, and Catch Up, *supra* note 33; Andrés Lopéz, *Innovation and IPR in a Catch-Up-Falling-Behind Process: The Argentine Case*, *in* Intellectual Property Rights, Development, and Catch Up, *supra* note 33. With Brazil, the most successful high-tech sectors are said to have succeeded due to factors other than the patent system, which was neither a help nor a hindrance. Roberto Mazzoleni & Luciano Martins Costa Póvoa, *Accumulation of Technological Capabilities and Economic Development: Did Brazil's IPR Regime Matter?*, *in* Intellectual Property Rights, Development, and Catch Up, *supra* note 33.

36. Shawn J. Kolitch, *The Proper Scope of Patentability in International Law*, 11 Marq. Intell. Prop. L. Rev. 149, 153 (2007).

37. Odagiri et al., *Introduction*, *supra* note 33.

38. Sanjaya Lall, *Indicators of the Relative Importance of IPRs in Developing Countries*, (UCCTAD-ICTSD Project on IPRs and Sustainable Dev. Series, Issue Paper 3, 2003) *available at* http://ictsd.org/i/publications/11770.

work of economist Keith Maskus,[39] he went on to identify another argument implying that opposing harmonization would be a rational stance for most developing countries:

> Many rich countries used weak IPR protection in their early stages of industrialization to develop local technological bases, increasing protection as they approached the leaders. Econometric cross-section evidence suggests that there is an inverted-U shaped relationship between the strength of IPRs and income levels. The intensity of IPRs first falls with rising incomes, as countries move to slack IPRs to build local capabilities by copying, then rises as they engage in more innovative effort.[40]

This is the basis for the differentiation argument: diversity is good for development, and uniformity is either bad or at best neutral. It is undeniable that the nations of the world are highly diverse in terms, for example, of economic potential, technological capacity, educational levels, and population size. Accordingly, logic suggests that patent systems designed to foster innovation-based economic growth would have to reflect this diversity. What makes differentiation more than a subset of the levels-of-development argument is its focus not just on domestic patent policy but also on foreign laws. Certain relatively advanced catchup country firms can take advantage of the differences between their domestic patent regimes and those in foreign markets. Thus, it is not just domestic patent policy that counts: the patent regimes of foreign countries may matter a great deal too, if not more, especially for countries seeking to grow by exporting high value products as did Germany in the late nineteenth century, and Japan and South Korea from the late twentieth. Having patent laws that were stronger in foreign markets was at least as important as having weak laws at home, if not more so. Thus, any harmonization scheme would, if implemented, be an obstacle to such strategies for industrial development.

In the late nineteenth century when Germany was a developing country seeking to catch up with Britain and France, it adopted a patent law that was at a level suitable for its stage of industrial development and consistent with its social welfare norms. But German chemical firms, which were soon to become globally dominant, benefited most of all from being able to take advantage of the higher levels of protection in these two countries and the United States. These companies not only grew rapidly as a result but were able to capture a significant share of these foreign

39. Keith E. Maskus, Intellectual Property Rights in the Global Economy (2000).

40. Lall, *supra* note 38, at 11.

markets.[41] In the same way, Switzerland was able around the same time to take advantage of the relatively higher levels of protection in Germany, which for Swiss firms was a key market for its chemicals, and initiate an era of sustained international success over many decades.

One can of course say that the world has changed a lot since those days, so these examples are not to be relied upon. And in any case, we are only talking about one industrial sector: fine chemicals, including pharmaceuticals. But some other industrial sectors seem to tell a similar story.[42] Besides, what was true for Germany and then Switzerland in the late nineteenth century became true for Japan, South Korea, and Israel[43] in the late twentieth century.

Japan, South Korea, and Taiwan were well-known for being less than respectful toward the intellectual property interests of foreign firms, and hardly suffered on account of it. Like South Korea and Taiwan, the Japanese patent system had, for almost all of its existence, been very much "TRIPS-minus." According to one study, "the meteoric rise from feudal serf to technological whiz-kid that the country has undergone in less than 150 years is little short of astounding, and poses the question of whether it could be repeated today."[44] The author concludes that "the homogenization of patent law, the claim implicit in TRIPS that one size can, and indeed should, fit all, does not adequately correspond with the picture of Japan's evolution."[45] Even so, useful as such a weak patent regime may have been for Japanese firms, and this is not entirely conclusive,[46] it is surely very significant in understanding their success that they were able to take advantage of the much stronger patent regimes of North America and Europe where most of their goods were exported.

In this century, China is another "pirate" nation that is upgrading its technological capacity in various sectors and is now increasing its patent filings in the main overseas markets. Indeed, in 2011, Chinese applicants were the fourth biggest source of PCT international patent filings, a rise of four places since

41. Graham Dutfield, Intellectual Property Rights and the Life Science Industries: Past, Present and Future (2009); Johann Peter Murmann, Knowledge and Competitive Advantage: The Coevolution of Firms, Technology and National Institutions (2003); May & Sell, *supra* note 27.

42. Dutfield & Suthersanen, *supra* note 3; May & Sell, *supra* note 27.

43. Meir Pugatch, Morris Teubal & Odeda Zlotnick, *Israel's High-Tech Catch-Up Process: The Role of IPR and Other Policies, in* Intellectual Property Rights, Development, and Catch Up, *supra* note 33.

44. Matthew Fisher, *Growth of the Japanese Patent System: A Lesson for Us All?*, 22 Intell. Prop. Q. 85, 113 (2004).

45. *Id.*

46. According to a recent empirical study, Odagiri et al. found it to be "extremely difficult to argue whether IPR helped or deterred Japan's catch-up." Hiroyuki Odagiri, Akira Goto

2006.[47] ZTE Corporation (China) came in first place for the number of PCT applications published in 2011 with Huawei Technologies (also China) following in third.[48] Overall, in 2011, China ranked first worldwide as the recipient of the most patent filings, overtaking the United States.[49] This is a reflection of how rapidly the geography of patent ownership is changing, marked primarily by a shift eastward. Incidentally, Philips, a company that started by copying Swan's and Edison's carbon filament lamps thanks to the abolition of patents in Holland in the late nineteenth century, and which was the number one PCT applicant cumulatively between 1978 and 2011, came in fifth in 2010 and ninth in 2011.[50] Although all developing countries have good reason to favor differentiation over harmonization, this does not make their interests identical. Indeed, it is largely because they are so diverse that a one-size-fits-all regime is such an unprecedentedly radical idea, and a potentially dangerous one at that.

C. The "Nobody-Can-Be-Sure-What's-Best-for-Any-Country" Argument

What does an optimal patent system look like? From a purely instrumentalist perspective, the issue can be framed like this: How much protection should be made available to inventors that is sufficient to induce them to invent *and* disclose, but that is not so extensive as to discourage or prevent other inventors from innovating themselves? That this is such a simplistic way to view the problem is itself a testament to the theoretical impossibility of creating an optimal system for any jurisdiction, let alone the whole world. Suffice it to say that trying to strike an optimal balance between the needs of patent holders, competitors, potential market entrants, and the public interest confronts policymakers with a host of imponderables that defy the efforts of economic theoreticians to model an optimally efficient and balanced system for *any* national economy.

It is not only countries that differ in their conditions: in-country industrial sectors are bound to also vary in their views of what a favorable patent regime

---

& Atsushi Sunami, *IPR and the Catch-Up Process in Japan*, *in* Intellectual Property Rights, Development, and Catch Up, *supra* note 33.

47. World Intellectual Prop. Org., World Intellectual Property Indicators 63 (2012), *available at* http://www.wipo.int/export/sites/www/freepublications/en/intproperty/941/wipo_pub_941_2012.pdf.

48. *Id.* at 65.

49. *Id.*

50. World Intellectual Prop. Org., PCT Yearly Review, The International Patent System 10, 36 (2012), *available at* http://www.wipo.int/export/sites/www/freepublications/en/patents/901/wipo_pub_901_2012.pdf.

should comprise. It is true that patent rules are sometimes tweaked to accommodate the specificities of individual industrial sectors.[51] Thus, the first German patent law of 1877 deliberately favored domestic chemical and engineering firms. Many countries today provide regulatory review (Bolar) exemptions for generic drug companies and term extensions for pharmaceutical innovator firms under certain circumstances.[52] But no jurisdiction provides a patent law for, say, pharmaceuticals, and then a very different one for software or for electrical appliance producers. Efforts to revise the US patent system were long hampered in part by different industrial sectors disagreeing on what is the optimal one-patent policy for their type of business.[53] There are good reasons for avoiding the confusion of having a diversity of patent rules operating in a single jurisdiction. But without doing so, there is no possibility of designing an optimal system for *any* sector or national economy. Shortly after SPLT talks recommenced at the start of this century, the nongovernmental organization (NGO) GRAIN warned that harmonization could conceivably "mean the end of patent policy as a tool for national development strategies."[54] Economic analyses and the historical record both suggest that although this view is a sweeping one and probably goes too far, it is far from being alarmist.

IV. Counter-Expertise and the Postponement of the SPLT Process

Developing country diplomats based in Geneva have, in the last ten years, improved their capacity to negotiate in pursuit of the collective interests of the countries they represent, and they deserve credit for this. But NGOs and so-called "counter-experts" have played a crucial role in offering alternative critical perspectives countering the propagandist fare typically served up, for example, in the guise of technical assistance programs.[55] Initiatives of the latter kind frequently

---

51. The US patent rules seem to go quite far in this regard. *See* Mark A. Lemley & Dan A. Burk, *Policy Levers in Patent Law*, 89 VA. L. REV. 1575 (2003).

52. *See, e.g.*, John A. Tessensohn, *Reversal of Fortune—Pharmaceutical Experimental Use and Patent Infringement in Japan*, 4 J. INT'L LEGAL STUD. 1 (1998).

53. Rebecca S. Eisenberg, *Wisdom of the Ages or Dead-Hand Control? Patentable Subject Matter for Diagnostic Methods After* In re Bilski, 3 CASE W. RESERVE J.L. TECH. & INTERNET 1, 48 (2012).

54. *WIPO Moves Toward "World" Patent System*, GENETIC RESOURCES ACTION INT'L (GRAIN) (July 2002), http://www.grain.org/system/old/docs/wipo-patent-2002-en.pdf.

55. DUNCAN MATTHEWS, INTELLECTUAL PROPERTY, HUMAN RIGHTS AND DEVELOPMENT: THE ROLE OF NGOs AND SOCIAL MOVEMENTS (2011); Carolyn Deere Birkbeck & Santiago Roca, *An External Review of WIPO Technical Assistance in the*

claim, with little evidence, that more intellectual property equals more investment, more research and development, and more economic growth, and that this is so regardless of whether the nation receiving this "assistance" could possibly be of more than passing interest to large patent-owning corporations except as a relatively minor market for their goods. These divergent "expert" views generate competition between ideas, leading to domestic interagency conflict and confusion. On the one hand, some developing country government departments, often those with responsibility for trade and intellectual property, are true believers in the notion that intellectual property regulations based on those of Europe or the United States are vital to attract vast amounts of "cargo" in the form of investment and high-tech goods.[56] Health and agriculture officials, and Geneva-based diplomats, though, often have a much more skeptical point of view. This is why US, EU, and Japanese officials sometime bypass the latter and go straight to the capitals to persuade developing country governments, and especially the departments responsible for trade, industry, and intellectual property, of the "rightness" of their views on patents and harmonization.

Indeed, the developing country representatives in Geneva were generally highly skeptical about the need for such a treaty, especially given that their countries had only just implemented TRIPS and some had still not fully done so. Knowing that open opposition might not be the most politically prudent approach, they decided instead to participate but to put forward some counter-proposals of their own, both within the text of the patent harmonization treaty and elsewhere at WIPO, and these effectively stopped the process in its tracks.[57]

Developing countries insisted that the treaty make reference to such issues as access to genetic resources and disclosure of origin, protection of traditional knowledge, conservation of biological diversity, public health, and ethics in scientific research.[58] By 2004, the process was grinding to a halt as a result of these counter-demands. The United States, the European Union, and Japan sought to

---

Area of Cooperation for Development, WIPO Doc. CDIP/8/INF/1 (Sept. 1, 2011); Carolyn Deere Birkbeck, *The Technical Assistance Principles of the WIPO Development Agenda and Their Practical Implementation* (ICTSD, Working Paper 28, 2010).

56. The author's own experiences in technical assistance suggest that some developing-country officials are indeed convinced that the "right" intellectual property rules are crucial to attract investment in R & D and in medium and high technology production facilities. In the case of most least-developed countries this is sadly unfounded. There is zero evidence to suppose that regulatory imitation by these countries such that their patent rules match those of the developed world will foster innovation.

57. Susan K. Sell, *TRIPS Was Never Enough: Vertical Forum Shifting, FTAS, ACTA, and TPP*, 18 J. Intell. Prop. L. 447, 455 (2011).

58. For a more detailed account of the SPLT negotiation process, see F. Sisule Musungu & Graham Dutfield, *Multilateral Agreement and a TRIPS Plus World: The World Intellectual*

streamline the agenda so as to get the negotiating process back on track and, with the collaboration of the WIPO Secretariat, attempted to isolate and then divide the two most influential and skeptical of the developing countries, India and Brazil.[59] These tactics did not succeed. Developing countries found it quite easy to oppose the harmonization process once they organized and worked out that they had shared interests in blocking progress until other matters important to them had been added to the agenda. The perceived danger for developing countries was that the treaty would erode their freedom to craft their patent systems according to their specific requirements. Most likely, for example, it would have prevented them from designing limitations and exceptions in accordance with public policy, social welfare, and national interests.

The activism and the literature produced by NGOs and counter-experts was highly effective, in part because it was so timely. The early 2000s was a time when WIPO's Secretariat (the International Bureau) was facing heavy criticism. Indeed, much of the debate about harmonization focused on WIPO itself. Some developing-country government representatives were becoming quite openly critical about what they saw as WIPO's alleged deeply rooted bias toward the interests of users and its overdependence on the direction of certain governments and organizations with a commercial or professional stake in capturing the international intellectual property regime.[60] This bias, which admittedly was not usually shared by heads of developing-country patent offices, was felt to be particularly inappropriate given WIPO's status as a specialized agency of the United Nations (UN). WIPO came under heavy pressure to become a more open institution that was welcoming to non-business and nonprofessional NGOs. In addition there were serious allegations of mismanagement and corruption at the highest level, and this was becoming a matter of diplomatic and public concern.[61]

Property Organization, QUNO & QIAP (2003), *available at* http://www.geneva.quno.info/pdf/WIPO(A4)final0304.pdf.

59. Carolyn Deere, The Implementation Game: The TRIPS Agreement and the Global Politics of Intellectual Property Reform in Developing Countries 130 (2008).

60. *See* Christopher May, The World Intellectual Property Organization: Resurgence and the Development Agenda (2007); Musungu & Dutfield, *supra* note 58. The AIPPI influence on WIPO and its predecessor organizations is nothing new, as it goes back a century. Along with the International Chamber of Commerce, AIPPI was an active participant in many, if not most, intergovernmental conferences on patent law norm-setting during the twentieth century, including revisions of the Paris Convention for the Protection of Industrial Property.

61. William New, *WIPO Faces Persistent Complaints, Allegations of Mismanagement*, Intellectual Property Watch (June 7, 2005), http://www.ip-watch.org/2005/06/07/wipo-faces-persistent-complaints-allegations-of-mismanagement.

Consequently, WIPO was put on the defensive, finding it necessary to behave as a true member-driven organization sensitive to the concerns of all members, but without giving offense to the United States, Europe, and the large corporations whose fees, paid into the PCT system administered by WIPO, make it one of the wealthiest UN agencies. This was obviously a difficult balancing act, but WIPO performed it quite skillfully. In 2006, shortly after an event held by WIPO called the Open Forum on the Draft Substantive Patent Law Treaty, at which several speakers nominated by WIPO member states, including the present author, presented papers strongly critical of the idea that patent law harmonization is beneficial for economic development,[62] the SPLT process was effectively halted, to the chagrin of the United States, the European Union, and Japan. One suspects that the WIPO Secretariat released a collective sigh of relief.

There is one more reason for the failure of the SPLT process that merits a mention. Unlike the WTO, the opportunity for demander countries to "trade" agreement on intellectual property for market access concessions (among other "carrots") is much more difficult when nothing but intellectual property is on the menu and nothing about anything else can be added to it.[63] Consequently, "buying off" more than a few developing countries proved to be unfeasible.

V. Counter-Counter Harmonization

Brazil had been especially active in opposing the SPLT and expressing concern about the political orientation of WIPO. It was keen to promote a more pro-development orientation inherently contrary to the ethos of harmonization. In 2004, the governments of Brazil and Argentina proposed that WIPO adopt a "Development Agenda."[64] This led to the formation of a grouping of fourteen developing countries called the Friends of Development,[65] which produced a follow-up submission to WIPO. This latter document elaborated on the first one and responded to some developed countries, especially the United States, that were opposed to any comprehensive initiative to explicitly incorporate

62. The program and papers are downloadable from World Intellectual Prop. Org., Open Forum on the Draft Substantive Patent Law, *available at* http://www.wipo.int/meetings/en/2006/scp_of_ge_06/scp_of_ge_06_inf1.html (last visited Sept. 4, 2013).

63. *See, e.g.*, Ruth L. Okediji, *Beyond Fragmentation: WIPO-WTO Relations and the Future of Global IP Norms*, Netherlands Y.B. Int'l L. (2008); *see also* José E. Alvarez, *International Organizations: Then and Now*, 100 Am. J. Int'l L. 324 (2006).

64. World Intellectual Prop. Org., *Proposal by Argentina and Brazil for the Establishment of a Development Agenda for WIPO*, WIPO Doc. WO/GA/31/11 (Aug. 27, 2004) (note prepared by the Secretariat).

65. The Friends of Development were Argentina, Bolivia, Brazil, Cuba, Dominican Republic, Ecuador, Egypt, Iran, Kenya, Peru, Sierra Leone, South Africa, Tanzania, and Venezuela.

development concerns into the mandate and activities of WIPO.[66] The four issues covered in the second submission, all of which were highly controversial, concerned WIPO's mandate and governance, norm-setting, technical cooperation, and transfer of technology.

With regard to WIPO's mandate, the Friends of Development noted that as a UN agency, WIPO should be guided by the UN's development goals.[67] Moreover, development concerns should be incorporated into all of WIPO's work.[68] After considerable debate, the submission of a vast range of proposals and counter-proposals by numerous governments, and a certain amount of resistance, even from WIPO, members of the Organization agreed in September 2007 to formally adopt the Development Agenda by establishing a new Committee on Development and Intellectual Property.[69] The Committee is charged with overseeing the implementation of forty-five proposals about which consensus was reached.[70] It is difficult to predict whether the Committee's work will lead to any outcomes of substance. But its existence, rather than forming part of a quid pro quo with the SPLT, should make substantive patent law harmonization even harder to achieve. This is because the Committee's mandate makes explicit reference to the levels-of-development argument[71] and is in any case premised on the presumption that national patent regimes can be different.

This muscle-flexing by certain developing countries has undoubtedly provoked a reaction. If substantive patent law harmonization cannot be achieved in the ways envisaged by the demanders, other ways must be sought to pursue the same practical ends. Bilateral and regional trade agreements, known as free trade or economic partnership agreements, have frequently been used to "raise"

---

66. World Intellectual Prop. Org., *Proposal to Establish a Development Agenda for the World Intellectual Property Organization (WIPO): An Elaboration of Issues Raised in Document WO/GA/31/11*, WIPO Doc. IIM/1/4 (Apr. 6, 2005) (submission by the Group of Friends of Development).

67. *Id.* at 2.

68. *Id.*

69. World Intellectual Prop. Org., Assemblies of the Member States of WIPO, Forty-Third Series of Meetings, *General Report Adopted by the Assemblies*, Geneva, Sept. 24 to Oct. 3, 2007, WIPO Doc. A/43/16 (Nov. 17, 2007).

70. World Intellectual Prop. Org., *Report of the Provisional Committee on Proposals Related to a WIPO Development Agenda*, WIPO Doc. A/43/13 Rev. (Sept. 17, 2007).

71. Recommendation 15 adopted as part of the WIPO Development Agenda states in part: "Norm-setting activities shall...be inclusive and member driven; [and] take into account different levels of development." World Intellectual Prop. Org., The 45 Adopted Recommendations Adopted under the WIPO Development Agenda, *available at* http://www.wipo.int/export/sites/www/ip-development/en/agenda/recommendations.pdf (last visited Sept. 4, 2013).

the standards of patent protection in developing countries that are parties to these deals to match those of the United States or Europe. The role of technical assistance programs has already been mentioned.[72] In addition, there is patent office cooperation and the recent emphasis on enforcement that culminated in the highly controversial plurilateral Anti-Counterfeiting Trade Agreement[73] and ongoing negotiations for a Trans-Pacific Partnership Agreement.[74]

Given the importance of patents for the pharmaceutical industry, drug companies have been especially active in pursuing their patent-related regulatory interests individually, collectively through associations such as the International Federation of Pharmaceutical Manufacturers Associations (IFPMA), and by recruiting governments to act on their behalf. Among their successes has been the promotion of extended data exclusivity terms, regulatory linkage measures, patent term extensions, and extension ofding patent protection to incremental and "old" inventions, such as new dosage regimes and new uses of known substances. All such measures are specifically intended to maintain monopolies for longer and longer periods, in part by holding up generic drug competition beyond the standard maximum term of twenty years. These tend to be achieved through trade agreements where the United States and/or the European Union comprise one of the parties.

Less visibly, considerable efforts have been expended on finding ways to enhance cooperation between different patent offices receiving applications on identical inventions, so as to reduce pendency times, achieve greater legal certainty, and reduce costs for the applicants and workloads for the offices. The US, European, and Japanese patent offices ("the Trilateral Offices") have been in close contact since 1983; they continue to cooperate with each other and, in recent years, also with patent granting offices throughout the world. In his detailed empirical study on patent offices, Peter Drahos shows that considerable progress has already been made toward deeper harmonization through cooperation between patent offices.[75]

Although the goals of patent office cooperation seem on their face to be perfectly reasonable, as long as nothing is done to undermine the territorial nature of patent rights, there are some reasons for concern. First, although a greater number

---

72. *See generally* Carolyn Deere Birkbeck & Santiago Roca, *An External Review of WIPO Technical Assistance in the Area of Cooperation for Development*, Final Report Submitted on 31 August 2011, *available at* http://www.wipo.int/edocs/mdocs/mdocs/en/cdip_8/cdip_8_inf_1-annex1.pdf.

73. Anti-Counterfeiting Trade Agreement, Dec. 3, 2010, 50 I.L.M. 243 (2011).

74. Ian F. Fergusson et al., The Trans-Pacific Partnership Negotiations and Issues for Congress (2013).

75. Drahos, *supra* note 11.

of businesses than before have the opportunity to benefit, there is little doubt that already established transnational corporations that are well resourced and experienced in the ways of acquiring and managing large patent portfolios in many countries stand to gain the most. The kinds of rules and interpretations convenient for entrenched elites with financial clout and political influence may differ greatly from those that might benefit up-and-coming firms having quite different business models and contrasting perspectives on the patent system. Second, these traditional businesses, with their patent professional allies, are already engaged heavily in overseeing the work of developed-world patent offices. It is not too cynical to suggest that patent office cooperation initiatives may constitute the means to extend the influence of these interests to the granting offices of the rest of the world without much if any public accountability or debate. Indeed, Peter Drahos refers to patent office networking as "invisible" harmonization.[76]

Conclusion

Despite recent renewed efforts on the SPLT at WIPO, substantive patent law harmonization seems doomed to failure. Given the inability of harmonization advocates to make a strong case that harmonization pitched at the level of the developed countries would benefit any country, least of all the developing ones, this is just as well. But harmonization efforts will not go away; to the contrary, such efforts move forward in ways that are often unaccountable due largely to their being kept away from the public gaze. Even so, given the diversity of interests involved, harmonization is never going to be a unidirectional process toward total uniformity. Unlike former British Prime Minister Tony Blair, patent regulation *does* have a reverse gear. The important thing is to use it.

---

76. *Id.*

II GLOBAL APPROACHES TO SUBJECT MATTER STANDARDS AND PATENTABILITY

6 PATENT BARBARIANS AT THE GATE: THE WHO, WHAT, WHEN, WHERE, WHY AND HOW OF US PATENT SUBJECT MATTER ELIGIBILITY DISPUTES

Margo A. Bagley*

Introduction

Shortly after the 1980 US Supreme Court decision in *Diamond v. Chakrabarty*[1] allowing patents on life forms, a US patent attorney attended an AIPPI[2] conference in a foreign country. After introducing himself to another attendee (a patent attorney from a different nation) and identifying himself as an American, the foreign patent attorney said disgustedly to the US patent attorney: "You Americans are patent barbarians!"[3] Perhaps such a statement is not the best way to make friends and influence people, but the speaker voiced a sentiment

---

* Professor of Law, University of Virginia School of Law. This chapter benefitted from comments received during presentations at the European Policy on Intellectual Property (EPIP) conference at the University of Lueven (Belgium), DePaul University School of Law's Manzo seminar, the University of Virginia School of Law faculty retreat, and the 2nd Annual Briggs & Morgan Patent Law Symposium at the University of Minnesota School of Law. Excellent research assistance was provided by Xiaoyan (Sabrina) Wang and the exceptional research librarians of the University of Virginia School of Law.

1. 447 U.S. 303 (1980). However, although patents on living organisms had been granted prior to this time (such as U.S. Patent No. 141,072 to Louis Pasteur for yeast in 1873) the patentability of such inventions had later come under question. See ROBERT P. MERGES & JOHN F. DUFFY, PATENT LAW & POLICY: CASES AND MATERIALS, 147 (6th ed. 2013) (citing P.J. Federico, *Louis Pasteur's Patents*, 86 SCI. 327 (1937), *reprinted in* 19 J. PAT. OFF. SOC'Y 966, 967 (1937)).

2. Association Internationale pour la Protection de la Propriété Intellectuelle or International Association for the Protection of Intellectual Property.

3. Special thanks to John White, Esq. for sharing this anecdote.

that probably has been shared over the years by many foreign observers of the US approach to patent subject matter eligibility.

Granting patents for "anything under the sun made by man"[4] even with the judicially defined limits concerning abstract ideas, laws of nature, and natural phenomena, certainly could be deemed evidence of a lack of refinement. Nevertheless, much of the rest of the world has followed America's lead, to differing degrees, in expanding the scope of patent-eligible subject matter, particularly with regard to living matter. *Diamond v. Chakrabarty*'s recognition of an oil-eating, living bacterium as patent-eligible subject matter may have appeared "barbaric" to some at the time (including, perhaps, the four dissenting Supreme Court justices in the case) but the patent laws of all EU countries, and most other countries with well-developed patent systems, routinely provide protection for microorganisms and have for quite some time. Many countries also have followed the United States in extending utility patent protection to isolated DNA and other genomic material, plants, animals, and more, while drawing the line at other "barbaric" American expansions such as human cloning methods, human embryonic stem cell cultures, and some diagnostic and therapeutic methods.[5]

However, patent-eligible subject matter determinations continue to present a moving target in the United States primarily for biotech, software, business methods, and diagnostic methods, with court decisions, patent office policies, and legislative initiatives sometimes expanding, and sometimes contracting, eligibility boundaries. And decisions in the United States continue to have repercussions for patenting abroad.

This chapter will explore, from a bird's-eye view, selected aspects of the who, what, when, where, why, and how behind some of the developments and forces fueling the expansion and contraction we see in US patent subject matter eligibility determinations,[6] and the sometimes surprisingly different approaches taken by other countries. Section I provides a brief background on the "what" with which patent eligible subject matter (PESM) disputes are concerned. Sections II

4. *Chakrabarty*, 447 U.S. at 309.

5. *See generally* Geertrui Van Overwalle, *Biotechnology and Patents: Global Standards, European Approaches and National Accents*, *in* GENETIC ENGINEERING AND THE WORLD TRADE SYSTEM (Daniel Wüger & Thomas Cottier eds., 2008); Deming Liu, *Now the Wolf Has Indeed Come! Perspective on the Patent Protection of Biotechnology Inventions in China*, 53 AM. J. COMP. L. 207, 209 (2005); Margo A. Bagley, *Patent First, Ask Questions Later: Morality and Biotechnology in Patent Law*, 45 WM. & MARY L. REV. 469, 475–76 (2003); Directive 98/44/EC of the European Parliament and of the Council on the Legal Protection of Biotechnological Inventions, 1998 O.J. (L 213), 13–21 (EC).

6. This chapter highlights issues relating to biotechnology, computer-implemented inventions, and methods. *See* Dan L. Burk, *Patent Law's Problem Children: Software and Biotechnology in Transatlantic Context*, *in* PATENT LAW IN GLOBAL PERSPECTIVE ch. 7 (Ruth L. Okediji & Margo Bagley eds., 2013), for a discussion of the underlying reasons for the controversies in these technological areas.

and III describe several changes in the players and venues ("who" and "where") associated with PESM inquiries that appear to be contributing to the disputes, as well as the relevancy of timing ("when"), while Sections IV and V highlight some of the possible reasons for, and complicating factors in, continuing PESM controversies ("why" and "how").

I. The "What" of US Patent-Eligible Subject Matter Challenges

The question of what types of inventions should be eligible for patent protection is an important, policy-laden inquiry, making it a difficult subject for harmonization efforts internationally and even within a given country. Article 27 of the Agreement on Trade-Related Aspects of Intellectual Property (TRIPS) requires member countries to provide patent protection to inventions "whether products or processes, in all fields of technology," but it also provides a laundry list of inventions members can exclude from patentability, such as plants and animals (other than microorganisms); diagnostic, therapeutic, and surgical methods for treating humans and animals; and inventions whose commercial exploitation might endanger public order or be contrary to morality. Although the patent laws of many countries take advantage of a number of these exclusions, the United States historically has not, choosing instead, as noted above, to grant patents on anything under the sun made by man, subject to judicially created exclusions for laws of nature, natural phenomena, and abstract ideas. Thus, for example, patents on human embryonic stem cells (which require the destruction of human embryos) have issued and survived in the United States, while being invalidated in the European Patent Office and the European Court of Justice due to conflicts with morality-based provisions of the European Union Biotechnology Directive.[7]

In *Diamond v. Chakrabarty,* the US Supreme Court gave a green light to biotech researchers and investors by confirming that a living organism, in that case a genetically modified bacterium, can comprise patent-eligible subject matter under 35 U.S.C. § 101. Section 101 of the US Patent Act provides that "whoever invents or discovers any new and useful process, machine, manufacture, or composition of matter or any improvement thereof, may obtain a patent therefor."[8] In the years that followed, the United States Patent and Trademark Office (USPTO) used the decision as the basis for granting patents on a wide variety

7. And providing another possible example of American "patent barbarianism." *See generally* Oliver Brüstle v. Greenpeace e.V., Case C-34/10 18, Oct. 2011; Case G2/06, WARF/stem cells, [2009] E.P.O.R. 15 (Enlarged B. App. 2008).

8. 35 U.S.C. § 101 (2012).

of controversial biotech inventions, including isolated DNA, human embryonic stem cell cultures, animal and human cloning methods, and transgenic plants and animals. The Supreme Court largely affirmed the USPTO's broad approach to biotech PESM in 2001 in *J.E.M. Ag Supply, Inc. v. Pioneer Hi-Bred International, Inc.,* which relied heavily on the *Chakrabarty* decision.[9] There, the Court held that sexually and asexually reproducible plants can be the subject of utility patents, despite Congress' enactment of more specific statutory protection schemes for both types of plants.[10]

Another contested PESM area characterized by considerable divergence across national approaches is computer-implemented inventions and business methods. Although some countries, such as signatories to the European Patent Convention, have explicit statutory exclusions from patentability for methods of doing business and computer programs as such,[11] the US Patent Act contains no such prohibitions. Thus, as with biotechnological inventions, the United States has been at the forefront of full recognition of both business methods and software as patent-eligible subject matter, although recent decisions indicate a retrenchment of sorts in this space.[12] One year after handing down *Diamond v. Chakrabarty*, the US Supreme Court decided *Diamond v. Diehr*,[13] directed to a computer-implemented rubber-curing process. After declining to assign patent-eligible subject matter status to computer-assisted inventions in two earlier cases, *Gottschalk v. Benson*[14] and *Parker v. Flook*,[15] the Court in *Diehr* held the claimed process to be patent-eligible. Full acceptance of the patent eligibility of such inventions proceeded gradually through a series of lower court decisions, with the Court of Appeals for the Federal Circuit explicitly expanding PESM to business methods (in dicta no less) in the 1998 *State Street Bank v. Signature*

---

9. J.E.M. Ag Supply, Inc. v. Pioneer Hi-Bred Int'l, Inc., 534 U.S. 124, 134 (2001).

10. *Id.* The more specific plant protection statutes were the Plant Patent Act of 1930, 35 U.S.C. §§ 161–64 (2000); and the Plant Variety Protection Act of 1970, 7 U.S.C. § 2422 *et seq.* (2000). The *J.E.M.* case affirmed the USPTO's longstanding practice of granting patents on plants, first approved by the USPTO Board of Patent Appeals and Interferences in *Ex parte* Hibberd, 227 U.S.P.Q. (BNA) 443, 444 (1985). The USPTO had been granting patents on animals, also a controversial category, since 1988. *See* U.S. Patent No. 4,736,866 (filed on June 22, 1984); and *Nonnaturally Occurring Non-Human Animals Are Patentable Under § 101*, 33 PAT. TRADEMARK & COPYRIGHT J. (BNA) (Apr. 23, 1987).

11. *See* Convention on the Grant of European Patents, art. 52, Oct. 5, 1973, 1065 U.N.T.S. 199 (entered into force Oct. 7, 1977) [hereinafter EPC].

12. *See, e.g.,* Bilski v. Kappos, 130 S. Ct. 3218 (2010); CLS Bank Inc. v. Alice Corp. Pty, Ltd., 717 F.3d 1269 (2013).

13. 450 U.S. 175 (1981).

14. 409 U.S. 63 (1972).

15. 437 U.S. 584 (1978).

Financial Group decision, holding that a transformation of data that provides a "useful, concrete, and tangible result" comprises PESM under Section 101. The decision very quickly resulted in a veritable flood into the USPTO of business method patent applications where previously there had been just a trickle. As business method patents were granted by the USPTO, infringement suits asserting these patents also proliferated, generating concerns regarding the appropriateness of such subject matter for patent protection and the ability of the USPTO to grant high quality patents in this area.

Signs of an impending judicial retreat from this exceedingly broad approach to PESM arguably first appeared in *Laboratory Corp. of America Holdings v. Metabolite Laboratories., Inc.*,[16] relating to a diagnostic method patent where the patentee was charged with seeking to "claim a monopoly over a basic scientific relationship."[17] The US Supreme Court dismissed the grant of certiorari in the case as improvidently granted because the patent subject matter question had not been properly raised in the lower court. However, in dissent, Justice Breyer sent a warning signal to the Federal Circuit that the breadth of the *State Street* PESM decision was in conflict with Supreme Court precedent.[18]

Heeding the warning, the Federal Circuit retreated from its broad test for business method patent eligibility *en banc* in *In re Bilski*,[19] a case involving a claim to a method of hedging risks in commodities trading. In a fractured decision that generated several dissenting opinions, the court articulated the "machine or transformation" test for determining the patent-eligibility of a process: a process must either be implemented on a machine or it must transform an article to a different state or thing to be eligible for patent protection.

The Supreme Court granted certiorari in *Bilski* and, in a 5–4 decision, declined to ratify the Federal Circuit's bright-line machine or transformation test as the only determinant of process patent eligibility. Instead, the Court called for a flexible approach, upholding the rejection of Bilski's application but on the

---

16. 548 U.S. 124 (2006).

17. *Id.* at 125.

18. He warned:

> Neither does the Federal Circuit's decision in *State Street Bank* help respondents. That case does say that a process is patentable if it produces a "useful, concrete, and tangible result." But this Court has never made such a statement and, if taken literally, the statement would cover instances where this Court has held the contrary.

Id. at 136.

19. 545 F.3d 943, 954 (Fed. Cir. 2008).

ground that Bilski's business method was an unpatentable abstract idea, without explaining how courts are to identify such abstract ideas in the future.[20]

Mayo Collaborative Services v. Prometheus[21] brought together business method and biotech-related PESM concerns with a patent directed to diagnostic methods, which tend to be relevant in personalized medicine, an important area for biotechnology innovation. The case also provided an opportunity for the Court to opine on the parameters of a second exclusion from patent eligibility, laws of nature, having taken on abstract ideas (with dubious success) in *Bilski*. The Court provided slightly more of a bright-line rule in *Mayo* by declaring all laws of nature patent-ineligible unless their application involves an inventive concept. However, how inventive additional steps have to be, and how far the notion of a law of nature reaches, is still unknown.

Most recently, the Court opined on the third exclusion from PESM, natural phenomena, in the *Ass'n for Molecular Pathology v. Myriad Genetics, Inc.*,[22] case where it had granted certiorari on the question "Are human genes patentable?"[23] In *Myriad Genetics*, a large group of plaintiffs (patients, researchers, organizations, and others, all represented by the American Civil Liberties Union and the Public Patent Foundation) challenged the validity of selected claims in several of Myriad's patents on breast cancer susceptibility genes (BRCA1 and BRCA2), their mutations, and related methods, as ineligible under § 101. The district court originally invalidated all of the contested claims, finding claims to isolated genomic DNA (gDNA) and complementary DNA (cDNA) to be unpatentable as lacking characteristics "markedly different" from sequences found in nature. It also invalidated claims to various methods, such as "detecting a germline alteration in a BRCA1 gene" by analyzing a sequence from a human sample, as abstract idea comparisons that failed the machine or transformation test.

---

20. Bilski v. Kappos, 130 S. Ct. 3218 (2010). In dissent, Justice Stevens disparaged the lack of guidance supplied by the majority opinion:

> The Court, in sum, never provides a satisfying account of what constitutes an unpatentable abstract idea. Indeed, the Court does not even explain if it is using the machine-or-transformation criteria. The Court essentially asserts its conclusion that petitioners' application claims an abstract idea. This mode of analysis (or lack thereof) may have led to the correct outcome in this case, but it also means that the Court's musings on this issue stand for very little.

Id. at 3236. *See also* Mark A. Lemley, Michael Risch, Ted Sichelman & R. Polk Wagner, *Life after* Bilski, 63 Stan. L. Rev. 1313 (2011).

21. 132 S. Ct. 1289, 1299 (2012).

22. Ass'n for Molecular Pathology v. Myriad Genetics, Inc., 133 S. Ct. 2107 (2013).

23. 132 S. Ct. 1794 (2012).

The Federal Circuit reversed (twice actually, as the Supreme Court vacated the first reversal decision and ordered reconsideration in light of *Mayo*, and the second Federal Circuit decision came to the same conclusion as the first) as to all except the method of comparison claims, concluding that the isolated gDNA and cDNA claims constituted patent-eligible subject matter.[24] Not surprisingly, the Supreme Court reversed, in a sense splitting the baby by holding claims to isolated gDNA not patent-eligible as directed to products of nature, while upholding as patent-eligible nonnaturally occurring complementary DNA (cDNA) claims.[25] Unfortunately, the reasoning underlying the Court's distinction between the patent eligibility of gDNA and cDNA could generously be called "opaque."

In the wake of this renewed interest in PESM by the Supreme Court, and for a variety of other reasons, some of which are discussed below, PESM disputes are persisting and growing in the United States.[26] And for a variety of reasons, most of which *cannot* be fully discussed below, patent developments in the United States often influence, or are at least considered in, patent developments in other countries.[27] Thus it is worth exploring the players, venues, timing, and other reasons behind these enduring PESM controversies to glean a better picture of what we may expect long-term in the United States and perhaps abroad on this topic.

II. "Who": The Players Driving Biotech Patent Eligibility Determinations

In the United States, patent eligibility cases historically have resulted from applicants appealing USPTO § 101 rejections up to the US Supreme Court.[28] However, although these cases are still an important source of challenges, other players are emerging whose interests and motivations exert new and

24. Ass'n for Molecular Pathology v. U.S. Patent & Trademark Office, 689 F.3d 1303, 1343–44 (Fed. Cir. 2012).

25. Ass'n for Molecular Pathology v. Myriad Genetics, Inc., 133 S. Ct. 2107 (2013).

26. *See* Classen Immunotherapies, Inc. v. Biogen IDEC, 659 F.3d 1057, 1074 (Fed. Cir. 2011) (Rader, C.J., additional views) ("In the last several years, this court has confronted a rising number of challenges under 35 U.S.C. §101.").

27. *See* Margo A. Bagley, Ruth L. Okediji & Jay Erstling, International Patent Law and Policy (2013) (discussing how patent developments in the United States provided impetus for the EU Biotech Directive); *see also* Cancer Voices Austr. v. Myriad Genetics, Inc., [2013] FCA 65, ¶¶ 129–35 (discussing the Federal Circuit *Myriad Genetics* decision and reaching the same result but declining to adopt its reasoning due to differences in the relevant statutory provisions in Australia and the United States).

28. *See, e.g.,* Bilski v. Kappos, 130 S. Ct. 3218, 3225 (2010); Diamond v. Diehr, 450 U.S. 175 (1981); Diamond v. Chakrabarty, 447 U.S. 303 (1980); Parker v. Flook, 437 U.S. 584 (1978); Gottschalk v. Benson, 409 U.S. 63 (1972).

different pressures on PESM determinations. Moreover, certain PESM-related decisions by traditional players also contribute to continued controversy in this area.

A. Non-Competitor Alleged Infringers and Civil Society

"Who" includes non-competitor alleged or potential infringers such as farmers, patients, and researchers. With the expansion in the scope of eligible subject matter has come an expansion in the kinds of entities and individuals that may find themselves charged with patent infringement. For example, Vernon Bowman and Percy Schmeiser of *Bowman v. Monsanto Co.*[29] and *Monsanto Canada Inc. v. Schmeiser*[30] fame, respectively, are both farmers, as are many of the hundreds of plaintiffs in the *Organic Seed Growers and Trade Association v. Monsanto Co.* suit.[31] In *J.E.M. Ag Supply v. Pioneer Hi-Bred*,[32] the defendant was a seed retailer; and in *Myriad Genetics*, many of the plaintiffs were researchers or patients. Such individuals or groups may see a threat from patents, to their livelihood or other interests, that either was not present in the past or was not an issue, due to the absence of aggressive patent enforcers in their particular field, and they now may be more inclined to raise PESM challenges.[33]

A sometimes related "who" are civil society groups representing public interest concerns. The involvement of such groups in biotech subject matter eligibility determinations is common in Europe and can be seen both in the negotiations that resulted in the European Union Biotechnology Directive as well as several opposition challenges to patents on plants, animals, human embryonic stem cells,

---

29. 133 S. Ct. 1761 (2013).

30. [2004] 1 S.C.R. 902 (Can.).

31. Organic Seed Growers & Trade Ass'n v. Monsanto Co., 718 F.3d 1350 (Fed. Cir. 2013).

32. J.E.M. Ag Supply, Inc. v. Pioneer Hi-Bred Int'l, Inc., 534 U.S. 124, 134 (2001).

33. In addition, the rise in patent infringement suits by "patent trolls" (or, less pejoratively, "patent assertion entities") likely is also resulting in an increase in PESM challenges. A recent study commissioned by the nonpartisan Government Accountability Office (GAO) found the following:

> [L]awsuits filed by monetizers [a.k.a trolls] increased from 22% of the cases filed five years ago to almost 40% of the cases filed in the most recent year. In addition . . . of the 5 parties in the sample who filed the greatest number of lawsuits during the period studied, 4 were monetizers. Only one was an operating company.

Sara Jeruss, Robin Feldman & Joshua Walker, *The America Invents Act 500: Effects of Patent Monetization Entities on US Litigation*, 11 DUKE L. & TECH. REV. 357 (2012).

and more at the European Patent Office and in national courts.[34] However, in the United States, the involvement of such groups in patent eligibility determinations historically has been marginal at best.

The Supreme Court arguably facilitated the increased involvement of the public in patent eligibility questions in the United States by relaxing, in its 2006 *MedImmune v. Genentech*[35] decision, the standing requirement for challenging a patent. According to the Federal Circuit, "[w]hether intended or not, the now more lenient legal standard facilitates or enhances the availability of declaratory judgment jurisdiction in patent cases."[36] It appears likely that *MedImmune* emboldened groups such as the Public Patent Foundation (PubPat) and the American Civil Liberties Union (ACLU) to challenge Myriad's patents on genes associated with an increased risk of breast and ovarian cancer (BRCA1 and BRCA2) on PESM grounds, and for PubPat and various farmers and organic trade groups to challenge Monsanto's RoundUp Ready® patents on moral utility and other grounds in the *Organic Seed Growers v. Monsanto* case.[37] These challenges are notable in part because the last such significant challenge occurred more than twenty years ago, even though many controversial patents have issued from the USPTO since that time.[38] After the USPTO announced in a 1987 Notice that it considered "non-naturally occurring, non-human multicellular living organisms, including animals, to be patentable subject matter" based on *Diamond v Chakrabarty*, opponents of patents on animals sought relief in court.[39] Nine plaintiffs, including the Animal Legal Defense Fund, the American

---

34. *See, e.g.,* Shobita Parthasarathy & Alexis Walker, *Observing the Patent System in Social and Political Perspective: A Case Study of Europe, in* PATENT LAW IN GLOBAL PERSPECTIVE (Ruth L. Okediji & Margo Bagley eds., 2013). Kali Murray & Esther van Zimmeren, *Dynamic Patent Governance in Europe and the United States: The* Myriad *Example,* 19 CARDOZO J. INT'L & COMP. L. 287, 321–25 (2011); Plant Genetic Sys. v. Greenpeace, Ltd., T 0356/93, 1995 OFF. J. EUR. PAT. OFF. 545 (1995); Oliver Brüstle v. Greenpeace e.V., Case C-34/10, Oct. 18, 2011. Australia also has seen its first such patent challenge in *Cancer Voices Australia v. Myriad Genetics Inc.*, [2013] FCA 65. Although standing is not an issue for such groups in Australia, the "loser pays" system does add a considerable level of risk of financial exposure that likely serves as a detriment to suit.

35. 127 S. Ct. 764 (2006).

36. Micron Tech. v. MOSAID Tech. Inc., 518 F.3d 897, 902 (Fed. Cir. 2008) (applying an "all circumstances" test instead of its pre-*MedImmune* "reasonable apprehension of suit" test).

37. *See* Ass'n for Molecular Pathology v. U.S. Patent & Trademark Office, 689 F.3d 1303 (Fed. Cir. 2012); Organic Seed Growers & Trade Ass'n v. Monsanto Co., 718 F.3d 1350 (Fed. Cir. 2013).

38. In fact, the main patents at issue in both cases issued well over a decade ago. *See, e.g.,* U.S. Patent Nos. 5,747,282; 5,693,473; 5,837,492; 5,352,605; 5,322,938.

39. *See* USPTO, Notice on Patentability of Animals, 1077 Off. Gaz. Pat. Off. 24 (Apr. 21, 1987) [hereinafter USPTO, 1987 Notice]; *see also Nonnaturally Occurring Non-Human Animals Are Patentable under § 101, supra* note 10, at 664.

Society for the Prevention of Cruelty to Animals, and the Humane Farming Association, filed suit alleging that the USPTO Commissioner had violated the Administrative Procedures Act in filing the Notice without complying with the required public notice and comment period, and sought an injunction against the issuance of any patents on multicellular living organisms, including animals, without explicit congressional authorization.[40] The plaintiffs also asked the court to hold that animals were not patentable subject matter under Section 101.

In affirming dismissal of the suit for lack of standing, the Court of Appeals for the Federal Circuit noted that none of the plaintiffs were able to show (1) "that he personally has suffered some actual or threatened injury as a result of the putatively illegal conduct" (personal injury); (2) that "the injury 'fairly can be traced to the challenged action'" (causation); and (3) that the injury "is likely to be redressed by a favorable decision" (effective relief). Although these requirements are still prerequisites to suit, *MedImmune* gave civil society groups a basis to argue (successfully in *Myriad Genetics*, less so as of this writing, in *Organic Seed Growers*, where the Federal Circuit has affirmed a denial of standing against the plaintiffs)[41] that these requirements were met outside of the normal patentee/competitor context. Such groups might be encouraged to bring future PESM challenges in light of the relative success of the *Myriad Genetics* suit; however, the difficulties encountered in the *Organic Seed Growers* case is likely to temper such challenges. Nevertheless, non-competitor plaintiffs facing infringement allegations can be expected to continue to have both the requisite standing and motivation to challenge PESM in the future.

B. Competitor Alleged Infringers

The "who" is starting to include competitors who are alleged or potential infringers. Historically, at least in the United States, competitors sued by a patentee for infringement have avoided challenging validity on § 101 grounds, focusing instead largely on § 102, § 103, inequitable conduct, or ownership issues.[42] This behavior is not surprising, as competitors may own, or seek to own, patents

---

40. *See* Animal Legal Def. Fund v. Quigg, 932 F.2d 920, 924 (Fed. Cir. 1991).

41. *Organic Seed Growers*, 718 F.3d at 1352.

42. *See, e.g.*, Ass'n for Molecular Pathology v. U.S. Patent & Trademark Office, 702 F. Supp. 2d 181, 221 (S.D.N.Y. 2011) ("[I]n the absence of a § 101 challenge to patent validity, the fact that courts have previously upheld the validity of patents directed to biological products in response to § 102 and/or § 103 challenges has no bearing on the present inquiry.... The Patent Act sets out patent invalidity as an issue to be raised by the parties, see 35 U.S.C. § 282, and it would be erroneous to treat a case involving DNA-related patents as holding that isolated human genes constitute patentable subject matter under § 101.") (citations omitted); *see also* Christopher M. Holman, *The Impact of Human Gene Patents on Innovation and Access: A Survey of Human*

on the same type of technology and would not want to categorically eliminate the subject matter from eligibility. As Professor Andrews explained more than a decade ago:

> There has yet to be a definitive legal case to address directly whether human genes are an appropriate subject matter for a patent in the first place.... [T]he court cases on gene patents are generally battles between two entities (such as a university and a biotech company) about who has rights to a particular patent. There is no incentive for either side to challenge whether a gene patent is an inappropriate patent on a product of nature because each side wants to reap the financial rewards of a gene patent.[43]

However, we now are seeing § 101 challenges by competitors such as in *Mayo Collaborative Services v. Prometheus Laboratories, Inc.*[44] There, the Mayo Clinic was sued for patent infringement after deciding to discontinue use of the kit sold by Prometheus Labs and to start producing and selling its own diagnostic test kit. To the extent a competitor simply wants freedom to operate, and has other, perhaps policy-based reasons for not intending to seek patent protection on a class of inventions, such suits can be the best hope for early, clarifying challenges to the patent eligibility of a new technology before judicial notions of "settled expectations" make it difficult to reach an objectively sound result.[45] Such notions appear to have been the deciding factor in Judge Moore joining Judge Lourie instead of Judge Bryson in the Court of Appeals for the Federal Circuit *Myriad Genetics* decision regarding the patent eligibility of isolated genomic DNA.[46]

C. The US Supreme Court and the Court of Appeals for the Federal Circuit

For good or for ill, the US Supreme Court has chosen to eschew bright-line rules in favor of murky standards, leaving further subject matter criteria development to

---

Gene Patent Litigation, 76 U. Mo.-Kan. City L. Rev. 295 (2007) (surveying patent infringement cases involving human gene patents).

43. Lori B. Andrews, *Genes and Patent Policy: Rethinking Intellectual Property Rights*, 3 Nature Revs.: Genetics 805 (2002).

44. Mayo Collaborative Servs. v. Prometheus Labs., Inc., 132 S. Ct. 1289 (2012).

45. *See infra* Section III.A.

46. Ass'n for Molecular Pathology v. U.S. Patent & Trademark Office, 689 F.3d 1303, 1343–44 (Fed. Cir. 2012), *aff'd in part, rev'd in part sub nom.* Ass'n for Molecular Pathology v. Myriad Genetics, Inc., 133 S. Ct. 2107 (2013).

the Federal Circuit and inviting continuing conflicts in lower court decisions regarding the proper boundaries of PESM.[47] For example, in *Diamond v. Chakrabarty*, the Court took an expansive view of patent-eligible biotech subject matter, essentially declaring anything under the sun made by man to be eligible for patent protection, animate or inanimate, as long as the invention is "markedly different" from what exists in nature.[48] But how different does something have to be to qualify as "markedly different?" In *Myriad Genetics*, the district court judge concluded that isolated gDNA, and cDNA, were not markedly different from native DNA in the body, and in fact their virtually identical nature was the sole basis for the value of isolated DNA.[49]

However, on appeal, the Court of Appeals for the Federal Circuit disagreed, concluding both types of DNA *were* markedly different from DNA in a human body because isolated gDNA "has been cleaved (i.e., had covalent bonds in its backbone chemically severed) or synthesized to consist of just a fraction of a naturally occurring DNA molecule)" and as such it exists "in a distinctive chemical form—as distinctive chemical molecules" from native DNA; and cDNA, with its introns removed, was clearly not naturally occurring.[50] By not providing parameters for determining a marked difference from a natural product (which admittedly would be no easy feat), the Supreme Court left lower courts to make policy-based decisions regarding where the patent-eligibility line should be drawn. Even after the Supreme Court decision in *Myriad Genetics* rejecting the patent eligibility of isolated gDNA claims and upholding the patentability of cDNA claims, the PESM boundaries in this area are still quite murky, raising perhaps as many questions as the decision

---

47. *See* Rochelle C. Dreyfuss, *Percolation, Uniformity, and Coherent Adjudication: The Federal Circuit Experience*, 6–7 S.M.U. L. Rev. (forthcoming 2013) ("The debate between the Supreme Court and the Federal Circuit over the use of bright line rules is another indicator that the sound administration of justice requires more than uniformity. The Federal Circuit is drawn to clear rules... and yet the Supreme Court has rejected every one of these rules."). *See generally* Tun-Jen Chiang, *The Rules and Standards of Patentable Subject-Matter*, 2010 Wis. L. Rev. 1353 (2010); John F. Duffy, *Rules and Standards on the Forefront of Patentability*, 51 Wm. & Mary L. Rev. 609, 622–23 (2009). *See also* Peter S. Menell, *Forty Years of Wandering in the Wilderness and No Closer to the Promised Land:* Bilski's *Superficial Textualism and the Missed Opportunity to Return Patent Law to Its Technology Mooring*, 63 Stan. L. Rev. 1289, 1291 (2011) ("[T]he failure of the *Bilski* majority to elucidate the basis—constitutional, statutory, and/or jurisprudential—for deciphering the boundaries of patentable subject matter leaves other important industries and decision-makers in the wilderness.").

48. Diamond v. Chakrabarty, 447 U.S. 303, 310 (1980).

49. Ass'n for Molecular Pathology v. U.S. Patent & Trademark Office, 702 F. Supp. 2d 181, 221 (S.D.N.Y. 2011).

50. *Ass'n for Molecular Pathology*, 689 F.3d at 1328.

answers and that lower courts will be struggling to apply to claimed inventions for years to come.[51]

Moreover, the Supreme Court has refused to allow lower courts to bypass the often thorny subject matter eligibility inquiry. In *Classen Immunotherapies v. Biogen*,[52] Chief Judge Rader of the Court of Appeals for the Federal Circuit opined on why the court should avoid § 101 subject matter determinations:

> In the last several years, this court has confronted a rising number of challenges under 35 U.S.C. § 101. The language of § 101 is very broad. Nevertheless, litigants continue to urge this court to impose limitations not present in the statute....
>
> This court should decline to accept invitations to restrict subject matter eligibility.... [E]ligibility restrictions usually engender a healthy dose of claim-drafting ingenuity, [evasions that] add to the cost and complexity of the patent system and may cause technology research to shift to countries where protection is not so difficult or expensive.[53]

Judge Rader is not alone in wanting fewer cases to be decided on patent eligibility grounds. Professors Crouch and Merges also have argued in favor of moving § 101 inquiries to the back of the line and allowing most patentability determinations to be based on novelty, non-obviousness, or description grounds, as did the US government in *Mayo Collaborative Services v. Prometheus Laboratories. Inc.*[54] However, the Supreme Court dismissed such an approach in no uncertain terms, noting:

> This approach, however, would make the "law of nature" exception to § 101 patentability a dead letter. The approach is therefore not consistent with prior law. The relevant cases rest their holdings upon section 101, not later sections.... [We] decline the Government's invitation to substitute

---

51. Ass'n for Molecular Pathology v. Myriad Genetics, Inc., 133 S. Ct. 2107 (2013). For example, the Court held that isolated gDNA is not PESM because it is naturally occurring, yet suggested that claims to the subject matter might have been treated differently if drafted in terms of chemical composition as opposed to genetic information. Also, its analysis of why cDNA was indeed PESM was cursory and conclusory, and the Court explicitly declined to opine on the PESM status of other "scientific alterations of the genetic code." *Id.* at 2120.

52. Classen Immunotherapies, Inc. v. Biogen IDEC, 659 F.3d 1057, 1074 (Fed. Cir. 2011).

53. *Id.* (Rader, C.J., additional views).

54. *See* Dennis D. Crouch & Robert P. Merges, *Operating Efficiently Post-*Bilski *by Ordering Patent Doctrine Decision-Making*, 25 Berkeley Tech. L.J. 1673 (2010); *see also* Mayo Collaborative Servs. v. Prometheus Labs., Inc., 132 S. Ct. 1289 (2012).

§§ 102, 103, and 112 inquiries for the better established inquiry under § 101.[55]

In response to this clear requirement *to* adjudicate PESM coupled with a lack of clear guidance on *how* to adjudicate it, the judges on the Federal Circuit are free to decide PESM questions through the prism of their particular view of appropriate patent policy. As one commentator explains in regards to the eligibility of method claims:

> The underlying philosophical divide in the Federal Circuit is evident in the increasingly divergent approaches to § 101 that different panels have applied over the last two decades.... On one side, some members of the Federal Circuit adhere to a "coarse filter" approach, characterizing § 101 as an initial filter meant to disqualify only the most bare abstract ideas that are not tied to any useful application.... On the other side of the divide are judges who favor a "limiting test" approach, seeking to replace the machine-or transformation test with a new formulation that would invalidate a significant number of overbroad patents under § 101 and the abstract idea exception.[56]

The 2013 *CLS Bank International v. Alice Corp. Pty. Ltd.*[57] decision provides a timely illustration of this conundrum. In *CLS Bank*, the ten member *en banc* Federal Circuit delivered a disturbingly fractured per curium opinion with a fifty-eight–word majority holding and seven separate opinions, spanning 135 pages. No single opinion, concurring or dissenting, garnered a majority of votes, thus the simple holding of patent ineligibility has no rationale supporting it on which lower courts, and later Federal Circuit panels, can rely. Although the *CLS Bank* case dealt with method, system, and computer-readable medium claims directed to lessening the settlement risk of financial instrument trades, the

---

55. *Id.* at 1303–04. *See also* Rebecca S. Eisenberg, *Wisdom of the Ages or Dead-Hand Control? Patentable Subject Matter for Diagnostic Methods After* In re Bilski, 3 CASE W. RES. J.L. TECH. & INTERNET 1, 50 (2012) ("Some scholars have suggested that patentable subject matter is redundant to other doctrinal limitations on patentability that would support the same outcomes, raising the question of whether it is necessary or appropriate to use patentable subject matter limitations to do this work. But these other doctrines may sometimes fail, leaving patentable subject matter limitations as a backstop. Doctrinal redundancy is a common feature of legal systems and may make sense if the interest at stake is important.").

56. Dina Roumiantseva, *The Eye of the Storm: Software Patents and the Abstract Idea Doctrine, in* CLS Bank v. Alice, 28 BERKELEY TECH. L.J. 569 (2013).

57. 717 F.3d 1269 (Fed. Cir. 2013) (en banc). The Supreme Court granted cert. on December 5, 2013.

problems with PESM determinations it evidences are not limited to a particular technological environment.

D. The US Congress

Congress recently waded back into the PESM waters with the 2011 Leahy–Smith America Invents Act (AIA),[58] which included three relatively new opportunities for PESM challenges: codification of the Weldon amendment in Section 33 of the Act,[59] the Tax Strategy Patent provision, and the transitional Covered Business Method Program. The Weldon amendment prevents the issuance of patents "encompassing a human organism" and was apparently designed to give effect to the USPTO's interpretation of the Thirteenth Amendment as banning patents on humans.[60] It is new to the patent statute, but not new to US law as it has been included in annual appropriations bills since 2004. The provision states "[n]otwithstanding any other provision of law, no patent may issue on a claim directed to or encompassing a human organism." Although § 33 of the AIA does not amend § 101 of the Patent Act, it is a type of PESM limitation and it seems ripe for litigation, as key terms such as "human" and "encompassing" are not defined in the statute.[61]

Section 14 of the AIA contains a limitation on patents directed to tax avoidance strategies, providing that "[f]or purposes of evaluating an invention under section 102 or 103 of title 35, . . . any strategy for reducing, avoiding, or deferring tax liability, whether known or unknown at the time of the invention or application for patent, shall be deemed insufficient to differentiate a claimed invention from the prior art." Section 14 thus serves as an indirect PESM limitation, as a tax strategy alone cannot qualify for a patent. Finally, Section 18 of the AIA creates an eight-year window for a party who has been sued for infringement of a covered business method (CBM) to challenge the patent in a special, speedy, post-grant review (PGR) proceeding and simultaneously stay the infringement

---

58. Pub. L. 112-29, 125 Stat. 325 (codified as amended in scattered sections of 35 U.S.C.)

59. *Id.* § 33. The provision is named after Rep. David Weldon, M.D., the congressman who introduced the bill.

60. *See* Comm'r of Patents and Trademarks, *Policy Statement on the Patentability of Animals*, 1077 Off. Gaz. Pat. Office 24 (1987); *see also* Wendy H. Schacht & John R. Thomas, Cong. Research Serv., The Leahy-Smith Act: Innovation Issues 19 (2013), *available at* http://www.fas.org/sgp/crs/misc/R42014.pdf.

61. The USPTO Manual of Patent Examining Procedure (MPEP) states: "If the broadest reasonable interpretation of the claimed invention as a whole encompasses a human being, then a rejection under 35 U.S.C. § 101 must be made indicating that the claimed invention is directed to non-statutory subject matter." USPTO, MPEP § 2105 (2012), *available at* http://www.uspto.gov/web/offices/pac/mpep/mpep-2100.pdf.

suit.[62] The statute defines a "covered business method patent" as "a patent that claims a method or corresponding apparatus for performing data processing or other operations used in the practice, administration, or management of a financial product or service, except that the term does not include patents for technological inventions."[63] The first challenge under § 18 was filed on September 16, 2012, the effective date of the provision, by SAP against Versata, with SAP arguing that the claims at issue were not directed to PESM.[64] The USPTO agreed with SAP and invalidated the claims under 35 U.S.C. § 101.[65] These three new provisions have the potential to generate an increasing number of PESM challenges in the future.

Of course, this is not a comprehensive list of the players whose actions are affecting PESM disputes, nor are their actions only affecting the § 101 inquiry.[66] But these selected examples do illustrate the important impact changing dynamics in system participants can have on the incidence and determination of patent eligibility questions.

III. "When" and "Where" Challenges Are Occurring (and Why It Matters)

In addition to developments in PESM participants, we are also seeing changes in where disputes are being adjudicated and possibly also when. Interestingly, both timing and location can affect the outcome and availability of PESM challenges.

---

62. *See* Pub. L. 112-29, § 18, 125 Stat. 345 (codified as amended in scattered sections of 35 U.S.C.). The regular PGR procedure created by the AIA only applies to patents including a claim with an effective filing date on or after March 16, 2013, the effective date of the AIA first-inventor-to-file provisions.

63. *Id.*

64. SAP Am., Inc. v. Versata Dev. Grp. Case CBM2012-00001 (MPT) (P.T.A.B. 2013).

65. *Id.*

66. For example, special interest groups who lobby Congress for the various statutory changes, amici who write persuasive briefs encouraging or discouraging review in a particular case, the Solicitor General when asked by the Supreme Court for input on whether to hear a case, and many more are all important players in the patent system generally and in PESM disputes more particularly. *See generally* John F. Duffy, *The Federal Circuit in the Shadow of the Solicitor General*, 78 GEO. WASH. L. REV. 518 (2010) (describing the Supreme Court's increasing reliance on the views of the Solicitor General in deciding whether to grant certiorari in patent cases); Nicholas Thompson, *Gene Blues*, WASH. MONTHLY 14, Apr. 2001 (discussing special interest groups); Colleen V. Chien, *Patent Amicus Briefs: What the Courts' Friends Can Teach Us about the Patent System*, 1 U.C. IRVINE L. REV. 395, 388 & n.21 (2011) (noting the filing of over one thousand briefs in Federal Circuit and Supreme Court patent cases over a twenty-year period).

A. Timing of PESM Disputes

In the United States, third-party PESM challenges sometimes are not brought until long after the USPTO has started granting patents in the area.[67] This is likely due, at least in part, to pre-AIA post-grant proceedings in the USPTO not allowing for PESM challenges and that competitors historically have been loath to use PESM deficiencies as an infringement defense. This is unlike the situation in the European Patent Office and elsewhere, where post-grant opposition proceedings allow patent eligibility questions to be raised by third parties very early in the life of a patent. The fact that challenges are often brought long after many patents have been granted in a biotech area appears to affect some court conclusions of patent subject matter eligibility due, at least partially, to concerns regarding the impact of late challenges on innovation incentives and the effects on long-standing property rights and settled expectations. In her concurring opinion in *Myriad Genetics*, Judge Moore explained the importance of settled expectations, citing Supreme Court precedent:

> If I were deciding this case on a blank canvas, I might conclude that an isolated DNA sequence that includes most or all of a gene is not patentable subject matter. The scope of the law of nature/manifestation of nature exception was certainly enlarged in *Prometheus*. But we do not decide this case on a blank canvas.... [T]he United States Patent Office has allowed patents on isolated DNA sequences for decades, and, more generally, has allowed patents on purified natural products for centuries. There are now thousands of patents with claims to isolated DNA, and some unknown (but certainly large) number of patents to purified natural products or fragments thereof.... I believe we must be particularly wary of expanding the judicial exception to patentable subject matter where both settled expectations and extensive property rights are involved.... The settled expectations of the biotechnology industry—not to mention the thousands of issued patents—cannot be taken lightly and deserve deference. This outpouring of scientific creativity, spurred by the patent system, reflects a substantial investment of time and money by the biotechnology industry to obtain property rights related to DNA sequences. The type of fundamental alteration in the scope of patentable subject matter argued in this case "risk[s] destroying the legitimate expectations of inventors in their property."... The Supreme Court has warned that "courts must be

67. *See, e.g.*, J.E.M. Ag Supply, Inc. v. Pioneer Hi-Bred Int'l, Inc., 534 U.S. 124, 134 (2001); Ass'n for Molecular Pathology v. U.S. Patent & Trademark Office, 689 F.3d 1303, 1343–44 (Fed. Cir. 2012).

> cautious before adopting changes that disrupt the settled expectations of the inventing community." ... "'To change so substantially the rules of the game now,'" after more than a century of practice, "'could very well subvert the various balances the PTO sought to strike when issuing the numerous patents which have not yet expired and which would be affected by our decision.'"[68]

Although Judge Moore's reasoning was not dispositive for the Supreme Court in the *Myriad Genetics* case, the Court has previously given much weight to settled expectations in some of its decisions, giving lower courts the basis to do so as well in future cases.[69] Consequently, *when* a PESM challenge is brought can affect the PESM decision, suggesting that challenges earlier in the life of a patent may be more likely to focus on the true merits of the case.[70] If we begin seeing challenges sooner, before rights and expectations are entrenched, such as through the new Post Grant Review (PGR) proceeding established by the AIA or as a result of the new players in the system, that could have an important effect on judicial determinations of eligibility.

B. "Where" Patent Eligibility Challenges Are (for now) Occurring

New venues and challenges are increasing the importance of where patent eligibility challenges can take place. Conversely, the increase in patent eligibility disputes and the lack of clear guidance on how to adjudicate the PESM question may be contributing to a move to eliminate some venues as a way to reduce the number of, or completely eliminate, such challenges.

---

68. *Ass'n for Molecular Pathology*, 689 F.3d at 1343–44.

69. *See* Ass'n for Molecular Pathology v. Myriad Genetics, Inc., 133 S. Ct. 2107 (2013) (refusing to defer to the USPTO's practice of granting isolated DNA patents in part due to change in governments position regarding eligibility of such claims). *See generally* Microsoft Corp. v. i4i Ltd. Partnership, 131 S. Ct. 2238 (2001) (relying in part on settled expectations in retaining clear and convincing evidence standard for presumption of validity); Festo Corp. v. Shoketsu Kinzoku Kogyo Kabushiki Co, 535 U.S. 722, 739 (2002) ("[C]ourts must be cautious before adopting changes that disrupt the settled expectations of the inventing community.") (citing Warner–Jenkinson Co. v. Hilton Davis Chem. Co., 520 U.S. 17, 28 (1997))); *J.E.M.*, 534 U.S. at 134 (relying in part on the USPTO's 16-plus-year practice of granting plant patents in deciding the case).

70. However, the converse is that a court may be hesitant to stifle a technology near its inception before harms and benefit have been meaningfully demonstrated. *See, e.g.,* Greenpeace v. Plant Genetic Systems NV, [1995] 8 OJ EPO 545 (refusing to revoke a patent under Article 53 of the EPC, because at the time of the decision sufficient evidence had not been adduced of a threat to the environment from commercialization of the invention.).

1. The USPTO: In many countries, patent office challenges to a granted patent may be brought in the patent office during a limited period of time after the grant; for example, nine months in the European Patent Office (EPO),[71] through an opposition proceeding. Post-grant opposition proceedings, such as those in the EPO, include the opportunity for oral hearings and significant participation by third parties. As earlier noted, the AIA gave the United States a post-grant opposition proceeding called Post Grant Review (PGR), which applies to patents obtained under the new first-inventor-to-file rules.[72] PGR challenges must be initiated within the first nine months after a patent grant, and can be based on most patentability criteria, including § 101 subject matter eligibility, unlike the other USPTO third-party challenge proceedings (inter partes reexamination, ex parte reexamination, and inter partes review), which were and are limited to issues of novelty and non-obviousness.[73] PGR thus creates a new venue, the patent office, for patent subject matter eligibility challenges in the United States. Although PGR applies only to post-AIA patents, the AIA also created a special PGR for certain pre-AIA "Covered Business Method" patents, as discussed above in relation to the *Versata v. SAP*[74] case.

2. District Court under the APA: In *Exela PharmaSciences, LLC v. Kappos*,[75] the district court approved a different strategy for patent infringement defendants to challenge the improper issuance of a patent later asserted against them. The court denied a motion by the USPTO to dismiss a declaratory judgment action under the Administrative Procedure Act (APA) against the USPTO for issuing the patent. The decision appeared to hold real promise as an additional avenue for parties seeking to challenge the issuance of a patent on subject matter eligibility grounds.

However, on December 6, 2012, the Federal Circuit held in a different case, *Pregis Corp. v. Kappos*,[76] that a competitor who has been sued as an infringer cannot lodge an APA challenge against the USPTO's issuance of a patent. This issue thus is likely to be the subject of a petition for certiorari to the US Supreme Court in the future and may, ultimately, provide an additional path to challenge USPTO patent eligibility determinations.

---

71. EPC, *supra* note 11, art. 99.

72. *See* 35 U.S.C. § 321 (2012).

73. *See id.* §§ 303, 311, 321 (2012).

74. 717 F.3d 1255 (Fed. Cir. 2013).

75. Civ. A. No. 1:12-cv-469, 2012 WL 3638552 (E.D. Va. Aug. 22, 2012).

76. 700 F.3d 1348 (Fed. Cir. 2012).

3. District Court Patent Infringement/Declaratory Judgment Proceedings: An interesting effort to eliminate subject matter eligibility determinations in US courts is the argument that courts are barred from adjudicating Section 101 PESM questions in patent infringement or declaratory judgment litigation. This argument was recently advanced in the *RMail v. Amazon.com* case in the Eastern District of Texas. As argued by Rmail in response to a motion for summary judgment on ineligible subject matter:

> The entire jurisprudential "murky morass" of subject matter eligibility need not arise again in any litigation. Rmail acknowledges that this argument is for the good faith extension or modification of existing caselaw. Courts until now have uniformly overlooked Congressional will on this question. Namely, while Section 101 analyses are appropriate in Patent Office application proceedings, this Court lacks any statutory basis for analyzing Section 101 issues as a litigation defense.... Under the Patent Act of 1952, only enumerated patent defenses exist. If an issue is not denominated an infringement defense within the Patent Act, then the Court lacks jurisdiction to address it....It does not matter that the statutory misinterpretation has lasted so long, or so pervades conventional thinking....*Prometheus*, admittedly arose within an infringement defense context. However, no one seems to have pointed out to the Supreme Court this important threshold issue of statutory construction, and statutory limitations on the powers of the federal courts. Prometheus thus does not bar this Court from issuing a correct ruling in the present adversarial context.[77]

35 U.S.C. § 282 (2012) specifies defenses that must be alleged in a civil action on a patent:

The following shall be defenses in any action involving the validity or infringement of a patent and shall be pleaded:

(1) Noninfringement, absence of liability for infringement or unenforceability,
(2) Invalidity of the patent or any claim in suit on any ground specified in part II of this title as a condition for patentability,
(3) Invalidity of the patent or any claim in suit for failure to comply with–
  (A) any requirement of section 112, except that the failure to disclose the best mode shall not be a basis on which any claim of a patent may be canceled or held invalid or otherwise unenforceable; or
  (B) any requirement of section 251.
(4) Any other fact or act made a defense by this title.

---

77. Plaintiff RMail's Limited Opposition to Defendant's Motion for Partial Summary Judgment of Invalidity, at 25–27, RMail Ltd. v. Amazon.com, Inc., No. 2:10-cv-00258-JRG (E.D. Tex. May 18, 2012) (citations omitted).

As Rmail further notes:

> Only parts (2) or (4) [of Section 282] may even arguably apply to subject matter ineligibility under Section 101. Part (2) does not apply because the "ground specified in part II of this title as a condition for patentability" applies solely to Sections 102 and 103 (as confirmed by their headings). Part (4) does not apply because statutory language must clearly demarcate a fact or act as a litigation defense for it to apply.[78]

Professors Hricik[79] and Crouch[80] also have opined on the availability of PESM as a litigation defense under § 282, and Chief Judge Rader argued in favor of such a reading of the statute in his additional reflections in the *CLS Bank* case, noting that "the Supreme Court long ago held that Section 101 is not a "condition of patentability" [and] the statute does not list Section 101 among invalidity defenses to infringement."[81] Actually, Supreme Court precedent appears rather murky on the issue;[82] thus, a court ultimately will need to confirm or refute the availability of § 101 challenges in USPTO and district court actions.

---

78. *Id.* at 26.

79. David Hricik, *Are the Courts Correct in Their Assumption That a Patent Issued on Non-patentable Subject Matter Is Invalid?*, Patently-O (Aug. 27, 2012), http://www.patentlyo.com/patent/2012/08/are-the-courts-correct-in-their-assumption-that-a-patent-issued-on-non-patentable-subject-matter-is-invalid.html. A counter proposition by Professor Mark Lemley in commenting on Professor Hricik's post is that "102, 103, 112, 251 are defenses that must be pled, but 101 are threshold issues that are not defenses at all, but pure questions of law, and therefore not things on which the defendant bears the burden of proof, and which can be raised even if not pled." *Id.* (reply posted Aug. 28, 2012, 12:49 PM).

80. Dennis Crouch, *Can a Third Party Challenge Section 101 Subject Matter Eligibility in the USPTO's New Post-Grant Review Procedure?*, Patently-O (Aug. 29, 2012), http://www.patentlyo.com/patent/2012/08/can-a-third-party-challenge-section-101-subject-matter-eligibility-in-the-usptos-new-post-grant-review-procedure.html.

81. CLS Bank Int'l Inc. v. Alice Corp. Pty. Ltd., 717 F.3d 1269, 1334 (Fed. Cir. 2013) (*en banc*) (Rader, C.J. additional reflections).

82. For example, in *Graham v. John Deere*, 383 U.S. 1, 12 (1966), the Court stated: "The [Patent] Act sets out the conditions of patentability in three sections. An analysis of the structure of these three sections indicates that patentability is dependent upon three explicit conditions: novelty and utility as articulated and defined in § 101 and § 102, and nonobviousness, the new statutory formulation, as set out in § 103." Although it does not mention subject matter, according to the Federal Circuit (in dicta) it is "beyond question" that subject matter is a condition of patentability. *See* Aristocrat Tech. v. Int'l Game Tech., 543 F.3d 657, 661–63 (Fed. Cir. 2008).

Because § 282 also governs challenges in the AIA PGR proceedings,[83] a determination that PESM challenges under § 101 are not allowable under § 282 seemingly would have the further troubling effect of eliminating the USPTO as a venue for such challenges, potentially leaving alleged infringers, civil society groups, and other interested parties with no opportunity to challenge patents on PESM grounds.[84]

IV. The "Why" of Continued Patent-Eligible Subject Matter (PESM) Controversies

There are many factors contributing to the enduring nature of patent subject matter eligibility inquiries and scholars have posited a variety of explanatory theories.[85] The few examples discussed below show interesting connections with several recent patent eligibility cases and illustrate some of the many reasons we are continuing, and will continue, to see patent eligibility challenges.

A. Technological Advances and the Patent Incentive

As the US Supreme Court noted in *Bilski*, "times change. Technology and other innovations progress in unexpected ways... Section 101 is a 'dynamic provision designed to encompass new and unforeseen inventions.'"[86] Despite § 101's ostensibly dynamic nature, as Professor Duffy notes, "law follow[s] technology."[87] That lag creates legal uncertainty during the period of time before the courts are able to adjudicate the subject matter issue (which, as noted above, may be a very long time indeed).

---

83. *See* 35 U.S.C. § 311 (2011).

84. It is theoretically possible that the USPTO could raise, *sua sponte*, the issue of patent eligibility as an extraordinary action in a reexamination, similar to the withdrawal from issue in *Blacklight Power Inc. v. Rogan*, 295 F.3d 1269 (Fed. Cir. 2002).

85. *See, e.g.,* Kevin E. Collins, Prometheus Laboratories, *Mental Steps, and Printed Matter*, 50 Hous. L. Rev. 391 (2012); Joshua D. Sarnoff, *Patent-Eligible Inventions after* Bilski*: History and Theory*, 63 Hastings L. J. 53 (2011); Efthimios Parasidis, *A Uniform Framework for Patent Eligibility*, 85 Tul. L. Rev. 323, 329 (2010); Arti Rai, *Addressing the Patent Gold Rush: The Role of Deference to PTO Patent Denials*, 2 Wash. U. J.L. & Pol'y 199 (2000).

86. Bilski v. Kappos, 130 S. Ct. 3218, 3227 (2010).

87. John F. Duffy, *Why Business Method Patents?*, 63 Stan. L. Rev. 1247, 1263 (2011) (arguing that patents on business methods arose due to technological and industrial changes that in effect transformed many business fields into branches of engineering, and that law followed and accommodated these developments).

According to Abraham Lincoln, patents added "the fuel of interest to the fire of genius."[88] Patents can be powerful weapons, allowing their owners to block off markets, charge monopoly pricing, gain leverage in negotiations, change the competitive landscape, and much more.[89] Thus the incentives to obtain patent protection are strong. And that creates pressure on patent subject matter boundaries. As Professor Sarnoff explains:

> Given the substantial incentives to seek patents on applications of newly discovered but categorically ineligible science, nature, and ideas at the forefront of medical and biotechnological research and development, we can continue to expect disputes over patent eligibility to arise. Such disputes will likely be hotly contested due to the long-standing historical normative commitments to protecting this public domain from piecemeal encroachment through wholesale or more limited retail patent claims to those discoveries. Furthermore, with the accelerating pace of scientific and technological discoveries, we can expect to confront these complex issues of eligibility in regard to a wide range of new and important products and processes, such as personalized medicine, computational genomics, synthetic agriculture, nanotechnology, and so on.[90]

Because knowledge will continue to increase, patent offices will continue to be faced with mapping § 101 and other statutes/treaties onto new and complex technological developments, resulting in continued controversies. And, as noted by Professor Dreyfuss, the "traditional tools for balancing public and private interests are largely broken"; thus "it is no wonder, then, that subject matter challenges have been on the upswing in US courts."[91]

---

88. Abraham Lincoln, Lecture on Discoveries and Inventions (1859), *cited in* MICHAEL NOVAK, THE FIRE OF INVENTION, THE FUEL OF INTEREST: ON INTELLECTUAL PROPERTY 6 (1996)

89. *See generally* Tom Ewing, *Introducing the Patent Privateers*, 45 INTELLECTUAL ASSET MGMT. MAG. 31 (2011).

90. Joshua D. Sarnoff, *Patent Eligible Medical and Biotechnology Inventions After* Bilski, Prometheus, *and* Myriad, 19 TEX. INTELL. PROP. L.J. 393 (2011).

91. Rochelle C. Dreyfuss, *Implications of the DNA Patenting Dispute: A U.S. Response to Dianne Nicol*, 22 J.L. INFO. & SCI. 1 (2012), *available at* http://www.jlisjournal.org/abstracts/dreyfuss.22.1.html (citing the basic research defense, compulsory licensing, and patent misuse as some of the broken tools).

B. Percolation and the Federal Circuit

Professor Dreyfuss traces some of the problems we see in patent law doctrinal development to creation of the Federal Circuit, which has resulted in a lack of percolation of patent law issues in the appellate courts,[92] and which may be a significant contributing factor in why we see so much dissatisfaction with US Supreme Court PESM decisions. When an issue is the subject of a regional circuit split and the Supreme Court grants certiorari, the issue likely has come before different courts in varied factual scenarios leading to diverse judgments, thus allowing the Supreme Court to see a variety of facets of the problem and adjudicate accordingly. When the Court grants certiorari in a Federal Circuit case, if it is a bad case on the facts, we may be more likely to see "bad law" resulting.[93] According to Professor Dreyfuss:

> [I]t is time to consider whether the proceduralists who founded the court were wrong to dismiss the role of percolation. After all, federal law is not typically the product of a dialogue between two courts hierarchically related to one another.... in a country as complex as the United States, the adjudicatory system is stronger when courts have overlapping jurisdiction and the independence to question one another, modify each other's outcomes, and resolve similar issues differently; that the Supreme Court benefits from—and depends upon—this activity; and that percolation provides important information to Congress.[94]

Applying this reasoning to a recent decision, one can view *Bowman v. Monsanto* as an unfortunate case for the Supreme Court to take on the question of patent exhaustion for self-replicating technologies because Mr. Bowman actually was trying to obtain, through his activities, the benefits of the patented RoundUp Ready® genetically modified seed technology.[95] A different set of facts before a different regional circuit court could have resulted in a different appellate result and a different set of considerations for contemplation and review by the Supreme Court. For example, in the *Organic Seed Growers and Trade Association v. Monsanto Co.* case, the farmer-plaintiffs do not want genetically modified seeds and pollen in their fields and fear being sued for infringement

92. Dreyfuss, *supra* note 47.

93. *Id.* (citing Frederick Schauer, *Do Cases Make Bad Law*?, 73 U. Chi. L. Rev. 883 (2006), for the notion that not only hard cases, but all cases can make bad law).

94. *Id.* at 15–16.

95. Bowman v. Monsanto Co., 133 S. Ct. 1761 (2013).

due to contamination they absolutely will not take advantage of yet cannot avoid.[96] A merits decision by a different appellate court in this case could, perhaps, have resulted in a more nuanced decision by the Court on this important issue.

Professor Dreyfuss notes that a lack of percolation also can be detrimental to patent law development on the way down from Supreme Court review.[97] Federal Circuit judges who have made public statements regarding a particular issue may not be as amenable to gracefully accepting the Court's decision (with its often opaque analysis) as judges on a different court that had not been reversed (which would be more likely to happen with certiorari granted to resolve a circuit split). As she explains, the Supreme Court

> often leaves the hard work of implementing its approach to further development below. In the context of the Federal Circuit, this is a real problem. First, it is difficult for any court to figure out how to implement a standard with which it does not agree, and that is especially true, as it often is with the Federal Circuit, when an en banc (rather than a circuit split) was the signal to grant certiorari. In such cases, every judge suffers from anchoring and framing effects; each has already debated the issue with everyone else on the court and staked out a public position. Second, because the Federal Circuit is more specialized that the Supreme Court, its judges can become intellectually complacent (some might call it arrogant) about whose resolution is correct.[98]

For example, in *CLS Bank v. Alice*, a post-*Mayo* PESM case, Chief Judge Rader, criticized the Court's broad characterization of a "law of nature" in *Mayo*, stating that "equating the personalized medicinal effect of a human-created pharmaceutical in patients of different metabolic rates and genetic makeups with the speed of light (or even gravity) is only possible in a netherworld of undefined judicial insights."[99] This is interesting, in part because, as noted above, the chief judge stated in *Classen* that he would prefer courts to avoid § 101 enquiries, an approach rejected in no uncertain terms by the Supreme Court in *Mayo*.[100]

---

96. *See* Organic Seed Growers & Trade Ass'n v. Monsanto Co., 718 F.3d 1350 (Fed. Cir. 2013); *see also* Margo A. Bagley, *The Wheat and the GMO Tares: Lessons from Plant Patent Litigation and the Parables of Christ*, 11 St. Thomas L.J. (forthcoming 2014).

97. Dreyfuss, *supra* note 47, at 18.

98. *Id.*

99. CLS Bank Int'l, Inc. v. Alice Corp. Pty., Ltd., 717 F.3d 1269, 1334 (Fed. Cir. 2013) (en banc) (Rader, C.J., additional reflections).

100. *See supra* Section II.C.

Unfortunately, we can expect the effects of this lack of percolation to manifest themselves in patent eligibility determinations and other patent disputes for the foreseeable future.

C. Patent Inflation

Another "why" underlying the continuing biotech subject matter eligibility controversies is the notion of "Patent Inflation" theorized by Professor Jonathan Masur.[101] According to Professor Masur, the asymmetrical relationship between the USPTO and the Federal Circuit causes, over time, a gradual expansion in patent subject matter boundaries. As he explains:

> If a patent applicant is denied a patent by the PTO, she can appeal that denial to the Federal Circuit. However, if the PTO grants the patent, no other party has the right to appeal that grant. Accordingly, the PTO can avoid appeals and reversals, both of which are costly in monetary and reputational terms, simply by granting any patent that the Federal Circuit might plausibly allow. Because the PTO will grant nearly any plausible patent, the vast majority of rejected applications that are appealed to the Federal Circuit will concern boundary-pushing inventions that are unpatentable under current law. Occasionally, a particularly patent-friendly panel of Federal Circuit judges will elect to reverse the PTO and grant a patent that the Agency has denied. The Federal Circuit's decision will create a new, inflationary precedent. The boundaries of patentability will expand slightly, as this new precedent exerts influence on the other circuit judges. And as the Federal Circuit's conception of what may be patented expands, the PTO will similarly inflate its own standards in order to maintain an adequate margin for error and avoid denying a patent that the Federal Circuit is likely to grant on appeal. Patent law will thus be subject to a natural inflationary pressure.

Although the patent inflation theory is not without its critics or flaws,[102] it does have explanatory power, particularly in the area of biotech patents. The

---

101. Jonathan Masur, *Patent Inflation*, 121 Yale L.J. 470 (2011). Professor Wasserman proposes a similar theory in Melissa F. Wasserman, *The PTO's Asymmetric Incentives: Pressure to Expand Substantive Patent Law*, 72 Ohio St. L.J. 379 (2011).

102. *See, e.g.*, Arti K. Rai, *Who's Afraid of the Federal Circuit?* 121 Yale L.J. Onl. 335 (2011) (responding to Masur's article); Lisa Larrimore Ouellette, *What Are the Sources of Patent Inflation? An Analysis of Federal Circuit Patentability Rulings*, 121 Yale L.J. Onl. 347 (2011) (same). Also, recent research indicates that patentability decisions vary considerably by examiner at the USPTO, making it exceedingly difficult to assess a particular policy approach by

overt[103] and covert[104] expansion in patent subject matter eligibility by the USPTO certainly contributed to cases such as *J.E.M. Ag Supply*[105] and *Myriad Genetics*[106] finally reaching the Supreme Court, and, as discussed above, there are enough differing views of the proper boundaries of PESM on the Federal Circuit to create the opportunity, with any given panel, for a further expansion. Moreover, although the new Post Grant Review procedure created by the AIA will allow concerned parties to attack the grant of a patent on § 101 grounds, the proceeding is costly, which may give pause to would-be challengers, and standing to appeal may be an issue.[107]

D. Chain Reaction

Another driver of biotech patent subject matter controversies appears to be the "tit for tat" chain reaction phenomenon identified by Sabrina Safrin. Professor Safrin posits that the expansion of patent subject matter eligibility requirements to grant property rights in genetic material, effectively begun in the *Diamond v. Chakrabarty* decision, initiated a "chain reaction" leading to a desire among biodiversity-rich developing countries for second-generation property rights over genetic resources used to create inventions. As she explains:

---

the agency. *See* Mark A. Lemley & Bhavan Sampat, *Examiner Characteristics and Patent Office Outcomes*, 94 Rev. of Econ. & Stats. 817 (2012) (pointing to the importance of human resource policies as patent system reform levers because "the most important decisions made by the patent office are significantly affected by the happenstance of which examiner gets an application.").

103. *See, e.g.*, USPTO, 1987 Notice, *supra* note 39.

104. Like the unannounced decision to stop denying patents on mammalian cloning methods, originally adopted because such patents, under 35 U.S.C. § 271(g), could "encompass" a human as the provision grants exclusionary power over products (e.g., cloned humans) to owners of patents on processes (e.g., human cloning). *See* Office Action, U.S. Pat. Application No. 08,935,052, Mar. 28, 1999 (issued as U.S. Pat. No. 6,235,970), discussed in Bagley, *supra* note 5, at 506–07.

105. J.E.M. Ag Supply, Inc. v. Pioneer Hi-Bred Int'l, Inc., 534 U.S. 124, 134 (2001).

106. *See* Ass'n for Molecular Pathology v. Myriad Genetics, Inc., 133 S. Ct. 2107 (2013).

107. *See also* Elizabeth A. Doherty, *IPR and PRG: They Are Not "American Oppositions,"* Finnegan—America Invents Act (AIA) Blog (Mar. 18, 2013), http://www.aiablog.com/global-harmonization/ipr-and-pgr-they-are-not-american-oppositions ("The current filing fee for an EPO opposition is €745. As of March 2013, the current filing fee for an IPR or PGR is $23,000 or $30,000, respectively, plus additional fees for challenging more than 20 claims. Moreover, because of the lack of discovery, an EPO opposition is a lower cost proceeding overall."). *See also* consumer Watchdog v. Wisconsin Alumni Research Foundation, No. 13-1377 (Fed. Circ. 2013)(facing standing issue).

> Developing countries harbor the greatest amount of the world's naturally occurring genetic material.... Why, these countries asked, should individuals and companies from gene-poor developed countries obtain genetic material free of charge from gene-rich developing countries when they then patent these genes and at times sell them back to the country where the genetic material originated? Moreover, developing countries faced increasing pressure to extend patent protection to man-made living organisms and their genetic material.... The key operating dynamic is that of a tit-for-tat. Namely, if developed countries assert and demand that developing countries recognize intellectual property rights over man-made living organisms and isolated and purified genetic sequences, then developing countries believe that they should also assert property interests over the raw genetic material that may contribute to the patented goods. Raw genetic material has contributed to pharmaceutical innovations and improved crops from time immemorial. Yet sovereigns only asserted ownership rights over this material after the patent system recognized private ownership rights over the material and internationalized these property rights.[108]

Many biodiversity-rich countries such as China, India, and Brazil have enacted biodiversity protection laws that regulate access to genetic resources and their use in creating inventions, so as to enable the sharing of the benefits garnered from the resources with the providing country. In view of Professor Safrin's theory, it is interesting and ironic to note that a biotech patent subject matter challenge (based on a biodiversity law that was itself enacted, per the Safrin theory, in response to the expansion of patent subject matter to include living things) has been lodged at the European Patent Office against European patent number EP 1962578 issued to Monsanto Invest B.V. on May 4, 2011, for closterovirus-resistant melon plants.

Two oppositions to the melon patent were filed before the opposition period ended in February 2012: one based on lack of novelty and inventive step filed by Nunhems, the vegetable seed-producing subsidiary of Bayer CropScience, the other from a coalition of nongovernmental organizations (NGOs) acting under the name "No Patents on Seeds."[109] The coalition has asserted technical

108. Sabrina Safrin, *Chain Reaction: How Property Begets Property*, 82 NOTRE DAME L. REV. 1917, 1928–31 (2007).

109. *See The Melon Patent Case FAQ*, EUR. PATENT OFFICE (EPO), http://www.epo.org/news-issues/issues/melon.html (last updated Feb. 14, 2012). The coalition apparently is supported by a diverse group of organizations, including the Berne Declaration (Switzerland), GeneWatch (UK), Greenpeace (Germany), Misereor (Germany), Development Fund (Norway), No Patents on Life (Germany), Reseau Semences Paysannes (France) and Swissaid

arguments against the patent, but the opposition papers also charge that the melon claimed "is not patentable under Article 53(a) as it is against order public for violating provisions of the Indian Biological Diversity Act, 2002." In other words, the group is arguing that because Monsanto did not comply with the biodiversity law in India in obtaining genetic material for its invention, that should be grounds for invalidating the patent under Article 53(a), which provides "European patents shall not be granted in respect of: (a) inventions the commercial exploitation of which would be contrary to 'ordre public' or morality...."[110] Although the chances that the opposition on Article 53(a) grounds will be successful seem slim, the novel use of biodiversity law violations as a basis to challenge PESM is another reason to expect biotech subject matter challenges to continue.

E. Patent First, Ask Questions Later

As biotechnological advances increase and create new opportunities, they often also create or revive questions regarding the role societal concerns (e.g., environmental, moral, legal, and ethical) should play in determining limits on patent subject matter. As noted earlier, the World Trade Organization's (WTO) Agreement on Trade-Related Aspects of Intellectual Property (TRIPS) allows member countries on the basis of morality, public order, environmental, and other concerns to exclude inventions from patentability, and many countries have incorporated such exclusions into their laws. Moreover, proposals (and in some cases, requirements) for patent applications to include additional information (which may impact patentability determinations) on issues such as the origin of genetic resources used in creating the invention are indicative of a more critical view of appropriate patent subject matter limits.[111] Such considerations are largely foreign to US patent law, which tends to view technological progress as synonymous with societal progress.[112]

(Switzerland). These organizations are seeking a revision of the EPC to exclude breeding material, plants and animals, and food from patentability. More than 250 organizations and over 50,000 individuals have signed a petition to this effect. *See also* Vandana Shiva, *Opposition to Monsanto's Patent on Indian Melon*, NAVDANYA (Feb. 3, 2012), http://www.navdanya.org/news/223-opposition-to-monsantoas-patent-on-indian-melon.

110. EPC, *supra* note 11, art. 53(a).

111. *See* QUEEN MARY INTELL. PROP. INST., REPORT ON DISCLOSURE OF ORIGIN IN PATENT APPLICATIONS 52–55 (2004), *available at* http://trade.ec.europa.eu/doclib/docs/2005/june/tradoc_123533.pdf (describing a variety of such requirements in national laws).

112. Complaint, ¶¶ 78, 92, Organic Seed Growers & Trade Ass'n v. Monsanto Co., No. 2012-1298 (Fed. Cir. 2013) [hereinafter Organic Seed Growers Complaint], *available at* http://www.pubpat.org/assets/files/seed/OSGATA-v-Monsanto-Complaint.pdf (arguing that patents on genetically modified plants lack moral utility due to the variety of ill-effects caused by such plants and related products).

Certainly from *Chakrabarty* to the present, the United States has taken the lead in pushing the boundaries of patent-eligible biotech subject matter. As a group, biotech inventions tend to be more likely than most to implicate moral considerations, either because of concerns that some research should not be conducted and patents provide government-granted incentives for that research, or concerns that the underlying research should be conducted and should be free from exclusive rights.[113] In addition, biotech patents may be more likely to raise expressive concerns that may negatively impact particular groups.[114]

Moreover, US law states that "[a] person shall be entitled to a patent unless" the USPTO can establish he is not so entitled, and due to the demise of the moral utility doctrine, there is no statutory basis for inquiring into the morality of an invention as a criterion for patentability.[115] The "moral utility" doctrine is a common law construct based on the idea that to be "useful" within the meaning of the patent statute, and thus eligible for patent protection, an invention has to meet certain standards of morality. For over 150 years, courts cited this requirement as the basis for rejecting a variety of morally controversial inventions, including gambling machines and fraudulent articles. Over time, however, courts and the USPTO grew uncomfortable with making *ad hoc* morality determinations and virtually cabined the doctrine out of existence.[116] Thus, patents routinely issue from the USPTO on useful, novel, non-obvious, yet morally controversial biotech inventions, and to the extent some segment of the population is disturbed by the patent's issuance, questions are asked *ex post* about whether the Office should be granting such patents. In a 2003 article, I discussed this US "Patent First, Ask Questions Later" regime and contrasted it with "Ask Questions First, Then Patent" systems in which moral concerns can be

---

113. Morally controversial biotech subject matter includes: transgenic animals (devaluing life), genetically modified plants (impact on farmers, the environment, food supply), genes/DNA (products of nature, building blocks of life/research), methods of cloning humans/cloned humans ("ownership" of humans), human embryonic stem cells (destroying human embryos), and mixed human/animal chimeras (human dignity, species integrity).

114. *See, e.g.*, Shubha Ghosh, *Race-Specific Patents, Commercialization, and Intellectual Property Policy*, 56 Buff. L. Rev. 409 (2008); Johnathan Kahn, *Race-ing Patents/Patenting Race: An Emerging Political Geography of Intellectual Property in Biotechnology*, 92 Iowa L. Rev. 353 (2007); Timothy R. Holbrook, *The Expressive Impact of Patents*, 84 Wash. U. L. Rev. 573 (2006).

115. The judicially created tool for assessing morality historically in the United States is the now largely defunct moral utility doctrine first articulated in *Lowell v. Lewis*, 15 F. Cas. 1018 (C.C.D. Mass. 1817) (stating that to be patentable, an invention must not be "injurious to the wellbeing, good policy, or sound morals of society.").

116. *See, e.g.*, Diamond v. Chakrabarty, 447 U.S. 303 (1980); Juicy Whip, Inc. v. Orange Bang, Inc., 185 F.3d 1364, 1367 (Fed. Cir. 1999); *Ex parte* Murphy, 200 U.S.P.Q. (BNA) 801, 802 (Bd. App. 1977).

considered during the patentability inquiry such as in Europe, China, Japan, and many other countries.[117] Without statutory limits on morally controversial subject matter, patents will continue to issue from the USPTO on inventions that will be challenged eventually, either in the United States or abroad as foreign offices are pressured by applicants to expand subject matter eligibility to be coextensive with that in the United States. As Lord Justice Jacob explained in the UK *Aerotel and Macrossan* decisions:

> [T]here is pressure from would-be patentees on patent offices.... This pressure in part stems from the fact that... people have been getting patents for these subject-matters in the USA. Since they can get them there, they must as a commercial necessity apply for them everywhere. If your competitors are getting or trying to get the weapons of... patents you must too. An arms race in which the weapons are patents has set in... [I]t by no means follows that because of pressure from applicants, the grant of patents for excluded categories should be allowed or that the excluded categories... should be construed narrowly. Just as with arms, merely because people want them is not sufficient reason for giving them.[118]

However, the possibility of moral limits on patent subject matter in the United States just refuses to go away. The most recent approving citation of the doctrine was by the USPTO in 1998 when, faced with a patent application on a mixed human/nonhuman creature, the Office issued an advisory that stated "inventions directed to human/non-human chimera could, under certain circumstances, not be patentable because, among other things, they would fail to meet the public policy and morality aspects of the utility requirement."[119] But now the doctrine is being revived in an argument against the patentability of

---

117. Bagley, *supra* note 5, at 475–76. *See also* Margo A. Bagley, *The New Invention Creation Boundary in Patent Law*, 51 Wm. & Mary L. Rev. 577 (2009).

118. Nevertheless, the pressure has resulted in the whittling away of patent subject matter exclusions in many countries, often through patent office claim drafting fictions such as Swiss-type claims in the EPO, which, for many years, allowed applicants to circumvent the statutory prohibition in EPC Article 53(c) against diagnostic and therapeutic methods by claiming, for example, "the use of composition A in the manufacture of a medicament for the treatment of disorder B," now replaced by claims to "Substance X for the treatment of disease Y." *See In re* Abbott Respiratory, G 0002/08 (2010), *available at* http://www.epo.org/law-practice/case-law-appeals/recent/g080002ex1.html.

119. Media Advisory, U.S. Patent & Trademark Office, *Facts on Patenting Life Forms Having a Relationship to Humans* (Apr. 1, 1998), http://www.uspto.gov/web/offices/com/speeches/98-06.htm.

genetically modified seeds by plaintiffs in the *Organic Seed Growers and Trade Association v. Monsanto Co.* case.[120] There the plaintiffs noted Justice Story's original articulation that to be patentable, an invention must not be "injurious to the wellbeing, good policy, or sound morals of society . . . a new invention to poison people . . . is not a patentable invention."[121] They then cited a variety of ill-effects associated with GM plants, including the potential loss of biological diversity as GM crops contaminate and overtake conventional crops, unknown toxic effects from the dramatically increased and widespread use of glyphosate, and harmful, poisonous effects on people.[122]

It is worth noting that even if the USPTO evaluated moral utility at the time the patents under contention in the *Organic Seed Growers* case were issued, it is unlikely the patents would have been denied on that basis, simply because the evidence being asserted now was not available when the applications were being prosecuted. Thus, under the Supreme Court's reasoning in *Diamond v. Chakrabarty*, the hypothetical possibility of harm likely would not have been deemed sufficient for a rejection. However, even the possibility to ask questions "much later" when relevant evidence is available, may not be available to concerned persons, as the case was dismissed for lack of standing, a holding affirmed by the Federal Circuit.[123] Nevertheless, the plaintiffs have petitioned for Supreme Court review, which ultimately may allow the merits of the case to be reached and the question of moral utility to be addressed.

In the meantime, the US "Patent First" regime, in conjunction with the other theories discussed in this section, and the pressures the US approach creates for other countries, also supports the notion that biotech subject matter eligibility challenges will continue in the United States and abroad.

V. The "How" of Biotech Patent Eligibility Challenges

As the cases above show, patent eligibility challenges are often brought as appeals of patent office rejections, defenses to charges of infringement, and requests for declaratory judgments of invalidity based on subject matter. But patent eligibility issues do not always appear as straightforward interpretations of the relevant subject matter statute; the presentation can be far more subtle than that and can

120. Organic Seed Growers & Trade Ass'n v. Monsanto Co., 718 F.3d 1350 (Fed. Cir. 2013).

121. *Lowell*, 15 F. Cas. at 1018.

122. Organic Seed Growers Complaint, *supra* note 112, ¶¶ 78–92.

123. *Organic Seed Growers*, 718 F.3d at 1352.

result in a policy-based interplay with other patent doctrines, as seen in the following non-US cases.

A. Claim Construction and Claim Scope: Broadening

Consider the Canadian *Monsanto v. Schmeiser* and the EU *Monsanto v. Cefetra* decisions.[124] The Monsanto Company owns numerous patents covering technology for genetically modifying plants to make them resistant to the herbicide glyphosate, sold under the trademark RoundUp®. Monsanto prohibits farmers planting the patented "RoundUp Ready®" seeds from saving any harvested seeds for replanting in a subsequent season.

In *Monsanto v. Schmeiser*, Monsanto did not have a patent on its genetically modified RoundUp Ready canola plants in Canada; the patents only covered the genes, cells, etc. used to modify and produce the plants.[125] Just two years earlier, the Canadian Supreme Court had ruled in *Harvard College v. Canada* that higher life forms, such as plants, were not patent-eligible subject matter, as they failed to qualify as either a manufacture or a composition of matter.[126] Nevertheless, the Court concluded that a claim to a plant cell (lower life form) offered the same patent infringement protection as a claim to the plant, making obviation of the higher life form patent prohibition a simple claim drafting exercise. In other words, interpreting a claim broadly can be equivalent to expanding patent subject matter eligibility.

B. Claim Construction and Claim Scope: Narrowing

The reverse is also true and evidenced in the Court of Justice of the European Union (CJEU) *Monsanto v. Cefetra* decision.[127] There Monsanto sued importers of soymeal prepared from RoundUp Ready soybeans grown in Argentina, where Monsanto lacks patent protection for the RoundUp Ready technology and was thus unable to collect royalties from the farmers who grew the soybeans that were converted to soymeal. However, Monsanto does have patent protection for the chimeric genes, cells, and plants throughout Europe, and sued the meal importers for infringement of those patents, on the basis that the meal still contained the chimeric patented gene.

---

124. Monsanto Canada, Inc. v. Schmeiser, [2004] 1 S.C.R. 902 (Can.); Monsanto Tech., LLC v. Cefetra BV, Case C 428/08 (Eur. Ct. Justice, 2010).

125. *Monsanto Canada*, 1 S.C.R. at 902.

126. Harvard College v. Canada (Commr. of Patents), [2002] 4 S.C.R. 45 (Can.).

127. Case C-428/08, Monsanto Tech. LLC v. Cefetra BV, 2010 E.C.R. I-6756.

The CJEU concluded that Article 9 of the EU Biotechnology Directive governed the case. Article 9 of the Directive, entitled "Scope of protection," provides:

> The protection conferred by a patent on a product containing or consisting of genetic information shall extend to all material... in which the product is incorporated and in which the genetic information is contained and performs its function.[128]

Because the function of the claimed gene was to protect the plant from glyphosate, which otherwise would kill the plant, the Court ruled that the patented gene was not performing its function in the dead soymeal; thus there was no patent infringement.

As the CJEU's ruling in *Monsanto v. Cefetra* shows, how broadly or narrowly a claim is construed affects the patent owner's ability to enforce the patent against third parties. It also illustrates a reaction to perceived patentee overreaching.[129] It is unlikely that Monsanto would have brought the case if it had patent protection on Roundup Ready® seed in Argentina. Thus, allowing Monsanto to sue dealers in downstream products appears not only unnecessary, but also could have had more far reaching consequences. As Dr. Christopher Heath of the EPO Board of Appeals notes:

> If Monsanto's position were correct, similar cases could arise for the importation of cotton, the [RoundUp Ready®] variety of which Monsanto has started selling in India without corresponding patent protection. Any trace of the protected DNA in, say, imported jeans or t-shirts would allow the patentee to raid the premises of any manufacturer or shop commercially manufacturing or selling these goods—an interesting, but somewhat worrying scenario.[130]

The CJEU decision does not mention such concerns; it focuses instead on interpretation of the Directive in light of its goals and purposes. Nevertheless, the decision certainly constrains the scope of gene patents by denying absolute product protection to this specific subject matter category, possibly because of the perception of patentee overreaching. Thus an exercise in claim construction

---

128. Directive 98/44/EC, *supra* note 5, art. 9.

129. *See* Elizabeth K. Rowe, *Patents, Genetically Modified Foods, and IP Overreaching*, 64 SMU L. Rev. 859, 866–67 (2011).

130. Christopher Heath, *The Scope of DNA Patents in Light of the Recent* Monsanto *Decisions*, 40 IIC 956 (2009).

for infringement purposes can have the effect of an eligibility limitation decision, especially in light of the fact that Article 9 was ostensibly designed to be a patent eligibility provision. Consider the following complaint:

> [T]here is a significant risk that countless biotechnology products are not protected by enforceable patent claims.... Furthermore, the opinion's impact may extend far beyond a narrow conception of the biotechnology industry. For example, the viability of patents claiming isolated DNA or RNA sequences used as reagents—including reagents used in diagnostic methods such as gene tests and DNA chips—are now in jeopardy.... The [CJEU] decision is surprising because *Article 9 of the Directive... was intended to define what constitutes patentable subject matter when the claims in question cover living and replicating organisms. Article 9 was not intended to define the scope of enforceable rights in the context of alleged patent infringement.*[131]

Whether or not Article 9 was intended to define the scope of enforceable rights in the patent infringement context, it clearly does now, a result that might not have occurred without Monsanto arguably overreaching in bringing this particular case. The case also raises the question of whether restricting patent enforcement (e.g., via compulsory licensing, research exemptions, etc.) might, in some cases, be preferable to or provide a safety valve for subject matter eligibility determinations.

The use of other patent doctrines as proxies or levers for indirectly narrowing or broadening patent eligibility is not a new phenomenon,[132] nor is it foreign to the United States. In fact, concerns about patent claim scope preempting research, justified or not, have been the subject of some debate in the United States.[133] As one group of scholars notes: "We don't exclude inventions from patentability because the invention is too abstract. We refuse to patent certain claims when those claims reach too broadly and thereby threaten downstream

---

131. Richard Peet et al., *The Future of Biotechnology Patents in the European Union: The Potential Impact of* Monsanto Technology LLC v. Cefetra et al. *on Patent Infringement*, IP-WATCH (Aug. 17, 2010), http://www.ip-watch.org/2010/08/17/the-future-of-biotechnology-patents-in-the-european-union/ (emphasis added).

132. For more on various patent doctrines as levers to accomplish particular policy objectives generally, see DAN L. BURK & MARK A. LEMLEY, THE PATENT CRISIS AND HOW COURTS CAN SOLVE IT (2009).

133. *See* Katherine Strandburg, *Much Ado About Preemption*, 50 HOUS. L. REV. 563 (2013) ("'Preemption' has emerged as the leading contender for conceptual grounding of the patentable subject matter doctrine's exclusion of abstract ideas and natural phenomena from patentability.").

innovation."[134] Moreover, the moral utility challenges against genetically modified plants brought in the *Organic Seed Growers* case can also be viewed as indirect attacks on PESM. In addition, as noted earlier in Section II, the new Weldon amendment and tax strategy patent provisions in the 2011 AIA, while not part of Section 101, both do implicate patent subject matter eligibility.

Conclusion

The American approach to patent-eligible subject matter determinations may not be barbaric, but it is in flux. After almost three decades of taking an exceptionally broad approach to patent-eligible subject matter the US Supreme Court has, in *Bilski*, *Mayo*, and *Myriad Genetics*, revisited the abstract idea, law of nature, and natural phenomena exclusions from subject matter eligibility in ways that appear to narrow eligibility in some respects while leaving it open or broadening it in others, all with the stated goal of ensuring both an open system and incentives for innovation.

In addition, patent subject matter eligibility issues have again captured the attention and interest of civil society groups to the extent of prompting legal action, as manifested in the *Myriad Genetics* and the *Organic Seed Growers* litigations, both of which involve numerous plaintiffs challenging, in declaratory judgment actions, patent validity in a way rarely seen before in the United States. Both cases also raise important standing questions regarding how open the patent system really is and should be to all stakeholders. These US movements are taking place concurrently with patent eligibility decisions in Europe, Australia, and beyond that sometimes limit, sometimes expand patent subject matter eligibility and patent scope in ways that are likely to generate more patent eligibility questions in the future.

A serious consideration of the who, what, when, where, why and how of patent subject matter determinations leads inevitably to the conclusion that there is no end in sight to such challenges and that we are likely to see a continued morphing of the types of parties, actions, and venues involved. How these disputes play out in the United States and abroad can be expected to impact innovation and to have other collateral effects. As the US Supreme Court has noted:

> Patent protection is, after all, a two-edged sword. On the one hand, the promise of exclusive rights provides monetary incentives that lead to creation, invention, and discovery. On the other hand, that very exclusivity

---

134. Mark A. Lemley, Michael Risch, Ted Sichelman & R. Polk Wagner, *Life after* Bilski, 63 Stan. L. Rev. 1313, 1346 (2011).

> can impede the flow of information that might permit, indeed spur, invention, ... At the same time, patent law's general rules must govern inventive activity in many different fields of human endeavor, with the result that the practical effects of rules that reflect a general effort to balance these considerations may differ from one field to another.[135]

Patent eligible subject matter questions are often complex and not amenable to easy or lasting resolution, due to a variety of factors both exogenous and endogenous to patent offices and courts, many of which have been discussed in this chapter. By recognizing some of the new players in the system and areas where additional pressure has been put on the PESM inquiry to do work previously done by other patent law doctrines or to address new problems, the courts, the USPTO, and/or Congress possibly could alleviate some of that pressure by reinvigorating or calibrating existing patent tools (e.g., the experimental use exception) to do the work they are more properly designed to do, and thus reduce the need and desire for recourse to the PESM inquiry. Ultimately, however, the patent barbarian label notwithstanding, accepting that patent subject matter boundaries will continue to be a moving target in the United States and abroad, and affirmatively recognizing the levers that affect them, may be our most viable option.

135. Mayo Collaborative Servs. v. Prometheus Labs, Inc., 132 S. Ct. 1289, 1305 (2012).

7 PATENT LAW'S PROBLEM CHILDREN: SOFTWARE AND BIOTECHNOLOGY IN TRANSATLANTIC CONTEXT

Dan L. Burk*

Introduction

Modern patent law applies across a vast range of developed, developing, and as yet undeveloped technologies. Indeed, the international TRIPS Agreement requires that, with only a few exceptions, patent protection is to be available in signatory nations without respect to technological field.[1] At the same time, modern technologies display a highly varied and even disparate set of technical features, economic characteristics, and innovation profiles.[2] In order to accommodate this diverse array of technologies, patent statutes incorporate flexible provisions allowing their application to be modulated to the needs of different subject matter categories.[3]

And yet, even with such statutory flexibility, some technologies fit only very uncomfortably within the ambit of current patent statutes. In particular, recombinant DNA technology and computer software have long been the problem children of modern patent law.[4] For at least the last thirty years, on both sides of the Atlantic, patent doctrine has been in a constant state of turmoil regarding these technologies,

---

* Chancellor's Professor of Law, University of California, Irvine.

1. Agreement on Trade-Related Aspects of Intellectual Property Rights, art. 27, Apr. 15, 1994, Marrakesh Agreement Establishing the World Trade Organization, Annex 1C, 33 I.L.M. 81 (1994) [hereinafter TRIPS Agreement].

2. Dan L. Burk & Mark. A. Lemley, The Patent Crisis and How the Courts Can Solve It (2009).

3. *Id.*

4. Dan L. Burk, *The Problem of Process in Biotechnology*, 43 Hous. L. Rev. 561, 588 (2006).

sometimes accepting them, sometimes rejecting them, and almost continually revising itself in unsuccessful attempts to accommodate them. This difficulty has been apparent across multiple patent doctrines, including disclosure, obviousness, and utility.

But perhaps the most pronounced difficulties have manifested in attempting to fit inventions from these high technology fields reliably, if at all, within the subject matter provisions of American and European patent law. Despite a lengthy period of consideration and reconsideration, patent authorities in these major jurisdictions have generally failed to settle on a solution to the subject matter problems concerning software and biotechnology inventions. The law on this question remains in constant flux, attempting new formulation every few years, leaving the industries founded on these technologies uncertain as to the criteria for patentability in their fields. No stable resolution to these issues is currently in sight.

However, it may be that the extended effort that has gone into considering the problem has not been entirely in vain, even if no firm criteria for determining software and biotechnology subject matter have yet emerged. Instead, what arguably has emerged from thirty years of judicial, legislative, and administrative struggle is the outline of the problem that bedevils these technologies. This outline is apparent in examining the common threads running not only between decisions regarding these two technologies, but among the law developed by adjudicators in different jurisdictions. A comparative study of such decisions shows that such institutions have consistently relied upon a similar pattern of doctrines, interpretive strategies, and jurisprudential constructs in their attempts to accommodate software and biotechnological developments within patentable subject matter.

In tracing the skein of such decisions we must of course be aware of certain differences in the contexts of the various decisions regarding patentable subject matter. Much of the development of this law in North America has been common law development, generated in the courts, either by judicial interpretation of statutes, or from entirely judicially created doctrine. The United States Patent Office in particular, has a relatively subordinate role in developing the meaning of the patent statute. In Europe, only Great Britain follows the common law approach; other national courts are largely creatures of civil law.

Perhaps more important, much of European patent law development emerges from the intersection of national laws and interpretation of international treaties. The European Patent Office is a treaty organization, with quasi-judicial functions in interpreting the language of the European Patent Convention. Separately and contiguously, the Court of Justice of the European Union, a different treaty organization, assumes responsibility for interpreting the language of European Union directives, as well as the compatibility of national laws with the functioning of a common market under the relevant community treaties. The language of certain directives has a bearing on the development of patent laws, whereas the functioning of national patent laws has a bearing on the common market. The interpretive latitude of such bodies is markedly different than that of common law courts.

Additionally, it is important in such comparisons to recognize that these treatments of patent law do not develop in isolation. Judges and patent examiners on both sides of the Atlantic are well aware of the decisions of their counterparts elsewhere; they are in frequent communication and familiar with doctrinal developments in other jurisdictions.[5] International treaties lend a measure of harmonization to the patent regime in different jurisdictions, or may impose constraints channeling local developments along similar courses. Consequently, to the extent that similar patterns of addressing patent issues constitute parallel development, it is often conscious parallelism rather than an independent convergence.

Nonetheless, even bearing such caveats in mind, tracing particular patterns across jurisdictions can help us triangulate on core issues that exist despite disparate institutional and jurisprudential contexts. In examining similar fact patterns from different jurisdictions, I will highlight common approaches to common problems, which I will argue reveal a fundamental deficiency to be corrected in modern patent law.

I. Categorizing Subject Matter

Perhaps the best place to begin tracing the commonalities of treatment in biotechnology and software patenting is in North America, with the landmark decision of the Canadian Supreme Court in *Harvard College v. Canada (Commissioner of Patents)*.[6] The invention at issue in the decision was the so-called "Oncomouse," a genetically modified rodent carrying genetic sequences, called oncogenes, that confer on the animal a greater susceptibility to cancer, making it a valuable laboratory model for biomedical research. The mouse was the subject of a lengthy patent battle in the European Patent Office; was celebrated in the United States as one of the first patents on a genetically modified animal; and was the subject of a patent rejection in the Canadian Patent Office, which contended that such living subject matter was not within the scope of the Canadian patent statute.

That decision was upheld by the Canadian Supreme Court. Over a vigorous dissent, the majority of the Court in *Harvard College* accepted the position of

---

5. *See* Peter Drahos, The Global Governance of Knowledge: Patent Offices and Their Clients (2010).

6. [2002] 4 S.C.R. 45 (Can.).

the patent office, holding that a living organism was not within the meaning of the subject matter provisions of the patent statute. The Canadian patent statute defines patentable subject matter as "any new and useful art, process, machine, manufacture, or composition of matter."[7] The Court concluded that a living, "higher" organism could not fit the definition for any of these categories. In particular, the majority held that the mouse could not be a "composition of matter" under the statute, reasoning that the ordinary meaning of this term included only inanimate material, not living organisms, or at least, not "higher" organisms such as a mouse. And, if the mouse could not be a composition of matter, then similarly it surely could not be an art, process, machine, or manufacture, placing it outside all the statutory categories.

One finds a surprising strain of vitalism in this opinion—in considering the proper definition of "composition of matter," the Court dwells on the transcendence of "higher" life forms, arguing that the common understanding of "matter" does not capture this transcendent quality.[8] The court all but argues that the mouse has a soul; at least it asserts that the common understanding of the term "matter" would be that it lacks whatever vital spark a mouse has. As a consequence, given that a "composition of matter" cannot be alive, the Court concluded that only a new act of the legislature—presumably, using a different terminology—could bring genetically modified higher organisms within the scope of the statute.

In other words, the ultimate rationale behind this conclusion was that a particular technology, such as genetically modified living subject matter—or at least, multicellular genetically modified subject matter—must be within the contemplation of the legislature when enacting the statute, rather than within the ambit of the language chosen by the legislature. The import of such a holding for patent policy is striking. It effectively means that the patentable subject matter canon is closed, reasoning that if the legislature did not envision a particular technology when enacting a statute, it can only mean that unanticipated innovation is not patentable. According to the majority in *Harvard College*, only if Parliament indicated acceptance of a new category could an invention such as a genetically modified mouse become eligible for patenting.

As I have pointed out in previous work,[9] this result is particularly striking because two decades earlier, the US Supreme Court considered essentially the same question, but reached a diametrically opposite result in *Diamond v. Chakrabarty*,[10] the first American case to squarely address the patenting of living

7. Patent Act, R.S.C. 1985, c. P-4, s. 2.

8. Dan L. Burk, *Reflections in a Darkling Glass: A Comparative Contemplation of the* Harvard College *Decision*, 40 CAN. BUS. L.J. 219 (2003).

9. *Id.*

10. 447 U.S. 303 (1980).

organisms. The *Chakrabarty* decision concerned a patent application covering a genetically engineered microorganism capable of digesting petroleum. Most of the arguments on either side of the *Chakrabarty* decision paralleled those that either supported or opposed the patent in *Harvard College*. But most surprisingly, the courts in each case were interpreting the identical statutory language, because the subject matter provisions of the Canadian patent statute were lifted almost verbatim from the US statute.

As with the Canadian statute, the US patent statute also states as its subject matter new and useful processes, machines, articles of manufacture, and compositions of matter.[11] But unlike the Canadian statute, the US understanding of these terms extends to living organisms. According to the US Supreme Court, the bacterium at issue in *Chakrabarty*, being composed of matter, clearly qualified as a composition of matter under the statute. The Court adopted the very straightforward position that bacteria, whatever their properties as organisms, are composed of matter, and that term as used in the statute includes technologies not specifically foreseen by the legislature. This conclusion was informed by a policy conclusion somewhat contrary to that of the Canadian court: that the US Congress in choosing the language of the statute had intended patentable subject matter to encompass "anything under the sun made by man," so consequently the language must be read expansively.[12] On the American view, the subject matter categories might be viewed as illustrative, or at least broadly inclusive.

In fact, unless one is possessed by the vitalism that seems to have gripped the Canadian Supreme Court in the *Harvard College* case, it is fairly difficult to identify items in the modern world that do *not* fit into one or more of the four categories of process, machine, composition of matter, or article of manufacture. For example, in addition to the *Chakrabarty* bacterium being a composition of matter, one could easily imagine drafting claims by which such a recombinant bacterium might also constitute an article of manufacture, given that it was the object of human processing, produced in a laboratory. Similarly, one might envision a creative set of claims under which the bacterium could be claimed as a machine—molecular biologists view various microbial structures as molecular machines, including nanoscale electrical motors, flywheels, and driveshafts that drive bacterial flagella.[13] Indeed, much of current nanotechnology is based on the mechanics of molecular assemblies originating in microorganisms.

11. 35 U.S.C. § 101 (2011)

12. *Chakrabarty*, 447 U.S. at 309 (quoting S. Rep. No. 1979, at 5 (1952)).

13. *See* Jared Diamond, *Transport Mechanisms: The Biology of the Wheel*, 302 Nature 572 (1983).

For that matter, claims to the *Chakrabarty* invention need not necessarily be confined to the "product" categories of machine, composition, or manufacture. One could imagine the bacterium being claimed in terms of its processes—although product patents and process patents are generally placed into separate categories, living organisms in fact constitute a discrete set of biological processes. On this view, Chakrabarty's bacterium comprised a large bag of biochemical reactions, if you will, and could likely be distinguished in terms of such characteristics. It may be that as a practical matter such a characterization would be too onerous as a claims-drafting exercise. But the implication of such a thought exercise is that the Supreme Court's reading of the statutory subject matter categories as illustrative and inclusive largely transforms the subject matter question into a descriptive problem.

There is of course a critical policy consideration animating the US Supreme Court's reading of the statute, which is a prescription remarkably different from the Canadian approach in *Harvard College*. Each court read essentially identical language, and purported to defer to the legislative meaning of that language, to reach diametrically opposite defaults: in one case that the categories of subject matter were intended to be open to unforeseen innovation, in the other that the categories were intended to be closed. In other words, the US Supreme Court held that the door to new technologies was open unless Congress explicitly shuts it; the Canadian Supreme Court held that the door to new technologies is shut unless Parliament explicitly opens it.[14] The implication for our analysis here is that the former position invites new readings of the subject matter categories, whereas the latter position precludes them.

A. Describing Biotechnology Inventions

But the *Harvard College* decision was short-lived, at least in its practical effect; within two years the Canadian Supreme Court negated its previous opinion, albeit without directly overruling it. In the subsequent *Monsanto Canada Inc. v. Schmeiser*[15] opinion, the subject matter restrictions of *Harvard College* were undermined and perhaps entirely undone, even though the particular holding of the opinion was retained. The *Schmeiser* case again concerned the patentability of a living, multicellular organism, this time in the form of a genetically modified plant. The plant in question was a patented recombinant canola plant that was resistant to the herbicide glyphosphate; the herbicide was marketed by

14. Burk, *supra* note 8, at 229.

15. [2004] 1 S.C.R. 902 (Can.).

Monsanto under the brand name "Round-Up," making the plants "Round-Up Ready." Monsanto alleged that a farmer, Percy Schmeiser, had been growing such "Round-Up Ready" plants without a license from Monsanto, in violation of the Monsanto patent.

Schmeiser argued at trial that he was unaware of the presence of the patented plants in his fields, and did not know how they came to be growing there. It is fairly clear that neither the trial court nor subsequent appellate courts believed this assertion, finding contrary evidence that the plants were placed in Schmeiser's fields intentionally. Schmeiser's level of intent or knowledge may have made little difference to the infringement analysis, as intentional or not, making using or selling a patented invention constitutes violation of the patent statute as a matter of strict liability, and requires no scienter. So before the Canadian Supreme Court, Schmeiser argued that Monsanto's patent could not be proper under the subject matter criteria articulated in the *Harvard College* case: the genetically modified plant was clearly a multicellular, "higher" organism with complexity equivalent to a genetically modified mouse. If the Canadian patent statute does not cover living organisms, so that it does not cover transgenic mice, surely it does not include transgenic plants either.

The Court rejected this argument, albeit in a somewhat circuitous fashion. In the interim between *Harvard College* and *Schmeiser*, the Court had undergone a shift in personnel; the *Schmeiser* opinion was decided by a new majority, written by a justice who had dissented in *Harvard College*, over the dissent of the justice who had written the majority opinion in *Harvard College*. The *Schmeiser* majority, while acknowledging that the court had previously held higher organisms to be unpatentable, observed that the claims in the Monsanto patent were directed to a particular recombinant DNA sequence that conferred herbicide resistance. The *Harvard College* opinion had said nothing about the exclusion of DNA molecules from patentable subject matter. Thus the court reasoned that the Monsanto patent was not precluded by the *Harvard College* decision.

By permitting the "Round Up Ready" plant to be patented as a practical matter by means of a claim to a constituent transgenic DNA sequence, the *Schmeiser* opinion avoided expressly overruling *Harvard College* but instead limited it to insignificance. True, the DNA claimed in the patent happened to be present in the plant, but it was not the plant itself that was claimed. This formalism propagates a version of the *Chakrabarty* description exercise above. After *Schmeiser*, clearly the path to patenting a plant, or a mouse, or any other "higher" organism in Canada is to draft the claims in terms of the genetic modification, not in terms of the host embodying the modification. The *Harvard College* prohibition thus becomes no more than a prohibition on terminology.

B. Expression and Function

Crossing now over the Atlantic, the conceptual thread that begins with *Harvard College* takes on yet an additional twist in the more recent opinion of the Court of Justice of the European Union in *Monsanto v. Cefetra*.[16] This case concerned, once again, Monsanto's herbicide resistance "Round-Up Ready" technology. However, this time the genetically modified plants were soy, rather than canola, and were grown in Argentina, a country where Monsanto held no patent on the herbicide-resistant plant. Patents are national in effect, and no patent owner holds patents to an invention in every possible jurisdiction. As the transgenic soy beans at issue in the case had been grown in Argentina, no patent infringement occurred in the use of the plant there.

The beans were then ground into soymeal, which was imported into the European Union through Rotterdam. Monsanto did hold a Dutch "Round Up Ready" patent, but of course it was not the plant that was being imported: it was a processed product derived from the plant.

However, as in the *Schmeiser* case, the Monsanto patent claims were drafted in terms of the recombinant DNA sequence. Monsanto asserted that the DNA sequence was present in the ground soymeal, so that importation of the meal constituted importation of the sequence, violating its patent. In particular, Monsanto argued that the patent was drawn to protect the DNA sequences *as such*, not in meal or in a plant or any other particularized context.

The importers alleged that this claim was improper under the European Biotechnology Directive, an EU requirement to which the patent law of member states—including the Netherlands—must conform. Article 9 of the European Biotechnology Directive provides:

> The protection conferred by a patent on a product containing or consisting of genetic information shall extend to all material, save as provided in Article 5(1), in which the product is incorporated and in which the genetic information is contained and performs its function.[17]

Recital 23 of the directive additionally provides that:

> . . . a mere DNA sequence without indication of a function does not contain any technical information and is therefore not a patentable invention.[18]

---

16. Case C-428/08, Monsanto Tech. LLC v. Cefetra BV, 2010 E.C.R. I-6765.

17. Directive 98/44/EC of the European Parliament and of the Council on the Legal Protection of Biotechnological Inventions, 1998 O.J. (L 213) 13 (EC).

18. *Id.* pmbl. ¶ 23.

Interpreting the Biotechnology Directive, the Court of Justice of the European Union rejected the assertion of the patent against the DNA in the soymeal. The Court took the inverse proposition of Article 9 as limiting biotechnology patents: protection shall *not* extend to material in which the genetic information does *not* perform its function. The DNA sequence, although present in the soymeal, was clearly not performing its function, because the material in which it was situated was dead. Even if the patent were properly read under Dutch law to cover the DNA sequence as found in the meal, this reading would conflict with the language of the Directive.

The opinion distinguishes between DNA as "genetic information" and as constituting "technical information," which is to say, as enabling biological function. It seems clear the language in the Directive serves to deter naked claims to DNA, by linking molecular claims to a biological function.[19] In the United States such claims have been similarly limited, under the utility requirement for patentability, by requiring disclosure of the biological function of the gene to which the patent is directed.[20] Because the rate of discovery for DNA sequences far outpaces their characterization, absent such limitations, claims to isolated DNA molecules could be advanced long before their biological function is known, preempting later beneficial applications of the molecule.

However, the *Cefetra* opinion extends this restriction to create a functional limitation on DNA *product* claims. This not only assumes that the biological function of the process is known and disclosed, it ties DNA product claims to a particular process implementation.[21] By limiting DNA claims to a functional milieu, nucleotides become patentable only as part of a larger biological apparatus. This mirrors the rationale of decisions concerning software in the context of digital processing apparatus,[22] a parallel line of precedent to which we now turn.

II. Subject Matter Exclusions

As we have seen above, nearly everything "under the sun made by man," if properly described, potentially fits into one or another of the American subject

19. Andreas Hübel, *General Issues of Biotech Patents, in* BIOPATENT LAW PATENT STRATEGIES AND PATENT MANAGEMENT 1, 10 (Ulrich Stolts ed., 2012).

20. *In re* Fisher, 421 F.3d 1365 (Fed. Cir. 2005).

21. *See* Sven J.R. Bostyn, *A Decade After the Birth of the Biotech Directive: Was It Worth the Trouble?, in* BIOTECHNOLOGY AND SOFTWARE LAW: A COMPARATIVE REVIEW OF NEW DEVELOPMENTS 221, 231 (Emanuela Arezzo & Gustavo Ghidini eds., 2011).

22. *See* Kambiz Fathi, *Scope of Protection of Gene Patents in Europe*, 48 (Dec. 9, 2011) (Master's Thesis, University of Stockholm), *available at* http://papers.ssrn.com/sol3/papers.cfm?abstract_id=1970209.

matter categories. This realization leaves one to wonder if there are items that, even if described so as to fit one or more of the statutory categories, are *not* "made by man" and so might be excluded from patentable subject matter. In fact patent law, both in Europe and America, deploys a series of doctrines intended to winnow out patent candidates that are not inventions by virtue of existing independent of, or perhaps prior to, human ingenuity. In the United States these doctrines have been primarily addressed in a series of cases dealing with the patentability of computer software.[23] This developmental path may seem odd, given that computer technology seems clearly a product of human engineering, and other technologies such as biotechnology may seem drawn from natural materials. However, as process-based inventions, digital processors implement general principles in a manner that tends to implicate the division between the natural and the artificial.

Additionally, such software decisions are tightly coupled to the biotechnology developments we have already explored. The relevant software cases begin with a trilogy of opinions from the US Supreme Court, beginning in the 1970s. These cases are contemporary with the *Chakrabarty* case, making their way through the US court system at approximately the same time that the *Chakrabarty* case was wending its way to the Supreme Court. *Chakrabarty* followed the first of the major software decisions, *Gottschalk v. Benson*,[24] and was initially remanded back to an intermediate court of appeal for reconsideration in light of the second decision in the software trilogy, *Parker v. Flook*.[25] The *Chakrabarty* opinion was in turn heavily relied upon in the subsequent decision on software in *Diamond v. Diehr*.[26] Notwithstanding the different subject matter, *Chakrabarty* was effectively part of the same contemporaneous discourse as the *Gottschalk, Flook,* and *Diehr* decisions.

The earliest of these Supreme Court decisions, the *Gottschalk* case, involved a method or "algorithm" for converting one type of numerical notation into a different type of numerical notation. The algorithm involved application of a particular mathematical procedure to convert the numerical notation; the court characterized patenting the algorithm as tantamount to patenting a mathematical procedure. As previous commentators have noted, *Gottschalk* lay at the confluence of several concatenated patent doctrines dealing with subject matter: the laws of nature doctrine, the abstract ideas doctrine, the mental steps doctrine, and the printed matter doctrine.[27]

---

23. Dan L. Burk & Mark A. Lemley, *Inherency*, 47 WM. & MARY L. REV. 371, 406 (2005).

24. 409 U.S. 63 (1972).

25. 437 U.S. 584 (1978).

26. 450 U.S. 175 (1981).

27. *See Pamela Samuelson*, Benson *Revisited: The Case Against Patent Protection for Algorithms and Other Computer Program-Related Inventions*, 39 EMORY L.J. 1025 (1992).

Each of these closely related, judicially created doctrines serves to restrict the subject matter categories articulated in the statute.

The first of these, figuring prominently in the opinion, is the prohibition against patenting laws or principles of nature. Among the possible exceptions to the very broad categories encompassed by the language of the American and Canadian patent statutes are items that exist independent of, or perhaps prior to human invention. If the Court in *Chakrabarty* held that anything under the sun made by man falls within patentable subject matter, this implies the inverse that items under the sun *not* made by man do *not* fall within patentable subject matter. Although there is no statutory language to this effect, the Court has at times held that laws and principles of nature, such as the theory of relativity, as well as naturally occurring items such as "a mineral discovered in the earth or a new plant found in the wild"[28] constitute unpatentable products of nature.

But this prohibition rests on the somewhat philosophically dubious proposition that what we label "natural law" originates, not from human invention or ingenuity, but as a feature of the material world, preceding human perception. The *Gottschalk* opinion equates mathematical procedures with laws of nature, extending the prohibition on patenting the latter to a prohibition on the former. One might of course question whether mathematics is in some sense constitutive of natural law, somehow embedded in the fabric of the universe, rather than constituting a human construct that is useful to describe the universe.[29] And regardless of one's view on the natural state of mathematics, one might question separately whether what we call natural law is not a product of human formulation. Nonetheless, US patent law since *Gottschalk* assumes that mathematics, as well as natural law, is somehow embedded in the fabric of the universe, waiting to be discovered, not invented.

Acknowledging the laws of nature issue leads quickly to the second and related doctrine at issue in the case: the prohibition on patenting abstract ideas. Black-letter patent law has long prohibited the patenting of ideas that have not been reduced to some specific instantiation.[30] The relationship to the principles of nature doctrine is almost immediately apparent: all patentable inventions are at some level based upon the operation of natural law or principles. A description of the invention at this abstract level frequently constitutes the idea or concept of the invention. Thus, claims that are not wedded to a fairly specific embodiment may be sufficiently imprecise to encompass the principle of the invention rather than its implementation, and so encompass laws of nature.

---

28. Diamond v. Chakrabarty, 447 U.S. 303, 309 (1980).

29. *See* Phillip J. Davis & Reuben Hersh, The Mathematical Experience 410 (1981).

30. Dan L. Burk, *Patenting Speech*, 79 Texas L. Rev. 99, 137 (2000).

The abstract idea prohibition in turn raises a third problem. Although perhaps not stated as explicitly as the other doctrines, the *Gottschalk* opinion indicated a concern with the mental steps doctrine.[31] Patent law has long resisted claims encompassing steps that could be carried out in the human mind. To some extent, such claims simply raise practical problems of enforcement: if mental activity were covered by exclusive legal rights, it would be difficult to detect, police, assess, or prove infringement when someone is thinking patented thoughts. The very concept of exclusive rights in mental processes raises in turn related concerns regarding encroachment on freedom of speech and thought. Such concerns might not seem implicated by processes carried out in a digital computer. But to the extent that computers were in the 1970s and early 1980s viewed as "electronic brains," some correspondence was seen between operations in carbon memory and operations in silicon memory. And, human cognitive processes might be implicated by broad undifferentiated claims including steps that could be carried out in either silicon or carbon memory, which is to say in a constructed data processor or in the human brain.

Fourth, following from the mental steps doctrine is an inevitable concern regarding the "printed matter doctrine."[32] Patent law has tended to reject as unpatentable any claims where the novelty of the invention lies in the arrangement of symbolic indicia. As a practical matter, this prohibition helps to police the boundary between patent and copyright, preventing patents on novel arrangements of symbols that constitute literary or artistic works. But symbolic indicia are of course largely constitutive of software: human-readable indicia at the level of source code, machine-readable indicia at the level of object code. Software is inscribed in magnetic and optical media, and encoded in electronic circuits. The power of computing technology is that it is quite literally inscribed with writing, so that anything that can be described can be modeled in a data processor.[33] Thus, patent claims involving software or recorded data tend to blur the line between function and expression; claims encompassing encoded functional operations could just as well read on recorded texts or music.[34]

---

31. *See* Samuelson, *supra* note 27, at 1056–57.

32. *See* Burk, *supra* note 30, at 141.

33. Philip E. Agre, *Internet Research: For and Against*, *in* INTERNET RESEARCH ANNUAL: SELECTED PAPERS FROM THE ASSOCIATION OF INTERNET RESEARCHERS CONFERENCES, 2000–2002, at 25, 27 (Mia Consalvo et al. eds., 2004).

34. *See* Burk, *supra* note 30, at 144–45.

A. Tangible Inventions

In a follow-on case, *Parker v. Flook*, the US Supreme Court held that software that was used to calculate the parameters for an industrial process was not patentable; a numerical "alarm limit" calculation, used to monitor the state of an industrial process, was held too abstract to meet the tests for subject matter.[35] But the outcome was decidedly different in the third case the Court took during its twentieth-century run of software decisions, *Diamond v. Diehr*.[36] This opinion, citing repeatedly to *Chakrabarty*, held that an industrial process for curing rubber *was* sufficiently concrete to garner a patent, even though the process involved the use of a chemical equation in a digital processor to perform calculations that determined the timing of the process. The difference between *Diehr* and *Flook* appeared to rest largely on the palpability of the end product of the claimed process. In *Flook* it was a number, but in *Diehr* the output was a concrete, tangible, and useful result—that is, cured rubber.

The approval of software claims grounded in "concrete" and "tangible" embodiments ushered in an era of what Professors Lemley and Cohen have called the "doctrine of the magic words."[37] For the next two decades, the Patent Office and lower courts in the United States struggled with the problem of when a claim constituted a claim to software, and when it constituted a *Diehr*-type claim to software implementation. Patent applicants and their attorneys drafted what were effectively software claims in terms of apparatus, engaging in much the same subject matter exercise illustrated above with regard to the *Chakrabarty* bacterium: How many statutory subject matter categories could claims for a software invention be drafted to fit?

The most obvious characterization of software is of course the statutory category of process, as computer code implements a series of electromechanical transformations with a specified output. However, given that "algorithm" is very nearly synonymous with "process," and *Gottschalk* largely disparaged the patenting of algorithms, the logical category for software became suspect. *Diehr* permitted claims to a type of algorithm, but seemingly only as embodied in mechanism, not pure process claims. But this of course points to a second statutory categorization; as the language of *Diehr* ostensibly valorized software in the context of apparatus claims, many software claims oriented toward characterization as a machine.

35. Parker v. Flook, 437 U.S. 584.

36. Diamond v. Diehr, 450 U.S. 175 (1981).

37. Mark A. Lemley & Julie E. Cohen, *Patent Scope and Innovation in the Software Industry*, 89 Cal. L. Rev. 1, 9 (2001).

Thus a potential alternative was to claim software as a machine. As some commentators struggling with software in the copyright context have noted, software constitutes a machine built of text.[38] But this is of course precisely the characteristic that runs afoul of the printed matter doctrine. Hence, machine-based claims were drafted in terms of the state of the hardware, as configured by software. In a similar vein, yet another alternative was to attempt to describe software as an article of manufacture: for example to claim semiconductor chips inscribed with novel code, or in the now-infamous patent application in *In re Beauregard*, to claim a floppy disc incorporating novel magnetic flux.[39] After some resistance, the Patent Office grudgingly agreed to accept such claims, although it is worth noting that recent decisions have limited the use of these claims by examining the patentability of the underlying encoded process.[40]

The overall consequence of the *Gottschalk* decision was that securing allowance of software claims became fundamentally a claims-drafting problem. Claims were couched in terminology either to obscure the presence of software entirely, or to describe it in a way palatable to the examiner's degree of adherence to the cases interpreting the subject matter provisions. This trend reached its apex with the now-infamous Federal Circuit decision in *State Street Bank*, holding that any invention that produced a "concrete, tangible, useful result" fell within patentable subject matter[41]—but "concrete" and "tangible" appear to have been metaphors, as the Federal Circuit approved as concrete and tangible results such things as numerical output, which was expressly disapproved by the Supreme Court in *Gottschalk* and *Flook*.

And it is critical to note, with regard to composition of software patents, that several of the limiting doctrines identified in *Gottschalk* might themselves be viewed as inchoate products of indefinite claims drafting and inadequate disclosure. For example, one view of the prohibition against claiming laws of nature is that such claims violate the requirements of enablement, commensurability, and written description. The patentee cannot claim multiple instantiations of an invention unless she has enabled and described those instantiations, but claiming abstractions or principles of nature inevitably encompasses embodiments that are not enabled or described. Conversely, claims that are insufficiently precise might encompass the principle of an invention rather than its implementation,

---

38. Pamela Samuelson et al., *A Manifesto Concerning the Legal Protection of Computer Programs*, 94 Colum. L. Rev. 2308, 2320 (1994).

39. *In re* Beauregard, 53 F.3d 1583 (Fed. Cir. 1995).

40. CyberSource Corp. v. Retail Decisions, Inc., 654 F.3d 1366, 1374 (Fed. Cir. 2011); Digital Vending Servs. Int'l LLC v. Univ. of Phoenix, Inc., 672 F.3d 1270 (Fed. Cir. 2012).

41. State Street Bank & Trust Co. v. Signature Fin. Grp., 149 F.3d 1368 (Fed. Cir. 1998).

and so encompass laws of nature. Likewise, mental steps are almost by definition difficult to describe and enable.

This view of descriptive inadequacy certainly was the position taken by the Federal Circuit in *State Street*, dealing with business methods.[42] American patent jurisprudence has displayed a fairly strong historical bias against business method patents, many of which have traditionally involved mental steps or printed matter. Business methods became a recurring issue in software cases; once patent software begins to be patented, it is inevitable that some software will be directed to accounting or investment or inventory or other business methods. But the Federal Circuit opined that there was in fact no prohibition or animus against business methods; rather, that the cases disfavoring business methods had simply invalidated specific patents that failed enablement and claim definiteness—which happened to involve business methods.

B. In Search of the Technical

Such issues were of course not confined to the United States; software patents are sought in other jurisdictions as well. On the other side of the Atlantic, the European patent systems were confronting much the same set of problems.[43] There has not been and is not at the time of this writing a pan-European patent; consequently a significant portion of this jurisprudence evolved in the context of diverse national court decisions.[44] Those from Germany and the United Kingdom are particularly notable in their distinctive approaches to software as patentable subject matter.[45]

Because a complete review of such decisions would exceed the limits imposed by this volume's editors (and perhaps the patience of the audience, as well), I will focus here on decisions rendered by the appellate boards of the European Patent Office. The EPO is charged with examining applications under the provisions of the multilateral European Patent Convention (EPC). The EPC contains explicit treaty prohibitions addressing most of the judicially created doctrines at issue in *Gottschalk*. Article 52(2) of the EPC excludes from patentable subject matter:

(a) discoveries, scientific theories and mathematical methods;
(b) aesthetic creations;

---

42. *Id.*

43. *See* Justine Pila, *Dispute Over the Meaning of the Word "Invention" in Article 52(2) EPC—The Patentability of Computer-Implemented Inventions in Europe*, 36 I.I.C. 173 (2005).

44. *See* Rosa Maria Ballardini, *Software Patents in Europe: The Technical Requirement Dilemma*, 9 J. Int'l Prop. L. & Prac. 563, 567–68 (2008).

45. Convention on the Grant of European Patents, *opened for signature* Oct. 5, 1973, 13 I.L.M. 268 (entered into force Oct. 7 1977).

(c) schemes, rules and methods for performing mental acts, playing games or doing business, and programs for computers;
(d) presentations of information.[46]

On its face the provision might seem to end the question of patents for computer programs, as the treaty is explicit about excluding them from patentable subject matter. However, Article 52(3) adds a troubling caveat: that the enumerated categories are only excluded from patentable subject matter "as such."[47] This leaves the troubling question as to what could be meant by a computer program "as such"—that is, a computer program as a computer program and not as something else.

Answering that question has produced, much as in the United States, two to three decades' worth of uncertainty and turmoil over software patenting. The EPO has focused on whether an invention as claimed has some type of "technical" nature, which the Article 52 categories such as business methods and computer programs "as such" have been said to lack. Various tests have evolved to assay the technical character of suspicious claims. Initially the EPO looked for inventions to make a "technical contribution" to the art, that is, to have some inventive aspect outside the software.[48] Later this was amended to look for inventions having a "technical effect," that is, some concrete output.[49] More recently even this approach was abandoned to allow any claim involving hardware to come within patentable subject matter.[50]

The parallels between these approaches and those in the contemporary *Gottschalk/Flook/Diehr* trilogy are apparent, as are the attendant pitfalls. As the suspect Article 52 categories frequently are implemented in computing apparatus, it becomes extremely difficult to determine when a computer program, business method, or even a game has a technical character. The result has been the EPO's version of the "doctrine of the magic words," by which software patents passed the scrutiny of the examining corps so long as certain suspect terminology was avoided. The attempt to distinguish technical character has produced largely rhetorical stratagems: decades of patent applications concerning software, whose claims were drafted in terms of hardware or apparatus so that the claims would not be drawn to a computer program as a computer program.

---

46. *Id.* art. 52(2).

47. *Id.* art. 52(3).

48. *See* Ballardini, *supra* note 44, at 565.

49. *Id.* at 566.

50. *Id.*

The issue is well bracketed by two decisions of the Technical Board of Appeal of the European Patent Office that concern business methods implemented in computer technologies. The first of these is the *Pension Benefits* decision, which considered an automated system of investment—that is to say, a business method.[51] The claims of the patent were drawn to both the method of investment and an apparatus implementing the method. Reviewing the application, the Board opined that a technical character is inherent in the EPC concept of invention, and that merely reciting computing means within a method claim does not confer a technical character on a claimed process invention. At the same time, an apparatus claim to a suitably programmed data processing device would constitute patentable subject matter, rendering the claimed invention patentable subject matter. However the method claims in the patent were rejected as lacking a technical nature.

A later decision, *Hitachi*,[52] in some respects supports and in others contradicts the *Pension Benefit* outcome. The patent application in *Hitachi* claimed an automated system of auctioning; again, from a categorical perspective, the invention constituted both a computer program and a business method, and was described in an application encompassing both apparatus and process claims. Following the *Pension Benefits* reasoning, the Board's decision again held that the implementation of the invention in hardware, by virtue of incorporating the apparatus, clearly presented a claim with a technical nature. However, unlike the *Pension Benefits* decision, the Board held the *method* claims as executed by means of a physical apparatus also entail a technical nature.

Each of these decisions takes the position that by reciting an apparatus claim implementing the method, the method characterized as hardware comes within patentable subject matter, and no longer constitutes a program "as such." This is essentially a version of the drafting techniques already discussed for the United States following *Diamond v. Diehr*. The opinions part ways with regard to the permissibility of method claims as method claims. But both cases then deny the applicant a patent for failure of an inventive step, holding that given the invention, implementation in a general purpose data processor is obvious.

This approach is logically somewhat suspect due to what American courts would term "hindsight reconstruction"; that is, taking the invention as given when assessing its own inventiveness. Such an approach comes very close to the discredited analysis of *Parker v. Flook*, which considered an abstract method as its

---

51. Case T 0931/195, *Controlling Pension Benefit System PBS Partnership*, [2001] OJEPO 441.

52. Case T 0258/03, *Auction Method/Hitachi*, [2004] OJEPO 575.

implementation's own prior art.[53] But notwithstanding the questionable logic of the inventive step analysis, the overall outcome of the cases may chart the most pragmatic and productive route for questions involving problematic technologies. Rather than struggle with intractable subject matter questions, the decisions concede that interminable, unwinnable fight; grant the method patentable subject matter status; and instead hold firm on the patentability questions of utility, novelty, and obviousness. This approach was unfortunately rejected by the US Supreme Court when recently urged by the Obama administration as a sensible resolution to subject matter confusion.[54] But the EPO decisions may chart a path to a more sensible subject matter jurisprudence.

Or perhaps they will not. It is worth noting that the president of the EPO, perhaps in response to the invitation of a British court attempting to fathom the EPO case decisions,[55] convened an enlarged Board to reconcile the disparity between the approach typified by *Hitachi* and that typified by *Pension Benefits*.[56] After spending considerable time with the problem, the Board declined to do so, issuing an opinion that some variation was inevitable in developing technologies. Given that the technology, and the problems associated with claiming, have been developing for a good thirty years, it is difficult to understand how much more development is needed before a coherent approach could be articulated.

III. Process and Apparatus

The subject matter themes found in biotechnology and software cases converge in a series of recent disputes over patents for diagnostic tests. These cases combine the recurring questions we have seen regarding principles of nature, mental steps, abstract ideas, and subject matter categories. Although these process inventions frequently involve some biological molecule, sometimes including DNA or another nucleic acid, and may be automated, employing some type of digital processor, the doctrinal convergence is due to more than a technological convergence. The method claims at issue in such cases involve biological rather than electronic processes, but implicate the same set of doctrines and policies.

---

53. *See* Diamond v. Diehr, 450 U.S. 175, 189 n.12 (rejecting the prior art approach of Parker v. Flook, 437 U.S. 584 (1978)).

54. Mayo Collaborative Servs. v. Prometheus Labs., Inc., 132 S. Ct. 1289, 1304 (2012).

55. Aerotel Ltd. v. Telco Holdings Ltd. [2006] EWCA (Civ) 1371 (Eng.).

56. *See* Justine Pila, *Software Patents, Separation of Powers, and Failed Syllogisms: A Cornucopia from the Enlarged Board of Appeal of the European Patent Office*, 70 CAMBRIDGE L.J. 203 (2011).

The commonality is largely a product of the form of the claims at issue more than of the particular technology involved. The most troublesome diagnostic claims have been directed to minimalist versions of test procedures, seemingly relatively low-tech processes centered on drawing a biological sample and performing simple chemical analysis, possibly by hand. Rather than reciting a specific apparatus they claim the steps of testing and diagnosis simpliciter. Consequently, as in the software controversy over mental steps, these claims do not distinguish whether the procedure is to be performed in silicon or carbon data processors, and they are concomitantly broad enough to encompass the idea or principle on which the test is based.

Claims of this type were at issue in *Labcorp v. Metabolite,*[57] where the diagnostic process claims at issue involved a "correlation" step—that is, testing for the presence of a certain substance in the blood of a patient, and deciding on the basis of the concentration of that substance whether the patient was ill. This case was originally accepted for review by the US Supreme Court, then dismissed as having been improvidently granted review.[58] Justice Breyer, dissenting from the dismissal order, indicated in a lengthy opinion his discomfort with such patents as possibly encompassing laws of nature—that is, the relationship between a certain metabolite and disease—and possibly encompassing a "mental step" in the doctor's recognition of disease from a test result.[59] Breyer urged review of the case or a similar case in order to reign in subject matter overreaching. Breyer's commentary signaled to lower courts the likelihood of Supreme Court action on the topic, and is widely credited with moving the inferior Court of Appeals for the Federal Circuit away from its permissive "useful concrete tangible result" test for patentability.

Nonetheless, similar patents continued to issue, and the Supreme Court took up this series of questions in the *Mayo v. Prometheus*[60] decision, another diagnostic method case. The method claims in the patent held by Prometheus were drawn to the process of measuring the levels of certain metabolites in a patient's blood after administering thiopurine drugs, and then adjusting the dosage to avoid either an ineffectively low or harmfully high dosage. As in *Metabolite*, the patent was challenged on the ground that its claims were essentially drawn to laws of nature, that is, to the relationship between the level of metabolites in the patient's blood and the efficacy of the drug dosage. Additionally, also as in

---

57. Lab. Corp. of Am. Holdings (LabCorp) v. Metabolite Labs., Inc., 548 U.S. 124 (2006) (per curiam).

58. *Id.*

59. *Id.* at 126–28 (Breyer, J., dissenting).

60. 132 S. Ct. 1289 (2012).

Metabolite, the claims included a "determining" step that could have been characterized as an impermissible mental step.

The court focused on the "principle of nature" argument, relying on the reasoning from the *Diehr* and *Flook* software decisions to emphasize that processes consisting largely or entirely of physical correlations, without some significant physical apparatus or application, lie outside patentable subject matter. The Federal Circuit had opined that the claims met this standard because they involved a "transformation" of matter in vivo, which is to say that the drug administered to the patient metabolized in the body to the substances measured. But the Supreme Court viewed metabolic transformation itself as merely the product of natural biological processes, and inconsequential in limiting claims to the process of gathering data from which physicians could draw an inference about the patient's treatment. The Court particularly emphasized that broad method claims of this type could preclude use of the underlying principle, potentially inhibiting future research or advancement based on the principle.

The commonalities between the software cases and the diagnostic cases are not limited to the appearance of the laws of nature or mental steps doctrines. The printed matter doctrine has reared its head in the diagnostic cases as well, as for example in the *King Pharmaceuticals v. Eon Labs*[61] decision from the Federal Circuit. The claims in the contested *King* patent concerned a method of administering a pharmaceutical orally with food, which was shown to increase the efficacy of the drug. There was nothing new about the drug or its use in treatment, or oral administration. For that matter, people had long taken the drug with food, but had done so to avoid an upset stomach, rather than to increase the efficacy of the drug. The only new aspect of the method was informing or encouraging the patient to take the drug with food for enhanced efficacy. Thus, the claims of the method of treatment patent at issue were drawn to either verbally informing the patient of the positive effects of administering the drug with food, or communicating this information to the patient by means of a printed label on the drug container.

The most recent chapter in this ongoing story is found in the American court decisions involving the so-called Myriad gene patents, *Ass'n for Molecular Pathology v. United States Patent and Trademark Office*,[62] a case that was recently decided by the US Supreme Court.[63] The lawsuit is contemporary with the *Bilski*, *Prometheus*, and *King* lawsuits, moving through the court system in

61. 616 F.3d 1267 (Fed. Cir. 2010).

62. Ass'n for Molecular Pathology v. U.S. Patent & Trademark Office, 689 F.3d 1303 (Fed. Cir. 2012), *aff'd in part, rev'd in part*, 133 S. Ct. 2107 (2013).

63. Ass'n for Molecular Pathology v. Myriad Genetics, Inc., 133 S. Ct. 2107 (2013).

parallel with those decisions. The *AMP* lawsuit concerned claims to genetic sequences and to diagnostic procedures employing the gene sequences, for BRCA-1 and BRCA-2, which confer on carriers of the genes a heightened risk of breast cancer.[64] Both the product patent claims directed to the nucleotides and the process patent claims directed to using the sequences were challenged as encompassing unpatentable laws or products of nature.[65]

The Myriad product claims included different types of molecules with different relationships to naturally occurring DNA. The patent claimed both "isolated and purified" genomic (gDNA) sequences, as well as complementary (cDNA) sequences, that are reverse transcribed from messenger RNA (mRNA) transcripts drawn from human cells. The genomic sequences were a particular target of the "products of nature" argument, as their nucleotide sequences were drawn from those found in the human body. However, relying on several decades of precedential practice from the USPTO, Myriad argued that the limitation in the claim to "isolated and purified" sequences set them apart from naturally occurring DNA. The DNA sequences as found in human somatic cells are surrounded by and bound to an array of other macromolecules, making the isolated sequence a product of human intervention that is not naturally occurring.

Myriad further argued that, even if the claims to isolated genomic DNA read on products of nature, the patent claims to cDNA did not. cDNA does not normally exist in human cells, being the product of a laboratory reverse transcription procedure using mRNA transcripts as a template, read by a special type of viral enzyme that will assemble a DNA strand complementary to the RNA.[66] As a result, cDNAs typically have structural and coding differences from genomic human DNA; in particular, they are missing long "intervening sequences" or "introns" that are not transcribed into mRNA, and so are not present in the reverse transcribed cDNA.[67] Transcription from RNA to DNA does not normally occur in human cells, so that the claimed cDNAs were fairly clearly the product of human intervention.

However, the trial court held both types of DNA molecules, cDNA and gDNA, unpatentable, reasoning that they contained the same "information" as a native DNA strand, which is to say that they contained essentially the same coding sequence.[68] As the "physical embodiment of genetic information," the

64. *Ass'n for Molecular Pathology*, 689 F.3d at 1309.

65. *Id.* at 1317.

66. James Watson et al., Molecular Biology of the Gene 745 (6th ed. 2008).

67. *Id.* at 415.

68. Ass'n for Molecular Pathology v. U.S. Patent & Trademark Office, 702 F. Supp. 2d 181, 230 (S.D.N.Y. 2010).

trial judge found them to constitute products of nature rather than products of human invention. The court additionally held Myriad's process claims, drawn to the use of the genes in diagnosis, to be unpatentable on much the same grounds as the Supreme Court later articulated for invalidating the *Prometheus* diagnostic claims. The process claims recited a method for assessing the relationship of particular nucleotide sequences to the risk of breast cancer, without specifying any particular apparatus or embodiment. This, the court reasoned was tantamount to patenting a natural principle: the correlation between the presence of certain DNA variants in the body, and the risk of developing breast cancer.[69]

On appeal to the Federal Circuit, and again subsequently on remand to the Federal Circuit from a brief sojourn at the Supreme Court, a three-judge panel unanimously upheld the trial judge's holdings with regard to the Myriad diagnostic claims, unanimously reversed the trial judge's holding regarding cDNA, and reversed by a majority the holding regarding gDNA.[70] The panel all agreed that the process claims encompassed abstract laws of nature, and all agreed that cDNAs are the patentable product of human intervention, but split regarding the "product of nature" claims for gDNA.

In fact, even the majority of the panel was in many senses a plurality, reflecting a three-way division of opinion as to how to think about the genomic DNA claims. Judge Bryson, in dissent, fully embraced the trial court's characterization of the genomic sequence DNA as an unpatentable natural product.[71] Judge Lourie, writing for the majority, reached his conclusions by focusing primarily on the structure of the gDNA molecules, noting that the nucleotide chains, having been separated from their native environment of the chromosome, have a different chemical structure than the same sequences as found in human cells.[72] Judge Moore, in concurrence, reached the same conclusions, but did so by focusing primarily on the function of the gDNA, noting that the DNAs claimed in the patent could be used for a variety of purposes that a native molecule could not.[73]

The interplay between the district and appellate courts in *AMP* echoes and reprises certain themes recognizable from the computer software cases. Aside from the familiar "principle of nature" treatment of the diagnostic method claims, much of the discussion hinges on the proper categorical treatment of information-based inventions. Indeed, the discourse among the appellate judges,

69. *Id.* at 233–34.

70. *Ass'n for Molecular Pathology*, 689 F.3d at 1303.

71. *Id.* at 1327–28.

72. *Id.* at 1350.

73. *Id.* at 1341–43.

and between the appellate and trial judges, can be seen as a disagreement over "technical character" in a different context. At the Supreme Court, a unanimous court held gDNA to be unpatentable, but preserved the patentability of cDNA. The Supreme Court majority opinion appeared to adopt the informational view of gDNA, rather than focusing on the structure and function of the gDNA molecule. The Court also explicitly declined to pass judgment on the application of gene sequences and on other manipulations of the genetic code.

Conclusion

It is impossible to trace the trajectory of these cases without being struck by the multiple intricate cross-connections. The issues raised by biotechnology and software patenting cross both technological and jurisdictional boundaries. It may come as something of a surprise that *Gottschalk* and *Diehr* can only be properly read in the context of *Chakrabarty*, or that *Hitachi* and *Pension Benefits* are best understood in tandem with *Monsanto v. Cefetra*. But what may be more surprising is that full comprehension of the ramifications of the *Cefetra* decision requires the context provided by *AMP* and *Schmeiser*, or that the reasoning of *Hitachi* and *Pension Benefits* tracks both *Gottschalk* and *AMP*.

Reading such cases together demonstrates the recurrent doctrinal pattern in both biotechnology and software subject matter decisions. First, in comparing the cases reviewed here, it should be apparent that the doctrines invoked in such cases—laws of nature, abstract ideas, mental steps, printed matter—not only substantially overlap with one another but are frequently indistinguishable from one another, blurring across a doctrinal gradient: patent claims that are too abstract run the risk of capturing within their ambit a law of nature; claims that encompass mental steps are too abstract; claims that encompass symbolic indicia may read on mental steps. Hence these doctrines travel in a pack, appearing en masse in the patent decisions addressing biotechnology and software.

The effect of these doctrines, as shown above, is to transform the subject matter problem in biotechnology and software patents into a drafting problem. Language that seems sufficiently concrete, avoiding abstractions or conceptual claiming, will tend to skirt the around the limiting doctrines. Conversely, the prohibitions against software or biotechnology in certain forms shunt claims drafting toward descriptions that encompass familiar and permissible statutory categories (machines, articles of manufacture, compositions of matter), and away from process-based claims.

These acceptable categories are all product categories, and so necessarily tied to a particular physical substrate. And of course process claims, being largely conceptual, are most acceptable when described in terms of some physical apparatus.

The appearance of these doctrinal patterns in contexts that cross technological and jurisdictional boundaries allows us to observe not merely doctrinal similarities, but to excavate the underpinnings of such similarities. A key insight is the presence of informational claims, as acknowledged in cases such as *AMP* and *Cefetra*.[74] At some level all technology involves the embodiment of information; to a greater or lesser extent all artifacts carry information encoded in technical structures.[75] This is sometimes easily apparent, as in the case of a mechanical lock and key, and sometimes less apparent. Even artifacts whose informational encoding is not immediately recognizable entail informational content, as thermodynamically defined.[76]

But where classic mechanical, electrical, or chemical inventions were concerned, this aspect of their character could largely be ignored or overlooked. Indeed, because description and specification of informational content is relatively recent and often imprecise, patent law proceeding out of the nineteenth century developed an antipathy toward informational characterizations, leaning instead toward definition of inventive artifacts in terms of more readily observable physical attributes.[77] We have seen in the reviewed cases above this tropism, across subject matter categories, toward definition by means of physical characteristics.

However, the informational character of process-based inventions is less amenable to being ignored than that of product inventions. Processes, whether chemical, mechanical, or electrical, necessarily implicate the transfer of information, as entropic changes between physical vectors.[78] Little wonder, then, that processes become suspect subject matter for patents unless limited to the objects or subjects of such transfer. And this antipathy extends to information-based technologies generally. As I have pointed out in previous work, informational encoding is eminent in both macromolecules and computer devices, where informational content is only lightly embodied, and the descriptive line between product and process is blurred.[79]

---

74. *Cf.* Justine Pila, *Patents for Genes and Methods of Analysis and Comparison*, 126 L.Q. Rev. 534, 538 (2010) (comparing the approaches to definition of invention in the *AMP* and *Ceftera* cases).

75. Burk, *supra* note 4, at 588–90.

76. *Id.*

77. *See* Dan L. Burk, *Method and Madness in Copyright Law*, 2007 Utah L. Rev. 587, 609–10.

78. *Cf.* John Collier, *Causation Is the Transfer of Information*, *in* Causation and Laws of Nature 215, 221–23 (Howard Sankey ed., 1999).

79. *See* Burk, *supra* note 4, at 587.

In sum, all the doctrines common to trans-Atlantic software and biotechnology cases concern the disposition of information within patentable subject matter. Broad abstract claims constitute claims to information or information transfer; mental steps implicate the organic recording of information, just as printed matter necessarily implicates the codification of information. Patent doctrine attempts to confine software claims to a particular apparatus for data processing, or limit a DNA claim to the context of a particular function, in an effort to avoid the hard question of how to situate these information technologies within a nineteenth-century legal framework. The cases deploying these doctrines, in any jurisdiction, are difficult to parse, but the deficiency lies not in the "problem" technologies they address. Rather the problem is in the capacity of patent law to function in the information age. Until patent law develops some way to accommodate information-based technologies, courts on both sides of the Atlantic will continue to struggle to apply subject matter limitations.

8 PATENTING PLANTS: A COMPARATIVE SYNTHESIS

Mark D. Janis*

Introduction

Both the TRIPS Agreement[1] and the 1991 Act of the UPOV Convention[2] leave it to their member states to decide whether to permit utility patent protection for plants. TRIPS Article 27(3)(b) permits members to exclude "plants... and essentially biological processes for the production of plants" from utility patent protection, on the condition that they "provide for the protection of plant varieties either by patents or by an effective *sui generis* system or by any combination thereof...."[3] UPOV(1991) omits a provision that had banned UPOV member states from providing both plant variety protection and patent protection for the same plants.[4]

---

* Robert A. Lucas Chair in Law and Director, Center for Intellectual Property Research, Indiana University Maurer School of Law, Bloomington, IN. Leslie Prill provided excellent research assistance.

1. Agreement on Trade-Related Aspects of Intellectual Property Rights, Apr. 15, 1994, Marrakesh Agreement Establishing the World Trade Organization, Annex 1C, 1869 U.N.T.S. 299, 33 I.L.M. 1197 (1994), *available at* http://www.wto.org/english/docs_e/legal_e/27-trips.pdf [hereinafter TRIPS Agreement].

2. International Convention for the Protection of New Varieties of Plants, Dec. 2, 1961, amended Mar. 19, 1991 [hereinafter UPOV(1991)], *available at* http://www.upov.int/en/publications/conventions/1991/pdf/act1991.pdf (English text of the 1991 Convention).

3. The TRIPS Agreement also permits members to exclude inventions from patentability as necessary to protect the *ordre public*, which includes the protection of "plant life" and the avoidance of "serious prejudice to the environment." TRIPS Agreement, *supra* note 1, art. 27(2).

4. Barry Greengrass, *The 1991 Act of the UPOV Convention*, 12 EUR. INTELL. PROP. REV. 466, 467 (1991). For the text of the dual protection ban, see, e.g., International Convention for the Protection of New Varieties of Plants, art. 2(1), Dec. 2, 1961, as amended Nov. 10, 1972 and Oct. 23, 1978 [hereinafter UPOV

No other language in TRIPS speaks specifically to plants. Member states that choose to offer utility patent protection for plants merely must adhere to TRIPS minimum standards applicable to any utility patent. In regards to patentability, this requires that inventions be "new, involve an inventive step and [be] capable of industrial application,"[5] and that they be disclosed "in a manner sufficiently clear and complete for the invention to be carried out by a person skilled in the art."[6] As to scope, TRIPS delineates the package of exclusive rights that patents must confer,[7] and provides blanket permission for members to provide "limited exceptions" to those exclusive rights, on the conditions that the exceptions "do not unreasonably conflict with a normal exploitation of the patent and do not unreasonably prejudice the legitimate interests of the patent owner, taking account of the legitimate interests of third parties."[8] The absence of any mandated exceptions from patent scope for plants in TRIPS is especially significant because the UPOV agreement requires its member states to recognize a number of exceptions from the scope of plant breeders' rights.[9]

The TRIPS Agreement's grant of such wide discretion on plant patenting practices presents some interesting questions about the merits of harmonization in patent law. Some scholars have suggested that in an un-harmonized legal environment, legal innovation should flourish.[10] Is there evidence that innovation is occurring in national legal rules governing plant patenting? To the extent that it is, which institutions are primarily responsible for it?[11] With these questions in

---

(1978)], *available at* http://www.upov.int/en/publications/conventions/1978/pdf/act1978.pdf (English text of the 1978 Convention).

5. TRIPS Agreement, *supra* note 1, art. 27(1) (specifying that "inventive step" and "capable of industrial application" may be understood as synonymous with "non-obvious" and "useful," respectively).

6. *Id.* art. 29(1) (also permitting members states to impose a best mode requirement).

7. *Id.* art. 28(1)(a) (applicable where the subject matter of the patent is "a product"); *id.* art. 28(1)(b) (applicable where the subject matter is a process). Both provisions are potentially relevant for plant patenting because patents on plant innovation might be directed to plants themselves (among other products) or to breeding processes (among other processes).

8. *Id.* art. 30. A separate article sets out the conditions under which a member state can permit "other uses" without the patent holder's authorization, such as by compulsory licensing. *Id.* art. 31.

9. UPOV(1991), *supra* note 2, art. 15(1) (compulsory exceptions for breeding, experimentation, and noncommercial activities); *id.* art. 15(2) (optional exception for seed saving).

10. *See, e.g.*, John Duffy, *Harmony and Diversity in Global Patent Law*, 17 Berkeley Tech. L.J. 685 (2002).

11. In this study, I am especially interested in evidence of judicial innovation, which has also been an area of primary focus for other patent scholars. *See, e.g.*, Dan L. Burk & Mark A. Lemley, The Patent Crisis and How the Courts Can Solve It (2009).

mind, in this chapter I survey the plant patenting practices of the major patenting jurisdictions (the United States, Europe, China, Japan, and South Korea), along with others where agriculture is important to the domestic economy (Canada, Australia, and India, for example). I begin by examining approaches to the threshold question of eligibility for plants and plant-related inventions.[12] I then consider other substantive patent law issues that are emerging, or are likely to emerge, in jurisdictions that permit some level of plant patenting.[13] I find abundant evidence of diversity in approaches to the question of patent protection for plants, but I conclude that it is doubtful—or at least difficult to tell—whether this diversity exemplifies robust and progressive innovation in legal thinking.

I. Patent Eligibility of Plants: Three Paradigms

Many of the debates about patent protection for plants have centered on the threshold issue of whether plants constitute patent-eligible subject matter. Reflecting the lack of consensus internationally over this issue, and the lack of any mandate in TRIPS, major patenting jurisdictions have developed over the past several years a diverse array of approaches to patent eligibility for plants. Three approaches have emerged, with some variations within each grouping: a permissive approach; a restrictive approach; and an approach or set of approaches that resist easy characterization, but may be referred to loosely as intermediate approaches.

A. Permissive Approaches

1. United States: Plants as Eligible Compositions of Matter

Notwithstanding the upheaval in the law of patent eligibility as applied to isolated gene claims[14] and claims to certain types of methods,[15] it remains well-established that claims to plants define patent-eligible subject matter under US law. The Supreme Court confirmed in *Diamond v. Chakrabarty* that biological subject matter could qualify as a patent-eligible "composition of matter" under 35 U.S.C. § 101, although it also reaffirmed that "laws of nature" and "physical phenomena" (along with abstract ideas) are excluded from § 101.[16] In *Ex parte*

---

12. *See infra* Section I.

13. *See infra* Section II.

14. Ass'n for Molecular Pathology v. Myriad Genetics, Inc., 133 S. Ct. 2107 (2013).

15. Bilski v. Kappos, 130 S. Ct. 3218, 3227–28 (2010); Mayo Collaborative Servs. v. Prometheus Labs, Inc., 132 S. Ct. 1289 (2012).

16. Diamond v. Chakrabarty, 447 U.S. 303 (1980).

Hibberd,[17] the Board of Patent Appeals and Interferences of the USPTO applied these standards to various claims to conventionally bred maize seed having specified characteristics. The Board held that the claims defined patent-eligible compositions of matter.[18] Thereafter, the USPTO issued numerous utility patents on plants based on *Hibberd*. Some years later, in 2001, the US Supreme Court ratified the USPTO's practice in *J.E.M. Ag Supply v. Pioneer Hi-Bred*.[19] United States' patents directed to plant innovation frequently include claims directed to the plant itself, the seed of the plant, and various plant parts.[20] All of these constitute patent-eligible compositions of matter under existing law; there is no serious challenge to eligibility.

A number of US patents also include claims to breeding processes. For example, a utility patent issued in 2011 to Monsanto includes the following claim:

> 23. A method of producing an inbred soybean plant derived from soybean variety A1016494, the method comprising the steps of—
> (a) preparing a progeny plant derived from soybean variety A1016494 by crossing a plant of the soybean variety A1016494 with a soybean plant of a second variety, representative seed of said soybean variety having been deposited under ATTC Accession No. PTA-11456;
> (b) crossing the progeny plant with itself or a second plant to produce a seed of a progeny plant of a subsequent generation;
> (c) growing a progeny plant of a subsequent generation from said seed and crossing the progeny plant of a subsequent generation with itself or a second plant; and
> (d) repeating steps (b) and (c) until an inbred soybean plant derived from the soybean variety A1016494 is produced.[21]

---

17. 227 U.S.P.Q. (BNA) 443 (BNA) (B.P.A.I 1985).

18. *Id.* 444.

19. J.E.M. Ag Supply, Inc. v. Pioneer Hi-Bred Int'l, Inc., 534 U.S. 124 (2001). For a more detailed nalysis, see Mark D. Janis, Herbert H. Jervis & Richard C. Peet, Intellectual Property Law of Plants ch. 7 (forthcoming 2014).

20. For example, US Patent No. 7,880,066 (issued Feb. 2011 and assigned to Monsanto) includes a claim to "A plant of soybean variety A1016494, representative seed of said soybean variety having been deposited under ATCC Accession No. PTA-11456," *id.* col. 25, claim 2, and corresponding independent claims to a "seed" of the same soybean variety and "a tissue culture of regenerable cells" of the soybean variety. *See id.* cols. 25–26, claims 1, 5. The patent also includes dependent claims to other specified plant parts. *See id.* col. 25, claim 4.

21. *Id.* col. 28.

Claims in this form fall squarely within the realm of Section 101–eligible "processes." The judicially developed exclusion for abstractness that has been invoked against some classes of process claims is not implicated. Traditionally, process claims that entail a transformation of matter will be deemed to avoid abstractness problems,[22] and claims to breeding processes invariably do entail a transformation of matter. In this regard, US practice differs sharply with that in Europe.[23]

The permissive approach of US utility patent law toward plants is especially significant because US law also offers additional protection options for plant innovation. In addition to offering plant patent protection for asexually reproducing plants,[24] US law offers plant variety protection for sexually reproducing varieties, and permits breeders to hold utility patent protection and plant variety protection concurrently on the same variety.[25] This suggests that in the United States, the key future issues on patenting plants will involve questions about regulating the interface between the utility patent regime and other regimes, especially the plant variety protection regimes.[26] Indeed, these issues may be more significant than questions internal to utility patent law about applying particular utility patent doctrines to plants.

2. Plants as Eligible "Inventions"

The United States is by no means alone in adopting a permissive approach to the grant of utility patents for plants. Although definitive judicial rulings on the matter are scarce, the practice of issuing utility patents on plants now appears to be well-established in Japan, South Korea, and Australia.

Japanese law extends patent-eligible subject matter to "inventions," and defines inventions as the "highly advanced creation of technical ideas utilizing the laws of nature."[27] Although the reliance on "invention" as the rubric for eligibility distinguishes Japanese law from that of the United States, the Japanese and US approaches to eligibility for plants are ultimately quite similar. In Japan,

---

22. Bilski v. Kappos, 130 S.Ct. 3218, 3226–27 (2010) (discussing the transformation of matter test while declining to adopt it as the exclusive measure of eligibility for process claims).

23. *See infra* Section I.B.1.

24. Plant patents have traditionally been used for nursery stock, for which asexual propagation has been the most convenient commercial option. *See generally* JANIS ET AL., *supra* note 19, ch. 6 (discussing the requirements of the plant patent regime).

25. *See id.* ch. 4 (discussing the US Plant Variety Protection Act, codified at 7 U.S.C. § 2321).

26. For one perspective on these interface issues, see Mark D. Janis, *Interfaces in Plant Intellectual Property*, *in* OVERLAPPING IP RIGHTS (Neil Wilkof & Shamnad Basheer eds., 2012).

27. Tokkyohō [Japan Patents Act], art. 2(1), Law No. 121 of 1959 [hereinafter Japan Patents Act].

as in the United States, the patent statute contains no express exclusionary provision directed toward plants or plant varieties. Accordingly, it is left to Japanese judicial and administrative bodies to determine whether the statutory concept of "invention" should encompass plants and plant innovation. No definitive judicial decision has been rendered to date, but the Examination Guidelines of the Japan Patent Office (JPO)[28] assume that plants can constitute patent-eligible subject matter.[29] A claim to a plant specifying the plant's allegedly novel and inventive characteristics defines eligible subject matter under the Guidelines.[30] Likewise, a claim to a plant in product-by-process form is acceptable, assuming that the claim recites the process steps and relevant conditions (such as environmental conditions or plant characteristics that influence the selection steps).[31] The Guidelines do specify that a claim to "[a] newly discovered plant per se" is regarded as failing to comply with the requirement for an industrially applicable invention.[32] But, apart from this caveat, the Guidelines establish a permissive approach to eligibility.

Likewise, in Australia, patent eligibility for plants has developed via judicial and administrative authority rather than explicit legislative directive. Australian patent law requires that subject matter be a "manner of manufacture within the meaning of section 6 of the Statute of Monopolies" in order to be patent-eligible,[33]

---

28. *Examination Guidelines for Patent and Utility Model in Japan* (June 2010), *available at* http://www.jpo.go.jp/tetuzuki_e/t_tokkyo_e/1312-002_e.htm [hereinafter *2010 Examination Guidelines*].

29. This is reminiscent of US law, where the practice of granting utility patents for plants emerged as a USPTO administrative initiative that courts eventually ratified. *See supra* notes 17–19 and accompanying text.

30. *2010 Examination Guidelines*, *supra* note 28, pt. VII, ch. 2, § 3.1.2 (providing the following example: "A plant belonging to *Castanea crenata* (Japanese chestnut) having the ATCC Accession No. xxxx whose bark contains catechol tannin and pyrogallol tannin in the ratio of X1-X2: Y1-Y2 and has a catechol tannin content of z1-z2 ppm (weight ratio), or its mutant having the said characteristics.").

31. The Guidelines include the following examples:

> A watermelon obtained by crossing a diploid watermelon with a tetraploid watermelon obtained by polyploidizing a diploid watermelon, whose somatic cell has 33 chromosomes.
>
> A process for creating a cabbage characterized by crossing a cabbage strain having the ATCC Accession No. xxxx as a seed parent with another cabbage as a pollen parent by having resistance for the herbicide X.

Id.

32. *Id.* § 3.2.1 (example 1); *see also* Japan Patents Act, *supra* note 27, art. 29(1) (requirement for industrially applicable inventions).

33. Patents Act 1990 (Cth), § 18(1)(a).

where "manner of manufacture" is not otherwise defined, but has been said to call for "a product, process, or method that arises through some form of 'human intervention with nature to bring about some physical change.'"[34] The leading Australian case interpreting this language, *National Research Development Corp. v. Commissioner of Patents*, advocated a flexible and generous approach to the language, although it recognized that mere scientific theories and laws of nature should fall outside the reach of the language.[35] This is comparable to the approach of US patent eligibility law as expressed in *Chakrabarty* and more recently in *Bilski v. Kappos*.[36] *National Research Development Corp.* opened the door for the patenting of plant varieties in Australia. In *Grain Pool v. Commonwealth*, a case on the constitutionality of the plant variety protection legislation, the Australian High Court acknowledged that "the effect of the decision in *National Research Development Corporation* is to confirm that there is no intrinsic impediment to the patentability of plant varieties."[37] The Australian patent statute lends further confirmation, specifying that plants are excluded from eligibility for innovation patents,[38] thus arguably confirming by negative implication that plants are within the range of eligible subject matter for standard patents.[39] Similarly, the Australian Manual of Patent Examination and Procedure reiterates the rule that a plant can be regarded as a manner of manufacture if it satisfies the requirements of *National Research Development Corp.* and progeny.[40]

The evolution of South Korean law on the patent eligibility of plants is more convoluted, but it likewise reflects a permissive attitude toward the patenting of plants. In the aftermath of World War II, South Korea incorporated many American patent law concepts into its 1946 Act, including four provisions

---

34. *See* Ctr. for Int'l Econ., Economic Analysis of the Impact of Isolated Human Gene Patents 22 (2013) (quoting Mark J. Davison, Anne L. Monotti & Leanne Wiseman, Australian Intellectual Property Law 377 (2008)), *available at* http://www.ipaustralia.gov.au/pdfs/IPA_Final_Report__Human_Gene_Patents_2013.pdf.

35. Nat'l Research Dev. Corp. v. Comm'r of Patents, (1959) 102 CLR 252.

36. *See supra* note 15.

37. Grain Pool of W. Austl. v. Commonwealth, [2000] HCA 14.

38. Patents Act 1990 (Cth), § 18(3).

39. On the differences between the standard and innovation patents in Australia, see Mark D. Janis, *Second Tier Patent Protection*, 40 Harv. J. Int'l L. 151 (1999).

40. Australian Patent Office Manual of Patent Practice and Procedure, ¶ 2.9.2.15 (2011), *available at* http://www.ipaustralia.gov.au/pdfs/patentsmanual/WebHelp/Patent_Examiners_Manual.htm (specifying that a new plant can be regarded as a manner of manufacture if it is not naturally occurring and if it "involves the technical intervention of man" and "is useful in economic affairs"); *see also* Cancer Voices Australia et al. v. Myriad Genetics Inc., [2013] FCA 65 (interpreting manner of manufacture requirement in context of isolated DNA).

corresponding loosely to 35 U.S.C. §§ 161–64, which implement the US plant patent regime directed at patent protection for asexually reproduced plants.[41] By 1961, in the course of revising its patent law to conform it more closely to that of Japan, Korea deleted three of the plant patent provisions, but retained the provision (Article 31) that established that asexually reproduced plants were eligible subject matter.[42] This reportedly gave rise to the argument that by providing for eligibility for asexually reproducing plants, the Act was implicitly rejecting eligibility for sexually reproducing plants.[43] Such an argument is reminiscent of the failed argument offered in the United States in *J.E.M.* that the existence of the US plant patent regime excluded plants from the utility patent regime. Perhaps responding to signals from the US Supreme Court's *J.E.M.* decision, Korea deleted Article 31 from its patent law in the course of large-scale revisions undertaken in 2006.[44] It is accepted that as a consequence of this change, all plants are patent-eligible under Korean law.[45]

B. Intermediate Approaches

Neither Europe nor Canada follows the permissive paradigm described above. However, patent protection for plants remains important in those jurisdictions. Both permit patent protection for plant innovation to some extent, and hence both may be described as following an intermediate approach. This may suggest a greater level of coherence than the law actually displays.

1. Europe: Eligible "Plants" v. Ineligible "Plant Varieties"; Ineligible "Essentially Biological Processes"

The European Patent Convention, Article 53(b), includes what appears on its face to be a robust exclusion of both "plant varieties" and "essentially biological processes for the production of plants."[46] Experience has demonstrated that the

---

41. *See* K.-S. Yoon, *Plant Variety Protection in the Republic of Korea*, *in* AGRICULTURAL BIOTECHNOLOGY AND INTELLECTUAL PROPERTY: SEEDS OF CHANGE 274 (Jay P. Kesan ed., 2007).

42. Thus, as of 1961, Article 31 of the Patent Act of South Korea, Act No. 9381, provided that "[a]ny person who invents a new and distinct variety of plant which reproduces itself asexually may obtain a plant patent therefor."

43. Yoon, *supra* note 41, at 275.

44. Patent Act of South Korea, Act No. 9381, Jan. 30, 2009, *translation available at* http://www.kipo.go.kr.

45. *See, e.g.*, JANIS ET AL., *supra* note 19, ch. 8.

46. Convention on the Grant of European Patents, art. 53(b), Oct. 5 1973, as revised Nov. 29, 2000, 1065 U.N.T.S. 254, *available at* http://www.epo.org/law-practice/legal-texts/html/epc/1973/e/ar53.html ("European patents shall not be granted in respect of...(b) plant or

European Patent Office has a more mixed normative commitment to excluding plant innovation from the utility patent system than this language might seem to project.

As for the "plant varieties" exclusion, the Enlarged Board's decision in the *Novartis* case remains the leading articulation.[47] Novartis had claimed, inter alia, a transgenic plant, employing claims that specified the nature of the exogenous DNA sequences but did not identify a particular plant variety as the transformed plant.[48] In response to a question about whether such a claim fell within the Article 53(b) exclusion, the Enlarged Board ruled that "[i]n the absence of the identification of a specific plant variety in a product claim, the subject-matter of the claimed invention is not directed to a plant variety or varieties" that is excluded by Article 53(b).[49] Under this ruling, a claim directed to a plant variety per se is excluded under Article 53(b), whereas a claim directed to a plant—which may encompass a multiplicity of excluded varieties—could still stand notwithstanding Article 53(b).

Addressing the argument that such a result was illogical, the Enlarged Board asserted that the question before it did not call for resort to "arithmetic logic," but instead was to be resolved as a policy matter.[50] In the Enlarged Board's view, protecting the products of plant biotechnology solely under plant variety protection regimes would not provide an adequate innovation incentive, because transgenic technology might apply across a wide range of "appropriate plants," whereas plant variety protection afforded rights only in individual plant varieties.[51] It helped that Article 53(b) seemed to distinguish between "plant *varieties*" (in the first exclusion) and *plants* (in the second exclusion for essentially biological processes for the production of plants).[52]

---

animal varieties or essentially biological processes for the production of plants or animals; this provision does not apply to microbiological processes or the products thereof.").

47. NOVARTIS II/Transgenic Plant, G 0001/98, [2000] E.P.O.R. 303 (Enlarged Bd. 1999).

48. In claim 17 of the patent-in-suit, Novartis claimed: "A transgenic plant comprising recombinant DNA sequences encoding one or more cell membrane-degrading components and one or more hydrolytic enzymes." Anti-pathogenically effective compositions comprising lytic peptides and hydrolytic enzymes, claim 17, EP 0448511 (published Sept. 25, 1991).

49. *Novartis*, [2000] E.P.O.R. at ¶ 3.10.

50. *Id.*

51. *Id.* ¶ 3.8.

52. *Id.* ¶ 3.3.1. The European Biotechnology Directive adopts the rule from *Novartis*. Directive 98/44/EC of the European Parliament and of the Council of 6 July 1998 on the Legal Protection of Biotechnological Inventions, 98/44/EC, art. 4(2), 1998 O.J. (L 213), 13 [hereinafter Biotech Directive 98/44/EC] ("[i]nventions which concern plants... shall be patentable if the technical feasibility of the invention is not confined to a particular plant... variety.").

As to the Article 53(b) exclusion for essentially biological processes for the production of plants, the EPO has arrived at a more straightforward position, although the "essentially biological" language remains somewhat troublesome. In consolidated decisions on applications relating, respectively, to tomato and broccoli breeding methods, the Enlarged Board appeared to conclude that method claims to many conventional breeding techniques would be excluded under Article 53(b).[53]

The broccoli case included claims for breeding broccoli by a series of selection, crossing, and backcrossing steps, although the claims also refer to the use of doubled haploid technology and molecular markers, as claim 1 below illustrates:

1. A method for the production of Brassica oleracea with elevated levels of 4-methylsulfinylbutyl glucosinolates, or 3-methylsulfinylpropyl glucosinolates, or both, which comprises:
   a) crossing wild Brassica oleracea species selected from the group consisting of Brassica villosa and Brassica drepanensis with broccoli double haploid breeding lines;
   b) selecting hybrids with levels of 4-methylsulfinylbutyl glucosinolates, or 3-methylsulfinylpropyl glucosinolates, or both, elevated above that initially found in broccoli double haploid breeding lines;
   c) backcrossing and selecting plants with the genetic combination encoding the expression of elevated levels of 4-methylsulfinylbutyl glucosinolates, or 3-methylsulfinylpropyl glucosinolates, or both; and
   d) selecting a broccoli line with elevated levels of 4-methylsulfinylbutyl glucosinolates, or 3-methylsulfinylpropyl [glucosinolates], or both, capable of causing a strong induction of phase II enzymes, wherein molecular markers are used in steps (b) and (c) to select hybrids with genetic combination encoding expression of elevated levels of 4-methylsulfinylbutyl glucosinolates, or 3-methylsulfinylpropyl glucosinolates, or both, capable of causing a strong induction of phase II enzymes.[54]

In an auxiliary request, the applicant added the step "deriving broccoli double haploid breeding lines" as the first step of the method.[55]

The Enlarged Board took up four questions regarding the challenged applications. Responding to the fourth, the Enlarged Board quite sensibly concluded that in applying Article 53(b)'s "essentially biological" criterion, "it is not relevant

---

53. State of Israel-Minister of Agriculture/Tomatoes, G 0001/08, (Enlarged Bd. Dec. 9, 2010); Syngenta Participations AG/Broccoli, G 0002/07 (Enlarged Bd. Dec. 9, 2010), *available at* http://www.epo.org/law-practice/case-law-appeals/pdf/g070002ex1.pdf.

54. *Id.* Referral (G02/07), ¶ 2.

55. *Id.*

whether a step of a technical nature is a new or known measure, whether it is trivial or a fundamental alteration of a known process, whether it does or could occur in nature or whether the essence of the invention lies in it."[56] Alluding to its evolving jurisprudence in the area of software-related processes, the Enlarged Board noted that a rule of eligibility that depended on an assessment of the state of the art would be flawed because it would conflate eligibility with novelty and inventive step doctrines.[57] As the Board pointed out, if the analysis of whether a plant breeding process was "essentially biological" depended upon what was known in the art, then the analysis would necessarily be subject to revision over time as plant breeding techniques advanced.[58]

The other three questions addressed the core concerns about the extent to which breeding processes should be excluded from patent protection. The steps of breeding processes invariably evidence human intervention, so under an approach that merely calls for the existence of a modicum of human intervention—the *Diamond v. Chakrabarty* approach of US law—breeding processes should pose no eligibility concerns. However, the Enlarged Board could not rationalize that principle with the existence of the Article 53(b) exclusion. Even "essentially biological" processes could involve some human intervention, the Enlarged Board reasoned, so it is "fundamentally correct . . . that not just any kind of human intervention can suffice to make an invention in this field escape the exclusion."[59]

The Enlarged Board proceeded to examine the nature of the human intervention that would render a process not "essentially biological." Reviewing the drafting history of the EPC, the Enlarged Board concluded that Article 53(b) must have been intended to exclude "plant breeding processes which were the conventional methods for the breeding of plant varieties" at the time when the UPOV Agreement and the EPC were being drafted.[60] The Enlarged Board observed that such processes "included in particular those . . . based on the sexual crossing of plants (i.e., of their whole genomes) deemed suitable for the purpose pursued and on the subsequent selection of the plants having the desired trait(s)."[61] This

56. *Id.* Order, question 4.

57. *Id.* Reasons, ¶ 6.4.1.

58. *Id.* (arguing that under such an approach, any plant breeding process initially deemed not essentially biological would become essentially biological "once it has become known and found entrance into plant breeders' sets of routine operations").

59. *Id.* Reasons, ¶ 6.4.2.1.

60. That is, the late 1950s and early 1960s. *Id.* Reasons, ¶ 6.4.2.3. So, notwithstanding the Enlarged Board's stated aversion to conflating prior art doctrines with eligibility doctrines, some understanding of the state of the art is necessary in order to apply Article 53(b).

61. *Id.*

observation informed the Enlarged Board's answer to the first referred question, and the main holding in the case, that "[a] non-microbiological process for the production of plants which contains or consists of the steps of sexually crossing the whole genomes of plants and of subsequently selecting plants is in principle excluded from patentability as being 'essentially biological' within the meaning of Article 53(b) EPC."[62]

Of course, the claims at issue arguably included more than just crossing and selection steps. The claims in the broccoli case referred to molecular markers, as noted, and the claim as amended in the auxiliary request included an initial step of deriving double haploid breeding lines, both of which were arguably steps of a "technical" (i.e., not "essentially biological") nature.[63] Here, the Enlarged Board confronted a familiar dilemma in patent eligibility jurisprudence: the freedom afforded in claim drafting frequently allows applicants the opportunity to sculpt claims closely around eligibility restrictions, resulting in disputes that tend to center predominantly on the form of the claims rather than directly on policy considerations. The Enlarged Board attempted to resolve the dilemma by again resorting to the drafting history of the EPC. Noting that "[t]he application of technical means or other forms of human intervention in [breeding] processes which helped to perform them was already common" when Article 53(b) was being drafted, the Enlarged Board concluded that the mere existence of a technical step in a breeding process claim would not necessarily save the claim from exclusion under Article 53(b).[64] According to the Enlarged Board, the critical distinction was whether the technical step merely serves to assist in the crossing or selection steps as opposed to itself introducing or modifying a trait.[65] If the

62. *Id.* Order, question 1.

63. The arguments for the existence of technical steps as to the tomato claims are perhaps less apparent. The illustrative claim from the tomato application reads:

> A method for breeding tomato plants that produce tomatoes with reduced fruit water content comprising the steps of: crossing at least one Lycopersicon esculentum plant with a Lycopersicon spp. to produce hybrid seed; collecting the first generation of hybrid seeds; growing plants from the first generation of hybrid seeds; pollinating the plants of the most recent hybrid generation; collecting the seeds produced by the most recent hybrid generation; growing plants from the seeds of the most recent hybrid generation; allowing fruit to remain on the vine past the point of normal ripening; and screening for reduced fruit water content as indicated by extended preservation of the ripe fruit and wrinkling of the fruit skin.

Id. Referral (G01/08), ¶ 2. In an auxiliary request, the applicant amended the claim to add, inter alia, the initial step of "selecting plants with tomato fruits having an increased dry weight percentage." *Id.*

64. *Id.* ¶ 6.4.2.3.

65. *Id.*

former, the claim would not escape the Article 53(b) exclusion;[66] if the latter, the claim would not be excluded under Article 53(b).[67] Predictably in light of eligibility jurisprudence from the software area, the Enlarged Board felt compelled to add that technical steps injected into a claim as extraneous upstream or downstream steps would not suffice to insulate the claim from an Article 53(b) exclusion.[68]

The Enlarged Board's decision is important because it highlights the divergence between European and US patent law on eligibility of breeding processes, and because it is likely to have an impact in other countries whose patent laws adopt European elements. However, it is difficult to discern whether the decision represents the sort of legal innovation that proponents of patent law diversity might endorse. If the Enlarged Board is to be taken at its word, its decision is essentially conservative, implementing policy choices (primarily a policy bias in favor of UPOV-style protection) made decades previously. In any event, alongside the *Novartis* decision, it places European law in an intermediate posture, tolerating some utility patenting of plant innovation while excluding a good deal.

2. Canada: Plants Per Se as Ineligible "Higher Life Forms"

The Canadian approach to utility patent eligibility of plants also may be characterized as an intermediate approach, more akin to European law than to American. This is ironic, because, like the US legislation, the Canadian Patent Act sets out categories of patent-eligible subject matter, among them "manufactures" and "compositions of matter."[69] Beyond that point, however, Canadian jurisprudence on patenting plants bears little resemblance to US jurisprudence. Canadian law has followed a path so circuitous as to make US law by comparison look like a model of clarity.

In 1989, in *Pioneer Hi-Bred Ltd. v. Canada (Commissioner of Patents)*,[70] the Supreme Court of Canada reviewed a judgment that claims to a soybean variety

66. *Id.* Order, question 2 (Enlarged Board's response to the second referred question).

67. *Id.* Order, question 3 (Enlarged Board's response to the third referred question). Specifically, the Enlarged Board called for a technical step that "by itself introduces a trait into the genome or modifies a trait in the genome of the plant produced, so that the introduction or modification of that trait is not the result of the mixing of the genes of the plants chosen for sexual crossing." *Id.*

68. *Id.* ¶ 6.4.2.3 (asserting that "upstream steps dealing with the preparation of the plant(s) to be crossed or downstream steps dealing with the further treatment of the plant resulting from such crossing and selection process" should be ignored when analyzing the Article 53(b) exclusion).

69. Patent Act, R.S.C. 1985, c. P-4, § 2 (defining "invention" as "any new and useful art, process, machine, manufacture or composition of matter, or any new and useful improvement in any art, process, machine, manufacture or composition of matter").

70. 1 S.C.R. 1623 (1989) (Can.).

did not fall within any of the categories of patent-eligible inventions. The Court appeared to express some sympathy for the argument that a breeder who was undertaking a conventional hybridization process was simply harnessing laws of nature rather than creating a patent-eligible invention, but the Court stopped short of ruling on those grounds, citing the complexity of the question and the lack of adequate technical evidence in the record.[71] Instead, the Court ruled that Pioneer's disclosure was insufficient to support the claim, even though Pioneer had deposited seed of the claimed variety.[72]

Within a year, Canada passed a Plant Breeders' Rights Act.[73] This set the stage for subsequent arguments about whether Parliament was implicitly acknowledging that plants—or other higher life forms—fell outside the utility patent regime.[74] Eventually, the Supreme Court of Canada addressed these arguments in *Harvard College v. Canada (Commissioner of Patents)*.[75] In a 5-4 decision, the Court ruled that a claim to a "higher life form," such as a genetically modified mouse, did not define a patent-eligible "manufacture" or "composition of matter."[76] The Court's opinion explicitly categorized plant varieties as "a subset of higher life forms" in the context of this rule.[77]

That ruling appeared to settle the question of the patent-eligibility of claims to plant varieties per se, but, as the *Novartis* case from Europe illustrates, there may be many ways to formulate commercially meaningful claims to plant innovation. That fact became evident in Canada in the infamous litigation between Monsanto and the canola farmer Percy Schmeiser.[78] Mr. Schmeiser's allegations about infringement—rejected by the courts—earned the litigation its notoriety, but the judicial determinations about patent eligibility are perhaps of more lasting significance in Canadian patent jurisprudence. The patent-in-suit[79] did not contain claims to a canola plant per se, or to its seeds. Rather, the claims were directed to the chimeric gene responsible for conferring the glyphosate tolerance,

---

71. *Id.* ¶ 19.

72. *Id.* ¶¶ 30–36.

73. Plant Breeders' Rights Act, S.C. 1990, c. 20; *see also infra* Section II.C. (discussing exemptions from protection under the Canadian act).

74. Analogous arguments were offered in the United States in *Chakrabarty*, 447 U.S. 303 (1980).

75. Comm'r of Patents v. Harvard Coll., [2002] 4 S.C.R. 45 (Can.).

76. *Id.* ¶ 120.

77. *Id.* ¶ 194.

78. Schmeiser v. Monsanto Can. Inc., [2004] 1 S.C.R. 902 (Can.).

79. Canadian Patent No. 1,313,830 to Monsanto Canada, Inc., entitled "Glyphosate Resistant Plants."

along with plant cells containing the gene and various expression and plant transformation vectors containing the gene.[80] Schmeiser had lost on validity and non-infringement in the lower courts, and continued to press his arguments at the Supreme Court of Canada. As to eligibility, Schmeiser argued that the practical result of enforcing these claims would be to exclude others from using plants or seeds containing the claimed gene or plant cells, and so the claims were effectively drawn to ineligible higher life forms.[81]

In another 5-4 decision, the Court rejected this argument en route to upholding the lower court's infringement determination. The opinion disposed of the eligibility issue in two brief paragraphs. In the *Harvard College* case, the Commissioner had allowed claims to a plasmid and a somatic cell culture without objection, and these claims were analogous to the claims at issue in *Schmeiser*, the Court reasoned.[82] Moreover, the Court in *Harvard College* had endorsed the proposition that "a fertilized, genetically altered oncomouse egg would be patentable subject matter, regardless of its ultimate anticipated development into a mouse," which disposed of Schmeiser's argument about Monsanto's claims effectively covering ineligible higher life forms.[83]

The dissenters found this outcome "hard to reconcile" with *Harvard College*,[84] at least as to the product claims.[85] To avoid this perceived conflict, the dissenters found it necessary to interpret the claims so as not to extend protection to "whole plants," thus preserving the claims' eligibility.[86] Under a purposive construction, it was "clear" that the gene claims did not "extend protection to the plant."[87] As to the plant cell claims, they could not be construed to "extend past the point where the genetically modified cell begins to multiply and differentiate into plant tissues," because at that point, the claim would cover "every cell in the plant, i.e.... the plant itself."[88] According to the dissenters, this was a construction

80. *Id.* (claims 1–28 and 43–51). The patent also included claims to a method for regenerating a glyphosate-tolerant plant, but these claims were not at issue in the case.

81. *Schmeiser*, [2004] 1 S.C.R. at ¶ 21 (reciting Schmeiser's argument).

82. *Id.* ¶ 22. The Court also seemed to find it relevant that a "mammal" was at issue in *Harvard College*, and not (of course) in *Schmeiser.*

83. *Id.* ¶ 23.

84. *Id.* ¶ 109.

85. Regarding the method claims for regenerating the plant cell into a plant, the dissenting opinion concluded that such a process entailed "substantial human intervention and does not follow the 'laws of nature' as would natural asexual or sexual reproduction," and thus would constitute an eligible invention. *Id.* ¶ 137.

86. *Id.* ¶ 111.

87. *Id.* ¶ 130.

88. *Id.* ¶ 138.

that allowed Monsanto a fair scope of protection against its competitors, because Monsanto could still preclude competing biotechnology companies from using the claimed gene "to create a glyphosate-resistant plant cell that can then be regenerated into a glyphosate-resistant plant."[89] That is, Monsanto's composition claims were "solely for genetically modified chimeric genes and cells in the laboratory prior to regeneration."[90]

C. Restrictive Approaches

The patent laws of China, India, and certain South American countries provide the best illustration of the restrictive approach to utility patent protection for plants. Although these countries adopt similar paradigms resisting patent protection for plants, they offer plant variety (plant breeders' rights) protection, and not necessarily on identical terms. Accordingly, it would go too far to suggest that the overall level of protection for plant innovation in these countries is uniform.

Let us turn first to Chinese patent law, where Article 25.1(4) prohibits the grant of an invention patent for "plant varieties,"[91] which might seem to set the stage for a Chinese version of the European *Novartis* dispute.[92] However, administrative guidelines in China raise a question about whether a *Novartis*-style claim directed to a "plant" having a specified genetic modification would fall outside the exclusion on the ground that it is not limited to a plant variety per se. The SIPO Guidelines for Examination apply Article 25.1(4)'s restrictive language vigorously, contending that transgenic plants "still belong[] to the category of... 'plant variety'" and therefore should be excluded from eligibility.[93] One may wonder whether this proscription applies only to claims to transgenic plant varieties per se, or also to claims in the style of *Novartis*, which encompass, but are not limited to, specific varieties.[94]

---

89. *Id.* ¶ 131.

90. *Id.* ¶ 139.

91. Patent Law of the People's Republic of China, *as revised by* Decision Regarding the Revision of Patent Law of the People's Republic of China (promulgated by the Standing Comm. Nat'l People's Cong., Aug. 25, 2000) [hereinafter China Patents Act], *available at* http://www.sipo.gov.cn.

92. *See supra* Section I.B.1.

93. SIPO GUIDELINES FOR EXAMINATION, pt. 2, ch. 1, § 9.1.2.4 (2010) [hereinafter SIPO GUIDELINES], *available at* http://www.gechengip.com/information/content2010_en.htm; *see also id.* § 4.4 (defining "plant variety").

94. Other references in the Guidelines to the plant variety exclusion may lend more confusion than clarification. Section 9.1.2.3 specifies that a "single plant" that survives by photosynthesis is an excluded plant variety, but "[i]f a cell, tissue and an organ of a plant do not possess the

The Guidelines also specify that claims to seeds would fall within the plant variety exclusion.[95]

Notably, Article 25 does not refer to utility models.[96] Presumably, utility model protection for plant varieties would be permissible.

China's Article 25.1 arguably departs from European practice concerning claims to plant breeding processes. Rather than excluding such processes from eligibility, Article 25.1 specifies that "for processes used in producing products referred to in item (4) ... patent right *may* be granted in accordance with the provisions of this law."[97] However, here again the SIPO Guidelines present some questions. According to the Guidelines, the processes referred to as patent-eligible are "non-biological processes, and do not include those for the production of animals or plants through essentially biological processes."[98] The distinction rests on the degree of "human technical involvement," per the Guidelines. If human technical involvement is "the controlling or decisive factor for achieving the result or effect of that process," then the process is not essentially biological and is patent-eligible.[99] It seems inevitable that the EPO Enlarged Board's decision in the broccoli/tomatoes cases[100] will prove important in Chinese law on the patent-eligibility of breeding processes, perhaps imparting a more robust exclusion into Chinese law than the current Guidelines may imply.

Among the other jurisdictions that have chosen a patent-restrictive approach, India's patent law is particularly notable. India's Patent Act departs from the European formulation of excluding "plant varieties." Instead, India's legislation excludes "plants ... in whole or any part thereof other than microorganisms but including seeds, varieties and species."[101] India's Patents Act also excludes

---

above-mentioned characteristic, they cannot be regarded as 'plant varieties' ... ." *Id.* Recitations of this sort highlight the familiar proposition that eligibility restrictions tend to mutate into restrictions on claiming format.

95. *Id.*

96. *Cf.* China Patents Act, *supra* note 91, art. 22 (referring to both patents and utility models in specifying the requirements for novelty, inventive step, and industrial applicability).

97. *Id.* art. 25(1)(4) (emphasis added).

98. SIPO GUIDELINES, *supra* note 93, § 4.4.

99. *Id.*

100. *See supra* Section I.B.

101. The Patents Act, § 3(j), No. 39 of 1970, INDIA CODE (2005) [hereinafter India Patents Act]. The plant exclusion was added in 2002. The Patents (Amendment) Act, § 4, No. 38 of 2002, INDIA CODE (2002). *See generally* Janice Mueller, *The Tiger Awakens: The Tumultuous Transformation of India's Patent System and the Rise of Indian Pharmaceutical Innovation*, 68 U. PITT. L. REV. 491, 559 (2007) (comprehensive analysis of the Indian patent system from a US scholar's perspective).

"essentially biological processes for production or propagation of plants... other than non-biological and microbiological processes."[102] However, Indian law also includes a unique plant breeders' rights system that is in some respects more generous than others.[103]

Some commercially significant South American countries also have adopted restrictive approaches to patent protection for plants. Argentina's patent law excludes from patent eligibility "all kinds of live material and pre-existing substances in nature."[104] Argentina does have a plant variety protection regime.[105]

The persistence in these countries of a restrictive approach to plant patenting comes as no surprise in view of attitudes that have existed in many parts of the world for decades. Is the restrictive approach indicative of legal innovation? This is a difficult question that prompts further questions about what legal innovation, or even legal innovation specifically in the area of patent eligibility rules, actually means. Does it mean that individual countries will be free to engage in adept customization that produces diverse rules that are calibrated for local conditions? Some may argue that the restrictive patent eligibility rules discussed here exemplify just such a process. Perhaps that is most true of India, where a non-UPOV plant variety protection scheme was created as part of a clear decision to channel plant variety innovation away from the utility patent system.

On the other hand, does legal innovation contemplate that individual countries will engage in a period of experimentation that eventually leads to the emergence of a "market dominant" standard that becomes the de facto international norm? There is no evidence of such a process occurring in the plant patenting eligibility jurisprudence in the major patent jurisdictions. To the contrary, a better case can be made that diverse rules on eligibility are hardening and reinforcing themselves in a number of jurisdictions, making ultimate convergence less likely. Perhaps more important, this process reduces the likelihood of creative experimentation of a sort that would seem to lie at the core of meaningful legal innovation.

---

102. India Patents Act, *supra* note 101, § 3(j).

103. The Protection of Plant Varieties and Farmers' Rights Act, No. 53 of 2001, Acts of Parliament, 2001, *available at* http://indiacode.nic.in/fullact1.asp?tfnm=200153. For background, see, e.g., Mrinalini Kochupillai, *India's Plant Variety Protection Law: Historical and Implementation Perspectives*, 16 J. INTELL. PROP. RTS. 88 (2011), *available at* http://papers.ssrn.com/sol3/papers.cfm?abstract_id=1780230; Prabhash Ranjan, *Recent Developments in India's Plant Variety Protection, Seed Regulation and Linkages with UPOV's Proposed Membership*, 12 J. WORLD INTELL. PROP. 219 (2009).

104. Law No. 24481, art. 6(g), May 23, 1995 [LV-D] A.D.L.A. 5635 [hereinafter Patent Law of Argentina].

105. Law. No. 20247, Mar. 30, 1973, B.O. [hereinafter Seeds Law of Argentina]; *see also* Law. No. 24376, Oct. 25, 1994, [LIV-D] A.D.L.A. 4376 [hereinafter Seeds Law and Regulatory Decree 2183/91], which adopts the 1978 version of the UPOV Convention.

II. Emerging Issues in Patent Protection for Plants

In many of the major patenting jurisdictions, the question of whether plant varieties constitute patent-eligible subject matter has reached a point of equilibrium. Although the durability of that equilibrium may be questioned in some countries (Canada, for example), on the whole, in countries where at least some utility patent protection for plants is available (under either a permissive or intermediate approach as outlined above), judicial attention in the near future is likely to be directed to a number of other substantive issues. A sampling of these issues is provided below.

A. Non-obviousness/Inventive Step

In the United States, numerous patents have been issued with claims to identified plant varieties (and their seeds, and other plant parts). Many of these patents also include claims to breeding methods that produce the identified plant variety. It seems inevitable, particularly in the post-*KSR* climate, that patent challengers in litigation or post-grant proceedings will attempt to press arguments to invalidate variety claims or breeding process claims on the basis of obviousness. There is some indication that there has been little or no explicit discussion of obviousness in the ex parte prosecution of such patents.[106]

Likewise, there is virtually no US case law exploring obviousness in the context of patented plants.[107] Chemical patent practice may supply the closest analog, yet the overriding theme here is that obviousness is highly fact-dependent. For example, one might be tempted to argue that a claim to a breeding method that produces a novel product, but uses standard breeding steps to do so, fails the obviousness criterion. However, in the context of chemical inventions, the Federal Circuit has asserted that although "[a] process yielding a novel and nonobvious

106. This assertion is based in part on anecdotal evidence. It is also based on an informal review of patents issued in Class 800, Subclass 260 over roughly a five-year period from 2006 to 2011. A sample of ten patents was chosen, and a review of their file histories showed no discussion of obviousness. Of course, a rigorous statistical study would need to be undertaken in order to validate this initial impression.

107. Two non-precedential opinions have some relevant discussion. *See In re* Pod-ners, L.L.C., 93 U.S.P.Q.2d (BNA) 1411 (Fed. Cir. 2009) (upholding an obviousness determination as to claims for bean plants and methods for breeding them, where a grower had brought yellow beans from Mexico and had "followed normal and well-established agricultural techniques" for reproducing the beans in a specified color range); *In re* Sigco Research, 36 U.S.P.Q.2d (BNA) 1380, 48 F.3d 1238 (Table) (Fed. Cir. 1995) (overturning an obviousness rejection as to claimed sunflower plants and seeds where there was insufficient evidence that applying conventional breeding practices to existing plants would have predictably yielded plants that were true breeding for the specified oleic acid content as claimed).

product may nonetheless be obvious," the fact remains that "there are no per se rules relating the patentability of product and process in either direction."[108]

Although the US law of obviousness as applied to patented plants has yet to be fleshed out, the Japanese approach has been more fully articulated, at least at the administrative level, in the JPO's Examination Guidelines.[109] The Guidelines articulate a general rule that a claim to a plant will be deemed to lack evidence of an inventive step "where characteristics of the plant created can be easily predicted from the characteristics of publicly known plants within the species to which the plant belongs and where the invention does not have advantageous effects that a person skilled in the art cannot foresee."[110] For example, a claimed plant produced by "merely crossing" publicly known parents A and B, where the claimed plant possesses the combined (known) characteristics of A and B, lacks an inventive step. In addition, a claimed plant "whose shape or color is similar to that of publicly known plants within the species to which the plant belongs" lacks an inventive step.[111]

The Japanese Guidelines also extend inventive step analysis to claims that are directed to breeding processes. According to the Guidelines,

> An invention of a process for creating a plant does not have an inventive step, where the selection of parent plants, means, conditions or the like is not considered to be difficult and where the created plant does not have advantageous effects that a person skilled in the art cannot foresee.[112]

Whether and to what extent the Japanese inventive step law mirrors the US post-*KSR* obviousness jurisprudence is beyond the scope of this discussion. Even allowing for differences, the JPO Guidelines may prove useful as a source of arguments for those who seek to raise questions about the obviousness of patented plant inventions in the United States.

---

108. TorPharm, Inc. v. Ranbaxy Pharms., Inc., 336 F.3d 1322, 1327 (Fed. Cir. 2003); *see also In re* Ochiai, 71 F.3d 1565, 1570 (Fed. Cir. 1995) (concluding that there is no per se rule "that a process claim is obvious if the prior art references disclose the same general process using 'similar' starting materials"). In patents on breeding processes, one may wonder whether the USPTO is essentially applying *Ochiai* as if it establishes a per se rule of non-obviousness.

109. *2010 Examination Guidelines, supra* note 28. For the source of the inventive step requirement in Japanese patent law, see Japan Patent Act, *supra* note 27, art. 29(2).

110. *2010 Examination Guidelines, supra* note 28, § 3.2.2(1).

111. *Id.* The reference to color is perhaps significant in light of the *Pod-ners* case in the United States. *See supra* note 107.

112. *2010 Examination Guidelines, supra* note 28, § 3.2.2(2).

B. Utility and Disclosure

In countries where patent protection of plants is permitted, once controversies over threshold eligibility questions are settled, the focus of subsequent disputes predictably turns to issues that are less exotic, but still may be difficult, such as whether the patentee has provided adequate disclosure in the patent specification.[113] Adequacy of disclosure has always been a difficult issue in plant intellectual property,[114] and, indeed, in intellectual property law for all sorts of biological subject matter.

In the United States, the disclosure requirements embodied in 35 U.S.C. § 112, para. 1, apply with full force to utility patents claiming plant subject matter.[115] Enablement challenges have been raised successfully in a number of US plant biotechnology cases. These cases have involved questions about the extent to which disclosed plant transformation techniques could be adapted to work in numerous diverse plant species.[116] In addition, as is the case with many other types of biological subject matter, the patentee may need to deposit seeds or other materials in order to comply with disclosure requirements.[117] The 2010 Japanese Examination Guidelines express similar expectations for utility patent disclosures involving plant innovation.[118]

The foregoing indicates that the law of utility patents for plants in permissive jurisdictions may be expected to proceed incrementally and perhaps even predictably, more akin to applications of patent law to other more settled areas. The litigation strategy in the *Organic Seed Growers and Trade Association v. Monsanto* case may be an exception.[119] There, the plaintiff organizations—mainly organic seed businesses and farmers—sought declaratory relief that would preclude

113. *See, e.g.*, Mark D. Janis, *Rules v. Standards for Patent Law in the Plant Sciences*, 24 L. Context 54 (2006) (identifying this trend).

114. J.E.M. Ag Supply, Inc. v. Pioneer Hi-Bred Int'l, Inc., 534 U.S. 124, 142–43 (2001).

115. By contrast, for US plant patents, the disclosure need only be "as complete as reasonably possible." 35 U.S.C. § 162 (2012); *see also* Janis et al., *supra* note 19, ch. 6 (discussing the disclosure requirements for plant patents).

116. Janis, *supra* note 113 (discussing the cases). The recurring question in these cases was whether *agrobacterium*-mediated transformation could be applied to monocots where the disclosure (and background knowledge in the art) demonstrated success only in dicots.

117. *See* 37 C.F.R. §§ 1.801–809 (2013) (regulations governing the deposit requirement).

118. For example, plant characteristics should be described in conventional quantitative terms, plant colors should be described by reference to standard references, and any special cultivation practices should be noted. *2010 Examination Guidelines*, *supra* note 28, § 3.1.3.1 (applying the enablement requirement to plants). Seeds should be deposited where it is not otherwise possible to enable a person of ordinary skill to reproduce the plant. *Id.*

119. Complaint, Organic Seed Growers and Trade Ass'n v. Monsanto Co., 851 F. Supp. 2d 544 (S.D.N.Y. 2012).

Monsanto from enforcing various patents on transgenic plants.[120] The case was dismissed by the district court for lack of standing, a decision that was affirmed by the Court of Appeals for the Federal Circuit; however, a petition for review by the Supreme Court has been filed.[121] Should the case ever proceed to the merits, a court may have an opportunity to opine on whether the utility doctrine of 35 U.S.C. § 101 may be used to invalidate virtually any transgenic plant patent on the grounds that transgenic technology is "injurious to the well-being, good policy, or sound morals of society."[122] Quite aside from whether the plaintiffs could muster facts sufficient to support this allegation, arguments for injecting an *ordre public*–style objection into US utility doctrine have been soundly rejected, and there is no present indication that the United States will change course.[123] Nonetheless, the argument deserves mention in light of the fact that the legal basis for *ordre public* arguments is better established in other jurisdictions.

C. Scope

Issues regarding enforceable scope are the least well-established, but arguably the most important emerging issues in the substantive law of patents for plants. Illustrative disputes are detailed below. In some cases, doctrines that are wholly internal to patent law, such as claim construction, infringement rules, and exhaustion, have been (and will continue to be) invoked. Other doctrines such as patent misuse, which inject external competition values into the patent law analysis, have also been important, as have contracts principles. The discussion below considers whether patents on plants or plant innovation are likely to cover transactions in downstream products such as processed grain; whether they are likely to cover the saving of seed for reuse on the farm; and whether they are likely to cover a competitor's breeding activities undertaken without permission.

1. Processed Grain

The existence of diverse patent eligibility rules for plants in various countries gives rise to the potential for cross-border flows of plant material, and these cross-border flows implicate patent rights, at least from the patent holder's

---

120. *Id.*

121. *See id.*, *aff'd*, 715 F.3d 1350 (Fed. Cir. 2013).

122. Complaint ¶ 111, Organic Seed Growers & Trade Ass'n v. Monsanto Co., 851 F. Supp. 2d 544 (S.D.N.Y. 2012) (No. 11 Civ. 2163). The complaint also sets out numerous other theories of invalidity, unenforceability, and noninfringement, some of which are closer to the mainstream of the law than the plaintiffs' utility argument. *See id.* ¶¶ 112–130.

123. *See generally* Margo A. Bagley, *Patent First, Ask Questions Later: Morality and Biotechnology in Patent Law*, 45 WM. & MARY L. REV. 469 (2003).

perspective. For example, Monsanto was unable to obtain patent protection for the Roundup Ready® gene under the laws of Argentina. Reportedly, a significant amount of Roundup Ready® soybean is grown in Argentina without Monsanto's authority, harvested and processed as ground soymeal, and exported. In one instance, an importer (Cefetra) sought to import the soymeal into Europe, where Monsanto holds European patent rights for a class of glyphosate-resistant enzymes that are produced by Monsanto's Roundup-Ready® soybeans, along with the gene sequence coding for those enzymes. The soymeal at issue tested positive for the presence of the patented enzyme and gene sequence, and Monsanto sued in a Netherlands court, the Rechtbank's-Gravenhage, for patent infringement.[124] The infringement dispute involved a novel question of patent scope: whether Monsanto's patent claiming the plant material extended to the processed grain.

The operative legal question on infringement concerned the proper interpretation of Article 9 of the European Biotechnology Directive, which provides that "protection conferred by a patent on a product containing or consisting of genetic information shall extend to all material...in which the product is incorporated and in which the genetic information is contained and performs its function."[125] Article 9 on its face may seem to establish a generous scope of protection for biotechnology patents, but Cefetra focused on the "performs its function" clause to craft an argument for restricting scope. According to Cefetra, the patented material in the soybean meal was not performing its function of conferring glyphosate resistance (even though it had previously performed that function when the plants were growing and might conceivably be extracted, and be capable of doing so again). The Netherlands court referred the question to the Court of Justice for the European Union (CJEU) for a preliminary ruling, and the Court issued a ruling agreeing with Cefetra's position:

> Article 9 of the Directive must be interpreted as not conferring patent right protection in circumstances such as those of the case in the main proceedings, in which the patented product is contained in the soy meal, where it does not perform the function for which it was patented, but did perform that function previously in the soy plant, of which the meal is a processed product, or would possibly again be able to perform that function after it had been extracted from the soy meal and inserted into the cell of a living organism.[126]

---

124. Case C-428/08, Monsanto Tech., L.L.C. v. Cefetra BV, 2010 E.C.R. I-6765, ¶¶ 15–21, *available at* http://curia.europa.eu/jurisp/cgi-bin/form.pl?lang=EN&Submit=Submit&numaff=C-428/08. Monsanto also brought similar challenges in Spain and the United Kingdom.

125. Biotech Directive 98/44/EC, *supra* note 52, art. 9.

126. *Cefetra*, 2010 E.C.R. I-6765, ¶ 50.

It seems safe to predict that the *Cefetra* ruling will inspire analogous arguments in other jurisdictions, irrespective of whether those jurisdictions have any legislative counterpart to Article 9. For example, in Canada, where plant genetic material is subject to patent protection even though plants per se are not, it has been argued that a *Cefetra* rule might emerge.[127] The argument is that Canadian courts could arrive at the *Cefetra* outcome by construing unauthorized "use" for infringement purposes (under Section 24 of the Canadian patent statute) to call for an examination into whether the patented invention played an important part in the alleged user's activities, or whether the alleged user had any foreseeable opportunity to take advantage of the claimed invention.[128]

It is not clear whether similar arguments would gain any foothold in US law. It is well understood in US law that claims to chemical compositions cover all uses of those compositions, a proposition that seems to be in some tension with the CJEU's construction of Biotech Directive Article 9 in *Cefetra*. In addition, it is generally accepted in the United States that the concept of use for infringement purposes should be construed broadly.[129] On the other hand, in the context of a claim to a system, the Federal Circuit has held that "to 'use' a system for purposes of infringement, a party must put the invention into service, i.e., control the system as a whole and obtain benefit from it."[130] To the extent that this ruling opens the door for litigation over nice questions about obtaining "benefit," it is an unfortunate ruling, especially if extended to composition claims. One might imagine a *Cefetra* argument that an unauthorized user of processed grain containing the Roundup Ready® gene is not obtaining "benefit" from the claimed invention and thus is not liable for infringement. Such an argument, if accepted, would take US patent law in an unwelcome new direction.

2. Farm-Saved Seed

In jurisdictions permitting patent protection for plants, it seems inevitable that another patent scope issue will arise: whether patent rights extend to the saving and replanting of patented seed. The issue has received considerable attention in the United States, primarily in cases in which Monsanto has sued farmers who

---

127. Nathan Fan, *Importing* Cefetra's *Soy Meal Case into Canada: How Well Will Patents for Genetically Modified Plant Genes and Cells Found in a Residual State Fare Under Canada's Patent Laws?*, 22 INTELL. PROP. J. 333 (2010).

128. *Id.* at nn.40–44 and accompanying text. The leading Canadian enunciation of the "use" analysis for infringement purposes is found in *Schmeiser*, discussed in Section I.B *supra*.

129. *See* NTP, Inc. v. Research in Motion, Ltd., 418 F.3d 1282, 1316 (Fed. Cir. 2005) (citing Bauer & Cie v. O'Donnell, 229 U.S. 1 (1913)).

130. Centillion Data Sys., LLC v. Qwest Comm. Int'l, Inc., 631 F.3d 1279, 1284 (Fed. Cir. 2011).

have allegedly saved and replanted soybean seed containing the Roundup Ready® gene. Complicating the issue—at least arguably—is the fact that the UPOV Agreement (1991) includes an optional seed-saving limitation on plant variety rights,[131] and jurisdictions that recognize patent protection for plants all have UPOV-style plant variety systems that adopt a seed-saving limitation,[132] with some variations, especially in Europe[133] and Canada.[134]

---

131. UPOV(1991), *supra* note 2, art. 15(2) ("Notwithstanding Article 14, each Contracting Party may, within reasonable limits and subject to the safeguarding of the legitimate interests of the breeder, restrict the breeder's right in relation to any variety in order to permit farmers to use for propagating purposes, on their own holdings, the product of the harvest which they have obtained by planting, on their own holdings, the protected variety....").

132. For US law, see 7 U.S.C. § 2543 (no infringement of US plant variety rights "for a person to save seed produced by the person from seed obtained, or descended from seed obtained, by authority of the owner of the variety for seeding purposes and use such saved seed in the production of a crop for use on the farm of the person, or for sale as provided in this section," and likewise no infringement to engage in "[a] bona fide sale for other than reproductive purposes, made in channels usual for such other purposes, of seed produced on a farm either from seed obtained by authority of the owner for seeding purposes or from seed produced by descent on such farm from seed obtained by authority of the owner for seeding purposes"); *see also* 7 U.S.C. § 2401(b)(1) (2012) (defining sale for nonreproductive purposes). For Japanese law, see Plant Variety Protection and Seed Act, art. 21(2), No. 83 of 1998, *amended by* No. 49 of 2007. For South Korean law, see Seed Industry Act, art. 58(2), Act. No. 5024, Dec. 6, 1995, *amended by* Act. No. 6374, Jan. 26, 2001 ("The Minister of the Ministry of Agriculture and Forestry may restrict a variety protection right for a variety, if a farmer collects the seeds of the variety for himself for the purpose of self-production."); *see also id.* art. 58(3) (providing that "[m] atters relating to the scope of restriction, procedures, method, etc. under Paragraph (2) shall be prescribed by a Presidential Decree."). For Australian law, see Plant Breeders' Rights Act 1994 (Cth), § 17.

133. In Europe, the Biotech Directive provides that when a patent holder sells, otherwise commercializes, or consents to the sale of plant-propagating material to a farmer for agricultural use, the farmer is impliedly authorized "to use the product of his harvest for propagation or multiplication by him on his own farm," to the extent permitted by the seed-saving provision of the Community Plant Variety Regulation (CPVR). Biotech Directive 98/44/EC, *supra* note 52, art. 11(1), *referring to* Council Regulation 2100/94, art. 14, Community Plant Variety Rights, 1994 O.J. (L 227) 1 (EC). The seed-saving provision in CPVR art. 14 is relatively limited as compared to the seed-saving exemption in US law. The CPVR seed-saving exemption does not apply to any "hybrid or synthetic variety," art. 14(1); can only be invoked in connection with specified varieties listed in the Regulation, art. 14(2) (which notably excludes soybeans); and requires the farmer to pay "equitable remuneration" to the plant variety protection owner, art. 14(3), under the terms of implementing rules found in the Agricultural Exemption Regulation. Commission Regulation 1768/95, Implementing Rules on the Agricultural Exemption Provided for in Article 14(3) of [Regulation 2100/94], 1995 O.J. (L 173) 14 (EC). The Implementing Rules are at issue in Case C-509/10, Geistbeck v. Saatgut-Truehandverwaltungs GmbH, O.J. C 30/15 (Jan. 29, 2011) (reference from the Bundesgerichstof for a preliminary ruling).

134. Canadian law provides a good reminder that the scope of protection for plant-related innovation depends both on utility patent rules and plant variety protection rules in many jurisdictions. Because current Canadian patent law prohibits patents on plants and seeds per

In the United States, the Court of Appeals for the Federal Circuit has rejected the argument that the existence of a seed-saving exemption in the US Plant Variety Protection Act indicates implicitly that there must be a seed-saving exemption from utility patent infringement.[135] The court reasoned that the plant variety and patent systems, respectively, impose different conditions for protection, and thus accord different "rights and privileges."[136]

The Federal Circuit has also rejected arguments that the doctrine of patent exhaustion shields various seed-saving practices.[137] In *Monsanto v. McFarling*, Mr. McFarling had made an authorized purchase of Roundup Ready® soybean seed, but the court held that this purchase did not exhaust Monsanto's patent rights in the 1,500 bushels of progeny seed that McFarling produced from the original seed, harvested, and replanted.[138] In *Scruggs*,[139] the Federal Circuit reached the same result on similar facts. Some have questioned whether these rulings survive the Supreme Court's decision on patent exhaustion in *Quanta*.[140]

---

se, it might be supposed that seed-saving practices in Canada are clearly insulated from intellectual property liability. In fact, however, the Canadian Plant Breeders' Rights Act does not include an explicit exemption for seed saving. The Plant Breeders' Rights Office Guidelines claim that farmers may save and reuse seed on their own land without liability under the Act, but there is no statutory authority for that assertion. *See Guide to Plant Breeder's Rights in Canada*, Can. Food Inspection Agency (Jan. 14, 2010), http://www.inspection.gc.ca/english/plaveg/pbrpov/guidee.shtml#a9Guide ("Farmers may save and plant their own seed of protected varieties on their own land without infringing on the holder's rights.").

135. Monsanto Co. v. McFarling, 302 F.3d 1291 (Fed. Cir. 2002). The Federal Circuit declined to reconsider the ruling in Monsanto Co. v. McFarling, 363 F.3d 1336, 1344 (Fed. Cir. 2004).

136. *McFarling*, 302 F.3d at 1299.

137. In addition, the Federal Circuit has turned back arguments that Monsanto's practices constituted patent misuse. *See* Herbert Hovenkamp, Mark D. Janis, Mark A. Lemley & Christopher Leslie, 1 IP and Antitrust ch. 3 (2d ed. 2010).

138. *McFarling*, 302 F.3d at 1299.

139. Monsanto Co. v. Scruggs, 459 F.3d 1328 (Fed. Cir. 2006). In *McFarling*, the initial seed purchase was made under Monsanto's standard license agreement, which expressly forbids seed saving, whereas in *Scruggs*, the farmer had not signed any agreement. The court in *Scruggs* considered this difference immaterial; in neither case had there been an authorized sale of the saved and replanted progeny seed. The court in *Scruggs* concluded that:

> Without the actual sale of the second generation seed to Scruggs, there can be no patent exhaustion. The fact that a patented technology can replicate itself does not give a purchaser the right to use replicated copies of the technology. Applying the first sale doctrine to subsequent generations of self-replicating technology would eviscerate the rights of the patent holder.

Id. at 1336.

140. Quanta Computer, Inc. v. LG Elecs., Inc., 553 U.S. 617 (2008); *see, e.g.*, Yee Wah Chin, *Inexhaustible: Patents on Self-Replicating Technologies*, 3 Landslide 13 (2011) (arguing that,

The US Supreme Court recently addressed the exhaustion question for self-replicating technologies in *Bowman v. Monsanto Co.*[141] Bowman had purchased commodity soybeans from a grain elevator rather than purchasing them from a soybean seed dealer. He planted the commodity soybeans, harvested the progeny, and saved some of the seed for replanting in the subsequent year, blending the saved progeny seed with seed from additional commodity purchases.[142] He repeated the process for some time. Some portion of the commodity seed (and the subsequent generations of seed produced by Bowman) contained the Roundup Ready® trait. In all likelihood, this occurred because other area farmers who sold their crops to the same grain elevator had grown Roundup Ready® soybeans, quite probably as a result of their authorized purchases of seed from Monsanto. Bowman sprayed his crops with RoundUp® herbicide to cull non-RoundUp Ready® plants, thus obtaining for himself a cheaper source of RoundUp Ready® seeds for planting in subsequent seasons. Bowman argued that the sale from the licensed farmers to the grain elevator constituted an unconditional sale of a patented good that exhausted Monsanto's patent rights, such that the subsequent sale from the elevator to Bowman could not trigger patent infringement liability, an argument that would seem to find some footing in *Quanta*.[143] On the other hand, *McFarland* held that even where patent rights are exhausted as a result of the authorized sale of a batch of seed, rights in the progeny of that batch are not exhausted because the progeny has never been sold. The trial court accepted the *McFarland* argument,[144] ruling for Monsanto, and the Federal Circuit affirmed. The US Supreme Court granted certiorari and affirmed Bowman's infringement liability, focusing on Bowman's affirmative efforts to reproduce the patented invention:

> Bowman devised and executed a novel way to harvest crops from Roundup Ready seeds without paying the usual premium. He purchased beans from a grain elevator anticipating that many would be Roundup Ready; applied a glyphosate-based herbicide in a way that culled any plants without the patented trait; and saved beans from the rest for the next season. He then planted those Roundup Ready beans at a chosen time; tended and treated them, including by exploiting their patented glyphosate-resistance; and

---

after *Quanta*, the exhaustion rule should apply to shield growers of progeny seed from patent infringement liability).

141. Bowman v. Monsanto Co., 133 S. Ct. 1761 (2013).

142. Monsanto Co. v. Bowman, 686 F. Supp. 2d 834, 835–36 (2011).

143. *Id.* at 836.

144. *Id.* at 839.

> harvested many more seeds, which he either marketed or saved to begin the next cycle. [I]t was Bowman...who controlled the reproduction (unto the eighth generation) of Monsanto's patented invention.[145]

Irrespective of this outcome, disputes over whether patent owners can regulate seed-saving practices are likely to persist. Had the Court found Monsanto's rights exhausted, patent holders presumably would have sought a remedy in contract law.[146] The Court's ruling that Monsanto's patent rights are not exhausted may give further stimulus to ongoing efforts to persuade Congress to intervene.[147]

3. Breeding Activities

Perhaps the most important issue concerning the scope of utility patent protection for plants has received relatively little attention outside academia: whether utility patent rights in plants preclude others from using the patented plant without permission as a source for breeding new (and potentially competing) plants. UPOV (1991) contains a mandatory breeders' exemption from plant variety rights,[148] but as the Federal Circuit's *McFarling* decision on seed-saving suggests, the mere existence of a statutory limitation on the scope of plant variety rights does not alone justify recognizing a comparable limitation on the scope of utility patent rights.[149]

The more intriguing question is whether breeding activities undertaken without a patent owner's permission qualify as exempt from infringement liability under the experimental use doctrine. In the United States, the breeder undoubtedly would lose an experimental use argument under current law. The United

---

145. Bowman v. Monsanto, 133 S. Ct. 1761, 1768–69 (2013).

146. Some have argued that the availability of contract remedies is further justification for a robust exhaustion rule. *See, e.g.*, Chin, *supra* note 140, at 16. In a case such as *Bowman*, one may wonder how a contract remedy would operate. Monsanto has no contract with the grain elevator or with Bowman, so the issue would seem to be whether Monsanto can bind its licensed growers to prevent downstream third parties such as the grain elevator from selling to other third parties for the purpose of replanting.

147. For example, the Seed Availability and Competition Act of 2011, H.R. 307, 112th Cong. (2011), proposed subjecting seed patents to a limited compulsory license for the replanting of saved seed, with license fees to be paid into a "Patented Seed Fund" administered by the US Department of Agriculture. The legislation also purported to nullify "any contractual limitation on retaining such seed, or any requirement to pay royalties or licensing or other fees, by reason of the patent, for retaining such seed." *Id.* at proposed section 2(f).

148. UPOV (1991), *supra* note 2, art. 15(1). For the implementation of this exemption in the US Plant Variety Protection Act, see 7 U.S.C. § 2544 (2012) ("The use and reproduction of a protected variety for plant breeding or other bona fide research shall not constitute an infringement of the protection provided under this Act.").

149. *See supra* notes 135–36 and accompanying text.

States has no general statutory experimental use defense, and the common law defense as currently construed would not extend to activities that bear even the slightest hint of commerciality.[150] Similarly, in Australia, the very existence of an experimental use defense is subject to question.[151]

In other countries that adhere to a permissive approach to patenting of plants—namely Japan and South Korea—a statutory experimental use defense does exist.[152] However, it is not clear that the statutory exemption would encompass commercial breeding activities. Indeed, even if it did, it is not clear that the exemption would extend to the activities themselves, or that it would likewise extend to the sale of plants produced as a result of the breeding activities.[153] Accordingly, a breeder may not be much more free to operate without license in these countries than in countries that rely on a seemingly more ephemeral common law experimental use doctrine. At all events, the scope of any breeders' exemption is likely to be dictated by the outcome of larger battles over the extent of permissible use of patented technologies generally, rather than by considerations specific to the seed industry.

Conclusion

In the major patenting jurisdictions, the approach to utility patent protection for plants is characterized by considerable diversity on critical issues of eligibility and scope. This diversity is permissible under TRIPS and UPOV. However, it is not

150. *See* Madey v. Duke Univ., 307 F.3d 1351 (Fed. Cir. 2002) (construing the doctrine narrowly); *see also* Esther van Zimmeren & Geertrui van Overwalle, *A False Sense of Security Offered by Zero-Price Liability Rules? Research Exceptions in the United States, Europe, and Japan in an Open Innovation Context*, *in* Patent Law in Global Perspective (Ruth L. Okediji & Margo Bagley eds., 2013).

151. *See, e.g.*, Senate Community Affairs References Comm., Gene Patents 2.2.2, 5.108 (Nov. 2010) (explaining that the Australian patent statute includes no explicit research exemption, and that there is ongoing debate in Australia about whether to incorporate one).

152. Japan Patent Act, *supra* note 27, art. 69(1) ("[T]he effects of the patent right shall not extend to the working of the patent right for the purposes of experiment or research"); Patent Act of South Korea, *supra* note 44, art. 96(1)(i) (specifying that "the effect of a patent right does not extend to...working a patented invention for research or experimental purposes").

153. For the claim that this exemption extends to the act of experimentation, but not to the act of selling products that result from the experimentation, see Y. Hiraki, *Reality and Problems of Plant Protection Under Patent Law and Seed and Seedlings Law in Japan*, *in* Agricultural Biotechnology and Intellectual Property: Seeds of Change 268, 271 (J.P. Kesan ed., 2007). For a European view of the question, see, e.g., Joseph Straus, *Measures Necessary for the Balanced Co-Existence of Patents and Plant Breeders' Rights—A Predominantly European View*, WIPO-UPOV/SYM/02/7 (Oct. 23, 2002) (arguing for the application of the experimental use exemption under some circumstances).

clear whether the diversity is the consequence of post-TRIPS legal innovation. It may as easily be attributed to lingering pre-TRIPS sentiment, coupled with stare decisis or simple inertia. It may be the product of differing conceptions of institutional roles. For example, judges in the major patenting jurisdictions are operating within markedly different conceptions of the judicial role. Or, indeed, it may be the product of steadfast beliefs that current approaches are well-tailored normatively for local conditions.

Regardless, without any mandate—or even any nudge—from international instruments, the diversity of approaches discussed here seems likely to persist. As judicial and administrative authorities reiterate these rules over time, it seems likely to become increasingly difficult for future adjudicators to innovate, and innovation may have to come primarily from a local legislative process.

9 ENABLEMENT AND WRITTEN DESCRIPTION

Matthew Fisher*

Introduction

A patent is a somewhat curious legal creation. It provides exclusivity in an information source—an invention—that is, in economic terms, both free and public. The patent system creates a framework within which this information can be controlled, and this, in turn, enables limitations to be placed upon its supply and use. Information is itself a wayward thing: difficult to tame and coax into a proprietary regimen. Economically speaking, it is a non-wasting asset: one whose utility is not lessened by widespread use. Consequently, arguing for the artificial imposition of personal control once information has transitioned from the secret to the public realm requires some forceful justification. As Hermann Rentzsch once noted, it is meaningless for a person to argue theft of an idea (or by extension, any other informational good) as this is to complain "that something has been stolen which he still possesses, and he wants something back which, if given to him a thousand times, would add nothing to his possession."[1] The only element that will have been eroded in such a situation is the exclusivity of the originator.

The perils of making an invention public (a necessary consequence of marketing the vast majority of products that could be covered by patents) while failing to contain the inventive information and reduce it to one's possession are further highlighted when one considers other economic consequences that flow from its definition as an information

---

* Senior Lecturer and Vice Dean (Education), University College London.

1. Subcomm. on Patents, Trademarks & Copyrights of the S. Comm. on the Judiciary, 85th Cong., 2d Sess., An Economic Review of the Patent System 22 (Comm. Print 1958) (quoting Hermann Rentzsch, *Geistiges Eigenthum*, *in* Handwörterbuch der Volkswirtschaft 334 (1866)) [hereinafter Patent Subcomm. Report].

good. Classical theory would explain that once information has been created there are no further economic costs associated with its use (and, accordingly, its misuse), beyond those of communication and learning.[2] Therefore, returning to the subject matter of the patent: once implemented or otherwise disclosed, the invention can (in the absence of legal protection) be freely, and often relatively inexpensively, copied by others. It is the job of the patent to control this secondary use of information, to create scarcity where there would otherwise be none, and thus to protect the originator from the incursions (whether intentional or innocent) of others.

Structurally speaking, therefore, the patent grant facilitates the formation of property rights in the subject matter that it protects. As such, it cannot be justified on the basis that invention is inherently someone's property: without the patent system, property in an invention simply does not exist. The invention is only property *because* it is protected by a patent, because this imposes a structure of scarcity, and not vice versa.

The patent is, accordingly, not to be thought of lightly. The grant of exclusivity takes from the public a freedom that they would otherwise be at liberty to enjoy. If nothing more, it removes the possibility of exploring and commercializing the selfsame pasture—developing the same prospect[3] —that the patentee now asserts as his own.[4] A price must therefore be extracted: the public must be compensated for the inconvenience that the grant of exclusivity imposes upon them. Within a framework that has been developed on an international basis over many hundreds of years, the agreed toll has been standardized. Ignoring the financial, essentially administrative, costs that the patentee must meet in order to secure the grant—bureaucratic costs not fundamentally

2. See further the discussion of free and public goods given in CHRISTOPHER T. TAYLOR & Z.A. SILBERSTON, THE ECONOMIC IMPACT OF THE PATENT SYSTEM 24 (1973) (quoting Kenneth Arrow, *Welfare Economics and Inventive Activity*, *in* THE RATE AND DIRECTION OF INVENTIVE ACTIVITY (1962)).

3. Utilizing the nomenclature of Edmund Kitch, *The Nature and Function of the Patent System*, 20 J.L. & ECON. 265 (1977).

4. History tells us that this cost is more important than one might otherwise presume. There are numerous examples of inventions, discoveries, and breakthroughs being made by independent parties at or around the same time. Probably the most famous example of this phenomenon concerns the tale of the electric telephone, with both Alexander Graham Bell and Elisha Gray independently applying to patent the device on the same day. However, there are many other similar coincidences: Ogburn and Thomas provided what is possibly the first relatively comprehensive study into such "multiples" in 1922—see W.F. Ogburn & D. Thomas, *Are Inventions Inevitable? A Note on Social Evolution*, 37 POL. SCI. Q. 83 (1922)—in which they list 148 cases distilled from the fields of mathematics, astronomy, chemistry, physics medicine, biology, psychology, and practical mechanics.

associated with public inconvenience—the real price of the patent is disclosure. This means informing the world of one's invention and, critically, providing it with instructions that enable one skilled in the art to replicate the invention without undue burden.

The remainder of this chapter considers the birth and development of the specification (the chosen vessel for communicating the invention to the public) and the interpretation of the standards to which is it now held. It examines the genesis of the English patent custom, a system whose roots extend further than any other extant patent regime,[5] and that of the United States—a system grown from an English seed, but that rapidly evolved from its anglicized rootstock. Finally, it discusses the more recent developments in the law concerning enablement, sufficiency, and written description concentrating on UK and US jurisprudence in this area.

First, however, the context.

I. The Framework

In order to gain patent protection in the United Kingdom, United States, or indeed any of the other national or international patent systems that comply with the TRIPS framework,[6] an invention must satisfy certain criteria relating to patentability: it must be new, possess an inventive step (i.e., be non-obvious) and be capable of industrial application (i.e., have utility).[7] In addition, Article 29 of the TRIPS Agreement requires that:

> [A]n applicant for a patent shall disclose the invention in a manner sufficiently clear and complete for the invention to be carried out by a person skilled in the art and [Members] may require the applicant to indicate the best mode for carrying out the invention known to the inventor at the filing date or, where priority is claimed, at the priority date of the application.

---

5. Readers might note that at the time of the genesis of the English patent custom both the potato and tobacco were new "inventions" within the contemporary understanding of this term. It is important to note, however, that the English custom was not the first patent system: it is generally accepted that the city-state of Venice can lay claim to this accolade. *See* G. Mandich, *Venetian Patents*, 30 J. Pat. Off. Soc'y 166 (1948).

6. Agreement on Trade-Related Aspects of Intellectual Property Rights, Apr. 15, 1994, 33 I.L.M. 1197 [hereinafter TRIPS Agreement], which came into effect on January 1, 1995 following the Uruguay Round of negotiations under GATT. Ratification of TRIPS is a prerequisite to membership in the World Trade Organization (WTO). TRIPS provides certain minimum standards that a country's intellectual property protection must attain.

7. *See id.* art. 27.

The overlaying and interweaving of the patentability requirements creates a rich blanket in which to wrap worthy creations, protecting them from the incursions of others. Together these elements form the foundations of the patent system, their combination underpinning the genesis of the patent *right*: the notion that the applicant is *entitled* to protection from unauthorized competition provided the invention objectively satisfies the agreed criteria and fees are paid. This concept of fair and objective judgment allows separation from the "discretionary and quixotically granted monopolistic privileges"[8] sometimes awarded in the past, and forms an element of the social contract into which the patentee may be notionally considered to enter with the state upon filing. Thus, bureaucratic property vests in the patentee's application[9] on the basis of the offer of new and inventive teaching, with a full (although geographically and temporally limited)[10] property right being conveyed upon the invention successfully passing formal examination.

Scarcity is maintained, and property policed, by the textual boundaries formed by the patent's claims—words of the inventor's choosing that demarcate the extent of the patent's protective sphere. In the thirty-eight member states of the European Patent Convention (the EPC) this is mandated by Article 69 EPC, which explains that the extent of protection shall be determined by the claims

---

8. Jessica M. Silbey, *The Mythical Beginnings of Intellectual Property*, ESCHOLARSHIP 14 (2007), http://www.escholarship.org/uc/item/3gf7h2nz.pdf. The published version, Jessica M. Silbey, *The Mythical Beginnings of Intellectual Property*, 15 GEO. MASON L. REV. 319, 331 (2008), omits "and quixotically granted" from this sentence, which is a shame: some were. *See also* Oren Bracha, *The Commodification of Patents 1600–1836: How Patents Became Rights and Why We Should Care*, 38 LOY. L.A. L. REV. 117 (2004).

9. In the United Kingdom, by virtue of Patents Act, 1977, § 30(1) [hereinafter PA 1977]. The European Patent Convention (EPC) deals with this concept within the Convention on the Grant of European Patents arts. 71–74, Oct. 5, 1973, 1065 U.N.T.S. 254, as revised Nov. 29, 2000 [hereinafter EPC 2000]. 35 U.S.C. § 261 makes similar provisions for US patents.

10. The patent right is evidently limited geographically by the territoriality of the state granting protection. Multiple applications must be made in order to gain protection in multiple states. The temporal limit on the patent is imposed by statute. Accordingly, TRIPS art. 33 provides that the "term of protection available shall not end before the expiration of a period of twenty years counted from the filing date." In the United Kingdom, and all other states contracting to the European Patent Convention, the term of protection is understood as being a *maximum* of twenty years from filing—see, e.g., PA 1977, *supra* note 9, § 25; EPC 2000, *supra* note 9, art. 63. Patents may, in practice, expire sooner due to the patent holder's failure to pay the relevant renewal fee. There is, however, the possibility for a limited extension of this term for medicinal or plant protection products if supplementary protection certificates have been issued. *See* Regulation (EC) No. 469/2009 of the European Parliament and of the Council of 6 May 2009, 2009 O.J. (L 152) (concerning the supplementary protection certificate for medicinal products (codified version)); Regulation (EC) No. 1610/96 of the European Parliament and of the Council of 23 July 1996, 1996 O.J. (L 198) (concerning the creation of a supplementary protection certificate for plant protection products). The US position is dealt with under 35 U.S.C. § 154 (term) and §§ 155–56 (extension).

as interpreted in light of the description and drawings. Article 84 then adds that the "claims shall define the matter for which protection is sought." In the United States, a similar provision is found within the second paragraph of 35 U.S.C. § 112, which requires that the "specification shall conclude with one or more claims particularly pointing out and distinctly claiming the subject matter which the applicant regards as his invention."[11]

In terms of the protective nature of the patent, then, the claims are everything. It is only permissible to enclose subject matter within their embrace that is both patentable, in the sense of being new, inventive, etc.,[12] and that is supported by the teaching of the broader specification, supplemented, as necessary, by the common general knowledge of the person skilled in the art (the notional skilled addressee, or person having ordinary skill in the art (PHOSITA)).[13] Thus, whereas sufficiency is a fundamentally internal, structural element of the patent, both novelty and inventive step are restrictions imposed upon the claims by external forces. The integrity of the patent, and the extent of the claims, will necessarily be influenced by the combination of all the elements taken together.

Hence the majority of patents will be judged from two distinct points of view: internally on the basis of the sufficiency of disclosure, and from an external perspective, on the interwoven elements of patentability. Judgments on both will be made in light of the scope of exclusivity adopted by the patentee in her choice of claim language.

II. The Rationale and Its Historical Development

The rationale behind the requirements of sufficiency and enablement, that the patentee must place the public in possession of the teaching that forms the core of the invention, is evidently the consummation of the patentee's bargain with the state. The specification fulfills this requirement by clothing the invention in words of instruction that one skilled in the art may follow. There is evidently an economic, perfectly rational, and natural disincentive to make such a disclosure if it is not explicitly demanded, as describing the invention in a manner clear

---

11. As of November 16, 2011, when the Leahy Smith America Invents Act was signed into law, the provision is now entitled 35 U.S.C. § 112(b) and reads: "[t]he specification shall conclude with one or more claims particularly pointing out and distinctly claiming the subject matter which the inventor or a joint inventor regards as the invention."

12. The intricacies of the exclusions from patentability are, for the purposes of this article, conveniently forgotten.

13. See, for example, EPC 2000, *supra* note 9, arts. 83 and 84, which in turn are based on Patent Cooperation Treaty, June 19, 1970, as amended, 28 US.T. 7645, 1160 U.N.T.S. 23. Similar provisions are found within PA 1977, *supra* note 9, § 14, and 35 U.S.C. § 112.

enough and complete enough for it to be performed will also effectively teach third parties how to compete. Disclosure must therefore be extracted by making it a condition of grant. Accordingly, when the patent's exclusivity expires, and thereby the patentee's monopoly comes to an end, this teaching becomes a legacy for future generations of technologists to enjoy and exploit for themselves. Such is the accepted mantra, arrived at following many hundreds of years of development.

A. Early English Development

The underlying rationale of the system—that the patentee must enrich the state in return for the monopoly—has a very long pedigree indeed. One of the earliest examples of such considerations impacting upon the formulation of state policy toward invention can be found in the approach adopted by the English Crown in the mid-1500s. When the institution of a concerted policy of stimulating domestic industry by the award of letters patent took root in England during the reign of Elizabeth I, the Crown's primary aim was to make the technologically backward state self-sufficient.[14] Central to the success of this plan was the acquisition of superior technology and know-how from England's Continental neighbors—it being a far more reliable prospect to simply steal another state's technological secrets than to develop one's own from scratch. Accordingly, those areas that had previously "figured most prominently on the list of imports—viz. alum, glass, soap, oils, salt, saltpetre, latten, etc"[15]—were targeted for "development" in this fashion. The Crown set about, through Elizabeth's First Minister, William Cecil (Lord Burghley),[16] to incite the importation of foreign teaching and expertise by the offer of monopoly power.

The instruction of a native workforce was considered critical to the success of such a plan, and the idea that consideration must move from the patentee to the state was also clearly ingrained within the nascent system. Accordingly, distinct expectations were placed on the patentee, which, although the specifics varied from grant to grant, were "nonetheless consonant with the basic premise of developing new trade and industry within the Realm."[17] Nowadays the price extracted

---

14. *See* L. Getz, *History of the Patentee's Obligations in Great Britain*, 46 J. PAT. OFF. SOC'Y 62, 69–71 (1964); Edward C. Walterscheid, *The Early Evolution of the United States Patent Law: Antecedents (Part 2)*, 76 J. PAT. & TRADEMARK OFF. SOC'Y 849, 855 (1994).

15. E.W. Hulme, *The History of the Patent System under the Prerogative and at Common Law*, 12 L.Q. REV. 141, 152 (1896); *see also* CHRISTINE MACLEOD, INVENTING THE INDUSTRIAL REVOLUTION: THE ENGLISH PATENT SYSTEM 1660–1800, at 12 (1988).

16. *See* MACLEOD, *supra* note 15, at 11; Walterscheid, *supra* note 14, at 855.

17. Walterscheid, *supra* note 14, at 856–57.

by the state for the provision of the monopoly is evidently the requirement for written instruction to teach the populace of the invention and thereby to demonstrate its merit. Elizabethan practice, by contrast, was generally far more direct.

1. "Working" and "Revocation" Clauses—A Doctrine of Introduction

Clauses were often worked into grants, especially those of foreign artisans, to ensure the enrichment of native industry following the grant of a patent monopoly.[18] The most basic were those that required the patentee to "work" the new art, trade, or industry within the kingdom.[19] Strict time limits were often imposed, with the grant being void for lack of consideration if they were not adhered to.[20]

As time went on and the policy became more established, the "working" clause was gradually phased out in favor of a general revocation clause. This allowed the Crown to revoke grants on grounds of their being "generally inconvenient," a simple but all-encompassing concept easily applicable to a failure to introduce the patented industry within the specified time period.[21] The power arose from the fact that the grant of a patent was, at this time, inherently a matter of royal grace—and the hand that gave could also take away. Despite the potential breadth of the clause, it was mostly exercised in cases of nonuse, cases where the grant was made on a false suggestion of novelty, and cases where the true inventor was discovered to be someone other than the patentee.[22] The clause was included in patents granted for the next two hundred years.

---

18. See discussion in E.W. Hulme, *On the Consideration of the Patent Grant, Past and Present*, 13 L.Q. REV. 313, 314 (1897). It was a well-known axiom of English law, relating to all forms of royal grant, that the "King must not be deceived in his grant," the penalty for contravention of this rule being annulment. For a number of authorities that state this rule, see D.S. Davies, *The Early History of the Patent Specification*, 50 L.Q. REV. 86, 100 (1934). Given the whole point of the patent grant was to see the institution of the relevant manufacture within England, it was taken as given that failure to work the invention within the realm would lead to annulment. However, as Walterscheid notes, foreigners were "apt to plead ignorance of the English common law!" Walterscheid, *supra* note 14, at 857. It was therefore considered expedient to explicitly work such a clause into the grant.

19. As opposed to merely introducing it.

20. Other clauses with much the same end took the form of requirements obligating the patentee to employ and train native artisans to practice the trade covered by the grant. *See* Walterscheid, *supra* note 14, at 857, explaining that this type of requirement was usually applied only to foreign patentees in order to safeguard the continuance of the industry should they abscond before the term of their patent expired.

21. Davies, *supra* note 18, at 102; *see also* Walterscheid, *supra* note 14, at 857.

22. Davies, *supra* note 18, at 103–04.

2. From "Introduction" to "Explanation" as the Price of Monopoly

The transition from introduction to explanation (or, more accurately, a written disclosure) as the price of monopoly occurred at some point during the eighteenth century in England. Despite a number of potential precursors,[23] it is generally accepted that the first enrollment of a true specification of an invention covered by a patent was performed by John Nasmith in 1711.[24]

Nasmith's precise motivation in providing a written description of his invention is the subject of conjecture. Some commentators have suggested that the most likely explanation was that he sought to make his grant more secure.[25] The fear of inventive theft was well grounded, and the path toward the specification had already been tentatively explored in the courts[26] and elsewhere in society.[27] However, others suggest that because its first mention appears in the report of the law officer dealing with Nasmith's petition, it may be deduced that the initiative came from the Crown; the officer required a better disclosure before he would issue a favorable report.[28] Nevertheless, it would

---

23. The earliest being an agreement between the Crown and Gerard Honricke, a German sea captain, in which he promised that in return for £300 he would "teache perfectlie by demonstraĉon and practice with the hand without all manner of decipte and falsehood the said arte of makinge salte peter in the moste perfecte sorte." This he did, and it is this document, reproduced in the State Papers for 1561 under the title of "The trew and perfecte arte of the making of Saltpeter to grow in Cellars, Barns, or in Lyme or Stone quarrees," CALENDAR OF STATE PAPERS, DOMESTIC SERIES 1547–1580, QUEEN ELIZABETH, VOL. 16, at 29–31 (1856), that Hulme contends is the first example of a specification. *See* Hulme, *supra* note 15, at 145; E.W. Hulme, *The Early History of the English Patent System*, *in* 3 COMM. ASS'N OF AM. LAW SCHS., SELECT ESSAYS IN ANGLO-AMERICAN LEGAL HISTORY, 117, 142 (1909). However, as Davies notes, Honricke's claim to the first specification is dealt a fatal blow by the fact that he was never granted a patent; indeed he never asked for one—he just offered to sell his knowledge. *See* Davies, *supra* note 18, at 263–64.

24. In disputes akin to Garill's case in the Privy Council in 1664. *See* H. DUTTON, THE PATENT SYSTEM AND INVENTIVE ACTIVITY DURING THE INDUSTRIAL REVOLUTION 1750–1852, at 75 (1984); A.A. GOMME, PATENTS OF INVENTION: ORIGIN AND GROWTH OF THE PATENT SYSTEM IN BRITAIN 27 (1946); MACLEOD, *supra* note 15, at 49; J.N. Adams & G. Averley, *The Patent Specification—The Role of* Liardet v Johnson, 7 J. LEGAL HIST. 156, 158 (1986); Davies, *supra* note 18, at 87; Edward C. Walterscheid, *The Early Evolution of the United States Patent Law: Antecedents (Part 3)*, 77 J. PAT. & TRADEMARK OFF. SOC'Y 771, 782 (1995).

25. *See* Hulme, *supra* note 18, at 317.

26. *See, e.g.*, *Garill's Case*, a dispute heard in the Privy Council in 1664. For a full account of Garill's dispute, see Davies, *supra* note 18, at 274.

27. In publications such as Sturtevant's *Treatise of Metallica*: S. STURTEVANT, METALLICA: OR THE TREATISE OF METALLICA (1612), *in* LETTERS PATENT AND SPECIFICATIONS OF LETTERS PATENT FOR INVENTIONS 140 (B. Woodcroft ed., 1858).

28. *See* GOMME, *supra* note 24, at 33; Walterscheid, *supra* note 24, at 788. The role played by the law officers in the grant at this time was significant. *See* E.W. Hulme, *The History of*

be incorrect to suggest that with Nasmith's patent came a modern approach to assessing the sufficiency of the disclosure. It did not. Indeed for the remainder of the eighteenth century most of the specifications submitted for enrollment were hopelessly vague by any, let alone modern, standards.[29] Accordingly, at this point in time, and for many years after "it is doubtful whether patentees had any clear idea what the function of a specification was or how full and accurate it ought to be."[30]

However, even though the quality of the disclosures contained in these early specifications was very much in doubt, the practice soon took root, becoming customary in about 1734.[31] Four decades later, at least by the time of the decision in *Liardet v. Johnson*[32] in 1778, the courts had clearly acknowledged the death of the doctrine of introduction, requiring instead the enrollment of a specification as the price of the patentee's monopoly.

Liardet v. Johnson is a case described by some as a "landmark in the history of English patent law"[33] and dismissed by others as insignificant.[34] Whichever view is held, it is certainly noteworthy as "one of the earliest statements by an English

---

the Patent System under the Prerogative and at Common Law—A Sequel, 16 L.Q. Rev. 44, 53 (1900). Their influence in deciding patent grant policy began around the beginning of the seventeenth century. MacLeod, *supra* note 15, at 48, suggests that in the century after 1660 all modifications in the patent system were made by the law officers in the course of reporting on inventors' petitions. Walterscheid, *supra* note 24, at 779, states that by the seventeenth century the participation of the law officers in deciding patent policy was becoming standard practice.

29. *See* Adams & Averley, *supra* note 24, at 161; *see also* MacLeod, *supra* note 15, at 49 (stating that the specification, at this time, could be as informative or evasive as the patentee saw fit).

30. MacLeod, *supra* note 15, at 50.

31. E.W. Hulme, *On the History of the Patent Law in the Seventeenth and Eighteenth Centuries*, 18 L.Q. Rev. 280, 283 (1902), states that the first requirement for a specification can be found in a patent of 1716, but that the practice was not uniform until about 1740. Davies, *supra* note 18, at 89, states that the practice was made customary in 1734 but that there are examples of the requirement being made in 1712 (three times in 1716, twice in 1717, twice in 1718) and that between 1720 and 1733 a further fifteen specifications were required. Gomme, *supra* note 24, at 34, states that between 1711 and 1734 (when the practice became customary) 29 of 158 grants had a specification enrolled.

32. Reports of the first trial were published in the *Morning Advertiser* and the *Daily Post* on February 23, 1778, and in the *London Chronicle* and the *Daily Advertiser* the following day. The second trial is reported in the *Morning Post and Daily Advertiser*, *The Gazetteer*, and the *New Daily Advertiser* on July 20, 1778. *See* Adams & Averley, *supra* note 24, at 174. The case is reported at Liardet v. Johnson, (1778) 1 Carp. P.C. 35 (K.B.).

33. Hulme, *supra* note 18, at 317.

34. Adams & Averley, *supra* note 24.

judge of the modern requirement that a specification must be enabling."[35] Lord Mansfield, charging the jury, famously explained that:

> [T]he condition of giving encouragement is this: that you must specify upon record your invention in such a way as shall teach an artist, when your term is out, to make it—and to make it as well by your directions: for then at the end of the term, *the public shall have benefit of it.* The inventor has the benefit during the term, and the public have the benefit after... [Where the invention is a composition] the specification must state... the proportions; so that any other artist may be able to make it, and it must be a lesson and direction to him by which to make it. If the invention be of any other sort, to be done by mechanism, they must describe it in a way that an artist must be able to do it.[36]

Irrespective, therefore, of whether the case was the first in which the importance of the disclosure was stressed—and as Walterscheid notes, there is at least one earlier case in which Mansfield discussed the adequacy of the specification[37] —*Liardet v. Johnson* is, at least, evidence of an important step having been taken in the development of patent law. It stresses the importance of consideration moving to the public and firmly reflects the grant's status as a contract with the state in which a temporary monopoly is exchanged for benefit accruing from the inventor's knowledge entering the public domain.

By the end of the eighteenth century, *Liardet v. Johnson* was settled law, and the patent could finally be said to have started its separation from grants of Crown favor. It had entered the market economy as an item of commerce, and the price demanded for its existence was disclosure.[38] By the dawn of the 1800s, it had become settled law that this meant the patent should teach the operation of the invention without further experimentation,[39] defective teaching being grounds for avoiding the grant.[40] Working of the invention was no longer enough.

---

35. Walterscheid, *supra* note 24, at 797. But see Adams & Averley, *supra* note 24, at 171, who suggest that the novelty of the case lies in its reliance on the testimony of expert witnesses.

36. Hulme, *supra* note 31, at 285 (emphasis added).

37. Walterscheid, *supra* note 24, at 797 (citing E.W. Hulme, *Privy Council Law and Practice of Letters Patent for Invention from the Restoration to 1794 (continued)*, 33 L.Q. Rev. 180, 192 (1917) for the proposition that Mansfield had previously held a Mr. Brand's patent invalid in 1771 owing to the fact that the patentee had omitted certain material information from his specification).

38. J. Buller was therefore able to definitively state in 1795 that the "specification is the price that the patentee is to pay for the monopoly." Boulton v. Bull, (1795) 126 Eng. Rep. 651, 654 (K.B.).

39. *See, e.g.*, Turner v. Winter, (1787) 99 Eng. Rep. 1274, 1276 (K.B.).

40. It should be noted that a specification could equally be defective if it included too much as if it disclosed too little. This was especially the case if the superfluous material was thought to

3. The Specification as a Mark of Evolution

Uncertainties concerning the scope and substance of patent rights in England had been evident since before the passage of the Statute of Monopolies in 1624, and were to continue for many decades after the advances marked by Nasmith's patent. They were the inevitable result of an evolution within both the patent sphere and society. When the system was conceived during Elizabeth I's reign, all grants were semi-contractual agreements between the patentee and the monarch whereby protection was offered in return for the introduction of new manufacture. The system was relatively parochial: certain industries were targeted by the Crown for development, and the link between petition, grant, and enforcement of the patent, and the working of the manufacture within the realm, was eminently clear. The overall small number of patents, minute by today's standards, and the clear connection that the patent maintained with the exercise of royal favor, meant that it was possible to keep an eye on individual patentees to ensure that they continued to uphold their end of the bargain.

By the mid-eighteenth century, however, the patenting process had evolved into something altogether different. Whereas "patents originally represented royal privileges issued under the royal prerogative to achieve royal policy goals,"[41] numerous legal advances, beginning with the decision of *Darcy v. Allein* in 1602[42] and following on through the Statute of Monopolies 1624,[43] had made significant steps toward curtailing the monarch's power in this area. The Civil Wars (1642–1651) had also obviously impacted upon the Crown's freedom to dispense favors under the exercise of the prerogative. The combined effect of these developments necessitated a shift away from patents being seen as royal privileges and toward their being viewed in the context of the common law and legal rights. The separation from the Crown was further enhanced by the increasing formalization of the disclosure requirement.

Accordingly, patentees in the post-Nasmith period began to take a far more active role in the definition of their own scope of protection because they were the ones in charge of the drafting of the specification. Suddenly they were the authors of their own fortunes, and this, in turn, allowed focus to be switched from the issuing body to the petitioners themselves. The specification thus assisted in the disengagement

---

be included for the purpose of misleading the public. *See* Walterscheid, *supra* note 24, at 802, where he discusses R v. Arkwright, (1785) 1 Web. P.C. 64 (C.P.).

41. A. Mossoff, *Rethinking the Development of Patents: An Intellectual History, 1550–1800*, 52 Hastings L.J. 1255, 1274 (2001).

42. Darcy v. Allein, (1603) 77 Eng. Rep. 1260 (K.B.). For a discussion of the case and its legacy, see M. Fisher, *The Case That Launched a Thousand Writs, or All That Is Dross? Re-conceiving* Darcy v Allen: *The Case of Monopolies*, 2010 Intell. Prop. Q. 356 (2010).

43. Statute of Monopolies, 1624, 21 Jac. 1, c. 3 (Eng.).

of the patent from its prerogative-steeped roots by grafting it onto a bureaucratic, proprietorial system the like of which we are all familiar. The increasing distance that this placed between the Crown and the grant is significant, as it enabled opponents of the system to begin to voice their complaints, and competitors to challenge the grants, without this being viewed as criticism of the sovereign per se. This, in turn, eventually led to wholesale reform of the UK patent system following a bitter struggle against a strong abolitionist movement in the mid-to-late–nineteenth century.[44]

Central to the anti-abolitionists' cause was the idea of the patent as a contract between the inventor and the state. Three of the main treatises extant at the beginning of the crisis, those of Carpmael,[45] Hindmarch,[46] and Spence,[47] went to lengths to emphasize the nature of this bargain. Carpmael, for example, states that the possibility of gaining a patent was:

> [A] great incentive to the exertion of ingenuity; as the... patentee[s] found themselves rewarded for their labour... and the public were ultimately benefited by being made acquainted with the means of producing the invention, which became public property at the expiration of the term of the grant, or earlier.[48]

As Coulter explains, this view of the patent as a bargain accorded with both the common law and the "exchange" arguments of the classical economists. "Any restrictions that the patent placed upon use of the new manufacture... were temporary ones acceded to by the public in return for the information contained in the written specification."[49]

Hindmarch reiterated this point, rationalizing that the only way in which the patentee could have exclusive property in his invention, once it was made public, was by the application of some positive law made with the actual or implied

---

44. *See* M. Fisher, *Classical Economics and Philosophy of the Patent System*, 2005 INTELL. PROP. Q. 1 (2005).

45. WILLIAM CARPMAEL, THE LAW OF PATENTS FOR INVENTIONS, FAMILIARLY EXPLAINED, FOR THE USE OF INVENTORS AND PATENTEES (1832).

46. W.M. HINDMARCH, TREATISE ON THE LAW RELATING TO PATENT PRIVILEGES FOR THE SOLE USE OF INVENTIONS (1846).

47. W. SPENCE, A TREATISE ON THE PRINCIPLES RELATING TO THE SPECIFICATION OF A PATENT FOR INVENTION (1847).

48. CARPMAEL, *supra* note 45, at 3 (quoting M. COULTER, PROPERTY IN IDEAS: THE PATENT QUESTION IN MID-VICTORIAN BRITAIN 78 (1991)).

49. COULTER, *supra* note 48, at 78. For a more modern appreciation of the information function patents, see F.K. Beier & J. Straus, *The Patent System and Its Information Function—Yesterday and Today*, 8 INT'L REV. INTELL. PROP. & COMPETITION L. 387 (1977); *see also Note, The Disclosure Function of the Patent System (or Lack Thereof)*, 118 HARV. L. REV. 2007 (2005).

consent of the whole community.[50] Such consent was deemed to exist because of the benefits that would accrue to society from the publication of the invention—information that, absent the patent system, might have remained secret.[51] In essence, therefore, the only relevant question to be asked of a particular specification to see if it was worthy was whether it was "such that a mechanist can make the machine from the description there given."[52]

In terms of justifications, this "exchange theory" enabled the pro-patent lobby to sidestep any difficult questions concerning the nature of invention and the inventive process and to claim they were supporting a system that encouraged the dissemination of knowledge. The patent was no longer a privilege meted out by a benevolent/capricious[53] monarch, it was a simple bargain whereby the inventor agreed to tell the world of her invention in return for a temporary monopoly.[54] As John Farey stated before the 1829 Select Committee charged with investigating the patent system: "The first applicant who is able, and willing, to make disclosure of the secret, ought to have the patent, that is to be given as the price of such disclosure."[55]

B. The Early Position in the United States

The development of patent law in the United States tells a similar story. Although early custom was heavily based on English practice, the constitutional authorization provided by Article 1, § 8, cl. 8 ("to promote the Progress of Science and useful Arts, by securing for limited Times to Authors and Inventors the Exclusive Right to their respective Writings and Discoveries"), and the creation of a nationwide patent system in the early days of the first Congress[56] soon took the system in its own direction.

---

50. HINDMARCH, *supra* note 46, was at pains to point out that patents, regardless of their social utility, were not a natural right of the inventor; rather they were a "matter of favour."

51. *See* HINDMARCH, *supra* note 46, at 1; *see also* COULTER, *supra* note 48, at 79.

52. See the arguments of Romily and Scott (for the plaintiff) in Harmer v. Plane, (1807) 14 Ves. 130, 131, who attribute the quote to Lord Eldon, "when Lord Chief Justice," in Cartwright v. Eamer, (1800) (unreported).

53. Depending upon which side of the fence one happened to be standing in respect of any given grant.

54. *See* PATENT SUBCOMM. REPORT, *supra* note 1, at 24; COULTER, *supra* note 48, at 94–95; DUTTON, *supra* note 24, at 22; Fritz Machlup & E.T. Penrose, *The Patent Controversy in the Nineteenth Century*, 10 J. ECON. HIST. 1, 26 (1950).

55. John Farey in his evidence before the 1829 Select Committee: REPORT OF THE SELECT COMMITTEE ON THE LAW RELATIVE TO PATENTS FOR INVENTIONS, 1829, H.C. 332, at 22.

56. Patent Act of 1790, ch. 7, 1 Stat. 109 (1790).

From the outset, the US legislation emphasized the need for a written specification detailing the invention. Section 2 of the Patents Act 1790 required a

> Specification in writing, containing a description ... of the thing or things, by him or them invented or discovered, and described as aforesaid, in the said patents; which specification shall be so particular ... as not only to distinguish the invention or discovery from other things before known and used, but also to enable a workman or other person skilled in the art or manufacture, whereof it is a branch, or wherewith it may be nearest connected, to make, construct, or use the same, to the end that the public may have the full benefit thereof, after the expiration of the patent term.[57]

Substantially identical requirements were subsequently made in both the 1793 Act and that of 1836. However, throughout the duration of these three enactments the specification was not only the source of the teaching; it also performed a public notice function warning third parties of the extent of the patent's protective influence. As Justice Story explained in *Evans v. Eaton*,[58]

> [The] specification ... has two objects: one is to make known the manner of constructing the machine ... so as to enable artizans to make and use it.... The other object of the specification is, to put the public in possession of what the party claims as his own invention, so as to ascertain if he claims anything that is in common use, or is already known, and guard against prejudice or injury from the use of an invention which the party may otherwise innocently suppose not to be patented....[59]

At this point in time there was not yet any requirement for the patentee to include separate claims within their application; these were only introduced with the Patents Act 1836.[60] Thus, when Justice Story was examining the issue in

57. *Id.*

58. The case reached the Supreme Court in 1818, Evans v. Eaton, 16 U.S. 454 (1818); it was the first time that a substantive patent issue had done so. The Court remanded the issue for a new trial, and it returned to the Supreme Court in 1822, 20 U.S. 356 (1822). The practice was formally established in this latter case.

59. *Evans*, 20 U.S. at 433.

60. Readers will note that it was to take another forty-seven years for a similar requirement to be made in the United Kingdom. See the Patents, Designs, and Trade Marks Act, 1883, 46 & 47 Vict., c. 57, §5 (5), which required for the first time that "a specification, whether provisional or complete, must commence with the title, and in the case of a complete specification must end with a distinct summary of the invention claimed."

Evans, the specification inevitably fulfilled this definitional function—short of directly examining the physical embodiment of the invention itself,[61] there was nothing else that could do so. Furthermore, even when the 1836 Act made claims a requirement,[62] they initially did little more than serve as guardians of novelty. Indeed it was not until almost thirty-five years later that they came to occupy their current position and to determine the patents' extent of protection. Accordingly, until the passage of the Patents Act 1870,[63] infringement was determined by reference to the description in the specification and the drawings alone—the patent being deemed to cover all forms of invention that embraced the principle or mode of operation disclosed in the patent documentation and that gave the same effect.[64] The claims were simply signposts utilized to indicate the invention's prominent features and to point out what was novel about the patentee's creation.[65] With no claims to police the boundaries of protection and constrain the invention, the concept of sufficiency encompassed what would nowadays be considered questions of patentability simpliciter—including whether the patentee had succeeded in distinguishing the old from the new.[66]

Nevertheless, throughout the life of the 1836 Act a slow but inexorable shift in the importance attributed to claim language became evident in the patent jurisprudence. Courts and patentees sought more certainty in the scope of the patent's reach, and claim language became the focus of much debate.[67] The final

---

61. A less than ideal situation given the invention's technical contribution would inevitably be broader than the precise embodiment created.

62. Patent Act of 1836, ch. 357, § 6, 5 Stat. 117 (1836), required that the patentee "particularly specify and point out the part, improvement, or combination, which he claims as his own invention or discovery." *See* R.D. Hantman, *Doctrine of Equivalents*, 70 J. PAT. & TRADEMARK OFF. SOC'Y 511, 517 (1988); *see also* T. TAKENAKA, INTERPRETING PATENT CLAIMS IN THE UNITED STATES, GERMANY AND JAPAN 6 (1995); W. Jessup, *The Doctrine of Equivalents*, 54 J. PAT. OFF. SOC'Y 248 (1972).

63. Patent Act of 1870, ch. 230, 16 Stat. 198 (1870).

64. *See, e.g.*, the comments of Justice Curtis, delivering the opinion of the court, in *Winans v. Denmead*, 56 U.S. 330, 342 (1853), when he stated: "It is generally true, when a patentee describes a machine, and then claims it as described, that he is understood to intend to claim, and does by law actually cover, not only the precise forms he has described, but all other forms which embody his invention."

65. K.B. Lutz, *Evolution of the Claims of U.S. Patents*, 20 J. PAT. OFF. SOC'Y 134, 147 (1938), explains that during the period from 1836 to 1870 the claims "rarely, if ever, received consideration on the question of infringement."

66. See further discussion in Lutz, *supra* note 65, at 147–56.

67. *See* F. Prager, *Trends and Developments in American Patent Law from Jefferson to Clifford (1790–1860)*, 6 AM. J. LEGAL HIST. 47 (1962) for a discussion of this trend. *See also* F. Prager, *The Changing Views of Justice Story on the Construction of Patents*, 4 AM. J. LEGAL HIST. 14 (1960).

act, which clearly signaled the claims' ascendance to their current status as the guardians of the patent's boundaries, occurred with a subtle change in wording introduced under the 1870 Statute, section 26 of which required that the inventor "shall particularly point out *and distinctly claim* the part, improvement, or combination which he claims as his invention or discovery."[68]

Subsequent Supreme Court decisions highlighted the importance of the claims as a separate and distinct part of the patent and emphasized the value of clear and concise language in order that the public interest not be impeded.[69] The 1870 Act also maintained the requirement that the patentee:

> File in the patent office a written description of the [invention]..., and of the manner and process of making, constructing, compounding, and using it, in such full, clear, concise, and exact terms as to enable any person skilled in the art or science to which it appertains, or with which it is most nearly connected, to make, construct, compound, and use the same; and in case of a machine, he shall explain the principle thereof, and the best mode in which he has contemplated applying that principle so as to distinguish it from other inventions.[70]

Although it may be noted that the language used to form the foundation of the requirement for an enabling disclosure had changed somewhat from that contained in the Act of 1790,[71] the essential principles were maintained. The specification had to be sufficient to enable a person skilled in the art to perform the invention; enablement was still the price of the patent. This said, with the claims now defining the invention for which protection was sought and determining the extent of the resultant patent's influence, it might have been thought that the need for the specification to lay claim to, and demonstrate possession of, the inventive concept should either be tied to enablement, or transferred to the claims. This is not, however, what happened.

---

68. Patent Act of 1870, ch. 230, 16 Stat. 198 (1870) (emphasis added). Section 6 of the 1836 Act simply required the patentee to "particularly specify and point out the part, improvement, or combination, which he claims as his own invention or discovery."

69. *See, e.g.*, Merrill v. Yeomans, 94 U.S. 568, 570 (1876), in which Justice Miller, giving the opinion of the Court, states that the language of the claim in the instant cases is "far from possessing that precision and clearness of statement with which one who proposes to secure a monopoly at the expense of the public ought to describe the thing which no one but himself can use or enjoy."

70. Patent Act of 1870, ch. 230, § 26, 16 Stat. 198 (1870).

71. Reproduced in text accompanying note 57, *supra*.

Notwithstanding the changes wrought by the ascendance of the claims as the definitional medium for the patent's exclusive territory, the idea that the specification contributes to determination of the patentee's possession of the invention has been maintained even under a system of peripheral claiming.[72] Notably, the Court of Appeals for the Federal Circuit (CAFC) declared in *Vas-Cath v. Mahurkur*,[73] that:

> The purpose of the "written description" requirement is broader than to merely explain how to "make and use"; the applicant must also convey with reasonable clarity to those skilled in the art that, as of the filing date sought, he or she was in possession of the invention. The invention is, for purposes of the "written description" inquiry, whatever is now claimed.[74]

As Lefstin notes, this might have made sense in the early stages of US patent law, but once concepts of "invention" and "claim" became synonymous, the logic of possessing the invention under the description should have evaporated.[75] That it did not is a point to which we return in due course.

III. The Modern Law of Sufficiency and Support in the United Kingdom

A. The Statutory Framework

Pre-grant issues of sufficiency and support are currently dealt with under Section 14 of the Patents Act 1977 (PA 1977). Accordingly, Section 14(3) requires that "[t]he specification of an application shall disclose the invention in a manner which is clear enough and complete enough for the invention to be performed by a person skilled in the art." Section 14(5)(c) then continues, requiring that the claim or claims shall be "supported by the description." In other words, an application should be rejected if the patentee attempts to claim more than she discloses in her specification. This evidently concurs with the notion that disclosure is the price the patentee pays for her patent. It would make a mockery of the pact between the inventor and the state if the former was able to gain broader

72. That is, a system where claims form the outer boundary of protection, as opposed to central claiming—as under the 1836 Act, where the claims simply point out what is new and inventive.

73. 935 F.2d 1555 (Fed. Cir. 1991).

74. *Id.* at 1563–64.

75. *See* J.A. Lefstin, *The Formal Structure of Patent Law and the Limits of Enablement*, 23 Berkeley Tech. L.J. 1141, 1197–98 (2008).

protection from the latter than the area that they would bequeath upon expiry of their monopoly.

Any analysis of the sections of the PA 1977 that relate to sufficiency and support must take into account the provisions of the European Patent Convention (EPC) upon which these sections are based. Like so much else of the Act, these sections are "so framed as to have, as nearly as practicable, the same effects in the United Kingdom as the corresponding provisions of the European Patent Convention."[76] The precise benefit of translating one overarching statutory provision into another, and then proclaiming that it has the same effect as the language that the draftsman eschewed in favor of his own, is somewhat of a mystery—a point often made by Lord Justice Jacob when sitting in the English Court of Appeal.[77] Nevertheless, Section 14(3) finds reflection in Article 83 EPC and Section 14(5) in Article 84.

Section 14(3) of PA 1977 is further reflected in the grounds for revocation of a patent found in Section 72—itself based on Article 138 EPC. Accordingly, Section 72(1)(c) (equivalent to Article 138(1)(c) EPC) refers to a patent being invalid if the "specification... does not disclose the invention clearly enough and completely enough for it to be performed by a person skilled in the art." Thus, the grant can be revoked if the patentee has not provided the consideration required as her side of the social bargain that the patent represents. The issue of over-broad claiming, or lack of support, found within Section 14(5) has no clear analog within the grounds of invalidity. This lies in stark contrast to the previous, pre-EPC, position wherein Section 32(1)(i) of the Patents Act 1949 (PA 1949) allowed objection on the basis that any claim was not fairly based on the matter disclosed in the complete specification. This was generally agreed to mean that the patentee was "not entitled to claim a monopoly more extensive than is necessary to protect that which he has himself (in his specification) said is his invention."[78] Blanco-White noted that as well as encompassing claims that were wider than what was new and inventive, this also extended to "any case in which... it appears that the inventor has not given adequate consideration for the grant."[79] We shall return to this subject in due course. However, for the present our concentration must rest on insufficiency proper.

76. *See* PA 1977, *supra* note 9, § 130(7).

77. Shütz (UK) Ltd. v. Werit (UK) Ltd., [2011] EWCA (Civ) 303, [39] (Jacob, L.J.) ("To this day I remain baffled, nay flabbergasted, by this convoluted and roundabout way of implementing the relevant provisions of the treaties.").

78. T.A. Blanco-White, Patents for Inventions 4-801 (5th ed. 1983).

79. *Id.*

B. Insufficiency Proper

Insufficiency proper, that is, the requirement that the disclosure in the specification be adequate to enable the invention as claimed to be performed, now found within PA 1977 Sections 14(3) and 72(1)(c), was also a ground of objection to the validity of a patent under the old UK legislation—that is, PA 1949 § 31(1)(h)—and had been explicitly so since the Patent and Designs Act 1932.[80] Prior to this point, the selfsame considerations that Lord Mansfield had laid down in *Liardet v Johnson* were applied directly by force of law; a failure to adequately disclose the invention could therefore result in the grant being revoked at common law.[81] The rule for insufficiency first adopted under a UK statutory framework is, accordingly, warmly familiar. In *No-Fume v. Pitchford*, Romer LJ explained that:

> Specifications very frequently contain mistakes; they also have omissions. But if a man skilled in the art can easily rectify the mistakes and can readily supply the omissions, the patent will not be held invalid. The test to be applied... [is] this: Can he rectify the mistakes and supply the omissions without the exercise of any inventive faculty? If he can then the description of the specification is sufficient.[82]

This approach was to continue practically unchanged under PA 1949—as Blanco-White explained: "To be proper and sufficient, the complete specification as a whole... must in the first place contain such instructions as will enable all those to whom the specification is addressed to produce something within each claim 'by following the directions of the specification, without any new inventions or additions of their own' and without 'prolonged study of matters which present some initial difficulty.'"[83] Disclosure of the best method of performing the invention was also required under PA 1949, although, as with invalidity for lack of support, this did not make it into the post-EPC statute.

The leading decision in the United Kingdom on sufficiency of disclosure is that of the Court of Appeal in *Mentor v. Hollister*,[84] decided in 1992. This was the first decision of a higher court under the 1977 Act in which it was specifically required to consider the degree of clarity and completeness required of a patent

---

80. *See* Patents and Designs Act, 1932, 22 & 23 Geo. 5, c. 32, § 25(2)(h) (Eng.) ("That the complete specification does not sufficiently and fairly describe and ascertain the nature of the invention, and the manner in which the invention is to be performed.").

81. *See, e.g.*, R v. Arkwright, (1785) 1 Web. P.C. 64 (C.P.) 66.

82. No-Fume v. Pitchford, (1935) 52 R.P.C. 231 (C.A.) 243.

83. Blanco-White, *supra* note 78, at 4-502 (citations omitted).

84. [1993] R.P.C 7.

specification. The facts are relatively simple. The patent in question related to a male urinary incontinence device—a one-piece apparatus designed to be rolled over and adhered to the penis and which then drained into a collection bag. The advantage that this had over the prior art was that it was simpler to apply and remove; the prior teaching all pointed toward two-part devices employing separate adhesive tape. Mentor had originally brought an action against Hollister for patent infringement. At first instance, the defendants denied infringement and counterclaimed that the grant was invalid on grounds on obviousness and insufficiency. The judge, Aldous J, held the patent valid and infringed,[85] and the defendants appealed.

Before the Court of Appeal, the question of sufficiency was the only live issue—the defendants had conceded that the relevant claim was inventive and infringed. Lord Justice Lloyd, giving the judgment of the Court, explained that the requirement that "the specification should sufficiently and fairly describe the invention and the method by which it [could be]...performed [was]...the price which the inventor pays for his twenty year monopoly."[86] The Court noted that this had "always been a requirement of our patent law;" adding that disclosure of the best method of performance was no longer demanded of the patentee, this having been abandoned with the passage of PA 1977.

Concentrating solely on the question of insufficiency proper, the Court cautioned against adopting too strict a requirement of disclosure: "Disclosure of an invention does not have to be complete in every detail, so that anyone, whether skilled or not, can perform it."[87] Lord Justice Lloyd explained that the patent's teaching was addressed to the person skilled in the art, who inevitably approached the question of sufficiency armed with her own understanding based on her common general knowledge. The patent does not, he explained, therefore have to disclose what would be "self-evident" to this person. Nevertheless, then comes the difficulty—where do the boundaries lie between those things that the addressee is expected to do for herself and those things that she must be told? Place the bar too low and the requirement of sufficient disclosure would be robbed of all its significance and the specification turned into an enigma for the addressee to decrypt and solve. Place it too high, however, and the burden placed upon the patentee to minutely explain and instruct others in the performance of the invention becomes too burdensome and potentially renders patents invalid where only minor, easily surmountable, errors or omissions remain.

85. Mentor v. Hollister, [1991] F.S.R. 557.

86. Mentor v. Hollister, [1993] R.P.C. 7, 9.

87. *Id.* at 10.

The Court cautioned itself over adopting any hard-and-fast rule on the matter, noting that the "language in which such a rule might be couched does not itself admit of precision."[88] Lord Justice Lloyd explicitly approved of the explanation of this point made by the judge at first instance when he had noted that:

> In each case, it is a question of fact, depending on the nature of the invention, as to whether the steps needed to perform the invention are ordinary steps of trial and error which a skilled man would realise would be necessary and normal to produce a practical result.[89]

Further, Lloyd L.J. explained that there were a great many different words that could be used to describe the steps that the addressee might be required to take once departing from those explicitly laid down in the teaching of the patent. A continuum stretched from "mere practice at one end of the scale, through enquiry, trial, experiment and research at the other, until one reaches the threshold of a fresh invention."[90] However, the Court acknowledged that the lack of precise boundaries between any of these descriptors—they "shade into each other"[91] —rendered their utility as definitional markers somewhat dubious.

Nevertheless, appreciating that a working definition was probably required, Lloyd L.J. turned to an earlier judgment of the Court of Appeal in *Valensi v. British Radio Corporation*,[92] a decision explicitly relied upon in the court below. In the *Valensi* decision, Buckley L.J. explained that the notional person skilled in the art, a hypothetical construction of the court, was "not a person of exceptional skill or knowledge," and moreover could not be expected to "exercise any invention no[r] any prolonged research, inquiry or experiment." However, that person must be "prepared to display a reasonable degree of skill and common knowledge of the art in making trials and to correct obvious errors in the specification if a means of correcting them can readily be found."[93]

Notwithstanding that *Valensi* was decided before the passage of PA 1977, Lord Justice Lloyd considered the test laid in the case was "as appropriate" under the current Act as it was before. Nevertheless, he noted that the judge at first instance's classification of the sort of experiments that the addressee should be

88. *Id.* at 11.

89. *Id.* at 11 (quoting from Aldous J., [1991] F.S.R. 557, 562).

90. *Id.* at 12.

91. *Id.*

92. [1973] R.P.C. 337.

93. *Mentor*, [1993] R.P.C. 7, 13.

expected to conduct in light of this test—"routine trials"—perhaps provided "a surer test of what is meant by "clearly enough and completely enough" in Section 72(1) of the Act than the negative test proposed in *Valensi*."[94] This approach has been summarized in subsequent decisions as essentially asking "whether the patent imposes an undue burden on the skilled reader to arrive at a workable prototype of the invention."[95]

On the facts of *Mentor*, the patentee's failure to provide (1) information concerning the adhesive used to affix the device to the body, and (2) how to select a suitable material to release the adhesive as the sheath was unrolled, were not considered to render the specification insufficient. Both omissions could be remedied by simple trial–and-error experimentation. The fact that it would have taken significantly more time to create a commercially saleable product that could compete with the plaintiff's was not a relevant consideration. Lord Justice Lloyd noted that it was "only the work involved, and the time taken, in reaching a workable prototype that is relevant for the purpose of testing sufficiency."[96]

Errors and omissions in the specification may result (to borrow the phraseology of Justice Arnold in a recent decision[97]) in the skilled team being:

> [V]ery surprised that the inventors had failed to explain such a basic matter, and rather cross at being presented with such a puzzle to solve.... [H] owever, the skilled team is deemed to read the specification with a mind willing to understand it. It follows that they would not throw up their hands when confronted with the problem, but would consider the specification with care to see if it was possible to work out what was meant....[98]

Accordingly, whether defects and/or omissions in the instructions contained in the specification will serve to render a disclosure insufficient is clearly a question of degree. As Floyd J. noted recently, in *Zipher v Markem*: "a specification may present the skilled team with such a combination of defects that, whereas individually no single defect would have stopped the team from being able to perform the invention, in combination they may do."[99] Furthermore, if the teaching in a document is such that a skilled person following it could find himself at one

94. *Id.* at 14.

95. Zipher v. Markem, [2009] F.S.R. 1, [425] (Floyd, J.).

96. *Mentor*, [1993] R.P.C 7, 15.

97. Generics (t/a Mylan) v. Yeda and Teva [2012] EWHC 1848 (Pat) (Arnold, J.)

98. *Id.* at [224].

99. [2009] F.S.R. 1, [380].

of a number of possible endpoints, some within the scope of the claim and some without, then this is evidently not enabling.[100]

C. The Broad Claim Problem in English Law

Although being a valid objection at the pre-grant stage, the requirement in Section 14(5) that the claims be adequately supported by the description has no direct analog in the post-grant arena. If mistakes are made, and an unsupported claim issues, what then is to happen following the grant? It is evident that PA 1977 § 72 contains a closed list of grounds—such is clear from the language of the provision on which it is based, and "so framed" as to have the same effect: Article 138 EPC. This states that a European Patent "may be revoked with effect for a Contracting State *only* [on the five grounds that follow]," and lack of support is not one. Nevertheless, this has not prevented parties, and on occasion members of the judiciary, from attempting to shoehorn the old grounds of revocation into the post-EPC regime. This generally has been met by a rather stern response from the courts. Accordingly, efforts by the judge at first instance to resurrect the "lack of fair basis" as a post-grant objection in *Genentech's Patent*[101] were wholeheartedly rejected by the Court of Appeal—Mustill L.J. noting, for example, that the judge had "held these claims ... to be bad on a ground which, as all members of this court agree, was beyond his jurisdiction."[102]

Notwithstanding these comments, the dissonance between the grounds of objection pre- and post-grant was a source of consternation for some time following the Court of Appeal's clear statement of principle in *Genentech*. This problem was compounded by a belief that the European Patent Office (EPO), in particular, was apt to grant claims of very broad scope, especially in the field of biotechnology.[103] How best, then, to deal with these? Attempts to draw the breadth argument into the objection concerning insufficient disclosure under Section 72(1)(c) of the Act met with resistance in *Chiron v. Organon (No. 3)*.[104] Following EPO decisions on this matter,[105] the court rejected arguments that the

100. *See* Evans Medical Ltd's Patent, [1998] R.P.C. 517, 536–37.

101. [1987] R.P.C. 553 (Whitford, J.).

102. Genentech's Patent, [1989] R.P.C. 147, 260 (Mustill, L.J.).

103. *See, e.g.*, the discussion in T. Roberts, *Broad Claims for Biotechnological Inventions*, 1994 Eur. Intell. Prop. Rev. 371, and J. Wibbelmann, *Broad Claims: A Nuisance?*, 1997 Eur. Intell. Prop. Rev. 515.

104. [1994] F.S.R. 202.

105. *See, e.g.*, Decision T 292/85, GENENTECH I/Polypeptide expression, 1989 O.J. 275, 1989 E.P.O.R. 1.

claimants' broad claims were insufficiently enabled by the description, as there was "no dispute that the skilled man could, using his general knowledge and the information in the specification make an embodiment within all the claims."[106] Mr. Justice Aldous therefore considered that the section could not be the "vehicle for the judicial massage that the defendants would wish."[107]

Further inventive attempts to circumvent the deficiencies in the grounds of invalidity also met with resistance. Thus, the argument that "if a patentee 'got away with' [to be understood as: "was mistakenly allowed"] a claim contrary to s14(5)(c), then that was a relevant matter or consideration when the court was exercising its discretion to allow amendment [of the patent],"[108] was roundly rejected in *Chiron v. Organon (No. 5).*[109] Mr. Justice Aldous considered it inappropriate to raise the issue of support for a broad claim when a narrower claim in the patent was to be deleted, explaining that it was not "relevant or right to consider matters which have no nexus with the amendment nor with the cause for the amendment."[110]

The broad claim problem therefore remained substantially unresolved until the decision of the House of Lords in *Biogen v. Medeva*[111] almost twenty years after the promulgation of PA 1977. By this time, the EPO Boards of Appeal had clarified their understanding of the requirement of sufficiency.[112] Instrumental to this clarification was the decision of the Technical Board of Appeal (TBA) in T_409/91 *EXXON/Fuel Oils*, described by the editor of the European Patent Office Reports as reflecting a "significant and (it is submitted) wholly desirable back-tracking from the excessively liberal view of Article 84 expressed previously by the Technical Board..."[113]

---

106. [1994] F.S.R. 202, 241. The quote continues with the words "except the vaccine and in vitro cell growth claims"; however, these were challenged on the basis of insufficiency proper and not due to the fact of their broad scope. There was no requirement that the skilled man should be able to make all embodiments falling with the claims—see the discussion in *Mölnlycke A.B. v. Procter & Gamble Ltd.*, [1992] FSR 549, 600.

107. [1994] F.S.R. 202, 242.

108. Chiron v. Organon (No. 5), [1994] F.S.R. 258, 266.

109. There were fourteen cases in this series in total.

110. *Chiron*, [1994] F.S.R. 258, 267.

111. [1997] R.P.C. 1.

112. For the reader not versed in the intricacies of the European patent system, the House of Lords has explained, for example in *Merrell Dow Pharmaceutical, Inc. v. H.N. Norton & Co. Ltd.*, [1996] R.P.C. 76, 82, that decisions of the EPO's Boards of Appeal are "of considerable persuasive authority."

113. [1994] E.P.O.R. 149, 150.

In the decision, the TBA held that in order to satisfy the requirements of Article 83 EPC (on which Section 14(3) PA 1977 is based), the specification must contain sufficient information to enable a person skilled in the art, utilizing her common general knowledge, to carry out the invention across the whole area claimed.[114]

Subsequent decisions of the TBA confirmed this approach; so by the time the House of Lords came to consider the issue of broad claims in *Biogen v Medeva*, it was settled that the Boards of Appeal had abandoned their more lenient position. The established view was that, in order to be sufficient, a claim needed to be enabled across its breadth. Lord Hoffmann, providing the leading judgment in the *Biogen* case, took this as his starting point; referring directly to *EXXON/Fuel Oils* he noted that the TBA had held that:

> Article 84 EPC also requires that the claims must be supported by the description, in other words, it is the definition of the invention in the claims that needs support. In the Board's judgment, this requirement reflects the general legal principle that the extent of the patent monopoly, as defined by the claims, should correspond to the technical contribution to the art in order for it to be supported, or justified.

The key, his lordship noted, was the generality of the claim and the disclosure. If the patentee was able to show that the teaching encompassed a beneficial effect shared in common by all of a class of products, then the patentee would not be barred (on grounds of insufficiency at least) from claiming all within the class, even though the patentee had made only one or two of them. On the other hand, if the patentee "cannot demonstrate that there is a common principle by which that effect will be shared by other products of the same class, [they]...will be entitled to a patent for that product but not for the class, even though some may subsequently turn out to have the same beneficial effect."[115] Accordingly, depending on the technical contribution made to the art by the patent, it is not necessarily enough to simply disclose just one way of performing the invention.

On the facts: having invented and disclosed a perfectly good method of producing the hepatitis "B" virus antigen, the patentee had claimed "any method of making a DNA molecule which would achieve the necessary expression."[116] Such a broad claim could not be left to stand as it clearly exceeded the patentee's technical contribution to the art. As Arnold J. lucidly explained in a recent

114. *Id.* at 152.

115. [1997] R.P.C. 1, 48.

116. *Id.* at 40 (Lord Hoffmann).

decision: "The breadth of the claim will exceed the technical contribution if the claim covers ways of achieving the desired result which owe nothing to the patent or any principle it discloses"[117] In *Biogen*, this is precisely what had occurred.

The House of Lords revisited the issue in *Generics (UK) v. Lundbeck* in 2009.[118] Lundbeck was the proprietor of a patent for escitalopram, the (+) enantiomer of citalopram (a known antidepressant). The isolated enantiomer had superior properties, and fewer side effects, than the known racemic mixture. The patent claimed the (+) form of the molecule however it was made, but disclosed only two routes to manufacture in the specification. At first instance the judge held that this was insufficient, citing *Biogen* as his authority. This conclusion was overturned in the Court of Appeal, a view subsequently endorsed by the House of Lords.

Their lordships considered it to be a question of identifying the correct technical contribution of the patent in question, and it was a mistake to equate this with the inventive concept. Here, the contribution made by the patent was the provision of the isolated enantiomer—a product—and not the process by which it was made. This was the case "even though the inventive step lay in finding a way to make the product."[119] Where the technical contribution was the product itself, it was settled law that in order to be sufficient, the patentee need only explain one route by which it could be made. The claim was accordingly not too broad.

Another recent case to discuss the interface between sufficiency and over-broad claims is that of *Novartis v. Johnson & Johnson*[120] in the Court of Appeal. The facts of the case are admirably summed up by Jacob L.J., giving the judgment of the Court:

> The reader might be forgiven for initially supposing that this apparently detailed list of elements would lead to a monopoly of reasonably defined

---

117. MedImmune v. Novartis, [2011] EWHC 1669, [469] (Arnold J.).

118. [2009] UKHL 12, [2009] R.P.C. 13. The case is not the easiest decision from which to extract a rational consensus. As Arnold J. has explained in a recent judgment of the patents court:

> The principal speeches were given by Lords Walker of Gestingthorpe, Mance and Neuberger of Abbotsbury. Lord Phillips of Worth Maravers (*sic*) said that they reached the same conclusion for the same reasons, and agreed with all of them. Lord Scott of Foscote agreed with Lord Neuberger. Lord Walker said he understood his reasons to be essentially the same as those of Lords Mance and Neuberger, and Lord Neuberger said that he understood that his reasons to be effectively the same as those of Lords Walker and Mance. In these circumstances, it is not easy to quote particular passages from just one opinion as representing the reasoning of at least a majority of the panel.

MedImmune v. Novartis, [2011] EWHC 1669, [474].

119. *MedImmune*, [2011] EWHC 1699, [475] (Arnold, J.).

120. [2010] EWCA (Civ) 1039.

> scope, that each of the elements actually meant something by way of delineating the monopoly. But the reader would be wrong. Upon analysis it turns out that the elements are mostly meaningless and what is left is no more than a claim to a [contact] lens made from two types of polymer, provided it works.[121]

Not only were the criteria for choosing the constituent components of the substance from which the lens was to be created "extremely wide," but each was also described as having "woolly limits" as well. There was, according to the Court, "a lot to be said for the view that the claim should never have been allowed as not complying with Article 84 of the EPC"—that is, the requirement for clarity and support. Nevertheless, as Jacob L.J. explained, such "[u]ndue width" could be remedied using either non-obviousness, or insufficiency, or a combination of the two.[122]

The judge at first instance had found that the skilled person would only know whether he had succeeded in creating something falling within the claims if he conducted "a small scale test on actual people." However, as Jacob L.J. explained, it was unclear what effect the failure of such a trial would have. Accordingly:

> [We] come to an astonishing conclusion. Although the claim has a number of elements, hardly any of them have any significance.... In substance the claim amounts to this: "if you try any pair of polymers, to see if they work (perhaps only after surface treatment) and find anything that does, we claim it."[123]

Moreover, the teaching of the patent was held to give the skilled reader hardly any assistance in determining whether any given combination of polymers would "work" at all—notwithstanding that it stretched to 422 paragraphs over 53 pages. "Something fishy" had gone on.[124]

Being unable to predict from the patent documentation whether any of the examples therein "worked" or not, the addressee was left with a vexing question: What was he to do? Trial and error would be a major enterprise: the patent suggested selection of two polymerizable materials from two vast classes, with little guidance on proportions and no advice on how to assess if the blend would be

121. *Id.* at [10].

122. *Id.* at [19].

123. *Id.* at [50].

124. *Id.* at [44].

successful short of actual trial. If a selection "worked" then this was all "well and good—but that would tell you nothing about the remainder of the vast ambit of the claim.... If it does not 'work' then the Patent does not help you as to what to do next."[125]

In essence, therefore, Novartis's patent "did no more than to invite the reader to perform a research program where, if he succeeded, the patent claimed the fruits of his research." This, according to Jacob L.J., was "a long way off from satisfying the sufficiency test."[126]

This case, perhaps more than any other in recent years, illustrates the difficulty of assessing the quality of the specification without intensively probing the facts and calling for expert evidence on the matter. Indeed, as Jacob L.J. noted,[127] this was a European Patent issued by the EPO and the selfsame specification had survived challenge in at least three other European states, as well as before the TBA. All assumed that the examples given in the patent did, in fact, work; however, this was not actually the case.

In summary, therefore, it would appear that the requirement for support under Section 14(5) PA 1977 can lay a foundation to an invalidity action on grounds of insufficiency where the breadth of the claim exceeds the invention's technical contribution to the art. As Arnold J explained after conducting an exhaustive review of the authorities in *MedImmune v. Novartis*, this may occur in at least two ways:

> [First] where the patent claims results which it does not enable, such as making a wider class of products when it enables only one and discloses no principle to enable the others to be made, and [second] where the patent claims every way of achieving a result when it enables only one way and it is possible to envisage other ways of achieving that result which make no use of the invention." [128]

This is the case notwithstanding that lack of support (or rather lack of fair basis) is no longer a freestanding ground of invalidity under the Act. Nevertheless, by adopting such an approach, the dissonance between pre- and post-grant objections to patentability in this respect has been ameliorated, preserving the integrity of the underlying justifications for the requirements while sidestepping the lapse in statutory wording. As such, it is a pragmatic, and eminently more palatable,

---

125. *Id.* at [70], [71].

126. *Id.* at [77], [72].

127. *Id.* at [62].

128. [2011] EWHC 1669, [469].

solution to the broad claim problem than the "judicial massage" suggested in the Act's formative years.

IV. The Modern Law of Enablement & Written Description in the United States

The provision within modern US patent law that governs issues of sufficiency and written description is Section 112 of the 1952 Act (as amended). Paragraph 1 of which requires that:

> (a) IN GENERAL—The specification shall contain a written description of the invention, and of the manner and process of making and using it, in such full, clear, concise, and exact terms as to enable any person skilled in the art to which it pertains, or with which it is most nearly connected, to make and use the same, and shall set forth the best mode contemplated by the inventor or joint inventor of carrying out the invention.

The jurisprudence of the Court of Appeals for the Federal Circuit (CAFC) and its predecessor, the Court of Customs and Patent Appeals (CCPA), has established[129] that there are three separate requirements contained within this section: (1) enablement; (2) written description; and (3) best mode.[130] Although there will inevitably be a degree of overlap among the three, the CAFC has nevertheless been at pains to note that each is distinct, and none are subsumed within the others.[131] In terms of their underlying rationales, the three requirements have been supported by the courts on different bases. Whereas the requirement for an enabling disclosure appears predicated upon provision of consideration for the patent grant,[132] written description has been stated to be a broader concept—requiring the patentee to "convey with reasonable clarity to those skilled in the art that, as of the filing date sought, he or she was in possession of the invention."[133] It

---

129. Although not without some significant dissent—see, for example, the judgment of Rader J., dissenting from the court's refusal to rehear the case en banc, in *Enzo Biochem v. Gen-Probe*, 323 F.3d 956 (Fed. Cir. 2002) (Rader, J., dissenting).

130. *See, e.g.*, Ariad Pharm. v. Eli Lilly, 598 F.3d 1336 (Fed. Cir. 2010); *In re* Ruschig, 379 F.2d 990 (C.C.P.A. 1967).

131. See, for example, the categorical statement of the majority in the en banc decision in *Ariad Pharmaceuticals v. Eli Lilly*, 598 F.3d 1336 (Fed. Cir. 2010), considered below.

132. See comments to this end in *Genentech, Inc. v. Novo Nordisk A/S*, 108 F.3d 1361, 1366 (Fed. Cir. 1997), where Judge Lourie, giving the judgment of the court, explained that "[p]atent protection is granted in return for an enabling disclosure of an invention, not for vague intimations of general ideas that may or may not be workable."

133. Vas-Cath v. Mahurkur, 935 F.2d 1555, 1563–64 (Fed. Cir. 1991).

has been noted that this rationale was once thought to apply only where the patent's claims had been amended after filing, where there were questions of entitlement to priority from an earlier application, or "where a party asserted that counts in an interference were supported in a specification."[134] However, in more recent years the standard has been found to apply to all claims, whether original or amended.[135] Best mode is relatively self-explanatory, and relates to a robust understanding of the bargain the patentee strikes with the state. Not only, therefore, must she disclose a method of performing the invention, she must disclose her best.

A. Enablement Proper

As may now be anticipated, the approach taken to the issue of enablement under modern US law is strikingly similar to the formulations adopted under previous statutes. Its core, that the patent is only justifiable if it contributes to society by teaching the invention to those skilled in the art, is immediately familiar. So too is the standard by which enablement is judged: the CAFC *In re Wands*,[136] for example, explained that "[e]nablement is not precluded by the necessity for some experimentation such as routine screening. However, experimentation needed to practice the invention must not be undue experimentation."[137]

Notwithstanding, therefore, that the term "undue experimentation" does not feature in the statutory language, it is now well established that this is the standard to be applied. What precisely amounts to undue experimentation will inevitably vary from case to case and will depend on the weighing of a number of factual considerations, having regard to both the nature of the invention and the state of the art.[138]

Judge Newman explained the rationale of enablement in her (part concurring, part dissenting) judgment *In re Wands* in the following terms:

> The premise of the patent system is that an inventor, having taught the world something it didn't know, is encouraged to make the product available for public and commercial benefit, by governmental grant of the right to exclude others from practice of that which the inventor has disclosed. The

---

134. *See* Mark D. Janis, *On Courts Herding Cats: Contending with the "Written Description" Requirement (and Other Unruly Patent Disclosure Doctrines)*, 2 WASH. U. J.L. & POL'Y 55, 59–60 (2000).

135. See comments to this end in *Regents of the University of California v. Eli Lilly & Co.*, 119 F.3d 1559, 1567 (Fed. Cir. 1997).

136. 858 F.2d 731 (Fed. Cir. 1988).

137. *Id.* at 736–37.

138. *Id.*

> boundary defining the excludable subject matter must be carefully set: it must protect the inventor, so that commercial development is encouraged; but the claims must be commensurate with the inventor's contribution. Thus the specification and claims must meet the requirements of 35 U.S.C. § 112.[139]

Thus we see once more that enablement is strongly tied to the scope of the claims, it being settled that one cannot claim more than one teaches.[140] The notion of avoiding undue experimentation also meshes well with the concept of only allowing routine trial and error experiments to supplement the specification's teaching under UK jurisprudence. The policy goals of each system are clearly cut from the same cloth. Both seek to enrich the state of the public domain with the patent's teaching following its expiry; both require instruction of the hypothetical person skilled in the art; and both allow leniency in respect of minor, easily surmountable, errors or omissions.

Nevertheless, the extent of the enablement doctrine's reach in the United States is worthy of further comment. The language found within § 112 is important; not only does it require instructions to "make" the patented subject matter, but it also demands instructions to "use." The nexus between enablement and utility[141] is accordingly very strong—a point made clear in *In re '318 Patent Infringement Litigation*.[142] The case concerned a patent for a method of treating Alzheimer's disease with a particular (known) compound, galanthamine. Judge Dyk, giving the judgment of the Court, explained the problem that had arisen: "The specification for the '318 patent was only just over one page in length, and it provided almost no basis for its stated conclusion that it was possible to administer 'an effective Alzheimer's disease cognitively-enhancing amount of galanthamine.' "

Although the specification did provide short summaries of a number of scientific papers in which galanthamine had been administered to humans or animals, in no case were these studies conducted on organisms that showed "physiological changes" similar to Alzheimer's disease. Experimental data suggesting that the compound could be a promising treatment for the disease were only obtained some two months after the patent had issued. It was settled that enablement was to be determined at the patent application filing date,[143] and therefore the

---

139. *Id.* at 741.

140. See similar conclusions in *In re Fisher*, 427 F.2d 833 (C.C.P.A. 1970).

141. The requirement for which is found in 35 U.S.C. § 101 (2012).

142. 583 F.3d 1317 (Fed. Cir. 2009).

143. Plant Genetic Sys, N.V. v. DeKalb Genetics Corp., 315 F.3d 1335, 1339 (Fed. Cir. 2003); *see also In re* Glass, 492 F.2d 1228, 1232 (C.C.P.A. 1974).

question before the Court was whether the specification contained sufficient evidence that the compound would work in the manner suggested. The issue of enablement thus hung on whether the patentee was in a position to prove utility at the filing date. Judge Dyk explained that: "[a]llowing ideas, research proposals, or objects only of research to be patented has the potential to give priority to the wrong party and to 'confer power to block off whole areas of scientific development, without compensating benefit to the public.'"[144]

In the Court's opinion, the specification provided by the patentee, even read in light of the knowledge of those skilled in the art at the filing date, did "no more than state a hypothesis and propose testing to determine the accuracy of that hypothesis. That is not sufficient."[145]

B. Best Mode

The robustness of the enablement requirement, and its emphasis as a central doctrine of patent validity, is ostensibly further enhanced by the requirement of "best mode."[146] The basic principle is straightforward—as well as requiring the invention to be taught to the person skilled in the art, the specification must also disclose the best mode of operation (if any) that the inventor possessed at the time of filing. As such, the information necessary to satisfy this requirement potentially goes far beyond that demanded under sufficiency proper. As the CCPA noted in *In re Gay*: "Manifestly, the sole purpose of [best mode]... is to restrain inventors from applying for patents while at the same time concealing from the public preferred embodiments of their inventions which they have in fact conceived."[147] Nevertheless, recent legislative changes,[148] although stopping just short of outright slaughter of the provision, have emasculated this justification (and indeed the requirement of best mode per se) by ripping from 35 U.S.C. the effective means for enforcing the requirement post-grant.

---

144. *In re '318 Patent Litigation*, 583 F.3d at 1324 (citations omitted).

145. *Id.* at 1327.

146. Found in the third paragraph of 35 U.S.C. § 112 (2012). Although it may be argued that this enhancement is now little more than notional—see the discussion of the changes wrought by the Leahy–Smith America Invents Act ("the AIA"), *infra*. The reader will recall that a similar requirement under UK law to disclose the "best method" of performing the invention, required under the Patents Act 1949, was abandoned with the passage of the 1977 Act. It is not a requirement made of the patentee under the TRIPs Agreement, and, prior to the AIA, had been the subject of a degree of criticism in the United States. *See, e.g.*, S.B. Walmsley, *Best Mode: A Plea to Repair or Sacrifice This Broken Requirement of United States Patent Law*, 9 MICH. TELECOMM. & TECH. L. REV. 125 (2002).

147. *In re* Gay, 309 F.2d 769, 772 (C.C.P.A. 1962).

148. Ushered in by the AIA, considered *infra*, in late 2011.

Prior to the passage of the Leahy–Smith America Invents Act[149] (the AIA) in late 2011, Section 112's insistence on the provision of this preferred embodiment was further reinforced by Section 282. This gave courts the power, notwithstanding a general presumption of validity, to declare a patent invalid or render it unenforceable for failure to disclose best mode. However, in much the same manner as the transition to a post-EPC statutory regime robbed the UK courts of the ability to hold a patent invalid for "lack of fair basis" while still preserving the pre-grant requirement that the claims must be supported by the description, so too has the AIA neutered best mode. Following the enactment of the statute, Section 282 has been amended. The relevant part now reads:

> "The following shall be defenses in any action involving the validity or infringement of a patent and shall be pleaded:
> ...(3) Invalidity of the patent or any claim in suit for failure to comply with—
> (A) any requirement of section 112, *except that the failure to disclose the best mode shall not be a basis on which any claim of a patent may be canceled or held invalid or otherwise unenforceable;*" (emphasis added)

Accordingly, in any proceedings commenced on or after September 16, 2011, the effective date for the amendments made by the AIA in this respect, the courts are no longer in a position to enforce Section 112(a)'s third mandate. Nevertheless, as the first paragraph of Section 112 remains substantially untouched (cosmetic titivation aside), the USPTO has confirmed that the AIA's changes do not, and will not, affect their examination practice in this area.[150]

Therefore, even following the AIA's amendments in this respect, best mode remains a pre-grant requirement; the test for which is essentially two-pronged. The first question is subjective:[151] Did the inventor possess a best mode of operation at the time the patent was filed?[152] Assuming that the answer to this first

149. "An Act to amend title 35, United States Code, to provide for patent reform," to provide its long title. Pub. L. No. 112-29, 125 Stat. 284 (2011) (codified as amended in scattered sections of 35 U.S.C.).

150. See the answers to "AIA Frequently Asked Questions," specifically questions BM2 and BM3, *AIA Frequently Asked Questions*, USPTO, http://www.uspto.gov/aia_implementation/faq.jsp (last visited Sept. 5, 2013).

151. As confirmed in N. Telecom Ltd. v. Samsung Elecs. Co., 215 F.3d 1281, 1286 (Fed. Cir. 2000).

152. *See* U.S. Gypsum Co. v. Nat'l Gypsum Co., 74 F.3d 1209, 1212 (Fed. Cir. 1996).

question is "yes," the second enquiry is objective: Has the inventor "concealed" the best mode from the public? As such, best mode stands distinct from enablement proper—which would not necessarily require the disclosure of anything approaching a marketable variant. Walmsley sums the position up with great clarity when he notes that:

> [T]he best mode requirement is a subset that represents the intersection of two larger sets—the written description and enablement requirements. Thus, the best mode requirement is not limited to just the "what" (description requirement) or the "how" (enablement requirement) of an invention. Rather, the best mode requirement may well be thought of as the "what best" and "how best" of an invention.[153]

Evidently, therefore, complete failure to set forth any mode at all would be equivalent to non-enablement,[154] whereas concealing the most advantageous method of operation known at the filing date of the application would fall foul only of this "sufficiency-plus" best mode requirement. However, as the AIA has ensured, once granted, lack of best mode can no longer be used as an explicit ground of challenge to an otherwise healthy patent—mistakes of issuance in this respect cannot be remedied *post hoc*.

The practical effects of the amendments made by the AIA in this context are therefore difficult to assess. Although best mode has evidently not been entirely scrubbed from the patent landscape, it is now but a shadow of its former self—an old soldier: shell shocked and broken. However, it is perhaps over-egging the pudding to claim, as some have done, that the changes wrought by the AIA have "effectively eliminated the best mode requirement from patent law."[155] As we have seen with similar changes instituted under English law, the pre-grant requirements for support under Section 14(5) PA 1977 have become subsumed within a general notion of sufficiency—effectively outflanking the deletion of "lack of support" as a freestanding ground of invalidity in the transition from the old law. It is therefore hard to accept that inventive lawyers will be willing to let best mode simply fade away. To return to the justification for the requirement given in *In re Gay,* its continued relevance to prosecution would suggest at the very least the possibility of advancing defensive arguments of inequitable conduct where the

---

153. Walmsley, *supra* note 146, at 132.

154. See comments to this end in *In re* Glass, 492 F.2d 1228, 1233 (C.C.P.A. 1974).

155. *See, e.g.*, L. Petherbridge & J. Rantanen, *In Memoriam Best Mode*, 64 STAN. L. REV. ONLINE 125, 126–27 (2012).

inventor has intentionally concealed her preferred embodiment. Accordingly, the old adage may well prove to be true: this old soldier might never die.

C. Written Description

1. A Separate Requirement for a Written Description

Finally we arrive at the requirement of "written description." This is undoubtedly one of the more controversial aspects of Section 112(a), having spawned a great deal of discussion, both within the courts and without, on the merits, or otherwise, of separating this requirement from that of enablement proper.[156] Battle lines are drawn between those who favor a clear divide, and those who believe written description should essentially be subsumed within enablement itself.

The former group believes that requiring a patentee to demonstrate possession of his invention, above and beyond teaching one skilled in the art how to make and use it, is a necessary demand with an extensive historical pedigree.[157] They explain that this function of the specification can be traced back at least to the Act of 1793, and point to Justice Story's classic statement in *Evans v. Eaton*, noted above, as supporting this fact. It will be recalled that Justice Story declared in that case that the specification had two objects. The first was to teach the invention. The other was:

> [T]o put the public in possession of what the party claims as his own invention, so as to ascertain if he claims anything that is in common use, or is already known, and guard against prejudice or injury from the use of an invention which the party may otherwise innocently suppose not to be patented....[158]

Although this statement would appear to support the view taken, it must be recalled that the patent system at this point in time was a very different animal to that which we see today. Most critically: in 1822 claims, our modern-day notice-givers, custodians of the patent's extent of protection, the means by which we now define the invention itself, were simply not required. The

---

156. Compare, for example, the comments of the majority and of those in dissent in *Ariad Pharmaceuticals v. Eli Lilly*, 598 F.3d 1336 (Fed. Cir. 2010) (*Ariad II*) and *Enzo Biochem v. Gen-Probe*, 323 F.3d 956 (Fed. Cir. 2002). *Compare* J.A. Schroeder, *Written Description: Protecting the Quid Pro Quo since 1793*, 21 FORDHAM INTELL. PROP. MEDIA & ENT. L.J. 63 (2010), *with* Janis, *supra* note 134.

157. *See, e.g.*, the comments of Miller, J., in the majority, in *In re* Barker, 559 F.2d 588, 592–93 (C.C.P.A. 1977); *see also* Vas-Cath v. Mahurkur, 935 F.2d 1555, 1560–61 (Fed. Cir. 1991).

158. Evans v. Eaton, 20 U.S. 356, 433 (1822).

task of demarcation, of definition, therefore fell to the specification. As Janis notes: "[r]ead in context, the "possession" language of *Evans* clearly is directed towards satisfying... [the] notice function, one which the modern description does not require."[159]

Sensing, perhaps, that putting too much weight on the historical leg might make the table collapse, proponents of a separate and distinct written description requirement under Section 112 also attempt to support their argument on broader policy grounds. They explain that although it might be imagined that there is little to differentiate describing an invention, on the one hand, and enabling someone skilled in the art to make and use it, on the other, this is not actually the case. There can be a vast difference, especially in the chemical arts. Circuit Judge Rich attempted to maintain this distinction in *In Re Hunter*, where he explained that "[a] patent specification... may fortuitously enable those of skill in the art to make and use an invention *that an applicant did not make* before filing the patent."[160] Accordingly, "[a]lthough a specification that meets the written description requirement always satisfies the enablement requirement, the converse is not always true."[161]

If this is the case, and enablement is always within written description, then one might wonder why we bother having two separate standards at all. Nevertheless, as Janis has noted, it is Judge Rich's reference to "fortuitous" enablement that is truly "startling," appearing, as it does, to be an invitation for "judges to distinguish between those inventors who deliberately provided information about how to make and use their invention, and those who 'got lucky.' "[162] Janis continues, asking "what justifies penalizing the inventor" in cases where the inventor's teaching enables one skilled in the art to perform an invention that falls within a claim, but has somehow not been demonstrated to have been possessed by the inventor at the date of filing.[163]

So there is controversy: but nevertheless the separate requirement of a written description now seems firmly ensconced within US patent law. Judge Linn made the point elegantly in his concurring opinion in *Ariad Pharmaceuticals v. Eli Lilly* (*Ariad I*), when he stated: "I join the opinion of the court because I concur that it is supported by our precedent. I write separately to emphasize, as I have before,

159. Janis, *supra* note 134, at 64.

160. *In re* Hunter, No. 94-1301, 1995 WL 364375, 59 F.3d 181 (Table), at *5 (Fed. Cir. June 19, 1995).

161. *Id.*

162. Janis, *supra* note 134, at 67.

163. *Id.* at 67–68.

my belief that our engrafting of a separate written description requirement onto section 112, paragraph 1 is misguided."[164]

2. Development of the Standard

Whichever side of the fence one happens to be on, it seems uncontentious to state that the modern revival of the separate written description doctrine (or its invention, if one happens to hold that view)[165] occurred in the CCPA's 1967 decision in *In re Ruschig*.[166] In this case, Judge Rich explicitly distinguished written description from enablement, noting:

> While we have no doubt a person so motivated would be enabled by the specification to make... [the compound in question], this is beside the point for the question is not whether he would be so enabled but whether the specification discloses the compound to him, specifically, as something appellants actually invented.[167]

The case itself concerned an amended claim, and so the court was faced with a question of priority. This might have been dealt with under Section 132 as an added matter issue,[168] but instead the Court considered written description best placed to deal with the shortcoming. Accordingly, although undisputedly enabled, there was still an issue of whether the patentee had shown that she was in possession of the invention at the filing date.

Cases subsequent to *Ruschig* not only heralded a tightening of the manner in which Section 132 was utilized—essentially relegating it to policing amendments to the specification; written description took a prominent place when the claims were redrafted[169] —but also saw the idea of a separate written description standard being fully embraced. Accordingly, by the time of the CAFC's en banc decision in *Ariad Pharmaceuticals v. Eli Lilly* (*Ariad II*),[170] Circuit Judge Lourie, speaking for

---

164. *Ariad I*, 560 F.3d 1366 (C.C.P.A. 1973).

165. See, for example, the comments of Judge Rader in *Enzo Biochem v. Gen-Probe*, 323 F.3d 956, 977 (Fed. Cir. 2002) (Rader, J., dissenting).

166. 379 F.2d 990 (C.C.P.A 1967).

167. *Id.* at 995.

168. Some cases following *In re Ruschig*, did just this. *See, e.g., In re* Barker, 559 F.2d 588 (C.C.P.A. 1977) (seeming to treat written description and added matter interchangeably). See further, Janis, *supra* note 134, at 64.

169. See comments to this end in *In re Rammussen*, 650 F.2d 1212, 1214–15 (C.C.P.A 1981). However, note the powerful dissent of Judge Rader in *Enzo Biochem*, 323 F.3d at 978, where he argues that the priority function of the written description requirement is redundant in the face of Section 132.

170. 598 F.3d 1336 (Fed. Cir. 2010).

the majority, was able to remark that "[s]ince its inception, this court has consistently held that § 112, first paragraph, contains a written description requirement separate from enablement, and we have articulated a 'fairly uniform standard,' which we now affirm."[171]

Furthermore, by the time of *Ariad II*, the written description requirement had ceased to be confined to cases in which the claims had been amended: it now applied to original claims as well. This transition to an "extended" written description standard, as some have called it,[172] occurred with the 1997 decision in *Regents of the University of California v. Eli Lilly*.[173] The case's facts are strongly redolent of those before the House of Lords in *Biogen*[174]: at the time of patent's filing, the patentee, the University of California (UC), had developed and fully enabled a perfectly good method for producing rat insulin cDNA, but had nevertheless claimed far more broadly. Their patent encompassed methods for producing vertebrate, mammalian, and human insulin cDNA within its scope.[175] The university charged Lilly with infringement. Lilly responded, denying that it infringed and alleging, in any case, that the claims to mammalian, vertebrate, and human cDNA were invalid for lack of an adequate written description.

The court agreed. It explained that:

> Contrary to UC's argument, a description of rat insulin cDNA is not a description of the broad classes of vertebrate or mammalian insulin cDNA. A written description of an invention involving a chemical genus, like a description of a chemical species, "requires a precise definition, such as by structure, formula, [or] chemical name," of the claimed subject matter sufficient to distinguish it from other materials.[176]

As some have noted, the "fact pattern in *Eli Lilly* is similar to past Federal Circuit enablement determinations,"[177] and it is therefore somewhat odd that the court had chosen to create an additional requirement when enablement

---

171. *Id.* at 1351.

172. *See, e.g.*, G.M. Whitley, *A Patent Doctrine without Bounds: The "Extended" Written Description Requirement*, 71 U. Chi. L. Rev. 617 (2004).

173. 119 F.3d 1559 (Fed. Cir. 1997).

174. [1997] R.P.C. 1.

175. *Regents*, 119 F.3d at 1562–63.

176. *Id.* at 1568 (citations omitted).

177. Whitley, *supra* note 172, at 621.

could have done the job just as well.[178] Nevertheless, the intention of the court is clear: whether or not a patent contains an enabling disclosure, it may still be invalidated for lack of written description.[179]

The decision in *Lilly* attracted significant criticism, both from academic commentators[180] and from within the ranks of the CAFC itself. Judge Rader notably proclaimed not only that "[a] straightforward reading of the text of section 112 suggests that the test for an adequate written description is whether it provides enough written information for others to make and use the invention,"[181] but also that the court in *Lilly* had erred because:

> Instead of invalidating under the statutory test for adequacy of disclosure, i.e., enablement, the Lilly court purported to create a new doctrine for adequacy of disclosure that it labeled incorrectly "written description." As noted, from its creation through thirty years of application, WD had never been a free-standing substitute for enablement.[182]

Nevertheless, with the en banc decision of the court in *Ariad II*, any hope of resigning the separate written description requirement to the trashcan of history must surely now have withered and died. The majority's support for the enhanced standard is unwavering:

> [A] separate requirement to describe one's invention is basic to patent law. Every patent must describe an invention. It is part of the quid pro quo of a patent; one describes an invention, and, if the law's other requirements are met, one obtains a patent. The specification must then, of course, describe how to make and use the invention (i.e., enable it), but that is a different task.[183]

---

178. Judge Rader makes the point succinctly in his dissent in *Enzo Biochem v. Gen-Probe*, 323 F.3d 956, 980 (Fed. Cir. 2002) (Rader, J., dissenting), when he notes that: "the patent claimed vertebrate insulin cDNA a category ranging from fish to humans again claims whose scope far exceeds the patent's enabling disclosures."

179. See comments to this end in *Regents*, 119 F.3d at 1567.

180. *See, e.g.*, Janis, *supra* note 134; J.M. Mueller, *The Evolving Application of the Written Description Requirement to Biotechnological Inventions*, 13 Berkeley Tech. L.J. 615 (1998); M. Sampson, *The Evolution of the Enablement and Written Description Requirements under 35 U.S.C. § 112 in the Area of Biotechnology*, 15 Berkeley Tech. L.J. 1233, (2000); Whitley, *supra* note 172; *see also* D.L. Burk & M.A. Lemley, *Is Patent Law Technology Specific?*, 17 Berkeley Tech. L.J. 1155 (2002) (providing a critical assessment of its application by the Federal Circuit, generally discuss the requirement in positive terms).

181. *Enzo Biochem*, 323 F.3d at 976 (Rader, J., dissenting).

182. *Id.* at 980.

183. Ariad Pharmaceuticals v. Eli Lilly, 598 F.3d 1336, 1345 (Fed. Cir. 2010). This point is repeated *id.* at 1353–54.

Conclusion

Parallels are easy to draw between the expanded written description standard under *Lilly* and the machinations undertaken by the House of Lords in *Biogen* in order to invalidate an overly broad claim. Both, if one is being unkind to the courts, effectively resurrect a doctrine from a bygone era and apply it to curtail the new upstart technology's running away with the show. The fact that *Lilly*, *Ariad II*, and *Biogen* all hail from the ranks of biotechnology is not insignificant, as it demonstrates the inherent concern that courts seem to have with breakthroughs at the frontier of new fields of technology. Despite the patent system being all about progress and technological advancement, the history of its own development is littered with concerns about providing overly broad monopolies in the formative years of any given art. Accordingly, Morse's broad claims were struck down using the saber of enablement.[184] So too was Arkwright's patent for his "water frame" (which essentially paved the way for the mechanization of the spinning industry) eventually felled,[185] the court citing the patentee's attempts at obfuscation as one of the grounds of invalidation (elements in the specification being "of no use but to be thrown in merely to puzzle").[186] Biotechnology is the new frontier, and the courts' caution in this area is understandable. Nevertheless, the perils of requiring too much disclosure of prospective patentees are difficult to ignore.

The most enduring theories that are said to underpin the modern patent system are those that rationalize it on the basis of incentive: one commentator has even noted that this "simple model" has "been the model for 200 years."[187] Therefore, whether economically sound or not,[188] the popular conception is that the patent system provides monopoly power by creating a zone of exclusion as an

---

184. O'Reilly v. Morse, 56 U.S. 62 (1854).

185. R v. Arkwright, (1785) 1 Web. P.C. 64 (C.P.).

186. *Id.* at 69

187. M.A. Lemley, *Reconceiving Patents in the Age of Venture Capital*, 4 J. Small & Emerging Bus. L. 137, 139 (2000).

188. Needless to say, the view of the patent as a monopoly does not go unchallenged. Despite the validity of Pretnar's assertion that "this view can easily be verified by opening almost any textbook on economics or industrial organization." Bojan Pretnar, *The Economic Impact of Patents in a Knowledge-Based Market Economy*, 34 Int'l Rev. Intell. Prop. & Competition L. 887, 887 (2003). Additionally, some writers have questioned whether conventional thinking can, in fact, be said to be reflective of the true situation. Kitch, for example, argues that the patent only rarely confers monopoly power; more usually it provides a simple property right that is subject to diverse competitive market pressures. *See* E.W. Kitch, *Patents: Monopolies or Property Rights?*, 8 Res. L. & Econ. 31 (1986); *see also* G.S. Rich, *Are Letters Patent Grants of Monopoly?*, 15 W. New Eng. L. Rev. 239 (1993) (concluding that patents are not monopolies).

incentive for investment in, or disclosure of, the results of innovative endeavor.[189] The system is predicated upon assumptions that technological progress is good for society and that the offer of a patent is justified as a means of extracting supra-normal levels of inventive teaching *from* the populace *for* the populace. This view is encapsulated in the comments of Sir Donald Nicholls VC when he famously noted that one of the golden threads of the jurisprudence relating to patents is that:

> [p]atents exist today to reward and thereby encourage inventors; they are not intended to make it possible to take out of public use processes or products already made available to the public.[190]

However, it is apparent that any such system for the reward and encouragement of innovative endeavor can only be legitimate if technological progress is actually obtained—in other words, if the knowledge the state gains as a result of extending legal protection to the patentee is something that its people did not have access to before. Novelty and non-obviousness patrol this requirement from the outside, making sure that the patentee does not monopolize things already in the state of the art (or minor, obvious, additions thereto). It is the task of sufficiency, and "best mode" where demanded, to police from within, ensuring not only that the public does not lose out from the bargain, but in fact gains something new and valuable from their toleration of the monopoly so granted.

Sufficiency—the requirement for teaching the invention—is the bedrock of the modern patent grant. As important as technological development is to the state, and as important is the need to offer inducement for innovators to invest in the process of invention, the patent system simply could not justifiably operate without payment of this price. Put simply, the patent system lives or dies by the quality of its disclosure: without disclosure there can be no bargain, and without a bargain there should be no patent. But bargain it is, and the scales of justice are finely balanced.

---

189. This is certainly the accepted justification for Elizabeth I's institution of a patent custom in the sixteenth century. *See supra* note 23.

190. Merrell Dow v. Norton, [1995] R.P.C. 233, 238.

III PATENTS, INSTITUTIONS, AND INNOVATION PATHWAYS

10 INDIGENOUS DEVELOPMENTAL NETWORKS AND THE NON-DEVELOPMENTAL STATE: MAKING INTELLECTUAL PROPERTY WORK FOR INDIGENOUS PEOPLE WITHOUT PATENTS

Peter Drahos*

Introduction

Economically successful states can be non-developmental for indigenous people. Australia, Canada, New Zealand, and the United States are examples of prosperous states, but all four have living within their borders indigenous people who do poorly on socioeconomic measures. For example, the UN Special Rapporteur in 2004 drew attention to the "human development gaps" existing for indigenous people in Canada in areas such as healthcare, housing, and education.[1] Australia has a long history of underdevelopment of its indigenous people. Faced by continuing poor outcomes on matters as basic as the mortality rates for indigenous children, the Australian federal and state governments in 2008 launched another long-term initiative called "Closing the Gap."[2] In New Zealand socioeconomic measures also paint a picture of disadvantage for indigenous people. So, for example, rates of suicide and incarceration among indigenous people

---

* Professor in the Regulatory Institutions Network, Australian National University; Chair in Intellectual Property, Centre for Commercial Law Studies, Queen Mary, University of London.

1. *See* UN Comm'n on Human Rights, Report of the Special Rapporteur on the Situation of Human Rights and Fundamental Freedoms of Indigenous People, Rodolfo Stavenhagen, U.N. Doc E/CN.4/2005/88/Add.3 (Dec. 2, 2004)

2. For the details, see *Closing the Gap in Indigenous Disadvantage*, COUNCIL OF AUSTL. GOV'TS, http://www.coag.gov.au/closing_the_gap_in_indigenous_disadvantage (last visited Sept. 4, 2013).

remain disturbingly high.[3] In the United States, the census data for 2010 shows that approximately 28 percent of American Indians and Alaska Natives were living in poverty compared to 15 percent for the nation as a whole.[4]

The economic and social problems of indigenous people are not confined to these four states. A UN special report on the position of indigenous people in Asia offers the following bleak summary:

> Indigenous peoples in Asia are among the most discriminated against, socially and economically marginalized, and politically subordinated parts of the society in the countries where they live.... [T]hey are victims of serious human rights violations as a consequence of the dispossession of their lands and natural resources, widespread violence and repression, and assimilation.[5]

The evidence points to a globally persistent pattern of indigenous peoples' disadvantage that occurs in both developed and developing countries. This persistence in developed countries is interesting because on indices of institutional integrity relating to matters such as rule of law and control of corruption, countries such as Australia, Canada, New Zealand, and the United States do very well.[6] In Section I of this chapter, drawing on work from institutional economics, I argue that a key variable in helping to explain the seemingly intractable problem of indigenous peoples' underdevelopment in rich states lies in the existence of extractive property institutions. Section I provides a conceptual account of an extractive property order, drawing on Australia for examples of the effects of such an order. The Australian experience broadly follows a pattern seen in other settler states such as Canada and the United States.

---

3. For the details, see UN Human Rights Council, Report of the Special Rapporteur on the Rights of Indigenous Peoples, James Anaya, The Situation of Maori people in New Zealand, U.N. Doc. A/HRC/18/35Add.4 (May 31, 2011), *available at* http://www.ohchr.org/Documents/Issues/IPeoples/SR/A-HRC-18-35-Add4_en.pdf.

4. United States Census Bureau, American Indian and Alaska Native Heritage Month: November 2011, CB11-FF.22 (Nov. 1, 2011), http://www.census.gov/newsroom/releases/archives/facts_for_features_special_editions/cb11-ff22.html.

5. United Nations (UN) Human Rights Council, Report of the Special Rapporteur on the Situation of Human Rights and Fundamental Freedoms of Indigenous People, Rodolfo Stavenhagen, U.N. Doc. A/HRC/6/15/Add.3, (Nov. 1, 2007).

6. Results of various indices for these countries can be found on the Transparency International's website at http://www.transparency.org as well as on the site of the Worldwide Governance Indicators Project, http://info.worldbank.org/governance/wgi/index.asp.

The central challenge for indigenous people is to find ways to overcome the disadvantages of a property order of the state that is largely non-developmental for them in terms of economic opportunities. From here the chapter narrows its focus to consider the case of indigenous knowledge assets and the use of intellectual property rights by indigenous people in economic enterprises of their own making. Assets are resources that can be used to generate a flow of income.[7] They can be divided into tangible assets such as real property or intangible assets such as reputation and knowledge. Many indigenous groups around the world have lost ownership of their traditional lands and territories, a fact recognized by state actors.[8] The knowledge assets of indigenous people form potentially useful economic resources for them in pursuing their own development agendas. The capacity of economic actors to obtain private returns from knowledge assets depends, at least on some views, on a property rights regime that confers some degree of excludability on those assets.[9] However, this is the beginning rather than the end of the story. The actual economic gains and losses of an intellectual property rights regime depend deeply on a state's administrative and organizational systems. For example, how a state administers its patent office or a multinational enterprise organizes its patent portfolio affects its economic success.[10] Patents, although clearly relevant to indigenous knowledge asset exploitation, are not well suited to providing returns to indigenous groups on such assets. I argue that indigenous groups need to adopt an organizational form I term the "indigenous developmental network." Faced with a non-developmental state property order, indigenous groups have to compensate by organizing into developmental networks. Indigenous developmental networks have to carry out two key tasks when it comes to exploiting knowledge assets. The first is to develop a strategy of protection, and the second is to build capacity into the network so that the full value of the asset can be understood for the purposes of exploiting it in non-indigenous systems of governance and regulation. We live in an era

---

7. Robert Cooter & Thomas Ulen, Law & Economics 39 (2000).

8. The sixth opening paragraph of the UN Declaration on the Rights of Indigenous Peoples (UNDRIP) states that "indigenous peoples have suffered from historic injustices as a result of, inter alia, their colonization and dispossession of their lands, territories and resources." Declaration on the Rights of Indigenous Peoples, pmbl. ¶ 6, G.A. Res 61/295, UN Doc. A/RES/61/295 (2007).

9. However, there are other ways to create incentives for private investment in the search for knowledge, such as the use of prizes. *See* Suzanne Scotchmer, Innovation and Incentives 40–41 (2004). There are also those who argue that markets will perform better without intellectual property monopolies. *See* Michele Boldrin & David K. Levine, Against Intellectual Monopoly (2008).

10. *See generally* Peter Drahos, The Global Governance of Knowledge: Patent Offices and Their Clients (2010).

of regulatory capitalism where all the products that we buy, ranging from food and medicines to the clothes that we wear, are the objects of multiple regulatory standards and systems.[11] This places huge capacity demands on those wishing to enter markets with new products and services. In the case of indigenous people, meeting this capacity demand is contingent upon indigenous developmental networks forging close relations with science (see Section III). Indigenous people have to find ways to fit their knowledge assets into existing regulatory systems (such as food and drug regulation) or economic governance systems (such as carbon markets or trade in ecosystem services). This requires their assets to have a lot of epistemic mobility, as it were, and can really only be achieved in cooperation with science.

For the purposes of developing the analysis, I refer to fieldwork interviews conducted as part of a three-year Australian Research Council–funded project that examined the relationship among indigenous knowledge, intellectual property, and biodiversity.[12] During the course of the project we interviewed members of indigenous communities in Western Australia, Northern Territory, Queensland, and South Australia; members of land councils representing indigenous groups; representatives from indigenous corporations formed by indigenous groups; relevant state and federal government departments; and nongovernmental actors such as fair trade organizations, as well as private sector actors such as mining corporations.

I. Extractive Property Rights

A. The Road to Riches

One influential idea within economic history is that it is institutions that set a country on the path to rags or riches.[13] Private property rights are generally the heroes in this story. Institutions that provide individuals with the opportunity to make secure investment decisions at the micro level lead to the macro success of a country. However, institutions can also play the role of villain. Some institutional

---

11. *See* THE RISE OF REGULATORY CAPITALISM: THE GLOBAL DIFFUSION OF A NEW ORDER 598 (David Levi-Faur & Jacint Jordana eds., 2005); JOHN BRAITHWAITE, REGULATORY CAPITALISM: HOW IT WORKS, IDEAS FOR MAKING IT WORK BETTER (2008).

12. Peter Drahos & Luigi Palombi, *The Sustainable Use of Australia's Biodiversity: Transfer of Traditional Knowledge and Intellectual Property*, Australian Research Council Project DP0880606, 2008–2010.

13. For examples of this line of argument, see D.C. NORTH, GROWTH & WELFARE IN THE AMERICAN PAST (1974); D.C. NORTH & R.T. THOMAS, THE RISE OF THE WESTERN WORLD (1973).

arrangements concentrate power in the hands of an elite few and allow these elites to prey on the economic efforts of the larger population. Such "extractive institutions" generally reduce the growth prospects of a country.[14]

The distinction between secure property institutions and extractive institutions has been used as part of a larger analysis aimed at explaining why the countries that were formerly linked to the flourishing economies of the Aztecs, Incas, and Mughals experienced a decline in economic prosperity while countries that began as colonies (Australia, Canada, New Zealand, and the United States) became much richer. How did colonies from poor beginnings grow to be rich and countries from rich empires become poor?

The critical factor in explaining why these colonies prospered lies not in their geographical endowments but rather in their institutional inheritance. When colonists created property rules that were broadly enfranchising of individual productivity they set a colony upon a long-run growth path. Where they established extractive institutions, as did the Spanish and Portuguese in the Americas, they created the conditions for long-term decline. The choice between establishing productive property rights and extractive ones was heavily affected by population density and whether Europeans decided to settle in the colony in large numbers.[15] In those colonies where Europeans began to settle in large numbers they were more likely to end up with property arrangements that were less discriminatory as among themselves than if they were an elite minority seeking to subjugate an indigenous population. As among themselves, the colonists progressively created an autonomous legal system. Such a system is characterized by a formal independence from politics and a commitment to rules as tools of governance, accountability, and procedure.[16] A property order embedded in autonomous law meets the economist's ideal of well-defined and secure property rights.

B. The Road to Poverty

As we will see in this section, at the same time as colonial authorities were creating the conditions for secure property rights to serve the interests of their ever-expanding populations they were engaged in the creation of a property order that was maximally extractive in terms of its effects on indigenous populations. The case of the Australian colonization provides an example of this broader

---

14. Daron Acemoglu, Simon Johnson & James A. Robinson, *Reversal of Fortune: Geography and Institutions in the Making of the Modern World Income Distribution*, 17 Q. J. ECON. 1231 (2002).

15. *Id.*

16. PHILIPPE NONET & PHILIP SELZNICK, LAW AND SOCIETY IN TRANSITION: TOWARD RESPONSIVE LAW 54 (1978).

pattern of setting indigenous people along a path of having to fight the effects of an extractive property order. For present purposes I use the concept of an extractive property order to apply to cases where the rules of the property order allow one group the benefit of large-scale, uncompensated transfers of assets belonging to another group.

The taking of indigenous land by colonists can be divided into a formal process of rule imposition and a practical process of dispossession. The practical process of dispossession consisted of fighting a war of invasion and then occupying conquered territory.[17] The formal process of imposition involved the application of the common law system to the territory of what became Australia. The common law operated on the principle that the Crown was the root of all individual titles held by the subjects of the Crown. Once the common law was applied to the colonies, the Crown became the owner of the territory.[18] The extended doctrine of *terra nullius* enabled the application of the common law system. In its extended version a state could, through an act of occupation, claim sovereignty over a territory that was already inhabited but where the inhabitants had very low levels of social and political organization and so could be considered primitive.

The pattern that prevailed after colonization was one in which the colonizers created property rules that allowed them to make productive use of the assets they had acquired through military and juridical conquest while at the same leaving indigenous people with virtually no assets to enable them to pursue economically productive activity. For example, responding to demand from the London wool market in the early nineteenth century, Australian colonial authorities in the 1820s began to grant larger tracts of land, much of which was turned to sheep grazing.[19] By 1850, colonial Australia was supplying almost 50 percent of Britain's wool needs, this expansion in wool production being financed by London commercial credit.[20] Changes in British industrial and financial capitalism created an aggressive hunger for pastoral property in Australia. Squatters supported by London financiers became a powerful interest group, and in 1847 were able to secure changes to the law granting them long-term leases at comparatively little

---

17. For details, see Amanda Nettlebeck & Robert Foster, In the Name of the Law: William Willshire and the Policing of the Australian Frontier (2007); Henry Reynolds, The Other Side of the Frontier (1981).

18. However, the Australian High Court in its famous *Mabo* decision found that the reception of British law into Australian colonies did not produce the chain of extinguishment supposed by colonial legal authorities. *See* Mabo v. Queensland (No. 2), (1992) 175 CLR 1.

19. Philip McMichael, Settlers and the Agrarian Question: Foundations of Capitalism in Colonial Australia 55 (1984).

20. *Id.* at 101–03.

cost.[21] Millions of acres, which had for tens of thousands of years been the subject of indigenous systems of ownership, fell into the hands of "shepherd-kings."[22]

At the same time as colonial authorities were creating property rules for productive investments by various groups within the colony (pastoralists, farmers, miners etc.) under the new property order they left indigenous people with almost no assets. The principal mechanism that colonial authorities settled on to deal with the pressure they were receiving from indigenous people as well as from politically influential antislavery interests in England was the creation of reserves. These were areas of Crown land set aside for use by aboriginal people. Thirty-five reserves were created in 1850 in New South Wales (NSW).[23] They were approximately one square mile each in area.

The reserve system did not deliver security of property. By way of example, the Weipa reserve in far North Queensland was cut by the government in 1958 from 354,828 hectares to 124 hectares in order to accommodate the bauxite leases of the mining company Comalco.[24] In NSW, reserve lands in 1938 amounted to about 15,000 acres. By 1964 this had been cut to about 1,500 acres, the rest being revoked or leased to white people.[25]

So far we have seen how after colonization indigenous people in Australia faced an extractive property order and continued to face one even as autonomous law delivered secure property institutions for non-indigenous Australians. Consistent with Acemoglu's broader analysis of economic development, this extractive property order consigned indigenous people to a path of poverty, one that has proven difficult to leave. For example, home ownership for most Australians represents their biggest investment decision and their most important source of wealth. On the measures surrounding home ownership indigenous Australians do badly. They are more likely to be homeless, less likely to be investing in their own home, and more likely to live in overcrowded households.[26]

---

21. Stephen H. Roberts, The Squatting Age in Australia: 1835–1847, at 271 (1975).

22. Roberts, *supra* note 21.

23. Heather Goodall, Invasion to Embassy: Land in Aboriginal Politics in New South Wales, 1770–1972, at 54–56 (1996).

24. Frank Brennan, Land Rights Queensland Style: The Struggle for Aboriginal Self-Management 8 (1992).

25. Goodall, *supra* note 23, at 321–22.

26. Indigenous Australians make up 2.4 percent of the population but account for 9 percent of the homeless population. Based on 2006 census data, 36 percent of the 166,700 indigenous households were investing in home ownership compared to 69 percent of seven million non-indigenous households. For these and other statistics, see Dep't for Families, Hous., Cmty. Servs. & Indigenous Affairs, Indigenous Home Ownership Issues (2010).

Over long periods of time indigenous people in settler states have, through persistent and resilient land rights activism, found ways to recover control of some land. For example, in Australia in 2008 indigenous-owned or controlled land amounted to 17 percent of the area of Australia.[27] Elsewhere I have argued that indigenous people have distinctive place-based forms of innovation.[28] As groups of indigenous people regain a degree of control over land, they have the opportunity to strengthen and reinvigorate their innovation systems. Their process of innovation is deeply integrated into a cosmological connectionist scheme in which all innovation has threads leading back to ancestors. The innovation system depends on the transmission of noncodified personal knowledge that has to be learned through personal forms of interaction (storytelling, individual training, ritual) rather than through the study of codified rules. Finally, the goals and expression of innovation have less to do with products and more to do with services to the land of one's ancestors.

The value of knowledge assets generated by indigenous people in Australia and other countries has received a level of symbolic recognition in treaties such as the Convention on Biological Diversity (CBD) of 1992.[29] But for the most part the dominant property rights systems for any potential exploitation of indigenous knowledge assets remain national systems of intellectual property rights. These systems have their origins in the economic policies of Europe's medieval kingdoms or the industrializing states of early European capitalism.[30] Not surprisingly, a study by WIPO observed that at some point "existing IP mechanisms cannot fully respond to the characteristics of certain forms of traditional knowledge, namely, their holistic nature, collective origination and oral transmission and preservation."[31] There are individual cases where indigenous people have

---

27. Steering Comm. for the Review of Gov't Serv. Provision, Overcoming Indigenous Disadvantage: Key Indicators 54 (2009), *available at* http://www.pc.gov.au/__data/assets/pdf_file/0003/90129/key-indicators-2009.pdf.

28. Peter Drahos, *When Cosmology Meets Property: Indigenous People's Innovation and Intellectual Property*, 29 Prometheus 233 (2011).

29. Article 8(j) of the Convention on Biological Diversity (CBD) is the most prominent example. *See* Convention on Biological Diversity, art. 8(j), June 5, 1992, 31 I.L.M. 818 [hereinafter CBD]. For a discussion of the CBD and other treaties that deal with indigenous knowledge and intellectual property, see Christoph Antons, *The International Debate About Traditional Knowledge and the Approaches in the Asia-Pacific Region, in* Traditional Knowledge, Traditional Cultural Expressions and Intellectual Property Law in the Asia-Pacific Region 39 (Christoph Antons ed., 2009).

30. *See* Christopher May & Susan K. Sell, Intellectual Property Rights: A Critical History (2006).

31. World Intellectual Prop. Org. (WIPO), Intellectual Property Needs and Expectations of Traditional Knowledge Holders 216 (2001), *available at* http://www.wipo.int/export/sites/www/freepublications/en/tk/768/wipo_pub_768.pdf.

successfully used systems such as copyright to protect a given asset.[32] Despite such individual successes, the structure of intellectual property systems allows for the possibility of uncompensated transfers of indigenous knowledge assets. In order to better understand the extractive effect of intellectual property, the next subsection draws a distinction between the public domain and the intellectual commons.

C. *The Extractive Public Domain*

In the context of intellectual property systems, a minimal conception of the public domain would be an information domain linked to a given intellectual property system (e.g., copyright, plant variety rights, patents) that a third party is not barred from accessing by virtue of a property rule. For example, when a patent expires "all the world may make and use the article [that is] the subject of the patent."[33] This minimal conception of the public domain ties a public domain to an individual system. Rather than there being a single public domain there is a public domain for the purposes of copyright law, one for patent law, one for trademark law, and so on. The commons, on the other hand, is an arrangement in which a defined group is linked to assets, and the members of the group have some of the incidences of ownership (for example, use rights) over those assets.[34]

When it comes to understanding how the ownership of knowledge assets works within indigenous communities in Australia, the concept of the intellectual commons is a more useful one than that of the public domain. Indigenous knowledge assets are generated as part of a cosmological framework linked to the deeds of ancestors who shaped the land to which indigenous people belong.[35] Most or all knowledge that is part of a group's system can be traced back to the acts of powerful ancestors in the Dreamtime.[36] Dreamtime stories are the threads that connect different parts of an indigenous knowledge system.

---

32. For a discussion of the copyright cases in Australia dealing with indigenous artwork, see Kathy Bowery, *Indigenous Culture, Knowledge and Intellectual Property: The Need for a New Category of Rights?*, *in* Emerging Challenges in Intellectual Property 46 (Kathy Bowery, Michael Handler & Dianne Nicole eds., 2011).

33. Cellular Clothing Co. v. Maxton & Murray (1899 AC 326), at 344 (cited with approval by Brennan, J. in Sanofi v. Parke Davis Ltd. (No. 2) [1983] HCA 32).

34. One can develop different types of commons using criteria such as inclusiveness and exclusiveness. For a discussion, see Peter Drahos, A Philosophy of Intellectual Property ch. 3 (1996).

35. Ian Keen, Aboriginal Economy and Society: Australia at the Threshold of Colonisation 211 (2004).

36. For examples of the power of ancestors to shape the land, see Howard Morphy, Ancestral Connections: Art and an Aboriginal System of Knowledge 220–22 (1991); David McKnight, People, Countries, and the Rainbow Serpent 78–79 (1999).

The knowledge assets of indigenous people can be thought of as being part of a commons because these assets circulate within a defined group or subgroup. Who gains access to what knowledge depends on a variety of things, but even if some knowledge is widely known within a group, it does not follow that all members of the group may use the knowledge. For example, Howard Morphy's work on the Yolgnu system of knowledge reveals a split level of operations.[37] At one level it operates in a clearly delineated way to allow women and uninitiated men access to public knowledge while initiated men have access to secret knowledge, with the most important secret knowledge being controlled by male elders. Probing beneath the surface reveals a murkier world in which much secret knowledge is in fact known by women, and secrecy is less about keeping knowledge hidden and more about who can work with the knowledge.[38] In the case of the Yolgnu one can say there is an intellectual commons made up of access and use rights to knowledge. But there is not a public domain in which strangers have unrestricted use rights. What is public to the group is not open to all those outside of the group to use.

Along similar lines, Ian Keen's case studies of indigenous land tenure in Australia show that there were a number of systems characterized by different mixes of usufructuary rights, individual rights of exclusion, and community title.[39] It was not so much a system of land tenure as a resource tenure system that was cosmologically anchored in a web of kinship relations, rights, and obligations. In all regions, individual or group ownership arrangements were integrated with a system that gave non-owners some use rights over the resources of the "Country."[40] In this networked system of resource governance the specific rights an individual had depended on where one stood in the web of kinship relations. McKnight's work with the Lardil People, for example, shows that the Country of one's father or father's sister would give individuals specific rights in the division of game whereas the Country of one's sister-in-law would allow for general hunting rights but with an obligation to give big game to the senior member of the Country.[41] Under this system of resource governance resource assets did not, as in the case of the patent public domain, fall under a property rule that allowed for unrestricted third-party access.

---

37. Howard Morphy, Ancestral Connections: Art and an Aboriginal System of Knowledge (1991).

38. *Id.* at 89.

39. The case studies are reported in Keen, *supra* note 35, ch. 9 (2004).

40. Keen uses the term "Country" to refer to the land and waters "accorded the identity of a *country group* that 'held' or 'owned' the country." *Id.* at 421. According to McKnight, "Country" is a place of belonging where one can truly "sit down." McKnight, *supra* note 36, at 81.

41. McKnight, *supra* note 36, at 84.

The public domain rules of intellectual property systems have an extractive effect precisely because they allow for transfers of assets that under indigenous systems of resource governance remain the subject of individual or group control. For example, plants for indigenous people have many dimensions. They may be ancestors and function as totems for particular groups.[42] A tree may have utilitarian functions such as providing shelter and being a source of medicine, but it may also be linked to a person by virtue of a kinship relation because it features in an ancestral story on that person's mother's side leading that person to say that "'this tree is my mother.'"[43] Knowledge about the plant will be multilayered with some levels of knowledge only open to some people, and then only after initiation. Indigenous people may be prepared to reveal some simple uses of plants to ethnobotanists, but it does not follow they are putting the plant into the public domain to be used by any third party.[44] Plants do not exist in the "wild," but rather are part of Countries, which, in their role as ancestors, they may have shaped. For the purposes of the patent system, a plant in the wild is a discovery that is open to anyone to use in the making of an invention. If, the plant turns out to be useful in a genetically based invention over which a patent is granted, the patentee enjoys a limited period of protection, after which third parties can make use of the invention. From an indigenous perspective, a plant with which a population has long and deep multilevel ties is under patent rules treated as open to anyone to use, manipulate, and turn into a private asset.

It might be argued that the extractive effect of intellectual property rules on indigenous resource systems has been modified in some sectors. The most obvious example of this modification is the recognition in the CBD of principles of informed consent and benefit sharing when it comes to the use of genetic resources.[45] These principles have received a more detailed expression in the Bonn Guidelines and the Nagoya Protocol on Access to Genetic Resources and the Fair and Equitable Sharing of Benefits Arising from Their Utilisation, to the Convention on Biological Diversity.[46] The linkages that the CBD regime creates

42. W.E.H. Stanner, *Religion, Totemism and Symbolism*, *in* W.E.H. Stanner, White Man Got No Dreaming, 106, 127–29 (1979).

43. Ian Keen, Knowledge and Secrecy in an Aboriginal Religion 107 (1994).

44. For an example of where an indigenous group has revealed some simple uses of plants, see L.G. Liddy et al., Wagiman Plants and Animals: Aboriginal Knowledge of Flora and Fauna from the Mid Daly River Basin, Northern Australia (2006). But as this publication makes clear, there is much knowledge about the plants that has not been revealed. *See id.* at 15.

45. *See* CBD, *supra* note 29, art. 15(5)–(6).

46. For example, a system of informed consent should also require the consent of indigenous communities where it is appropriate. *See* Secretariat of the Convention on Biological Diversity, Bonn Guidelines on Access to Genetic Resources and Fair and Equitable Sharing of the

among the principles of informed consent, benefit sharing, and the interests of indigenous people are important. However, they do not necessarily stop the extractive effect of intellectual property rules.

The CBD itself expressly recognizes the sovereignty of states over their natural resources.[47] The principles within the CBD regime that potentially favor indigenous people represent negotiating gains that may or may not be turned into real economic gains at the level of state law. If we accept that states are rational egotists, acting in accordance with what they perceive to be their national interest, then we have to accept the possibility that they may symbolically comply with an international agreement while finding ways to avoid the import of its obligations.[48] The complexity of national property systems means that a state potentially has many ways to devise rules that pay symbolic attention to the needs of indigenous people while in fact depriving them of concrete economic benefits.[49]

The issue of whether the extractive effects of intellectual property rules have been modified by CBD principles depends on a fine-grained examination of a given state's property rights regime. It is important to emphasize that in this analysis the entire property regime of a state needs to be evaluated. It is unlikely, I suggest, that the extractive effects of a state's property order upon indigenous people will depend on just one rule. It is the combination of rules that determines whether the institution of property within a state is extractive for a given indigenous group and therefore whether the state is setting its indigenous people on the path to prosperity or poverty. In order to illustrate this claim the next two sections briefly examine the cases of Queensland's Biodiscovery Act of 2004 and India's Traditional Knowledge Digital Library (TKDL) initiative.

D. Queensland's Biodiscovery Regime

One of the main purposes of the Queensland Biodiscovery Act 2004 is to allow for the collecting of "minimal quantities of native biological resources on or in State land or Queensland waters."[50] Collecting on private land is not covered. The

---

Benefits Arising out of their Utilization (2002), ¶¶ 26(d), 31, *available at* https://www.cbd.int/doc/publications/cbd-bonn-gdls-en.pdf.

47. *See* CBD, *supra* note 29, arts. 3, 15(1).

48. On states as rational egotists, see ROBERT O. KEOHANE, AFTER HEGEMONY: COOPERATION AND DISCORD IN THE WORLD POLITICAL ECONOMY (1984).

49. The classic work on the symbolic use of politics to reassure people while delivering real benefits to special interest groups is MURRAY EDELMAN, THE SYMBOLIC USES OF POLITICS (1964).

50. *See* Queensland Biodiscovery Act, § 3(1)(a) (2004), *available at* http://www.legislation.qld.gov.au/LEGISLTN/CURRENT/B/BiodiscovA04.pdf.

Biodiscovery Act makes no provision concerning any interests that indigenous people may have in the collection of native biological material. Although it does refer to Article 15 of the CBD it does not refer to Article 8(j) of the CBD, the central animating principle of the CBD on the recognition of indigenous peoples' interests in the use of their innovation systems as they relate to biodiversity.[51]

Queensland does have the Queensland Biotechnology Code of Ethics[52], a code that was first released in 2001. Within the Code there are guidelines requiring those in the biodiscovery business to negotiate benefit-sharing arrangements with indigenous people in cases where they have used their traditional knowledge.[53] There also is a guideline prohibiting both biopiracy and assisting a third party to commit biopiracy.[54] The Code is mandatory for Queensland government organizations, organizations receiving biotechnology funding from the government, or research organizations that have government officials as part of their membership.[55] The principal sanction for an organization failing to comply with the Code is a review of its government funding.

The Biodiscovery Act and the Code offer lots of scope for a biotechnology organization to steer away from having to deal with an indigenous group if it so wishes. Under the Biodiscovery Act, indigenous people are not involved in the permission process to gain access to state land or Queensland waters. An organization can also choose to collect on private land not owned by indigenous people.[56] The Code does apply to private land.[57] The obligation to enter into a benefit-sharing agreement under the Code applies only to knowledge obtained during the course of biodiscovery.[58] No obligation arises for a company under the guidelines if it uses ethnobotanical information in the public domain (as defined under the rules of the settler state) as a guide to searching.

---

51. *See id.* § 4(3)–(4).

52. Queensland Gov't., Queensland Biotechnology Code of Ethics (2006), *available at* http://www.qld.gov.au/dsitia/assets/documents/qld-biotechnology-code-of-ethics.pdf [hereinafter the Code].

53. *Id.* at 8, ¶ 10.

54. *Id.*

55. *Id.* at 4.

56. As Natalie Stoianoff observes, the exemption of private land provides considerable scope for circumventing the operation of the Biodiscovery Act. *See* Natalie P. Stoianoff, *The Recognition of Traditional Knowledge Under Australian Biodiscovery Regimes: Why Bother with Intellectual Property Rights?*, *in* TRADITIONAL KNOWLEDGE, TRADITIONAL CULTURAL EXPRESSIONS AND INTELLECTUAL PROPERTY LAW IN THE ASIA-PACIFIC REGION 293, 308 (Christoph Antons ed., 2009).

57. *See* the Code, *supra* note 52, at 8, ¶ 10.

58. *See id.*

More revealing, however, is the history of collecting in Queensland prior to the release of the Code in 2001 and the enactment of the Biodiscovery Act in 2004. In 1993, Griffith University forged what was to become a long-term partnership with the pharmaceutical company AstraZeneca (then simply Astra).[59] During the 1990s, Griffiths's laboratory became a node of world excellence in high-throughput screening. From 1994 until 1997 continuous screening was used, allowing about 20,000 extracts to be evaluated against target screens. In 1997 the laboratory shifted into campaign screening with 30,000 extracts being tested in four weeks, and by 2000 the first 100,000-extract campaign was completed.[60]

Crucial to the success of this approach in terms of drug discovery is a large extract library. The more extracts, the more opportunities there are to find something that might react with the target. Griffith's extract library tripled in size between 1997 and 2000.[61] Two public bodies, the Queensland Museum and the Queensland Herbarium, were central to the effort of collecting the thousands of samples that formed the basis of the extract library.[62] At first collecting in relatively small areas reaped rich results, as one might expect in a region of high biodiversity. But over time the collectors from these institutions found that they were traveling farther and getting less and less.[63]

For most of the Griffith–AstraZeneca partnership, collecting took place in the absence of specific access and benefit-sharing legislation. High-throughput screening of the AstraZeneca–Griffith collection began in 1994.[64] Australia ratified the CBD in 1993. The Queensland government published its Code in 2001, and its Biodiscovery Act came into operation in 2004. Collecting, as we have seen, was particularly intense in the late 1990s when the size of the extract library tripled. Also, some vagueness surrounds the areas of collecting. There were allegations that some collecting took place on indigenous peoples' land, but this has been denied by the Museum and the Herbarium.[65] But even if collecting did

59. This partnership has been the subject of a detailed case study. *See* SARAH LAIRD, CATHERINE MONAGLE & SAM JOHNSTON, QUEENSLAND BIODIVERSITY COLLABORATION, THE GRIFFITH UNIVERSITY PARTNERSHIP FOR NATIONAL PRODUCT DISCOVERY: AN ACCESS & BENEFIT SHARING CASE STUDY (2008).

60. *See* Ronald J. Quinn, Priscila de Almeida Leone, Gordon Guymeyer & John N.A. Hooper, *Australian Biodiversity via Its Plants and Marine Organisms: A High-Throughput Screening Approach to Drug Discovery*, 74 PURE APPLIED CHEM. 520 (2002).

61. *Id.*

62. Author interviews with the Queensland Museum and Herbarium, 2009.

63. Author interview with the Queensland Herbarium, 2009.

64. Quinn et al., *supra* note 60.

65. LAIRD ET AL., *supra* note 59, at 26.

take place on indigenous land (as legally defined by the settler state) this may not matter much because, as some officials at the Queensland Department of Environment and Resource Management pointed out, "most people got what they wanted before the legislation kicked in."[66]

This case is instructive because it shows how the extractive nature of a property order for indigenous people does not lie in one or two rules but in the combined operational effect of the property institution, including when rules are actually brought into operation and how they are brought into operation (as law or ethical guidelines). As I have explained, indigenous peoples' knowledge assets are place-related assets. Rules that place plants with which indigenous people have had a working relationship for thousands of years in the public domain make it difficult for indigenous people to protect the asset value of the plant, especially if they are combined with a real property regime that allows organizations to steer away from having to deal with indigenous groups.

From the perspective of indigenous people, the settler property institution has allowed for uncompensated transfers of indigenous peoples' assets to take place. Queensland's mega biodiversity, they would argue, is not just a random event, but the product of ecosystems innovation by indigenous people using management tools such as fire.[67] As explained earlier, no resource assets under their system of resource tenure went unallocated but rather were the subject of networked governance based on a mix of property arrangements involving the commons, individual ownership, and use rights. The settler state has relocated most of these resource assets to different parts of its property order (e.g., land law, environmental law, intellectual property law, biodiscovery law), leaving indigenous people with the challenge of managing whatever assets they have or can reclaim under a property order that is fundamentally alien to them.

---

66. Author interview with the Queensland Department of Environment and Resource Management, 2009.

67. On the links between indigenous fire management and its effects on the Australian ecological systems, see D. Yibarbuk et al., *Fire Ecology and Aboriginal Land Management in Central Arnhem Land, Northern Australia: A Tradition of Ecosystem Management*, 28 J. BIOGEOG. 325 (2001); Jeremy Russell-Smith et al., *Contemporary Fire Regimes of Northern Australia, 1997–2001: Change Since Aboriginal Occupancy, Challenges for Sustainable Management*, 12 INT'L J. WILDLAND FIRE 283 (2003), *available at* http://www.southwestnrm.org.au/sites/default/files/uploads/ihub/russell-smith-j-2003-contemporary-fire-regimes-northern-australia-1997-2001-change.pdf; CULTURE, ECOLOGY AND ECONOMY OF FIRE MANAGEMENT IN NORTH AUSTRALIAN SAVANNAS: REKINDLING THE WURRK TRADITION (Jeremy Russell-Smith, Peter Whitehead & Peter Cooke eds., 2009).

E. An Iron Law of Extraction?

My hypothesis that the property order of states is extractive for indigenous people is not necessarily confined to settler states in which colonizers conquered indigenous people and established their own property regime. It may also be true of states where the colonizer left or was thrown out, and it may be true in states where the political elite is itself indigenous. In other words, there is the possibility of an iron law of extraction that operates when states begin to define property orders for indigenous groups. The composition of the state's political elite may not matter to the operation of this iron law.[68] The testing of this hypothesis depends on an assessment of the operational effect of the whole property order and not just some part of it. An example that shows why systems testing of a property order is needed rather than rule testing is India's TKDL initiative.

In response to European patent claims relating to the neem plant and US patent claims concerning turmeric, the Indian government set up a task force in 2000 to track the extent of misappropriation of Indian traditional medicinal knowledge.[69] The task force studied the patent databases of the major patent offices and produced numbers suggesting that tens of thousands of patents were being granted over materials with medicinal uses that had been known about for a long time in Indian knowledge systems such as Ayurveda, Unani, and Siddha. India was able to push for reforms of the International Patent Classification (IPC) system, the system of hieroglyphic complexity used by patent offices to classify inventions. Among other things, India designed a traditional knowledge resource classification system that shares structural similarities with the IPC. As a result, India's traditional knowledge systems are much better integrated into the patent-searching systems used by the world's major patent offices.[70]

On the surface it appears that India has taken steps to prevent uncompensated transfers of knowledge assets from taking place. But as noted earlier, an assessment of the extractive effects of property upon indigenous peoples' resources has to be evaluated by examining the interactional effects of the rules that make

68. For a study of the protection of traditional knowledge in Samoa and Vanuatu that lends support to this iron law, see Miranda Forsyth, *Lifting the Lid on "The Community": Who Has the Right to Control Access to Traditional Knowledge and Expressions of Culture?*, 19 INT'L J. CULT. PROP. 1 (2012).

69. For a brief description of the history of the task force as well as the data produced by it, see Dep't of Ayurveda, Yoga & Naturopathy, Unani, Siddha and Homoeopathy, TRADITIONAL KNOWLEDGE DIGITAL LIBR., http://indianmedicine.nic.in/showfile.asp?lid=316 (last visited Sept. 4, 2013).

70. For a description of the cooperation between WIPO and India on the TKDL, see Press Release, WIPO, WIPO and India Partner to Protect Traditional Knowledge from Misappropriation, WIPO Doc. PR/2011/682 (Mar. 22, 2011), *available at* http://www.wipo.int/pressroom/en/articles/2011/article_0008.html.

up a state's property institution. If we subject India's property institution to this broader empirical test it may be that we would find extractive effects.

By way of example it is worth focusing on what happens to the land rights of indigenous people in India when they have to confront mining companies interested in the mineral resources on their land. The State of Orissa in India is a state rich in tribes and indigenous knowledge. It is also a state rich in natural resources such as water and forests as well as minerals.[71] One of its minerals is bauxite, the high quality of which leads to lower refining costs. In the south, where this better grade of bauxite is to be found, live tribes such as the Kondhs, for whom the mountainous hills are the sacred centerpiece of their cosmological scheme.[72] Like indigenous people in Australia, the Dongaria Kondh have used fire over the generations to cultivate a habitat rich in biodiversity.[73] There is a deep knowledge of plant medicines to treat a range of things including "scorpion and snake bites, stomach disorders, arthritis, tuberculosis, paralysis, cholera, acidity, eczema, tumours, menstrual disorders, wounds and sores, diarrhoea, dysentery, bone fractures, rheumatism, asthma, malaria."[74]

As documented by the anthropologist Felix Padel, confrontations between tribal people and mining companies in India frequently turn tribal people into landless laborers who enter a spiral of ill health and lack of income, a spiral that propels them down into a netherworld of destitution and desperate acts.[75] Benefit-sharing obligations tend to drift away on the wind in these kinds of real world situations. Once indigenous people have been turned into landless wanderers they disappear into India's long night of poverty. There is nobody with whom to sign a benefit-sharing agreement. One arrives at a mining site, not a thriving ecosystem. The landscape has been transformed beyond recognition. The top of the hill has disappeared. The giant neem trees have died. The indigenous people have vanished. Who had custody of this plant? Who was responsible for it? Nobody knows. There are just mining company roads, giant trucks, noise, and a lot of dust.

---

71. For a discussion of mining in Orissa and its impacts on tribes, see Leah Temper & Joan Martinez-Alier, *Is India Too Poor to Be Green?* 42 Econ. & Pol. Weekly 1489 (2007).

72. For a detailed report, see N.C. Saxena, S. Parasuraman, Promode Kant & Amita Baviskar, Report of the Four Member Committee for Investigation into the Proposals Submitted by the Orissa Mining Company for Bauxite Mining in Niyamgiri, Submitted to the Ministry of Environment and Forests, Government of India (2010), *available at* http://moef.nic.in/downloads/public-information/Saxena_Vedanta.pdf.

73. *Id.* at 17.

74. *Id.* at 18.

75. Felix Padel & Samarendra Das, Out of This Earth: East India Adivasis and the Aluminium Cartel (2010).

Thus, before we can conclude that India's TKDL database is not extractive in its effects we need to know more about it. It is a proprietary database that the Indian government makes available to other patent offices on the condition that these offices use it only for patent searching and they only disclose so much as is essential as part of their reporting processes.[76] The aim is to preserve the secrecy of the information in the TKDL. The TKDL is a regulatory tool that is being used in a defensive strategy to stop the foreign patenting of medicines related to traditional knowledge and uses of plants originating from India. There is some evidence that the TKDL is being used by patent offices in decisions about the grant of a patent as well as by applicants to modify their claims.[77]

Indigenous people are also interested in positive economic agendas that liberate them from poverty and dependency on the state. The role of the TKDL in positive agendas for indigenous people is much less clear. How does the TKDL help indigenous tribes in India, especially if they lose their land? The loss of land represents a potentially deadly threat to the survival of a place-based cosmology and the generation of new knowledge. There may be few real benefits to indigenous groups in initiatives such as the TKDL. Once an indigenous group discloses its knowledge, that knowledge may well end up in codified form in a database such as the TKDL. In a world where an indigenous group has lost its land and has faded away after its knowledge assets have been extracted and relocated to an electronic cloud, the real beneficiaries may turn out to be the companies and researchers who use the database as leads to patentable compounds and genes. The TKDL does not necessarily stop a patent application related to the traditional use of a plant. It may instead improve the quality of the claims over the invention and produce a stronger patent.[78]

---

76. A copy of the agreement allowing the European Patent Office access to the TKDL is available at http://www.spicyip.com/docs/TKDL-EPO.pdf (last visited Sept. 4, 2013).

77. For a list of patent applications at the European Patent Office involving the TKDL, see *Examples of EPO Applications Set Aside, Modified or Withdrawn Based on TKDL Evidence*, WIPO, http://www.wipo.int/meetings/en/2011/wipo_tkdl_del_11/epo_examples.html (last visited Sept. 4, 2013).

78. An example of where this occurred is patent EP 1849473, granted by the European Patent Office to Livzon Pharmaceutical Group, a Chinese pharmaceutical company. The invention concerned a traditional medicine composition for the treatment of avian influenza. The Council of Scientific and Industrial Research, which administers the TKDL in India, filed relevant prior art under the third-party observation procedure at the EPO. This led to an amendment of some of the claims. All the relevant documents are available at https://register.epo.org/espacenet/application?number=EP07001215&lng=en&tab=doclist (last visited Sept. 4, 2013).

II. Indigenous Developmental Networks

A. Developmental Networks and Adaptive Governance

The upshot of the previous part is that state property orders are likely to be extractive and therefore non-developmental for indigenous people. As the cases of Australia, Canada, New Zealand, and the United States suggest, overcoming the effects of an extractive property regime is a long-term process. These are four rich states yet indigenous people in these countries remain on the bottom rungs of the socioeconomic ladder. Extractive property regimes would seem to generate long-term path dependencies.

In the remainder of the chapter I will argue that where indigenous groups have managed to obtain some degree of control over their land they are in a position to build an indigenous developmental network for the purpose of making economic gains from their knowledge assets. An assumption is that indigenous groups will probably continue to face an extractive property regime, including extractive rules of intellectual property. Their challenge is to manage the effects of this regime in ways that meet their development objectives. Changing the rules of the regime is desirable, but such change does face political feasibility constraints. In the twentieth century, intellectual property standard-setting was dominated by coalitions of developed states, industry peak bodies, and multinational companies.[79] It is true that the politics of intellectual property standard-setting has seen the entrance of more actors, including social movement actors.[80] As a result the alliance politics of intellectual property standard-setting has become more complex. That said, states and multinational actors remain dominant players. In the short to medium term at least, indigenous people will have to work with intellectual property rules as they find them rather than as they would like them to be.

My claim that groups of indigenous people should invest in the creation of indigenous developmental networks draws on a body of organizational literature that recognizes how successful knowledge development within a firm depends on an informal network operating with a "sense of shared destiny" and mutual trust.[81] The firms that do best in terms of managing knowledge assets rely on techniques that build the common identity needed to create the trust that makes employees ready to reveal knowledge to others. Using networks to manage resource assets is not an unfamiliar way of working for indigenous people. For example, local

---

79. For an account, see PETER DRAHOS & JOHN BRAITHWAITE, INFORMATION FEUDALISM: WHO OWNS THE KNOWLEDGE ECONOMY? (2002).

80. For a comprehensive analysis, see ACCESS TO KNOWLEDGE IN THE AGE OF INTELLECTUAL PROPERTY (Gaelle Krikorian & Amy Kapczynski eds., 2010).

81. For an overview of this literature, see Paul S. Adler, *Market, Hierarchy, and Trust: The Knowledge Economy and the Future of Capitalism*, 12 ORG. SCI. 215, 220 (2001).

networks have been crucial to the successful community management of marine resources in many Pacific Island states.[82] A similar kind of local network philosophy can be applied to the management of knowledge assets.

The purpose of an indigenous developmental network is to do what the non-developmental state has failed to do—to promote the development goals of the indigenous group that the network represents. Based on the fieldwork carried out in Australia we see there are two basic conditions that have to be fulfilled in order for an indigenous developmental network to emerge. The first is that an indigenous group must find a path back to the control of land, and the second is that the network must be self-determining. The second condition simply means that the network has to be led by indigenous leaders. This does not mean that there cannot be non-indigenous participants in the network. In fact, as will become clear, it is almost certain that non-indigenous participants will need to be part of the network as one indigenous group is unlikely to have all the capacity it needs to develop its knowledge assets.

My description of an indigenous developmental network is not an *a priori* analytical construct but rather an inductively arrived at category based on fieldwork with indigenous communities in Australia. As mentioned in the introduction, the fieldwork covered parts of Central Australia, Eastern Australia, Northern Australia, and Southern Australia. Over a period of three years we came across two indigenous groups that were being represented by what I have called a developmental network. One group was located in the Kimberley region in Western Australia (the Jarlmadangah Burru Aboriginal Community) and the other in Central Cape York Peninsula in Queensland (a group of families from the Kuuku I'yu Northern Kaanju). In both these cases, indigenous groups had found a way back to Country and were, among other things, using intellectual property rights as part of a plan to exploit their knowledge assets.[83] Both groups had a clear development focus. Both groups had strong and creative leadership prepared to take risks. They wanted to create enterprises around their knowledge assets that would lead to further investment, as well as employment and education opportunities, especially for their children. Everywhere we went in Australia

82. For a survey and discussion, see H. Govan, *Status and Potential of Locally-Managed Marine Areas in the South Pacific: Meeting Nature Conservation and Sustainable Livelihood Targets through Wide-Spread Implementation of LMMAs,* Coral Reefs Initiative for the Pacific Study Report (2009), http://www.crisponline.net/Portals/1/PDF/0904_C3A_Govan_MMAs.pdf.

83. For details, see Drahos, *supra* note 28; David J. Claudie, Susan J. Semple, Nicholas M. Smith & Bradley S. Simpson, *Ancient but New: Developing Locally Driven Enterprises Based on Traditional Medicines in Kuuku I'yu Northern Kaanju Homelands, Cape York, Queensland, Australia, in* Indigenous Peoples' Innovation: Intellectual Property Pathways to Development 29 (Peter Drahos & Susy Frankel eds., 2012).

indigenous people spoke about their hopes for a better future for their children. It is an overriding concern as several researchers note:

> We envisage that culturally appropriate development of medicinal plant products will contribute to improved opportunities for Kuuku I'yu people to live and work on homelands. This will also allow younger people to engage with and learn about natural resources on their homelands, and provide an alternative to life in centralised townships.[84]

Indigenous developmental networks are locally led entities, and in order to be effective as agents of development they have to remain so. Information asymmetry helps to explain why. Clearly, each indigenous group will have the best information about what knowledge assets exist in the group, which ones can be used for economic purposes, and under what circumstances. Each indigenous group will be situated within a highly local context with a different set of assets in terms of knowledge, land, and leadership. For example, some indigenous groups will have to work in landscapes that have been highly degraded because of overexploitation through farming practices, while other groups will have the advantage of less-degraded landscapes. The level of knowledge assets will be different for each group, much depending on the extent to which the group's authority structures and methods of knowledge transmission have survived the long, hard processes of colonization. Once some successful indigenous developmental networks emerge we might expect a modeling effect as other groups learn from the success of these networks.[85] If a number of indigenous developmental networks spring up within the same region there may also be gains from coordinating on projects that generate joint benefits.

The difficult task that faces an indigenous developmental network is the adaptive management of the intellectual property tools that relate to the assets of the network. Adaptive management is an approach that has evolved in response to the uncertainties of dealing with natural resource problems. It is characterized by an emphasis on mechanisms of monitoring and evaluation of interventions into systems so as to gather more knowledge in order to improve the quality of the intervention.[86] One cycles through a process of doing and learning that leads to more knowledgeable doing, this cycling being repeated and always informed

---

84. Claudie at al., *supra* note 83, at 29, 30.

85. Modeling here is used in the sense of observational learning. *See* John Braithwaite & Peter Drahos, Global Business Regulation 580 (2000).

86. *See* George Stankey & Catherine Allan, *Introduction*, *in* Adaptive Environmental Management: A Practitioner's Guide 3, 5 (Catherine Allan & George H. Stankey eds., 2009).

by data gathering, evaluations, and explanatory model building.[87] In these cycles there is a systematized approach to working out precisely what one has learned from the doing. One can, of course, make the process of adaptive management more or less rigorous.

The failure of the settler state to act developmentally for indigenous people means that the task of managing the property institution falls to indigenous developmental networks. Adaptive management becomes the process through which indigenous networks journey in order to find responsive uses of systems that remain part of a fundamentally alien, autonomous, and extractive legal order. There really is no choice for indigenous developmental networks but to approach the intellectual property issues in the spirit of adaptive management. If they do not then there will be lost opportunities when it comes to building businesses using indigenous knowledge assets, or non-indigenous people will become the primary beneficiaries of those assets.

B. The Secret of the Mars Bar

By definition one cannot specify in advance of experience the steps in an adaptive management strategy. That said, there are reasons for thinking that in the case of managing indigenous knowledge assets, the core of an adaptive strategy lies in secrecy. Keeping knowledge secret is a highly effective form of protection.[88] For indigenous people secrecy is, unlike the patent system, a familiar form of protection. The trade in pituri by indigenous people in Australia, which covered some 550,000 square kilometers, provides an illustration.[89] Pituri is a plant drug that is a source of nicotine. The plant from which the drug was made, *Duboisia hopwoodii*, was found to grow widely in Australia's arid interiors, but it was pituri made from plants growing around a river system in the southwestern part of Queensland that was most desirable. An important part of the explanation for this lies in the curing techniques developed by the clans in this region. Members of these clans had

---

87. *See* Robert M. Argent, *Components of Adaptive Management*, *in* ADAPTIVE ENVIRONMENTAL MANAGEMENT, *supra* note 86, at 11–30.

88. For a discussion of surveys showing that many firms prefer to rely on first entry advantages and secrecy, see David Encauoua, Dominque Guellec & Catalina Martinez, *Patent Systems for Encouraging Innovation: Lessons from Economic Analysis*, 35 RES. POL'Y 1423 (2006). In a survey of 1478 R&D labs in the United States, Cohen, Nelson and Walsh found that secrecy, lead time, and complementary capabilities were the major mechanisms of appropriation, with secrecy being a very important mechanism. Patents were seen as a less-important mechanism. Wesley M. Cohen, Richard R. Nelson & John P. Walsh, *Protecting Their Intellectual Assets: Appropriability Conditions and Why U.S. Manufacturing Firms Patent (or Not)*, (Nat'l Bur. of Econ. Research Working Paper 7552, 2000).

89. Pamela Watson, *This Precious Foliage*, OCEANIA MONOGRAPH 26 (1983).

discovered that the strength of pituri varied with drying time, and that methods of drying by means of the sun weakened the potency of the drug. These clans developed an artificial method of drying the pituri by cooking it in hot sand. The great secret lay in the cooking time, which is consistent with scientific knowledge about the curing of plant alkaloids.[90] As one researcher explains, this knowledge was tightly controlled:

> The preparation and wholesale distribution of the narcotic *pituri* or pitcheri was in the hands of the descendants of the man who first discovered its narcotic properties. The preparation was a close monopoly and any child born to the horde or tribe who belonged to the pitcheri Moora automatically succeeded to all rights and privileges in the distribution.[91]

Whenever trade secret protection is being compared with patents, the risk of trade secret protection is said to be that trade secrets do not protect against the possibility of independent invention.[92] Although this is a risk, it is a risk that companies have learned to manage. For example, the Mars company limits knowledge of its recipes to as few employees as possible and allows very few visitors to its factories.[93] Those who make it past the gates have to sign ferocious-sounding confidentiality agreements. But anyone can buy a Mars product, read its ingredients on the wrapper, and subject the bar to a chemical analysis in order to learn about it.[94] The calculation that Mars is making is that a competitor may learn a lot about the chocolate bar through analysis, but not everything. Studying the bar will not reveal the exact source of the cocoa beans, the grinding machinery and techniques employed to process the beans, other processes used on the other ingredients in the bar, and so on.

Even if a competitor was able to produce a bar that a consumer in a blind taste test could not pick from the Mars Bar, the competitor could not use the Mars trademark and packaging to sell the bar. The combination of secrecy and

---

90. *Id.* at 49.

91. George Aiston, *The Aboriginal Narcotic Pitcheri*, 7 OCEANIA 372 (1937).

92. *See, e.g.*, MICHAEL A. GOLLIN, DRIVING INNOVATION: INTELLECTUAL PROPERTY STRATEGIES FOR A DYNAMIC WORLD 179 (2008); Michael Risch, *Trade Secret Law and Information Development Incentives*, *in* THE LAW AND THEORY OF TRADE SECRECY: A HANDBOOK OF CONTEMPORARY RESEARCH 152, 168 (Rochelle C. Dreyfuss & Katherine J. Standburg eds., 2011).

93. For an account of secrecy in the chocolate world, see JOËL GLENN BRENNER, THE EMPERORS OF CHOCOLATE: INSIDE THE SECRET WORLD OF HERSHEY AND MARS (1999). For an analysis of trade secrets in the context of the candy industry, see Jeanne C. Fromer, *Trade Secrecy in Willy Wonka's Chocolate Factory*, *in* THE LAW AND THEORY OF TRADE SECRECY, *supra* note 92.

94. BRENNER, *supra* note 93.

the brand image of Mars along with legal protection of these two things through trade secret law, trademark law, and jurisdiction-specific actions relating to the notion of unfair competition form a hugely powerful legal shield for Mars's products.

Secrecy and controlling access to land (just as Mars controls access to its factories) form the core of an adaptive management strategy for indigenous groups. The Hopi in the United States have been described as "fanatically committed to secrecy," not allowing tourists the use of media equipment and sketch pads.[95] It is a fanaticism that large corporations such as Mars understand and practice when it comes to protecting their most valuable secrets. Beyond the embrace of secrecy, an indigenous developmental network has to make calculations based on the products or processes it wishes to commercialize and the strengths and weaknesses of particular intellectual property systems.[96] At the same time the network has to keep in mind its own capacities. For example, it might be thought that in the case of plant knowledge, patent protection will be more important for indigenous groups than trade secrets. Although this might be true for pharmaceutical companies, it is probably not true for indigenous groups, or at least the cost-benefit calculation is different for indigenous groups than it is for pharmaceutical companies. One of the indigenous communities we interviewed had obtained patents over a class of compounds derived from a plant that was known to them.[97] Their experience with the patent system has thus far been disappointing. Obtaining venture capital to commercialize the compound has proven very difficult. They described the cost of paying patent fees as a burden on their community. When we asked them whether they would go down this path again they replied that they probably would not. One alternative in this situation is to use the plant in the preparation of a complementary medicine and invest in marketing the unique properties of the medicine. Trademarks rather than patents become important in this kind of commercialization strategy.

---

95. Michael F. Brown, Who Owns Native Culture? 13 (2003).

96. As Michael Gollin argues, it is all about becoming strategic. Michael A. Gollin, Driving Innovation: Intellectual Property Strategies for a Dynamic World 131 (2008). Not surprisingly, there are different views about the utility of other forms of intellectual property protection for indigenous knowledge. For a positive case for geographical indications, see Daniel Gervais, *Traditional Innovation and the Ongoing Debate on the Protection of Geographical Indications*, *in* Drahos & Frankel, *supra* note 83, at 121. For the negative case, see Susy Frankel, *The Mismatch of Geographical Indications and Innovative Traditional Knowledge*, 29 Prometheus 253 (2011).

97. *See* Drahos, *supra* note 28.

C. Canada as the Developmental State: The "Igloo Tag"

An example of where trademark protection has produced long-run benefits for an indigenous group is the case of the Inuit art market in Canada. Critical to the commercial emergence of Inuit art was James Houston, an artist who in 1948 traveled to Northern Canada and encouraged the Inuit to make traditional carvings out of soapstone.[98] A successful exhibition in Montreal led to government support for more carving. Within a decade the Inuit went from a situation of "near starvation" to benefiting from a "multimillion-dollar enterprise."[99]

For several decades, beginning in the 1950s, the Canadian government has played the role of the developmental state, assisting in organizing village cooperatives, buying Inuit art when sales went through a low period, and helping to organize international exhibitions to promote the art abroad.[100] Success always attracts imitators. In the case of the Inuit art market the imitators mass-produced plastic igloos, polar bears, and the like. The Canadian Department of Indian and Northern Affairs responded by registering in 1958 the "Igloo Tag."[101] Only Inuit artists and their marketing agencies could use this certification mark on their artwork. The Department was responsible for authorizing the printing of labels for use by marketing agencies.

Over the last several decades the Canadian government has become less interventionist than it was in the developmental decades of the 1950s and 1960s, and there are concerns about the future of the Inuit art industry.[102] But to the outside eye at least there are also impressive successes with networks of galleries, co-operatives, festivals, and associations working in different ways to support Inuit art.[103] This world of Inuit art production is very different from the impoverished communities that James Houston visited in the summer of 1948 at Port Harrison in Northern Quebec.

The lesson of this Canadian story is not so much about the power of trademarks, but the evolution of a network that ended up delivering greater capacity to the Inuit to commercialize their art, a network that made good use of trademark law. Houston's energy and leadership were obviously critical. He introduced

---

98. Leanne Stuart Pupchek, *True North: Inuit Art and the Canadian Imagination*, 31 AM. REV. CAN. STUD. 191, 199 (2001).

99. Nelson H.H. Graburn, *Inuit Art and Canadian Nationalism: Why Eskimos? Why Canada?*, 1 INUIT ART Q. 5, 6 (1986).

100. *See id.*

101. For a brief description of the mark, see *The Igloo Tag*, 5 INUIT ART Q. 57 (1991).

102. *See The Support System for Inuit Artists*, INUIT ART Q. 19, 86 (2004).

103. *See id.*

new ideas and materials to the Inuit and set up a payment system that worked. Also important were the promotion and lobbying activities of the Canadian Handicrafts Guild (formerly the Women's Art Association of Montreal).[104] The Canadian government, attracted by the Inuit success, increased the scale of that success through an international public relations campaign, as well as sending arts-and-craft specialists to help the Inuit with the quality of production.

D. Commodity Innovation, Systems Innovation

The previous two subsections suggested the potential usefulness of trade secrets and trademarks for the protection of indigenous knowledge assets. Neither of these systems of protection creates a public domain through fixed duration rules. However my basic point remains. Intellectual property orders are likely to remain extractive for indigenous groups. The distinctive feature of indigenous peoples' innovation lies in their capacity to innovate at the level of ecological services.[105] In other words, it is a capacity to devise systems for the maintenance of ecological systems. For reasons having to do with the commodity nature of capitalism, intellectual property rights are more suited to commodity-based forms of innovation (for example, an invention, a plant variety right, a semiconductor chip).[106] The collective labor of groups of indigenous people using a system of fire burning to manage Country (an ecological system) does not in any obvious way produce a single commodity. The improvements to an ecological system have important use values, but as we saw from the discussion of the extractive public domain, it is hard for indigenous people to capture private benefits from these use values.

One form of regulation that has enabled indigenous people to make some economic gains from their contribution to ecological services has been "regulation by information" in which ways are found to release information about goods or services to purchasers of those goods or services.[107] The social and environmental fields have seen the rise and globalization of certification organizations such as the Forest Stewardship Council and the Fairtrade Labelling Organizations International relying on labeling as part of their certification processes.[108] These organizations use the trademark system to communicate information to investors

---

104. Pupchek, *supra* note 98, at 200.

105. Drahos, *supra* note 28.

106. Marx's theory of capitalism links it to the fetishism of commodities. *See* Ben Fine, Marx's Capital 25 (1984).

107. Marc Schneiberg & Tim Bartley, *Organizations, Regulation, and Economic Behaviour: Regulatory Dynamics and Forms from the Nineteenth to the Twenty-First Century*, Ann. Rev. L. & Soc. Sci. 4, 31, 43 (2008).

108. *See* Margaret Chon, *Marks of Rectitude*, 77 Fordham L. Rev. 2311 (2009).

and consumers about the origins and quality of their standard-setting services. The standards of some organizations expressly recognize the role of indigenous knowledge in the maintenance of ecological systems.[109] These voluntary standard-setting bodies are also beginning to develop standards for specific ecological services markets. For example, in 2012, the Rainforest Alliance and its partner Imaflora validated a forest carbon project led by the Paiter Suruí, an indigenous tribe in the Brazilian Amazon.[110] Among the standards used in this process were the Climate, Community, and Biodiversity Standards, standards developed by an alliance of voluntary standards bodies, NGOs, and private companies.[111] Among other things, the standards require project developers wishing to use the standards to draw on the knowledge of indigenous groups when it comes to identifying the impacts of a project on natural resources and ecosystem services.[112]

By providing certification and labeling services, these voluntary standard systems help indigenous groups sell their products into consumer markets willing to pay a premium for ethical and sustainable production.[113] As the example of the Paiter Suruí shows, models using certification services can be devised that to a limited extent will support private investment in a function of an ecosystem. We are still a long way from being able to value the uses of ecological systems.[114] But the field of regulation by information does offer some opportunities for indigenous groups to benefit from the systems of innovation they have used over long periods of time to maintain ecosystems. It seems reasonably clear that the future of capitalism, at least in the context of ecosystems, lies less in Schumpeterian creative destruction and more in creative preservation. Over time we may be able to

---

109. Principle 3.4 of the Forest Stewardship Council's principles recognizes the rights of indigenous people and requires, among other things, that they be "compensated for the application of their traditional knowledge regarding the use of forest species or management systems in forest operations." Forest Stewardship Council (FSC), FSC Principles and Criteria for Forrest Stewardship, FSC-STD-01-001 (version 4-0), 1996, *available at* http://www.fsc.org/principles-and-criteria.34.htm.

110. *See* Rainforest Alliance, Suruí Forest Carbon Project Achieves Climate Validation, Jun. 4, 2012, *available at* http://www.rainforest-alliance.org/newsroom/news/surui-forest-carbon-project.

111. *See* The Climate, Community & Biodiversity Alliance, Climate, Community & Biodiversity Project Design Standards (2d ed. 2008), *available at* https://s3.amazonaws.com/CCBA/Upload/ccb_standards_second_edition_december_2008+(1).pdf.

112. *See id.* at 25.

113. *See* Anna Hutchens, Changing Big Business: The Globalisation of the Fair Trade Movement 114–19 (2009).

114. Nick Hanley & Edward B. Barbier, Pricing Nature: Cost-Benefit Analysis and Environmental Policy 205 (2009).

devise better regulatory models for maintaining ecological services.[115] For indigenous people these regulatory models may offer greater opportunities to make gains from their knowledge assets.

III. The New *Voyage of the Beagle*: Indigenous People and Scientists

A. Ecosystems: Crisis and Opportunity

Over time, science has helped to reveal the importance of ecological systems and the services that they provide. In 2001, the Millennium Ecosystem Assessment (MA) was launched under the auspices of the United Nations.[116] Involving more than 1,360 scientists from 95 countries it produced a series of technical studies and reports that give us the best understanding we have so far of changes in ecosystems and the likely consequences for human well-being. The picture that emerges from the MA is one in which our current state of development has been supported freely by streams of services in the form of clean water, food, disease control, climate regulation, and many others emanating, often in hidden ways, from mountain, polar, inland water, coastal, marine, forest, desert and other ecosystems. Economic growth has impacted these ecosystems to the point where some two-thirds of the services provided by these systems are in decline around the world.[117]

The MA and its sister global assessment project, the International Panel on Climate Change, are, much like Darwin's *Voyage of the Beagle* (1839), examples of scientific journeys. They are not the work of one scientist. They sweep across the scientific disciplines, piecing together a picture of many ecosystems operating at different levels and with high degrees of integration. The interdisciplinary science behind the MA has, in a sense, caught up with a basic truth known to many indigenous people. In Australia this truth is captured in the phrase "Caring for Country."[118] For indigenous people it is a truth that should be stated as a

---

115. For an example of a major initiative, see THE ECON. OF ECOSYSTEMS & BIODIVERSITY (TEEB), THE ECONOMICS OF ECOSYSTEMS AND BIODIVERSITY (TEEB), MAINSTREAMING THE ECONOMICS OF NATURE: A SYNTHESIS OF THE APPROACH, CONCLUSIONS AND RECOMMENDATIONS OF TEEB (2010), *available at* http://www.teebweb.org/publications/teeb-study-reports/synthesis.

116. *See Millennium Ecosystem Assessment*, U.N. ENV'T PROGRAMME, http://www.maweb.org/en/index.aspx (last visited Oct. 12, 2013).

117. MILLENNIUM ECOSYSTEM ASSESSMENT, ECOSYSTEMS AND HUMAN WELL-BEING: SYNTHESIS 1 (2005), *available at* http://www.millenniumassessment.org/documents/document.356.aspx.pdf.

118. The Australian government funds natural resource management projects under its "Caring for our Country" program. *See* Austl. Gov't, Dep't of Agric., Fisheries & Forestry, *Caring for*

categorical imperative—"Care for Country." But it also has a conditional form: If you want Country to Care for you then Care for Country. A truth known and lived locally by many indigenous groups now has a scientific macro-equivalent. If we are to continue to enjoy well-being we will have to care for the ecosystems on which we have depended for so long.

The global state of ecosystem crisis documented by the MA is also a time of opportunity for indigenous people. Behind the MA is a science that is more integrated and discursively open than perhaps at any time in history. The smokestack capitalism of the nineteenth century began, as Marx correctly foresaw, the process of bending science to its industrial and military needs, culminating in the twentieth century in a science that labored with some autonomy, but heeded the calls of its military and industrial paymasters.[119] The prospect of cooperation between what Thomas Kuhn called "normal science," an inward-looking science in which its practitioners do not lift their head above the paradigm they have internalized, and indigenous people is slim.[120] The prospects are better now for deeper epistemic cooperation between science and indigenous people than at any other time. For example, the MA did involve a multi-stakeholder board that included indigenous people and produced a report that examined how local and traditional knowledge systems could be incorporated into managing ecosystems.[121]

For indigenous people, the benefit of cooperation with scientists lies in the global recognition that science can bring to the achievements of indigenous people in creating systems for managing ecosystems. For example, Australia's colonizers had no understanding of the importance of indigenous fire methods to managing Australia's ecosystems.[122] Over the last few decades, as a result of the scientific study of these methods, their utility has come to be much better understood.[123] Science's techniques for analyzing, measuring, and quantifying the role

---

Our Country, 2008–2013, available at http://www.nrm.gov.au/about/caring/index.html (last visited Oct. 12, 2013). For indigenous people, the phrase refers to specific duties and responsibilities they have toward particular places. *See* Julian Gorman & Sivaram Vemuri, *Social Implications of Bridging the Gap through "Caring for Country" in Remote Indigenous Communities of the Northern Territory, Australia*, 34 RANGELAND JOURNAL 63, 66 (2012).

119. DRAHOS, *supra* note 34, at 112.

120. For criticisms of Kuhn's conception of "normal science," see CRITICISM AND THE GROWTH OF KNOWLEDGE (Imre Lakatos & Alan Musgrave eds.,1970).

121. *See* Walter V. Reid, Fikret Berkes, Thomas J. Wilbanks & Doris Capistrano, *Introduction, in* BRIDGING SCALES AND KNOWLEDGE SYSTEMS: CONCEPTS AND APPLICATIONS IN ECOSYSTEM ASSESSMENT 1, 3 (Walter V. Reid, Fikret Berkes, Thomas Wilbanks & Doris Capistrano eds., 2006).

122. Drahos, *supra* note 28, at 241.

123. *See* references cited *supra* note 67.

of ecological services are crucial to understanding the historical contribution that indigenous people have made to the maintenance of those services. Science, in other words, is needed in order for us to be able to understand fully the social value of indigenous innovation.

At a practical level, partnerships between scientists and indigenous developmental networks carry risks. I argued earlier that secrecy was a natural starting point of protection for indigenous developmental networks. Partnerships between indigenous developmental networks and scientists mean that indigenous people will have to reveal something of their knowledge assets. In revealing assets to outsiders indigenous groups run the risk that these outsiders will behave opportunistically with those assets. In the next subsection I argue that the only real way in which this problem can be dealt with is through building a trusted network.

B. The Trusted Network

Trust is seen by economists as socially efficient.[124] This instrumental version of trust is not the kind of trust that matters to indigenous people. When indigenous people enter a research relationship with scientists over plants there will be all sorts of issues that unfold over time. These may relate to restrictions about where the plants can be collected from, how the plants are to be treated, whether they can be taken off country, under what conditions, who can handle them, what can be done to them, and so on. In the words of one interviewee "plants are ancestral beings—they can't just be thrown around." Writing a contract to cover every contingency that might occur over a period of years is difficult, as the theorizing around the problem of incomplete contracting shows.[125]

Trust and respect can resolve problems in ways that contractual rules cannot. If one group deeply trusts a second group, it also believes that the trust relationship will guide the second in making decisions about problems that affect the interests of the first. There is no need to prescribe a rule for every contingency because when the contingency arises, the trust relationship will drive the search for the best possible solution. Trust completes a relationship that contract cannot. But this is not any trust. The trust that allows an indigenous group to place in the hands of a scientist an ancestral being has to be something more than the trust one has in one's bank or insurance company. The former trust has to be one between people. The psychological literature draws a distinction between

124. K. Arrow, The Limits of Organization 23 (1974).

125. Padmashree Gehl Sampath, Regulating Bioprospecting: Institutions for Drug Research, Access and Benefit-Sharing 71 (2005).

instrumental/exchange trust and social identity trust.[126] In the latter type of trust there is a preexisting process in which the emergence of individual identity is linked to membership of social groups. When individual identity draws on group identity, trust emerges as a byproduct: "trusting others in the group to pursue the group's interests is little different from trusting oneself to do so."[127]

When an indigenous group enrolls a scientist into its network it creates the opportunity for a process of trust-building to take place. The outcome of that process may be a version of group identity trust if the scientist comes to care for and, in some respects, identify with the indigenous group that makes up the core of the network. Working out the factors that lead to this kind of trust is a matter of social psychological investigation, but one can be reasonably confident that time will be a crucial factor, time in which the indigenous group will test in various ways whether the scientist can be trusted with knowledge. It will most likely be a staged process of trust-building in which the scientist progressively steps into a community as opposed to simply being a member of a network.

The claim here is not that membership of an indigenous group in the fullest sense is a prerequisite for a phenomenon akin to social identity trust to occur. Rather, enough connections have to be created between the scientist and the indigenous group so that the scientist begins thinking particularistically and diffusely about the indigenous group. Borrowing the language of Parsonian action theory, the scientist as a social actor moves away from orientations of universalism and specificity toward particularism and diffuseness.[128] The relationship between the scientist and the indigenous network ceases to be just a standard research business relationship tightly focused on the one pharmacological task (universalism and specificity). Behind the network the scientist discovers and comes to know and feel affection for a fragile human community with goals and ambitions for its children and Country (particularism replaces universalism). A process of empathetic connection with the community and its goals shapes the scientist's decisions about the directions in which to take the research. The written words of the research contract are supplemented by a broad-ranging concern for how research decisions affect the many aspects of community life (diffuseness replaces specificity). For example, more research effort might be put into examining the types of oil in the plant because a business based on extracting oils from plants is something within the capacity of the community.[129]

---

126. Valerie Braithwaite, *Communal and Exchange Trust Norms: Their Value Base and Relevance to Institutional Trust*, in TRUST & GOVERNANCE 46, 52–53 (Valerie Braithwaite & Margaret Levi eds., 1998).

127. *Id.* at 52.

128. The discussion that follows is based on Parsons's analysis of pattern variables in T. PARSONS, THE SOCIAL SYSTEM (1951).

129. *See* Claudie at al., *supra* note 83, at 44.

Under conditions of trust, the indigenous developmental network becomes a collective that searches for an appropriate research agenda that can help it to fulfill its broader cultural goals. The research agenda is no longer dominated by a global pharmaceutical agenda based on the improbable success of finding a magic molecule, an agenda that relegates indigenous people to the status of knowledge tipsters, that does not draw on their expertise in an ongoing way and does not create jobs for them. It is by planning together as a trusted network that indigenous people and scientists can find the research agendas that will lead to Schumacher's goal of sustainable development based on appropriate or intermediate technologies.[130]

There are advantages for both the scientist and the indigenous developmental network once the scientist becomes the trusted insider. For the indigenous group the risk of opportunistic behavior lessens because the scientist begins to think about the project as a group project (in Parsonian terms collectivity-orientation replaces self-orientation). Consultation ceases to be a formal act required by an organization's protocol and more a natural process of persons who respect each other conferring about important decisions to be taken. For the scientist there are, of course, the inherent benefits of being trusted and of being connected to a new group.

But social identity trust may also bring instrumental benefits for scientists. Once an indigenous group begins to trust a scientist, they may reveal more about the plant, its connections to them, and the land. Not every piece of information will make sense, but just as the strokes of a brush on a canvas eventually cohere into a painting, so pieces of information revealed over time come together to reveal a system of plant knowledge. Some of this information will be vital to understanding a plant's therapeutic efficacy. Traditional medicine often uses plants together when treating disease, something that poses challenges for clinical research into plant medicines.[131] If the causal power of the plant as a treatment lies in a combination of agents and methods of preparation that remain undisclosed to scientists, then high-throughput screening methods may miss what is important. Williamson, for example, points out that if synergistic causality is at work then a bioassay-led search for the single compound "is doomed to failure."[132] There is mounting scientific evidence that studying whole plants along

130. Appropriate and intermediate technologies are argued for in E.F. SCHUMACHER, SMALL IS BEAUTIFUL: ECONOMICS AS IF PEOPLE MATTERED (1973).

131. Ranjit Roy Chauhury, Urmila Thatte & Jianping Liu, *Clinical Trial Methodology*, *in* TRADITIONAL MEDICINAL PLANTS AND MALARIA 389, 391 (Merlin Willcox, Gerard Bodeker & Philippe Rasoanairo eds., 2005).

132. E. M. Williamson, *Synergy and Other Interactions in Phytomedicines*, 8 PHYTOMEDICINE 401, 404 (2001).

with their traditional methods of preparation will lead to models of synergistic causality in drug development and treatment, such models being important for the treatment of diseases such as malaria where resistance to existing drugs such as chloroquine has developed.[133]

Through building a trusted network, indigenous groups can reveal more of their knowledge assets and create a greater networked capacity to explore how those assets can be used to bring benefits to their communities. Without the trusted network in place, indigenous people may never move beyond a defensive secrecy about their assets. With a trusted network in place the sharing of knowledge becomes possible, and this in turn opens the door to planning and investment.

Conclusion

The knowledge that indigenous people have built up over long periods of time are potential assets that could help them improve their socioeconomic position. The evidence from wealthy settler states suggests that it is difficult for indigenous groups to move up the socioeconomic ladder. One hypothesis that might explain this is that the property orders of states are extractive for indigenous people. Essentially this means that uncompensated asset transfers take place from indigenous people to other groups and sectors within the state. In the case of indigenous knowledge assets this extractive effect occurs because indigenous knowledge assets are placed into an open public domain and indigenous innovation, which is centrally aimed at maintaining Country through systems, fits poorly with innovation rules that are aimed at creating exclusivity over commodities. The upshot is that the property order of states makes them non-developmental for indigenous people.

The question of whether a state's property order is extractive for an indigenous group can only be answered by looking at the overall effect of the rules that make up the order. State initiatives that aim to protect indigenous knowledge may nevertheless end up functioning more to protect state interests (see the section on India's TKDL initiative), or they may simply be examples of symbolic regulation that disguise the overall extractive nature of a state's property order for indigenous people (see the section on Queensland's biodiscovery regime). Although the chapter posits an iron law of extraction that governs state behavior, this law may not hold under all historical conditions (see the section on Canada and Inuit art).

---

133. For a comprehensive assessment of the evidence in the case of plant-based treatments for malaria, see Traditional Medicinal Plants and Malaria, *supra* note 131.

Faced by a property order that is non-developmental, indigenous groups have little choice but to form developmental networks. Modifying the property order of a state is a politically difficult task for indigenous groups and, in the case of intellectual property rules, faces geopolitical constraints. In the face of slow rule reform, indigenous groups have to find ways to work within existing intellectual property systems. Obtaining benefits from existing intellectual property rules depends crucially upon indigenous groups creating a distinctive organizational form—the indigenous developmental network. Two conditions for the emergence of such a network are strong indigenous leadership and control over the land to which the knowledge assets relate. Secrecy is a powerful, natural, and low-cost starting point when it comes to these networks adaptively managing their assets. Patents may have some role for some networks, but generally trademarks are likely to offer the most feasible long-term forms of protection, especially for products and ecosystem services that can potentially be the subject of voluntary certification standards.

Indigenous knowledge assets are, in the first instance, best protected by secrecy, but if they are to be used in market contexts they will, at least to some extent, have to be revealed to outsiders who possess the relevant expertise to help develop those assets. The global crisis in the degradation of ecosystems and the relevance of indigenous innovation to the management of ecosystems creates an important opportunity for indigenous developmental networks to work with scientists in particular. The resolution to the problem of revealing knowledge assets to an outsider lies in making the outsider an insider within the trusted network that works for indigenous development. A trusted network is the best organizational form to encourage sharing of knowledge. Within a trusted network an indigenous group and scientists can begin to share information so that the indigenous group can make the long-term planning and investment decisions that will help realize their hopes for their Country and their children.

11 OBSERVING THE PATENT SYSTEM IN SOCIAL AND POLITICAL PERSPECTIVE: A CASE STUDY OF EUROPE

Shobita Parthasarathy and Alexis Walker*

Introduction

The major patent systems across the world seem substantially similar. They all reward inventors with intellectual property rights for a limited period of time, in exchange for disclosure about the methods of making the invention. They all bring scientific and legal knowledge to bear on the assessment of patent applications and the grant of patents. Although they may exhibit minor national or regional quirks, their governing laws and approach to patentability are also substantially the same. In fact, many of them work actively to ensure harmonization of their laws and practices. Finally, in most places, patent systems are thought to play a central role in encouraging innovation and therefore economic growth.

But this is not the full story. In this chapter, we argue that when patent systems are understood through a social and political lens, they exhibit many differences that are important for understanding both their role in contemporary societies and their possible futures. In order to make this argument, we use the European patent system as a case study. Although distinguished scholars have argued that it is largely similar to its counterpart in the United States,[1] we argue that the European patent system's distinct history, and social and political context, shape its policies and practices in important ways.

---

* Shobita Parthasarathy is Associate Professor at the Ford School of Public Policy, University of Michigan. Alexis Walker is a Ph.D candidate in the Department of Science and Technology Studies, Cornell University.

1. Peter Drahos, The Global Governance of Knowledge: Patent Offices and Their Clients (2010).

In particular, we demonstrate that it has developed a distinctive response to the activist challenges that have plagued patent systems over the past three decades. We hope that our approach and findings will encourage social scientific analysis of legal institutions, and of patent systems in particular. Although some historians and anthropologists have examined aspects of intellectual property closely[2], this area of scholarship is still extremely limited.

We begin the chapter by exploring the political history of the European patent system. We demonstrate that it was developed as part of a *European innovation policy* apparatus in the wake of World War II. This scope and orientation shaped its practices and policies. We then briefly describe the rise of activist challenges to the system since the 1980s. European activists have rather similar concerns as those fighting other patent systems across the world: they demand that intellectual property regimes change to take into account the broad implications of their decisions, including the consequences of trade liberalization for the developing world, the rights of citizens across the world to affordable healthcare, the ethical boundaries of markets, and the appropriate limits of scientific inquiry. The European patent system's history and social and political position have made it particularly sympathetic, we argue, to these challengers and led it to shift its culture and practices. These shifts may lead to long-term changes in the system that would be difficult to predict if we focused solely on the law.

These conclusions are based on qualitative research conducted from 2007 to 2012. This included over one hundred interviews with US and European patent office and other government officials, advocacy groups, patent lawyers, inventors, and other stakeholders. We also analyzed a variety of documents including: patent office publications; individual patents (and associated documents, including challenges and oppositions); archival records on the development of the US and European patent offices (held, for example, at the archives for the European Commission, European Parliament, Council of Europe, European Free Trade Association, and the European Economic Community); transcripts of congressional, parliamentary, and patent office hearings (from the eighteenth century to the present); and US Supreme Court cases from the eighteenth century to the present. Finally, participant observations were conducted at multiple patent office hearings and meetings in both the United States and Europe.

---

2. *See, e.g.*, B. Zorina Khan, The Democratization of Invention: Patents and Copyrights in American Economic Development: 1790–1920 (2009); Cori Hayden, When Nature Goes Public (2003); Christopher May & Susan K. Sell, Intellectual Property Rights: A Critical History (2006); Ingrid Schneider, Das Europäische Patentsystem. Wandel von Governance durch Parlamente und Zivilgesellschaft (2010); Daniel Kevles, Diamond v. Chakrabarty *and Beyond: The Political Economy of Patenting Life*, *in* Private Science: Biotechnology and the Rise of the Molecular Sciences (Arnold Thackray ed., 1998).

I. The History of the European Patent System

Creators of the European patent system built it with explicit policy goals in mind. After decades of war on the continent, they hoped that it would stitch European countries together to create a common, democratic future of economic prosperity and peace. Building the system, however, was not easy. The process lasted almost thirty years, and required extensive negotiations. Through the construction of this system, negotiators not only took an important step toward the creation of European law and a European economy, they also took steps toward creating the ideas of the European public and public interest.

The first significant steps toward a European patent system occurred in 1947, when France and the "Benelux" countries (Belgium, the Netherlands, and Luxembourg) established the International Patent Institute (IPI) in The Hague (Netherlands) to coordinate their efforts to determine a patent application's novelty.[3] Although most European countries had patent systems by this time, they differed in many respects, including what each considered patentable subject matter, how applications were reviewed, who was eligible for a patent, and what the grant of a patent actually meant. The IPI was the first effort among European countries to establish a unified approach to the search process and the novelty determination.[4] IPI personnel would investigate the prior art (relevant patents and scientific publications) and produce a "search report," which was then used by the national patent offices. Soon afterwards in 1949, the new Council of Europe (CoE) took up the idea of a unified European patent system as one of its first tasks. A unified patent system, according to the CoE, would help it ensure its goal: "to achieve a greater unity between its members for...facilitating their economic and social progress."[5] The Council's Committee on Legal and Administrative Questions described its interest by noting "that technical progress and development of inventions make increasingly imperative the pooling of the means and resources of each of the Member States of the Council for the protection of the inventor and of national industries which benefit from his activities."[6] At the time, Council membership comprised Belgium, Denmark, France, Ireland, Italy, Luxembourg, the Netherlands, Norway, Sweden, and the United Kingdom. From the outset, proponents saw the European patent

3. D. Guellec & B.V.P. de La Potterie, The Economics of the European Patent System: IP Policy for Innovation and Competition (2007).

4. Eda Kranakis, *Patents and Power: European Patent-System Integration in the Context of Globalization*, 48 Tech. & Cult. 689 (2007).

5. Neill Nugent, The Government and Politics of the European Union (2010).

6. Committee on Legal and Administrative Questions, Council of Europe, Recommendation 23—European Patents Office (Doc 110) (1949), *available at* http://assembly.coe.int/Main.asp?link=/Documents/AdoptedText/ta49/EREC23.htm.

system as a policy instrument that, by providing industrial rights to the inventor, would provide public benefit through industrial, and therefore economic, growth.

The first proposal, offered by French senator Henri Longchambon, suggested a European "Patents Office" that would issue inventor's certificates that would establish the novelty and patentability of an invention.[7] In this scheme, national patent offices would still grant and administer the actual patents. The Council established a Committee of Experts on Patents, composed of each member country's highest patent officials, to study Longchambon's draft proposal. Members of this committee soon agreed that although the goals of the proposal were good, they were not yet "practicable," and then began to work on initial steps toward the idea of a European patent system. Work toward the creation of a European patent system can be divided into two projects. First, negotiators needed to develop a unified approach to the processes of search and examination and to the definition of patentable subject matter. Second, they needed to consider how to create appropriate governance, including a unified European patent office. Most of the countries involved in the negotiations already had a patent office that employed technical personnel and that had established its own practices and policies. What would become of them? Would they become part of a European patent system? How? How would their responsibilities change, and how would they be able to handle these changes? How would the governance of a patent system work? How would it be organized?

A. A European Approach to Patent Law

The CoE's Committee of Experts on Patents focused on the first challenge, composing a Preliminary Comparative Study on Novelty and Patentability, in April 1951.[8] This report identified the similarities and differences between member countries on a variety of issues including how they treated prior use of an invention for purposes of determining novelty, and whether and how they defined exclusions from patentability. The Committee's treatment of these exclusions demonstrates the centrality of public policy concerns in developing the European patent system.

Most contentious were discussions about a unified approach to the patentability of "chemical products," "foodstuffs," and "horticultural products." Although few countries allowed patents on both pharmaceutical products and

7. Committee on Economic Questions, Consultative Assembly of the Council of Europe, Creation of a European Patents Office (Doc 75) (1949).

8. Committee of Experts on Patents, Council of Europe, Criteria of Novelty and Patentability (1951).

the processes of making them, most either prohibited "product" patents because they facilitated monopolies that would be detrimental to the public's health, or allowed patents on pharmaceutical products but also allowed compulsory licensing under certain circumstances in order to deal with the potential problems that a monopoly could create. These approaches demonstrated that European countries saw some limits to patentability as necessary to achieve public interest goals. The Austrian delegation described its position in the negotiation: "For the Austrian economy, the protection of chemical substances is unacceptable. Furthermore, it seems entirely inadmissible that foodstuffs, stimulants, pharmaceuticals, and disinfectants necessary for the preservation of the life and health of the general public should be subject to exclusive rights for the benefit of individuals."[9] Rather than assuming that the economic benefits that accrued to the patent holder would automatically achieve social benefit, some countries argued that the patent system had to take an active role in *balancing* public and private interests to achieve the public good.

Negotiators also had to contend with the "*ordre public*" exclusion contained in many national patent laws. In general, this exclusion prevented patents on inventions deemed contrary to public policy or morality, but there was considerable diversity in how different countries articulated the concept. Italy, for example, used the exception to disallow patents "detrimental to health," whereas Ireland disallowed "inventions liable to cause an increase in the price of commodities, a hindrance to the freedom of commerce or any other public inconvenience."[10] Nevertheless, all of the negotiating countries agreed quickly that the European system should include a clause that prevented patents on inventions that were contrary to public policy or morality. In fact, there was almost no discussion of the matter. This suggests that the relevance of the public policy and morality clause was uncontroversial and obvious. Patent law was relevant to public policy, and therefore required a clause in the pan-European treaty to address potential conflicts.

The idea that patents should not be allowed on inventions that are contrary to public policy or morality can be traced historically in three directions. The first comes directly from the history of patent law, as the seventeenth-century English Statute of Monopolies prevents the issuance of patents that are "contrary to law," "mischievous to the state," or "generally inconvenient."[11] It was then incorporated into the 1883 Paris Convention, one of the first international treaties regarding

9. Committee of Experts on Patents, Appendix II: Declaration by the Austrian Delegation CM (63) 101 (Strasbourg, France, 1963), at 16.

10. Criteria of Novelty and Patentability, *supra* note 8, at 19.

11. Chris Dent, *"Generally Inconvenient": The 1624 Statute of Monopolies as Political Compromise*, 33 MELBOURNE UNIV. L. REV. 16 (2009).

intellectual property. The Convention stated that signatories could not allow patents that "mislead the public."[12] The United States, however, does not have such an exception in its governing statute. Although Supreme Court Justice Joseph Story did suggest the existence of a "moral utility" exception in the nineteenth century, it appears rarely in modern US patent jurisprudence.[13]

The second origin of the public policy exception developed in the wake of the French Revolution in the eighteenth century, in which newly written administrative laws of all kinds referred to the concept of *ordre public* to demonstrate explicit attention to the public good.[14] The third can be traced through international treaties: starting in the nineteenth century, international treaties (particularly those in Europe) began to include an "*ordre public* exception" to demonstrate recognition of national sovereignty in particularly sensitive areas.[15] The TRIPS Agreement, a significant international treaty on intellectual property, contains the clause.[16] Given the clause's historical origins and meaning, it is likely that negotiators of the European patent system saw it simply as a necessary part of patent law, especially one that involved international negotiation and consideration of national sovereignty. As we will demonstrate later in the chapter, it is only in the wake of activist challenges that it gained the association with the ethical dimensions of patent law that it has now.

As the CoE's Committee of Experts on Patents tried to develop agreement on a variety of issues including definitions of "industrial character," "novelty," "technical progress and creative effort," "influence of prior rights and applications," and "scopes and roles of the description and the claims,"[17] they sought advice from stakeholders at both national and international levels. Each national delegation consulted with inventor, patent law, patent agent, and industrial associations in their country, and the Committee as a whole heard comments

12. S. Thambisetty, *Understanding Morality as a Ground for Exclusion from Patentability under European Law*, 12 EUBIOS J. ASIAN & INT'L BIOETHICS 48 (2002).

13. Andrew R. Smith, *Monsters at the Patent Office: The Inconsistent Conclusions of Moral Utility and the Controversy of Human Cloning*, 53 DEPAUL L. REV. 159 (2003); B.D. Enerson, *Protecting Society from Patently Offensive Inventions: The Risk of Reviving the Moral Utility Doctrine*, 89 CORNELL L. REV. 685 (2003).

14. M.J.B. Bernier, Droit Public *and* Ordre Public, 15 TRANS. GROTIUS SOC. 83 (1929).

15. M. Forde, *The "*Ordre Public*" Exception and Adjudicative Jurisdiction Conventions*, 29 INT'L & COMP. L.Q. 259 (1980); Gerhart Husserl, *Public Policy and* Ordre Public, 25 VA. L. REV. 37 (1938).

16. C.M. CORREA, INTELLECTUAL PROPERTY RIGHTS, THE WTO, AND DEVELOPING COUNTRIES: THE TRIPS AGREEMENT AND POLICY OPTIONS (2000).

17. Lodovico Benvenuti, Written Question No. 73 to the Committee of Ministers by Mr. Heckscher (CM/AS (60)Quest73FinalE) (1960).

from the International Association for the Protection of Intellectual Property, the International Federation of Patent Agents, the International Chamber of Commerce, and the European Industrial Research Management Association.

Eventually, these negotiations led to the 1963 Strasbourg Convention on the Unification of Certain Points of Substantive Law on Patents for Invention, which was signed by all Council countries (which by this point included Belgium, France, Italy, Luxembourg, the Netherlands, Ireland, Denmark, Norway, Sweden, the United Kingdom, Greece, Turkey, Austria, Cyprus, and Switzerland).[18] It disallowed patents on plant and animal varieties, and on "inventions the publication or exploitation of which would be contrary to ordre public or morality, provided that the exploitation shall not be deemed to be so contrary merely because it is prohibited by a law or regulation."[19] Despite the earlier controversy, it allowed patents on food and pharmaceutical products, and agricultural and horticultural processes, but provided signatories with a ten-year grace period to bring their policies into agreement with the Convention.

The 1973 European Patent Convention, which established the European patent system, relied heavily on the Strasbourg Convention. It incorporated the Strasbourg Convention's two categorical exclusions (on plant and animal varieties and on inventions that violated *ordre public* or morality) and added a third: "methods of treatment of the human or animal body by surgery or therapy and diagnostic methods practised on the human or animal body."[20] In identifying these exceptions, the creators of the European patent system were clearly thinking about it in terms of public policy, and were concerned that an unfettered patent system may not always achieve social benefit. These exclusions also suggested a broad approach to what might be counted as relevant knowledge and expertise in the patent system. Traditionally, participants in the patent system focused on industrial and patent law, and to a lesser extent economic and scientific expertise. Consider, for example, the stakeholders who participated in the CoE discussion described above. However, excluding methods of treatment and diagnostic methods "in the interests of public health" [21] raised the possibility that public health and medical experts might be able to claim relevance and legitimacy in patent-related discussions, and that these discussions would consider

18. C. Wadlow, *Strasbourg, the Forgotten Patent Convention, and the Origins of the European Patents Jurisdiction*, 41 Int'l Rev. Intell. Prop. & Compet. L. 123 (2010).

19. Council of Europe, Convention on the Unification of Certain Points of Substantive Law on Patents for Invention, opened for signature Nov. 11, 1963, C.E.T.S. No. 047 (1980).

20. Convention on the Grant of European Patents, opened for signature 5 October 1973, 13 I.L.M. 268 (entered into force Oct. 7, 1977).

21. European Patent Office, Revision of the European Patent Convention (EPC 2000), Synoptic presentation EPC 1973/2000—Part I: The Articles, at 50, 2007.

public policy and public interest broadly. In addition, the *ordre public* exclusion could create an opportunity for a wide variety of individuals and groups to claim knowledge and expertise relevant to the relationship between morality and patent law. These openings, however, were not intentional. Negotiators never mentioned the possibility that such figures might be relevant to patent law. Rather, this broader approach was connected to negotiators' vision of the European patent system as a policy instrument, and as we will see it became quite useful to public interest groups later on.

B. Organizing the European Patent System

As the Council negotiated the substantive patent issues, the new European Economic Community and European Free Trade Association (which were both eventually absorbed into the European Union) considered the potential governance structure of a European patent regime. The EEC, which was created in 1957 among Italy, Germany, France, and the Benelux countries to facilitate European economic integration, saw unification of the patent system as an early priority.[22] In 1962, its Committee of Experts (which, similar to the one convened by the Council of Europe, was made up of the heads of patent offices from the member states) proposed a European Patent Law that would create a single European patent that would coexist with existing national patents. A central European Patent Office would perform examination and granting functions, and the European Court of Justice and an Administrative Council made up of representatives from the EEC member countries would oversee its activities. Non-EEC countries would be allowed to participate in the system if the Administrative Council approved them. The structure envisioned by this proposal echoed the American system, with a central bureaucratic apparatus and external legislative and judicial oversight. However, this system would create a new layer of pan-European decision-making authority and expertise, which would coexist with the expertise and authority of national systems. Thus, it would create new challenges. What would European expertise in patent searching and examination mean? Would it be the same as national expertise? Would non-EEC countries accept decisions by the Administrative Council when they were not represented in it? Would the European patent office operate in all of the languages of its member states?

22. D. Thompson, *The Draft Convention for a European Patent*, 22 Int'l & Comp. L.Q. 51 (1973); F. Froschmaier, *Some Aspects of the Draft Convention Relating to a European Patent Law*, 12 Int'l & Comp. L.Q. 886 (1963).

Because EEC negotiators could not agree on whether, and how, non-EEC countries could participate in the proposed system, their draft proposal did not make much progress. In the meantime, the heads of patent offices from the member countries of the EFTA, which was then made up of Austria, Denmark, Norway, Sweden, Switzerland, and the United Kingdom, made another proposal: a "two-part arrangement," in which there would be a unified system up to and including the grant of the patent, but then policies regarding the effects and validity of the granted patent would be determined separately for EEC and non-EEC member countries. In this proposal, the European office had fewer responsibilities and therefore, more limited political exposure.

The EFTA Memorandum attracted a great deal of interest. The French government, primarily at the request of French industry, advocated this proposal inside the EEC and soon the Chairman of the EEC's expert committee incorporated it into a draft Memorandum for review and discussion. The EEC then used the EFTA proposal as the basis for discussion when it convened an Inter-Governmental Conference made up of the directors of the patent offices of its own member countries. The Conference included observers from the Council of Europe, the International Patent Institute, the United International Bureaux for the Protection of Intellectual Property (the predecessor of the World Intellectual Property Organization), the EEC Commission, patent agents, lawyers, and industrialists. By late 1969, the Inter-Governmental Conference had published its first preliminary draft with the aim of bringing "interested circles," defined as "the private persons concerned, the inventors, the companies that will exploit the patents and produce the products, the patent agents, and the patent lawyers," more actively into the discussion.[23] Thus, the parties began to discuss a framework that would include EEC and non-EEC member states in a system existing outside of the EEC. Discussion continued over the next few years, and in 1973, sixteen countries met in Munich to sign what became known as the European Patent Convention.[24] This Convention guides the European patent system today, and its basic organization is quite similar to the EFTA proposal: a common system for the examination and grant of a patent, with policy on effects and validity determined independently by each member state.

---

23. Thompson, *supra* note 22.

24. Twenty countries participated in the discussions. By 1977, seven had become full-fledged members. Today, there are thirty-eight member countries. *See* European Patent Office, Member States of the European Patent Organisation, http://www.epo.org/about-us/organisation/member-states.html (last visited Sept. 5, 2013).

C. Governing the European Patent System

The EPC established a European Patent Organization, composed of a European Patent Office and an Administrative Council, to grant patents.[25] The European Patent Office performs both bureaucratic and judicial functions, including receiving patent applications, conducting searches of prior art and examination of the application, and resolving legal questions through boards of appeals. The Administrative Council, composed of representatives from the member countries, performs supervisory and legislative functions. It meets periodically to review the rules and regulations governing the Patent Office and its personnel, to appoint adjudicators to the appeals boards, and to elect the Organization's president. It also determines whether and when to review the European Patent Convention itself. In most cases, each member country can cast one vote, and most decisions are made through a simple majority. The European Patent Organization is located in Munich, with additional branches in The Hague[26], Berlin, and Vienna, and its official languages are English, French, and German.

Because the European Patent Organization predates the 1993 Treaty of Maastricht establishing the European Union and because it has always included non-EU countries, it is a pan-governmental organization that operates independently. As a result, it is disconnected from European Union institutions, including the Parliament and European Court of Justice. This has important implications for how the system has developed and maintained its legitimacy.[27] While the policies of the US Patent and Trademark Office (USPTO) are guided by the US Congress and the federal court system, for example, the EPO's policies are largely its own, with the Administrative Council—composed of representatives from member countries—playing an important formal and informal role. Even if they are not translated directly into policy, EPO employees quickly learn about the concerns of Administrative Council representatives and do their best to address them.[28] Thus, in contrast to the USPTO, the EPO has enormous power to shape patent policy. As a result, it also plays a significant role in shaping the definition of European public interest, and the meaning of Europe itself. However, the EPO

---

25. *See* European Patent Office (EPO), The Administrative Council of the European Patent Organisation, EPO, http://www.epo.org/about-us/epo/administrative-council.html (last visited Sept. 5, 2013).

26. Initially, The Hague office did all searches (because it was previously the home of the International Patents Institute) and the Munich office did most of the examination (some examiners were in Berlin). Today, however, all examiners perform both search and examination.

27. J. Karanović, *Contentious Europeanization: The Paradox of Becoming European Through Anti-Patent Activism*, 75 Ethnos 252 (2010).

28. Interview with Employee A, EPO, in Munich, Ger. (June 25, 2008).

has chosen to comply voluntarily with EU law, particularly when dealing with controversial issues. This not only ensures smooth harmonization between EU member states and EPO practices, but it also provides external legitimacy for the institution.

Although there are many similarities between the patent application process in Europe and in other major patent systems, a few differences are worth mentioning. After the EPO Receiving Section classifies a patent application according to its subject matter, it is sent to an Examining Division made up of three examiners. By contrast, only a single examiner reviews patents at the USPTO. Although one examiner plays the primary role at the EPO, all three review and sign off on the interactions with the applicant and the final decision. This team approach encourages examiners to seek advice and learn from one another, and also to develop a more uniform approach to search and examination. European examiners have similar backgrounds to their counterparts in other systems. They must have scientific or engineering training and learn the relevant patent law through a dedicated patent academy and through an apprenticeship. However, they must also be competent in at least two (and preferably all) of the Organization's official languages. Examiners tend to make a career at the EPO, in large part because it offers excellent benefits (including a substantial pension). This means that the examining corps has not only formal training but also considerable experiential expertise. It also means that they identify themselves with the goals of the institution, and often refer to themselves as "civil servants"[29] even though they are not part of a national government or the European Union. This contrasts with the United States, where retention of examiners has been a serious problem for over a century.[30]

Once a patent is granted by the EPO, it becomes a bundle of national patents each of which can be independently governed in each country (which countries are a part of the bundle depends on the applicant's choice and the fees paid). Within nine months of a patent's issue, it can be opposed at the EPO on the grounds that the invention did not meet the criteria for patentability or that it was not sufficiently described, or that the granted patent goes beyond the initial application. By contrast, until recently, the United States had a re-examination process that only allowed patent challenges on "prior art" grounds.[31] Before the creation of the European patent system, patent offices across Europe had developed a pre-grant opposition procedure to

---

29. Interview with Employee B, EPO, in Munich, Ger. (June 25, 2008).

30. U.S. Gov't Accountability Office, GAO-07-1102, USPTO: Hiring Efforts Are Not Sufficient to Reduce the Patent Application Backlog (2007).

31. Although the recently passed America Invents Act (2011) permits post-grant review of patents by the USPTO, it is as yet unclear how this will be interpreted in practice. Pub. L. No. 112-29 § 1, 125 Stat. 284 (2011).

correct mistakes in their search, examination, and granting processes quickly and cheaply rather than forcing inventors to deal with lengthy litigation. This approach could be the result of the difficulty of negotiating litigation across member countries, or it could be the result of Europe's relative comfort with bureaucratic power (the United States, by contrast, is characterized by trust in the fairness of its litigation processes to achieve a fair and just result). Because of the challenges of granting patents that would be valid across Europe, including the costs of translation, the EPC changed it into a post-grant process.[32] If a third party opposes a patent, an Opposition Division consisting of three technical examiners—two of whom were not involved in the initial examination—review the case. Approximately one-third of oppositions result in a patent's revocation, while another one-third results in narrowing the claims.[33]

Applicants and third parties can appeal the decisions of the Receiving Division, Examining Division, or Opposition Division to a Technical Board of Appeal (similar to the USPTO's Board of Patent Appeals and Interferences). Technical Boards of Appeal are made up of either legally qualified members, technically qualified members, or a mixture of both, depending on the type of appeal. The final level of judicial review is the Enlarged Board of Appeal, which analyzes broad legal questions that are either referred by a Technical Board or by the president of the European Patent Office. This is in contrast to the United States, where the federal court system performs this function. In Europe, members of the appeals boards are full-time employees of the Office for a five-year fixed term, but they work in a separate Directorate-General from the examining corps. Cases cannot be appealed to the European Court of Justice because the Organization is not part of the European Union.

II. Challenging the European Patent System

Like its counterparts across the world, the European patent system has experienced growing challenges to its decisions, policies, and practices. Over the last thirty years, a wide variety of groups, including environmentalists, animal rights activists, patient advocates, scientists, and development organizations, have argued that it needs to develop a broader understanding of its responsibility to the public interest that would include ethical, social, environmental, and health

32. R. Singer, *The European Patent Enters a New Phase*, STAFF UNION EUR. PATENT OFFICE (1970), *available at* www.suepo.org/public/background/singer.pdf.

33. Dietmar Harhoff & Markus Reitzig, *Determinants of Opposition Against EPO Patent Grants—The Case of Biotechnology and Pharmaceuticals*, 22 INT'L J. IND. ORG. 443 (2004).

considerations.[34] Some also argue explicitly that human rights need to be balanced against the rights of the patent holder.

In Europe, these challengers have used two strategies. First, they try to influence the European patent system through the European Parliament. In the late 1980s, for example, challengers developed a large, global network of opponents when Parliament considered a Directive that would harmonize patent law regarding biotechnology across member states.[35] The biotechnology industry had strongly advocated this legislation (known as the EU Biotech Patent Directive), arguing that it would facilitate innovation and economic growth. The EPO also agreed to follow whatever legislation the European Parliament ultimately passed, as acceptance would facilitate relationships with patent offices in EU countries (most of whom sat on the EPO's Administrative Council), and it would also provide some political legitimacy for its decision-making in an area that was becoming controversial.

The Biotech Patent Directive generated controversy soon after the European Commission introduced it into the European Parliament in 1998. While the Commission saw the Directive as merely a legal and economic initiative, most Parliamentarians disagreed and urged attention to the ethical, social, ecological, and economic implications of patents.[36] Buoyed by support from Greenpeace and other groups, they refused to pass the law without some consideration of these issues. Many argued that the European Patent Convention's *ordre public* clause disallowed many biotechnology patents, particularly those involving life forms. They demanded that the Directive articulate this and clarify the clause's relevance to the implications of patents. Despite its original meaning allowing for sensitivity to national differences, this language had now created an opportunity for a vigorous debate about the role of social, ethical, and ecological concerns in discussions about patent law. By 1995, critics had won some concessions, but they

---

34. S. Parthasarathy, *Breaking the Expertise Barrier: Understanding Activist Strategies in Science and Technology Policy Domains*, 37 SCI. & PUB. POL'Y 355 (2010), *available at* http://www-personal.umich.edu/~shobita/files/breaking%20the%20expertise%20barrier.pdf; S. Parthasarathy, *Whose Knowledge? What Values? The Comparative Politics of Patenting Life Forms in the United States and Europe*, 44 POL'Y SCIENCES 267 (2011), *available at* http://www-personal.umich.edu/~shobita/files/Parthasarathy%20policy%20sciences%20preprint.pdf.

35. Directive 98/44/EC of the European Parliament and of the Council of 6 July 1998 on the Legal Protection of Biotechnological Inventions, 1998 O.J. (L 213), 13; Steve Emmott, *No Patents on Life: The Incredible Ten-Year Campaign Against the European Patent Directive*, *in* REDESIGNING LIFE? THE WORLDWIDE CHALLENGE TO GENETIC ENGINEERING (Brian Tokar ed., 2001).

36. Emmott, *supra* note 35.

rejected the Directive in order to force the Commission to revise the language in line with their concerns.[37]

The European Commission introduced an amended version in December 1995, which again mobilized challengers both inside and outside of Parliament.[38] The European Parliament debated the legislation for another two years and passed a revised version in July 1998.[39] The final Directive included some attention to critics' concerns. It specified the EPC's *ordre public* clause and prohibited, among other things, inventions that involved "commercialization or exploitation" of the human embryo (e.g., stem cells and methods of isolating them, embryos and methods of creating them). It outlawed patents on plant and animal varieties. It also required examiners to perform a risk/benefit "balancing" test when considering patents on animals, to ensure that the benefits to humanity outweighed the risks to the animal involved.[40] It gave the *ordre public* clause bioethical meaning, and also made patent law an important political issue ripe for widespread public engagement. In fact, it was the first time that the European Parliament had deliberated for so long on an issue, and had such an impact on policy.

In the years since, challengers have continued to lobby the European Parliament to address their concerns regarding the patent system. In the early 2000s, they persuaded it to reject a directive designed to harmonize laws governing software patents. Various groups including the EuroLinux Alliance and the Foundation for a Free Information Infrastructure argued that such patents would stifle innovation.[41] Challengers have also convinced the European Parliament to issue non-binding "resolutions" condemning decisions by the EPO to grant specific patents including those covering genes linked to breast and ovarian cancer

---

37. Interview with Rep., Women's Environmental Network, in London, UK (2007).

38. Global 2000 & Kein Patent auf Leben, Directive: Legal Protection of Biotechnological Inventions (Munich/Brussels, May 25, 1997); Church of Scotland Science, Religion, and Technology Project, Churches Challenge European Parliament on Patenting Life, *available at* http://webzone1.co.uk/www/srtproject/meppat.htm (1997); Dinham, B, Pesticides Trust, et al. to P Stevenson, Compassion in World Farming et al., Suggested Plans for Meeting with John Battle MP (July 24, 1997).

39. Directive 98/44/EC, *supra* note 35.

40. I. Schneider, *Can Patent Legislation Make a Difference? Bringing Parliaments and Civil Society into Patent Governance, in* POLITICS OF INTELLECTUAL PROPERTY: CONTESTATION OVER THE OWNERSHIP, USE, AND CONTROL OF KNOWLEDGE AND INFORMATION 129 (S. Haunss & Kenneth C. Shadlen eds., 2009); Peter Drahos, *Biotechnology Patents, Markets, and Morality*, 9 E.I.P.R. 441 (1999).

41. Karanović, *supra* note 27.

susceptibility (known as the BRCA genes), the cryopreservation of germ cells, and stem cells.[42]

In addition to challenging patent policy at the European Parliament, many challengers have mobilized against specific patents issued by the EPO. If they deem a patent to be problematic, they both launch a public relations effort (including press conferences, press releases, and street protests) and use the EPO's official opposition process. As discussed above, the EPC designed the EPO's opposition process to reduce post-grant litigation and enhance patent quality. Challengers, however, have used it to force the EPO to consider the consequences of its actions, often by arguing that a problematic patent violates the *ordre public* clause. They have opposed dozens of patents this way, including those that cover life forms, traditional knowledge, and important drugs.

In one of the largest challenges that proved to be particularly important for the EPO, twenty-one groups (including Greenpeace; Belgian and Swiss chapters of No Patents on Life; and British, German, and Austrian animal rights organizations) opposed a patent on a mammal genetically engineered to contract cancer (known worldwide as the Harvard "Oncomouse" patent). The patent was controversial even before it was granted: multiple groups submitted third-party observations to the EPO, urging it not to issue the patent because it violated the *ordre public* clause. EPO granted the patent in 1992, arguing that it passed a "weighing up" test in which EPO compared the benefits to humankind with the animal's suffering. Challengers responded to the EPO by filing twenty-one oppositions, and engaged the bureaucracy in a debate over what constituted relevant knowledge for determining a violation of the *ordre public* clause.

Ultimately, the Opposition Division asked the inventors to rewrite the patent to cover only rodents. If we focus on the legal consequences of the Oncomouse decision, we may observe only the challengers' impact on the individual case. However, interviews with EPO officials suggest that it was a major turning point in the culture and orientation of the office. One high-level official recalled that period:

> [W]hen the Oncomouse Harvard patent was granted in 1992, a big storm arose that nearly swept away the EPO, because people [here] were so strongly inward looking at this place that they didn't realize what was going on. That the work that they were doing was not just considered to

---

42. *See, e.g.*, European Parliament, European Parliament Resolution on the Patenting of BRCA1 and BRCA2 ("Breast Cancer") Genes, (Oct. 3, 2001), *available at* http://www.europarl.europa.eu/sides/getDoc.do?pubRef=-//EP//TEXT+MOTION+P5-RC-2001-0633+0+DOC+XML+V0//EN; European Parliament, Resolution on the Decision by the European Patent Office with Regard to Patent No. EP 695351 Granted on 8 December 1999, O.J. C378 (Dec. 29, 2000), at 96–97.

be of relevance to pharmaceutical companies and other innovators, but there was a strongly social component as well. And this is the bitter lesson they had to learn. . . .[43]

During and after the Oncomouse controversy, challengers involved in the case and other like-minded groups filed oppositions against dozens of other patents.[44] They challenged patents on genes, stem cells, genetically modified plants and animals, and methods of gene and embryo manipulation, among many other things. In each case, challengers used evidence and expertise and invoked civil rights that the EPO had not traditionally considered. As they had to contend with these issues, EPO officials began to believe that they had to find a way to incorporate such concerns into their decision-making. One noted:

In general, I think that it's important...for people dealing with patents to keep in mind where the system comes from. What its legitimacy and its social basis are. So to really keep in mind the social contract of private benefit and public benefit and incentive for innovation on the one hand but disclosure and promoting innovation for the public good on the other hand. I think that probably has been lost. . . .[45]

Both the debates at the European Parliament and the individual oppositions at the EPO attacked the practices, policies, and laws of the European Patent Office. Challengers argued that the EPO was not considering the public interest adequately, and they used a variety of tactics to demonstrate this. The EPO could have responded as the USPTO did, for example, by arguing that such concerns were irrelevant. But its personnel did not, in part because of the system's history and political context. It had always been part of the emerging policy apparatus of Europe, and its employees saw themselves as civil servants. Therefore, the EPO could not ignore the cries of challengers who claimed to represent the public interest. Furthermore, its structure and position as a pan-governmental organization outside the European Union connected it directly to the European citizenry. While the USPTO could look to the US Congress to deal with political concerns

43. Interview with Employee A, EPO, in Munich, Ger. (June 25, 2008).

44. *See, e.g.*, Greenpeace, Liste Der Einsprechenden Gegen Patent EP 695 351/ Sammeleinspruch Greenpeace Stand 17.4.00 (Apr. 18, 2000); Kein Patent Auf Leben, Sehr Geehrter Herr Wolinski, Beilegend Möchte Ich Nochmal Acht Dokumente in Das Beschwerdeverfahren Einbringen (Nov. 4, 2004); *see also* S. Parthasarathy, Building Genetic Medicine: Breast Cancer, Technology, and the Comparative Politics of Health Care (2007).

45. Interview with Employee B, EPO, in Munich, Ger. (June 25, 2008).

and represent the will of the public, the EPO stood alone and therefore had to think more strategically about maintaining its legitimacy.

III. Systemic Responses

High-level officials at the EPO tried to address these issues through changes to its organization and culture. One set of initiatives focuses on educating its employees, including examiners, about the concerns of activists, national governments, and the media. In the early 1990s, *The Gazette*, the EPO's internal newsletter that is distributed to all employees, began to publish original articles about ethics and biotechnology, reprint relevant EPO press releases announcing the activist oppositions, and display photographs of the protests occurring outside of its buildings.[46] Previously, it had published only internal announcements, including listings of job openings, promotions, and new employees; short articles about foreign visitors to the EPO; and announcements about and reviews of (mostly social) EPO events. In 1992, for example, the *Gazette* announced an essay competition on "Patents and Ethics in the Context of Modern Technology" among employees at the EPO. Two winners received 15000 and 10000 German marks (about $10,000 and $6,500, respectively) with the essays published not only in the *Gazette*, but also in major European newspapers (*Financial Times*, *Le Monde*, *Neue Züricher Zeitung*, and *Süddeutsche Zeitung*).[47] The EPO's decision to engage its employees in discussions about the social and ethical dimensions of patenting is quite interesting. Officials could have decided that coverage of these issues would be distracting or demoralizing, or perhaps worst of all, interrupt the objectivity of their employees' work. However, they decided the opposite. Thinking more about the issues of controversy, high-level officials seemed to believe, would make all employees do a better job.

Internal discussions about the social and ethical implications of patenting grew considerably in the 2000s, when Alain Pompidou, a trained biomedical scientist and doctor, became president of the EPO. Pompidou had long been involved in discussions about the appropriate development of biomedicine and biotechnology, and sought to encourage such discussions in the Office. For example, he asked the EPO's Learning and Development Directorate (which is in charge of all personnel training efforts) to organize a series of lectures entitled "Ethics and Science: Are They Connected?" The purpose of the nine lectures,

---

46. B. Karet, *Patents and Ethics in the Context of Modern Technology*, EPO GAZETTE (Newsletter) (Munich) 1992, at 13–17.

47. *Patente auf Lebewesen und Gene*, FRANKFURTER ALLGEMEINE ZEITUNG (July 14, 1993), at N1.

which were simulcast in the EPO's Munich and Hague offices and held over the course of a year and a half, was to "increase staff awareness of the ethical aspects of novel technologies."[48] Pompidou suggested the list of speakers for the series, most of whom were European professors of philosophy or bioethics. He sought to use the series to teach the EPO's personnel, and examiners in particular, "that patents were no longer limited to technical and legal subject matter. It should be looked at in broader context."[49]

This training effort also emphasized the relevance of social and ethical expertise to better work output, although it was not explicitly connected to the patent decision-making process. Perhaps the grandest attempt to change the EPO's practices was the "Scenarios" project, a multi-year effort initiated in 2004 to encourage strategic thinking about the future direction of the Office and the role of the patent system in global society. These efforts began with interviews of over one hundred scholars and stakeholders from all over the world, who were asked to predict and discuss the different social, political, and economic pressures that the Office might face by the year 2025. A small group of personnel from the EPO—including people from all departments and levels—interviewed both traditional stakeholders and experts (e.g., from inventor organizations, patent law firms, high-technology companies) and new players in this arena (e.g., Greenpeace, philosophers, political scientists, international development organizations). The effort was meant to expose the EPO's personnel and the public to new sources of expertise about the role of the patent system in society, while also providing external players with an opportunity to shape the future of the organization.

The EPO's team then used this interview data to construct four possible future scenarios. Although all of them touched on the activist challenges in some way, the "Trees of Knowledge" scenario addressed social and ethical concerns directly. In this scenario, the EPO's personnel tried to predict the kinds of dilemmas that might arise if the grassroots movements continued to grow and public trust in government, industry, science, and technology continued to erode. In particular, authors of this scenario organized the discussion around three "key" questions: "How can public and private interest in IP [intellectual property] be reconciled for the benefit of society? How are the ethical and moral dilemmas raised by technology reflected by the patent system? Where should the limits of patentability be drawn? By whom?"[50] These were large and complicated questions, which required contemplation of issues that were quite new. Until

---

48. S. Bayrak, *Ethical Issues in Engineering and Science*, EPO GAZETTE (Newsletter) (Munich), July–Aug. 2007, at 15.

49. Interview with Employee D, EPO, in Munich, Ger. (Feb. 9, 2009).

50. EUR. PATENT OFFICE, SCENARIOS FOR THE FUTURE 83 (2007).

relatively recently, they might have seemed inappropriate for deliberation by a technical bureaucracy such as the EPO.

The final report of the Scenarios project was unveiled in April 2007 at a large event in the EPO's Munich Office.[51] Major figures in European politics and policy—including German chancellor Angela Merkel—participated, and not surprisingly, the event was well-attended by the EPO's employees.[52] All EPO employees received copies of both the final report and a book of the interview transcripts, to expose them to new ideas. Although most examiners have day-to-day contact with inventors and patent lawyers through the examination process, they rarely hear, for example, the arguments of the Church of Scotland regarding the patentability of living organisms.

Since the report's release, and as new presidents have taken over, the EPO has gotten involved in a variety of policy discussions. In 2009, it hosted a workshop designed to enhance its relationship with civil society groups, including the nongovernmental organizations that had attacked its work for decades. And in 2010, concerned about whether the monopolies created by patents were hurting climate change mitigation efforts by stifling access to innovation, the EPO (with the United Nations Environment Programme and the International Centre for Trade and Sustainable Development) issued: *Patents and Clean Energy: Bridging Evidence and Policy*.[53] It had clearly extended its responsibility to cover the implications of patents, and no longer focused simply on the "technical" job of issuing them. Furthermore, it seemed to acknowledge that simply granting intellectual property rights might not achieve the public good.

European Patent Office officials have also instituted changes to the patent examination process. The EPO has developed a Sensitive Cases, or SeCa, system, to identify and deal with applications it deems controversial. The SeCa system begins at the classification stage. The EPO's Receiving Section marks with a 3x5-inch sticker applications assigned to technical departments (known as "Clusters") that the Communication department has identified as particularly controversial (e.g., biotechnology, nanotechnology). The sticker provides a series of prompts meant to guide examiners through an assessment of an application's sensitivity, and procedures for additional review. Two aspects of this system

---

51. Press Release, European Patent Forum (Apr. 18, 2007)

52. *Id.*

53. U.N. Env't Programme, Eur. Patent Office & Int'l Ctr. for Trade & Sustainable Dev., Patents and Clean Energy: Bridging Evidence and Policy - Final Report (2010), *available at* http://www.unep.ch/etb/events/UNEP%20EPO%20ICTSD%20Event%2030%20Sept%202010%20Brussels/Study%20Patents%20and%20clean%20energy_15.9.10.pdf (last visited Sept. 5, 2013).

deserve special note. First, it is formalized through preprinted stickers and a systematic change in the administrative routine of the classification department and specific Clusters. Second, it focuses on applications in technical fields that personnel in the Communication department recognize as publicly controversial. These two elements thus serve as a constant reminder to examiners of the consequences of their work and the gravity of their responsibility to the European public.

When an examiner receives an application with a SeCa sticker, she must conduct an additional determination of sensitivity in addition to her evaluation of scientific and legal patentability. This determination also involves an assessment of whether the application might be in violation of the *ordre public* clause of the EPC and its elaboration in the EU Biotech Patent Directive. Usually, the examiner judges the sensitivity of an application using a set of examples provided by supervisors and the examiner's own intuition.[54] The only exception is the "balancing" test that examiners must perform when considering the patentability of animals. Even in these cases, however, examiners receive no training to help them perform this systematic ethical analysis; they must rely on their own judgments. One examiner explained his approach,

> It's a bit in the discretion of the examiner to decide what he thinks is [sensitive.] ... But it's a balancing act between the patient and the suffering animal. And of course in some cases, for a cat if you have cosmetic purposes and if that causes suffering. But in other cases it might be more difficult for you to decide if it has a small medical benefit and you have a lot of suffering....[55]

Although high-level officials at the EPO have sponsored many initiatives to emphasize the relevance and legitimacy of social and ethical knowledge, they have not trained examiners to deploy this knowledge in a systematic or rational way in patent decision-making. Even when they are required by law to engage in ethical reasoning, examiners have had to rely on their individual judgments. Upon request, they can also receive input from the EPO's patent law experts to facilitate their decision-making.

If the examiner ultimately makes a determination of sensitivity, the application must be reviewed and certified by Team and Directorate supervisors and the director of the Cluster. If they agree with the examiner's assessment, the application is registered in the SeCa database, which is available to all of the EPO's

54. Interview with Employee F, EPO, Munich, Ger. (2009).

55. Interview with Employee E, Munich, Ger. (Aug. 11, 2008).

employees. A specially designated SeCa Advisory Board, made up of examiners, patent lawyers, and public relations personnel inside the EPO periodically reviews the database so that they can be aware of the progress of these applications, and prepare if the examiner decides to issue a patent based on that application. Meanwhile, the examiner proceeds with the review process. If she made the sensitivity determination because she believes that the application is contrary to the *ordre public* clause, she will reject it and advise the inventor about the problem. The inventor may decide to revise the application to avoid the problem, or simply challenge the rejection. As with all other applications, this process continues until the patent is either issued or rejected. Overall, officials hope that the SeCa system will encourage examiners to scrutinize their work more carefully, and ask themselves: "What's happening here? Do we have a quality problem? Are you granting patents you shouldn't be granting?"[56]

The SeCa system, however, is only part of a multi-pronged crisis management system. The EPO issues press releases and fact sheets when it grants a patent that is likely to become controversial, after controversies have erupted, and in response to general areas of controversy (e.g., patents on living organisms or software)—describing the issues of public concern and the reasons behind its decision-making.[57] These external relations efforts are overseen not only by the Communications department, but also by newly created Issue Management Groups (IMGs), composed of examiners, patent lawyers, and public relations specialists, all from inside the organization.[58] Each IMG specializes in one controversial technical area (e.g., computer-implemented inventions, biotechnology), and helps communications officials develop strategies to deal with patent applications that have been identified through the SeCa system while also helping them interact generally with stakeholders on issues that have become controversial. Issue Management Group members are often sent out as envoys to patent-related events throughout Europe, so that they can listen to the concerns of various groups and also explain the position of the Office. They also might review drafts of press releases, or speak to members of the EPO's Administrative Council or the EU's Parliament to alert them to the impending issuance of a controversial patent. The IMGs and increased contacts with the press were created to better communicate with the public and explain the EPO's decisions, while also gathering information about issues of public concern for further discussion inside the EPO.

---

56. Interview with Employee A, EPO, Munich, Ger. (June 25, 2008).

57. *See, e.g.*, Press Release, European Patent Office, Opposition Hearing on Genetic Stem-Cell Patent (2002); European Patent Office, Hintergrundinfromationen zum "Öl-Mais"-Patent, Fact Sheet (2003).

58. Interview with Employee A, EPO, in Munich, Ger. (June 25, 2008).

Conclusion

This chapter has demonstrated major changes taking place in the European patent system, in response to the activist challenges that have emerged over the past three decades. If we had focused our analysis on the European patent system's response through case law, the picture would have looked quite different. First, we might have focused on the debate over the EPC's *ordre public* clause, rather than understanding this debate as part of a large, global challenge to the way that patent systems understand their responsibility to the public interest. Second, we would have probably observed the uneven and limited legal outcomes achieved by challengers and concluded that these activists are little more than a minor and temporary irritation to the European patent system.

When we explore these challenges and the responses by the European patent system in historical, social, and political context, not only do we see a much more comprehensive challenge, but also a more comprehensive response. The European patent system has responded to challengers legislatively, through non-binding resolutions and binding Directives. It has also responded organizationally and culturally. The European Patent Office has informed its employees about the nature of these challenges and changed its administrative routines through the SeCa system and the IMGs. It has also begun to take an active role in policy debates, issuing research reports focused on the implications of patents in controversial areas including climate change. These changes serve to teach EPO employees about the complex consequences of their work, and their responsibility to keep the public benefit at the forefront of their minds. These initiatives suggest that the EPO sees itself in a balancing role, rather than assuming that assigning rights to the patent holders will automatically benefit the public interest. Examiners learn, for example, to keep in mind the costs and benefits of patents and to understand that these costs and benefits extend beyond the economic realm. They are, however, provided with only limited guidance regarding how, exactly, to incorporate this understanding into the everyday practice of patent examination.

The European patent system's response seems less surprising when considered in historical context. Born out of the rubble of two world wars, it has always been more than a special legal domain. It was, and is, an important policy instrument designed to strengthen the idea of a unified Europe. Furthermore, even as its creators debated its structure and orientation, they considered its implications for public health, public policy, and even public morality. As a result, when challengers demanded that the system do a better job of considering the public interest, its insiders could not argue that such issues were irrelevant. Rather, they struggled to determine *how* to do a better job fulfilling the system's responsibilities.

But will these organizational and cultural changes have a long-term impact on decision-making by the European patent system? It is impossible to say. There are signs that it has already had an impact, and that the system is only increasing its attention to the complex implications of patents. A representative of Greenpeace noted: "The Patent Office has completely changed its attitude. . . . I think there are more internal quality checks and internal meetings and people following these discussions. So it has changed definitely."[59] But, there remains little guidance as to how, exactly, examiners should consider the broad implications of patents in their decision-making. Regardless, as we forecast the future path of the European patent system, we must consider its efforts in this area. As we have demonstrated, to focus solely on its legal decision-making would hamper our ability to understand what it is and how it might evolve.

Overall, we have demonstrated the importance of understanding legal institutions—and particularly patent systems—in their social and political context. This kind of analysis can be difficult and time-consuming, requiring data-gathering from personnel and records that do not seem directly connected to decision-making or legal precedent. However, this approach not only puts legal analysis in perspective, but it also demonstrates how factors outside of the law shape a legal domain's past, present, and future.

59. E-mail from Greenpeace Rep. #1 to authors (Aug. 8, 2008) (on file with author).

12 TOWARD A THEORY OF REGULATORY EXCLUSIVITIES

John R. Thomas*

Introduction

No intellectual property rights impact our lives more than those that concern our health. Yet scant policy discussion has been directed toward regulatory exclusivities, those proprietary interests established specifically to protect the innovative pharmaceutical industry from generic competition.[1] In particular, with the notable exception of Rebecca Eisenberg,[2] few members of the professorate have provided a sustained account of the role played by regulatory exclusivities with respect to the research, development, and distribution of medical treatments.

At a time when Congress has significantly enhanced the role of regulatory exclusivities—and is giving serious consideration to augmenting these rights even further—the paucity of guiding principles has been keenly felt. Congress recently enacted legislation establishing lengthy regulatory exclusivities for so-called biologics—a category of

---

* Professor of Law, Georgetown University. I gratefully acknowledge Bruce Artim, Scott Hemphill, Becky Eisenberg, Mark Lemley, Jessica Litman, Peggy Radin, Arti Rai, Bill Schultz, Hans Sauer, and participants in the University of Michigan Law School Intellectual Property Workshop for helpful comments and criticisms.

1. The regime described here in large measure applies to medical devices as well. *See* Paul D. Levin, *On the Inappropriate Inclusion of Medical Devices Within the Hatch–Waxman Act*, 89 J. Pat. & Trademark Off. Soc'y 456 (2007). For purposes of brevity, this chapter employs the term "pharmaceutical industry" to refer to both.

2. *See* Rebecca S. Eisenberg, *The Role of the FDA in Innovation Policy*, 13 Mich. Telecomm. & Tech. L. Rev. 345 (2007) [hereinafter Eisenberg, *Innovation Policy*]; Rebecca S. Eisenberg, *Patents, Product Exclusivity, and Information Dissemination: How Law Directs Biopharmaceutical Research and Development*, 72 Fordham L. Rev. 477 (2003) [hereinafter Eisenberg, *Biopharmaceutical R&D*].

products including vaccines, gene therapies, recombinant therapeutic proteins, and other products of biotechnology.[3] Policymakers have asserted that similar rights should be established for terrorism countermeasures, personalized medicine, and race-based medicine.[4] And one more extreme proposal would essentially abolish the patent system within the pharmaceutical field, with regulatory exclusivities taking their place altogether.[5] These measures create or call for terms of protection that approach that accorded to patents, but absent the usual requisites needed to secure the patent right. As the costs of both healthcare and medical innovation increase, future proposals that would establish new forms of regulatory exclusivities, as well as expand the reach of current proprietary rights, seem likely to proliferate.[6]

This contribution makes a modest effort to fill this policy gap by asserting that regulatory exclusivities should be framed as sui generis rights that complement, rather than supplement the patent system. In Section I, this chapter sets the stage by describing the regulatory framework in which regulatory exclusivities arise, as well as the seven sorts of regulatory exclusivities available under current law. Section II then contrasts the varying regulatory exclusivities with the patent right. In Section III, this chapter asserts that the incentive of a regulatory exclusivity is most appropriate where the patent framework is incapable of promoting activity that lies in the interest of public health. In contrast, the case for a regulatory exclusivity is weakest when it merely promotes the sorts of innovation the patent system was designed to protect, or, perhaps more important, secures protection for potentially patentable innovation that falls short of core patentability requirements. Section IV analyzes recent legislative proposals in terms of this framework. The chapter concludes with an argument that to the extent the patent system is deemed insufficient to encourage the development of technologies of central national concern, modest reform to the patent system presents a superior alternative to the creation of new forms of regulatory exclusivity.

3. Biologics Price Competition and Innovation Act of 2009 (BPCIA), §§ 3139 and 7001–7003 of the Patient Protection and Affordable Care Act, Pub. L. No. 111-148, as modified by the Heath Care and Education Affordability Reconciliation Act of 2010, Pub. L. No. 111-152.

4. *See, e.g.*, (BB 1.2(a)) S. 975, Project Bioshield II Act (109th Congress).

5. *See* H.R. 3947, MODDERN Cures Act of 2011 (112th Congress).

6. *See* Mark D. Janis, *Second Tier Patent Protection*, 40 HARV. INT'L L.J. 151, 177 (1999) (observing that "[i]t takes little imagination to extend this reasoning to justify the creation of a multitude of additional second tier patent regimes having specialized subject matter requirements").

I. Regulatory Exclusivities in Pharmaceutical Innovation

The term "regulatory exclusivity" refers generally to a period of time during which the Food and Drug Administration (FDA) affords an approved drug protection from competing applications for marketing approval.[7] Elevated to the status of an intellectual property right by the World Trade Organization (WTO),[8] regulatory exclusivities have become even more firmly entrenched through Free Trade Agreements championed by the United States.[9] Yet what is fairly described as the newest form of intellectual property remains poorly understood. This chapter next reviews these exclusivities within the regulatory setting in which they arise, first for traditional, small-molecule pharmaceuticals and then for biologics.

A. The Hatch–Waxman Act

Current law requires sponsors of new drugs to demonstrate that their product is safe and effective before it can be distributed to the public. This showing obligates a sponsor to conduct both preclinical and clinical investigations of drugs that have not been previously tested.[10] While deciding whether to issue marketing approval, the FDA evaluates the test data that the sponsor submits in a so-called New Drug Application (NDA).

Although the governing statute states that safety and effectiveness data disclosed in an approved NDA should be made available to the public absent "extraordinary circumstances,"[11] the routine practice of the FDA is to maintain this data in confidence.[12] In addition, because the required test data is quite costly to generate, sponsors

7. John R. Thomas, Pharmaceutical Patent Law (2d ed. 2010).

8. *See* Aaron Xavier Fellmeth, *Secrecy, Monopoly, and Access to Pharmaceuticals in International Trade Law: Protection of Marketing Approval Data Under the TRIPS Agreement*, 45 Harv. Int'l L.J. 443 (2004).

9. *See* Robert Galantucci, *Data Protection in a U.S.–Malaysia Free Trade Agreement: New Barriers to Market Access for Generic Drug Manufacturers*, 17 Fordham Intell. Prop. Media & Ent. L.J. 1083 (2007).

10. Thomas, *supra* note 7, at 302–03.

11. 21 U.S.C. § 355(*l*) (2012). In addition, public disclosure may not be made until the first date a generic application could first be approved—in other words, at such time that all relevant patents and regulatory exclusivities had expired. 21 U.S.C. § 355(*l*)(5).

12. The Food and Drug Administration (FDA) routinely considers a number of commonplace occurrences, including that the New Drug Application (NDA) incorporates trade secrets or contains information that could be used to justify an award of marketing approval by foreign health authorities, to rise to the level of extraordinary circumstances. The effective result is that the FDA can be expected to maintain the test data incorporated into each and every NDA in confidence. *See* James T. O'Reilly, *Implications of International Drug Approval Systems on Confidentiality of Business Secrets in the U.S. Pharmaceutical Industry*, 53 Food & Drug L.J. 123 (1998).

of new pharmaceuticals ordinarily do not disclose it to the public. Otherwise their competitors could file their own NDAs using that test data, and thereby avoid the expenses of developing it on their own.[13]

Although generic firms cannot access the information disclosed in a sponsor's NDA in support of their own applications for marketing approval, the FDA has long allowed them to rely upon such data in a less direct manner. The FDA established a second sort of marketing approval application, termed an "Abbreviated New Drug Application" (ANDA), which applied to generic firms in certain circumstances.[14] Congress subsequently confirmed these FDA practices with the Hatch–Waxman Act, for the first time codifying an abridged approval pathway for generic pharmaceuticals.[15]

The Hatch–Waxman Act allows a generic firm to file an ANDA if the active ingredient of its proposed product is the same as the approved drug.[16] An ANDA essentially permits a generic drug manufacturer to rely upon the safety and efficacy data of the original manufacturer by indirection, even if the generic firm cannot consult the data directly. It also allows a generic manufacturer to avoid the costs and delays associated with filing a full-fledged NDA.[17]

The Hatch–Waxman Act additionally incorporated features that benefitted innovative pharmaceutical firms. Congress observed that the lengthy FDA approval process usually diminishes the effective term of relevant patents. The Hatch–Waxman Act therefore allowed for extension of pharmaceutical patents to compensate for a portion of the lost term. The period of extension, which is applied to a single patent held by the innovative firm, is capped at five years. Further, the remaining term of the restored patent following FDA approval of the NDA may not exceed fourteen years.[18]

Congress further recognized that ANDA filings may implicate patents held by innovative firms. The Hatch–Waxman Act therefore established a complex

13. Rebecca S. Eisenberg, *Pharmaceutical Innovation and Cost: An American Dilemma*, 5 Yale J. Health Pol'y, L. & Ethics 717 (2005).

14. *See* Alan H. Kaplan, *Fifty Years of Drug Amendments Revisited: In Easy-to-Swallow Capsule Form*, 50 Food & Drug L.J. 179, 182–83 (describing pre-Hatch–Waxman Act Abbreviated New Drug Applications (ANDAs) established under the Drug Efficacy Study Implementation (DESI) program for pre-1962 drugs and proposed for newer drugs as well).

15. Drug Price Competition and Patent Term Restoration Act of 1984, Pub. L. No. 84-417, 98 Stat. 1585 (1984).

16. 21 U.S.C. § 355(j)(1) (2012).

17. Federal Trade Commission (FTC), Generic Drug Entry Prior to Patent Expiration: An FTC Study 5 (July 2002), *available at* http://www.ftc.gov/os/2002/07/genericdrugstudy.pdf.

18. 35 U.S.C. § 156 (2012).

patent dispute resolution regime. The filing of an ANDA may be deemed a "somewhat artificial" act of patent infringement,[19] allowing the patent owner to assert a charge of infringement immediately. If the innovative firm brings suit, the generic firm is subject to a thirty-month stay of FDA marketing approval—in essence an automatic preliminary injunction in favor of the innovative firm.[20]

Finally, the Hatch–Waxman Act established a regime of regulatory exclusivities. These proprietary rights consist of a period of time during which a competitor's ability to obtain FDA permission to sell a previously approved brand-name drug is restricted. The Hatch–Waxman Act itself established three forms of regulatory exclusivities that operate to reward, respectively, the sponsor of a drug that qualifies as a new chemical entity (NCE), the sponsor of a new clinical study concerning a drug that does not qualify as an NCE, and a generic firm that challenges a patent held by an innovative firm. Congress has also created two additional sorts of regulatory exclusivities pertaining to orphan drugs[21] and to the completion of pediatric studies.[22]

Coherence of terminology is, sadly, not a hallmark of legal commentators within the pharmaceutical field. Some writers employ the term "marketing exclusivity" to describe each of these entitlements. Others distinguish between "marketing exclusivities" and "data protection." Within this framework, a marketing exclusivity entirely prevents competitors from filing applications for FDA marketing approval. Data protection provides more limited rights. This entitlement blocks firms that wish to rely upon the clinical data presented by the brand-name firm. Unlike a marketing exclusivity, data protection would not block a firm that had prepared its own data package from seeking FDA approval.[23]

In practice, the distinction between marketing exclusivities and data protection may be more apparent than real. Due to the high costs, generic drug companies simply do not undertake repetitive versions of the clinical trials conducted by brand-name firms. That an application for FDA marketing approval could be filed with an independently generated data package is therefore of small comfort to generic firms. In any event, this chapter will employ the term "regulatory exclusivity" to describe FDA-administered proprietary rights generally, and will distinguish between marketing exclusivity and data protection as the need arises.

The first and arguably most significant of the regulatory exclusivities provides for a five-year period of regulatory exclusivity for NCEs. NCE exclusivity

19. Eli Lilly & Co. v. Medtronic, Inc., 496 U.S. 661, 676 (1990).

20. 21 U.S.C. § 355(j)(5)(B)(iii).

21. 21 U.S.C. § 360cc (2012).

22. 21 U.S.C. § 355a(b) (2012).

23. THOMAS, *supra* note 7.

prevents a generic applicant from submitting an ANDA for five years after the approval of a drug that qualifies as an NCE.[24] A drug is judged to be an NCE if the FDA has not previously approved that drug's active moiety.[25] NCE exclusivity does not preclude the FDA from accepting an application submitted by an entity that has performed all the required preclinical and clinical studies itself, or that has a right of reference to such data.[26]

NCE exclusivity effectively prevents generic manufacturers from bringing a product to market for five years plus the length of the FDA review of the generic application.[27] If, for example, the FDA requires two years to approve a particular generic application, the real-world impact of the NCE exclusivity has been seven years of protection. In this respect NCE exclusivity operates differently from other forms of FDA-administered exclusivities, which generally prevent the FDA from approving applications, rather than accepting them in the first instance.[28]

The Hatch–Waxman Act also provides for a three-year new clinical study exclusivity period. New clinical study exclusivity may be awarded with respect to a marketing application that contains reports of new clinical studies conducted by the sponsor that are essential to FDA approval. The FDA has granted new clinical study exclusivity for such changes as new dosage forms or new indications, or for a switch from prescription to over-the-counter status for the drug.[29] In contrast to NCE exclusivity, new clinical study exclusivity does not prevent the FDA from accepting a generic application with respect to the drug. If the new clinical study exclusivity continues to bar the issuance of marketing approval at the close of FDA review, the FDA will issue a tentative approval for the generic product that will become effective once the new clinical study exclusivity has run its course.[30]

---

24. The Hatch–Waxman Act allows the five-year term of new chemical entity (NCE) exclusivity to be decreased to four years in one circumstance. If the NDA holder owns a patent that the generic applicant believes is invalid or not infringed, then the generic applicant is allowed to file its application one year early—upon the expiration of four, rather than five years from the date the NDA was approved. 21 U.S.C. § 355(j)(5)(F)(ii). The apparent purpose of this provision is to encourage early resolution of patent disputes, in turn minimizing delay of generic market entry.

25. *See* 21 C.F.R. § 314.108(a) (2012).

26. *See* Elizabeth H. Dickinson, *FDA's Role in Making Exclusivity Determinations*, 54 Food & Drug L.J. 195, 200 (1999).

27. *See* Marvin M. Goldenberg, *Medicare and the New Generic Drug Legislation*, 29 Pharmacy & Therapeutics 89, 90 (Feb. 2004).

28. *See* Dickinson, *supra* note 26, at 200.

29. *Id.* at 201.

30. *Id.*

In addition, new clinical study exclusivity only applies to the use of the product that was supported by the new clinical study.[31] If, for example, the new studies support a new indication or dosage form of the previously approved ingredient, then the three-year exclusivity applies only to that particular use or dosage form. The FDA is not barred from approving generic drugs for other indications or dosage forms.[32] As with NCE exclusivity, new clinical study exclusivity does not preclude the FDA from approving a full NDA. If the sponsor of that subsequent NDA has performed all the required preclinical and clinical studies itself, or has a right of reference to such data, the FDA may approve the NDA without regard to the new clinical trial exclusivity.[33]

A drug may be subject both to NCE exclusivity and new clinical study exclusivity during its life cycle. A new drug will commonly enjoy a five-year NCE exclusivity. The sponsor of the drug may subsequently perform additional clinical trials to qualify changes in the drug for additional three-year exclusivities.[34]

Finally, the Hatch–Waxman Act also established a regulatory exclusivity that operates in favor of generic firms. Prospective manufacturers of generic pharmaceuticals are afforded exclusivity in the event they challenge a patent associated with an approved pharmaceutical. The reward consists of a 180-day generic drug exclusivity period awarded to the first ANDA applicant that challenges the validity or scope of a patent. During this 180-day period, the FDA may not approve another patent-challenging ANDA with respect to the same drug.[35]

In crafting these three regulatory exclusivities within the Hatch–Waxman Act, Congress looked to legislation it had enacted a year earlier.[36] The Orphan Drug Act was intended to encourage firms to develop pharmaceuticals to treat rare diseases and conditions.[37] These medicines are called "orphan drugs" because firms may lack the financial incentives to sponsor products to treat small patient

---

31. *See* Valerie Junod, *Drug Marketing Exclusivity under United States and European Union Law*, 59 Food & Drug L.J. 479, 500 (2004).

32. *See* Thomas J. Parker et al., *FDA Marketing Exclusivity for Single Enantiomers of Previously Approved Racemates*, 15(1) J. Proprietary Rights 8 (Jan. 2003).

33. Dickinson, *supra* note 26, at 200.

34. *See* Junod, *supra* note 31, at 489.

35. 21 U.S.C. § 355(j)(5)(B)(iv) (2012).

36. Orphan Drug Act, Pub. L. No. 97-414, 96 Stat. 2049 (1982) (codified as amended at 21 U.S.C. §§ 360aa-360ee (2012), 26 U.S.C. § 45C (2012), 42 U.S.C. § 236 (2012)).

37. *See, e.g.*, Gary A. Pulsinelli, *The Orphan Drug Act: What's Right with It*, 15 Santa Clara Computer & High Tech. L.J. 299, 301 (1999); Li-Hsien Rin-laures, *Recent Developments Concerning the Orphan Drug Act*, 4 Harv. J. L. & Tech. 269, 269–70 (1991). The Orphan Drug Act also applies to varying degrees to biologics and medical devices, which, along with drugs, are commonly termed "orphan products."

populations and therefore be unwilling to "adopt" them.[38] Congress therefore established a seven-year term of orphan drug regulatory exclusivity.[39] That exclusivity applies only to the indication for which the drug is approved, however, with the result that the FDA could approve a second application for the same drug for a different use.[40]

Congress established an additional regulatory exclusivity in 1997 in order to promote the availability of medicines for children.[41] This exclusivity reflects concerns that many marketed drugs had not yet been clinically tested upon children. Investigations upon a pediatric population raise complex issues of informed consent and the inability of children to describe the effect of a medication accurately. In addition, sponsors often have scant incentive to conduct pediatric studies when, even in their absence, many drugs are nonetheless prescribed for use by children.

By establishing a pediatric regulatory exclusivity, Congress hoped to encourage additional pediatric testing, which in turn could allow medications to be labeled for use by children.[42] Under this system, the FDA requests NDA applicants and holders to perform pediatric studies with respect to the drug. Response to this written request is wholly voluntary. If the innovative drug company conducts the requested study or studies and submits a report to the satisfaction of the FDA, however, then it will be awarded a six-month pediatric regulatory exclusivity that augments any existing patents or exclusivities held by that entity.[43]

Finally, Congressional concern over the spread of antibiotic-resistant "superbugs" led to the enactment of the Generating Antibiotic Incentives Now (GAIN) Act, enacted as Title VIII of the FDA Safety and Innovation Act.[44] The 2012 statute allows the FDA to designate a drug as a "qualified infectious disease product" (QIDP) if it consists of an antibacterial or antifungal drug intended to treat serious or life-threatening infections.[45] The GAIN Act stipulates that QIDPs

---

38. *See, e.g.*, David B. Clissold, *Prescription for the Orphan Drug Act: The Impact of the FDA's 1992 Regulations and the Latest Congressional Proposals for Reform*, 50 Food & Drug L.J. 125, 125 (1995).

39. 21 U.S.C. § 360cc (2012).

40. *Id.* § 360cc(b).

41. Food and Drug Administration Modernization Act of 1997, § 11, Pub. L. No. 105-115, 111 Stat. 2296 (1997) (codified at 21 U.S.C. § 355a (2012)).

42. *See, e.g.*, Lisa Jerles, *The Best Pharmaceuticals for Children Act and the Pediatric Research Equity Act—Helping or Hurting America's Children?*, 6 Cardozo Pub. L. Pol'y & Ethics 515 (2008).

43. 21 U.S.C. § 355a(b) (2012).

44. Pub. L. No. 112-144, § 801 (introducing 21 U.S.C. § 505E (2012)).

45. 21 U.S.C. § 505E(g) (2012).

include drugs that address drug-resistant tuberculosis, gram negative bacteria, and *Staphylococcus aureus*.[46]

Along with other measures intended to provide pharmaceutical and biotechnology companies with incentives to develop innovative antibiotics,[47] the GAIN Act adds five years to the term of the new chemical entity, new clinical study, and orphan exclusivities for any QIDP.[48] The statute stipulates that the five-year QIDP extension is cumulative with the pediatric exclusivity.[49] As a result, a QIDP that qualified as a new chemical entity, and was also awarded a pediatric exclusivity, would be entitled to a data exclusivity period of ten years and six months.

B. Biologics

A new category of products termed "biologics" has begun to play an increasingly important role in US healthcare.[50] Vaccines, blood components, monoclonal antibodies, gene therapies, enzymes, hormones, and other products that are or were derived from living organisms are considered biologics. Not only are sales of biologics growing rapidly, some experts estimate that in coming years nearly half of all newly approved drugs will result from biotechnology.[51]

Public health advocates have observed that competing versions of biologic products have been slow to enter the market following patent expirations. In particular, observers believe that the ANDA process does not comfortably apply to biologics.[52] Biologics differ significantly from traditional small-molecule pharmaceuticals in their size, structural complexity, and method of manufacture. These distinctions are so substantial that biologics are ordinarily governed by a different statute[53] and a different division of the FDA[54] than small-molecule

---

46. *Id.* § 505E(f).

47. See *Antibiotics Resistance Rising: Can New Drugs Keep Pace?*, BioWorld Insight (Sept. 17, 2012).

48. 21 U.S.C. § 505E(a).

49. *Id.* § 505E(b).

50. *See* Kathleen R. Kelleher, *FDA Approval of Generic Biologics: Finding a Regulatory Pathway*, 14 Mich. Telecomm. Tech. L. Rev. 245, 247 (2007).

51. Bryan A. Liang, *Regulating Follow-On Biologics*, 44 Harv. J. on Legis. 363, 364 (2007).

52. *See* Alan J. Morrison, *Biosimilars in the United States: A Brief Look at Where We Are and the Road Ahead*, 26 Biotech. L. Rep. 463, 465 (2007).

53. The Public Health Service (PHS) Act most often applies to biologics, as compared to the Federal Food, Drug, and Cosmetic Act. The Hatch–Waxman Act modifies the latter legislation.

54. The Center for Biologics Evaluation and Research (CBER) regulates biologics, whereas the Center for Drug Evaluation and Research (CDER) addresses small-molecule pharmaceuticals.

pharmaceuticals. Competitors who wish to develop their own version of known biologics may face considerable difficulties in demonstrating that their product is equivalent to a particular brand-name product.

Congress turned to these concerns when it enacted the Biologics Price Competition and Innovation Act (BPCIA) of 2009. The BPCIA was incorporated into Title VII of the Patient Protection and Affordable Care Act, the omnibus healthcare legislation that recently survived constitutional challenge. The BPCIA included three significant components. First, the BPCIA established a licensure pathway for competing versions of previously marketed biologics. In particular, the legislation established a regulatory regime for two sorts of follow-on biologics, termed "biosimilar" and "interchangeable" biologics. The FDA was afforded a prominent role in determining the particular standards for biosimilarity and interchangeability for individual products. Second, the BPCIA created a patent dispute resolution procedure for use by brand-name and follow-on biologic manufacturers.

Finally, the BPCIA created FDA-administered periods of regulatory exclusivity for certain brand-name drugs and follow-on products. Under this system, no application for a follow-on biologic—either biosimilar or interchangeable—may be filed with the FDA for four years from the date the reference product was licensed.[55] In addition, the FDA may not approve an application for a follow-on product—either biosimilar or interchangeable—for twelve years from the reference product's licensure date.[56] The BPCIA also provides for a term of regulatory exclusivity for the applicant that is the first to establish that its product is interchangeable with the brand-name product. The period of follow-on regulatory exclusivity is ordinarily one year but may run for a longer or shorter term depending upon the existence or outcome of patent litigation.[57]

Although the twelve-year biologics exclusivity is barely out of the gate, it has already been subject to considerable controversy. In particular, the Obama administration has proposed that the twelve-year term should be diminished

55. 42 U.S.C. § 262(k)(7)(B) (2012).

56. *Id.*

57. More particularly, the follow-on exclusivity expires on the date that is the earlier of (1) one year after the first commercial marketing of the first interchangeable biologic to be approved as interchangeable with that reference product; (2) eighteen months after either a final court judgment in patent infringement litigation or the dismissal of such litigation against the first applicant; (3) forty-two months after the approval of the first interchangeable biologic if patent litigation remains pending; or (4) eighteen months after approval of the first interchangeable biologic if the applicant has not been sued for patent infringement. 42 U.S.C. § 262(k)(6).

by five years. According to the President's Deficit Reduction Plan,[58] the suggested reduction in the exclusivity period "increases the availability of generic biologics to encourage faster development of generic biologics while retaining appropriate incentives for research and development for the innovation of breakthrough products." The president forecasts a savings of $3.5 billion over ten years to Medicare, Medicaid, and other federal health programs. After a hard-fought campaign before Congress to establish a twelve-year period of exclusivity, the brand-name pharmaceutical industry seems unlikely to capitulate on this matter.

The BPCIA arguably marks a significant turning point in the thinking of the legislative and executive branches concerning regulatory exclusivities. First, the twelve-year term of the biologics exclusivity endures substantially longer than those of earlier proprietary rights afforded by the Hatch–Waxman Act and related legislation. Second, although the legislative history of these statutes provides few insights as to the rationale for regulatory exclusivities, the Obama administration has explicitly viewed the twelve-year biologics exclusivity as a R & D incentive. This characterization holds significant consequences for healthcare innovation and deserves further exploration, an effort this chapter undertakes next.

II. Regulatory Exclusivities versus Patents

The seven regulatory exclusivities are proprietary rights that have been described as "quasi-" or "pseudo-patents."[59] This characterization is apt at a level of rough approximation. Yet viewing regulatory exclusivities as equivalents to patents would be very much a mistake. The twelve-year biologics exclusivity and five-year NCE exclusivity in particular provide significantly more robust exclusionary rights than do patents. And although at times dismissed, the three-year new clinical study exclusivity too can compare favorably with a patent from the perspective of its proprietor.

Regulatory exclusivities offer the extraordinary advantage of pairing perfectly with product markets.[60] For example, the NCE exclusivity effectively prevents generic drug companies from selling their products altogether even following

58. Office of Management and Budget (OMB), Living Within Our Means and Investing in the Future: The President's Plan for Economic Growth and Deficit Reduction (Sept. 2011), *available at* http://www.whitehouse.gov/sites/default/files/omb/budget/fy2012/assets/jointcommitteereport.pdf.

59. Eisenberg, *Biopharmaceutical R&D*, *supra* note 2, at 482; James T. O'Reilly, *Elders, Surgeons, Regulators, Jurors: Are Medical Experimentation's Mistakes Too Easily Buried?*, 31 Loy. U. Chi. L.J. 317, 324 (2000).

60. Eisenberg, *Innovation Policy*, *supra* note 2, at 355 (appreciating that "[p]atents cover inventions, and inventions do not necessarily correspond to product markets").

the expiration of its five-year term. In contrast, generic firms are often able to design around individual patents, particularly when they concern such inventions as formulations, methods of manufacture, and chemical intermediates.[61] The courts have appreciated that the patent system promotes innovation by encouraging others to "invent around" the proprietary rights of their competitors.[62] Such an effort is impossible with respect to the NCE exclusivity.

The marketplace value of the new clinical study exclusivity may also be of considerable value, provided that the new study brings changes that are not reflected solely through labeling. When the study results in changes to the finished dosage form—such as new release characteristics or chemical modifications—existing generic products cannot be substituted for the innovator drug. On the other hand, the new clinical study is of more limited practical importance in circumstances where the results of the new study do not result in changes to the dosage form itself.[63] If the study reveals a new indication for the drug, for example, generic forms may continue to sell the drug for its original use. Physicians are then free to prescribe the generic drug off-label for the new indication. In many circumstances the new clinical study exclusivity can nonetheless provide an effective barrier against generic competition during its three-year term.

As with each of the regulatory exclusivities, the new clinical study exclusivity is also superior to patents in that it is self-enforcing. The FDA renders the usually routine judgment about whether the sponsor should enjoy a regulatory exclusivity, and then upholds that right during subsequent marketing approval determinations.[64] In contrast, a patent must be enforced by its proprietor via litigation in the federal courts. Even if such a venture is ultimately successful, exclusion of competition during the pendency of the proceedings may require the difficult showing of demonstrating entitlement to a preliminary injunction.[65]

In addition, as Mark Lemley and Carl Shapiro have aptly phrased the concept, patent rights are notoriously probabilistic.[66] Put differently, patents provide not so

61. *See* Gary J. Elman et al., *FDA Approval of Generic Drugs: Instituting a First Successful Defense Requirement for Generic Exclusivity*, 22 BIOTECH. L. REP. 97, 99–100 (2003).

62. *See, e.g.*, Intel Corp. v. VIA Techs., Inc., 319 F.3d 1357, 1367 (Fed. Cir. 2003) (observing that "patent law encourages competitors to design around").

63. Eisenberg, *Innovation Policy*, *supra* note 2, at 360–61.

64. Junod, *supra* note 31, at 492 ("Marketing exclusivity confers a very strong protection to the pioneer company because competitors cannot really challenge it.").

65. *See* Christopher Holman, *Do Reverse Payment Settlements Violate the Antitrust Laws?*, 23 SANTA CLARA COMP. & HIGH TECH. L.J. 489, 514 (2007).

66. Mark A. Lemley & Carl Shapiro, *Probabilistic Patents*, 19 J. ECON. PERSP. 75 (2005).

much the right to exclude but the right to try to exclude.[67] Courts have held about half of the patents that are litigated to have been improvidently granted and therefore invalid.[68] Potential shortcomings with respect to patentability criteria are simply irrelevant to regulatory exclusivity determinations, however. Once the FDA makes the typically straightforward determination that a firm is entitled to a regulatory exclusivity, this entitlement is almost never contested.

The virtual incontestability, self-enforcing nature, and scope of coverage of many of the regulatory exclusivities provide them with considerable advantages over patents. Yet, at least prior to the enactment of the BPCIA, these benefits traditionally accrued for relatively short periods of time. For example, in comparison to the twenty-year term awarded to patents, the generic and pediatric exclusivities last only about six months. As a result, the robust title conferred by regulatory exclusivities has historically been balanced by the comparatively brief periods of protection that they confer.

Long experience suggests that legislatures face persistent pressures to extend the terms of established intellectual property rights,[69] however, and regulatory exclusivities provide no exception. The new twelve-year biologics exclusivity provides the latest example. Some commentators have proposed the prolongation of the NCE exclusivity to durations found elsewhere,[70] while other legislative proposals would create lengthy new exclusivities for terrorism countermeasures and public health innovations. Policy discussion of these reform proposals has been hampered by the paucity of theoretical exploration concerning regulatory exclusivities. Scant scholarship has endeavored to identify the rationale for these proprietary interests, situate them in relationship to other intellectual properties, or provide a mechanism for setting their terms. This chapter next seeks to make a modest contribution to this debate by developing a theoretical account of regulatory exclusivities.

---

67. 35 U.S.C. § 282 (2012) (listing "defenses in any action involving the validity or infringement of a patent").

68. *See* Lemley & Shapiro, *supra* note 66, at 75.

69. *See* Arlen W. Langvardt, *The Beat Should Not Go On: Resisting Early Calls for Further Extension of Copyright Duration*, 112 Penn State L. Rev. 783 (2008).

70. *See* Junod, *supra* note 31, at 501 (noting these assertions).

III. A Normative Account of Regulatory Exclusivities

A. The Role of Regulatory Exclusivities

The constellation of regulatory exclusivities awards dissimilar proprietary rights in distinct situations. Each was developed through intensely pragmatic compromises between two contending factions within one specific industry. One theme can nonetheless be developed that unifies these diverse rights. Simply put, regulatory exclusivities are best seen as a means of promoting activities that the patent system, through its essential doctrinal framework, is unable to encourage.

The patent system has proved a durable construct for promoting technological innovation throughout the nation's history. Its core doctrines, developed in view of the industrial, mechanical, and manual arts of the late eighteenth century, have proved remarkably capable of addressing biotechnology, nanotechnology, information technology, and other recent advancements.[71] Yet the patent law has never been viewed as a mechanism for encouraging all manner of technological activity. Rather, its designers purposefully denied the patent incentive to some advances in order to promote the overall purpose of promoting innovation.[72]

Among the doctrines that leave some technological activity without the scope of the patent system is the statutory subject matter requirement. The Patent Act calls for the patenting of any new and useful machine, manufacture, composition of matter, or process.[73] This statutory language, although broadly encompassing, has not been understood to encompass all discoveries.[74] In particular, abstract ideas and scientific principles cannot be patented even where that information was costly to obtain and of unquestioned value.[75] The courts have instead reasoned that such patents would ultimately discourage research by hampering the exchange and use of fundamental knowledge.[76]

This long-standing rule renders the production of the enormous quantity of data that drug sponsors generate for the FDA outside the scope of patenting. Sponsors of new drugs are required to conduct multiple stages of preclinical

71. *See* Dan L. Burk & Mark A. Lemley, *Policy Levers in Patent Law*, 89 VA. L. REV. 1575 (2003).

72. Bilski v. Kappos, 130 S. Ct. 3218 (2010).

73. 35 U.S.C. § 101 (2012).

74. *See* Diamond v. Diehr, 450 U.S. 175, 185 (1981) ("The Court has undoubtedly recognized limits to §101 and every discovery is not embraced within the statutory terms.").

75. *See Bilski*, 130 S. Ct. at 3225 (observing that laws of nature, physical phenomena, and abstract ideas are not patentable subject matter).

76. Lab. Corp. v. Metabolite Labs., Inc., 548 U.S. 124, 126–27 (2006) (Breyer, J., dissenting from dismissals as improvidently granted).

and clinical trials upon new drugs in order to submit an NDA.[77] This effort is supplemented by extensive nonclinical data, including chemical, manufacturing, and statistical studies.[78] The typical NDA comprises hundreds of thousands of pages (or more) that will ultimately be scrutinized by multiple teams of FDA reviewers.[79]

The value of a new drug depends entirely upon this detailed, hard-won knowledge about its effect upon the human body.[80] But this quantity of information is far greater than the Patent Act requires. Applicants need only identify a plausible use for their inventions in order to obtain a patent. In vitro testing, by itself or accompanied by limited in vivo testing in laboratory animals, often leads to the reasonable inference that a particular compound would be an effective medical treatment when used in humans.[81] Such a showing is enough for patenting, yet it falls far short of fulfilling the requirements of safety and efficacy required to obtain the FDA's permission to market a drug.[82]

Other forms of intellectual property are incapable of making up this shortfall in the patent system. Firms in many industries rely upon the trade secret regime to protect valuable information, such as manufacturing specifications or lists of customers, that is not patentable.[83] The Hatch–Waxman Act renders this alternative unavailable for approved drugs due to the ANDA mechanism, however, under which generic firms rely upon the data generated by the drug's original sponsor.[84] Nor does the copyright law protect compiled information per se.[85] Regulatory exclusivities may therefore be seen as filling a gap among the mainstream forms of

77. *See* Michael A. Carrier, *Two Puzzles Resolved: Of the Schumpeter-Arrow Paradox and Pharmaceutical Innovation Markets*, 93 IOWA L. REV. 393, 417–18 (2008).

78. *See* W. Christopher Matton & F. Scott Thomas, *The Continuing Balance: Federal Regulation of Biotechnology*, 44 JURIMETRICS J. 283 (2004).

79. *See* Michael D. Green, *Safety as an Element of Pharmaceutical Quality: The Respective Roles of Regulation and Tort Law*, 42 ST. LOUIS U. L.J. 163,173 (1998).

80. *See* Eisenberg, *Innovation Policy*, *supra* note 2, at 353.

81. *See, e.g.*, *Ex parte* Aggarwal, 23 U.S.P.Q.2d 1334, 1338 n.8 (B.P.A.I. 1992). The phrase "in vitro" literally means "in glass," and refers to testing an artificial environment such as a test tube or culture media. STEADMAN'S MEDICAL DICTIONARY 798 (25th ed. 1990). The phrase "in vivo" refers to a live subject and ordinarily implies human or animal trials. *Id.*

82. *In re* Brana, 51 F.3d 1560, 1568 (Fed. Cir. 1995) ("The stage at which an invention in this field becomes useful is well before it is ready to be administered to humans.").

83. *See* Andrew Wasson, *Taking Biologics for Granted? Takings, Trade Secrets, and Off-Patent Biological Products*, 2005 DUKE L. & TECH. REV. 4 (2005).

84. *See* Jacqueline Fox, *Reinvigorating the Concept of Benefit: The Failure of Drug Company-Sponsored Research on Human Subjects*, 38 SETON HALL L. REV. 605, 636–37 (2008).

85. *See* Feist Publ'ns, Inc. v. Rural Tel. Serv. Co., 499 U.S. 340 (1991).

intellectual property rights, offering incentives for generating the data necessary for the FDA to make its marketing approval determinations.

The simplicity of this account of regulatory exclusivities is one of its advantages. Yet, even among the scant scholarly thinking that has so far accompanied the rise of this newest form of intellectual property, it is not the dominant view. Other commentators have viewed regulatory exclusivities as an innovation incentive that competes with the patent system in stimulating pharmaceutical research and development. For Professor Eisenberg, regulatory exclusivities are best viewed as "an economic measure designed to promote costly investments in innovation."[86] Professor Roin also views regulatory exclusivities as an innovation incentive.[87] He views the patent system as possessing considerable shortcomings with respect to the pharmaceutical industry. Roin advocates the significant expansion of regulatory exclusivities in order to provide the optimal mechanism for supporting the development of innovative drugs.[88] The Obama administration has apparently sided with this camp, recently suggesting that the twelve-year biologics exclusivity also serves as a R & D promotion mechanism.

The account of regulatory exclusivity as a patent-like innovation promoter has both shortcomings and descriptive flaws. Viewing regulatory exclusivities as innovation incentives implies that the patent system necessarily possesses failings that are unique to pharmaceutical research and development. This conclusion contrasts starkly with the usual account that patents have provided more value to the pharmaceutical industry than to other industries.[89] It also suggests that perceived deficiencies in the patent system remain unaddressed.[90] The failure to make the patent system responsive to one of the nation's most science-based industries threatens to antiquate the regime that has long served as America's engine of innovation. Such an approach also maintains deficiencies in patent doctrine for the great majority of innovative industries that do not benefit from regulatory exclusivities.

As well, the conception that regulatory exclusivities promote innovation in the same manner as patents necessarily calls for conflicts between the two regimes. An award of regulatory exclusivity does not depend upon any of the

86. *See* Eisenberg, *Innovation Policy*, *supra* note 2, at 361.

87. Benjamin N. Roin, *Unpatentable Drugs and the Standards of Patentability*, 87 TEX. L. REV. 503 (2008).

88. *Id.*

89. *See, e.g.*, JAMES BESSEN & MICHAEL J. MEURER, PATENT FAILURE: HOW JUDGES, BUREAUCRATS, AND LAWYERS PUT INNOVATION AT RISK 16 (2008).

90. Roin, *supra* note 87.

usual patentability requirements.[91] In particular, a regulatory exclusivity may be granted even with respect to a pharmaceutical that is well-known to the state of the art. This result conflicts with bedrock patent doctrine stipulating that proprietary rights may be awarded only to an invention that advances the art.[92] This rule reflects a deliberate policy choice that, as stated by the Supreme Court, "otherwise patents might stifle, rather than promote, the progress of useful arts."[93] Deeming regulatory exclusivities as serving patent-like functions effectively refutes this central patent policy.

To some, pharmaceuticals appear to be worthy candidates for regulatory exclusivities because they remain particularly costly to bring to market relative to other sorts of patentable inventions, but at the same time potentially provide substantial social benefits. That the appropriate legislative reaction to inefficient innovation is the award of more expansive proprietary rights seems a dubious proposition. Competition surely provides at least as much of a spur to innovation as exclusion does. And as the award of regulatory exclusivities does not depend on an assessment of the marginal benefit to the pharmacopeia provided, this assertion too seems somewhat beside the point.

Others observe that patients are utterly indifferent as to whether a cure is patented or not. They seek an effective treatment regardless of whether it is novel, nonobvious, or otherwise in keeping with the requirements of the Patent Act. Regulatory exclusivities, which are awarded whether or not these patentability requirements are fulfilled, potentially reflect this reality. Yet this response is not unique to the pharmaceutical field. No consumer likely bases purchasing decisions upon whether a particular product, be it athletic footwear, a tablet PC, or automobile, embodies a patented invention or not. This assertion does little to support a sui generis, subsidiary intellectual property rights regime for a particular category of patentable invention.

Another response to this line of argument is that it is entirely backward. Pharmaceutical firms commonly commission patent lawyers to determine whether nascent products would be eligible for patent protection. A firm may choose to pursue certain products only when it recognizes that robust intellectual property rights will be available for them. Under this account, we will never know what drugs might be on the market had more potent forms of intellectual property protection been available. Yet this assertion too is overexpansive: surely firms in all industries weigh their intellectual property options prior to investing

---

91. THOMAS, *supra* note 7, at 349 (observing that "[m]arketing exclusivities and patent protection are separate entitlements that are administered by different federal agencies and that depend upon different criteria").

92. 35 U.S.C. § 103(a) (2012).

93. KSR Int'l Co. v. Teleflex, Inc., 550 U.S. 398, 427 (2007).

in particular projects. Nothing sets the pharmaceutical industry apart in this regard.

B. Theory versus Practice

Prior to the enactment of the BPCIA, the notion that regulatory exclusivities ought to complement, rather than compete with, the patent system held up fairly well when viewed against the governing statutes. The terms of regulatory exclusivities could well be seen as representing an armchair judgment pertaining to the length, complexity, and risk associated with clinical and nonclinical trials: five years for NCEs, three years for new clinical studies on a known compound, and seven years for orphan drugs, where the development of rigorous testing protocols may prove difficult due to smaller patient populations.[94]

The pragmatic function of the regulatory exclusivities also suggested they were not appropriately seen as innovation incentives. Most commonly the five-year NCE exclusivity acts merely to regulate the timing of patent litigation.[95] Along with the three-year new clinical study exclusivity, it serves as the sole source of proprietary rights only in those few cases when its term extends beyond the life of any relevant patents, typically when the FDA regulatory review process has consumed many years.[96] Similarly, the six-month pediatric exclusivity period may be obtained whether or not the study successfully demonstrates safety and effectiveness in children.[97] This right simply establishes incentives for drug sponsors to conduct research and submit their results to the FDA.

The 180-day generic exclusivity period is also not aptly viewed as an incentive to innovate. Congress established this proprietary right in order to ameliorate collective action problems that may arise with regard to pharmaceutical patent challenges.[98] A generic firm that challenges a patent must bear the expensive, upfront cost of litigation. If the generic firm is successful, however, the challenged patent is declared invalid with regard to all the world. Any firm—not just the one who challenged the patent—could then introduce a competing product to the

---

94. Roin follows this chain of reasoning, for although he describes regulatory exclusivities as R & D incentives, he views them as ineffective. Under this view, rational firms would pursue FDA approval only for patented drugs. *See* Roin, *supra* note 87.

95. *See* THOMAS, *supra* note 7, at 354 ("NCE marketing exclusivity has a significant role to play in the timing of patent challenges and the entry of generic drugs into the U.S. marketplace.").

96. *See* Eisenberg, *Innovation Policy*, *supra* note 2, at 352.

97. *See* THOMAS, *supra* note 7, at 370.

98. *See* Jeremiah Helm, *The Patent End Game: Evaluating Generic Entry into a Blockbuster Pharmaceutical Market in the Absence of FDA Incentives*, 14 MICH. TELECOMM. & TECH. L. REV. 175,180 (2007).

marketplace. Understandably, this forced sharing may undermine the incentives any one generic firm would possess to challenge a brand-name firm's patent. The award of 180 days of generic exclusivity is therefore intended to allow a successful patent challenger to capture an individual benefit for its effort, in turn encouraging such challenges in the first instance.[99]

Of the five traditional regulatory exclusivities, the one pertaining to orphan drugs might have been fairly characterized as an innovation incentive at one time. Under the original terms of the 1983 Orphan Drug Act, the seven-year exclusivity was unavailable if the drug was or had been patented.[100] The practical effect of this provision was to establish intellectual property rights in products that patent doctrine would judge to fall within the public domain. This requirement was eliminated in 1985.[101] In so doing, Congress allowed the orphan drug exclusivity to operate not as a patent substitute, but as an incentive for conducting further research upon known compounds that were potential treatments for orphan diseases.

The newly enacted biologics exclusivity is far more problematic. The BPCIA arguably marks a significant shift in legislative thinking toward regulatory exclusivity on several fronts. The twelve-year term of the biologics exclusivity begins to approach the duration of a patent. Indeed, supporters of the BPCIA cited perceived failings of the patent system as the basis for this extent of protection. They asserted that essentially the entire array of patent doctrines combine to provide the biopharmaceutical industry with inferior rights to those possessed by other industries.[102] The Obama administration has arguably backed this view, explicitly identifying this regulatory exclusivity as an R & D incentive.

This chapter has attempted to make the case that regulatory exclusivities are best viewed not as an innovation-promoting patent substitute, but rather as a reward for gathering sufficient information about a drug to obtain marketing approval. Under this view, characterizing the regulatory exclusivity as a

99. *See* Erika Lietzan & David E. Korn, *Issues in the Interpretation of the 180-Day Exclusivity*, 62 Food & Drug L.J. 49, 49 (2007). The same analysis potentially applies to rulings of non-infringement as other firms might be able to develop generic products previously judged to fall outside the scope of the brand-name firm's patents.

100. Pub. L. No. 97-414, 96 Stat. 2049 (1983).

101. Orphan Drug Amendments of 1985, Pub. L. No. 99-91, 99 Stat. 387 (1985).

102. *See, e.g.*, *Schools, Investors Push Heavy Protection for Brand Biologics*, FDA Week (May 30, 2008); Meghan McCarthy, *Biotech Industry Emphasizes Need for Longer Data Exclusivity Period for Follow-On Biologics*, CQ HealthBeat (May 16, 2008); Donna Young, *Follow-on Biologics Called Key Issue for Biotech This Year*, BioWorld Today (Feb. 12, 2008).

mechanism for remedying defects in the patent system is a mistake from the start. Yet the contention that the biologics industry deserves a lengthy patent substitute merits review on its own terms.

From the outset, the validity of the assertion that the patent system does a poor job of protecting biopharmaceuticals is not obvious on its face. Patent protection has long been available for biotechnologies in the United States.[103] The 1980 opinion in *Diamond v. Chakrabarty*,[104] in which the Supreme Court held a genetically modified microorganism constituted patentable subject matter, has led to the grant of thousands of patents in every segment of the biotechnology industry, including biopharmaceuticals.[105] The availability of US patents for biotechnology has commonly been trumpeted as a principal reason for the dominance of domestic firms in the global marketplace.[106] Although the Supreme Court has recently curtailed the availability of patents for diagnostic methods[107] and isolated genomic DNA,[108] the reality remains that US patents are broadly available for biotechnology innovation.

Despite this favorable judicial treatment, one of the claimed defects of the patent system is that product patents are often unavailable for biologics because they constitute so-called "products of nature."[109] Surely patents may not be obtained on naturally occurring substances per se.[110] Yet for over thirty years,

---

103. *See* Dawn Willow, *The Regulation of Biologic Medicine: Innovators' Rights and Access to Healthcare*, 6 CHI.-KENT J. INTELL. PROP. 32 (2006) ("By far, the U.S. is the most mature market for biologics with the largest number of products available, many of which have long track records.").

104. 447 U.S. 303 (1980).

105. *See* Roger D. Klein & Maurice J. Mahoney, LabCorp v. Metabolite Laboratories: *The Supreme Court Listens, But Declines to Speak*, 36 J.L. MED. & ETHICS 141, 143 (2008).

106. *See* Haley Stein, *Intellectual Property and Genetically Modified Seeds: The United States, Trade, and the Developing World*, 3 NW. J. TECH. & INTELL. PROP. 160, 166 (2005) ("In *Diamond v. Chakrabarty*, the Supreme Court laid the legal foundation that would establish the United States as the global biotech patent leader.").

107. *See* Mayo Collaborative Servs. v. Prometheus Labs., Inc., 132 S. Ct. 1289 (2012).

108. *See* Ass'n for Molecular Pathology v. Myriad Genetics, Inc., 133 S. Ct. 2107 (2013).

109. *See* Biotechnology Industry Organization, *A Follow-On Biologics Regime Without Strong Data Exclusivity Will Stifle the Development of New Medicines* 2–3 [hereinafter BIO Fact Sheet], *available at* http://www.bio.org/sites/default/files/FOBSData_exclusivity_20070926_0.pdf (last visited Sept. 10, 2013) ("[B]ecause of current limitations of patentability of naturally occurring substances, many biologics are protected only by process patents that may be easier to 'design around.'").

110. *See* Diamond v. Diehr, 450 U.S. 175, 185 (1981) ("Excluded from such patent protection are laws of nature, natural phenomena, and abstract ideas.").

Supreme Court precedent has held that patenting is appropriate once significant artificial changes are made to a natural substance. The Supreme Court explained in *Diamond v. Chakrabarty* that in such instances the inventor's "claim is not to a hitherto unknown natural phenomenon, but to a nonnaturally occurring manufacture or composition of matter—a product of human ingenuity 'having a distinctive name, character [and] use.'"[111]

Erythropoietin, or EPO, provides one example of a patented biologic product. EPO, sold by Amgen under the trademark Epogen®, is a hormone that regulates red blood cell production. The best-selling biotech drug ever,[112] Epogen® is used by most of the country's kidney dialysis patients.[113] Its sales have exceeded $2 billion per year since 2001, and in recent years the Medicare and Medicaid programs have spent more on Epogen® than any other drug.[114]

According to Amgen, the last of its patents covering Epogen® will not expire until 2016.[115] As Epogen received FDA approval in 1989,[116] the net result is that this drug will have enjoyed over a quarter-century of effective patent protection in the United States. Amgen's first generation of product patents have expired,[117] but Epogen® also enjoys protection under composition of matter and other patents that will remain extant through 2015.[118] It should also be appreciated that Epogen® was originally subject to orphan drug exclusivity.[119]

Epogen® is not an isolated instance of acquisition and enforcement of product patents within the biotechnology industry. The best-selling biologics in the United

111. Diamond v. Chakrabarty, 447 U.S. 303, 309–10 (1980).

112. Andrew Pollack, *FDA Warning Is Issued on Anemia Drugs' Overuse,* N.Y. TIMES, Mar. 10, 2007, at A8.

113. Andrew Pollack, *Amgen Defends Its Turf as Competition Looms for Anemia Drug,* N.Y. TIMES, Oct. 17, 2007.

114. *Id.*

115. Amgen, Inc. Annual Report (Form 10-K) at 25 n.8 (Feb. 27, 2007).

116. Andrew Pollack, *Amgen Wins Patent Battle over Roche's Anemia Drug,* N.Y. TIMES, Oct. 24, 2007, at C2.

117. U.S. Patent No. 5,955,422 (claiming, in part, "[a] pharmaceutical composition comprising a therapeutically effective amount of human erythropoietin and a pharmaceutically acceptable diluent, adjuvant or carrier, wherein said erythropoietin is purified from mammalian cells grown in culture."); *see also* Amgen, Inc. v. Hoechst Marion Roussel, Inc., 457 F.3d 1293 (Fed. Cir. 2006).

118. *See* Karen Pihl-Carey, *Amgen Suing Roche for Infringing on EPO Patents,* 16 BIOWORLD TODAY (Nov. 10, 2005).

119. Carolyne R. Hathaway & John R. Manthei, *Market Exclusivity Options Under the FFDCA: Strategic Opportunities for Innovators of Pharmaceutical Patents,* 2 PHARM. L. & INDUS. REP. 184 (Feb. 13, 2004).

States—a list that includes Epogen® along with Aranesp®,[120] Enbrel®,[121] Neulasta®,[122] Rituxan®,[123] Avastin®,[124] Herceptin®,[125] Lucentis®,[126] and Humira®[127]—have enjoyed or continue to enjoy protection from one or more product patents.[128] The case that product patents are often unavailable for biologics or are not enforceable against competition appears to be a difficult one to make.

Alongside product patents, biopharmaceutical firms have also commonly obtained process patents, typically on methods of making biologics.[129] Process patents are sometimes deemed to be weaker than product patents because competitors may be able to develop different methods of manufacturing the same product.[130] However, manufacturing processes have been judged to be critical to the nature of biologics. Because the characteristics of biologics depend significantly on the complex techniques used to manufacture them,[131] one observer goes so far as to say that the "production process is 90 percent of the intellectual property related to the product."[132] As a result, process patents will likely prove more valuable with respect to biologics than they are sometimes regarded in other fields of patentable technology—again calling into question the need for a lengthy regulatory exclusivity.

The case for a lengthy regulatory exclusivity in exchange for expedited review of follow-on biologics has been founded on other assertions besides the statutory subject matter requirement. Virtually every other provision of the Patent Act has

---

120. U.S. Patent No. 5,955,422.

121. U.S. Patent No. 5,712,155.

122. U.S. Patent No. 5,824,784.

123. U.S. Patent No. 5,677,180.

124. U.S. Patent No. 6,054,297.

125. U.S. Patent No. 6,054,561.

126. U.S. Patent No. 6,407,213.

127. U.S. Patent No. 5,580,755.

128. This list was assembled from 2010 sales figures.

129. *See* BIO Fact Sheet, *supra* note 109, at 2–3 ("[B]ecause of current limitations of patentability of naturally occurring substances, many biologics are protected only by process patents that may be easier to 'design around.'").

130. *See* Dennis S. Karjala, *Biotech Patents and Indigenous Peoples*, 7 MINN. J.L. SCI. & TECH. 483, 504 (2006) ("Process patents are generally considered weaker than product patents because it is often possible to invent around a process patent.").

131. *See* Henry Grabowski et al., *The Market for Follow-On Biologics: How Will It Evolve?*, 25 HEALTH AFFAIRS 1292 (Sept./Oct. 2006).

132. Debra Weintraub, *Next Generation of Biopharmaceuticals*, 9 J. MANAGED CARE MED. 35 (2006).

been claimed to have been imposed unfairly against biologics. Among these are the doctrines of utility, claim definiteness, written description, and enablement. The result, it is claimed, is the unavailability of patent protection or patents that are difficult to enforce successfully in court.[133]

These statements can only be described as puzzling to anyone who has observed recent legislative activity regarding the patent system. The recent enactment of the Leahy–Smith America Invents Act was the culmination of a lengthy debate about virtually every aspect of patent law. Yet if any of these concerns about the unfairness of Patent Act requirements with respect to biologics were ever given voice during this lengthy discussion, the record does not reveal it.

Assertions that lengthy exclusivity for biologics is needed to make up for patent weaknesses also contrast strongly with statements innovative biopharmaceutical firms have made about particular patents when the US Patent and Trademark Office allows them to issue,[134] or when a court finds them not invalid and infringed.[135] Such statements also are overbroad. Each of these legal requirements applies to all fields of technology. Failure to satisfy the utility requirement has been the cause of the invalidity of patents claiming a steroid,[136] an energy generation system,[137] and beverage-flavoring method,[138] for example, whereas indefinite claims have caused the demise of patents claiming such inventions as an information broadcast system,[139] a method of drilling for oil,[140] and a system for dispensing prepaid debit cards.[141] Failure to satisfy the written description requirement has invalidated patents on alkaline

133. *See* BIO Fact Sheet, *supra* note 109, at 2 n.2 ("[T]he so-called 'utility,' 'written description,' and 'enablement' requirements of the Patent Act are interpreted more stringently for biotechnology inventions than most other technologies."); *id.* at 3 ("Under rules of patentability specific to biotechnology inventions, patent claims on biologics must often be narrowly drawn to the specific innovative aspect (e.g., a specific protein or nucleotide sequence) to be allowable.").

134. *See, e.g.*, *Amgen Obtains Broader Patent Protection for NEUPOGEN*, Bus. Wire (Dec. 4, 1996) (quoting Amgen's chairman and CEO as stating that "[t]his new patent provides additional, broad patent protection and strengthens our intellectual property position on NEUPOGEN").

135. *See, e.g.*, *Judge Blocks Anemia Drug Sales by an Amgen Rival*, N.Y. Times, Feb. 29, 2008.

136. Brenner v. Manson, 383 U.S. 519 (1966).

137. Newman v. Quigg, 877 F.2d 1575 (Fed. Cir. 1989).

138. Fregeau v. Mossinghoff, 776 F.2d 1034 (Fed. Cir. 1985).

139. Finisar Corp. v. DirecTV Grp., Inc., 523 F.3d 1323 (Fed. Cir. 2008).

140. Halliburton Energy Servs., Inc. v. M-I LLC, 514 F.3d 1244 (Fed. Cir. 2008).

141. Default Proof Credit Card Sys., Inc. v. Home Depot U.S.A., Inc., 412 F.3d 1291 (Fed. Cir. 2005).

batteries,[142] digital image technology,[143] and even household furniture.[144] And the lack of an enabling disclosure has in recent years doomed patents on inventions such as a syringe,[145] airbag sensor,[146] and computer software.[147] To the extent there are concerns about the manner in which Congress has framed these requirements, or the ways courts have interpreted them, these critiques apply across many industries.

A lengthy regulatory exclusivity has also been justified on the grounds that follow-on biologics may be approved under the strict "sameness" criterion that applies under the Hatch–Waxman Act,[148] but also under a more lenient standard of interchangeability. Innovative biopharmaceutical firms have claimed that this approval standard will create a loophole that does not exist for traditional pharmaceuticals—namely, the possibility that a follow-on firm will claim that its product is similar enough to the reference product to merit marketing approval based on an abbreviated application, but sufficiently distinct from the sponsor's patents to warrant a judgment of noninfringement.[149] It should be appreciated, however, that the Hatch–Waxman Act expressly allows generic firms to assert a non-infringement defense as well.[150] Generic drug firms not uncommonly prevail in these assertions.[151] Indeed, a recent FTC study found, at least for the court decisions issued during its period of study, that determinations of non-infringement outnumbered invalidity decisions in cases where the generic firm prevailed.[152]

---

142. Energizer Holdings, Inc. v. ITC, 275 F. App'x 969 (Fed. Cir. 2008).

143. LizardTech, Inc. v. Earth Resource Mapping, Inc., 424 F.3d 1336 (Fed. Cir. 2005)

144. Gentry Gallery, Inc. v. Berkline Corp., 134 F.3d 1473 (Fed. Cir. 1998).

145. Liebel-Flarsheim Co. v. Medrad, Inc., 481 F.3d 1371 (Fed. Cir. 2007).

146. Automotive Techs. Int'l, Inc. v. BMW of N. Am., Inc., 501 F.3d 1274 (Fed. Cir. 2007).

147. Sitrick v. Dreamworks, 516 F.3d 993 (Fed. Cir. 2008).

148. 21 U.S.C. § 355 (2012).

149. BIO Fact Sheet, *supra* note 109, at 2 (asserting that under the Hatch–Waxman Act, "the manufacturer of a generic drug may not have it both ways—it cannot gain FDA approval of its product by demonstrating that the active ingredient is the same as the innovator product and then turn around and claim in the patent context that its product is different from the innovator's drug").

150. 21 U.S.C. § 355(j)(2)(A)(viii)(IV).

151. *See, e.g.*, Schwarz Pharma, Inc. v. Paddock Labs, Inc., 504 F.3d 1371 (Fed. Cir. 2007); Abbott Labs. v. Novopharm Ltd., 323 F.3d 1324 (Fed. Cir. 2003); Warner-Lambert Co. v. Apotex Corp., 316 F.3d 1348 (Fed. Cir. 2003); Bayer AG v. Elan Pharm. Research Corp., 212 F.3d 1241 (Fed. Cir. 2000); MedImmune Oncology, Inc. v. Sun Pharm. Indus., Ltd., No. MJG-04-2612, 2007 WL 6137013 (D. Md. Oct. 29, 2007).

152. FTC, *supra* note 17, at 13 ("Of the decisions favoring the generic applicant, there were slightly more non-infringement decisions (14) than patent invalidity decisions (11).").

This argument further supposes that biopharmaceutical firms are unable to secure patents of sufficient breadth to protect their products from similar or comparable products. Yet the case seems difficult to make that claims in product patents for biologic drugs cannot be drafted broadly enough to cover minor variations in chemical structure of the kind expected in follow-on biologics. Biotechnology products may commonly be defined through multiple techniques, including their structure, chemical or physical characteristics, and method of preparation, that in combination are capable of providing a potent shield against would-be competitors.[153] And as with other sorts of inventions, biologics may be covered by as many patent claims as the inventor is willing to prepare. Further, the fundamental professional obligation of patent attorneys is to secure the broadest claims allowable by law. It seems extremely unlikely that biopharmaceutical firms will be unable to procure the services of artful claim drafters in their dealings with the Patent Office.

The biologics market itself provides further evidence of the value of biologics patent portfolios. Sufficient incentives arguably exist for competitors to conduct their own clinical trials for a number of blockbuster biologics, including EPO, even in the absence of an expedited approval pathway specifically devised for this category of drugs. Yet competitors have not entered the US market even as they continue to sell products such as EPO overseas. The strength of patent rights surely has contributed to the lack of domestic competition in biopharmaceuticals.[154]

Finally, the successful licensing and litigation track record enjoyed by numerous patented biologics does not support the assertion that these intellectual property rights are somehow inferior. In addition to the EPO patents discussed previously, US Patent No. 4,399,216 provides another notable example. Columbia University's co-transformation patent reportedly generated some $400 million in royalty revenue, paid by such firms as Abbott, Amgen, Biogen, Genentech, Genzyme, and Wyeth.[155] Far from being of cabined scope, the seventy-three claims of the '216 patent reportedly covered twenty-nine different protein-based drugs.[156] Plainly stated, concerns about the lack of availability of robust patent protection for biologics were not realized in this instance.

---

153. *See* Jay P. Lessler, *Drafting Claims for Chemical, Pharmaceutical, and Biotechnology Patent Applications*, 909 PLI/PAT 289 (July–Aug. 2007).

154. *See, e.g.*, *Biogenerics to Help Save Nearly $100 bn over 10 Years*, FIN. EXPRESS (Aug. 6, 2008), http://www.financialexpress.com/news/Biogenerics-to-help-save-nearly-100-bn-over-10-yrs/345537.

155. *See* Recent Development, *Columbia, Co-Transformation, Commercialization & Controversy: The Axel Patent Litigation*, 17 HARV. J. L. & TECH. 583, 594 (2004).

156. *Id.* at 591–92.

C. Policy Implications

Regulatory exclusivities and patents are likely the only two intellectual property rights that Congress expressly designed to operate in tandem. This synergy does not imply that these two regimes grant duplicative proprietary rights, however. As a normative and, for the most part, descriptive matter, regulatory exclusivities are best viewed as a means for encouraging socially beneficial acts that the patent system cannot protect under its own terms.

This account holds implications for the design of regulatory exclusivities for additional categories of drugs. These rights should preferably be tailored to the difficulty, cost, and risks associated with the clinical trials associated with that sort of drug. The term and scope of rights associated with the regulatory exclusivity are among the parameters that may be manipulated in order to provide private actors with a fair return on investment. A regulatory exclusivity deviates from this account when it is intended to finance the costs of preliminary research, subsidize production expenditures, or address perceived shortcomings in the patent system. In these circumstances, regulatory exclusivities begin to substitute for patents as a vehicle for promoting innovation—a measure that leaves shortfalls in the patent system unaddressed and holds significant consequences for public health.

IV. Proposed Regulatory Exclusivity Reforms

In addition to advocating changes to existing proprietary rights, members of the pharmaceutical industry have again begun to call for the creation of new forms of regulatory exclusivity in order to protect their costly investments in research and development. Current discussion has focused on the establishment of regulatory exclusivities for the creation of terrorism countermeasures and biologics. However, regulatory exclusivities could in theory be applied to virtually any regulated product. Ethnic medicines,[157] personalized medicines,[158] and treatments for any particular condition that stimulates legislative interest would all be possible candidates for new sorts of regulatory exclusivities.

In contrast to reform of the patent system, alterations to the regulatory exclusivity regime remain relatively convenient for Congress. The Leahy–Smith America Invents Act was the product of years of divisive debate among fragmented

---

157. *See* Jay N. Cohn, *The Use of Race and Ethnicity in Medicine: Lessons from the African-American Heart Failure Trial*, 34 J.L. Med. & Ethics 552 (2006); Shubba Ghosh, *Race-Specific Patents, Commercialization, and Intellectual Property Policy*, 56 Buff. L. Rev. 209 (2008).

158. *See* Scott Sasjack, *Demanding Individually Safe Drugs Today: Overcoming the Cross-Labeling Legal Hurdle to Pharmacogenomics*, 34 Am. J.L. & Med. 7 (2008).

patent stakeholders. Further, international agreements to some degree cabin the ability of Congress to alter the patent system in industry-specific ways. One component of the Final Act forming the WTO is the Agreement on Trade-Related Aspects of Intellectual Property Rights (TRIPS Agreement).[159] Under Article 27, "patents shall be available and patent rights enjoyable without discrimination as to the place of invention, the field of technology and whether products are imported or locally produced."[160] Although the TRIPS Agreement allows for some exceptions to this general principle of technological neutrality,[161] they are of very narrow application.[162] As a result, the possibilities for partitioning the patent system to account for particular public health needs are limited.

The TRIPS Agreement has less to say about regulatory exclusivities.[163] Indeed, this international agreement arguably does not even require WTO member states to establish them.[164] Rather, protection against "unfair commercial use" must be provided to new chemical entities under certain conditions.[165] WTO member states appear free to adopt additional forms of data protection under circumstances they choose. In particular, no requirement of technological neutrality pertains and no minimum or maximum term of protection is specified. As a result, few constraints limit the ability of Congress to frame regulatory exclusivities as it desires.

What is most convenient is not always the most desirable option, however. Notable about the recent proposals is that they advocate the expansion of regulatory exclusivity terms to patent-like lengths. Further, they are being advocated on the basis that the patent system cannot protect the fruits of the desired labors of the pharmaceutical industry. These proposed regulatory exclusivities are therefore expressly designed to serve as innovation-promoting patent substitutes. To this extent they cause a noteworthy expansion of the concept of regulatory exclusivities, a proposition this chapter discusses next.

---

159. *See* Agreement on Trade-Related Aspects of Intellectual Property Rights, Apr. 15, 1994, Annex 1C, 33 I.L.M. 1197 (1994) [hereinafter TRIPS Agreement].

160. *Id.* art. 27, ¶1.

161. *Id.* art. 30.

162. *See* Maria Victoria Stout, *Crossing the TRIPS Nondiscrimination Line: How CAFTA Pharmaceutical Patent Provisions Violate TRIPS Article 27.1*, 14 Bos. U. J. Sci. & Tech. L. 177 (2008).

163. *See* G. Lee Skillington & Eric M. Solovy, *The Protection of Test and Other Data Required by Article 39.3 of the TRIPS Agreement*, 24 Nw. J. Int'l L. & Bus. 1 (2003).

164. *See* Carlos Maria Correa, *Unfair Competition Under the TRIPS Agreement: Protection of Data Submitted for the Registration of Pharmaceuticals*, 3 Chi. J. Int'l L. 69 (2002).

165. TRIPS Agreement, *supra* note 160, art. 33(b).

A. Terrorism Countermeasures

In October 2001, an unknown person or persons mailed a refined form of the bacteria *Bacillus anthracis*, commonly known as anthrax, to members of Congress and the media.[166] The US government treated affected individuals with a pharmaceutical compound known as "ciprofloxacin hydrochloride."[167] Ciprofloxacin is sold under the trademark Cipro by the German firm Bayer AG. Although widely prescribed for a number of medical indications, Cipro was at that time the only antibiotic approved for inhalation anthrax.[168] Public accessibility of Cipro and other potential treatments for inhalation anthrax immediately became a matter of great public concern.[169] The FDA responded by approving alternative antibiotics for the treatment of certain strains of anthrax, including the one used in the attacks.[170]

In the wake of the anthrax incident, Congress considered enacting legislation intended to promote the development and accessibility of terrorism countermeasures. In the 109th Congress, the Project Bioshield II Act, S. 975, would have awarded qualifying countermeasures expanded regulatory exclusivity terms. In particular, the five-year new chemical entity exclusivity would have been doubled to ten years; the three-year new clinical study exclusivity would have been doubled to six years; and the seven-year orphan drug exclusivity would have been extended to ten years.[171] The Biodefense and Pandemic Vaccine and Drug Development Act, S. 1873, offered an alternative approach. It would have allowed countermeasures to qualify for orphan drug exclusivity. This bill would have established a period of orphan drug exclusivity of ten years with respect to countermeasures.[172] Neither bill was enacted, however, nor has legislative interest in this project remained at a high level.

In terms of existing regulatory exclusivity paradigms, both of these bills sought to encourage drug companies to review libraries of known chemical compounds

---

166. *See* Stephen Engelberg & Judith Miller, *Sign of Escalating Threat*, N.Y. TIMES, Oct. 17, 2001, at A1.

167. Kathleen Pender, "Cipro Had Big Boost from U.S.," S.F. CHRON., Oct. 25, 2001, at D1.

168. *See* FDA, Approval Letter (Aug. 30, 2000), *available at* http://www.fda.gov/downloads/Drugs/EmergencyPreparedness/BioterrorismandDrugPreparedness/UCM130759.pdf.

169. *See* Erika Mullenbach, *The Influence of Disease on the Evolution of U.S. Patent Law and Policy towards Foreign Patent Laws in the Late Twentieth to Early Twenty-First Century*, 7 TUL. J. TECH. & INTELL. PROP. 227, 239–42 (2005).

170. *See* FDA, Cipro (Ciprofloxacin Hydrochloride) for Inhalation Anthrax: Information on Cipro for Consumers: Questions and Answers (Nov. 14, 2001), *available at* http://www.fda.gov/Drugs/EmergencyPreparedness/BioterrorismandDrugPreparedness/ucm130711.htm.

171. S. 975, 109th Cong. § 331 (2005).

172. S. 1873, 109th Cong. § 5 (2005).

with an eye toward bringing them to market.[173] The drafter of S. 1873 realized that this perceived shortfall of private incentives is similar to that associated with orphan drugs. As with diseases with small patient populations, countermeasures may never garner extensive sales figures.

Despite these similarities to the circumstances that inspired the Orphan Drug Act, the failure to enact either piece of legislation may suggest congressional discomfort with the creation of regulatory exclusivities with durations that begin to approach the patent term. The impetus for the longer proposed terms may have arisen from the recognition that the federal government will likely be the primary purchaser of terrorism countermeasures in circumstances of close public scrutiny. This situation may not prove to be especially favorable for the sponsor's profits or public image.[174]

Ameliorating the concerns of potential countermeasure sponsors through a regulatory exclusivity would not only impose new duties upon this class of proprietary right, but would also raise concerns over the ability of the federal government to manufacture its own supplies of protected drugs during public emergencies. Current law allows the federal government to use subject matter protected by patents, copyrights, plant variety protection certificates, and semiconductor mask works whether the right holder consents or not.[175] The federal compulsory license statute does not account for regulatory exclusivities, however, and no legislation has been proposed that would call it to do so. Although circumstances are hard to imagine where sovereign immunity concepts would not allow the government to manufacture countermeasures in appropriate circumstances, or authorize others to do so,[176] potential sources of delay should preferably be avoided. Addressing this situation would simply be a matter of a housekeeping amendment to the compulsory licensing statute. Yet doing so would seemingly defeat the purpose of the countermeasure exclusivity.

In view of these circumstances, regulatory exclusivities do not provide the most appropriate mechanism for stimulating the development of terrorism countermeasures. Other incentive structures beyond direct funding are viable should interest revive in this initiative, however. The Orphan Drug Act may once more prove exemplary, as this legislation provides sponsors with such benefits as FDA

173. *See* Arti K. Rai et al., *Pathways Across the Valley of Death: Novel Intellectual Property Strategies for Accelerated Drug Discovery*, 8 YALE J. HEALTH POL'Y, L. & ETHICS 1, 4 (2008) (observing that innovative pharmaceutical firms commonly hold libraries of potentially useful small molecules that they maintain as trade secrets).

174. Mullenbach, *supra* note 170, at 239–40.

175. 28 U.S.C. § 1498 (2012).

176. *See* W.L. Gore & Assocs., Inc. v. Garlock, 842 F.2d 1275, 1282–83 (Fed. Cir. 1988).

protocol assistance,[177] tax breaks,[178] and a program through which researchers may compete for grants to conduct clinical trials.[179]

B. Hatch–Waxman Term Extension

As suggested by the recent efforts of the Obama administration to recalibrate the term of the biologics exclusivity from twelve to seven years, active discussion is ongoing concerning the term of the regulatory exclusivities. Most of the argument is on behalf of brand-name pharmaceutical firms that seek to align the NCE exclusivity—currently set to five years—to match the ten-year term found in Europe. These efforts would benefit from a focus on complementing the patent system, rather than competing with it as a source of innovation incentives. In so doing, policymakers would do well to reflect upon the recent incident involving colchicine.

Colchicine is an ancient remedy for gout that has been marketed in the United States since the nineteenth century. For many years a number of firms sold colchicine as a generic drug; due to its grandfathered status under the 1938 Food, Drug, and Cosmetic Act, no conclusive safety and efficacy trials had been conducted. In 2006, as part of its Unapproved Drugs Initiative, the FDA required these firms to conduct clinical trials and other studies if they wished to continue to sell the drug.[180] Only one firm complied; it reportedly spent $100 million on this effort, of which $45 million constituted the FDA application fee.

The FDA subsequently awarded three-year regulatory exclusivity for gout and seven-year orphan drug exclusivity for another indication, familial Mediterranean fever, in keeping with the Hatch–Waxman Act. The NDA holder immediately increased the price of its colchicine product, Colcrys®, from $0.09 to $4.85 per tablet. Annual Medicaid expenditures on colchicine could reportedly rise as much as $50 million per year. And as of September 2010, no generic colchicine products have been available in the United States.[181]

The colchicine episode should cause us to question whether the current terms of regulatory exclusivities provide rewards appropriate for the required investment. Even as some observers push for longer terms of NCE exclusivity, it may be

177. 21 U.S.C. § 360aa (2012).

178. 26 U.S.C. § 45C (2012).

179. 21 U.S.C. § 360ee (2012).

180. Aaron S. Kesselheim & Daniel H. Solomon, *Incentives for Drug Development—The Curious Case of Colchicine*, 362 N. Eng. J. Med. 2045 (2010).

181. *See* FDA, FDA Orders Halt to Marketing of Single-Ingredient Oral Colchicine (Sept. 30, 2010).

that the term is already overly generous. A more firm sense of the policy goals regulatory exclusivities are meant to achieve—as well as a commitment to tailoring the exclusivities in pursuit of those goals—would allow more appropriate choices in terms of both intellectual property and public health policy.

C. The MODDERN Cures Act

Serious thought has been given to jettisoning the patent system altogether in the area of pharmaceutical innovation in favor of a regime solely focused on regulatory exclusivities. The perhaps too creatively titled Modernizing Our Drug & Diagnostics Evaluation and Regulatory Network Cures Act (MODDERN Cures Act) provides the most notable example of this work. Placed before the 112th Congress as H.R. 3947, the bill would require a drug sponsor to identify a therapy that fulfills an "unmet medical need" to the FDA. As part of this identification, the sponsor would indicate the patents associated with that "dormant therapy" and assert that they will provide fourteen or fewer years of effective protection. The sponsor would also file a waiver of those rights.

If the FDA agrees with the sponsor and ultimately grants marketing approval, then the sponsor would obtain fifteen years of marketing exclusivity. All of the identified patents would be granted an extended term of up to fifteen years after the product is approved. The waiver would then kick in; any patent term after the fifteen years would then be disclaimed, paving the way for generic competition.

Whether the MODDERN Cures Act will have much traction in Congress remains to be seen. The legislation nonetheless proposes a significant paradigm shift for patents and regulatory exclusivities. It would accelerate the trend toward a rivalrous relationship between patents and regulatory exclusivities, confuse the distinctions between and appropriate role of these separate entitlements, remove a leading science-based industry from the patent system that has traditionally sustained it, and continue to fragment US intellectual property law into a diverse collection of specialized and complex proprietary rights. As Congress contemplates changes of this magnitude, it would do well to reflect upon the appropriate roles of arguably the most significant intellectual property rights of them all.

Conclusion

Regulatory exclusivities should be granted only in circumstances where the patent system, under its own terms, cannot protect activities that promote public health. In particular, regulatory exclusivities are best seen as mechanisms for promoting the acquisition of detailed knowledge about the effect of known products upon the human body. They are least justified when intended to promote innovation in the same manner as the patent system, or simply to mitigate perceived shortcomings in patent doctrine. Patent law reform provides a more appropriate response to concerns over particular innovation environments than the creation of a gallery of additional regulatory exclusivities. Consideration of the distinct role of both patents and regulatory exclusivities in the research, development, and distribution of new healthcare products may yet promote worthy statutory successors to the Hatch–Waxman Act.

IV EXCEPTIONS AND LIMITS TO PATENT PROTECTION

13 A FALSE SENSE OF SECURITY OFFERED BY ZERO-PRICE LIABILITY RULES? RESEARCH EXCEPTIONS IN THE UNITED STATES, EUROPE, AND JAPAN IN AN OPEN INNOVATION CONTEXT

Esther van Zimmeren and Geertrui van Overwalle*

Introduction

Innovation is considered an important means to trigger economic activity and to sustain economic growth.[1] "Open innovation" in general, and Public Private Partnerships (PPPs) in particular, are widely regarded as effective vehicles for the long-term structural development of infrastructure and services, bringing together distinct advantages of the private and the public sector.[2] Two international large-scale PPPs

---

* Esther van Zimmeren is Assistant Professor of Law at the University of Liège and post-doctoral research fellow of the Research Foundation-Flanders (FWO), Centre for Intellectual Property Rights, University of Leuven. Geertrui van Overwalle is Professor of IP Law at University of Leuven and Professor of Patent Law and New Technologies at the Tilburg Institute for Law, Technology, and Society (TILT), University of Tilburg. This chapter is partially based on Esther van Zimmeren's PhD Thesis: Esther van Zimmeren, Towards a New Patent Paradigm in the Biomedical Sector? Facilitating Access, Open Innovation and Social Responsibility in Patent Law in the U.S., Europe, and Japan (May 16, 2011) (unpublished manuscript on file with the authors and the library of the University of Leuven, Belgium). The authors gratefully acknowledge the support of the Flemish Fund for Scientific Research (FWO, Belgium), the Research Fund of the Leuven University (Grant number IMPH006/3H110067 relating to the project "Open Innovation: The Interaction between Legal Architectures and Innovation Strategies"), and the Vancraesbeeck Fund (KU Leuven, Belgium).

1. Organization for Economic Cooperation and Development (OECD), *Ministerial Report on the OECD Innovation Strategy: Key Findings* (May 2010), *available at* http://www.oecd.org/dataoecd/51/28/45326349.pdf.

2. *Communication from the Commission to the European Parliament, the Council, the European Economic and Social Committee and the Committee of the Regions*

in the health sector have recently caught wide attention: the Innovative Medicines Initiative (IMI),[3] based in the European Union, and the Critical Path Institute (C-Path),[4] based in the United States. Both initiatives share similar objectives and aim to accelerate the development of safer and more effective medicines for patients. Both organizations bring together scientists from the pharmaceutical industry and academic institutions, as well as representatives from patient organizations, small and medium-sized enterprises, and regulatory agencies to form precompetitive PPPs. By combining their knowledge and expertise in innovative projects, the partners in these projects are able to tackle challenges that are too big and complex for individual research teams to handle on their own.

As valuable as these "open innovation" initiatives may be, it remains to be seen whether the current legal framework is capable of accommodating such global collaborations. In particular, it is doubtful whether the research exception will shield research activities carried out within these public/private consortia from patent infringement allegations. IMI funds research projects involving both public and private researchers, based all over Europe, and even though these research projects are precompetitive, the ultimate aim of these projects may differ for researchers working for a public entity and those operating within a company. The partnership with C-Path[5] may open up further opportunities for collaboration between public and private researchers from both sides of the Atlantic, but may further complicate the applicability of the research exception. Due to their large scale and global nature, these type of PPPs likely encounter challenges incommensurable with the preexisting issues that have emerged with respect to research exceptions.

It is widely accepted that research exceptions are one of the legitimate derogations from the exclusivity granted by patent rights.[6] In principle, a robust research

---

of 19 November 2009, COM (2009) 615 final (Nov. 19, 2009), *available at* http://eur-lex.europa.eu/LexUriServ/LexUriServ.do?uri=CELEX:52009DC0615:EN:NOT.

3. *See* Innovative Medicines Initiative (IMI), *available at* http://www.imi.europa.eu (last visited Sept. 2, 2013).

4. *See* Critical Path Institute (C-Path), *available at* http://www.c-path.org (last visited Sept. 2, 2013).

5. IMI and C-Path have now signed a memorandum to bundle their efforts. *See* Press Release, IMI, EU's Innovative Medicines Initiative Announces Formal Collaboration with US Critical Path Institute (June 14, 2011), *available at* http://www.imi.europa.eu/sites/default/files/uploads/documents/Press%20Releases/IMIandC-PathPressReleaseFINAL.pdf.

6. *See, e.g.*, World Intellectual Prop. Org. (WIPO), Standing Comm. on the Law of Patents (SCLP), *Exclusions from Patentable Subject Matter and Exceptions and Limitations to the Rights*, WIPO Doc. SCP/13/3 (Feb. 4, 2009) [hereinafter *Exclusions*], *available at* http://www.wipo.int/edocs/mdocs/scp/en/scp_13/scp_13_3.pdf; Chris Dent et al., *Research Use of Patented Knowledge: A Review* (OECD Directorate for Science, Tech. & Industry, STI Working Paper No. 2006/2, 2006), *available at* http://www.oecd.org/dataoecd/15/16/36311146.pdf; *cf.*

exception can fuel follow-on innovation,[7] ward off "patent thicket" problems,[8] and mitigate refusals to license, as researchers are free to use the patented inventions within certain carefully designed margins. Yet a comparative analysis of the research exception in the United States, Europe, and Japan shows that, as far as a general research exception does exist, its boundaries often remain unclear and confusing. In the past, numerous papers and reports by scholars and advisory committees[9] have

---

Agreement on Trade-Related Aspects of Intellectual Property Rights, art. 30, Apr. 15, 1994, 33 I.L.M. 1197 (1994) [hereinafter TRIPS Agreement or TRIPS] ("Members may provide limited exceptions to the exclusive rights conferred by a patent, provided that such exceptions do not unreasonably conflict with a normal exploitation of the patent and do not unreasonably prejudice the legitimate interests of the patent owner, taking account of the legitimate interests of third parties."). In one of the drafts of the TRIPS Agreement, dated July 23, 1990, a non-exhaustive list of specific exceptions to patent rights was suggested, including a specific exception for experimental purposes. However, this approach was not supported by the United States. Therefore, the final text adopts a more general formulation. *See* Carlos M. Correa, *The International Dimension of the Research Exemption*, AM. ASS'N FOR ADVANCEMENT SCI. 7–8 (Jan. 2005), *available at* http://sippi.aaas.org/Pubs/Correa_International%20Exception.pdf. An exception for research or experimentation is likely to fulfill the three-step test established by Article 30 of TRIPS. The panel in the EC-Canada case already hinted at the TRIPS consistency of such exceptions. *See* Panel Report, *Canada—Patent Protection of Pharmaceutical Products*, WT/DS114/R (Mar. 17, 2000), *available at* http://www.wto.org/english/tratop_e/dispu_e/7428d.pdf. Consistency is likely to be confirmed if the issue were specifically raised in a WTO dispute.

7. *See, e.g.*, Katherine J. Strandburg, *What Does the Public Get? Experimental Use and the Patent Bargain*, 2004 WIS. L. REV. 81 (2004).

8. *See* Carl Shapiro, *Navigating the Patent Thicket: Cross Licenses, Patent Pools, and Standard-Setting*, *in* INNOVATION POLICY AND THE ECONOMY 119 (Adam B. Jaffe, Josh Lerner & Scott Stern eds., 2001). Patent thickets relate to the fear that the upsurge in patenting—caused by the race on patents by private and public entities in various sectors—would lead to a growing presence of patents inhibiting further R & D. In many sectors, the essence of innovation is cumulative innovation: each invention builds on many previous findings. Public and private entities must thus gain permission of each person that previously contributed to the development of a certain product or process in order to exploit it and to pursue further R & D. If there are many patents in the hands of different owners, this may ultimately lead to difficulties in bargaining licenses. This phenomenon has been labeled a "patent thicket." Every licensor will require the payment of royalties or upfront license fees, which ultimately may lead to "royalty stacking."

9. *See, e.g.*, JOHN R. THOMAS, CONG. RESEARCH SERV., RL 32651, SCIENTIFIC RESEARCH AND THE EXPERIMENTAL USE PRIVILEGE IN PATENT LAW (2004); SEC'Y'S ADVISORY COMM. ON GENETICS, HEALTH & SOC'Y (SACGHS), REVISED DRAFT REPORT ON GENE PATENTS AND LICENSING PRACTICES AND THEIR IMPACT ON PATIENT ACCESS TO GENETIC TESTS 82 (2010) [hereinafter *Gene Patents*], *available at* http://oba.od.nih.gov/oba/sacghs/reports/SACGHS_patents_report_2010.pdf; Peter Chrocziel, *Die Benutzung patentierter Erfindungen zu Versuchs- und Forschungszwecken* (Carl Heymanns Verlag, Schriftenreihe zum gewerblichen Rechtsschutz, No. 67, 1986); Trevor Cook, *A European Perspective as to the Extent to Which Experimental Use, and Certain Other Defences to Patent Infringement, Apply to Differing Types of Research* (Intellectual Prop. Inst., No. 109, 2006); Geertrui van Overwalle, *Patent Law and Medicines*, *in* OCTROOIRECHT EN GENEESMIDDELEN 65 (Geertrui van

been published expressing concerns about the uncertainties regarding the availability and scope of the research exception.[10] It has frequently been recommended to introduce a statutory exception in the United States and to develop guidelines in Europe, including a common understanding on the application of the research exception in uncontroversial areas of research and experimentation. Despite the vast number of publications and suggestions, not much has happened yet. The failure of legislators to codify a new exception in the United States or create guidelines for Europe may well indicate that research is not hampered by the absence of a clear-cut research exception or the lack of clarifying guidelines.[11]

Researchers in a wide variety of sectors seem to be happily continuing their research projects without fears of patent infringement liability, presuming that their work is covered by a research exception.[12] However, the research exception

---

Overwalle ed., 2000); THE ROYAL SOCIETY, KEEPING SCIENCE OPEN: THE EFFECTS OF INTELLECTUAL PROPERTY POLICY ON THE CONDUCT OF SCIENCE (2003); William R. Cornish, *Experimental Use of Patented Inventions in European Community States*, 29 INT'L REV. INTELL. PROP. & COMPET. L. 735 (1998); Rebecca S. Eisenberg, *Patents and the Progress of Science: Exclusive Rights and Experimental Use*, 56 U. CHI. L. REV. 1017 (1989); E. Richard Gold, Yann Joly & Timothy Caulfield, *Genetic Research Tools: The Research Exception and Open Science*, 3 GENEDIT 1 (2005), *available at* http://www.cipp.mcgill.ca/data/publications/00000040.pdf; Janice M. Mueller, *No "Dilettante Affair": Rethinking the Experimental Use Exception to Patent Infringement for Biomedical Research Tools*, 76 WASH. L. REV. 1 (2001); Maureen A. O'Rourke, *Toward a Doctrine of Fair Use in Patent Law*, 100 COLUM. L. REV. 1177 (2000); Strandburg, *supra* note 7; CTR. FOR INTELLECTUAL PROP. POLICY (CIPP), THE RESEARCH OR EXPERIMENTAL USE EXEMPTION: A COMPARATIVE ANALYSIS (2004), http://www.cipp.mcgill.ca/data/newsletters/00000050.pdf; Nuffield Council on Bioethics, The Ethics of Patenting DNA (2002), *available at* http://www.nuffieldbioethics.org/sites/default/files/The%20ethics%20of%20patenting%20DNA%20a%20discussion%20paper.pdf.

10. Most of these publications especially express concerns about the potential hindering effects of patents on biomedical research and the applicability of research exceptions to so-called research tools. The National Institute of Health (NIH) defines research tools very broadly as to encompass all "tools that scientists use in the laboratory, including cell lines, monoclonal antibodies, reagents, animal models, growth factors, combinational chemistry and DNA libraries, clones and cloning tools (such as PCR), methods, laboratory equipment and machines." Sharing Biomedical Research Resources: Principles and Guidelines for Recipients of NIH Research Grants and Contracts, 64 Fed. Reg. 72,090 (Dec. 23, 1999). We do not intend to repeat this discussion here or add a detailed and critical analysis of our own to this debate, both of which would go beyond the aim of the current chapter. In this respect, we refer to the literature mentioned above.

11. However, Australia, for long one of the few other countries apart from the United States without a statutory research exception, has recently adopted legislation introducing an exemption for "experimental purposes." *See* Australian Government, Intellectual Property Laws Amendment (Raising the Bar) Bill 2011, § 119C, Act No. 35 (2012) [hereinafter Australian Amendment], *available at* http://parlinfo.aph.gov.au/parlInfo/search/display/display.w3p;query=Id%3A%22legislation%2Fbillhome%2Fs837%22.

12. *See, e.g.*, ESTHER VAN ZIMMEREN ET AL., PATENT LICENSING IN MEDICAL BIOTECHNOLOGY IN EUROPE: A ROLE FOR COLLABORATIVE LICENSING STRATEGIES? (2011).

might not be the "safe harbor" they think it is. First, not all countries have a research exception. In the United States, for example, there is no general (statutory) research exception. Second, in jurisdictions where a (de jure) research exception exists, both the text and the judicial interpretation of the text may differ. In Europe, for example, the wording of the research exception varies substantially from country to country, and, in most countries, case law does not always clearly define the scope of the exception. The same is true for Japan. Third, in countries where a de facto research exemption applies, "rational forbearance" may come to a stop. For quite some time patent owners in many countries who wanted to maintain goodwill and ensure access to future inventions, in particular if the damages were likely to be very small, have turned a blind eye to infringement by scientists, resulting in a de facto research exemption.[13] It is doubtful, however, whether this is a stable and sustainable basis for R & D, in particular in view of the fact that in some sectors an increasing number of lawsuits are based on inventions resulting from academic research licensed out exclusively.[14] Patent owners in those sectors might not remain so tolerant in the future toward academic researchers who ignore patents.[15]

In this chapter, we do not intend to replicate past discussions on the research exception. Rather, we would like to revisit the role and the impact of the research exception in a specific context, namely in an "open innovation" setting. The core research question this chapter seeks to address is how open and collaborative innovation challenges the concept, rationale, and scope of the research exception, and whether current national research exceptions can effectively deal with regional and global, large-scale collaborations. At present, innovation is often the result of a collaborative process including a wide variety of stakeholders, such as customers, suppliers, competitors, firms in unrelated industries, investors, universities, public research organizations, consultants, and intermediaries all over the globe. This collaborative process has been labeled "open innovation" when firms engage in combining internal and external knowledge to accelerate internal innovation and expand the markets for external use of innovation. Such collaborative research requires a dynamic exchange of ideas, knowledge, technology, and people. It often takes place with mixed teams of public and private researchers, or when companies open their laboratories to public researchers.

---

13. Lori Pressman et al., *The Licensing of DNA Patents by US Academic Institutions: An Empirical Survey*, 24 Nature Biotech. 31 (2006).

14. For the biomedical sector, see Chris M. Holman, *Trends in Human Gene Patent Litigation*, 322 Science 198 (2008).

15. Amy Yancey & C. Neal Stewart, *Are University Researchers at Risk for Patent Infringement?*, 25 Nature Biotech. 1225 (2007).

In our attempt to reexamine the research exception in an open innovation context, we will apply a two-step approach. First, we will describe and analyze the content and scope of the current research exceptions in the United States, Europe, and Japan. Next, we will assess and evaluate the current regimes using the property rights/liability rules dichotomy as an analytical framework, enabling us to detect deficiencies and suggest routes for improvement to the current regime. With respect to open innovation, we do not examine whether, to what extent, or in which circumstances open innovation is indeed a useful and efficient innovation strategy for firms. We start from the (normative) assumption that firms are increasingly interested in open innovation and are using this strategy more and more often. However, this does not imply that we believe that open innovation would be the optimal approach to investigate or commercialize a new idea or technology in all circumstances.

The setup of this chapter is as follows. We first provide a definition and rationale for the research exception (Section I). This conceptualization of the research exception is followed by an attempt to revisit the research exception in view of the property rights/liability rules dichotomy (Subsection II.A) and the open innovation paradigm (Subsection II.B). We then investigate the current research exceptions in the United States, Europe, and Japan (Section III), followed by an evaluation of so-called "zero-price liability rules" (Subsection IV.A) in an open innovation context (Subsection IV.B). We conclude with some thoughts on how the shortcomings, uncovered during the assessment exercise, can be overcome.

I. Conceptualizing the Research Exception

A. Definition

Under a research exception researchers are free—within certain limits—to carry out their research in spite of the existence of relevant patents. The terms "research exception," "research exemption," "experimental use exception," and "experimental use defense" are often used as synonyms. However, there are significant conceptual differences among these different terms. Let us first focus on the distinction between "exception" and "exemption."[16] The term "exception" refers to a certain activity that does *not* infringe a patent that is claiming the subject matter of the activity although it normally would be considered an infringement. The term "exemption" denotes that the activity *constitutes* an infringement of a patent,

---

16. *See also, e.g.*, Sean O'Connor, *Enabling Research or Unfair Competition? De Jure and De Facto Research Use Exceptions in Major Technology Countries*, *in* PATENT LAW AND THEORY: A HANDBOOK OF CONTEMPORARY RESEARCH 519, 521–22 (Toshiko Takenaka ed., 2008).

but that it is "excused" from actionable infringement liability. Most contemporary statutory limitations are exceptions.[17]

As to the second distinction, the one between "experimental use" and "research use," it has to be noted that "experimental use" typically relates to activities of a truly scientific nature, focusing on the testing of hypotheses and adding to the body of scientific knowledge rather than promoting commercial interests. In contrast, "research use" is conceptually broader and could in principle also comprise commercial research and development (R & D).[18] In the present contribution we will apply the term "research exception" as a general term that encompasses the various concepts. We will use the other terms only if we wish to emphasize the particular nature or parameters of the exception/exemption.

B. Rationale

A research exception can be justified from different perspectives. First, a research exception can be legitimized from a *public interest* point of view. Unlimited protection and enforcement of patents is not considered justified if access and further development would be hindered to the detriment of the public interest. The knowledge generated by the stimuli offered by patents is not only output, but also operates as input for further research and experimentation for the sake of scientific and technological progress. A research exception facilitates the advancement of technological knowledge and follow-on innovation.[19] In doing so, the exception goes beyond the duty to disclose that is inherent in patent law, as the research exception also allows the free *use* of the patented knowledge.

A research exception can also be based on an *institutional economics* rationale. Institutional economics offers a transaction cost argument in favor of a research exception. It would be very costly and probably impossible for patent owners to monitor infringements and enforce their patents against every single researcher. At the same time, not knowing which research line will be successful, and negotiating and concluding multiple patent licenses before every research project, would involve significant transaction costs for users, as well. In theory, a research exception could thus provide legal certainty for both patent owners and users of a patented technology and represent a benefit in terms of transaction costs. This rationale appears even more important in an era where collaborative research[20] is

17. *Id.*

18. *Id.*

19. Strandburg, *supra* note 7.

20. We note that this argument is equally valid for patent owner/user collaborations and user/user collaborations.

promoted as the key to many scientific challenges, limitations in R & D productivity, and the high risks and costs of R & D.

C. Parameters

States have lots of leeway to design their research exception to be narrow or broad. Four important parameters can be distinguished in the way a research exception is drafted. A first parameter relates to the *nature* of the research: the exception may concern research *on*, and/or research *with*, the patented invention. Research *on* the patented invention includes experiments that verify whether the patented invention works as described in the patent, whether the invention is indeed novel and inventive, and where unknown effects are explored or the invention is further developed.[21] Research *with* refers to experiments in which the patented invention is used to investigate something else: the patented invention is used as an instrument, a tool, or an apparatus.[22] The latter is particularly important with respect to research tools. Many biotechnology companies produce research tools as some or all of their primary product lines. A broad research exception that covers research *with* the patented subject matter could destroy a sizeable portion of their market.[23]

A second parameter relates to the final *goal* of the research: the research exception may include research with noncommercial purposes or mixed noncommercial and commercial purposes. Research with a purely noncommercial, scientific goal includes research that is carried out to satisfy the eagerness to learn, and gratify scientific inquisitiveness and pleasure, thereby contributing to the body of knowledge. Research with a mixed noncommercial/commercial goal usually encompasses research set up to develop new applications, improve methods of making, or enhance effectiveness, with the ultimate goal of developing and exploiting new products and processes in the marketplace. Research with a purely

---

21. *See, e.g.*, *Exclusions*, *supra* note 6, at 29; Geertrui van Overwalle, *The Implementation of the Biotechnology Directive in Belgium and Its After-Effects: The Introduction of a New Research Exemption and a Compulsory License for Public Health*, 37 Int'l Rev. Intell. Prop. & Compet. L. 889, 907–08 (2006) [hereinafter van Overwalle, *The Implementation of the Biotechnology Directive in Belgium and Its After-Effects*]; Geertrui van Overwalle, *Klinische proeven en volksgezondheid. Naar een herijking van het algemeen belang in het octrooirecht* [Clincial Trials and Public Health. Reassessing the Notion of Public Interest in Patent Law], Tijdschrift voor Privaatrecht (TPR) [Journal for Private Law], 899–968 (2000), *available at* http://ssrn.com/abstract=1718619.

22. *Exclusions*, *supra* note 6, at 29; van Overwalle, *The Implementation of the Biotechnology Directive in Belgium and Its After-Effects*, *supra* note 21, at 908.

23. O'Connor, *supra* note 16, at 532; Strandburg, *supra* note 7.

commercial goal encompasses research that solely aims at investigating market potential or price acceptance.

A third parameter revolves around the *focus* of the research: the research exception may relate to research at large or research performed in view of the regulatory review process.[24] The latter is also called a "regulatory review exception" or a "Bolar exception." This exception permits the use of a patented invention without the consent of the patent holder for testing required for the submission of data to obtain marketing authorization for pharmaceutical or other products.

The fourth parameter concerns the institutional *mode* by which the exception is implemented: the exception may be a de jure (statutory or case law) or a de facto exception. Additional parameters that have been put forward in scholarly literature and case law relate to the type of the experiment (process or product), the place of the experiment (public laboratory, hospital, university versus private company), or the scale of the experiment (small or large scale).[25] In our opinion, these additional parameters should not be determinative as far as the applicability of the research exception is concerned. In our analysis we will examine different combinations of the four major parameters.

II. Assessing the Research Exception

A. The Research Exception as a "Zero-Price Liability Rule"

The property–liability dichotomy can serve as an analytical framework to understand and assess the research exception. This dichotomy is based on Calabresi and Melamed's frequently quoted seminal article of 1972 *Property Rules, Liability Rules and Inalienability: One View of the Cathedral*.[26] The value of the theory developed by Calabresi and Melamed has been acknowledged by a large body of scholarly literature that has applied its main tenets to different fields of law, including to patent law.[27]

---

24. *See* O'Connor, *supra* note 16, at 532.

25. *See* van Overwalle, *supra* note 9, and the references cited there.

26. G. Calabresi & A.G. Melamed, *Property Rules, Liability Rules and Inalienability: One View of the Cathedral*, 85 HARV. L. REV. 1089 (1972).

27. Most of the literature on this topic is from US scholars, including both legal scholars and economists. *See, e.g.*, Daniel A. Crane, *Intellectual Liability*, 88 TEX. L. REV. 253 (2009); Mark A. Lemley & Phil Weiser, *Should Property or Liability Rules Govern Information*, 85 TEX. L. REV. 783 (2007); Robert P. Merges, *Contracting into Liability Rules: Intellectual Property Rights and Collective Rights Organizations*, 84 CAL. L. REV. 1293 (1996); Jerome Reichman, *Legal Hybrids Between the Patent and Copyright Paradigms*, 94 COLUM. L. REV. 2432 (1994). However, recently also some European scholars have started working with the theory. *See, e.g.*, NICOLAS BRAHY, THE PROPERTY REGIME OF BIODIVERSITY AND TRADITIONAL KNOWLEDGE (2008); ANNETTE KUR & JENS SCHOVSBO, EXPROPRIATION OR FAIR GAME

Calabresi and Melamed's theory revolves around two questions. The first is to which party a certain "entitlement" should be allocated. With the "entitlement" concept the authors refer to the basic question a legislator faces when drafting new legislation: In the case of conflicting interests between two or more people, or two or more groups of people, which of the conflicting parties should prevail? Translated into the context of inventions, it is clear that the states tend to grant the entitlement to inventors. One may say that there is a distributional bias in favor of inventors.[28]

The second question focuses on how the entitlement should be *enforced*. Once the initial choice has been made as to the allocation of the entitlement to a certain party, the legislator must decide how to enforce its choice. Enforcement of entitlements can take place in three ways: by way of a property rule, a liability rule, or an inalienability rule.[29]

If the legislator prefers a property enforcement regime, someone who wishes to use the entitlement must buy the entitlement from its owner in a voluntary transaction. Professor Merges refers to these rules as "absolute permission rules": one cannot take these entitlements without prior permission from the owner.[30] The essence of the property rule is that the owner has "a right to exclude." Appropriate relief against someone who uses the entitlement without the owner's permission is an injunction granted by the courts that bars the use of the property at stake. The property owner, hence, has significant *control* over his entitlement and the valuation. The value of the entitlement is negotiated ex ante between the buyer and the seller.[31] The state will not intervene in the decision on the value of the entitlement. Parties will negotiate and the seller will have a veto right if the buyer does not offer enough. A property rule regime, thus, requires only *limited* government intervention. Translation of the property rule approach into an innovation context implies that the main mechanism to enforce the entitlement is a property rule, an exclusive right granted to the inventor for a certain territory and for a limited period (generally twenty years), which allows the owner to prevent others

---

FOR ALL? THE GRADUAL DISMANTLING OF THE IP EXCLUSIVITY PARADIGM (Max Planck Inst. for Intellectual Prop., Competition and Tax Law Research Paper Series, Research Paper No. 09-14, 2009), *available at* http://papers.ssrn.com/sol3/papers.cfm?abstract_id=1508330.

28. *See, e.g.*, Bd. of Trustees v. Roche Molecular Sys., Inc., 131 S. Ct. 2188 (2011), *available at* http://www.supremecourt.gov/opinions/10pdf/09-1159.pdf.

29. An entitlement is *inalienable* to the extent that parties are not allowed to transfer the entitlement. Hence, the state intervenes to determine the initial entitlement and the level of compensation in case the entitlement is taken or destroyed, but also to prohibit the sale as such or under specific circumstances. Calabresi & Melamed, *supra* note 26, at 1092–93. The present chapter will focus only on the property rule and the liability rule.

30. Merges, *supra* note 27, at 1302.

31. Calabresi & Melamed, *supra* note 26, at 1092.

from making, using, offering for sale, selling, and importing the invention.[32] This exclusive right is known as a patent.

Calabresi and Melamed define some market characteristics that would be in favor of establishing a property rule regime: (1) few parties, (2) low transaction costs, and (3) difficult valuation of the entitlement.[33] Enforcement by way of a property rule has some advantages and inadequacies. The first advantage of a strong property rule combined with injunctive relief is that it will deter risk-averse free riders. Second, the possibility for owners to seek an injunction offers a "platform" for the parties to bargain over the property and to negotiate a deal.

However, the importance of injunctions for property rule regimes also points to two inadequacies of property rules, in particular with respect to patents. The first shortcoming relates to the uncertain scope of patents.[34] The patent claims serve to describe the metes and bounds of the protected invention. In practice, it is impossible, however, to perfectly define an invention in words.[35] This makes it difficult to know in advance whether a particular use is infringing and whether one needs to negotiate a license. The fact that a property right cannot be clearly defined means that injunctive relief is a remedy with significant error costs and a strong risk of overprotection.[36]

The second shortcoming concerns the extent of "control" granted to patent owners by way of a property rule. Property rules rely on the assumption that patent owners are best placed to evaluate the relevant unknown circumstances of the bargaining process with downstream users.[37] However, patent owners may not always be in the best position to control and "coordinate" the downstream use of the invention. The early-stage inventor cannot always foresee all possible

---

32. *See, e.g.*, TRIPS Agreement, *supra* note 6, arts. 28, 33.

33. Calabresi & Melamed, *supra* note 26, at 1106–08; Robert P. Merges, *Of Property Rules, Coase, and Intellectual Property*, 94 Colum. L. Rev. 2655, 2664 (1994).

34. *See* Mark A. Lemley & Carl Shapiro, *Probabilistic Patents*, 19 J. Econ. Persp. 75 (2005).

35. "Unfortunately, the nature of language makes it impossible to capture the essence of a thing in a patent application." Festo Corp. v. Shoketsu Kinzoku Kogyo Kabushiki Co., 535 U.S. 722, 731 (2002). Moreover, in most countries, the scope of a patent is not limited to its literal terms but instead embraces all *equivalents* to the claims to prevent others from exploiting the limitations of language to the disadvantage of the patent owner.

36. *See, e.g.*, Carl Shapiro, Injunctions, Hold-Up, and Patent Royalties 24 (Competition Policy Center working paper) (Aug. 1, 2006), *available at* http://works.bepress.com/carl_shapiro/10/ ("[G]ranting patent holders the nearly automatic and immediate right to obtain permanent injunctions after they prevail in patent cases over-rewards the owners of patents who license rather than practice their patents. The over-reward is especially great for the owners of patents on minor features used in high-margin products.").

37. Patent owners are better placed than, say, courts or agencies in a liability regime. *See, e.g.*, Henry E. Smith, *Property and Property Rules*, 79 N.Y.U. L. Rev. 1719, 1725–56 (2004).

exploitations or may even deliberately slow down or hinder innovation to promote the sales of her own products. Where the inventor is clearly entitled to some compensation for use, but where control would violate the objectives of patent law, it might be better to separate control and compensation.[38] For such cases, the establishment of liability rules as "safety valves" for the shortcomings of the property rule regime is the natural solution.

An entitlement is protected by a liability rule whenever one may take and use the entitlement without prior permission provided one is willing to pay an objectively determined value for it. The appropriate value is set in an ex post proceeding.[39] The value will be what it is believed the original holder of the entitlement would have sold it for. Professor Merges terms these rules as "take now, pay later rules."[40] In contrast with a property rule regime, there is an additional layer of state intervention in a liability rule regime, as the objective evaluation will likely be undertaken by some state agency rather than by the parties concerned.[41] Examples of areas where liability rules are commonly used are eminent domain and torts.

Similar to the property rule regime, Calabresi and Melamed define some characteristics that would favor a liability rule approach for entitlements in a particular market: (1) many parties; (2) high transaction costs; and (3) easy, objective valuation of the entitlement.[42] These characteristics also indicate the major advantage of a liability rule regime in that they operate quite efficiently in situations where there are many parties. If many parties are involved, there is a high likelihood of strategic bargaining and transaction costs are generally high. In theory, each of those parties could hold up the bargaining process under a property rule in order to get a higher price.[43] If the government or a judge intervenes to set the price, the holdup problems are gone. So, in such situations liability rules are the enforcement mechanism of choice.

Liability rules also have a number of inadequacies. The first one is related to the difficulty of objectively determining the value of the entitlement. The

---

38. *See* Lemley & Weiser, *supra* note 27, at 791 (with respect to copyright).

39. Calabresi & Melamed, *supra* note 26, at 1092.

40. Merges, *supra* note 27, at 1302.

41. Calabresi & Melamed, *supra* note 26, at 1092.

42. Besides economic efficiency considerations, such as transaction costs, distributive preferences, and other justice considerations may also plead in favor of liability rules. Although Calabresi and Melamed warn that there are some limitations to the role of distributional preferences. Such preferences should not be used as a hidden way of accruing benefits for a group whom we would not otherwise wish to benefit. *Id.* at 1115.

43. *Id.* at 1106–08; Merges, *supra* note 33, at 2664.

specialized nature of many products and complex business environments make it difficult to properly value the right holders' loss. What is the true value of the entitlement? Who should decide this, and on the basis of what criteria? Courts or government agencies do not seem well-situated to make complex valuations of this sort.[44] They do not normally have the expert knowledge on the underlying technology, the final product, the industry, and the industry norms. Judges also may feel quite uncomfortable in their role as rate regulators.[45]

A further shortcoming of liability rules that seems to be particularly onerous with respect to patented inventions is the "chasm" created between the inventor/patent owner and the user. Inventors generally perceive the loss of control with respect to their patented invention associated with liability rules as quite unfair. In spite of the labor they put into the R & D process, they lose control over the exploitation of the invention. Another, but major, shortcoming of liability rules for inventions concerns the impact on the incentive to innovate. By setting a fixed rate, courts or agencies place a ceiling on the value of the patented technology. This rate may be lower than what the patent owner would have negotiated under a property rule. Inventors/patent owners will, thus, be undercompensated for their efforts, even where these efforts have a clear social benefit.

To operate as useful safety valves in a property rule–dominated regime, Professors Lemley and Weiser posit that liability rule mechanisms should comply with a number of additional requirements to overcome these shortcomings.[46] Namely, (1) liability rules should be clearly defined; (2) the scope of application of the liability rule should be limited to very specific circumstances so that they do not undermine the incentives to innovate; and (3) liability rules should be complemented by a sound governance scheme to limit the costs of administration, the risk of undercompensation, and to facilitate enforcement.[47]

To be effective, the governance scheme associated with the liability rule should: (1) appoint actors with the necessary expertise and capabilities, (2) be able to deal with multiple demands for the application of the rule, (3) have information at its disposal to design workable access arrangements, (4) be able to deal with the dynamic nature of the business relationships in the particular sector and of the technology (e.g., limited term, periodic reviews, possibility of a petition for

---

44. *See, e.g.*, Makan Delharim, Dep't of Justice, Forcing Firms to Share the Sandbox: Compulsory Licensing of Intellectual Property Rights and Antitrust 12 (2004), *available at* http://www.justice.gov/atr/public/speeches/203627.pdf.

45. *See, e.g.*, Pac. Bell Tel. Co. v. Linkline Commc'ns, Inc., 555 U.S. 438, 452–53 (2009).

46. Lemley & Weiser, *supra* note 27, at 809–40.

47. *Id.*

review), and (5) provide for necessary oversight.[48] In reality it is extremely rare to find "pure" forms of the property and liability rule regimes. It is quite uncommon to come across a "pure" property regime in the sense of exclusivity. Property rules are typically subject to a host of exceptions. In a patent context, one may encounter private use exceptions, research exceptions, and compulsory licenses. Similarly, "pure" liability rules also are extremely rare. They tend to be limited in time and/or space and particular circumstances. Entitlements to assets are almost always subject to "mixed" regimes split among different stakeholders. [49] Patents are an example par excellence of a property rule regime (exclusive right) mixed with liability rule mechanisms (see Figure 1). Lately, the practical relevance of liability rule mechanisms in the patent context seems to be increasing through various court decisions,[50] a growing commitment of patent owners to voluntarily constructed liability rules (for example by way of patent pools),[51] and the introduction of new compulsory license statutes.[52]

Lemley and Weiser also identify a subcategory of liability rules, so-called "zero-price liability rules."[53] Under such a rule, access to the asset is provided for free. Transcribing this sub-scheme to the patent arena leads to the insight that research exceptions can be regarded as "zero-price liability rules," as researchers can freely use patented inventions under such an exception.

The virtue of zero-price liability rules in comparison to regular liability rules is that the former are easy to "calculate" and to administer. However, zero-price liability rules also reflect some of the same shortcomings of regular liability rules: the risk of the creation of a "chasm" between the inventor and the user and

---

48. *Id.*

49. *See* Dan Burk, *Critical Analysis: Property Rules, Liability Rules and Molecular Futures—Bargaining in the Shadow of the Cathedral, in* GENE PATENTS AND COLLABORATIVE LICENSING MODELS: PATENT POOLS, CLEARINGHOUSES, OPEN SOURCE MODELS AND LIABILITY REGIMES 294, 295–96 (G. van Overwalle ed., 2009).

50. In the United States, see, e.g., eBay Inc. v. MercExchange, L.L.C., 547 U.S. 388 (2006); Paice L.L.C. v. Toyota Motor Corp., 85 U.S.P.Q.2d 1001 (Fed. Cir. 2007).

51. *See, e.g.*, Merges, *supra* note 27; Esther van Zimmeren et al., *Patent Pools and Clearinghouses in the Life Sciences*, 29 TRENDS IN BIOTECH. 569 (2011).

52. *See Gene Patents*, *supra* note 9, at 87; Esther van Zimmeren & Geertrui van Overwalle, *A Paper Tiger? Compulsory License Regimes for Public Health in Europe*, 42 INT'L REV. INTELL. PROP. & COMPETITION L. 4 (2011). Compulsory licenses oblige patent owners to license their inventions to a certain technology user, provided particular circumstances are present. Hence, the patent owner loses control, as he cannot prevent this licensee from using the invention. The license fee is set ex post by a government official or a judge. Compulsory licenses are the typical example of a liability rule.

53. Lemley & Weiser, *supra* note 27, at 786, 815.

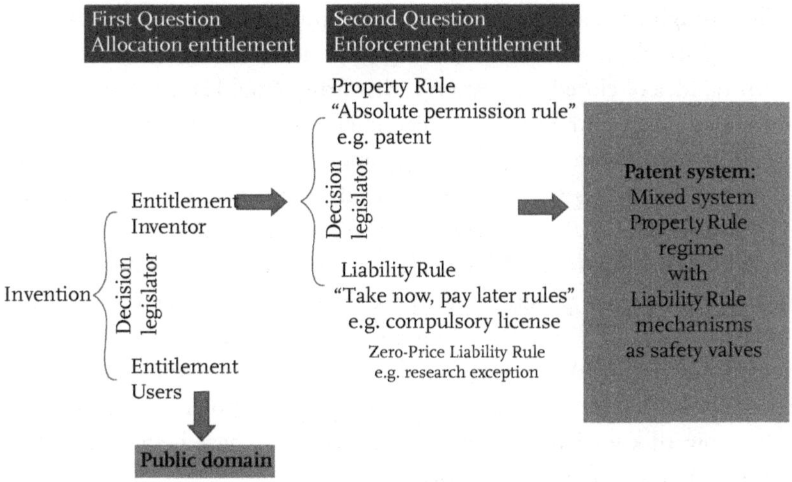

FIGURE I Property rule/liability rule dichotomy translated to patent law[54]

the potential impact on incentives to innovate. We now turn to explore such a research exception in an "open innovation" context.

B. The Research Exception in an Open Innovation Context

A trend toward "open" and collaborative innovation seems to be emerging. Some critical observers claim that open innovation is "old wine in new bottles,"[55] and argue that the main idea and discussion about the consequences of inter-organizational cooperation in R & D date back at least to the 1960s. Others regard open innovation as a new paradigm for fostering innovation, for example, in the biomedical sector.[56] Regardless of the merits of these claims, the term "open innovation" was coined for the first time in 2003 by Henry Chesbrough, a professor and executive director at the Center for Open Innovation at the University of California, Berkeley.[57]

---

54. This figure is in line with Geertrui van Overwalle, *Individualism, Collectivism and Openness in Patent Law: Promoting Access through Exclusion*, *in* INDIVIDUALISM AND COLLECTIVENESS IN INTELLECTUAL PROPERTY LAW 71, 77 (Jan Rósen ed., 2012).

55. Paul Trott & Dap Hartmann, *Why Open Innovation Is Old Wine in New Bottles*, 13 INT'L J. INNOVATION MGMT. 715 (2009).

56. Teri Melese et al., *Open Innovation Networks between Academia and Industry: An Imperative for Breakthrough Therapies*, 15 NATURE MED. 502 (2009).

57. HENRY W. CHESBROUGH, OPEN INNOVATION: THE NEW IMPERATIVE FOR CREATING AND PROFITING FROM TECHNOLOGY (2003).

Chesbrough explains the concept of open innovation by contrasting it with its counterpart, closed innovation. Closed innovation is characterized by control. The central idea of closed innovation is that firms should control the generation of their own ideas, knowledge, and technology, as well as the production, marketing, distribution, servicing, financing, and customer support. When universities and governments were not as involved in the commercial application of science, some firms took complete control of the production and supply chain. They created their own R & D departments to be able to control the whole development cycle of a new product within the company and became vertically integrated companies. Over the years a number of factors have emerged, such as the increased mobility of skilled workers, the growth of the venture capital market, the decreasing shelf life of particular technologies, the growing competition, and the shift toward knowledge-based societies, which all have helped pave the way towards greater use of open innovation.[58]

The central idea behind open innovation is that in a world of widely distributed knowledge, companies cannot afford to rely entirely on their own research, but should instead source ideas, knowledge and technology from other entities. In addition, internal ideas, knowledge, and technology that are not being used in a firm's business should be taken outside the firm. Thus, the boundaries between a firm and its environment become more permeable and innovations can easily transfer, resulting in a new market of knowledge.[59] Open innovation is thus defined as the use of inflows and outflows of knowledge to stimulate and accelerate innovation within firms and organizations, and to expand the markets for the exploitation of innovation by others.[60] The fundamental principle of open innovation is that firms can and should use external as well as internal ideas, knowledge, and technology, and employ internal and external paths to market as they advance their technology.[61] Open innovation is regarded as a "two-way street": it covers both "inbound open innovation" and "outbound open innovation." Inbound open innovation is an outside-in process that leverages ideas, knowledge, and technology acquired from others to complement the firm's internal R & D. Outbound open innovation is an inside-out process by which the firm is actively pursuing external exploitation of ideas, knowledge, and technology to

---

58. *Id.* at xxii.

59. *Id.*

60. Open Innovation: Researching a New Paradigm 1 (Henry W. Chesbrough et al. eds., 2008); Henry W. Chesbrough, Open Business Models: How to Thrive in the New Innovation Landscape (2006); Chesbrough, *supra* note 57.

61. Chesbrough et al., *supra* note 60; Chesbrough, *supra* note 60; Chesbrough, *supra* note 57.

enable commercialization exclusively or in addition to an internal application.[62] To enable these processes, various organizational modes are blossoming, such as R & D alliances, Public Private Partnerships (PPPs), consortia, networks, clusters, mergers and acquisitions, in-licensing agreements, out-licensing agreements, spin-offs, and R & D joint ventures.[63]

Although firms have been licensing in and acquiring external technologies for decades, these activities have lately been gaining more attention and are becoming part of firms' broader innovation strategies. At present, open innovation is becoming more accepted as a useful approach for innovation strategies not only for large firms, but also for small and medium-sized enterprises and universities. Governments and funding agencies are also gradually recognizing the role open innovation might play as a driver for economic growth and social welfare.[64]

Now that the analytical tool has been expounded (the property/liability rule dichotomy) and the peculiar ecosystem has been set forth (open innovation), we turn to the very essence of our interrogation and describe how research exceptions function at present in the United States, Europe, and Japan. We will then examine them through the prism of the property/liability rule dichotomy and the open innovation paradigm.

III. Analyzing Current Research Exceptions in the United States, Europe, and Japan

Many jurisdictions around the world provide a *research* exception. There is quite some variation in the wording of the exception[65] and in the interpretation of its scope in the United States, Europe, and Japan. Moreover, apart from the general

---

62. Chesbrough et al., *supra* note 60; CHESBROUGH, *supra* note 60; CHESBROUGH, *supra* note 57.

63. Esther van Zimmeren et al., *Patent Pools and Clearinghouses in the Life Sciences*, 29 TRENDS IN BIOTECH. 569 (2011).

64. *See, e.g.*, Communication from the Commission to the European Parliament, the Council, the European Economic and Social Committee and the Committee of the Regions—Europe 2020 Flagship Initiative—Innovation Union, COM (2010) 546 final (Oct. 6, 2010), *available at* http://ec.europa.eu/commission_2010-2014/geoghegan-quinn/headlines/documents/com-2010-546-final_en.pdf; Communication from the Commission to the Council, the European Parliament, the European Economic and Social Committee and the Committee of the Regions Improving Knowledge Transfer between Research Institutions and Industry across Europe: Embracing Open Innovation—Implementing the Lisbon Agenda, COM (2007) 182 final, *available at* http://ec.europa.eu/invest-in-research/pdf/com2007182_en.pdf; Japan Intellectual Property Strategy Headquarters, Intellectual Property Strategic Program 2009 (2009), *available at* http://www.kantei.go.jp/jp/singi/titeki2/keikaku2009_e.pdf.

65. The general nature of Article 30 of TRIPS has resulted in a broad variety of formulations of research exceptions among different states. At the moment of formulating the exceptions,

research exception, *regulatory* exceptions have been created, and *clinical* exemptions have been considered, in the United States, Europe, and Japan as well. Although, the regulatory exception and proposed clinical exemption pursue different objectives, these provisions may have an *indirect* effect on the interpretation of the scope of the general research exceptions and vice versa. Moreover, an extremely narrow interpretation of the research exception (or experimental use doctrine) may be complemented by a more moderate interpretation of the regulatory exception, giving more leeway to researchers in the biomedical sector.

A. Research Exceptions in the United States

Most major technology-oriented countries provide a statutory research exception. The United States and several other common law countries are an anomaly in this respect. In the United States, the research exception exists as a judicially created doctrine.

1. Experimental Use Doctrine

In *Whittemore v. Cutter*,[66] Justice Story introduced an extremely narrowly defined judicial experimental use exception. This exception shielded conduct relating to the patented technology "merely for philosophical experiments, or for the purpose of ascertaining the sufficiency of the machine to produce its described effects."[67] In *Sawin v. Guild*, Story defined infringement as making the patented invention "with an intent to use for profit, and not for the mere purpose of philosophical experiment, or to ascertain the verity and exactness of the specification."[68] Over the years, courts applied the exception if the purpose of the research was to gain knowledge and could not be tied directly to some profit-making

---

most countries had little guidance as to how they could maximize the benefits of the TRIPS Agreement in view of their level of development. At present, countries such as China, India, and Brazil have huge scientific and technological research potential, but it is not clear to what extent research exceptions will be beneficial in this respect. In fact, this was the motivating factor behind Brazil's proposal to the SCLP to establish a framework whereby detailed information on exceptions to patent rights in all jurisdictions, as well as information on experiences in implementing such exceptions and the impacts on development, would be compiled and made available as an important step in the implementation of the Development Agenda. WIPO-SCLP, Proposal by Brazil—Document Prepared by the Secretariat, WIPO Doc. SCP/14/7 (Jan. 20, 2010), *available at* http://www.wipo.int/edocs/mdocs/patent_policy/en/scp_14/scp_14_7.pdf.

66. Whittemore v. Cutter, 29 F. Cas. 1120 (C.C.D. Mass. 1813).

67. *Id.* at 1121.

68. Sawin v. Guild, 21 F. Cas. 554, 555 (C.C.D. Mass. 1813).

activity. The exception was widely assumed by scientists and university administrators to shield scientific investigation at universities from lawsuits.[69]

In a trilogy of cases, the US Court of Appeals for the Federal Circuit (CAFC) called into question the scope of the exception. The first case in this trilogy was *Roche Products, Inc. v. Bolar Pharmaceuticals* in 1984,[70] followed by *Embrex v. Service Engineering*[71] in 2000, and culminating in the well-known and controversial *Madey v. Duke* case in 2002.[72] In *Roche*, the defendant, a generic manufacturer, infringed Roche's drug patent to obtain test data necessary for product regulatory approval by the FDA. The CAFC ruled that tests, demonstrations, and experiments, which are in keeping with the legitimate business of the alleged infringer, are infringements for which the experimental use defense could not be invoked.[73]

In *Embrex*, the defendant had hired a university professor to invent around a patent held by Embrex. In the course of the experiment, the professor infringed the patent. The CAFC upheld the lower court ruling in favor of Embrex, stating that the commercial nature of the experimentation made the use neither experimental nor de minimis,[74] whether or not it led to the sale of infringing machines. According to the court, use of the patented technology in testing conducted to invent around the patent constitutes patent infringement.[75]

Madey appears to be the most controversial of the three cases holding that the noncommercial, non-profit status of university research does not constitute an automatic defense to patent infringement. In its decision, the CAFC articulated the modern conceptualization of universities as substantial economic players that have business objectives similar to other economic actors.[76] The CAFC first recalled that:

> [Regardless] of whether a particular institution or entity is engaged in an endeavor for commercial gain, so long as the act is in furtherance of the alleged infringer's legitimate business and is not solely for amusement, to

69. Stephen A. Merrill, Richard C. Levin & Mark B. Myers, A Patent System for the 21st Century 109 (2004); O'Connor, *supra* note 16, at 525.

70. Roche Prods., Inc. v. Bolar Pharm. Co., 733 F.2d 858 (Fed. Cir. 1984).

71. Embrex, Inc. v. Serv. Eng'g Corp., 216 F.3d 1343 (Fed. Cir. 2000).

72. Madey v. Duke Univ., 307 F.3d 1351 (Fed. Cir. 2002).

73. *Roche*, 733 F.2d at 863.

74. De minimis use concerns "use so inconsequential as to warrant being excused from infringement." *See* O'Connor, *supra* note 16, at 526.

75. *Embrex*, 216 F.3d at 1349.

76. O'Connor, *supra* note 16, at 528.

satisfy idle curiosity, or for strict philosophical inquiry, the act does not qualify for the very narrow and strictly limited experimental use defense.[77]

Furthermore, the profit or nonprofit status of the user is not decisive for the applicability of the doctrine. Research projects that are financed by major research universities but that have no prospect of commercialization advance the institution's legitimate business objectives, including "educating and enlightening students and faculty participating in these projects." Moreover, they "serve to increase the status of the institution and lure lucrative research grants, students and faculty."[78] And finally, the court noted, the university was not "shy in pursuing an aggressive patent licensing program from which it derives a not insubstantial revenue stream."[79] Concerns have been raised about the extremely limited interpretation of the experimental use doctrine in *Madey*.[80] Some scholars and practitioners believe that the doctrine no longer has any practical value. They fear that the decision could have a chilling effect on university research, as universities would be obliged to spend considerable resources on patent searches and licensing negotiations. However, in practice a kind of de facto research exemption has developed because companies are reluctant to sue universities for patent infringement—a practice of "rational forbearance."

2. De Facto Research Exemption

In addition to the experimental use doctrine and the practice of "rational forbearance," state sovereign immunity under the US Constitution has created another de facto research *exemption* protecting researchers working at state universities and the state universities themselves. Under the Eleventh Amendment of the US Constitution, federal courts may not be used by citizens to sue a state without the state's consent. Patent infringement suits may only be brought in federal courts. So, patent owners cannot directly sue states or state agencies for patent infringement.[81] However, after enjoying nearly a decade of protection on this ground,

---

77. *Madey*, 307 F.3d at 1355.

78. *Id.* at 1362.

79. *Id.*

80. *See, e.g.*, Tom Saunders, *Renting Space on the Shoulders of Giants:* Madey *and the Future of the Experimental Use Exception*, 13 YALE L.J. 261 (2003); Lawrence M. Sung & Claire M. Maisano, *Piercing the Academic Veil: Disaffecting the Common Law Exception to Patent Infringement Liability and the Future of a Bona Fide Research Use Exception After* Madey v. Duke University, 6 J. HEALTH CARE L. POL'Y 256 (2003).

81. *See* Seminole Tribe of Fla. v. Fla., 517 U.S. 44 (1996); *see also* three later decisions from the US Supreme Court on the same day dealing with this issue: Fla. Prepaid Postsecondary

states' immunity to intellectual property lawsuits was recently the subject of a court challenge.[82] In *Biomedical Patent Mgmt. Corp. v. California Department of Health Services*, the plaintiff, Biomedical Patent Management Corporation, claimed that sovereign immunity laws unfairly shield states—including state universities and research institutions—from patent infringement liability while at the same time allowing them to enforce their own patent rights. Biomedical Patent Management Corporation alleged that between 1990 and 2008, six patent actions were brought against the state of California, and in each case the state raised its immunity, whereas in the same period, the University of California filed at least fourteen patent infringement suits against various alleged infringers.[83] A change in case law would have had broad implications for biotechnology companies whose efforts to enforce their patents have frequently been thwarted by courts upholding states' immunity laws, as was the case in *Biomedical Patent Management*. The Supreme Court denied certiorari in *Biomedical Patent Management v. California* and, hence, for the moment state universities can continue to benefit from this escape route.

Given this de facto research exemption and the limited enforcement actions by industry against universities and individual scientists, it is interesting to see that state universities in the United States have continued to advocate for the creation of a statutory research exception.[84] Perhaps, this is an indication that universities are more concerned about the impact of patents on research than are individual scientists.[85] Universities may no longer feel confident that the past attitude of "rational forbearance" will last. Whereas US state universities still enjoy sovereign immunity from patent infringement actions, experts point to signs that the US Supreme Court may retreat from its "prior robust concept of state sovereign immunity."[86] Moreover, universities are probably more attractive financial targets for infringement lawsuits than are individual scientists. Universities' patent awareness and risk-averse nature may lead them to control liability risks.[87] The difficulty for university administrators is to exercise such control in the face

---

Educ. Expense Bd. v. Coll. Sav. Bank, 527 U.S. 627 (1999); Coll. Sav. Bank v. Fla. Prepaid Postsecondary Educ. Expense Bd., 527 U.S. 666 (1999); Alden v. Maine, 527 U.S. 706 (1999).

82. Biomedical Patent Mgmt. Corp. v. Cal. Dep't of Health Servs., 505 F.3d 1328 (Fed. Cir. 2007).

83. Emily Waltz, *University Patents Probed*, 26 NATURE BIOTECH. 596, 596 (2008).

84. Association of American Universities et al., Comments on H.R. 1908 and S. 1145, The Patent Reform Act of 2007, at 5, *available at* http://www.nacua.org/documents/PatentReformAct_Comments.pdf (last visited Sept. 2, 2013).

85. Rebecca S. Eisenberg, *Noncompliance, Nonenforcement, Nonproblem? Rethinking the Anticommons in Biomedical Research*, 45 HOUS. L. REV. 1059,1092 (2008).

86. *Id.*; *see also* O'Connor, *supra* note 16, at 548–52.

87. Eisenberg, *supra* note 85, at 1093.

of strong traditions of academic freedom, and the expectation that scientists have an obligation to investigate within their chosen fields. This suggests that relying on non-enforcement is less desirable than a statutory research exception, which we argue may be the best way forward.[88]

3. Statutory Research Exception?

In 2002, the proposed Genomic Research and Diagnostic Accessibility Act sought to introduce a very specific statutory research exception.[89] This research exception would allow researchers and medical practitioners to use patented gene sequences for noncommercial research purposes. "Research" was defined as "a systematic investigation, including research development, testing, and evaluation, designed to develop or contribute to generalizable knowledge."[90] The exception would not have been applicable to "any individual or entity that is *directly* engaged in the commercial manufacture, commercial sale, or commercial offer for sale of a drug, medical device, process, or other product using such patent for or patented use of genetic sequence information."[91] Hence, this proposed exception was restricted to a specific group of biomedical researchers, but was slightly broader than the common law exception in *Madey*. Nonetheless, in comparison to the European and Japanese research exceptions described below, the scope of this exception is still extremely limited and the Act was not passed into law by the US Congress.

Recently, the *Secretary's Advisory Committee on Genetics, Health, and Society* (SACGHS) has also recommended changing the US Patent Act by including a research exception.[92] Similar to the proposal in the Genomic Research and Diagnostic Accessibility Act, the research exception would be narrowly focused on those who use patent-protected genes in the pursuit of their research.[93] However, to the extent that it only addresses research regarding genetic tests this proposal is even more restrictive than that contained in the Genomic Research and Diagnostic Accessibility Act.

---

88. *Id.*

89. Genomic Research and Diagnostic Accessibility Act of 2002, H.R. 3967, 107th Cong. (2002), *available at* http://frwebgate.access.gpo.gov/cgi-bin/getdoc.cgi?dbname=107_cong_bills&docid=f:h3967ih.txt.pdf.

90. *See id.* § 2(j)(2)(E).

91. *Id.* § 2(j)(1).

92. *Gene Patents*, *supra* note 9, at 87–90.

93. *Id.* at 89.

4. Bolar Exception

The above-mentioned *Roche* case spurred the US Congress to include a "safe harbor" in the patent act, namely the so-called "Bolar exception." Congress enacted this exception in 1984 by way of the Hatch–Waxman Act, (35 U.S.C. § 271(e)(1)) "solely for uses reasonably related to the development and submission of information under a federal law which regulated the manufacture, use, or sale of drugs." This exception was added to permit clinical testing of generic versions of patented medicines during the patent term in order to facilitate quick market entry after expiry of the patent. Strictly speaking this is not a research exception, but it generally covers research carried out in the framework of a regulatory approval. Moreover, recent developments in this area may influence the way the common law research exception is perceived.

In 2005, the US Supreme Court gave an expansive interpretation of the Bolar exception in the *Merck v. Integra*[94] ruling. According to the Court, the use of patented inventions in preclinical research, the results of which are not ultimately included in a submission to the FDA, falls within the scope of the Bolar exception.[95] Recently, different panels of the Court of Appeals for the Federal Circuit issued conflicting decisions on whether the "safe harbor" embodied in 35 U.S.C. § 271(e)(1) extends to post-approval activities as well.[96] Overall, the trend in these cases seems to be toward a rather broad interpretation in comparison with the restrictive approach of the CAFC to the common law exception in *Madey*. Nevertheless, one cannot deny the limited scope of application of the statutory Bolar exception: the exclusion applies only to research that is related to drug development and submission of information to the FDA. It remains to be seen to what extent, if any, these cases will have an impact on the interpretation of the common law experimental use doctrine.

5. Clinical Exemption?

The existing exceptions in the United States are limited to a research and/or regulatory context and do not provide a safe harbor for clinicians who infringe patents when they perform genetic tests. However, there may be a close link between (for instance) clinical diagnostic testing and research. The patient's blood or tissue

94. 545 U.S. 193 (2005).

95. The data may concern safety, but also the pharmacological, toxicological, pharmakinetic, and biological qualities of a drug. Moreover, in footnote 7 of its opinion, the Court claimed that it made no ruling as to the status of biotechnology research tools. *Id.* at 205 n.7.

96. Whereas in *Classen Immunotherapies, Inc. v. Biogen IDEC*, 304 F. App'x 866 (Fed. Cir. 2011), the majority ruled against an extension of the safe harbor to post-approval activities; in *Momenta Pharmaceuticals, Inc., v. Amphastar Pharmaceuticals Inc.*, 686 F.3d 1348 (Fed. Cir. 2012), the majority decided otherwise, leading to a vigorous dissent by Chief Justice Rader.

samples on which tests are performed are often also necessary for doing research. In light of this, there have been several attempts to introduce a statutory safe harbor for clinicians. In 1996, the US Patent Act was amended to *exempt* medical practitioners from infringement liability when using patented medical or surgical techniques in medical or surgical procedures on a body.[97] However, this exemption was not extended to clinical laboratory services.[98]

To overcome this gap, the Genomic Research and Diagnostic Accessibility Act of 2002 also proposed a clinical exemption,[99] which would have extended the exemption for medical practitioners from patent infringement liability to clinicians performing genetic diagnostic, prognostic, or predictive tests.[100] This exemption would have allowed clinicians to carry out tests, creating effective competition with patent owners that offer the tests, which would presumably lead to lower prices for these tests, higher-quality testing, and opportunities for second opinions. As noted earlier, the Act did not become law.

More recently, the SACGHS has also recommended changing the US Patent Act by including a clinical exemption.[101] Moreover, in 2011, a kind of clinical exemption was proposed in the then-ongoing US patent reform debates. In the debates in the House regarding the "America Invents Act,"[102] a proposal was made to include a provision exempting "confirming genetic diagnostic test activity" from infringement liability where "an independent confirmation of the prior test is not available from another test provider under a license from the patent owner."[103] The American Civil Liberties Union (ACLU) opposed this proposal

97. Ganske-Frist Medical Procedure Exemption Act, 35 U.S.C. § 287(c) (2006). According to this amendment to the infringement statute in US patent law, courts can decide that a physician has infringed a medical process patent, but cannot order him or her to pay damages or enjoin him or her from using the technology.

98. *Id.*

99. Genomic Research and Diagnostic Accessibility Act of 2002, H.R. 3967, 107th Cong. (2002), *available at* http://frwebgate.access.gpo.gov/cgi-bin/getdoc.cgi?dbname=107_cong_bills&docid=f:h3967ih.txt.pdf.

100. The term "genetic diagnostic, prognostic, or predictive test" is defined as "any test, designed to detect disease, to predict the potential for a medical disorder, or to predict the effectiveness of therapeutics, which uses either an ordered listing of nucleotides comprising a portion of a human or human pathogen genetic code or the proteins encoded by such nucleotides." Genomic Research and Diagnostic Accessibility Act, § 3(a)(3)(F), H.R. 3967, 107th Cong. (2002).

101. *Gene Patents*, *supra* note 9, at 87–90.

102. Leahy–Smith America Invents Act, Pub. L. No. 112-29 (2011) (codified in scattered sections of 35 U.S.C.). The House version of the bill, Leahy–Smith America Invents Act, H.R. 1249, 112th Cong. (as passed by the House, June 23, 2011), can be found at http://www.gpo.gov/fdsys/pkg/BILLS-112hr1249ih/pdf/BILLS-112hr1249ih.pdf.

103. The exemption was proposed by Rep. Debbie Wasserman Schultz (D-FL). This proposal was made as an amendment to the Draft Manager's Amendment. For the original version of

on the ground that it "risks allowing gene patent holders to argue that Congress implicitly endorses the validity of such [gene] patents."[104] In light of the *Myriad* case,[105] such an endorsement would undermine the objectives pursued by the ACLU and the other plaintiffs in this case who challenged the patent eligibility of DNA sequences. In the end, the version of the bill enacted into law as the Leahy–Smith America Invents Act (AIA) of 2011 did not contain the proposed exemption.[106] Instead, Section 27 of the AIA obliges the director of the USPTO to conduct a study on effective ways to provide independent, confirming genetic diagnostic test activity where gene patents and exclusive licensing for primary genetic diagnostic tests exist.[107]

B. Research Exceptions in Europe

Most European countries provide a statutory research exception in their patent acts. More recently, a research exemption also has been created at the EU level.

1. Statutory Research Exceptions

Most national research exceptions in Europe have been modeled after Article 31(B) of the Convention for the European Patent for the common market. This convention, also called the Luxembourg "Community Patent Convention" (CPC),[108] was signed at Luxembourg on December 15, 1975, by the then nine member states of the European Economic Community. However, the CPC never entered into force, as it was not ratified by a sufficient number of countries. Fourteen years later, the Agreement relating to Community Patents (CP

---

the Manager's Amendment introduced in April 2011, see *Amendment to H.R. 1249 Offered by Mr. Smith of Texas*, Inventive Step (Apr. 12, 2011), *available at* http://inventivestep.files.wordpress.com/2011/04/managers-amendment-to-hr1249.pdf.

104. American Civil Liberties Union (ACLU), ACLU Opposes Inclusion of Flawed Second Opinion Testing Language in Manager's Amendment to H.R. 1249, the America Invents Act, June 15, 2011, *available at* http://www.aclu.org/files/assets/final_ltr_re_opposing_wasserman_schultz_amendment_with_coalition_letter.pdf.

105. Ass'n for Molecular Pathology v. U.S. Patent and Trademark Office, 702 F. Supp. 2d 181 (S.D.N.Y. 2010), *aff'd in part, rev'd in part*, 653 F.3d 1329 (Fed. Cir. 2011), *aff'd in part, rev'd in part sub nom.*, Ass'n for Molecular Pathology v. Myriad Genetics, Inc., 133 S. Ct. 2107 (2013). It is interesting to note that the US Department of Justice filed a brief in the lawsuit opposing the issuance of patents on isolated DNA, which was in conflict with the day-to-day granting practice of the USPTO, a view that was adopted by the Supreme Court.

106. *See* Leahy-Smith America Invents Act § 18, Pub. L. 112-29, 125 Stat. 284 (2011) (codified as amended in scattered sections of 35 U.S.C.).

107. *Id.* § 27.

108. Convention for the European Patent for the Common Market, 76/76/EEC, 1976 O.J. (L 17) 1 [hereinafter CPC].

Agreement), signed at Luxembourg on December 15, 1989,[109] was an attempt to revive the project. This Agreement consisted of an amended version of the original CPC.[110] Unfortunately, this attempt also failed.

As the provisions in the laws of the various European member states have the same roots, they are all quite similar[111] and some consensus seems to have emerged. It has been suggested that the various research exceptions apply irrespective of the way the patented subject matter has been put into operation (as a product or as a process) and irrespective of the place of the experiment (be it a public laboratory, hospital, university, or private company).

Even though the various research exceptions go back to the same "mother provision," the exact scope of the exception may differ from country to country due to imprecise wording,[112] subtle differences in the statutory text, and subsequent differing interpretations by national courts. One point of dissension relates to the nature of the research exception. Although the research exception generally tends to cover only research *on* the patented invention, the Belgian wording is different. With the implementation of the EU-Biotechnology Directive, the Belgian legislature decided to extend the exception in the Belgian Patent Act

---

109. Agreement Relating to the Community Patents, 89/695/EEC, 1989 O.J. (L 401) 1 [hereinafter CP Agreement].

110. Article 27(B) of the CP Agreement, *supra* note 109, is substantially the same as Article 31(B) of the CPC, *supra* note 108.

111. *See, e.g.*, U.K. Patent Act, § 60(5), c. 37, 1977 (U.K.), *available at* http://www.ipo.gov.uk/patentsact1977.pdf ("An Act which, apart from this subsection, would constitute an infringement of a patent for an invention shall not do so if (a) it is done privately and for purposes which are not commercial; (b) is done for experimental purposes relating to the subject matter of the invention . . . ."). Similarly, the German Patent Act provides that "Die Wirkung des Patents erstreckt sich nicht auf (1) Handlungen, die im privaten Bereich zu nichtgewerblichen Zwecken vorgenommen werden; (2) Handlungen zu Versuchszwecken, die sich auf den Gegenstand der patentierten Erfindung beziehen . . . ." Patentgesetz [PatG] [Patent Law] May 5, 1936, § 11 (Ger.), *available at* http://www.gesetze-im-internet.de/bundesrecht/patg/gesamt.pdf. Compare also the French Patent Act: "Les droits conférés par le brevet ne s'étendent pas: a) Aux actes accomplis dans un cadre privé et à des fins non commerciales; b) Aux actes accomplis à titre expérimental qui portent sur l'objet de l'invention brevetée . . . ." Loi 2008-516 du 3 juin 2008 code de la propriété intellectuelle [Law 2008-516 of June 3, 2008 on the Code of Intellectual Property], art. L613-5 (Fr.). The Swiss Patent Act says that "Die Wirkung des Patents erstreckt sich nicht auf: a. Handlungen, die im privaten Bereich zu nicht gewerblichen Zwecken vorgenommen werden; b. Handlungen zu Forschungs- und Versuchszwecken, die der Gewinnung von Erkenntnissen über den Gegenstand der Erfindung einschliesslich seiner Verwendungen dienen; insbesondere ist jede wissenschaftliche Forschung am Gegenstand der Erfindung frei . . . ." PATENTGESETZ [PATG] [PATENT LAW], June 25, 1954, SR 232.14, art. 9 (Switz.).

112. In its recent Intellectual Property Laws Amendment (Raising the Bar) Bill, Australia tried to overcome these problems by stating more explicitly what is meant by experimental purposes: "For the purposes of this section, *experimental purposes* relating to the subject matter of the invention include, but are not limited to, the following: (a) determining the properties

(BPA) to acts accomplished for scientific purposes *on* and/or *with* the subject matter of the patented invention.[113] The aim of this modification was to establish clarity and legal certainty, but the new wording actually led to new questions.[114] An important one was how the term "scientific purposes" had to be interpreted. The Belgian Minister for Economic Affairs supported a broad scope of interpretation encompassing both acts with a strict scientific purpose and acts with a mixed scientific/commercial aim.[115] Another point of discord relates to the final goal of the exception. For instance, a UK court clarified that only experiments that generate genuinely new information are covered by the exception and not experiments designed to verify existing knowledge, for instance for customers or regulatory agencies.[116] In contrast, according to the German federal court the research exception covers the verification of statements made in the patent or the development of information for regulatory review.[117] Thus, disagreement remains about the nature (experiments *on* versus experiments *with* the patented subject matter) and the final purpose of the experiment (commercial versus noncommercial goal).[118]

The introduction of a unitary patent in Europe[119] could have put an end to the disagreements. In one of the proposals of the European Commission for a

---

of the invention; (b) determining the scope of a claim relating to the invention; (c) improving or modifying the invention; (d) determining the validity of the patent or of a claim relating to the invention; (e) determining whether the patent for the invention would be, or has been, infringed by the doing of an act." *See* Australian Amendment, *supra* note 11. However, the list is not exhaustive, and, hence, still involves a risk of interpretation problems.

113. Belgian Patent Act, art. 28(1)(b); *see also* van Overwalle, *supra* note 21, at 889–919.

114. Van Overwalle, *supra* note 21, at 906.

115. Statement of the Minister for Economic Affairs, Verwilghen, On the Scope of the Research Exception, Parl. Doc., Chamber, 2004–05, 3-1088/3, at 2, 59–60. Moreover, the modified Belgian research exception did not incorporate a Bolar-type exception.

116. Monsanto Co. v. Stauffer Chem. Co., [1985] R.P.C. 515.

117. Bundesgerichtshof [BGH] [Federal Court of Justice], Jul. 11, 1995, 1996 Neue Juristische Wochenschrift [NJW] 782 (Klinische Versuche I [*Clinical Trials I*]), *translated in* [1997] R.P.C. 623; BGH Apr. 17, 1997, 1997 NJW 3092 (Klinische Versuche II [*Clinical Trials II*]), *translated in* [1998] R.P.C. 423.

118. Geertrui van Overwalle et al., *Models for Facilitating Access to Patents on Genetic Inventions*, 7 Nature Rev. Genetics 143, 143 (2006). For more detailed information, see also CIPP, *supra* note 9; Chrocziel, *supra* note 9; Cook, *supra* note 9; van Overwalle, *supra* note 9; Cornish, *supra* note 9.

119. Work on a unitary EU patent, formerly known as Community patent, started in the 1970s with the CPC followed by the CP Agreement in 1989 and various proposals for regulations by the European Commission in the beginning of the twenty-first century. All these attempts failed. The current proposal had a rather turbulent and unusual course. In fact, twenty-five out of the twenty-seven member states decided to create a unitary EU patent using the procedure of so-called "enhanced cooperation." Because Spain and Italy could not

unitary EU patent, a research exception was incorporated in Article 8(b), which adopted the same wording[120] as the CPC without, however, further clarifying how the provision should be interpreted in practice. It was most unlikely that the research exception would lead to radical changes if incorporated into the unitary patent regime. However, the provision was not included in the regulation adopted by the EU Parliament. Thus harmonization in Europe on the research exception must await another day.

2. The EU "Bolar" Exception

The regulatory review exception has been harmonized at the EU level by way of a directive.[121] This directive has to be implemented at the national level.

---

agree on the proposed translation arrangements for the EU patent, it was impossible to build the necessary consensus within a reasonable period. Enhanced cooperation is a special procedure enshrined in the EU treaty allowing a group of countries to adopt new common rules when a unanimous EU-wide agreement cannot be reached within a reasonable period of time. (Enhanced cooperation is regulated by Article 20 of the Treaty on European Union (TEU) and Articles 326 to 334 of the Treaty on the Functioning of the European Union (TFEU). In principle, at least nine states must be involved in enhanced cooperation, but it remains open to any state that wishes to participate. It may not constitute discrimination between those participating and the other states. Any acts that are adopted within the framework of such cooperation are binding only on the participating member states and do not constitute a part of the so-called *acquis*. Enhanced cooperation must further the objectives, protect the interests, and reinforce the integration process of the Union.). The procedure is used only in special circumstances and as a last resort. It is only the second time that it has been employed (The only other area where enhanced cooperation has been utilized is cross-border divorce law. *See, e.g.*, Renata Goldirova, *Two-Speed Europe May Emerge over Divorce Rules*, EUOBSERVER.COM (July 25, 2008), *at* http://euobserver.com/9/26546/?rk=1. The Heads of State recently established a compromise on the so-called "patent package," consisting of a proposal concerning the unitary patent, a proposal on the translation arrangements, and a proposal for a unified patent court. *See* Conclusions European Council, June 29 2012, EUCO 76/12, CO EUR 4, Concl. 2. The unitary patent and translation regulations were adopted by the EU Parliament on Decemer 11, 2012 and entered into force January 20, 2013. However, they will only apply from the later of January 1, 2014 or the date of entry into force of the Agreement on a Unified Patent Court. *See* http://www.epo.org/law-practice/unitary/unitary-patent.html. Twenty-five EU member states signed the Agreement on a Unified Patent Court on February 19, 2013, but it will not enter into force until ratified by at least thirteen member states.

120. "[A]cts done for experimental purposes relating to the subject matter of the patented invention."

121. Council Directive 2004/27/EC of the European Parliament and of the Council of 31 March Amending Directive 2001/83/EC on the Community Code Relating to Medicinal Products for Human Use, art. 1(8), 2004 O.J. (L 136) 34.

3. Clinical Exception?

At present, it is unclear to what extent the European research exceptions cover clinical diagnostic testing. On the one hand, diagnostic testing may fall within the scope of the research exception, because patient blood or tissue sampling is often necessary for doing research. On the other hand, one could argue that diagnostic testing cannot fall within the exception because the diagnosis could be defined as an act of providing the referring medical doctor with an opinion as to whether the patient carries a deleterious mutation. In this view, the performance of the test would be a purely commercial act.[122] Indeed, the European Society for Human Genetics (ESHG) believes that diagnostic use does not fall within the scope of a research exception "even if the test is performed within the public health sector and, notably, irrespective of whether money is exchanged or not." [123] Therefore, the ESHG recommends genetic laboratories not rely on research exceptions for the practice of offering genetic tests.[124]

C. Research Exceptions in Japan

1. Statutory Research Exception

Similar to most European patent laws, the Japanese Patent Act (JPA) stipulates in Article 69(1) that "a patent right shall not be effective against the working of the patented invention for experimental or research purposes."[125] There is no case law available clarifying the *general* scope of the exception. There was one notable case where Hamamatsu Medical School was sued by AntiCancer, a US biotechnology company, for allegedly infringing on the company's patent for mouse models. The Medical School invoked the research exception as a defense, but unfortunately the court did not rule on the research exception because it decided that Hamamatsu Medical School's mouse did not infringe on the patented mouse as described in the claims.[126] Hence, one must rely on doctrine for the interpretation of the scope of the exception.

According to prevailing doctrine, whether an experiment or research falls within the scope of the exception should be decided "depending on its subject

---

122. van Overwalle et al., *supra* note 118, at 143.

123. Eur. Soc'y Human Genetics (ESHG), *Patenting and Licensing in Genetic Testing—Recommendations of the European Society of Human Genetics*, 16 EUR. J. HUM. GENETICS, Suppl. 1, S3, at S7 (2008).

124. *Id.*

125. Patent Act, Act No. 121 of 1959.

126. Tōkyō Chihō Saibansho [Tōkyō Dist. Ct.] Oct. 10, 2002, 2004 (ne) no. 675 (Japan).

and purpose."[127] An experiment or research falls within the scope of the exception only when its subject is limited to the patented invention itself and when its purpose is limited to "technological progress."[128] The Japanese Council for Science and Technology Policy (CSTP) has clarified that the concept "experimental or research purposes" is "limited to experiments improving or upgrading patented inventions per se."[129] According to a study by Someno, these experiments include: (1) tests of patentability, (2) tests of function, and (3) experiments for the purposes of improvement and/or development.[130] Similar to Europe, the JPA does not give any weight to whether the experiment or research is conducted by a profit-making entity. In contrast to Europe, whether the ultimate purpose of the research is commercial, noncommercial, or mixed does not seem relevant for the applicability of the research exception in Japan.

In Japan, the availability of, and access to, research tools has generated much concern among practitioners and legal scholars. Use of research tools does not appear to fall within any of the above-mentioned categories.[131] Research activities for which a patented research tool is used constitute patent infringement regardless of whether they are conducted by universities or companies, or for commercial or noncommercial purposes. There have been various economic and legal (comparative) studies[132] with regard to the need for and utility of expanding the scope of the Japanese research exception or introducing a separate research exception for research tools. To our knowledge, these studies have not led to a change of the JPA.

---

127. International Association for the Protection of Intellectual Property (AIPPI), Q202 Working Committee Report, The Impact of Public Health Issues on Exclusive Patent Rights, at 2 (2008), *available at* https://www.aippi.org/download/commitees/202/GR202japan.pdf.

128. *Id.*

129. Council for Science and Technology Policy (CSTP), Guidelines for Research Licenses for Intellectual Property Rights Stemming from Government-Funded Research and Development at Universities, at 7 (May 23, 2006), *available at* http://www8.cao.go.jp/cstp/siryo/haihu55/siryo3-2-e.pdf.

130. Koichi Sumikura, *A Consortium for Enhanced Access to Patented Research Tools: Japanese Policy and Proposal of a Novel Scheme*, 2007 PICMET 696, 697; *see also* AIPPI Q105, *Experimental Use as a Defence to a Claim of Patent Infringement*, Resolution (1992), *available at* https://www.aippi.org/download/commitees/105/RS105English.pdf.

131. AIPPI Q202 Working Committee Report, *supra* note 127, at 3; Sumikura, *supra* note 130, at 697 (referring to Working Group on Patent Strategy Plan, Industrial Structure Council, Intellectual Property Committee, Patent System Subcommittee, Issues Concerning Smooth Use of Patented Inventions (2004) (in Japanese)).

132. *See, e.g.*, Institute of Intellectual Property (IIP), *Facilitation of the Use of Patented Inventions*, IIP Bulletin (2007), *available at* http://www.iip.or.jp/e/summary/pdf/detail2006/e18_10.pdf; IIP, *Measures for Smooth Use of Patented Inventions*, IIP Bulletin (2006), *available at* http://www.iip.or.jp/e/summary/pdf/detail2005/e17_09.pdf; IIP, *Exceptions to and Limitation of the Effect of Patent Right*, IIP Bulletin (2005), *available*

2. The Japanese "Bolar" Exception

With respect to regulatory review, the Japanese Supreme Court has ruled that the working of a patented invention in view of clinical trials and the filing of an application for approval of a generic falls within the scope of the exception.[133] Thus, the court created a Bolar-type exception allowing both commercial and noncommercial use under the exception.

IV. Evaluating Current Research Exceptions in the United States, Europe, and Japan

A. The Research Exception as a Zero-Price Liability Rule

Having sketched the current research exception regimes in the United States, Europe, and Japan, we now examine to what extent these research exceptions meet the requirements of a zero-price liability rule regime. We will do so first by testing whether these research exceptions meet the three basic characteristics listed by Calabresi and Melamed: (1) many parties; (2) high transaction costs; and (3) easy, objective valuation of the entitlement.

The research exceptions in the United States, Europe, and Japan all extend to hundreds, if not, thousands of researchers, albeit from different backgrounds, In the absence of a research exception, researchers would be obliged to negotiate licenses with all relevant patent owners, leading to considerable transaction costs—thus, the research exception does remedy high transaction costs. Last but not least, the research exceptions in the United States, Europe, and Japan operate at no cost. So, (objective) valuation is easy and will not create any problems.

All in all, the current research exceptions that exist in the United States, Europe, and Japan in principle fit the characteristics for liability rules set forth in Calabresi and Melamed's seminal paper. Nonetheless, it appears that some of the weaknesses of liability rules remain, as well. The chasm between the inventor/patent owner and users may also arise in a research context. For example, users that call upon the extremely broad research exception in Belgium, covering research

---

at http://www.iip.or.jp/e/summary/pdf/detail2004/e16_05.pdf; IIP, *Issues Affecting Smooth Use of Intellectual Property: Applicability of the Statutory Exception for "Experiment or Research,"* IIP BULLETIN (2004), *available at* http://www.iip.or.jp/e/summary/pdf/detail2003/e15_10.pdf; Sadao Nagaoka & Reiko Aoki, *Economics of Research Exemption* (Hitotsubashi Univ., Working Paper Series of the Institute of Innovation Research of Hitotsubashi University, IIR Working Paper WP#06-04, 2006), *available at* http://www.iir.hit-u.ac.jp/iir-w3/file/WP06-04nagaoka.pdf.

133. Saikō Saibansho [Sup. Ct.] 1999, 1998 (ju) no. 153, 53 SAIKŌ SAIBANSHO MINJI HANREISHŪ [MINSHU] 627 (Japan); *see also* AIPPI Q202 Working Committee Report, *supra* note 127, at 3.

with the patented invention, will likely irritate owners of patented research tools. To prevent such nuisance, legislators and courts should be aware of the need to tailor the research exception carefully in order to prevent such irritation, to create an open environment for further collaboration, to strike the right balance between legitimate interests of patent owners and of third parties, and to prevent any harm to incentives to innovate.[134]

When assessing whether current research exceptions fit in a liability rule regime, we further consider the three requirements for liability rules, set forth by Lemley and Weiser: (1) clear definition: liability rules should be clearly defined; (2) limited scope: the scope of application of the liability rule should be limited to very specific circumstances so that they do not undermine the incentive to innovate; (3) sound governance scheme: liability rules should be complemented by a sound governance scheme to limit the costs of administration and the risk of undercompensation, and to facilitate the enforcement. We will now examine these requirements.

The various US research exceptions and proposals for new exceptions seem to meet the requirements pretty well: they are defined quite clearly and limited to specific circumstances. The common law experimental use doctrine is limited to pure scientific research *on* patented inventions, and research is restricted to noncommercial purposes. "Noncommercial" is interpreted very narrowly, to the extent that even universities would have a commercial objective: educating and enlightening students and faculty participating in research projects is regarded as a research institution's legitimate business objective. Only researchers that carry out their research solely for their amusement, to satisfy idle curiosity, or for strict philosophical inquiry are covered. The US experience shows that meeting the requirements of clarity and limited scope does not always provide an optimal result, as it may lead to a total erosion of the experimental use exception. Even though the scope should be limited to specific circumstances, the scope should not be *too* narrow. An exception that is too narrow does not seem to be particularly useful for achieving the public policy rationale and transaction cost savings provided by research exceptions. We recognize, however, that the narrow interpretation by the court in *Madey* might be the logical consequence of universities and research institutes starting to act more and more like companies filing for patents, actively promoting licensing, and fiercely engaging in patent litigation. Whereas these practices are not so common yet in Europe and Japan, we cannot exclude the possibility that the increased interests in open innovation and, in particular university–industry collaborations, may result in similar changes to the image of universities in Europe and Japan. If so, this may also have some

---

134. Lemley & Weiser, *supra* note 27, at 815.

tightening effects on the interpretation of research exceptions in Europe and Japan. However, because the US experimental use doctrine was formulated more narrowly from its inception than the European and Japanese statutory exceptions, we do not expect the consequences to be as harsh as in the United States.

The research exception proposed in the Genomic Research and Diagnostic Accessibility Act and by the SACGHS is also clear and limited in scope. Not only would it be restricted to the use of patented gene sequences, but, to our understanding, it even would be limited to research in view of the development of genetic tests. Further, the exception would not be applicable to any individual or entity that is *directly* engaged in the commercial manufacture, commercial sale, or commercial offer for sale of a drug, medical device, process, or other product using such patent for or patented use of genetic sequence information. The "direct" engagement with commercial activities is, however, defined more permissively than the hard line set by the CAFC in *Madey*.

The Bolar exception is clearly defined and limited to research that is related to drug development and submission of information to the FDA for approval. However, by extending the exception to the use of patented inventions in preclinical research, the results of which are not ultimately included in a submission to the FDA, the Supreme Court has blurred the boundaries of this seemingly clearly defined exception.[135] The current debate among the members of the Federal Circuit[136] regarding the extension of the safe harbor to post-approval activities evidently exacerbates this situation, which may ultimately require another intervention by the Supreme Court.

The proposed clinical exemption is perspicuously outlined and limited to diagnostic use of gene patents in the context of patient care. Thus, the prospect of gene patents for therapeutic uses could still serve to stimulate investment in this field of research. Moreover, patents related to particular test kits or platform technologies remain untouched. Whereas follow-on innovation for therapeutics, kits, and platform technologies is not hampered, patients would have better access to genetic tests as they would have various providers offering these services. The narrow scope and tailored specific use or application is in line with the first two requirements. However, three members of the SACGHS did not agree with these well-tailored recommendations of the majority.[137] They believe that "statutorily modifying the gene patents system, including the creation of exemptions from liability for infringement upon such patents . . . would be more harmful than

135. *See* Merck KGaA v. Integra Lifesciences I, Ltd., 545 U.S. 193 (2005).

136. *See supra* text accompanying note 96.

137. *Gene Patents*, *supra* note 9, Statement of Dissent from Ms. Aspinall, Dr. Billings, and Ms. Walcoff, 1–2.

helpful to patient access and to the quality of innovative genetic diagnostics."[138] According to the dissenters, a variety of other factors heavily impact which tests are performed and under what conditions. Therefore, they would rather encourage the Department of Health and Human Services to critically evaluate laws, regulations, and policies regarding quality standards, reimbursement policies, and the criteria and requirements of public health programs to ensure that every patient has reasonable, good quality, and timely access to genetic tests.[139]

In Europe, the text of the research exception is similar in most countries (with the exception of Belgium), but its interpretation shows some divergence. Hence, the European exceptions do not seem to fulfill the requirements for a clear definition and a limited scope of application. In Belgium, the modified patent act broadened the nature of the shielded research activities, including research with the patented subject matter. This departure from the common approach is likely to create more uncertainty in Belgium and beyond. Furthermore, the lack of consensus on the final goal (noncommercial and commercial) of protected research may become more and more problematic with the increase of mixed noncommercial/commercial research collaborations.

In Japan, the main problem is that there is no case law on the interpretation of the research exception. Whereas the Japanese Supreme Court has ruled on the applicability of the exception in the light of regulatory procedures, it has never applied or interpreted the exception beyond this regulatory context. The doctrine provides some insights into the interpretation of the first parameter, which seems to be restricted to research *with*, but the role of the second parameter noncommercial/commercial research has not explicitly been touched upon. Therefore, the exact scope and circumstances are unknown.

To summarize, in theory, the research exception seems to meet the core characteristics set forth by Calabresi and Melamed to be recognized as a (zero-price) liability rule. The research exceptions are a powerful model to accommodate follow-on research and innovation and to circumvent high transaction costs. However, the research exceptions sensu stricto do not always work optimally in practice, which may well be because they fail to fit the requirements introduced by Lemley and Weiser, suffering as they do from an unclear (Europe) or too narrow (US) scope. Legal uncertainty, in one way or another, thus seems to stand in the way of an effective and operational research exception.

The third Lemley and Weiser requirement for liability rules concerns a sound governance scheme. The nature of a zero-price liability rule would normally require only a limited institutional governance framework. There is no need

---

138. *Id.* at 1.

139. *Id.* at 2.

to set a price, and the exception simply operates by patent owners remaining inactive. However, in case of doubt the enforcement and interpretation of the research exception is a task left to the courts. In contrast with "regular" liability rules, courts will not have to act as rate regulators.[140]

Lemley and Weiser also give some suggestions for a sound governance scheme: (1) actors should have the necessary expertise and capabilities, (2) actors should be apt to deal with multiple demands for the application of the rule, (3) actors should have the necessary information to design workable access arrangements, (4) actors should be equipped to deal with the dynamic nature of the business relationships, and (5) the necessary oversight of the access arrangement should be provided. The interpretation of the research exception may indeed require some expertise regarding the sector, the technology concerned, and the industry norms. The lack of case law seems to indicate that the governance scheme is failing and that courts are not yet engaged with the application of the zero-price liability rule. Either stakeholders do not invoke the exception or, if the exception is invoked, courts decide the case on different grounds. However, there is no firm evidence that courts and stakeholders are avoiding the research exception. Certainly, it is quite uncommon for patent owners to take researchers or universities to court. Nonetheless, it is also possible that people fear a similar erosion as has occurred in the United States and that research exceptions will be challenged in court in view of the changing environment for innovation and collaboration. Moreover, if courts face multiple demands for the application of the research exception, long litigation procedures may hamper the progress of the research in question.

It is doubtful whether courts have all the necessary information to design workable access arrangements, and whether they are the most appropriate institutional actors to deal with the dynamic nature of business relationships. Cases where many parties are involved require detailed access arrangements, which may be more properly handled in a contract between the parties than by a court. With respect to the requirement for oversight of the access arrangement, a court does not seem to be the most appropriate actor for this role either.

140. However, because the patent owner receives no compensation with this type of liability rule, a careful interpretation of the boundaries of the exception seems even more important than for other liability rule mechanisms. Zero-price liability rules can only operate effectively if the rules of the game are clear, that is, if the scope of the exception is clear. This seems to be the major deficit of all research exceptions described here.

B. The Research Exception in an Open Innovation Context

Research exceptions, crafted as zero-price liability rules, may work relatively well as such. But how do they operate in an open innovation ecosystem? Open innovation in general, PPPs, R & D alliances, consortia, networks, clusters, and R & D joint ventures require close collaboration between researchers at public and private entities. This may complicate the application of the research exception, particularly in countries where the applicability of the exception is based on the final goal (noncommercial/commercial) of the research. In PPPs or university–industry collaborations with a mixed noncommercial/commercial goal, it seems unnecessary to distinguish between researchers associated with a public entity who have engaged in the project with a primarily noncommercial purpose, and researchers associated with a private entity that presumably mainly pursue a commercial purpose. It seems undesirable to exclude the "private firm" researchers from the collaboration and the workings of the research exception. And even though the university researchers may primarily aim at furthering their research, their work often has a mixed noncommercial/commercial purpose, and this is all the more so in an open innovation context.

An ideal research exception in the context of open innovation may well be an exception embracing mixed noncommercial/commercial research. However, such a widely crafted exception entails the risk of the liability rule becoming too broad. An alternative option might be to add an additional criterion, such as the length of time before the innovation is brought to market (distance to market). With such a criterion, precompetitive research in very early stages and with a broad window of opportunities, such as the research carried out under the responsibility of IMI and C-Path, likely would not result in patent infringement whether it is performed by public or private researchers, or whether the ultimate purpose is commercial. The "distance to market" criterion may have disadvantages as well as it requires an indication as to where the line will be drawn. Further reflection is needed on this question.

Conclusion

Our comparative analysis of the research exception in the United States, Europe, and Japan shows that this exception is usually designed as a zero-price liability rule, meeting the essential prerequisites set forth by Calabresi and Melamed. However, having been thoughtfully conceived as a zero-price liability rule, the research exception often fails in practice. This failure is mainly due to the fact that the research exception does not provide the necessary legal certainty, and stops short of an adequate governance scheme. In some exceptional cases, the problem may result from a too-narrow scope as well.

An optimal solution would be to adopt a clear and well-balanced *statutory* exception at the international level. The introduction of such a research exception could be taken up by the Trilateral Co-operation among the USPTO, JPO, and EPO, as each of the three jurisdictions is struggling with this issue in one way or another. However, the priorities of the Trilateral Co-operation are currently focused on other important projects[141] and prompt harmonization in this field from the side of the Trilateral Co-operation is quite unlikely.

Therefore, the introduction of a national statutory exception[142] or the harmonization of existing national research exceptions seems like a more realistic solution.[143] The situation in the United States could indeed be improved by introducing an explicit statutory research exception limited to specific circumstances, but crafted more broadly than the extremely restrictive common law experimental exception provided in *Madey*. The situation in Europe and Japan could be improved by policy choices extending the research exception to include mixed noncommercial/commercial research activities, and carefully defining the delicate borderline between noncommercial and commercial research with respect to clinical services and research tools. It would be useful if the European Commission and the Japanese CSTP would issue guidelines in this regard. However, the research exception has been "a supposititious child" and has not received a lot of attention by policymakers in the United States, Europe, or Japan. Legislators and policymakers have generally ignored suggestions from advisory committees. The present chapter invites policymakers to renew their interest in this topic in the context of the Trilateral Co-operation, or in the process of patent reforms in the three different jurisdictions

If international and national lawmaking is unsuccessful, legal uncertainties related to the scope of the research exceptions may be overcome on the initiative of patent owners. Patent holders may settle the lack of clarity in the legislative framework in day-to-day "open innovation" practice and management of

141. In particular projects to address workload problems through maximizing reuse of work results from the other offices. *See* The Trilateral Co-operation, Achievements of the Trilateral Offices in 2010 (2011), *available at* http://www.trilateral.net/news/TrilateralConference2010/trilateralachievements.pdf.

142. Recently, Australia indeed adopted a statutory research exemption. *See Australian Amendment*, *supra* note 11.

143. Several organizations have repeatedly insisted on the need to further define and potentially harmonize the scope of research exceptions in national patent laws. *See, e.g.*, ESHG, *supra* note 123, at S7.

international R & D alliances and PPPs by including specific research use provisions in their project agreements; indeed they are already doing so in some sectors and jurisdictions.[144] If patent holders agree to apply such provisions both to project partners and third parties, they will establish a *contractual* zero-price liability rule.

Extending research use immunity to third parties, and thus voluntarily constructing a zero-price liability mechanism, seems particularly justified in cases where the project is funded by public resources. Patent owners may even prefer this approach to the statutory zero-price liability rule, because they will be able to better *control* the scope of the research use umbrella. In doing so, they bring a voluntarily constructed zero-price liability rule closer to a property rule. However, having to negotiate such research use licenses would be a setback from the transaction cost savings offered by a statutory research exception. Moreover, such research use provisions would only cover patents held by owners involved in the project concerned. It would not cover patents held by entities who do not participate in such projects. Subsequently, blocking patents held by outsiders still could hold up an entire research project. Model license clauses for research use of inventions held by third parties may be used by PPPs to settle research use terms for its participants. This might solve the problem partially, but patent thickets and refusals to license will remain potential "deal breakers."[145] In addition, negotiating such research use licenses may require quite some work from the PPP administrator. Such an input will generally only be possible for PPPs with a separate executive office.

In conclusion, open innovation initiatives would benefit from an international *statutory* research exception, covering mixed noncommercial/commercial

144. A similar approach has also been suggested by the NIH and AUTM. "[NIH, Public Health Service (PHS)] believes that it is important for funding recipients and the intramural technology transfer community to reserve in their license agreements the right to use the licensed technologies for their own research and educational uses, and to allow other institutions to do the same, consistent with the Research Tools Guidelines." NIH PHS, Best Practices for the Licensing of Genomic Inventions, 70 Fed. Reg. 18,410, 18,415 (2005), *available at* http://www.ott.nih.gov/pdfs/70FR18413.pdf; NIH PHS, Principles and Guidelines for Recipients of NIH Research Grants and Contracts on Obtaining and Disseminating Biomedical Research Resources, 64 Fed. Reg. 72,090, 72,095 (1999), *available at* http://ott.od.nih.gov/pdfs/64FR72090.pdf; AUTM, In the Public Interest: Nine Points to Consider in Licensing University Technology, Point 1, at 2 (Mar. 6, 2007), *available at* http://www.autm.net/AM/Template.cfm?Section=Nine_Points_to_Consider; *see also* European Commission, Guide to Intellectual Property Rules for FP7 Projects, Version 3, 21, *available at* ftp://ftp.cordis.europa.eu/pub/fp7/docs/ipr_en.pdf.

145. Moreover, the recent decision of June 6 from the U.S. Supreme Court in *Stanford v. Roche*, 131 S. Ct. 2188 (2011), *available at* http://www.supremecourt.gov/opinions/10pdf/09-1159.pdf, shows that a small variation in an agreement can make a major difference in terms of patent ownership. So, parties should be very careful in drafting licenses for research use.

research, but restricted to early stage ("long distance to market") research. In the absence of such an international exception, a statutory or *contractually* constructed zero-price liability rule, encompassing mixed early stage research, could also help to accommodate collaborative research. Alternatively, and far less adequately, research partners may make use of bilateral research use licenses. The least attractive solution consists of trusting the present policy of "rational forbearance" by patent owners.

For the time being, it is quite likely that most researchers participating in regional or international collaboration projects stretching across Europe, the United States, and Japan will have to take refuge in the defective statutory research exceptions in Europe and Japan, and the de facto research exemption in the United States. These exceptions, which are largely based on "rational forbearance," unfortunately offer only a false sense of security to open innovation collaborations. The time has come to shape an international patent law exception to provide clear and robust statutory immunity for research, and to spark contractually constructed zero-price liability rule mechanisms shielding open and collaborative research from infringement liability.

14 EXHAUSTION AND PATENT RIGHTS

Christopher Heath*

Introduction

The issue of exhaustion of patent rights can be approached from various angles—legal, economic, or political. My approach in this chapter is primarily a legal one, while I will conclude with some remarks on the economic and political issues.

The legal issues I intend to touch upon in this chapter are:

I. Terminology
II. Doctrinal concepts
III. Principles of domestic exhaustion
IV. Exhaustion and reproduction
V. Exhaustion and repair
VI. Exhaustion and contract, and
VII. International exhaustion

The above structure merits a little explanation. Often, and particularly in the context of international trade, exhaustion and parallel imports are mentioned in one breath. Yet the possibility of parallel importation of patented products is a consequence of the domestic exhaustion regime a country has adopted. Thus, prior to touching upon the issue of exhaustion in the international context, it needs to be established to what the term "exhaustion" is meant to refer (Section I), and what different doctrinal concepts are behind it (Section II). This essentially lays the ground for the more complicated issues to follow. Section III explains the three doctrinal concepts of "implied license," "exhaustion," and "first sale" in greater detail and in the context of how these concepts historically developed. Section IV

* Judge, Boards of Appeal, European Patent Office; Former Head of the Asian Department, Max Planck Institute, Munich.

discusses exhaustion in the context of self-reproducing material. Once this basic framework has been explained, I then move on to the two issues that have been the most important ones in connection with the domestic exhaustion of patent rights.

The first is the issue of repair versus reconstruction. Exhaustion allows for an article to be repaired, but not to be reconstructed. Section V therefore discusses the different approaches a number of national courts have taken to the repair/reconstruction question, and to what extent "exhaustion" is a helpful concept here in the first place. Interestingly enough, the English courts—otherwise following the implied license approach—borrowed the concept of exhaustion in order to distinguish repair from reconstruction, whereas the German courts found some sympathy for the implied license doctrine in order to determine to what extent a manufacturer might have consented to certain acts of repair. Finally, the US courts have approached the question from a competition law perspective that helps us to understand that often enough, by prohibiting acts of repair, manufacturers try to extend their patent monopoly beyond what they have contributed to society.

The second issue, addressed in Section VI, has become controversial in connection with issues of domestic exhaustion: the possibility of abrogating exhaustion by contractual means. Conceptually, this possibility is part of the implied license doctrine, yet English courts have been cautious, accepting contractual limitations only once they have been brought home to the actual user, proof of which mostly failed. As the exhaustion doctrine would not allow contract to override the limits of patent law, cases of contractual limitations have been few and far between. Most cases on contractual limitations have thus been argued before US courts, which, interestingly enough, have little hesitation in accepting such limiting clauses despite anticompetitive effects. Finally, Section VII deals with issues of international exhaustion, discussion about which has been particularly vivid after the conclusion of the Agreement on Trade-Related Aspects of Intellectual Property Rights (TRIPS Agreement).[1]

In setting out the various angles of the exhaustion issue, I make extensive use of quotes from German and Japanese decisions, and from Dr. Josef Kohler's book on patent law. A good many of these quotes have not been published before in English, and would only be accessible in German or Japanese. Even where they are in English, they could not be easily accessed on the Internet. I thus deem it appropriate to quote these sources as firsthand research material that may be more useful than my own, inevitably biased, reading of these sources.

1. Agreement on Trade-Related Aspects of Intellectual Property Rights, Apr. 15, 1994, 33 I.L.M. 1197 (1994) [hereinafter TRIPS Agreement].

I. Terminology

The term "exhaustion," when referred to in the context of intellectual property rights, has nowadays been adopted in many languages. The French speak of *épuisement*, the Spanish of *esgotamiento*, the Portuguese of *esaustão*, the Japanese of *shomō riron* or *shojinron*, and the Germans of *Erschöpfung*. Dr. Josef Kohler, who might have been the first to lay the doctrinal foundations of the exhaustion concept, did not come up with a proper term, yet in a 1902 decision the German Imperial Supreme Court (Reichsgericht) used the term "*Konsumtion*,"[2] meaning consumption. The term—and this is sometimes not well understood—does not refer to the existence of the intellectual property right as such, but to the limits in exercising such right for a specific product (not the right as such is exhausted, but the possibilities of exercising it with respect to certain products).

II. Doctrinal Concepts

The concept of exhaustion relates to a limitation in exercising intellectual property rights. Developing such a limitation became necessary in order to reconcile the concept of exclusivity granted under patent and copyright law (much less so for trademarks) with common sense and the requirements of modern commerce. Patent and copyright laws grant the owner the exclusive rights of commercial exploitation. These relate not only to the production, but also to the commercial distribution of protected products. Taken literally, the right owner could not only control the production and first sale of protected products, but also all subsequent acts of resale, lending, rental, export, or import of the protected products. Common sense tells us that this can hardly be so, as it would suffocate trade in these products.

In determining and justifying the limits of such exclusive rights, a number of concepts have been developed. English common law developed the doctrine of "implied license," making products protected under intellectual property rights subject to the same rules as products protected under property rights. In the latter case, sale of a product by its owner transferred all proprietary rights to the purchaser, unless certain rights were withheld (for example, retention of title). The same applies to products protected under intellectual property rights: the purchaser is entitled to deal with these products as he sees fit, unless the seller has retained any rights. This concept is based on the principles for the transfer of title and thus on principles developed under contract law. It has the advantage of applying the same rules to the transfer of real and intellectual property, of making the buyer immediately aware of any restrictions regarding the disposal of the

2. Entscheidungen des Reichsgerichts in Zivilsachen [RGZ] 50, 362 ("Duotal").

purchased goods, and of putting the onus of "bringing home" such restrictions on the seller.[3]

Josef Kohler, in his book on patent law published in 1900, takes a different approach. Kohler found it unsatisfactory that the seller of protected goods should be in a position to limit their subsequent commercial distribution at all. To him, the free flow of goods required a notion of absolute, inherent limits of patent law. Kohler's ideas were developed against a different background. For one, until 1900, German law could not draw upon common principles of contract law, as these simply did not exist. Only the introduction of the German Civil Code harmonized and finally superseded the legal rules applied by the multitude of German Länder: the French Civil Code, the Prussian Common Country Law, and the principles of Roman and common Germanic law. Furthermore, impediment to free trade was one of the main arguments of those who opposed the introduction of a common patent law. All too fresh were the memories of the various obstacles in cross-border commerce between the German Länder, unified only in 1871.[4] To Kohler, only a doctrine that based free trade on absolute limits of patent law could provide a reasonable degree of certainty.

Kohler thus came up with two theses and one conclusion. His first thesis concerned the rationale of patent law. Patent law was meant to grant the patentee a reward for his inventive efforts by the allocation of a monopolistic right that allowed the exclusion of all others from the commercial exploitation of the patented invention. Kohler's second thesis related to the connection between the forms of commercial use. The different forms of patent use (production, advertisement, sale, rental, lending, export, import, etc.) could not be viewed in isolation, but were intrinsically linked with each other. In fact, they could be viewed as different expressions of one common right.

Based on these two theses, Kohler's conclusion was this: to the extent that the patent gave the patentee the opportunity for a reward by excluding others, this reward could be obtained once for each product. Once the patentee himself had used the patented invention with respect to a certain product, he could no longer exercise his patent right for this specific product. This limitation was not confined to the form of use the patentee had chosen, but, because the different forms of use flowed from each other, this had an effect for all subsequent acts of commercial use of the product. If the patentee had sold a patented article, he could no longer exclude third parties from making any sort of commercial use of the article (e.g., by its sale, rental, or export). This connection between the forms of

---

3. For details, see Section III.A *infra*.

4. This is particularly mentioned by F.K. Beier, *Wettbewerbsfreiheit und Patentschutz*, 1978 GEWERBLICHER RECHTSSCHUTZ UND URHEBERRECHT [GRUR] 123, 129 (Ger.).

use was justified by Kohler's underlying thesis of the patent's function: if the patentee was granted one chance of obtaining a reward for each patented product, the first act of commercial use satisfied this requirement, and the patentee could not be entitled to exercise his patent rights with respect to this product, whatever form of use was made. The above rules are not only relevant in determining the lawfulness of certain acts of use, but have also gained some importance in making a distinction between repair and reconstruction. Although the patentee cannot prevent any third party from prolonging the life of a product by repair, the rules of exhaustion can never justify a reconstruction that in effect equals the production of a patented article without permission.[5]

Kohler's concept also indicates a fact that is mostly disregarded nowadays:[6] if exhaustion is viewed as an inherent and often unwritten limit of the patent right, it cannot be regarded as an exception to the patentee's overall rights. This has wide-ranging implications when it comes to the burden of proof. An exemption has to be proven by the person relying thereupon. If, on the other hand, subsequent acts of commercial sale are regarded as lawful per se and covered by the patent's inherent limitations, it is the patentee who has to argue that there should in fact be no exhaustion. In other words, third parties are allowed to deal in patented products as they see fit unless the patentee can prove that she had no chance of obtaining a monopolistic reward from the patent. The existence of the patented products on the market are prima facie evidence that she had, and that the products are in lawful circulation; one should not forget that there is no good faith rule when dealing with counterfeit products, and that this may disproportionately affect parties with limited information and limited bargaining power.

Case law in the United States is neither here nor there on the subject of exhaustion. The so-called "first-sale-doctrine" seems to imply that the first sale of a patented article exhausts the patentee's right to exclude for this very article, but differing from Kohler's approach, the case law finds it permissible to invoke the patent right if the sale has been made conditional, an issue further discussed below. To that extent, case law in the United States comes closer to the doctrine of implied license[7] that in fact is also terminologically used in some decisions,[8]

---

5. It is interesting to note that even English courts recognize the doctrinal importance of exhaustion in this field. It is discussed in extenso in *United Wire Ltd. v. Screen Repair Services (Scotland) Ltd.*, (2000) 4 All E.R. 353 (H.L.), *available at* http://www.publications.parliament.uk/pa/ld199900/ldjudgmt/jd000720/wire.htm. For details of Kohler's exhaustion doctrine, see Section III.B *infra*.

6. *See, e.g.*, *Karate*, Case X ZR 61/98, Bundesgerichtshof [BGH] [Federal Court of Justice] Dec. 14, 1999, Gewerblicher Rechtsschutz und Urheberrecht, Internationaler Teil [GRUR Int] 635, 2000 (Ger.).

7. United States v. Univis Lens Co., 316 U.S. 241 (1942).

8. Wang Labs. v. Mitsubishi Elecs., 103 F.3d 1571 (Fed. Cir. 1997).

but not, as in the United Kingdom, a rule,[9] but rather as an exception under the doctrine of equitable estoppel.[10] And in contrast to the implied license doctrine, the US doctrine stops at the national border.[11]

III. Principles of Domestic Exhaustion

A. The English Common Law Approach

As stated in *National Phonograph Co. of Australia v. Menck*, under English common law—

> [I]t is open to the patentee, by virtue of his statutory monopoly, to make a sale *sub modo*, or accompanied by restrictive conditions which would not apply in the case of ordinary chattels; . . . the imposition of these conditions in the case of sale is not presumed, but, on the contrary, a sale having occurred, the presumption is that the full right of ownership was meant to be vested in the purchaser while . . . the owner's rights in a patented chattel would be limited, if there is brought home to him the knowledge of conditions imposed, by the patentee or those representing the patentee, upon him at the time of sale.[12]

The *Menck* decision in the first two instances before the Australian courts also provided a thorough review of the reasons for such doctrine—

> The foundation of patent law is the Statute of Monopolies passed in 1623, which, after declaring grants of monopolies to be void, made an exception in favour of grants of privilege for 14 years "of the sole working or making of any manner of new manufacture" to the true and first inventors. For nearly three centuries after this Act was passed it appears to have been the accepted law that when once the patentee by himself or his licensees had by putting the invention in practice produced an article and disposed of it to some member of the public, the making use of the article so disposed of, by anyone whomsoever, was not an infringement of the monopoly of "working or making" the new manufacture.[13]

---

9. Betts v. Willmott, (1871) 6 L.R. Ch. 239, 245 (Eng.).

10. Bandag, Inc. v. Al Bolster's Tire Stores, 750 F.2d 903 (Fed. Cir. 1984).

11. It does not for copyright, though: Kirtsaeng v. John Wiley, 132 S. Ct. 1905 (2013). For details, see Section III.C *infra*.

12. Nat'l Phonograph Co. of Austl. Ltd. v. Menck, (1911) 28 R.P.C. 229 (P.C.) 248 (appeal taken from Austl.).

13. *Menck*, 28 R.P.C. at 234 (internal citations omitted).

In another case from the United Kingdom, the reason for exhaustion was explained in the following terms—

> If a patentee sells the patented article to a purchaser, and the purchaser uses it, he, of course, does not infringe. But why? By reason of the fact that the law implies from the sale of a licence given by the patentee to the purchaser to use that which he has bought. In the absence of condition this implied licence is a licence to use or sell or deal with the goods as the purchaser pleases. If the patentee sells, imposing no restriction or condition upon his purchaser at the time of sale, he cannot impose a condition subsequently by delivery of the goods with a condition indorsed upon them or upon the package in which they are contained. Unless the purchaser knows of the condition at the time of the purchase, and buys subject to the condition, he has the benefit of an implied licence to use free from condition. But suppose the purchaser buys, not from the patentee, but from a licensee, the patentee may have attached to his licence any conditions he pleased, and if he did attach a condition, then upon principle it seems to me that nothing (so far as licence as distinguished from estoppel is concerned) can turn on the question whether the purchaser from the licensee knew of the condition or not. If a person innocently uses a patented invention, not knowing there is a Patent, he is none the less an infringer, and if a person innocently buys a patented invention from a licensee, and uses it, not knowing that there are limits on the licence, I conceive that he is equally in infringer.[14]

In other words, it was clear that the patent right should be treated no differently than the property right in the object as such. If title was passed unconditionally, the purchaser could do as he saw fit. If not, the purchaser had to abide by the terms of the contract, or else would infringe the patent.

Finally, English case law is unequivocal on the burden of proof:—

> In a suit to restrain the sale of a patented article, it is incumbent on the Plaintiff not only to prove the sale, but to prove that the article was not made by himself or his agents.... The onus is thrown on Mr. Betts of proving two things. One of them, it appears to me (though I have not heard the argument for the Defendant upon it), he has proved, namely, that there has been a use of his invention. But there is another question, namely, whether it has been an unauthorized use. It appears to me that, the onus being thrown on Mr. Betts to prove that, he has not discharged

14. Badische Anilin und Soda Fabrik v. Isler, (1906) 1 L.R. Ch. 605, 610 (Eng.).

himself of it. He has to show that this thing which the Defendant is using is not manufactured by him.[15]

B. Josef Kohler's Doctrine of Exhaustion

The doctrine of exhaustion, as applied in many jurisdictions, has been developed by Josef Kohler in his magnum opus "Deutsches Patentrecht." The Master deserves to be quoted himself—

> 5. Connection Between Forms of Use
> a. General
> § 178.
> The different forms in which an invention can be used enjoy a certain independence. Yet they do not lack any connection, and the connections manifest themselves in important legal précis.
> In particular, the following basic principles apply:
> I. Every patent grants all forms of exploitation rights to the extent possible according to the type of invention: It is improper to grant a patent with the mere right of use, another with the mere right of manufacture. It is equally impossible to subsequently relegate the patent to such step. It is only possible to grant third parties a licence for one or the other form within the geographical limits the patent may be used.
> II. The connections between the forms of use have certain consequences:
> A. The use and distribution of patented products, be it that the product or the manufacture is patented, is always in connection with the manufacture; it is a continuation of the latter, it is a further economic development of the patent exploitation that lies in the manufacture.
> B. The domestic use and distribution of goods can be the continuation of a previous act of domestic distribution; it is so if the use and subsequent distribution is a legal consequence of a previous distribution.
>
> From A. follows this:
> 1. Had the object been domestically produced in a legitimate manner, then it can also be domestically marketed and domestically used. Thus:
> a. Had it been produced by a licensee, the object can be sold and used by third parties even once the licence ceases to exist; the same applies in cases of manufacture by a person temporarily entitled to use the patent, if the patent is subsequently transferred to another;

15. Betts v. Wilmott, (1871) 6 Ch. App. 239, 245 (Eng.).

b. If the goods were legitimately produced by a prior user, the prior user also has the right to market the goods and third parties have the right to use them;
c. Once the product has been legitimately produced, any subsequent purchaser can use it, even a purchaser who cannot trace back his right to a person deriving his right from the patentee; in particular, the purchaser in good faith who obtained ownership due to the principle that "hand must keep hand." It applies to the possessor in good faith, yet also to the possessor in general; also the possessor in bad faith who uses a machine may commit an offence against the property, yet does not do so against the patent;
d. The purchaser of goods lawfully produced is covered by patent law even in cases where the purchaser has sold the goods upon certain restrictions; if the goods come into commerce, such restrictions can be imposed on the purchaser as contractual obligations, and they can also be imposed on subsequent purchasers to the extent that upon purchase, he has taken on the same obligations or to the extent that he has purchased in bad faith, e.g., by exploiting the ignorance of the owner's representative who consequently does not impose such restrictions; such restrictions, however, cannot be imposed under patent law (unlike the American view that assumes that such restriction has effect under patent law, at least towards the purchaser in bad faith).
e. Once the goods have been manufactured in accordance with patent law, these can be garnished by creditors and be subsequently sold on the basis of a lawfully obtained right of garnishment. And the same applies once the patentee goes bankrupt and the patented goods are sold in bankruptcy proceedings: the more so since creditorship not only affects property law, but also patent law; but even if creditors renounced obtaining the patent right, they would still have the right to dispose of the patented goods already existing, and any subsequent purchaser may use them. The same applies if the landlord uses his statutory right of garnishment: sale of the bond by the landlord not only has the effect of disposing the real object as such, but also in its quality as an expression of the invention.
f. Once the goods have been legitimately produced by the person licensed for the certain geographical area within the country, they can be sold and used beyond this circuit, provided that use and distribution occurs domestically, in other words in the country that covers the area of licensing.
g. If the goods have initially not been produced in accordance with patent law, they may still be further distributed and used once the

patentee has received complete compensation from such unlawful manufacture, and thus has been compensated with respect to all consequences from such manufacture and distribution.

2. The above principle cannot be abrogated by the patentee: a condition to the extent that a lawfully manufactured object may only be distributed in a certain way, or only be used by a certain person, is without effect; it would lead to an unbearable burden of trade. Once the entitled person has brought the object into commerce, it is open to free distribution and life would not tolerate any form of interference. Certainly it is possible to impose contractual limits; yet such a thing only has contractual effects, as has been explained above. This even applies further once the entitled person produces patented objects, but has not yet distributed them. If the goods are stolen, the laws of theft apply: the possessor of the goods may be responsible as a thief or receiver of stolen property, yet not as a patent infringer, even though he uses and distributes the goods. And if the goods become garnished, then the right of garnishment applies and the patentee may not object to such execution. This requires that the goods have already been produced, i.e., made ready.

   It can thus be a contravention against obligations under a contract if the inventor agrees with wholesaler A that A shall sell the goods to a number of persons indicated, and A acts contrary to such orders. But such a contravention can never be contrary to patent law; the same applies if a person acted as a retailer in order to indirectly supply the patented goods to those whom the patentee would like to withhold them from.

   The whole principle A only applies to the patent country. It does not apply in relation to two different patent countries, since the patent rights in two different countries are different patent rights, they do not stream from one and the same right.

Thus:

a. Goods manufactured lawfully under the patent laws of one country may thus not be distributed or used in another country, or only if distribution and use is permitted according to the patent relations of this other country.
b. This applies even in the case where the patents in these two countries are granted to the same person. It applies if they have been granted to one and the same person and subsequently end up with different owners; it also applies if they remain in the hands of the same person. Thus, if the patent of countries A and B belong to the

same patentee, and goods manufactured in country A are distributed in country B, then they may not be distributed in country B on the basis of the rights that stream from the patent in country A and the manufacture under this patent, but only on the basis of rights that flow from the patent in country B.

Thus follows this:

α. If the entitled person has granted a licence for country A, due to such licence the products manufactured in country A may be used and distributed in country A without restriction, yet not in country B; in country B only based on a special licence granted for country B.

β. It can thus happen that if the same inventor holds patents in countries A and B, and licenses these in country A to one person, and in country B to another, the first person may produce and distribute in country A, the other person in country B. In no way may products from the other country be imported and distributed or used, unless it were that the licensee in country A also has the licence to import and distribute these to country B, and licensee in country B had the same right regarding products from country A, and that importation and distribution are based on such licence. Without such licence or permission of the licensee, goods produced in country A may not be imported and distributed in country B, and vice versa, even if the patent rights in A and B belong to the same person. However, the licence can indeed be granted in such a way that every purchaser of goods produced in country A should have the licence to import them to country B, and vice versa.

γ. What has been developed above applies with particular force once the patents in A and B no longer belong to the same person, but due to sales or other real obligations have been separated or at least split in different spheres of interest.

This far the consequences from principle A.

Principle B has to be considered in particular when products manufactured abroad are imported based on a licence. Once imported, every further development of the product's legal fate is covered by the first lawful act of distribution to the extent that such act of distribution legally comprises the subsequent use and distribution; it is thus covered if the first lawful distribution was a distribution for sale. It is not covered if the goods are imported based on a licence, but only a licence for use, not a licence for sale. In this case, it is not even covered if someone obtains ownership of the product in good faith. Similarly, it is not covered in the case when someone obtains infringing products in good faith that the goods were manufactured in accordance with patent law. If, however, goods have been imported for sale and in accordance with the patent right, then their subsequent use and distribution is permissible under patent law even if these goods change hands, be it in good or bad faith. The tort of the bad faith possessor is a tort, but certainly only of a real right, not a contravention against patent law.

Previously, the consequences A and B based on the connection of forms of use were explained by an implied licence, in other words, by the fact that the first distribution of a product also granted a licence for everything else; I also was previously of this opinion, yet had later on reason to distance myself from some of the consequences of this doctrine.

From the above, it becomes self-explanatory that the whole doctrine of the granting of an implied licence in this connection is incorrect, unnecessary and insufficient. The right of distribution and use based on the above principles becomes self-explanatory; it stands to reason without the assistance of a licence, and it applies even in cases where a licence would not be sufficient, especially in the case of purchase by third parties in good faith; also in cases of compulsory sales of products and in cases where goods produced on the basis of a licence are distributed beyond the licensed territory after the licence has been revoked. The questions that were raised in connection with the licensing doctrine thus become moot.

Still, the doctrine of implied licence is important and correct in another context, namely to the extent that one form of use does not already comprise another under the principles developed above. This happens frequently: The lawful manufacture of a machine covers distribution and use that naturally follow from such manufacture. If, however, the machine is used as part of a patented process, the use of the machine, lawful in principle, of course does not cover the process of which the use of the machine forms a part. In other words, the purchase of the machine only covers the use of the machine *in abstracto*; it does not cover the use of the machine to the extent that this would form part of a patented process. Yet in this context, the question occurs whether the distribution of a machine by the person who concurrently owns the process patent also comprises a licence for the process to the extent that the machine mainly or exclusively is used for this certain process. This question now should be dealt with under the principles of licence, both for the machine and for other tools or apparatus which are preferably used for certain purposes. Such a licence would

> be contained within the limits specified. While it could be assumed that the transfer of the licence is an undefined and general one, and should apply to anyone obtaining property or good faith possession of the machine or the tool, yet the grant of the licence would amount to a partial transfer of the invention rather than an exploitation of the latter. It would thus only go as far as the right of the licensor goes.

From this follows that:

> a. If the prior user has a right of manufacture, yet no right for the process, he cannot grant a licence for the process;
> b. If the transferor of the machine has a right for the process within certain geographical limits, he can transfer the licence only within these limits;
> c. If the transferor of the machine has only a temporarily limited right for the process, he can transfer the right only within these timely limits.[16]

As one can perceive from Kohler's diction, he does not write *on* law, he simply writes law. He had a complete legal system in his head, and as such was not only the archetype of a German professor, but also the opposite of a common law lawyer; the latter develops law in a problem-oriented manner, whereas Kohler is system-oriented, apart from the fact that his knowledge of foreign jurisdictions was legendary. Kohler's construction of patent exhaustion mainly has three repercussions.

First, the patent right cannot be invoked in cases of a *sub modo* or conditional sale, nor can the patentee avoid exhaustion by, for example, renting instead of selling, or otherwise by retaining title. Kohler's is a doctrine of first marketing rather than first sale. Contravening contractual limitations may lead to a cause under contract law, but not under patent law; the goods used in contravention of a limitation would not be considered infringing goods.

Second, and due to the connection of the different forms of use, secondary acts of use such as offering for sale, advertising, etc. can only infringe if they relate to the preparation or exercise of an infringing sale. This is of importance where, for example, the offer for sale relates to sales abroad. According to Kohler, a domestic offer for sale of goods abroad to a purchaser abroad cannot infringe.

Third, as mentioned above, Kohler does not perceive patent exhaustion as an exception, but as an in-built limitation of the patent right.

16. Josef Kohler, Handbuch des deutschen Patentrechts [Handbook of German Patent Law] 452–59 (1900) (internal citations ommited).

An example of the application of Kohler's doctrine is the Japanese Supreme Court decision "BBS Car Wheels"—

> According to Sec. 68 Japanese Patent Act, a patentee shall have the exclusive right to work the patented invention commercially. In the case of a product invention, "working" means the acts of manufacturing, using, assigning, leasing, importing, or offering for assignment or lease of the product (Sec. 2(3)(i) Patent Act). Any person who has purchased products covered by a patented invention, either from the patentee or from a licensee with consent of the patentee, commits an act of use by re-selling these goods to a third party. Equally, such a third party who has obtained the patented products in such a way is, at least formally, working the patented invention when further leasing it, and would thus be liable for patent infringement. However, if patented products are sold domestically, either by the patentee or with his consent, the patent is deemed exhausted because it has fulfilled its purpose. The patent does not give rights to subsequent use of the patented product by acts of transfer or lease. First, patent law has to be understood as balancing the interests of invention protection, and the public benefit of society at large. Next, if a tangible object is transferred, the rights are obtained by the transferee, and the transferee obtains those rights that were originally vested in the transferor. Also insofar as patented products are distributed on the market, the transferee obtains an object from the patentee whose exercise of the right suggests that the right in further acts of re-sale has been transferred as well. If with respect to any acts of marketing patented products, the patentee's consent were necessary each time a transfer occurs, the free movement of goods would be seriously impeded, the smooth distribution of patented products would be hampered and as a result, the interests of the patentee himself would suffer. This would run counter to the purpose of the Patent Act "to encourage inventions by promoting their protection and utilisation so as to contribute to the development of industry" (Sec. 1 Patent Act). Finally, by making the invention available to the public, the patentee will have the opportunity to obtain a reward from selling the product or granting a licence for the use of the patent and thereby obtain a licensing fee. In order to protect the financial interests of the patentee who has made his invention public, it would not seem necessary to give the patentee or the licensee rights beyond the first act of marketing, as the patentee would then obtain an unnecessary double reward in the course of further distribution.[17]

---

17. *BBS Car Wheels*, Saikō Saiansho [Sup. Ct.] July 1, 1997 (Japan), *translated in* 29 Int'l Rev. Intell. Prop. & Compet. L. 331 (1998).

It is also irrelevant that the patentee had never contemplated a certain use of its product, as a subsequent Japanese decision has clarified. In the Japanese *Acycrobil* case,[18] the defendant had obtained marketing approval for a generic drug containing the same active formula as the one described in the plaintiff's (Glaxo Wellcome) patent. Marketing approval would expire unless the products were actually brought on the market within six months from the date of marketing approval. The defendant thus purchased the plaintiff's pharmaceuticals containing the active ingredient, distilled the latter, and used this basic patented substance for producing the generic drug. As the patentee had received the economic benefits deriving from the patent by the act of first marketing, it could no longer control further acts of commercial exploitation by the purchaser. The action was thus dismissed. The court applied the remuneration doctrine and cited the Supreme Court's *BBS* decision in this respect.

C. The US First Sale Doctrine

The principle of the first sale doctrine was laid down by the US Supreme Court in *Adams v. Burke* (1873).[19] As observed above, the doctrine partly rests on contractual considerations of the implied license doctrine, but also invokes issues of public policy as stressed by Josef Kohler. This legal muddle-puddle (three judges were dissenting, voicing serious doubts about public policy considerations other than the grant of monopolies as such) continues until this day and seems to give rise to considerable headaches when applied in practice.[20] Subsequent decisions on contractual limitations (see below Section VI) seem to indicate that, at least in the recent past, the courts do not view the first sale doctrine as a cornerstone of public policy, but rather of contractual arrangements.

The relevant passage of *Adams v. Burke* reads as follows:

> We have repeatedly held that where a person had purchased a patented machine of the patentee or his assignee, this purchase carried with it the right to the use of that machine so long as it was capable of use, and that the expiration and renewal of the patent, whether in favor of the original patentee or of his assignee, did not affect this right. The true ground on

---

18. *Acycrobil*, Tōkyō Chihō Saibansho [Tōkyō Dist. Ct.] Jan. 18, 2001, *aff'd*, Tōkyō Kōtō Saibansho [Tōkyō High Ct.] Nov. 29, 2011, *reprinted in* 15 L. & TECH. 83 (2002).

19. Adams v. Burke, 84 U.S. 453 (1873).

20. In the context of copyright, the US Supreme Court on Dec. 13, 2010 rendered an inconclusive 4-4 verdict in *Costco Wholesale Corp. v. Omega*, 541 F.3d 982 (9th Cir. 2008), *aff'd by an equally divided court*, 131 S. Ct. 565 (2010) (mem.) (per curiam).

which these decisions rest is that the sale by a person who has the full right to make, sell, and use such a machine carries with it the right to the use of that machine to the full extent to which it can be used in point of time.

The right to manufacture, the right to sell, and the right to use are each substantive rights, and may be granted or conferred separately by the patentee.

But in the essential nature of things, when the patentee, or the person having his rights, sells a machine or instrument whose sole value is in its use, he receives the consideration for its use and he parts with the right to restrict that use. The article, in the language of the Court, passes without the limit of the monopoly. That is to say the patentee or his assignee having in the act of sale received all the royalty or consideration which he claims for the use of his invention in that particular machine or instrument, it is open to the use of the purchaser without further restriction on account of the monopoly of the patentees.[21]

D. Comments

It is hard to say which of the above doctrines is "better" or "worse." One should first of all be aware of the different consequences of these theories, and the different scope a patent is given thereby. As we can see below, the doctrine of patent exhaustion has fewer problems when dealing with conflicting contractual clauses because patent limits take precedence over contractual arrangements. On the other hand, in international trade, the implied license doctrine as applied by the English courts may be politically less controversial and easier to apply.[22]

IV. Exhaustion and Reproduction

The above theories on exhaustion are unanimous in the view that exhaustion applies to a specific product that has been marketed, yet that the effects of exhaustion for this one product cannot have an effect on the marketing of other products that fall under the claims of the patent. In other words, exhaustion of the right to exclude applies and is limited to the specific product that has been marketed. One may wonder, however, what should happen in cases where the exhausted product is capable of multiplying itself, thereby producing progeny that also fall

---

21. *Burke*, 84 U.S. at 455–56.

22. Such a solution is therefore advocated by Hermann Cohen Jehoram, *Prohibition of Parallel Imports Through Intellectual Property Rights*, 30 INT'L REV. INTELL. PROP. & COMPET. L. 495 (1999), in connection with international trade.

under the patent claim. This issue is of relevance for patents on genetically modified seeds or animals. If, say, a claim reads: "A glyphosate-tolerant plant cell comprising a DNA-molecule [with the following specifications]"[23], or "A transgenic human animal that carries in the genome of its somatic or germ cells a nucleic acid sequence..."[24], then the most common (if not the only) way of obtaining such patented product would be by the natural breeding of those products already on the market. Ordinarily, the sale of patented seeds or animals carries with it the (implied or explicit) license to use the product for one-time breeding purposes and thereby produce first-generation offspring, but does not extend to the production of a further, second generation. If exhaustion rules were to apply to all further offspring, this would amount to a license for the production of otherwise infringing products in direct competition with those marketed by the patentee (at least to the extent that third and further generation products would be stable enough to still possess the advantageous characteristics of the first-generation product).

In Europe, the Biotech Directive[25] provides for a specific rule in Article 9. According to this provision, the protection extends (or is limited to, whatever the perspective) to all offspring where the genetic information subject to the exclusive right "performs its function," in other words still possesses its advantageous characteristics for which the patent was granted. The only case law on this provision in Europe relates to the limitations (to a product where the genetic information no longer performed its function), but not the extent (for third-generation offspring).[26]

In the United States, most cases related to modified seeds were decided on the basis of contractual restrictions on the buyer,[27] yet in the first case that went up to the Supreme Court, *Bowman v. Monsanto*, there was no contractual restriction due to the fact that Mr. Bowman had purchased the seed from a so-called "grain elevator" that had obtained a variety of different seeds without any contractual obligations or limitations as to the sale or use of such seeds. The Court of Appeals

---

23. EP no. 0546090 for glyphosate tolerant plants.

24. EP publication no. 1170994 ("Enviro Pig").

25. Directive 98/44/EC of the European Parliament and of the Council of 6 July 1998 on the Legal Protection of Biotechnological Inventions, art. 9, 1998 O.J. (L 213), 13 ("The protection conferred by a patent on a product containing or consisting of genetic information shall extend to all material, save as provided in Article 5(1), in which the product is incorporated and in which the genetic information is contained and performs its function.").

26. Case C 428/08, Monsanto Tech, LLC v. Cefetra BV, 2010 E.C.R. I-6765, *available at* http://curia.europa.eu/juris/liste.jsf?language=en&num=C-428/08.

27. Monsanto v. Scruggs, 459 F.3d 1328 (Fed. Cir. 2006); Monsanto v. McFarling, 302 F.3d 1291 (Fed. Cir. 2002).

for the Federal Circuit had affirmed that there were no specific exhaustion rules for self-replicating technologies.[28] It is indeed difficult to see why the (purposeful) reconstruction of a patented machine would amount to infringement (as reconstruction can never be placed under the rules of exhaustion), whereas the (mostly purposeful) planting of patented seeds with the aim of obtaining progeny would not. In both cases, the aim is to obtain a product that undoubtedly falls under the patent claim and that would be in direct competition with those products sold by the patentee under the monopolistic right. In its decision of May 2013,[29] the US Supreme Court dismissed the appeal and stressed that "the patent holder has 'received his reward'" only for the actual article sold, and not for subsequent recreations of it.... If the purchaser could make and sell endless copies, the patent would effectively protect the invention for just a single sale [, thus] the exhaustion doctrine does not enable Bowman to make *additional* patented soybeans without Monsanto's permission (either express or implied)... Because Bowman thus reproduced Monsanto's patented invention, the exhaustion doctrine does not protect him." This is indeed a result that should logically follow from the exhaustion doctrine as set out above.

V. Exhaustion and Repair

A. *Issues and Approaches*

"Ending is better than mending" is not only one of the famous mottos of Aldous Huxley's *Brave New World*, but also a philosophy of those who offer single-use items as a business model. Recycling, on the other hand, has become of increasing importance in a world with limited natural resources. It is thus no coincidence that cases of recycling patented products have increasingly come before the courts in recent years.[30] Typically in these cases, some parts of a complex article have a longer life than others, making it commercially viable to replace the worn-out parts so that the article may be reused. The approaches of the courts toward this problem are not uniform. The first tries to distinguish between repair and reconstruction, and thus to define the issue of product identity.[31] In order to

28. Monsanto v. Bowman, 657 F.3d 1341 (Fed. Cir. 2011).

29. Bowman v. Monsanto, 133 S. Ct. 1761 (2013).

30. In the United Kingdom: United Wire Ltd. v. Screen Repair Servs., [2001] R.P.C. 24 (H.L.). In Japan: *Fuji Camera*, Tōkyō Chihō Saibansho [Tōkyō Dist. Ct.] Jun. 6, 2000, 1712 Hanrei Jihô [Hanji] 175. In the United States: Jazz Photo Corp. v. Int'l Trade Comm'n, 264 F.3d 1094 (Fed. Cir. 2001); Fuji Photo Film Co. v. Jazz Photo Corp., 249 F. Supp. 2d 434 (D.N.J. 2003) The *Fuji Photo Film v. Jazz Photo* case was appealed and brought before the Court of Appeals for the Federal Circuit, 394 F.3d 1368 (Fed. Cir. 2005).

31. *See infra* Section V.B.

do so, some courts have looked at the question whether the exchanged part was an "essential element" of the invention. If so, there would no longer be product identity and therefore manufacture rather than repair.[32] Some courts have tried an approach that takes the expectations of the purchaser or society as a starting point.[33]

B. Repair and Product Identity

According to Section 40 of the 1844 French Patent Act, the patentee merely had the right to produce the protected product or process. Under Section 41, secondary acts of exploitation, such as possession, sale, or importation of protected products, constituted an infringement only if use was made of products infringing the patent. Thus, secondary acts of exploitation were unlawful only if there was an infringing primary use. This incorporates the exhaustion doctrine in its purest form: acts other than the manufacture of a patented product cannot infringe unless the manufacture itself was infringing. Or, put into the context of the repair cases: a repair is impermissible only if it amounts to the manufacture of a patented article. Determination if this is so requires a comparison between the relevant acts of repair and the wording of the patent. If a repair amounts to a new manufacture, it is infringing; if the product remains the same, it is not. This is the approach taken by a number of courts when dealing with repair.

1. The *United Wire* Case

The English House of Lords held in *United Wire*:

> The sale of a patented article cannot confer an implied licence to make another or exhaust the right of the patentee to prevent others from being made. A repair of the patented product is by definition an act which does not amount to making it.... Repair is one of the concepts (like modifying or adapting) which shares a boundary with "making" but does not trespass upon its territory.... As a matter of ordinary language, the notions of making and repair may well overlap. But for the purposes of the statute, they are mutually exclusive. The owner's right to repair is not an independent right conferred upon him by licence, express or implied. It is a residual right, forming part of the right to do whatever does not amount

32. *See infra* Section V.C.

33. *See infra* Section V.D. These approaches are dealt with in further detail by C. Heath & M. Mōri, *Ending Is Better than Mending*, 37 Int'l Rev. Intell. Prop. & Compet. L. 856 (2006).

> to making the product.... [The first instance judge] said that the frame was an important part of the assembly and that the defendants had prolonged "the screen's useful life." It is quite true that the defendants prolonged the useful life of the *frame*. It would otherwise presumably have been scrapped. But the *screen* was the combination of frame and meshes pre-tensioned by attachment with adhesive according to the invention. That product ceased to exist when the meshes were removed and the frame stripped down to the bare metal. What remained at that stage was merely an important component, a skeleton or chassis, from which a new screen could be made.[34]

In other words, the distinction is between manufacture and repair, and in the case at issue, the court held that exchanging the screens amounted to a reconstruction of the patented article.

This reasoning was subsequently applied to a 2011 case of the UK Court of Appeal[35] where the work undertaken by the defendant was a replacement of a plastic container of a (patented) assembly of container and protecting cage:

Patented assembly of container and protecting cage
Source: Schütz (UK) Ltd. v Werit (UK) Ltd., [2011] EWCA (Civ) 303, [5] (Eng.).

---

34. United Wire Ltd. v. Screen Repair Servs. (Scot.) Ltd., (2000) 4 All E.R. 353 (H.L.), *available at* http://www.publications.parliament.uk/pa/ld199900/ldjudgmt/jd000720/wire.htm.

35. Schütz (UK) Ltd. v Werit. (UK) Ltd., [2011] EWCA (Civ) 303 (Eng.), *available at* http://www.bailii.org/ew/cases/EWCA/Civ/2011/303.html.

According to the court, "The upshot is simply this: does the act of putting a bottle into a Schütz cage, constitute 'making a product which is the subject-matter of the patent' within the meaning of Art. 25 of the CPC...." The court arrived at the conclusion that "what Delta did is to recondition or repair (where necessary) a cage made by Schütz and re-use it to make an assembly as claimed in claim 1."

This approach must be contrasted to the first instance case of *Schütz v. Werit* where the judge denied infringement:

> In my judgment the correct approach is to ask whether, when the part in question is removed, what is left embodies the whole of the inventive concept of the claim. I think that Mr Meade's approach, with its emphasis on the form rather than the substance of the claim, is too mechanical, and leads to results which are contrary to principle. I think that if one applies this test to the facts of *United Wire* one reaches the same conclusion because the inventions did not reside wholly in the frame or wholly in the mesh. It was a true combination....I have come to the conclusion that the activity of replacing the inner container of a Schütz IBC with a Werit bottle does not amount to making the patented product of claim 1 of either 307 or 967. It seems to me that the inventive concept of each of these claims is wholly embodied in the Schütz cage. Thus when the bottle is removed, the part retained embodies the whole of the inventive concept.[36]

The UK Supreme Court (formerly House of Lords) confirmed the approach taken by the first instance and denied infringement, based on a multifactor test that the Court of Appeal refused to apply to cases of repair/reconstruction:

> Deciding whether a particular activity involves "making" the patented article involves, as Lord Bingham said, is an exercise in judgment, or, in Lord Hoffmann's words, it is a matter of fact and degree. In some such cases, one can say that the answer is clear; in other cases, one can identify a single clinching factor. However, in this case, it appears to me that it is a classic example of identifying the various factors which apply on the particular facts, and, after weighing them all up, concluding, as a matter of judgment, whether the alleged infringer does or does not "make" the patented article. In the present case, given that (a) the bottle (i) is a

---

36. Schütz v. Werit, [2010] EWHC (Pat) 660, [197], [206] (Eng.).

> freestanding, replaceable component of the patented article, (ii) has no connection with the claimed inventive concept, (iii) has a much shorter life expectancy than the other, inventive, component, (iv) cannot be described as the main component of the article, and (b) apart from replacing it, Delta does no additional work to the article beyond routine repairs, I am of the view that, in carrying out this work, Delta does not "make" the patented article.[37]

Quite a lengthy part of this decision is dedicated to a comparison with the corresponding opinion of the German Federal Supreme Court that had to decide the same case of infringement between the same parties (and that is dealt with below in Section V.C.2). The German Court had criticized the UK Court of Appeal "for the court's opinion that a reconstruction in the corresponding case was affirmed inter alia for the reason that the defendants in their advertisements had used the term 're-manufacturing.'"[38] The only difference between the approaches of the German and UK Supreme Court now seems to be the question of whether habits of the consumer in replacing the bottles should be taken into account (which the German Supreme Court answers in the affirmative, whereas the UK Supreme Court says it should not).

2. The Fuji Camera Cases

The US and Japanese cases regarding single-use cameras offer an interesting comparison in this respect, as they concerned the same patents and the same acts of recycling, dealt with in two different jurisdictions. The various patents held by Fuji concerned parts of the camera (film-winding mechanism; mechanism for smoothly advancing the film; film chambers and film path; flash unit; push button; shutter mechanism; winding wheel), but also the whole of the camera ("A lens-fitted photographic film package having an externally operable member for effective exposure"). The defendants were basically opening the (used) cameras, replacing the winding wheel, resetting the film counter, replacing the battery, winding new film into the camera, resealing the camera, and putting a new outside wrapping on it.

A. THE JAPANESE APPROACH

In Japan, Fuji held several utility models and designs related to "units of film and camera." The claim of the patent at issue (Japanese Patent No. 1,875,901)

---

37. Schütz v. Werit, [2013] UKSC 16, [78] (appeal taken from Eng.), *available at* http://www.supremecourt.gov.uk/decided-cases/docs/UKSC_2011_0266_Judgment.pdf.

38. Palettenbehälter II [Pallet Container II], Case X ZR 97/11, Bundesgerichtshof [BGH] [Federal Court of Justice] July 17, 2012 (Ger.), *reprinted in* 8 J. INTELL. PROP. L. & PRAC. 82 (2013). For the approach of the Court toward repair and reconstruction, see below.

reads: "Film unit with lens that has an unexposed film in camera body beforehand, and that gives light exposure to films that are ejected after consumers have taken all the pictures." The defendant in this case had recycled these cameras by inserting a new film and a new battery. In view of its utility models, the plaintiff thought this to be an infringing act. In this respect, the plaintiff could not rely on the concept of a single-use camera being novel as such, but argued that by putting in new film, the defendant had made use of thc protccted camera, more specifically the winding mechanism. It was not argued that the replaced parts (battery, film) were protected as such. The Tokyo District Court agreed with the plaintiff and tried to draw the line between repair and reconstruction by arguing that the defendant's acts commenced at the very stage where the patented product's useful life had come to an end, and that invoking the patent right here would not amount to a double remuneration of the patentee. It further reasoned that the defendant was replacing an important part of the principal item essential to the invention. After the defendant's refill, one could no longer speak of the same product. The reasoning of the court merits a closer look:

> Once a patented product has fulfilled its function, the patentee is allowed to re-enforce its patent rights over these used patented products. First, this will not affect the free market even if the patentee is allowed to claim patent infringement after the patented products are used up, and second, in such case the patentee does not make a double profit. If patented products are commercially reprocessed, used and sold, such act deprives the patentee of marketing opportunities and diminishes the patentee's profits[;] whether or not the used patented products have fulfilled their function shall be determined by a comprehensive examination such as function, structure, material, application, manner of usage, circumstances of marketing of the patented products, etc....
>
> Once the defendant exchanges the main component which comprises a substantial part of patented invention, and thus manufactures new products, the patentee can enforce its patent rights over such products to the extent that these are products identical to the original patented products the patentee has sold. However, exchanging minor components of patented products (e.g. batteries or filters for electric equipment), or components that have a shorter life than the products themselves (such as light bulbs for electronic equipment, or water-proof packing for underwater instruments), or repairing patented products by exchanging the damaged parts does not equal the manufacture of different products. The repaired products thus do not lose their identity as patented products... [In the case of equipping disposable cameras with a new film and battery],... it

> is difficult for consumers who use preinstalled film for photographing to eject the used film without exposure to light....
>
> With regard to patented products, the connection part such as the hook is damaged when the user removes the film, and the hook and weld supersonic wave parts are damaged when persons remove the back cover from the camera body, and therefore the recycled products that contain new film and whose back cover has been reaffixed will be of lesser quality in terms of light-blocking effect and general function when compared to the plaintiff's patented products.... The plaintiff's patented product, the disposable camera's body itself, is not supposed to be returned to consumers after these have brought the camera to a photo developing shop.... Thus, when a consumer has taken all the pictures using the built-in film, brings the camera to a photo shop, and the photo shop sends the disposable camera to a photo processing station where the built-in film is ejected, the patented products have done their duty under the normal conditions of use. Therefore, in this case, the patentee's patent rights over the patented products are not exhausted, be it nationally or internationally.[39]

The decision already contains the two elements that the IP High Court in the *Canon* ink tank case identified as crucial in order to distinguish recycling from repair: the fulfilment of function and the exchange of a "substantial part of the invention."[40]

B. THE US APPROACH

On the other hand, the US Federal Circuit found that the activities were not infringing, as:

> [T]he replacement of unpatented parts, having a shorter life than is available from the combination as a whole, is characteristic of repair, not reconstruction.... The ruling of reconstruction as to these patents is incorrect, because the remanufacturing processes simply reuse the original component, such that there is not issue of replacing parts that were separately patented. If the claimed component is not replaced, but simply is reused, this component is neither repaired nor reconstructed.[41]

---

39. Tōkyō Chihō Saibansho [Tōkyō Dist. Ct.] June 6, 2000, 1712 Hanrei Jihô [Hanji] 175 (Japan) (ordering an interim injunction), *upheld by the same court in the main action*, Aug. 31, 2000 (unreported).

40. *See infra* Sections V.B.3, V.D.

41. Jazz Photo Corp. v. Int'l Trade Comm'n, 264 F.3d 1094, 1107 (Fed. Cir. 2001).

3. Toner Cartridges ("Ink Tanks") in Japan

In a subsequent case on the recycling of patented toner cartridges, the facts were the following: the plaintiff was a Japanese corporation well-known for the manufacture of printers and sale of ink cartridges used with these. The plaintiff was the owner of Japanese patent JP3278410 for "LIQUID-CONTAINING VESSEL, MANUFACTURE THEREOF, PACKAGE THEREOF, INK JET HEAD CARTRIDGE INTEGRATED WITH VESSEL AND RECORDING HEAD, AND LIQUID JET RECORDER." The defendant was a Japanese corporation importing and selling recycled ink cartridges sold by the plaintiff in the following way: the defendant and its affiliates collected the empty cartridges from customers of the plaintiff, exported them to factories abroad, cleaned and refilled them with new ink, and imported them back to Japan. The plaintiff found the import and sale of the cartridges an infringement of its patent rights under Section 68 and 100(1) of the Patent Act, and sought the suspension of the import and the sale of the recycled ink tanks, and the disposal of the products. The defendant argued that the plaintiff's patent rights over the respective products were exhausted upon their first sale in Japan.

The Japanese courts were divided both as to the result and the reasoning. Although the first-instance court found that there was infringement (because there was no product identity), the second-instance court found infringement based on a variety of factors (see below Section V.D.2), and the Japanese Supreme Court found infringement based on the doctrine of product identity, but with some modifications:[42]

> However, exhaustion and thus a restriction of enforcement of the patent right is limited to the patented article assigned by the patentee. Therefore in the case where the patented article assigned by the patentee, etc., is modified or part of the article is replaced, and hereby a new patented article different from the original article is manufactured, the patentee should be permitted to exercise the patent right with respect to this new article. In order to determine whether a new patented article has been manufactured, it is appropriate to consider the totality of the circumstances including the attributes of the patented article, the contents of the patented invention, the manner of the modification and the parts replaced, as well as other conditions of trade. The court should consider the function, structure, material, intended usage, life span and actual usage of the article as the attributes of the patented article, the condition of the patented

---

42. *Ink Tank*, Saikō Saiansho [Sup. Ct.] Nov. 8, 2007 (Japan), *reprinted in* 39 Int'l Rev. Intell. Prop. & Compet. L. 982 (2008).

article when it was modified, etc., the contents and the degree of the modification, the life span of the replaced part, the technical function and the economic value of the parts within the patented article as the manner of the modification and replacement. . . .

In the present case, according to the aforementioned facts, when the respondent's [Canon] ink tanks are filled with new ink and reused, this could cause quality loss of the print-outs and the breakdown of the printer, for which reason the respondent intended its ink tanks to be single-use and afterwards to be replaced with new ones. Therefore, the respondent's article has no openings for adding new ink, and it is necessary to drill a hole in the ink tank to refill it. For the process of manufacturing the appellant's article, a hole is thus drilled into the ink tank that is closed after the addition of new ink. By such modification during the manufacturing process of the appellant's article, it is not appropriate to state that only consumable ink has been added to the ink tanks. Rather, the ink tanks have been transformed so that it is possible to refill them with new ink.

In addition, according to the above-mentioned facts, since the ink in the respondent's article has the function of a barrier that prevents any air movement at the compressed interface, some or all of the compressed interface no longer maintains the ink after the ink has been used up to some extent. And when the respondent's used articles are removed from the printer, the remaining ink hardens in about one week to ten days. Therefore, if the ink tanks in such condition would be filled with new ink, the compressed interface could not fulfil the function to act as a barrier to prevent the air movement at the boundary, even if the new ink would be refilled to as much as the whole liquid storage room permits and the storage of the negative pressure-generating component would be also filled to the extent that the ink exceeds the boundary of the component.

Moreover, when the appellant's articles are produced, the fixed ink inside the ink tanks is flushed away by refurbishment of the inside of the ink tanks, and the appellant hereby intends to recover the function to make a barrier to prevent the air movement at the compressed interface. The appellant's articles maintain the ink at the whole compressed interface part once they are refilled with the same amount of ink as the respondent's new articles. Thus, by such modification in the course of manufacturing the appellant's articles, it is not appropriate to regard the ink tanks as merely having been refilled with new ink. Rather, it should be deemed that the used ink tanks are reused and the essential composition of the patented invention . . . inside the ink tanks has been restored, where such characteristic had been lost before. Therefore, the appellant has restored the actual value of the patented invention by stopping the ink from escaping from the ink tank before the package is opened.

> Considering the totality of the circumstances including the ones described above, as well as other above-mentioned facts such as the actual conditions of trade, the appellant's articles should be considered as new articles different from the respondent's unmodified articles. Therefore, the enforcement of the patent right should not be restricted with respect to the appellant's articles produced from the used respondent's articles assigned by the Japanese patentee, etc., abroad. The respondent who is the patentee of the patent right should be permitted to seek the suspension of the import and sale, etc., and the disposal of the articles.

4. Comments

Applying the doctrine of product identity may lead to an undue focus on the product rather than the patent. This was particularly manifest in the Japanese camera cases, where the courts looked at the camera as if this had been the patented entity (which would have been correct had the patentee actually invented the disposable camera as such). But the patented entity was the winding mechanism, and the mechanism was not exchanged—it was merely rewound, and of course such use is covered by the exhaustion doctrine. It is also a bit odd for the Japanese Supreme Court (in the toner cartridge case) to look at such issues as whether the cartridge was difficult or easy to disassemble, whether a hole had to be drilled, or whether consumers would get a recycled product of good quality—none has any bearing on the patent (that in this case related to an ink tank, and a method of producing such).

C. Repair, Exhaustion, and the Exchange of an "Essential Element"

1. The Dutch Approach

In several jurisdictions, namely the Netherlands, Germany, and Belgium, a case was argued that concerned a patented coffee-making apparatus (manufactured by Senseo). The apparatus claim also covered the coffee pad, although the patentee could not protect the coffee pad as an independent claim due to prior art. The question arose whether exchange of the pad amounted to an infringement. Although the first-instance court had found such exchange infringing, as the coffee pads should be considered an essential element of the invention, the Dutch Court of Appeal disagreed and likened the coffee pads to hot water that was equally necessary to make coffee, but whose use could hardly qualify as an infringing act. The Supreme Court agreed with the Court of Appeal:

> The mere circumstance that a fitting coffee pod is necessary to apply the patented invention does not entail outright that this pod is a means that regards an essential part of the invention. Apparently, and in light of the interpretation of the patent by the Court of Appeal, understandably, the

> Court of Appeal concluded that the coffee pod that fit the container was not an element that, according to the patent specification, distinguished the teaching of the patent from the state of the art. That conclusion does not reflect an inaccurate legal opinion.[43]

Thus, the court interpreted the "essential element" as one that must contribute to the inventive merit of the apparatus.[44]

2. The German Approach

The leading case in this respect is *Flügelradzähler/Impeller Flow Meter*[45] where the measuring unit of a patented impeller flow meter was manufactured and sold by the defendant. The case concerned an issue of indirect infringement, because the action was raised against the supplier of the measuring unit. But the same question would have arisen had the action been against the person refitting the measuring unit to the impeller flow meter. In both cases, the court needed to determine whether an "essential element" of the invention had been replaced. The court thus had to decide first whether the measuring capsule was an "essential element," and second, whether replacement thereof was not covered by the rules of exhaustion. The court held as follows:

> 1. A means within the meaning of §10 of the Patent Act refers to an essential element of the invention if this is capable of interacting functionally with one or more features of the patent claim to implement the protected invention. A feature that is completely subordinate for the technical teaching of the invention can be ignored as a non-essential element of the invention.
> 2. The answer to the question when the replacement of parts of a device is the equivalent of making the device requires a balancing of the protectable interests of the patent holder in the commercial exploitation of the invention on the one hand, and those of the purchaser in the unrestricted use of the patented device put into circulation on the other hand, while taking into account the nature of the patented product.

---

43. Sara Lee v. Vomar Voordeelmarkt, Case C02/227HR, Hoge Raad [HR] [Supreme Court] Oct. 31, 2003 (Neth.), *available at* http://deeplink.rechtspraak.nl/uitspraak?id=ECLI:NL:HR:2003:AI0346.

44. This was also held in the (parallel) proceedings in Belgium on the same issue of supplying coffee pads for the Senseo machines. *See* Tribunal de Première Instance [Civ.] [Court of First Instance], Antwerp, Feb. 20, 2004 (Belg.), *summarized in* 2004 Eur. Intell. Prop. Rev. N-105.

45. *Flügelradzähler* [Impeller Flow Meter], Case X ZR 48/03, Bundesgerichtshof [BGH] [Federal Court of Justice] May 4, 2004 (Ger.), *reprinted in* 36 Int'l Rev. Intell. Prop. & Compet. L. 963 (2005).

Whereas the second issue concerns the distinction between permissible repair and impermissible reconstruction, the first issue is a specific interpretation of what constitutes an "essential element" of the invention. With respect to the exchange of spare parts, both of the above issues were further defined in subsequent decisions.

The *Laufkranz*[46] decision dealt with the question of permissible repair in the case of a patented rail vehicle wheel where the flanged tire was replaced. The tire was part of the patent claim, yet represented a part that would typically wear out quickly and could therefore qualify as a "replacement part." In these circumstances, the court held:

> The rights of a purchaser of a patented apparatus to use the latter in accordance with its function in principle also extend to the exchange of a part that during the ordinary lifespan of the apparatus will regularly be replaced due to wear and tear or for other reasons, unless the technical effect manifests itself especially in the very part that is to be exchanged.

The whole issue was revisited in *Pipettensystem* case,[47] where the headnotes read—

1. A feature of a claim may be regarded as non-essential in the sense of §10 Patent Act if it does not contribute to the achievement of the invention, that is, to the solution proposed by the patent in order to solve its underlying technical problem.
2. In order to distinguish between purpose-related use and reconstruction of a patented product, it is decisive whether the measures that were undertaken preserve the identity of the patented product already put into circulation or whether this would equal the manufacture of a new product, taking account of the specific characteristics, effects, and advantages of the invention.
3. The exchange of parts that is common in such type of apparatus may amount to a reconstruction of the same if the technical effect of the invention manifests itself particularly in the part that is exchanged, because the invention has an influence on its manner of use or lifespan.
4. It would not be correct to assume a reconstruction where the exchanged part is only an object of the improved functioning of the whole apparatus.

---

46. *Laufkranz* [Wheel Tread], Case X ZR 45/05, Bundesgerichtshof [BGH] [Federal Court of Justice] May 3, 2006, Gewerblicher Rechtsschutz und Urheberrecht [GRUR] 837, 2006 (Ger.).

47. *Pipettensystem* [Pipette System], Case X ZR 38/06, Bundesgerichtshof [BGH] [Federal Court of Justice] Feb. 27, 2007, Gewerblicher Rechtsschutz und Urheberrecht [GRUR] 769, 2007 (Ger.).

This multifactor test was further elaborated in the *Palettenbehälter II* case that corresponds to the above-mentioned UK case *Schütz v. Werit* (the patent was the same, and so was the allegedly infringing device). Whereas the UK High Court and the Court of Appeal had rejected the German approach as unhelpful,[48] the German Federal Supreme Court for its part criticized the approach taken by the UK Court of Appeal (and should now take comfort in that the UK Supreme Court's approach comes much closer to the German Supreme Court's line of reasoning). The official headnotes of *Palettenbehälter II* read as follows—

1. When the exchange of certain parts belongs to the intended use of a patented product, such exchange may be performed on a machine that has been marketed by the patentee by competitors that have purchased the machine in a state that requires repair, and may sell the repaired machine to third parties.
2. In determining whether by the exchange of parts the identity of the product is maintained or whether such exchange amounts to a new manufacture, the question of whether the exchange of such part could be reasonably expected during the lifespan of the apparatus, and whether the technical effect of the invention manifested itself particularly in the exchanged parts, should equally be asked where a direct patent infringement is claimed.
3. Whether the inventive concept is particularly embodied in the replaced parts is usually only of importance if their exchange would have been anticipated during the lifespan of the products. In order to determine this, it must be ascertained whether the interested circles in commerce regard such replacement as

---

48. According to the UK High Court:

> I do not think any of the ways in which the test is formulated in the German cases really solve the difficulty. To ask whether the advantages of the invention are reflected in the part replaced is a test which is difficult to apply. If an inventive propeller is claimed in combination with a boat, no doubt the advantages are enjoyed by the boat, but it cannot seriously be suggested that an owner of a patented propeller could not rebuild the boat to which it was attached, or transfer it to a new one. I would reject any reliance on a balancing exercise involving the respective interests of the patentee and the user: this would be to revert to considerations of implied licence, rejected by the House of Lords in both *United Wire* and *Canon*. A monopoly right should have clear boundaries, and the boundaries should not be forced backward by considerations derived from implied licence. Likewise the notion of whether the part is one which is likely to be replaced in the lifetime of the device: this again seems to me to rely on the notion of an implied licence and, as the German cases recognise, does not answer the question of whether there is "making" when the part itself embodies the inventive concept. Furthermore I do not find it useful to ask whether the identity of the patented product is preserved, unless by that one means whether parts embodying the whole of the inventive concept are retained.

Schütz v. Werit, [2010] EWHC (Pat) 660, [199] (Eng.).

> a usual measure of maintenance that does not call into question the identity of the assembly as an object of commerce.

Thus, it seems that the only difference in approach between the UK and the German Supreme Court relates to the question of whether users of a particular product would regard the replacement of a specific part as a usual measure of maintenance. The German Supreme Court places importance on this question in cases where the inventive concept is not embodied in the replaced part. For the UK Supreme Court, this kind of consumer expectation is of no importance.

Under the above case law, the definition of an "essential part" almost amounts to squaring the circle: The part need not be in any way different from the state of the art (as Section 10 of the Patent Act may also apply where all parts individually may form the state of the art, yet the inventive merit lies in the new and inventive assembly); it needs to make a technically important contribution and yet must not qualify as a staple product. When we look at the case law, it seems that the decisions in *Laufkranz* and *Pipettensystem* were attempts to contain a definition of the essential element that in *Flügelradzähler* was construed far too broadly.[49]

And although it is frequently stressed that the provision does not broaden the scope of patent protection,[50] it cannot be denied that when compared to the case law prior to the enactment of Section 10,[51] the scope of permissible repair/refill seems to have shrunk considerably.[52]

The *Coffee Pads* decision rendered by the Düsseldorf Appeal Court[53] (under the same facts as in the above Dutch case) is a good example to highlight the above two difficulties. Differing from the courts in the Netherlands and Belgium, both the District and the Appeal Court in Düsseldorf found the sale of coffee pads an infringement of the apparatus patent. The Appeal Court in *Coffee Pads* held—

> If a filter pad and a coffee filtering machine in their interaction make out the apparatus protected as an invention, and such apparatus is designed for the exchange of filter pads, the economic core of the invention lies in

---

49. In essence, essential parts seem to be all those that relate to elements of the claim.

50. *E.g.*, Keukenschrijver, *Flügelradzähler, Kaffeetüte und Drehzahlermittlung—neue Entwicklungen bei der mittelbaren Patentverletzung*, Festschrift 50 Jahre VPP, 331 at 336 (2005).

51. *See* Tatjana Kowal-Wolk & Reinhard Schuster, *Patentverletzung im Reparatur-, Ersatzteil- und Altteilgeschäft,* Festschrift F. K. Beier, 87 (1996).

52. Apparently in the same vein, see Neils Holder, 2005 GEWERBLICHER RECHTSSCHUTZ UND URHEBERRECHT [GRUR] 20, 22 (Ger.).

53. *Kaffee-Filterpads* [Coffee Pads], Case 2U 35/04, Oberlandesgericht [OLG] [Higher Regional Court], Düsseldorf, Nov. 17, 2005, GEWERBLICHER RECHTSSCHUTZ UND URHEBERRECHT RECHTSPRECHUNGS-REPORT [GRUR-RR] 39, 2006 (Ger.).

particular in providing the supply of replacement filter pads. Where each time a filter pad is exchanged the apparatus is newly reconstructed, there is no room for the assumption that the benefits allocated to the patentee could already be reaped by the initial supply of the apparatus.

The "essential part" is defined by the court in economic terms, an approach that (at least in Europe) is usually avoided when interpreting patent claims. It is also inconsistent with the subsequent decisions *Laufkranz* and *Pipettensystem* (as above), and would run the risk of essentially protecting nontechnical features of technical rights.[54] But even under the more narrow definition of the *Pipettensystem* case, the coffee pad may qualify as an essential part of the claim: although not distinguishable over prior art, it is essential for making what the apparatus is meant to do. But because the apparatus has to be "fuelled" with the pads, they may well qualify as staple products (this angle finds no mention in the decision), leading to the second question of exhaustion. And in the *Coffee Pads* case, the court held that "every time a user puts a new pad into the apparatus, the latter is reconstructed." This is an interesting idea. Imagine getting up in the morning, putting a new coffee pad into the Senseo machine, pouring in some water, pushing a button—and, thereby, though unaware of it at this ungainly morning hour, becoming the maker of a new machine. The Federal Supreme Court had no opportunity to review these findings in light of its later decisions, due to the fact that Sara Lee's patent was revoked by the European Patent Office's Boards of Appeal in August 2005.

At least in one case, the German Federal Supreme Court held that even for technically essential acts of wear and tear, the patentee might have given an implied consent to exchange,[55] a solution that one could well apply to the *Coffee Pads* case as above.

It is reassuring for the position taken here that in interim proceedings concerning yet another coffee pad (or, this time, a Nespresso coffee capsule), the Düsseldorf District Court explicitly rejected the previously advocated "reconstruction doctrine" when a purchaser inserts a new pad, and instead held that the act of putting a new capsule into a patented machine was covered by the exhaustion doctrine:

The rights of the applicant to its invention are exhausted by the sale of the Nespresso machine... In principle, it must be assumed that the exchange

54. Niels Hölder likens this to the protection of a business model. *See* Niels Hölder, 2007 GEWERBLICHER RECHTSSCHUTZ UND URHEBERRECHT [GRUR] 96, 97 (Ger.).

55. This was so held in *Rohrschweißverfahren* [Pipe Welding], Case X ZR 113/04, Bundesgerichtshof [BGH] [Federal Court of Justice] Feb. 27, 2007 (Ger.), *reprinted in* 39 INT'L REV. INTELL. PROP. & COMPET. L. 106 (2008), for a process covering several steps where the patentee himself had furnished data that allowed the recipient to initialize the process. The court found that this was an implied consent to use the protected process as intended.

> of a part that frequently needs replacement during the lifespan of the machine cannot be considered a reconstruction of the machine... while the capsule acts in functional conjunction with the other elements of the patented invention... the inventive achievement is not particularly mirrored in the capsules. Neither is the technical advantage of the invention realised in the capsules..., and the advantage of the invention is not realised in the sense that the invention would positively influence the function or duration of the capsules... Finally, in order to balance the conflicting interests involved it cannot be decisive that the capsules, rather than the coffee machine, possibly represent the economic focus of the invention by the expected need to replace them. After all, such intention of commercial use can only be protected if the technical advantages of the invention are particularly mirrored in the capsules, which is not the case....[56]

3. The UK Approach—*Nestec S.A. v. Dualit Ltd.*

The supply of capsules fit for Nespresso machines was also the object of a lawsuit in the United Kingdom[57] (as of this writing, further proceedings are pending in Antwerp (Belgium) and the patent court in St. Gallen, Switzerland). The court denied infringement under both the doctrines of implied licence and exhaustion:

> The purpose of a Nespresso machine is to make coffee. In order to use the machine for its intended purpose, the purchaser must insert capsules into the machine. It follows that the purchaser must be impliedly licensed to obtain and use capsules with the machine. Otherwise, it would be useless. In the absence of any restriction upon the purchaser preventing him from obtaining capsules from third parties, the purchaser is entitled to do so. In my view, it makes no difference even if it is the case that, as a result of the way the claim is drafted, the purchaser makes a system falling within claim 1 as soon as he acquires the capsule....
>
> [T]he position is perhaps even clearer if it is analysed in terms of exhaustion of rights. By consenting to the manufacture and sale of Nespresso machines, Nestec have exhausted their rights under the patent to restrict purchasers' freedom to use such machines in accordance with their normal function....

---

56. Nestec S.A. v. Ethical Coffee, Case 4b O 82/12, Landgericht [LG] [Regional Court], Düsseldorf, Aug. 16, 2012 (Ger.), *available at* http://eplawpatentblog.com/2012/August/DE%20-%20capsule.pdf. This decision was upheld in *Nespressokapseln* [Nespresso Capsules], Case I-2 U 72/12, Oberlandesgericht [OLG] [Higher Regional Court], Düsseldorf, Feb. 21, 2013, Gewerblicher Rechtsschutz und Urheberrecht Rechtsprechungs-Report [GRUR-RR] 185, 2013 (Ger.), reflecting a significant convergence between the German and the English approach.

57. Nestec S.A. v. Dualit Ltd., [2013] EWHC (Pat) 923, [164]–[167] (Eng.).

The court also affirmed that the capsules were an essential element of the invention (following the German rather than the Dutch approach), yet denied a new manufacture in applying the above-mentioned multifactor test set out in the UK Supreme Court's *Schütz v. Werit* decision.

4. The US Approach—*Aro I* and *Aro II*

The leading cases for distinguishing permissible repair from impermissible reconstruction are two Supreme Court cases, *Aro I*[58] and *Aro II*.[59] *Aro I* is most interesting in that it clearly highlights the relationship between direct and contributory infringement for spare parts, and in its message that parts without any inventive merit can be replaced ad lib. Here, the patent covered the combination, in an automobile body, of a flexible top fabric, supporting structures, and a mechanism for sealing the fabric against the side of the automobile body to keep out the rain. The defendant had replaced the fabric only. According to the Supreme Court, the patent did not confer a monopoly over the fabric or its shape, which the court considered but an unpatented element. Further, a car owner would not infringe the combination patent by replacing the worn-out fabric of the patented convertible top on his car, as such a replacement by the car owner is a permissible "repair" and not an infringing "reconstruction," regardless of whether the intended use was known to the sellers or not. The Supreme Court thereby established the rule that unless one element of a combination patent was separately patented, its replacement could not amount to infringement, however essential the element may be.[60]

When compared to the approach taken by the German court in the *Coffee Pads* case, the difference is particularly striking as regards the relationship between a component and a protected entity:

> For if anything is settled in the patent law, it is that the combination patent covers only the totality of the elements in the claim, and that no element, separately viewed, is within the grant. The basic fallacy in respondent's position is that it requires the ascribing to one element of the patented combination the status of the patented invention itself.... No element, not itself separately patented, that constitutes one of the elements of a combination patent is entitled to patent monopoly, however essential it may be to the patented combination and no matter how costly or difficult replacement may be.[61]

---

58. Aro Mfg. Co. v. Convertible Top Replacement Co., 365 U.S. 336 (1961) (*Aro I*).

59. Aro Mfg. Co. v. Convertible Top Replacement Co., 377 U.S. 476 (1964) (*Aro II*).

60. *Aro I*, 365 U.S. at 336.

61. *Id.* at 345.

In a concurring opinion, Justice Brennan held that it would indeed be useful to find a formula for distinguishing repair from reconstruction, and that could not depend on whether the replaced part was independently patentable, but rather on the totality of factors:

> The life of the top was approximately three years, in contrast to the several times longer life of the other components of the combination. The top was replaceable at a cost of from $30 to $70, depending on the fabric; in contrast, the cost of other elements of the combination was approximately $400. These considerations, of themselves, suggest that the replacement was mere "repair" of the worn component, and not "reconstruction" of the patented combination.... And particularly persuasive, I think, that this replacement was mere "repair" is the role of the top relative to other components in the inventive concept.... The wiper arm, rather than the shape of the material alone, accomplished the inventive purposes of providing a top which was weathertight and prevented unauthorized access. The shape of the fabric was thus not the essence of the device, and, in all the circumstances, it seems reasonable and sensible to treat the replacement of the top as "repair."[62]

D. Repair, Exhaustion, and Social Perception

Some courts have tried to find a distinction between permissible repair and impermissible reconstruction on the basis of, either, an implied license in the reuse (as a legitimate expectation of the purchaser), or expectations of society at large regarding the reuse or recycle of a certain object.

1. Implied Licence

In the English *Dunlop* case, the court held:

> I should add that I do not think the defendant is justified in saying that the mere fact that these wires had been used in an article made by the plaintiffs is a licence to use them for any article in the same kind with which they could be used. In my opinion, the only licence or authority given by the sale to the plaintiffs was that the tyre, including the wires, might be used till it should be worn out. Any simple repairs might be done by any person without licence from the manufacture. By when a person takes the whole things and makes and sells that which is really a

62. *Id.* at 367–68 (Brennan, J. concurring).

new tyre, having merely the old wires on it, there is no licence from the plaintiffs to use the old wires for the purpose of putting them into a new article making up precisely the combination which is the subject of the letters patent.[63]

2. Multifactor Test

Also in the Japanese *Canon Ink Tank* case, where the plaintiff owned two patents—an apparatus patent (claim 1) on an ink tank for inkjet printers ("patented products") and a method patent (claim 10) for the manufacture of the patented products—reuse was made by drilling a vent on the top surface of the liquid ink storage room of the ink tank bodies, rinsing them, drying them, refilling ink through the vent, plugging the vent, putting labels on the products, and thus producing commercialized recycled products by reutilizing the used-up patented ink tank bodies.

According to the Tokyo High Court (appeal decision in the *Canon Ink Tank* case, subsequently overturned by the Supreme Court), a patent right could not be considered exhausted and the patentee could—

> [E]nforce its right in the following situations:
>
> **Scenario 1**: When a patented product is re-used or recycled after the patented product has fulfilled its original service life and thus fulfilled its function, or
>
> **Scenario 2**: When a third party adds or exchanges part or all of the substantial element of the patented invention of the patented products....
>
> Scenario 1 depends on whether or not the patented products have fulfilled their function (determination is made based on the patented products). Scenario 2 depends on whether or not the addition or exchange has been made to part or all of the substantial element of the patented invention (determination is made based on the patented invention).
>
> So, when part or all of the substantial element of the patented invention has been damaged or used up, and thus has fulfilled its function, additions or exchanges to such substantial elements fall both within type one and two. Where additions or exchanges are made to parts other than the substantial element of the patented invention, such act may fall within scenario 1, but not 2.[64]

---

63. Dunlop Pneumatic Tyre Co. v. Neal, (1899) 16 R.P.C. 247 (Eng.).

64. *Canon Inkjet Printer*, Tōkyō Kōtō Saibansho [Tōkyō High Ct.] Jan. 31, 2006 (Japan), *reprinted in* 37 INT'L REV. INTELL. PROP. & COMPET. L. 867 (2006). The Supreme Court took a different view, as I mention in Section V.B *supra*.

E. Comments

None of the above approaches is entirely satisfactory, be it for lack of coherence in terms of doctrinal thinking, or for the fact that recycling is becoming a socially desired behavior. I will only make two brief comments.

1. Reconstruction and Essential Element

In order to approach the issue of product identity, some courts have tried to determine whether an "essential element" was exchanged. Whereas the Dutch and US Supreme Courts have defined this as patent-specific (essential is what distinguishes the invention from prior art), the German Supreme Court attempts a definition that is product-specific, yet ends up including all parts that are not completely irrelevant for the protected product in question. As mentioned above, it appears preferable to interpret "essential" in terms of patent law rather than product specification. "Essential" should be what distinguishes the invention over prior art and what determines the contribution of the invention to society.

One other argument may also speak in favor of this interpretation: the actual product the patentee brings on the market may not always conform to the patent specification. The danger is thus that the courts may interpret the essentiality of an element in light of a product that may not be identical to the patent teaching, or may not even have been around at the time of the patent application and should therefore not be used to interpret the patent. The same problem occurs when the courts try to determine whether consumers regard the product as one that may be refilled or recycled, an approach taken by the Tokyo High Court in the *Canon Ink Tank* decision.

Even for the approach that is favored here, the distinction is not always clear-cut. In the *Coffee Pads* case mentioned above, the coffee pads as such were state of the art and therefore did not directly contribute to the advantages of the invention over the prior art. Yet, according to the Düsseldorf Appeals Court, the coffee pad in its concrete dimensions did contribute to the realization of the advantage that the shortened grooves could confer. Only a tight-fitting coffee pad in its interaction with the modified bottom could prevent hot water from flowing around the pad and directly into the cup instead of being pressed through the pouch. At the end of the day, this does make the pad essential in the patent sense: also a particular fuel (diesel, kerosene) may be essential for realizing the advantages of an improved injection system without making the fuel "essential" in the patent sense.

At least the more recent decisions in Germany and the United Kingdom, and the Japan IP High Court decision in *Canon Ink Tank* have applied a multifactor test that, although not necessarily perfect in the issues to be considered, may be the most promising approach to distinguishing permissible repair from impermissible reconstruction. Less promising, as rather tautological, appears the approach taken by the UK House of Lords in *United Wire*, and the Japanese Supreme Court in *Canon Ink Tank*.

2. Exhaustion and Implied Permission

One doctrinal issue in connection with repair is the interesting fusion between the doctrines of exhaustion and implied licence. The distinction between repair and new manufacture is important in connection with exhaustion: the act of manufacture can never exhaust, and in this respect, the English courts have identified a concept that was originally developed by Josef Kohler for German law. On the other hand, the concept of exhaustion may not be sufficient to determine permissibility of repair due to the fact that in some cases, the purchaser of products may buy these in the legitimate expectation that the aftermarket is not tied to the manufacturer of the original product, and that certain acts of repair or recycle have implicitly been consented to by the patentee.

The purchase of a coffee machine may be such a case: if, as was decided by the Düsseldorf courts in the *Coffee Pads* case, the replacement of a coffee pad would indeed amount to a reconstruction of the machine, with the consequence that the patentee could prohibit such exchange (leaving aside the fact that when undertaken by a consumer, this would be an act of private use. However, quite a number of these machines can now be found in coffee bars), one may legitimately ask if this conforms to the legitimate expectation of the purchaser of such a product. In other words, would anyone be willing to purchase a coffee machine while knowing that it could only be used once?. In such case, one should reasonably infer that the patentee has consented to the replacement of the pads, and it is questionable whether such implied consent for replacement can be limited to replacement parts by the manufacturer if there is otherwise free competition in the market for replacement parts. Thus, in some cases of repair or replacement, implied consent by the manufacturer may prevent the latter from arguing infringement, although the exhaustion doctrine would not apply. This is perhaps what the German Federal Supreme Court in *Palettenbehälter II* tried to emphasize when remitting the case back to the first instance for further fact-finding as to whether consumers were actually exchanging the bottles without the cage, or whether they would replace both elements together.

This brings us to the next topic, namely exhaustion and contract.

VI. Exhaustion and Contract

The above-mentioned exhaustion doctrines of course have repercussions on the domestic scope of patent rights. This is particularly so for conditional sales or sales *sub modo*. Here, the doctrines of exhaustion and implied license arrive at different solutions.

A. Conditional Sales under the Implied License Doctrine

1. Scope of Contractual Limitations

The best example of the consequences of a conditional sale is the above-mentioned decision *National Phonograph v. Menck*. Here, the relevant contractual clause read as follows:

> Exchanging or tendering *Edison* phonographs or parts, records or blanks, in whole or part payment for privileges of any character, or for advertising, or for goods or some other make or nature, or the exchange or acceptance of merchandise of other make or nature in whole or part payment for *Edison* phonographs or parts, records or blanks is not permitted.[65]

The issue not clarified prior to the *Menck* decision was the question of patent infringement by subsequent purchasers who had no clear knowledge of the restriction. Regarding these conditions, the Privy Council held as follows:

> The sale of a patented article carries with it the right to use it in any way that the purchaser chooses to use it, unless he knows of restrictions. Of course, if he knows of restrictions, and they are brought to his mind at the time of the sale, he is bound by them. He is bound by them on this principle: the patentee has the sole right of using and selling the articles, and he may prevent anybody from dealing with them at all. Inasmuch as he has the right to prevent people from using them or dealing in them at all, he has the right to do the lesser thing, that is to say, to impose his own conditions. It does not matter how unreasonable or how absurd the conditions are.[66]

In the above-mentioned *Isler* case,[67] it was suggested that even the innocent purchaser could not be absolved from patent infringement. This was rejected in the *Menck* decision. The restriction was only valid if the purchaser bought the goods "with the knowledge brought home to him of the limitation of his rights."

---

65. Nat'l Phonograph Co. of Austl. Ltd. v. Menck, (1911) 28 R.P.C. 229 (P.C.) 247 (appeal taken from Austl.).

66. *Id.*

67. Badische Anilin und Soda Fabrik v. Isler, (1906) 1 L.R. Ch. 605 (Eng.).

2. Restriction to be "Brought Home"

This position was then reiterated in the 1995 *Roussel Uclaf* case:

> [The plaintiff] does say that the product as supplied from France was in drums bearing the label "For use in PRC only—re-export forbidden." The defendants answered that by an express challenge concerning the labelling of the drums. They had bought a drum which did not bear that legend... and I must take it that some drums were so marked but others were not so marked. I could well understand how somebody receiving two kinds of drums might think that they were entirely free positively to export those which are not so labelled, even in a joint venture company. So it is not established that it was brought home to the Chinese joint venture company by the labels on the drums that there should be not export whatever of deltamethrin technical grade.[68]

B. Conditional Sales and Patent Exhaustion under US Law

The most instructive, if not leading case concerning restrictions of use of patented goods, is *Mallinckrodt v. Medipart*:

> This action for patent infringement and inducement to infringe relates to the use of a patented medical device in violation of a "single use only" notice that accompanied the sale of the device. Mallinckrodt sold its patented device to hospitals, which after initial use of the devices sent them to Medipart for servicing that enabled the hospitals to use the device again. Mallinckrodt claimed that Medipart thus induced infringement by the hospitals and itself infringed the patent....
>
> The device is marked with the appropriate patent numbers, and bears the trademarks "Mallinckrodt" and "UltraVent" and the inscription "Single Use Only." The package insert provided with each unit states "For Single Patient Use Only" and instructs that the entire contaminated apparatus be disposed of in accordance with procedures for the disposal of biohazardous waste. The hospital is instructed to seal the used apparatus in the radiation-shielded container prior to proper disposal. The hospitals whose activities led to this action do not dispose of the UltraVent apparatus, or limit it to a single use.
>
> Instead, the hospitals ship the used manifold/nebulizer assemblies to Medipart, Inc. Medipart in turn packages the assemblies and sends them

68. Roussel Uclaf S.A. v. Hockley Int'l Ltd., (1996) 113 R.P.C. 441 (P.C.) 444.

> to Radiation Sterilizers Inc., who exposes the packages to at least 2.5 megarads of gamma radiation, and returns them to Medipart....
>
> The enforceability of restrictions on the use of patented goods derives from the patent grant, which is in classical terms of property: the right to exclude....
>
> This right to exclude may be waived in whole or in part. The conditions of such waiver are subject to patent, contract, antitrust, and any other applicable law, as well as equitable considerations such as are reflected in the law of patent misuse. As in other areas of commerce, private parties may contract as they choose, provided that no law is violated thereby:
>
> [T]he rule is, with few exceptions, that any conditions which are not in their very nature illegal with regard to this kind of property, imposed by the patentee and agreed to by the licensee for the right to manufacture or use or sell the [patented] article, will be upheld by the courts precedent.[69]

Similar restrictions have been held valid in a number of subsequent cases, be it under copyright or patent law.[70]

C. Conditional Sales and Patent Exhaustion in Civil Law Countries

Under Continental patent law, the patentee cannot enjoin acts that take place after the first lawful sale, in other words, after rights over a certain product have been exhausted. It is irrelevant in this context how the patentee wishes his products to be handled or not. A useful example in this respect would be the Japanese *Acycrobil* decision already mentioned above in Section III.B.[71] Here, the defendant had obtained marketing approval for a generic drug containing the same active formula as the one described in the plaintiff's (Glaxo Wellcome) patent. Marketing approval would expire unless the products were actually brought on the market within six months from the date of marketing approval. The defendant thus purchased the plaintiff's pharmaceuticals containing the active ingredient, distilled the latter, and used this basic patented substance for producing a generic drug. As the patentee had received the economic benefits deriving from the patent by the act of first marketing, he could no longer control further acts of commercial exploitation by the purchaser, however unwanted these might have

69. Mallinckrodt v. Medipart, 976 F.2d 700 (Fed. Cir. 1992)

70. *See* Ariz. Cartridge Remanufacturers Ass'n v. Lexmark Int'l, 421 F.3d 981 (9th Cir. 2005); Ioneer Hi-Bred Int'l v. Ottawa Plant Food, 283 F. Supp. 2d 1018 (E.D. Mo. 2003); *see also* Monsanto v. Scruggs, 459 F.3d 1328 (Fed. Cir. 2006); Quanta Computer v. LG Elecs., 128 S. Ct. 2109 (2008).

71. *See supra* note 18.

been. The action was thus dismissed by both the district and high courts in applying the remuneration doctrine as established by the Japanese Supreme Court in its *BBS Car Wheels III* decision.

What now would be the result if the patentee inserted certain provisions restricting the purchaser in the sales agreement? Although under common law breach of such provisions would amount to a patent infringement, under Kohler's exhaustion doctrine, the effect could only be inter partes. The patentee's possibility of broadening his monopoly by conditional sales was the very reason for Kohler to develop his doctrine.

Already in 1971, Ullrich Schatz could state: "thus, it is clear the principle of exhaustion is jus strictum in the member states of the EEC. Contractual limitations of the exploitation rights of buyers are without patent law effect."[72]

An example to prove the case could be the Japanese tissue paper case, slightly modified: here the plaintiff and patentee had supplied hospitals with tissue paper in patented dispensers. Upon concluding contracts with hospitals, it obliged them to have the dispensers refilled with tissue paper only through the patentee. The hospitals asked third parties to refill the dispensers. Under these circumstances, it is clear that an action for patent infringement must fail both against the hospitals and against third parties: against third parties because the rights over the patented dispensers are exhausted upon the first commercial marketing; against the hospitals, because of the same reason. This would be different under the implied license doctrine. Here, the hospitals would be liable for patent infringement, and the case against third parties would depend on proper notification of the conditions of the sale. The hospitals, nota bene, might still be liable for a breach of contract to the extent that one allows contractual obligations to supersede patent exhaustion.

In fact, the Japanese case was more difficult.[73] The patentee had not only imposed certain conditions, but had even retained title in the patented dispensers and under this retention obliged hospitals to have the dispensers refilled with the patentee's tissue paper. As the hospitals had asked third parties to refill the dispensers, the patentee sued only these third parties for patent infringement. The Osaka High Court found that the retention of title was invalid, as it was not clearly expressed, and refilling was permissible under the exhaustion doctrine as the paper as such was not patented.

---

72. Ullrich Schatz, *The Exhaustion of Patent Rights in the Common Market*, 2 INT'L REV. INTELL. PROP. & COMPET. L. 1, 11 (1971). This position was reiterated in *United Wire Ltd. v. Screen Repair Services (Scotland) Ltd.*, (2001) 118 4 All E.R. 353 (H.L.), available at http://www.publications.parliament.uk/pa/ld199900/ldjudgmt/jd000720/wire.htm ("The difference in the two theories is that an implied licence may be excluded by express contrary agreement or made subject to conditions while the exhaustion doctrine leaves no patent rights to be enforced.").

73. *Tissue Paper*, Osaka Kōtō Saibansho [Osaka High Ct.] Dec. 1, 2000, Heisei 10 (wa) no. 11089 (Japan).

It is easy to see that the court was uncomfortable with the patentee's tie-in obligation. Yet its legal reasoning is not convincing: retention of title is nothing other than a failure to pass property. Even if the retention of title was ambiguous or ambiguously worded, the question must have been to what extent the plaintiff wanted title to pass. If this was not so, property was not transferred even though the purchaser was not aware of this, or may even have a claim for such transfer to be made. In other words, unless the proprietor wanted title to pass, property did not pass. And in such event, the patented dispensers would still be the property of the patentee, regardless of whether the failure to transfer property was "brought home" to the third parties, or was known to them.

The correct result, however, does not depend upon the retention of title. To that extent, the "first sale doctrine" is misleading: exhaustion is not necessarily triggered by the first act of *sale*, but rather by the first *putting into commerce* by the patentee. It is the patentee's choice if he wants to put the article into commerce by sale, by rental, by lending, by giving it as a free gift or in whatever other fashion. All else would run counter to the exhaustion doctrine's purpose of protecting commerce in trading goods on the open market against unforeseeable acts by a patentee.

D. Further Remarks

Patents and contracts are based on different principles. A patent is essentially a contract between society and the inventor and thereby, one could argue, not open to arrangements between individual buyers and sellers, which is the nature of a private contract. The US courts (as above) have argued that the patentee's right of exclusion must allow for a gradual waiver of the right: retaining some rights must be part of the bundle of all rights of exclusion. Continental courts would start from free-market conditions and argue that the patentee has complete freedom of setting conditions under which the goods are first marketed, and no harm can occur if first marketing triggers a nonnegotiable exhaustion of rights for such goods. Some authors have viewed contractual arrangements as a solution for otherwise difficult cases of repair or recycle,[74] while others are more critical.[75] My concern is that a contract rule that can broaden the inbuilt limits of

---

74. P. Carstensen, *Post-Sale Restraints via Patent Licensing: A "Seedcentric" Perspective*, 16 Fordham Intell. Prop. Media & Ent. L.J. 1053 (2006); M.D. Janis, *A Tale of the Apocryphal Axe: Repair, Reconstruction, and the Implied License in Intellectual Property Law*, 58 Md. L. Rev. 423, 426 (1999); Mineko Mohri, *Patents, Repair and Recycling from a Comparative Perspective*, 2010 Int'l Rev. Intell. Prop. & Compet. L. 779; M.R. Patterson, *Contractual Expansion of the Scope of Patent Infringement through Field-of-Use Licensing*, 49 Wm. & Mary L. Rev. 157 (2007); E.I. Winston, *Why Sell What You Can License? Contracting Around Statutory Protection of Intellectual Property*, 14 Geo. Mason L. Rev. 93 (2006).

75. Estelle Derclaye, *Repair and Recycle Between IP Rights, End User License Agreements and Encryption*, *in* Spares, Repairs and Intellectual Property Rights 21 (Christopher Heath & Anselm Kamperman Sanders eds., 2009).

patent law may yet be another avenue to enforce patents against often weaker parties such as consumers, repair shops, etc., and that such contractual arrangements may contravene public policy in other fields such as environmental concerns where recycling should become the norm rather than the exception. In addition, one should also observe that the US courts seem to be much more generous in affirming the validity of limiting conditions on shrink-wrap or click-wrap licensing arrangements, thereby further weakening the purchaser's position.

VII. International Exhaustion

A. General Remarks

Just as there is no uniform rule on (domestic) exhaustion, there is no uniform rule on international exhaustion, either. International exhaustion becomes a relevant question where goods protected by an IP right in the country of importation are imported from abroad by a person who is not the domestic IP owner or licensee. In these cases, such goods would be imported "without consent," a situation commonly referred to as "parallel importation," an importation outside the channels controlled or approved by the domestic patentee. As acts of importation (and, of course, subsequent sale) belong to the exclusive scope of rights granted to an IP owner, the latter could prevent such importation without consent unless there are reasons the IP owner, and in the context of this chapter, the patentee, should be barred from exercising his rights. Such reasons may be found in the absence of a notice to the contrary (English common law), considerations of patent policy (the patentee may have received a sufficient reward already for the product that is now being imported), or some sort of abuse doctrine (the patentee himself marketed the goods abroad and should not be able to turn around and bar their importation on the basis that the goods allegedly infringe). All these theories are examined in further detail below.[76] Subsequently, the question of international exhaustion, territoriality and international trade in light of the Paris Convention, the TRIPS Agreement, and GATT are examined.[77]

The issue of international exhaustion, be it for patents or for other IP rights, has been far more politicized than issues of national exhaustion, mostly due to the fact that the possibility of separating domestic markets often allows for lucrative price discriminations. However, as is explained below, the legal case against parallel imports is rather weak barring any specific legislative rule to limit exhaustion to domestic or regional acts of marketing.

---

76. *See infra* Section VII.B–E.

77. *See infra* Section VII.F.

The subsequent remarks and case law refer to patents, and the doctrine of "international exhaustion" for trademarks and copyrights may be as different as for domestic exhaustion. Namely, trademarks serve a completely different function, that is, the indication of a commercial origin. It may thus be wrong altogether to speak of "exhaustion" in the case of trademarks, as the right to prevent confusion in trade can never exhaust (a case in point may be the alteration of trademarked goods post sale). After all, a trademark in its original conception is not a monopolistic right, and exhaustion is a concept that is appropriate for monopolistic rights only.

Finally, it should be pointed out that the lawfulness of importation must be determined according to the scope of the patent right in the country of importation. In case there is no patent right in the country of importation, importation can of course not be prevented on the grounds of alleged patent infringement. And, as is outlined below when dealing with territoriality, the fact that goods have been lawfully marketed in the country of exportation may be relevant for determining the lawfulness of importation (in other words, the scope of a domestic patent right may under some circumstances not allow for an action against the importation of goods that have been lawfully marketed abroad), but cannot impose any automatic consequences on domestic patent law. Thus, parallel importation of patented products may be lawful not "because the patentee marketed these goods abroad," but "because domestic patent law follows an exhaustion rule that is triggered by the goods being marketed by the patentee or with his consent" (to take the ECJ rule as an example).

B. International Exhaustion Rules under the Implied Licence Doctrine

The issue of patent rights combined with importation was first dealt with in *Betts v. Wilmott*:[78]

> The point is this: Supposing a man to have a patent in *France* and a patent in *England,* and he establishes manufactories in each country, it is contended that if he sells the patented article in *France* it is for the French market, and it does not justify a person buying the article in France, and bringing it over to *England,* and using it here; and Mr. *Grove* put the case of his having sold the English patent to an assignee, and still continuing the manufacture in *France*, and asked if the importer of any article sold in *France* would not come within the doctrine of *Caldwell* v. *Vanvlissengen*.

---

78. Betts v. Wilmott, (1871) 6 Ch. App. 239 (Eng.).

> No doubt, in the case so put, the importer would be restrained, because the license to sell, which belonged originally to the patentee, would then be vested in his assignee; and therefore no license in *England* given by the original patentee after he had sold the patent could authorize the use of the article, so as to defeat the right of the assignees in *England*. And, of course, in exactly the same way, if the original patentee assigned his patent in *France*, no sale by him in *England* would be allowed to defeat the rights of the assignee in *France*. In other words, it comes within the doctrine of leave and license, and there could be no leave and license in such a case. But where a man carries on the two manufactories himself, and himself disposes of the article abroad, unless it can be shown, not that there is some clear injunction to his agents, but that there is some clear communication to the party to whom the article is sold, I apprehend that, inasmuch as he has the right of vending the goods in *France* or *Belgium* or *England*, or in any other quarter of the globe, he transfers with the goods necessarily the license to use them wherever the purchaser pleases. When a man has purchased an article he expects to have the control of it, and there must be some clear and explicit agreement to the contrary to justify the vendor in saying that he has not given the purchaser his license to sell the article, or to use it wherever he pleases as against himself. He cannot use it against a previous assignee of the patent, but he can use it against the person who himself is a proprietor of the patent, and has the power of conferring a complete right on him by the sale of the article.
>
> I am therefore of the opinion that, if the Plaintiff cannot show that he has not himself sold the very article the use of which he now seeks to prohibit, he cannot succeed by way of injunction in this Court any more than he would in an action at law. Therefore, on that ground, the decision of the Vice-Chancellor being right, the appeal will be dismissed with costs.[79]

The issue came up again more than one hundred years later in a case that more particularly dealt with the question of what kind of notice should be necessary to enjoin an importer from importing patented goods put on the market abroad by the patentee or with his consent[80] —

> In this case, therefore, it is critical to Roussel's position that they establish that in the course of their original supply, in the words of Lord Shaw of Dunfermline in *National Phonograph Company of Australia Ltd. v. Menck,*

79. *Id.* at 245.

80. Roussel Uclaf S.A. v. Hockley Int'l Ltd., (1996) 113 R.P.C. 441 (P.C.).

(1911) 28 R.P.C. 229 at 248, they "brought home" to the relevant Party, whether the Chinese or the defendants, the alleged limited licence.

First then, the Chinese, the original supplyee.... In his evidence in chief M. De Seze said in relation to China that a joint venture company was set up, that Roussel supplied technical grade deltamethrin to the company in China for formulation. He said nothing about the terms of supply to the Chinese and in particular does not say that the Chinese had it brought home to them that the technical grade deltamethrin supplied to them should not be resold. He does say that the product as supplied from France was in drums bearing the label "'for use in PRC only, re-export forbidden.'" The defendants answered that by an express challenge concerning the labelling of the drums. They had bought a drum which did not bear that legend. The express challenge was in the following language: "'I note M. De Seze does not say when Roussel first began marking the kegs in this way nor has he sworn every single drum exported to China had those words stamped on it.'" The reply from M. De Seze does not take up the challenge, and I must take it that some drums were so marked but others were not so marked. I could well understand how somebody receiving two kinds of drum might think that they were entirely free positively to export those which are not so labelled, even in a joint venture company. So it is not established that it was brought home to the Chinese joint venture company by the labels on the drums that there should be no export whatever of deltamethrin technical grade.

It is of course not necessary that the label be on the product provided the message is brought home somehow. However closely one looks at the evidence one cannot see any such bringing home. The high water mark seems to be M. De Seze in reply in a sentence which says: "Roussel do not sell such products for purchasers either to use for their own purposes or to sell on to a third party as they see fit." That does not say that it is brought home to purchasers that they may not do those activities....

Next I turn to the position of the defendants as subsequent supplyees. Had it been brought home to them that there are such conditions? In a sense it is virtually impossible to do this once it has not been established that it was brought home to the original purchasers.... The evidence certainly establishes that people in Florida, Singapore, Holland, England, India, Taiwan and Jordan at least have been advertising the material freely for sale. Mr. Fysh for Roussel says that is a "grey market." I do not really understand the relevance of that observation. The question is not whether the market is grey or any other colour or shade, but whether or not conditions of a limited licence were imposed when the product was first supplied by Roussel.... Once the goods are sold without a limited licence then the purchaser buys them free of any patent restriction.

> In my judgment this case is far removed from a case strong enough for judgment under Order 14, and the application falls.[81]

This so-called "common law exhaustion" basically leaves the matter to contract law and the proper notification of commerce. Should the patentee not wish her patented products manufactured abroad to be imported and sold, she has to impose a contractual restriction at the point of sale. In order for such restriction to survive further acts of sale, proper notification has to be given also to subsequent purchasers. In the absence thereof, such purchasers can assume to obtain the same rights as with the purchase of other, non-patented articles: the right to commercialize the purchased item by export, import, sale, or distribution. The same rules now apply to Japan according to the Supreme Court's decision in *BBS Car Wheels III*[82] —

> If the patented products were marketed abroad, then it can be naturally expected that such goods may be imported into Japan if the patentee puts such goods into circulation abroad without any reservations at the time of transfer. The transferee or any other subsequent purchaser is understood to have purchased the product without any restrictions that might apply to such products in Japan.[83]

This approach has the benefit of simplicity and may therefore be particularly attractive to developing countries whose judiciary might not yet be so familiar with intellectual property rights. The decision of lawfulness of importation depends on the presence or absence of a proper notice on the goods, in other words something easy to detect both for purchasers and judges.

C. International Exhaustion under US Law

The legal muddle-puddle for domestic exhaustion rules continues for the issue of international exhaustion. *Boesch v. Graff*, decided by the US Supreme Court in 1890,[84] is considered the leading case in this respect:

> Letters-patent had been granted to the original patentees for the invention by the government of Germany in 1879 and 1880. A portion of the

81. *Id.* at 445.

82. *BBS Car Wheels III*, Saikō Saiansho [Sup. Ct.] July 1, 1997 (Japan), *reprinted in* 29 INT'L REV. INTELL. PROP. & COMPET. L. 331 (1998).

83. *Id.* at 334.

84. Boesch v. Graeff, 133 U.S. 697 (1890).

> burners in question were purchased in Germany from one Hecht, who had the right to make and sell them there....
>
> It appears that appellants received two invoices from Germany, the burners in one of which were not purchased from Hecht, but, in the view which we take of the case, that circumstance becomes immaterial. The exact question presented is whether a dealer residing in the United States can purchase in another country articles patented there, from a person authorized to sell them, and import them to and sell them in the United States, without the license or consent of the owners of the United States patent....
>
> The right which Hecht had to make and sell the burners in Germany was allowed him under the laws of that country, and purchasers from him could not be thereby authorized to sell the articles in the United States in defiance of the rights of patentees under a United States patent. A prior foreign patent operates under our law to limit the duration of the subsequent patent here, but that is all. The sale of articles in the United States under a United States patent cannot be controlled by foreign laws. This disposes of the second error relied on.[85]

The decision is certainly correct, but the approach is not. Domestic patent rights in the importing country are not automatically exhausted because they have been exhausted abroad (a position that, due to its automatic effect, would even run afoul of Article 4*bis*(1) of the Paris Convention). Rather, marketing abroad may be a circumstance that could trigger domestic exhaustion. It is not that domestic patent rights must follow what the foreign law of the exporting country decrees, but rather that domestic law must be interpreted in order to determine what acts abroad, if any, can trigger domestic exhaustion. Domestic law is and must be independent of any limitations of the law abroad if and because acts of foreign marketing may be taken into account in order to determine domestic exhaustion. In the US decision, the goods had been marketed in Germany under a prior user right, the conditions of which did not apply to the US market. Under these circumstances, the patentee had not consented to the marketing of the goods, and this could well be a reason for denying domestic exhaustion. The US court, however, did not interpret domestic law, but (correctly) ruled that foreign exhaustion could not impose a domestic one. But it is a doubtful precedent to argue that no acts abroad can trigger domestic exhaustion of patent rights, although exactly this was held in the subsequent *Fuji v. Jazz Photo* case discussed

---

85. *Id.* at 702–03.

above.[86] Apart from the issue of repair, the case also concerned the importation of cameras manufactured and marketed abroad:

> United States patent rights are not exhausted by products of foreign provenance. To invoke the protection of the first sale doctrine, the authorized first sale must have occurred under the United States patent. Our decision applies only to LFFPs [cameras] for which the United States patent right has been exhausted by first sale in the United States. Imported LFFPs of solely foreign provenance are not immunized from infringement of United States patents by the nature of their refurbishment.[87]

And, in the second *Jazz* case, the court further elaborated:

> This court does not construe the "solely foreign provenance" language or the *Boesch* citation to dictate a narrow application of the exhaustion principle. Specifically, this court does not read *Boesch* or the above language to limit the exhaustion principle to unauthorized sales. Jazz therefore does not escape application of the exhaustion principle because Fuji or its licensees authorized the international first sales of these LFFPs. The patentee's authorization of an international first sale does not affect exhaustion of that patentee's rights in the United States. Moreover, the "solely foreign provenance" language does not negate the exhaustion doctrine when either the patentee or its licensee sells the patented article abroad. Read in full context, this court in *Jazz* stated that only LFFPs sold within the United States under a United States patent qualify for the repair defense under the exhaustion doctrine. *Id.* Moreover, Fuji's foreign sales can never occur under a United States patent because the United States patent system does not provide for extraterritorial effect.[88]

As mentioned above, the *Boesch*[89] court was under the impression that to apply the exhaustion doctrine would mean to apply foreign law. This weakness was exposed in *Jazz*, but the only reason the court could find for why foreign lawful marketing should not be taken into account is that patent law should not have

---

86. *See supra* note 30.

87. Jazz Photo Corp. v. Int'l Trade Comm'n, 264 F.3d 1094, 1105 (Fed. Cir. 2001) (internal citations omitted).

88. Fuji Photo Film v. Jazz Photo Corp., 394 F.3d 1368, 1376 (Fed. Cir. 2005) (internal citations omitted).

89. *See supra* note 84.

extraterritorial effect. There are two flaws in this approach. One, for a court that has heavily relied on the implied license doctrine, one wonders where this doctrine featured in the above decision. Two, acts performed abroad may very well have an effect on US patent law—one only needs to think of foreign prior art, or reduction to practice abroad. It should be added that similar considerations of alleged extraterritorial effect were rejected by the US Supreme Court in the field of copyright.[90] On patents, the US Supreme Court recently declined to give a ruling on the issue.[91]

D. The ECJ and Regional Exhaustion Based on Consent

As is well-known, no Community patent law exists that would specify exhaustion rules within the Community or with respect to third countries. Still, national exhaustion rules may be superseded by the principles of intra-Community free trade according to Articles 28 and 30 of the EC Treaty. According to European case law, intra-Community parallel imports of patented products can only be objected to where the "specific object" of patent law would so require. Or, worded differently, the patent existing in the country of importation can only be invoked against parallel imports from other member states in situations that would merit a derogation from the principle of free intra-Community trade. This is the case where the specific object of the patent is concerned. The latter is defined as:

> Inter alia to ensure to the holder, so as to recompense the creative effort of the inventor, the exclusive right to utilise an invention with a view to the manufacture and first putting into circulation of industrial products, either directly or by the grant of licences to third parties as well as the right to oppose any infringement.[92]

The specific object thus seemed to refer to two conditions: compensation and consent. It was thus uncontroversial that the patentee could object to the parallel importation of patented products in cases where these had been produced in another member state under a compulsory licence:[93] the patentee had not consented, nor could it be said that he had obtained an equitable reward under market conditions. The issue was far more controversial in cases

90. Kirtsaeng v. John Wiley & Sons, Inc., 133 S. Ct. 1351 (2013).

91. Ninestar Tech. Co. v. Int'l Trade Comm'n, 667 F.3d 1373 (Fed. Cir. 2012), *cert. denied*, 133 S. Ct. 1656 (2013).

92. Case 15/74, Centrafarm v. Sterling Drug, 1974 E.C.R. 1147, *available at* http://curia.europa.eu/juris/showPdf.jsf?text=&docid=88859&doclang=EN.

93. Case 19/84, Pharmon v. Hoechst, 1985 E.C.R. 2281, *available at* http://curia.europa.eu/juris/showPdf.jsf?text=&docid=92992&doclang=EN.

of parallel imports from Community countries where no patent protection (at least for this type of invention) was available. The first case concerned the parallel importation of pharmaceuticals from Italy to the Netherlands. The products had been manufactured in Italy by the Dutch patentee who, however, was unable to obtain an Italian patent for these, as patent protection for pharmaceuticals did not exist in Italy at that time. The court decided that *consent* was enough and held that

> It is for the proprietor of the patent to decide, in the light of all the circumstances, under what conditions he will market his product, including the possibility of marketing it in a member state where the law does not provide patent protection for the product in question. If he decides to do so he must then accept the consequences of his choice as regards the free movement of the product within the common market, which is a fundamental principle forming part of the legal and economic circumstances which must be taken into account by the proprietor of the patent in determining the manner in which his exclusive right will be exercised.[94]

This approach of course disregards the function of securing the patentee a market-based reward for every product sold *under the monopolistic conditions of a patent*. After all, a product sold in the absence of a patent can never give the producer a reward under monopolistic conditions, and thus never trigger exhaustion—

> The doctrine of exhaustion was hardly appropriate to the *Merck* case: there was no Italian right to be exhausted by Merck selling a drug in Italy. The result was that to some extent Merck was deprived of any benefit from their patent rights in Holland. The parallel imported drugs were free of royalty.... I think these matters are a good reason for referring the question to the court.[95]

The subsequent case referred to the ECJ by the English High Court concerned the parallel importation of pharmaceutical products from Spain and Portugal into the United Kingdom. Although the products had been placed on the market in those countries by the patentee or with his consent, no pharmaceutical

---

94. Case 187/80, Merck & Co. v. Stephar, 1981 E.C.R. 2063, *available at* http://curia.europa.eu/juris/showPdf.jsf?text=&docid=91009&doclang=EN.

95. Merck v. Primecrown, [1995] F.S.R. 909, 913, 914 (Eng.).

patent protection could have been obtained in these countries at that time. The Advocate General was in favor of overturning *Merck v. Stephar*:

> The strongest argument in favour of the pharmaceutical companies' interpretation of *Centrafarm v. Sterling Drug* is that, since the specific subject matter consists of the exclusive right of first marketing the patented product, a rule permitting parallel imports of such products marketed by the patentee in a member state where no patent protection exists and where, consequently, the patentee was subject to potential competition at the first marketing stage, would empty that exclusive right of much of its significance, i.e. the patentee must at least have had the opportunity of obtaining monopoly profits in the exporting member state before its national rights in the importing member state can be said to have been exhausted.[96]

Also some academics, including this author, took the view that consent of marketing was insufficient to describe the specific object of a patent.[97]

The ECJ, however, remained unconvinced and affirmed the principle of *Merck v. Stephar*:

> The court held, finally, in paragraphs 11 and 13 of *Merck* that it was for the holder of the patent to decide, in the light of all the circumstances, under what conditions he would market his product, including the possibility of marketing it in a member state where the law did not provide patent protection for the product in question. If he decides to do so, he must then accept the consequences of his choice as regards free movement of the product within the common market. This being a fundamental principle forming part of the legal and economic circumstances which the holder of the patent must take into account in determining how to exercise his exclusive right. Under those conditions, to permit an inventor to invoke a patent held by him in one member state in order to prevent the importation of the product freely marketed by him in another member

96. *Primecrown*, [1995] F.S.R. at 909, 913, 914; Beecham v. Europharm, Dec. 5, 1996 (Eng.), *reprinted in* 29 INT'L REV. INTELL. PROP. & COMPET. L. 184 (1998).

97. Paul Demaret, *Industrial Property Rights, Compulsory Licenses and the Free Movement of Goods under Community Law*, 18 INT'L REV. INTELL. PROP. & COMPET. L. 161, 176 (1987); Christopher Heath, *Zur Paralleleinfuhr patentierter Erzeugnisse*, 1997 RIW 541, 544. *But see* Friedrich-Karl Beier, *Gewerblicher Rechtsschutz und freier Warenverkehr im europäischen Binnenmarkt und im Verkehr mit Drittstaaten*, 1989 GEWERBLICHER RECHTSSCHUTZ UND URHEBERRECHT, INTERNATIONALER TEIL [GRUR INT]. 603, 613.

> state where the product was not patentable would cause a partitioning of national markets contrary to the aims of the treaty.[98]

Based on the above case law, patentees would thus be unable to prevent the parallel importation of products from newly acceding member states even where patent protection was not available in the country of exportation.

The ECJ's approach of consented marketing has also been adopted by a number of other countries that permit *international* exhaustion on such condition, for example, Thailand,[99] Argentina,[100] and South Africa.[101]

E. International Exhaustion Based on Remuneration

1. Kohler's Exhaustion Doctrine and International Trade

Kohler's doctrine as described above has been applied by many countries in the context of domestic trade. Perhaps, the implications on the free flow of domestic trade would be too severe to assume that a patentee can monopolize not only the marketing of patented products, but also subsequent sales.

In the context of international trade, however, the exhaustion doctrine is faced with problems that differ from the English theory of common law exhaustion. Under the latter doctrine, the sole condition is sufficient notice of the limits set by the patentee. Under the exhaustion doctrine, it is considered to what extent the first marketing of products abroad has the same effect as it has within the context of domestic trade. If the first marketing abroad had such effect, any objection by the patentee would be irrelevant. If, on the contrary, marketing abroad had no such result, the patentee could object to the importation even without proper notice to the public.

---

98. Case C-267/95, Merck v. Primecrown and Beecham v. Europharm, 1996 E.C.R. I-6285, *available at* http://curia.europa.eu/juris/showPdf.jsf?text=&docid=100592&doclang=EN.

99. *See* Thailand Patent Act, § 36(3)(vii), B.E. 2522 (1979), as amended.

100. See Argentina Law on Patents and Utility Models, § 36, Law No. 24.481, Mar. 30, 1995, as amended ("The right granted by any patent shall not have effect against: ... (c) any person that acquires, uses, imports or commercialises in any way the product patented or obtained by the patented process, once such product has been lawfully marketed in any country."). This provision, of course, goes way beyond the above consent rule, as it basically obliges any patentee to obtain patents in all countries of the world to obtain proper protection. In the author's opinion, this is difficult to justify within the framework of the TRIPS Agreement.

101. The relevant provision cannot be found in South Africa's Patent Act, but in Section 15C(a) of the Medicines and Related Substances Control Act of 1965, Act No. 101, June 19, 1965, as amended in 1997. The amendment has not yet come into force due to questions of the compatibility of compulsory licensing provisions with the TRIPS requirements. The explicit purpose of the 1997 amendment was to ensure the proper supply of the South African population with pharmaceuticals, particularly against the HIV virus.

There are some twists in the argumentation, however. Under the exhaustion doctrine in the classical sense, it would of course be required that a foreign patent be exhausted upon the first sale. In other words, if products have been marketed abroad, the domestic patent right of the patentee can only be "exhausted" if the products were marketed abroad under an exclusive patent right, and if such patent right belonged to the domestic patentee. The—highly controversial—question is then how far marketing of patented products abroad under the above conditions can indeed provide the patentee with those benefits the domestic patent right was intended to confer. Interestingly enough, Josef Kohler, just as some national courts (see below), has denied that foreign marketing can trigger domestic exhaustion, and I will deal with the reasoning of these decisions below.

2. Exhaustion and Remuneration

The patentee should receive one chance to market her product under monopolistic conditions, thereby exhausting the patent's potential value with respect to such product. This is Kohler's exhaustion doctrine. Taking this to the realm of international trade, the court of importation would have to ask whether the patentee by the marketing of the goods abroad has received a chance to market the product that is now going to be imported under monopolistic conditions. This was the starting point of the Tokyo High Court's decision on international exhaustion (subsequently overturned by the Supreme Court that used the doctrine of implied consent):

> There is no difference to the case of national exhaustion where the patent holder puts goods into circulation abroad and can determine the prices for the patented products of his own free will as a remuneration for the disclosure of his invention. This means, the opportunity of the patent holder to receive compensation for the disclosure of his invention is limited to one opportunity. Under the material aspect of the above-cited doctrine of national exhaustion in accordance with economic development, it does not make any particular difference whether the putting into circulation takes place within this country or abroad.[102]

Strictly speaking, the remuneration doctrine does not necessarily require the patentee's consent in marketing the products in order to exhaust the right. If, say, the patentee has decided to commercialize his patent by selling it to an unrelated third party, he has received one form of compensation for the right in

---

102. *BBS Car Wheels II*, Tōkyō Kōtō Saibansho [Tōkyō High Ct.] Mar. 23, 1995 (Japan), *reprinted in* 27 INT'L REV. INTELL. PROP. & COMPET. L. 550 (1996).

toto, and therefore also for all goods subsequently produced under such right. When compared to the theory of consented marketing, the remuneration theory is preferable for the fact that it does not require the patentee to apply for patent protection in every country where future marketing is likely—often, the commercial potential of an invention cannot be properly assessed at the filing date, or even within the priority period prescribed by the Paris Convention. The remuneration approach has found some following in academic circles, particularly in Japan[103] and Germany,[104] though it has not been endorsed by legislation in any country yet.

3. Discussion

Although there are also political reasons that have weighed against the adoption of a theory of international exhaustion, the legal ones are essentially these:

(1) The costs of a patentee to register the patent right in different countries;[105]
(2) The patentee's legitimate right to obtain maximum profits for his right by separating national markets;[106]
(3) The further argument of the Japanese Supreme Court that a domestic exhaustion would not necessarily correspond to an international one because "it cannot be argued that all those selling the product under a parallel patent would receive a double reward, or that it may qualify as an act of use of the domestic patent right";[107]
(4) The finding of the Appeal Court of Paris under French law that the absence of any contractual limit on the patented product did not indicate any form of consent, and further that the doctrine of international exhaustion did not have any clear basis in national or community law;[108]

---

103. *See, e.g.*, Nobuhiro Nakayama, *Tokkyo seihin no heikô yunyû*, 1064 Jurisuto 31 (1995); Yoshiyuki Tamura, *Heikô yunyû to chiteki zaisanken*, 1064 Jurisuto 45 (1995); Naohiko Tatsumi, *Shôhin ryûtsû to chiteki zaisanken no hôteki kôsei*, 21 Tokkyo Kenkyû 52 (1996).

104. Christopher Heath, *Zur Paralleleinfuhr patentierter Erzeugnisse,* 1997 RIW, 541; Hans-Georg Koppensteiner, *Urheber- und Erfinderrechte beim Parallelimport geschützter Waren*, 1971 RIW/AWD 357.

105. Dietrich Reimer, *Der Erschöpfungsgrundsatz im Urheberrecht und gewerblichen Rechtsschutz unter Berücksichtigung der Rechtsprechung des Europäischen Gerichtshofs*, 1972 Gewerblicher Rechtsschutz und Urheberrecht, Internationaler Teil [GRUR Int]. 221.

106. Friedrich-Karl Beier, *Zur Zulässigkeit von Parallelimporten patentierter Erzeugnisse,* 1996 Gewerblicher Rechtsschutz und Urheberrecht, Internationaler Teil [GRUR Int] 1.

107. *BBS Wheels III*, Saikō Saiansho [Sup. Ct.] July 1, 1997 (Japan), *reprinted in* 29 Int'l Rev. Intell. Prop. & Compet. L. 334 (1998).

108. *Phyteron,* Cour d'Appel [CA] [Regional Court of Appeal] Paris, May 29, 1999 (Fr.).

(5) Further, the majority decision of the Swiss Supreme Court that allowing for parallel imports in the field of patents would require an explicit legal provision;[109]
(6) The concept that a marketing abroad would not give the domestic patentee the same reward as a domestic marketing.[110]

The argument under (1) would be convincing if patent rights in other countries were rendered worthless by the possibility of parallel imports. Yet this is not the case. Even when allowing parallel imports, patent rights in other countries would still serve their main—and perhaps only—purpose of granting absolute rights against the use, manufacture, and sale of infringing products originating from other sources. Yet it is questionable if it is the function of patent law—and perhaps intellectual property rights in general—to protect the patentee against his own dealings, which would amount to the protection of economic self-denial.

Argument (2) above either stands to reason (any patentee should be allowed to fully enjoy those rights granted under patent law) or is unconvincing. As patents have always been a compromise between the inventor and the public at large for the purpose of stimulating inventive activity, the system and structure of patent rights has never been "unlimited." It suffices to note that patent rights are limited in time—twenty years from the filing date. Therefore, interpreting patent law in conformity with its underlying purpose would be more consistent with the system of patent rights than an interpretation that disregarded the patent system as a proper compromise between the patentee and the public.

The decisions of the Japanese and French courts are certainly clear in stating that the rules for national exhaustion could not be applied to goods marketed abroad. Yet the material question is *why* this should be so, and this question is left unanswered by these two courts.[111] Also the German Federal Supreme Court was clear in rejecting international exhaustion, but could not give more convincing arguments than that there was no international rule mandating international exhaustion:

> On the other hand, the marketing of a patented product by the patentee or with his consent outside the geographical area of the European Union does not entail exhaustion of the German patent or the German part of a European patent. The principle of national exhaustion was already

109. Case 4C.24/99, Bundesgericht [BGer] [Federal Supreme Court] Dec. 7, 1999 (Switz.). The decision was reached by a verdict of 3-2.

110. Argued particularly by WOLFGANG BERNHARDT & RUDOLF KRASSER, LEHRBUCH DES PATENTRECHTS 582 (4th ed. 1986).

111. Interestingly enough, it is also left unanswered by many academics. HERMANN ISAY, PATENTGESETZ 221 (4th ed. 1926); JOSEF KOHLER, *supra* note 16, at 455.

endorsed by the Imperial Supreme Court. The Federal Supreme Court has endorsed this in various decisions. In addition, this reflects the majority view among academic writings. The principle of national exhaustion is further applied in most other Member States of the European Community.

The opposite view was taken by the Tokyo High Court in its decision of 1995, when it held that the parallel importation of patented products marketed by the patentee in the country of exportation did not infringe the patentee's exclusive right of use. The Japanese Supreme Court has narrowed this point of view to the extent that parallel imports of patented products may be prohibited by the patentee once an unambiguous indication of a prohibition of importation is attached to the goods. In addition, the commercial court Zurich in its decision of 23 November 1998 has taken the view that on the basis of Sec. 8 Swiss Patent Act, the marketing of patented products by the patentee or a related company would entail exhaustion even if undertaken abroad. On the other hand, other Swiss courts took the view that only the doctrine of national exhaustion should be applied, while a decision of the Swiss Federal Supreme Court in this matter has not yet been rendered. These latest decisions from Japan and Switzerland are no reason to change the practice of this court. The Federal Republic of Germany has not entered into any international agreement that obliges its member states to apply the principle of international exhaustion. Neither the Agreement on Trade Related Aspects of Intellectual Property Rights (TRIPs)... nor the Paris Convention for the Protection of Industrial Property of 20 March 1883... can be interpreted in that way. Art. 6 TRIPs Agreement is mostly interpreted in a way that the exhaustion of intellectual property rights, apart from not being actionable under the TRIPs Dispute Settlement Rules, has indeed left exhaustion rules at the discretion of the individual contracting states. According to a different view, Art. 6 in connection with Arts. 27(1) and 28(1) mandates the principle of national exhaustion. Whichever opinion is preferable can be left open. Neither the provision of Art. 6 TRIPs Agreement concerning exhaustion nor the provisions on patent law in Arts. 27 and 28 TRIPs Agreement require the introduction of the principle of international exhaustion of patent rights. The same applies to Art. 4*bis* (1) and (2) Paris Convention. Also for these provisions, it is disputed whether contracting Member States are required to prohibit parallel imports, or if the matter is left to the national legislation of Member States. In no case can an obligation to introduce the principle of international exhaustion for patents be concluded therefrom. Accordingly, neither the Tokyo High Court, nor the Japanese Supreme Court or the Commercial Court Zurich, in reaching their verdict of international exhaustion for patents,

> have based this on the TRIPs Agreement, the Paris Convention or any other international agreement, but have reached their interpretation based on domestic patent law.
>
> Within the scope of German patent law, apart from national exhaustion, only EU/EEA exhaustion has been recognised by the highest courts. None of the amendments to the Patent Act has concerned a change from national (EU/EA) exhaustion towards international exhaustion. On the contrary, current practice is confirmed by the commemorative writings on the Community Patent of 15 December 1975. It is stated there that the principle of national exhaustion recognised in Germany for more than 70 years could also apply to the Community Patent. The right of a Community Patent should be exhausted once the patented product has been marketed in any Member State of the European Community by the patentee or with his explicit consent.[112]

Finally, the Swiss Supreme Court is certainly correct in pointing out that a clear provision in the law would help the interpretation of the exhaustion question. Still, the verdict reached raises two questions. First, why should the lack of any clear provision in the law not lead to the assumption that parallel imports are *permitted* rather than prohibited? Only this would be in accordance with the principle that intellectual property rights as exceptions to the rules of free competition cannot be interpreted beyond the scope granted by law. In the absence of any provision to the contrary, and in the absence of any result that can be deduced from the rationale of patent law as such, the rules of free competition should be applied.

As to the argument of insufficient reward, the law is only interested in the opportunity for reward as such, not the actual sum of money. For that reason it is not important if the patentee could make a better or worse profit by the sale in another country, but that the patentee had the chance to market the product under the same conditions as domestically, that is, under the same monopolistic right. This is not the case if the product was sold abroad by the patentee or with her consent, yet without the existence of a patent right there. And, in addition, it is not the case where the patented goods have been sold abroad under a scheme of subsidized pricing, for example, where drugs are offered to developing countries under a specific agreement entered into with the national government.[113]

---

112. *Karate*, Case X ZR 61/98, Bundesgerichtshof [BGH] [Federal Court of Justice] Dec. 14, 1999, Gewerblicher Rechtsschutz und Urheberrecht, Internationaler Teil [GRUR Int] 635, 2000 (Ger.). (internal citations omitted).

113. The EU's solution to the problem was Council Regulation 953/2003 of 26 May 2003 to Avoid Trade Diversification into the European Union of Certain Key Medicines, O.J. (L 135) 5 (EC). The Regulation regards pharmaceuticals sold in developing countries below

F. Exhaustion and International Law

1. The Paris Convention

The implied license doctrine under English law only attaches importance to proper notification of distribution limits to all subsequent users of the patented product. The classical exhaustion doctrine suggests, however, that limits to a patent right are inherent rather than dependent upon a patentee's clearly expressed intentions. Applying the principle of international exhaustion, as, for example, the Tokyo High Court and Supreme Court have done, has sometimes been objected to by invoking the principle of "territoriality of patents" as expressed in Article 4*bis* of the Paris Convention.[114] Domestic patent rights, so the argument runs, because of their territorial scope cannot be limited by acts committed outside such scope. In other words, a Japanese patent could not become exhausted because patented products were marketed in Germany, which is outside the scope of the Japanese patent right. Such an argument, however, misinterprets the intention and wording of Art. 4*bis* of the Paris Convention.[115]

Historically, some countries—particularly France—made the existence of a domestic patent right obtained under the priority of a foreign patent right dependent upon the existence of the latter.[116] Other countries refused to grant a subsequently filed patent a longer term of protection than that of the original one (Brazil, France, the United States, Belgium, Italy, and Spain). This principle of dependence of patents, also applied to trademarks under the Madrid Agreement, was found undesirable and indeed in contravention of the original spirit of the Paris Convention. For this reason, Article 4*bis* of the Paris Convention was inserted at the Brussels Conference in 1901, and subsequently clarified at the Washington Conference in 1911.[117] The present wording makes clear that the independence of patents concerns "grounds for invalidation and forfeiture and as regards their normal duration." However, there is nothing in the provision to suggest that developments abroad cannot influence patent rights at all. It is now standard practice that patents are only granted on condition of absolute novelty.

---

market price and under a certain symbol mark as counterfeit products when imported to the European Union.

114. This has been argued by the German Federal Supreme Court in the *Tylosin* decision, *reprinted in* 8 Int'l Rev. Intell. Prop. & Compet. L. 64 (1977), and has been explicitly rejected by the Japanese Supreme Court in *BBS*.

115. Friedrich-Karl Beier, *Territoriality and International Trade*, 1 Int'l Rev. Intell. Prop. & Compet. L. 70 (1970).

116. Stephan Ladas, Patents Trade Marks and Related Rights 505 (1975).

117. Actes de la Conference de Bruxelles 311 (1901); Actes de la Conference de Washington 22, 249 (1911).

Absolute novelty, however, requires taking into account the worldwide state of the art, not only the national one. In a similar fashion, national patent law may decree that foreign acts of marketing may have an effect on the exercise of the patent right with regard to particular goods marketed abroad. Article *4bis* of the Paris Convention is concerned with the existence of a domestic patent right, whereas the exhaustion doctrine is concerned with acts that "exhaust" further economic exploitation with regard to specific goods marketed under a patent. Under the exhaustion doctrine, the limits of economic exploitation are defined, and the Paris Convention in fact never dealt with this problem in the first place.

The main, or perhaps the only, remaining proponent of the thesis that a regime of international exhaustion (at least for patents) would contravene the Paris Convention is Nuno Pires de Carvalho. He writes: "The fact that patent rights are exhausted in the territory of one country should have no impact whatsoever on the patent rights valid in the territory of another country, given that those rights are entirely independent."[118]

This view is a misunderstanding of the nature of the principle of independence. Stripped of all disguises, independence guarantees sovereignty. It does not allow for any legal mechanisms to tie the fate of a domestic patent to the fate of a foreign one. In other words, it would not be permissible for a domestic court to accept exhaustion of a patent right merely because there was exhaustion in the country of exportation. The US case *Boesch v. Graff* would be a good example, and the court correctly held that there could be no such automatism. A patent cannot be revoked for the simple fact that it has been revoked in the country of first filing. But that does not mean that a domestic patent cannot be revoked on the same grounds for which the patent in the country of first filing was revoked, for example, a public prior use or a novelty-destroying document.

It is of course permissible to take foreign circumstances into account when determining the validity and scope of a domestic patent. A European patent may be revoked because of a public prior use in India; a US patent may be revoked because the invention was first reduced to practice in Indonesia by an inventor there. Or, to return to exhaustion, a domestic court may hold that the patentee has received a sufficient reward for a certain patented item by marketing such item abroad under monopolistic conditions. Territoriality is only at stake when certain consequences that have occurred in another jurisdiction must be adopted *telle quelle*. But this is not the solution that is argued here, nor is it the solution of those who advocate exhaustion due to first voluntary marketing or those who favor the doctrine of implied license.[119]

---

118. Nuno Pires de Carvalho, The TRIPS Regime of Patent Rights 115 (3d ed. 2005).

119. Carvalho's argument has another weakness when considering the case that the patentee markets a product in a third country without patent protection (country B), and the product

2. The TRIPS Agreement

Although the Paris Convention may be silent on the issue of parallel importation, other international treaties may influence domestic law on this point. The most important one in the field of intellectual property is the TRIPS Agreement concluded in 1994 as a package together with the GATT/WTO Agreement. Indeed, it would be expected from a treaty covering all aspects of intellectual property rights that the matter of parallel importation be also included. Not so. Although it was recognized that parallel importation would indeed fit nicely within the objective of international free trade advocated by GATT,[120] agreement could not be reached to allow generally for parallel importation.

In order to overcome this stalemate situation, Article 6 of the TRIPS Agreement now provides that "for the purposes of dispute settlement under this Agreement,... nothing... shall be used to address the issue of exhaustion of intellectual property rights." The dispute settlement mechanism in general allows every member to bring an action against another state if there is insufficient compliance with the principles of the GATT/WTO Agreement in general. Yet according to Article 6, whatever national stance is taken on the matter of exhaustion, no complaint can be heard in this respect. Although this certainly means that no country can be put in the dock for deciding for or against international exhaustion, it does not necessarily mean that the TRIPS Agreement as such would not favor either one or the other position.[121]

As this exception relates to procedural matters, it means only that members of the GATT/WTO Agreement cannot be made subject to sanctions, no matter how they decide on international exhaustion. Nevertheless, the agreement may favor explicitly or implicitly a certain solution to the subject matter of international exhaustion and parallel imports.

One aspect that has been particularly mentioned in this respect is the obligation of members to grant patentees a specific right of importation along with other exclusive rights such as for production and sale. An importation right is certainly useful once it comes to preventing counterfeit products from entering

---

is subsequently imported by a third party in a country where the patentee owns a patent (country A). Here, Article 4*bis* can undoubtedly not be applied, because there was no corresponding patent in the country of first sale (country B). Nonetheless, in determining infringement, domestic patent law (of country A) could place emphasis on the fact that it was the patentee himself who marketed the product abroad (which the European Court of Justice affirmed for the common market in *Primecrown*, [1996] ECR I-6285).

120. Thomas Cottier, *The Prospect for Intellectual Property in GATT*, 28 COMMON MARKET L. REV. 383, 401 (1991).

121. Marco Bronckers, *The Impact of TRIPS: Intellectual Property Protection in Developing Countries*, 31 COMMON MARKET L. REV. 1245 (1994); Joseph Straus, *Implications of the TRIPS Agreement in the Field of Patent Law*, *in* FROM GATT TO TRIPS 191 (F.K. Beier & Gerhard Schricker eds., 1996).

the country. Without an importation right, the patentee would have to wait until the counterfeit products are put on the market in order to obtain relief. This is certainly undesirable and inadequate. However, it is difficult to argue that the right of importation should follow different rules from the rights of production and sale.

The importation right concerns an aspect of economic exploitation just as much as production and sale. If, under the classical doctrine of exhaustion, further rights in commercial exploitation are exhausted upon the first sale of a patented article, and if such exhaustion is also assumed when such patented article is marketed abroad, then the exhaustion relates to all aspects of other commercial exploitation including importation. This becomes particularly obvious in the case of reimports. If a patented article is put on the market in, say, Japan, by the patentee or with his consent, then further acts of economic exploitation are "exhausted." If the patentee therefore would not be able to prevent further acts of sale and distribution, then it is difficult to see how and why the patentee should be able to exert any influence over this article once it has been exported into another country and subsequently reimported. If a patentee is granted a bundle of rights under her patent, such as production, sale, and importation, then upon the act of first sale, the whole bundle becomes "exhausted" once and for all. Consequently, no importation right can be invoked later on for the very article that has already been marketed previously.

Concerning the doctrine of common law exhaustion as outlined by the above-mentioned English and Japanese decisions, there is nothing in the TRIPS Agreement to suggest that the importation right cannot be made subject to certain conditions such as giving proper notice to the public about any restrictions in this respect.

The above analysis would only merit a different evaluation once national patent rights were rendered worthless by permitting parallel importation. Such might be the case if the patentee could not object to the importation of products produced in third countries where no patent rights were obtained, because, in theory and practice, this would require a patentee to apply for patents in all possible countries in order to receive at least once proper compensation for putting the goods on the market. However, as yet, no country has permitted parallel imports under these circumstances.

The subsequent Doha Declaration on Public Health of 14 November 2001 has clarified that "the effect of the provisions in the TRIPS Agreement that are relevant to the exhaustion of intellectual property rights is to leave each member free to establish its own regime for such exhaustion without challenge, subject to the MFN and national treatment provisions."[122]

---

122. World Trade Organization (WTO), Doha Declaration on the TRIPS Agreement and Public Health, art. 5(d), WT/MIN(01)/DEC/W/2, Nov. 14, 2001, *available at* www.who.int/medicines/areas/policy/tripshealth.pdf.

3. The GATT Agreement

There is no doubt that agreements such as GATT and TRIPS, meant to erase barriers to legitimate trade, would be the first to introduce a rule of international exhaustion after having made sure that there is a level standard of intellectual property protection. Not so. That a consistently applied rule of domestic exhaustion would make a mockery of international trade has been highlighted in a decision of the Zurich Commercial Court:

> The first marketing allows the patentee to realise the possibility of making the profit that is due to him. Allowing him to prevent or influence the further commercial distribution would impede economic commerce in an insupportable manner. In this respect, one has to bear in mind that a decision in favour of national exhaustion would not only mean the prohibition of parallel imports, but also of any other commercial use of the invention according to Sec. 8 Patent Act, including the commercial use. As the Judicial Commission of the High Court Lucerne in its decision of 11 December 1987 has stated, this would lead to untenable consequences. "An international businessman could, e.g., use a patented writing tool in the course of his business only in the country where it was bought." The example is harmless. Indeed, international trade would collapse. E.g., every foreign merchant travelling to Switzerland on business would not be allowed to use his technical appliances (if patented here and there) and would thus be unable to work. If, to take an example based on the subject matter of this case, an English commercial photographer bought the Kodak films in question in England, he would not be allowed to use them for shooting in Switzerland. In order not to commit a patent infringement, he would have to buy the films here. Then, however, he would not be allowed to export them for development to England, as he would have to do this here. The same for further copies. He could only sell the negatives in the country where the film was bought. And if one further imagines that such photographer was following, e.g., the Tour de France that for one etappe comes to Switzerland and that all patented products (helmets, glasses, gloves, tools, wireless devices, film and photo cameras, medicinal products, doping products, etc.) that were not purchased in Switzerland and do not come under Art. $5^{ter}(2)$ Paris Convention would have to be left at the border and purchased again in Switzerland to be used, it becomes clear that such a scenario is not in conformity with the purposes of patent law. Patent law should serve promotion of the economy, not its paralysis. Also in industrial production where products in the course of their manufacture are transported from one country to another, it would hardly be thinkable that the patentee having sold a patented device in country A to be built into a product that is further processed in country B and finalised in country C, in order to be sold in countries D and E, would claim

> licensing fees for countries B and C and seizure of the infringing products in countries D and E. Obviously, the interests of the patentee have to give way in favour of those of commerce and consumers interested in the free circulation of products once sold by the entitled person...' In addition, there is no public interest, let alone an overwhelming public interest in preventing parallel Imports. The Supreme Court in the "Chanel" decision held as follows: "All in all, a co-existence between contractual distributors and parallel vendors seems to be the solution that guarantees the optimal results that can be expected within the framework of free competition for the benefit of competitors, customers and the public at large" (BGE 122 III 469 E. 10b). In the case at issue, it should be added that the defendant was supplied with original products via the plaintiff's distribution system and thus meets the requirements that according to the plaintiff should be expected from retailers. In other words, it is not one of the usual constellations where the trade mark or patent owner regularly argues that parallel retailers did not meet the requirements and thus contributed to a negative image of the right owner. Here, the sole and only issue is the fact that the plaintiff or the parent company would achieve a higher profit if the defendant could not obtain Kodak products via English, but rather via Swiss channels. This may be an interest, but not a public one. Thus, international exhaustion would be the only interpretation in conformity with the Swiss Constitution.[123]

Unfortunately, common sense such as displayed above is in short supply when it comes to the drafting of international agreements that more often than not represent a political compromise between differing trade interests.

Some academic writings have also taken the point of view that GATT does not permit a rule of limited exhaustion,[124] but rather mandates a regime of international exhaustion.

Conclusion

The above represents an analysis of the legal issues surrounding patent exhaustion. Economic and political considerations have mostly gone into the issue of international exhaustion and parallel importation rather than the difficult relationship between repair and exhaustion, or contract and exhaustion.

---

123. *Kodak*, Handelsgericht [Commercial Court] Zurich, Oct. 23, 1988 (Switz.), GEWERBLICHER RECHTSSCHUTZ UND URHEBERRECHT, INTERNATIONALER TEIL [GRUR INT] 555, 1995 (Ger.) (overturned by the Swiss Federal Supreme Court).

124. *See, e.g.*, CHRISTIANE FREYTAG, PARALLELIMPORTE NACH EG- UND WTO-RECHT 282 (2001) (based on Articles XI and XX of GATT); DANIEL KRAUS, LES IMPORTATIONS PARALLÈLES DE PRODUITS BREVETÉS 178 (2004) (based on Article XI of GATT).

As to economic studies, there seems to be a significant "empirical uncertainty" surrounding the issue, as costs and benefits depend on the size of the country, the product sector one is looking at, the distribution system, etc.[125] However, some economists have been advocating a regime of international exhaustion.[126]

Also in the political arena, issues of international exhaustion have featured prominently, namely in the context of bilateral free trade agreements. Although attempts to pressure—particularly developing—countries into adopting a certain exhaustion regime have become less visible after the Doha Declaration, the US government still appears interested in limiting exhaustion rules via such bilateral trade agreements.[127]

To summarize my own position on exhaustion and patent rights: courts have often been too generous to patentees in allowing them to protect business schemes rather than inventions (in connection with repair), to adopt policies that directly contravene national policies of avoiding waste and encouraging recycling (in connection with recycling), or to extend patent monopolies by contract beyond a scope that would be commensurate to the patentee's contribution to society. In many cases, courts have not clearly started from the basic equation of patent law that monopolies are an exception and need justification. Not least for this reason, exhaustion should not be regarded as an exception, but as the norm for all goods that have legitimately been put on the market. It should be for the patentee to prove that circulation of such goods is not legitimate, and it should be the patentee (and ultimately the courts) to come up with a good reason as to why the importation of goods that were marketed abroad with the patentee's consent, or for which the patentee could obtain a proper reward, should not be permissible. On the level of GATT and TRIPS, "it is astonishing that the WTO Agreement should still consider it to be compatible with fair world trade for consumers in different countries to be played off against each other."[128]

---

125. *E.g.*, Kerrin Vautier, *The Economics of Parallel Imports*, *in* Parallel Imports in Asia 10 (Christopher Heath ed., 2003).

126. Frederick Abbott, *First Report to the Committee on International Trade Law of the International Law Association on the Subject of Parallel Importation*, 1998 J. Int'l Econ. L. 607.

127. An example is Singapore, who prior to the US–Singapore FTA had a regime of international exhaustion, whereas the FTA imposes certain limitations for pharmaceutical products. Other provisions can be found in the US–Australia and US–Morocco FTAs. Ng Loy Wee Loon, *Parallel Imports of Pharmaceuticals: Doha v. Free Trade Agreements*, *in* Intellectual Property and Free Trade Agreements 166 (Christopher Heath & Anselm Kamperman Sanders eds., 2007).

128. Hanns Ullrich, *Technology Protection According to TRIPS: Principles and Problems*, *in* From GATT to TRIPS 357, 385 (Friedrich-Karl Beier & Gerhard Schricker eds., 1996).

15 A NEW APPROACH TO THE COMPULSORY LICENSE CONUNDRUM

Cynthia M. Ho*

Introduction

This chapter addresses the conundrum of compulsory licenses on drug patents—why are compulsory licenses controversial, even though they are permissible under domestic and international laws? In a post-TRIPS world where many nations have revised their patent laws to now provide patents on drugs, compulsory licenses are considered by some to be one of the few remaining tools to help promote access to affordable medicines.[1] However, when some countries exercise their ability to issue compulsory licenses, they are accused of stealing or other improper action. In addition, there is widespread confusion or misrepresentation concerning what kinds of compulsory licenses are permissible under TRIPS. Notably, some of the misrepresentations are made by scholars who would not seem likely to be confused. Accordingly, a more fundamental phenomenon may be at issue than simple rhetoric and issue framing, although these can be very powerful in shaping the creation and interpretation of law.[2]

---

* Clifford E. Vickrey Research Professor and Director, Intellectual Property and Technology Program, Loyola University Chicago School of Law.

1. TRIPS requires all WTO members (except least-developed countries) to provide patents on all inventions without discrimination based on subject matter, such that patents on drug compounds are now required. Agreement on Trade-Related Aspects of Intellectual Property Rights, art. 27, Apr. 15, 1994, Marrakesh Agreement Establishing the World Trade Organization, Annex 1C, 33 I.L.M. 1143 (1994) [hereinafter TRIPS Agreement or TRIPS].

2. The power of rhetoric and issue framing is widely recognized with respect to the development of international laws that promote patent rights. *See, e.g.*, SUSAN K. SELL, PRIVATE POWER, PUBLIC LAW (2003); Amy Kapczynski, *The Access to Knowledge Mobilization and the New Politics of Intellectual Property*, 117 YALE L.J. 804 (2008). On the other hand, the patent perspectives suggested here might

This chapter provides insights, based upon social science research, into why compulsory licenses have inspired distortions and misunderstanding. As is well-documented in the field of social science, all individuals perceive information based on preexisting "schemas" through which they view the world.[3] These schemas are known to be powerful lenses that can result in dramatically different perceptions of the same factual information.

This chapter suggests that there are different perspectives of patent policies that serve as schemas. In particular, views of patent policy exist along a continuum, benchmarked by two divergent views that tend to dominate discussions. On one end of the spectrum, there is a view of patents as a privilege granted by the state and inherently subject to limitations—especially if necessary to promote other social policies, such as access to medicines; this is referred to as the "privilege view." On the other end of the spectrum, patents are considered essential to promoting innovation, with more protection considered to be necessarily beneficial to promoting more innovation; this is referred to as the "uber-right" view, to reflect the view of patents as a very strong right.[4]

Although these patent perspectives may bear some similarity to existing patent theories about whether rights should be maximized or minimized, the important new insight is that the existence of these views powerfully impacts how facts and laws are interpreted. As will be explained, some clear requirements under TRIPS, such as what subject matter may be compulsorily licensed, have been distorted to reflect a desired perspective of patents. The competing perspectives thus provide a vehicle for an enriched understanding of compulsory licensing controversies.

This chapter begins by discussing the basis for patent perspectives in social science theory and how the patent perspectives complement existing patent theories. The chapter also provides a more in-depth explanation of the privilege and uber-right perspectives before examining how each perspective has influenced interpretation of TRIPS requirements concerning compulsory licenses. Before discussing how the perspectives influence interpretations, this chapter briefly reviews basic compulsory license requirements under TRIPS Article 31. In addition, the complicated process for waiving one of the usual TRIPS requirements to

---

be the backdrop for the development of rhetoric and issue framing; unlike those issues, the perspectives are suggested to operate largely on a subconscious level.

3. *See, e.g.*, SUSAN T. FISKE & SHELLEY E. TAYLOR, SOCIAL COGNITION 97–99 (2d ed. 1991); ZIVA KUNDA, SOCIAL COGNITION: MAKING SENSE OF PEOPLE 17–19 (1999); Ronald Chen & Jon Hanson, *Categorically Biased: The Influence of Knowledge Structures on Law and Legal Theory*, 77 S. CAL. L. REV. 1103, 1125 (2004).

4. The term "uber-right" is used to emphasize a view of patent rights as very strong. However, there is no intent to imply a connection with a particular political ideology.

help provide low-cost drugs to developing countries is also explained. The complex process was first proposed in a decision by the TRIPS Council on August 30, 2003, and formally proposed as an amendment to TRIPS as Article 31*bis* in 2005.[5] The August 30 Decision remains in effect until two-thirds of WTO members formally ratify the amendment.

I. Background

A. Social Science Support

The perspectives presented here are supported by social science literature. In particular, research suggests that people rely on prior "schemas" that are essentially unconscious biases in receiving and understanding new information. A schema has been defined as a "mental structure which contains general expectations and knowledge of the world."[6] Although schemas are recognized as essential to enable individuals to cope with the large amounts of information present in everyday life, they also have been identified as a possible source of bias. Indeed, one example of a schema is a stereotype.[7] Schemas are used by all and can result in cognitive biases and imperfect decision-making regardless of education or sociocultural background.[8] They are often discussed in connection with fundamental beliefs or stereotypes, and have provided a useful method for gaining new insights into the operation of existing law and policy.[9]

Schemas have a powerful effect on how people process new information. Research shows a tendency for people to maintain preexisting beliefs out of

---

5. WTO, Decision of the General Council, *Implementation of Paragraph 6 of the Doha Declaration on the TRIPS Agreement and Public Health*, WT/L/540 (Aug. 30, 2003) [hereinafter 2003 WTO Council Decision]; WTO, Decision of the General Council, *Amendment of the TRIPS Agreement*, Annex to the Protocol Amending the TRIPS Agreement (Art. 31*bis*), WT/L/641 (Dec. 8, 2005). The 2005 document actually comprises proposed Article 31*bis*, as well as a proposed Annex to TRIPS that would appear after Article 73. WT/L/641 ¶ 1. These will hereinafter be referred to as Proposed Article 31*bis* and Proposed TRIPS Annex.

6. Martha Augoustinos & Iain Walker, Social Cognition: An Integrated Introduction 32 (1995).

7. *See, e.g.*, Chen & Hanson, *supra* note 3, at 1125–26.

8. *See, e.g.*, Cass Sunstein, *Behavioral Analysis of the Law*, 64 U. Chic. L. Rev. 1175, 1183 (1997).

9. *See, e.g.*, Jerry Kang, *Trojan Horses of Race*, 118 Harv. L. Rev. 1489, 1494–95 (2005) (showing how relaxation of ownership requirements by the Federal Communications Commission (FCC) may exacerbate implicit racial bias); L. Song Richardson, *When Human Experimentation Is Criminal*, 99 J. Crim. L. & Criminology 89, 94 (2008) (suggesting that implicit bias results in more lenient treatment of doctors who commit criminal acts in the course of scientific study).

proportion with actual correctness, including subsequent evidence revealing that the initial information was incorrect.[10] The reason this occurs is that that people have an inherent need to ignore, discredit, or rationalize inconsistent information to avoid the discomfort of cognitive dissonance.[11]

Moreover, individual schemas are reinforced by group identities. For example, in one famous experiment a film of a football game in which officials made a series of controversial decisions was viewed differently by students of two different schools.[12] Students who attended the offending team's college reported seeing half as many illegal plays as did students from the opposing institution. In other words, group ties had an unconscious, yet powerful effect in people viewing the facts to favor their group affiliation.

As a result of all of the above, those with different cultural views may perceive the same set of facts very differently. In particular, worldviews impact perceptions of facts to result in perceiving those facts in a manner that best reflects, rather than challenges those views. Studies have repeatedly shown this to occur in a number of areas of divergent cultural values, such as abortion, gun control, and global warming.[13]

This chapter builds upon these existing studies. The patent perspectives presented here can be viewed as schemas through which facts and laws are interpreted. Although social science research thus far has not studied patent perspectives, patent policy—especially with regard to its impact on affordable access to medicines—shares characteristics with the topics that have been studied thus far. In particular, whether patent rights should be promoted over short-term access to drugs is a topic that is just as controversial as abortion or gun control. Although all agree that strong patent rights with limited exceptions will reduce short-term access to drugs, there are strong disagreements on whether this is a problem.

---

10. *See, e.g.*, RICHARD NISBETT & LEE ROSS, HUMAN INFERENCE 169 (1980); Craig A. Anderson et al., *Perseverance of Social Theories: The Role of Explanation in the Persistence of Discredited Information*, 39 J. PERSONALITY & SOC. PSYCHOL. 1037, 1041–42 (1980); Kari Edwards & Edward E. Smith, *A Disconfirmation Bias in the Evaluation of Arguments*, 71 J. PERSONALITY & SOC. PSYCHOL. 5, 18 (1996); Arthur G. Miller et al., *The Attitude Polarization Phenomenon: Role of Response Measure, Attitude Extremity, and Behavioral Consequences of Reported Attitude Change*, 64 J. PERSONALITY & SOC. PSYCHOL. 561, 564 (1993).

11. LEON FESTINGER, A THEORY OF COGNITIVE DISSONANCE 2–3 (1957).

12. Albert Hastorf & Hadley Cantril, *They Saw a Game: A Case Study*, 49 J. ABNORM. SOC. PSCYHOL. 129 (1954).

13. Dan Kahan, *The Cognitively Illiberal State*, 60 STAN. L. REV. 115 (2007) (documenting this phenomena in political debates concerning a variety of controversial topics, such as sodomy, guns, smoking, nuclear energy, and global warming).

The polarized discussions of access to medicines that include accusations of bad faith are also consistent with social science. Social science evidence documents that individuals are unaware of their own biases, yet have a heightened sensitivity to the presumed biases of others.[14] Applying this insight to the arena of patents would suggest that holders of a particular patent perspective are vulnerable to believing that those with different views are necessarily distorting the facts, while simultaneously ignoring their own similar conduct.

Problematically, social science shows that this phenomenon can create resentment toward those with opposing views that are perceived as the only group skewing facts—resentment that in turn further reinforces not only distrust, but also further adherence to worldviews.[15] Accordingly, identifying this problem helps to explain the discord as a first step to potentially diffusing the distrust and resentment.

Although social science research thus far has focused primarily on the impact of preexisting worldviews on how facts are interpreted, this chapter shows the impact of patent perspectives on legal interpretations that do not turn on factual questions. This is a powerful new extension of existing theories on social science. After all, although actions may be taken to immunize people from existing biases concerning facts, those actions do not easily translate to a process that, thus far, has been presumed to be neutral.

B. Complement to Existing Patent Theories

Although there are a variety of existing theoretical bases supporting the existence of patent rights, those theories predominantly focus on why we should have patent rights in general, but do not address how to balance patent rights against access to medicines.[16] For example, although many may agree on the utilitarian theory of patents that suggests that patents are granted as a utilitarian tool to promote innovation,[17] that theory only supports the existence of patent rights in

---

14. This concept has been referred to as "naïve realism." *See, e.g.*, Robert J. Robinson et al., *Actual versus Assumed Differences in Construal: Naïve Realism in Intergroup Perception and Conflict*, 68 J. PERSONALITY & SOC. PSYCHOL. 404 (1995).

15. *See, e.g.*, Kahan, *supra* note 13 (documenting this phenomena in political debates concerning a variety of controversial topics, such as sodomy, guns, smoking, nuclear energy, and global warming).

16. Although some have suggested that patent theories might be used to predict outcomes in patent litigation, such predictions would still not be helpful to the question of the proper delineation between patent rights and access to medicine. *See* Samuel Oddi, *Un-unified Economic Theories of Patents—The Not Quite Holy Grail*, 71 NOTRE DAME L. REV. 267 (1995) (attempting to apply various patent theories to patent requirements, as well as scope of protection).

17. This is sometimes framed as a "reward" theory; namely, that patents provide a reward in the form of exclusivity over the patented invention to promote useful innovation.

general; it does not adequately explain how to fine-tune a patent system to consider other social considerations.

In addition, traditional patent theories—from whether to grant patents to determining the proper scope of patent rights—seem out of touch with the reality that nations today are largely shaping patent rights not based on these patent theories, but because of external realities. In particular, as noted earlier, many nations today provide patents on drugs not because they believe that this will promote greater innovation, but because they are required to under TRIPS. Although TRIPS arguably may be more consistent with certain patent theories, that does not mean that nations bound by TRIPS necessarily agree with those theories.[18] Rather it is widely understood that TRIPS was concluded *despite* its patent requirements so that developing countries could gain other benefits, such as greater access to the markets of wealthy WTO country members.[19]

The perspectives approach outlined here bears some similarities to existing intellectual property theories, yet is nonetheless distinct. In particular, patent rights and exceptions are often referred to as promoting either a "liability" or "property" rule.[20] A compulsory license is seen as a liability rule in that the patent right is taken and the owner compensated afterward; the person granted the license incurs a liability, but the patent owner has no control in protecting against an unauthorized use of his property. A property regime, on the other hand, would suggest that a patent right be more absolute, subject to cases of extreme necessity; in other words, the patent owner can prevent unauthorized use of her right under a property regime. The privilege perspective view thus bears some similarity to a liability regime whereas the uber-right perspective bears similarity to the property regime. However, an important distinction is that debates about liability versus property rights are generally made in the context of discussing what is economically efficient and usually promotes a *single* view, without considering the possibility that there are valid reasons to support the opposite view.

---

18. A "natural rights" theory of patents might explain why there should be increased global protection of patent rights—based on the assumption that this extends from the "natural rights" of the inventor. *See* Samuel Oddi, *TRIPS—Natural Rights and a "Polite Form of Economic Imperialism,"* 29 VAND. J. TRANSNAT'L L. 415 (1996).

19. In fact, some suggest that greater market access was so important to developing countries that they essentially did not willingly agree to give up sovereignty over patent rights in exchange. *See, e.g.*, Donald Harris, *TRIPS and Treaties of Adhesion Part II: Back to the Past or a Small Step Forward*, 2007 MICH. ST. L. REV. 185 (suggesting that TRIPS was a treaty of adhesion).

20. *See, e.g.*, Robert Merges, *Of Property Rules, Coase, and Intellectual Property*, 94 COLUM. L. REV. 2655 (1994).

II. The Competing Patent Perspectives

Now that the theoretical basis for a new approach to patents has been established, this section provides a fuller description of the two views of patent policy. Importantly and in contrast to prior patent theories, this section does not aim to suggest that either view is necessarily correct. Rather, the goal of the patent perspectives theory is to provide a realistic understanding that each of these views may be reasonable to some. Although both the different views may be difficult to reconcile in a global economy, particularly one where TRIPS imposes some requirements, this section aims to first establish that there are legitimate, different views of patent policy. The impact of these policies on interpreting laws and facts are later addressed in Section IV.

A. Patents as Privilege

On one end of the spectrum of views is the concept that patents are inherently subject to limitations. Although many—including those with an uber-right view—may concur that patents are a utilitarian tool to promote innovation by providing an incentive or reward for new innovations, those who view patents as a privilege do not necessarily place innovation as a priority over all other social interests. Rather, promoting innovation is only one goal among other competing societal goals that inherently contemplate the need for balance. Patents may be viewed as a special type of property to those who see patents as a privilege. As stated by Professor Baker (who is also associated with the Health Global Access Project, which is dedicated to eliminating barriers to access to HIV treatment), "[P]atents are not 'property' in the traditional sense—they are government granted rights that are intended to balance the interests of innovators and the public at large, and which are granted by governments with many express and implied conditions. . . ."[21]

The UN Commissioner of Human Rights supports the idea of patents as a privilege that must be "subject to limitations in the public interest."[22] In particular, the UN has suggested that certain human rights, such as the right to health, are "inalienable and universal" ones that must be recognized over state-granted rights, such as patents.[23] Moreover, the UN Commissioner has suggested that to

---

21. Brook K. Baker, *Pharma's Seven Deadly Lies About Thai Compulsory Licenses*, IP DISPUTES IN MED. (Feb. 1, 2007), *available at* http://www.cptech.org/blogs/ipdisputesinmedicine/2007/02/pharmas-seven-deadly-lies-about-thai.html.

22. United Nations (UN) Office of High Comm'r for Human Rights, Sub-Comm'n on the Promotion & Prot. of Human Rights, *Intellectual Property Rights and Human Rights*, ¶ 1, UN Doc. E/CN.4/Sub.2/RES/2000/7 (Aug. 17, 2000) [hereinafter Resolution 2000/7].

23. UN Econ. & Soc. Council, Sub Comm'n on Promotion and Prot. of Human Rights, *The Impact of the Agreement on TRIPS on Human Rights* ¶ 14, UN Doc. E/CN.4/Sub.2/2001/13

the extent there is an "actual or potential conflict," patent rights should yield to the right to public health.[24]

If patents were conceived as a tool to promote innovation as one among many societal goals, exceptions to ensure the patent purpose is served would seem reasonable. In particular, although patents are assumed to provide an incentive to innovate, to the extent they fail to provide appropriate incentives (or actually interfere with additional innovation), modifications are necessary. Accordingly, under the view of patents as privilege, a nation would limit or craft exceptions to typical patent remedies if doing so would promote greater innovation, such as an exception for use of patented inventions by researchers.

In the area of healthcare and access to medicines, the idea that patents are a privilege has particularly dramatic consequences. One possible view is that "the lives of patients have to come before the patents of drug companies,"[25] such that substantial modifications of patents are appropriate—particularly if a nation places a premium on the right to health through programs such as universal access to essential drugs. Accordingly, possible limitations on a patent right to accommodate public health interests might include limits on the scope of patentability, shorter patent terms, and issuance of compulsory licenses.

Holders of the view of patents as a privileged tool do not necessarily advocate abolishing all patents. However, consideration of patents as a privilege impacts the view of what *should* be patented. For example, India's pre-TRIPS law provides one example of patents as a privilege. Previously, India permitted patents on methods of making drugs, but not on drugs themselves. This distinction was intended to provide an incentive to create new methods of making drugs that could promote competition that would reduce prices on drugs. This approach seems to fit with a view of patents as a tool to promote progress: one to be balanced against other social interests, such as broad access to low-cost drugs.

The view of patents as a privilege is also consistent with the historical requirement of many nations that a patent owner "work" the patent locally. Originally, patents were considered a privilege granted by a nation to help promote technology

---

(June 27, 2001) (suggesting that rights under TRIPS, which are state-granted rather than inalienable, should, where appropriate, bow to the more universal human rights, such as the right to health); *accord* Resolution 2000/7, *supra* note 22, ¶ 3 (reminding "all governments of the primacy of human rights obligations over economic policies and agreements").

24. *See* Resolution 2000/7, *supra* note 22, pmbl. ¶ 3.

25. Press Release, Campaign for Access to Essential Medicines, MSF Welcomes Move to Overcome Patent on AIDS Drugs in Thailand (Nov. 30, 2006), *available at* http://www.msfaccess.org/about-us/media-room/press-releases/msf-welcomes-move-overcome-patent-aids-drug-thailand.

transfer by the patent owner to that nation.[26] Accordingly, patent owners that failed to make the patented invention in the patent-granting nation forfeited their patent rights.[27] Although countries moved away from automatic forfeiture of patents for failure to work patents locally, they continued to permit penalties for lack of local working: in particular, compulsory licensing, a less onerous punishment than complete patent forfeiture.[28] Non-working remains a ground for compulsory license in the laws of many countries and is explicitly endorsed by the Paris Convention, subject to the caveat that no compulsory license be issued based on this ground until three years after the patent grant.[29]

Holders of a privilege view would take a sceptical approach toward requiring minimum patent standards for most nations. In particular, they would advocate allowing each nation to decide when and whether to grant patents, as well as to determine the scope of such patents. In other words, the resistance of developing countries to including patent rights within the World Trade Organization agreements reflects the view that nations should retain their traditional sovereign rights to determine how best to balance patent rights against other societal interests. However, to the extent an international agreement was required, those who view patents as privilege would likely be more amenable to one that permitted national discretion to recognize competing interests. Indeed, some of the language in TRIPS can be seen as the handiwork of those who view patents as privilege: Articles 7 and 8 talk about balancing patent rights, and Article 30 explicitly considers the rights of third parties in limiting the rights of the patent holder.

26. *See, e.g.*, Paul Champ & Amir Attaran, *Patent Rights and Local Working under the WTO TRIPS Agreement: An Analysis of the US–Brazil Patent Dispute*, 27 YALE J. INT'L L. 365, 370–71 (2002).

27. *See, e.g.*, Statute of Monopolies, 1623, 23 Jam. 1 c.3 (Eng.), *in* 1 Statues Revised (Britain) Henry III to James II, 1253–1685, at 693–96 (1870) (requiring local working as a condition to retain patent rights); Michael Halewood, *Regulating Patent Holders: Local Working Requirements and Compulsory Licenses at International Law*, 35 OSGOOD HALL L.J. 243, 251–52 (1997) (noting that in the eighteenth century, the United States and France required foreign inventors to work the patent locally to maintain patent rights).

28. *See* Halewood, *supra* note 27, at 285, 266–67. *Compare* Paris Convention for the Protection of Industrial Property, art. 5, Mar. 20, 1883, 25 Stat. 1372 (imposing local working requirement), *with* Paris Convention for the Protection of Intellectual Property art. 5, *as revised at* The Hague, Nov. 6, 1925, 47 Stat. 1789 (permitting revocation for non-working if compulsory license was inadequate).

29. Paris Convention for the Protection of Industrial Property, art. 5A(4), *as revised at* Stockholm, July 14, 1967, 828 U.N.T.S. 305.

B. Patents as an Uber-Right

The alternative perspective views patents as a very strong property right that should in fact be stronger than most property rights, such that it will be referred to as an "uber-right." For example, Professors Epstein and Keiff have stated, "patents are praised as a spur to innovation, which is only made possible with the predictable enforcement of rights of exclusion for the patented technology."[30] In other words, the uber-right perspective similarly views patents as a special type of property right, but in a very different manner than the privilege view. Patents are considered special in that they are much more limited in time; thus the strength of patent rights during that limited term is considered paramount.

According to the uber-right view, patents should be given more protection than other rights. For example, although injunctions are equitable remedies that typically require a court to consider and balance a variety of factors (including harm to the plaintiff, defendant, and public interest), an uber-right perspective may take a more extreme approach. In particular, the uber-right view would hold that if a patent is found to be infringed, a permanent injunction should be nearly automatic.[31]

Holders of the uber-right view also have invoked human rights norms to support their position. In particular, they argue, based upon language in the Universal Declaration of Human Rights as well as the International Covenant on Economic, Social and Cultural Rights (ICESCR), that inventors have a right to benefit from the fruits of their inventions.[32] Although neither document refers explicitly to either patents or intellectual property rights, both include a clause about how everyone should enjoy the benefits of scientific progress and benefit from protection of interests from any scientific production of which he or she is

---

30. Richard A. Epstein & F. Scott Kieff, *Questioning the Frequency and Wisdom of Compulsory Licensing for Pharmaceutical Patents*, 78 U. Chi. L. Rev. 71, 73 (2011).

31. This view has been espoused for years by the Federal Circuit, presumed in part on the finite term of the patent court. *See, e.g.*, W.L. Gore & Assocs., Inc. v. Garlock, Inc., 842 F.2d 1275, 1281 (Fed. Cir. 1988); H.H. Robertson Co. v. United Steel Deck, Inc., 820 F.2d 384, 390 (Fed. Cir. 1987). The US Supreme Court, however, has held that this is an incorrect reading of the law. Ebay, Inc. v. MercExchange, LLC, 547 U.S. 388, 392–93 (2006). Even so, that may not necessarily change existing perspectives.

32. *See* Universal Declaration of Human Rights, art. 27(2), G.A. Res. 217A (III), UN GAOR, 3d Sess., 1st plen. mtg. at 71, UN Doc. A/810 (Dec. 12, 1948) [hereinafter UDHR] ("Everyone has the right to the protection of the moral and material interests resulting from any scientific, literary or artistic production of which he is the author."); International Covenant on Economic, Social and Cultural Rights, art. 15(1), G.A. Res. 2200A (XXI), UN GAOR, 21st Sess., Supp. No. 16, UN Doc. A/6316 (Dec. 16, 1966) [hereinafter ICESCR].

the author.[33] The basic argument is that if patent rights are minimized, the author (i.e. inventor) is deprived of the protection of his or her interests.[34]

If patents are considered an uber-right that should seldom be subject to exceptions, strong patent rights may be considered a reasonable priority to emphasize, rather than short-term access issues. Exceptions to patent rights would be considered to threaten crucial incentives needed, for example, to develop new drugs that involve "huge, lengthy, and risky investments" of over a billion dollars per drug.[35] To the extent that high costs of patented drugs may impede access, the uber-right view would suggest the problem is one of poverty, not of patent protection.[36] The key issue from the uber-right perspective is that short-term access issues can and should be resolved outside the patent system because reduced patent protection necessarily compromises long-term innovation.

The uber-right perspective would want to limit any and all exceptions to patent rights. Compulsory licenses would be considered anathema to an uber-right perspective and only permissible in the narrowest of circumstances.[37] In particular, a compulsory license would likely be considered valid only if the patent owner was incapable of providing an adequate supply of the patented product.[38] Any other compulsory license would be considered not only as improper, but in fact

33. UDHR, *supra* note 32, art. 27(2); ICESCR, *supra* note 32, art. 15(1).

34. However, the same article has been read to support the perception of patents as privilege—that consumers are entitled to enjoy the results of scientific progress in drug discovery such that they have actual access to medicines, not merely theoretical access based upon economic conditions beyond their control.

35. Epstein & Kieff, *supra* note 30, at 78. Others with an uber-right view similarly acknowledge the high cost of patented drugs, but tend to emphasize that drug discovery is lengthy and expensive. *See, e.g.*, Amir Attaran & Lee Gillespie-White, *Do Patents for Antiretroviral Drugs Constrain Access to AIDS Treatment in Africa?*, 286 J. Am. Med. Ass'n 1886 (2001); Robert C. Bird, *Developing Nations and the Compulsory License*, J.L. Med. & Ethics 209, 216 (2009).

36. *See, e.g.*, Richard P. Rozek, *The Effects of Compulsory Licensing on Innovation and Access to Health Care*, J. World Intell. Prop. 889, 896–99 (2000) (pointing to other barriers to access).

37. *See, e.g.*, *id.* at 889, 890 ("[C]ompulsory licensing is a policy that should be used rarely, if at all."); *id.* at 904 (stating that compulsory licensing "is not intended as a general component for a system that protects IPRs").

38. *See, e.g.*, Martin J. Adelman, *Compulsory Licensing of Drugs: TRIPS Context*, Paper Presented at ATRIP Annual Meeting in Tokyo, Japan (Aug. 4, 2003) (suggesting TRIPS requires "adequate remuneration" that is equivalent to actual damages such that compulsory licensing "is only useful when the patent owner is unwilling or unable to provide a sufficient supply of a needed patented drug").

"stealing" or "piracy"[39] and those that impose compulsory licenses as the "free lunch crowd"[40] or "anti-property hooligans."[41]

An uber-right view would also be consistent with the genesis of TRIPS. In contrast with the early view of patents as privileges, the inclusion of intellectual property rights in the WTO framework was suggested as necessary to protect "rights" of inventors and creators.[42] In addition, the uber-right view was reflected in arguments in support of mandatory minimum patent rights under TRIPS. For example, some suggested that requiring countries to increase patent rights under TRIPS would promote global innovation and prosperity.[43]

III. TRIPS

This section provides a brief overview of the TRIPS rules for compulsory licenses under Article 31 as a foundation to later explore how patent perspectives impact legal interpretations.

A. Article 31

Member states can issue compulsory licenses so long as they comply with the procedures in TRIPS Article 31. Any country can issue a compulsory license under this provision; it is not limited to the poorest countries. In addition, TRIPS permits nations to decide for themselves what subject matter may be subject to compulsory licensing under TRIPS. This is implicit in TRIPS and also is reinforced by the unanimously concluded Doha Public Health Declaration.[44] Although some

---

39. *See, e.g.*, Ronald Cass, *Drug Patent Piracy*, WALL ST. J., May 7, 2007, at 15; Christopher Horner, *Thailand Stealing out of WTO?*, WASH. TIMES, May 17, 2007, at 1.

40. Ronald A. Cass, Op-Ed., *Thai Patent Turmoil*, WALL ST. J. ASIA, Mar. 13, 2007, at 13 [hereinafter Cass, *Thai Patent Turmoil*].

41. Ronald A. Cass, Op-Ed., *Patent Remedy*, WALL ST. J. ASIA, Aug. 28, 2007, at 13; *Bangkok's Drug War Goes Global*, WALL ST. J. ASIA, Mar. 7, 2007, at 13.

42. *See, e.g.*, PETER DRAHOS & JOHN BRAITHWAITE, INFORMATION FEUDALISM (2002); SUSAN SELL, PRIVATE POWER, PUBLIC LAW (2003).

43. *See, e.g.*, Negotiating Group of Trade-Related Aspects of Intellectual Property Rights, Including Trade in Counterfeit Goods, Meeting of 24 Mar. 1987, ¶ 4, MTN.GNG/NG11/1 (noting that greater protection of intellectual property rights was necessary to provide incentives to innovate); Martin Adelman & Sonia Baldia, *Prospects and Limits of the Patent Provisions in the TRIPS Agreement: The Case of India*, 29 VAND. J. TRANSNAT'L L. 507, 517, 530 (1996) (suggesting that TRIPS patent provisions would spur India's innovation).

44. TRIPS, *supra* note 1, art. 31 (containing no explicit limitations on what inventions may be subject to compulsory licenses); WTO, Ministerial Declaration on the TRIPS Agreement and Public Health of 14 November 2001, ¶ 5(b), WT/MIN(01)/DEC/W/2, 41 I.L.M. 755 (2002) [hereinafter Doha Public Health Declaration]; *see also* Frederick M. Abbott & Jerome

patent owners (and their sympathizers) tend to discount the Declaration as a political statement of no consequence,[45] most scholars consider it relevant to interpreting TRIPS.[46] In addition, even if the Declaration were not utilized, the interpretation of TRIPS Article 31 should be the same, as the Declaration simply makes explicit the principles already set forth under TRIPS.

TRIPS focuses on ensuring that proper procedures are followed, rather than dictating what subject matter may be licensed. The Doha Public Health Declaration confirms this by stating that "[e]ach member has the right to grant compulsory licences and the freedom to determine the grounds upon which such licences are granted."[47] However, there is a caveat to this freedom—nations are to consider the "individual merits" of each compulsory license; in other words, they cannot grant blanket licenses for entire categories of subject matter.[48] Other than that, nations are free to grant licenses on any patented products—if they comply with the other procedural requirements of TRIPS.

The usual TRIPS requirements for compulsory licenses envisage that before such a license is issued, there will be an attempt to first negotiate with the patent owner, but that this attempt fails to result in a voluntary license.[49] However, TRIPS clearly sets forth three separate conditions when prior negotiation may be waived: a national emergency, situation of extreme urgency, or public noncommercial use.[50] Although TRIPS does not define these terms, the Doha Public Health Declaration

---

H. Reichmann, *The Doha Round's Public Health Legacy*, 10 J. INT'L ECON. L. 921, 956 (2007); Cynthia M. Ho, *Patent Breaking or Balancing?: Separating Strands of Fact from Fiction Under TRIPS*, 34 N.C. J. INT'L L. & COMM. REG. 371, 424–25 (2009).

45. *See, e.g.*, US Government Accountability Office (GAO), Intellectual Property, US Trade Policy Guidance on WTO Declaration on Access to Medicines May Need Clarification (2007), at 3 (noting that the United States considers the Doha Declaration to be a political statement that does not modify TRIPS); Press Release, Pharmaceutical Research and Mfrs. of America, WTO Doha Declaration Reaffirms Value of Intellectual Property Protection (Nov. 14, 2001), *available at* http://ww1.aegis.org/news/pr/2001/PR011126.html (stressing that the Declaration was a "political statement").

46. *See, e.g.*, Denis Borges Barbosa et al., *Slouching Toward Development in International Intellectual Property*, 2007 MICH. ST. L. REV. 71, 131–32 (2007) (viewing the Doha Public Health Declaration as not only a subsequent interpretive agreement, but one that establishes the right to health as an important right not to be trumped by provisions of TRIPS); CARLOS CORREA, IMPLICATIONS OF THE DOHA DECLARATION ON TRIPS AGREEMENT AND PUBLIC HEALTH 45 (2002), *available at* http://www.who.int/medicines/areas/policy/WHO_EDM_PAR_2002.3.pdf (considering Doha relevant interpretative document).

47. Doha Public Health Declaration, *supra* note 44, ¶ 1.

48. *Compare* TRIPS, *supra* note 1, art. 31(a) (requiring licenses be considered on "individual merits"), *with* Patent Act, § 187, S.C., ch. 23 (1923) (Can.) [hereinafter Canada Patent Act].

49. TRIPS, *supra* note 1, art. 31(b).

50. *Id.* If negotiation is waived, the patent owner must nonetheless be notified as promptly as possible. *Id.*

clarifies that nations have discretion to determine what constitutes an emergency.[51] The Declaration states that "public health crises" can constitute either a national emergency or a situation of extreme urgency.[52] Further, the Declaration states that public health crises *include* HIV/AIDS, tuberculosis, malaria and other epidemics.[53]

The scope and duration of the compulsory license must be limited to the "authorized purpose."[54] Notably, TRIPS does not state that compulsory licenses in general should be limited—in contrast to one proposal.[55] Rather, the limitation of the scope and duration of the license is only restricted to the "authorized purpose," according to the license-granting nation.[56] In other words, the restriction on the scope and duration is a minimal one.[57] However, a compulsory license is nonetheless given some limitations. For example, a license is subject to termination if the circumstances leading to the license cease to exist and are unlikely to reoccur—as determined by the nation granting the license.[58]

A compulsory license consistent with TRIPS must have certain other characteristics. A license must be nonexclusive and non-assignable (unless assigned with an entire enterprise).[59] In addition, there is a general requirement that a license be authorized "predominantly" to supply a domestic market.[60] This is generally understood to mean that at least 50 percent of products made pursuant to a license must be for the domestic market, subject to two caveats. First, this requirement does not exist at all if a license is issued as a remedy to address an anticompetitive practice.[61] Second, this requirement can be waived under a complicated procedure as discussed in the next section.

---

51. Doha Public Health Declaration, *supra* note 44, ¶ 5(c) ("Each member has the right to determine what constitutes a national emergency or other circumstance of extreme urgency. . . .").

52. *Id.*

53. *Id.*

54. TRIPS, *supra* note 1, art. 31(c) ("[T]he scope and duration of such use shall be limited to the purpose for which it was authorized.").

55. For example, a 1990 draft suggested that "parties shall minimize the grant of compulsory licenses in order not to impede adequate protection of patent rights." Draft of July 23, 1990, W/76, reprinted in Daniel Gervais, The TRIPS Agreement: Drafting History and Analysis 387 (3d ed. 2008).

56. TRIPS, *supra* note 1, art. 31(c) ("[T]he scope and duration of such use shall be limited to the purpose for which it was authorized.").

57. *See* Ho, *supra* note 44, at 432–38.

58. TRIPS, *supra* note 1, art. 31(g).

59. *Id.* art. 31(d)–(e).

60. *Id.* art. 31(f).

61. *Id.* art. 31(k).

TRIPS also requires that certain procedures exist that benefit the patent owner. The patent owner is always entitled to "adequate remuneration."[62] The patent owner is also entitled to challenge both the issuance of a compulsory license and the amount of the remuneration provided.[63] Notably, any such challenge is within the jurisdiction of the nation granting the compulsory license.

*B. 2003 Decision; Proposed TRIPS Article 31*bis

There is an important caveat to the usual TRIPS requirement that compulsory licenses be "predominantly" for domestic use. In particular, since 2003, this requirement can be waived in limited circumstances pursuant to a decision by the TRIPS Council.[64] The waiver of this requirement was intended to address the fact that compulsory licenses were of no utility to countries that lacked manufacturing capacity to make use of licenses.[65]

A country that wants to import drugs pursuant to this process can do so only if a series of procedural hurdles are overcome in two separate countries. However, before outlining the procedural requirements, it is important to note what countries can import drugs made under this procedure, as well as what drugs are covered.

First, not all nations can import drugs made under this procedure. All least-developed countries may clearly use the procedure. Technically, any other country can use the procedure if it lacks adequate manufacturing capacity to make a particular patented product.[66] However, some countries promised not to use the procedure to import drugs, or to only do so in case of an emergency or extreme urgency.[67]

---

62. *Id.* art. 31(h).

63. *Id.* art. 31(i)–(j).

64. WTO, Decision of the General Council, *Implementation of Paragraph 6 of the Doha Declaration on the TRIPS Agreement and Public Health*, WT/L/540/Corr. 1 (Aug. 30, 2003).

65. Doha Public Health Declaration, *supra* note 44, ¶ 6. Of course, countries are not limited to using this procedure to obtain low-cost drugs. For example, they can alternatively secure lower cost drugs through donations. In addition, for drugs made as generics in India before 2005, they can continue to be made as generics.

66. Proposed TRIPS Annex, *supra* note 5, ¶ 1(b).

67. WTO General Council, General Council Chairperson's Statement, WT/GC/M82 (Nov. 13, 2003); WTO, Statement by the Chairperson to the General Council (Dec. 6, 2005), *available at* http://www.wto.org/english/news_e/news05_e/trips_319_e.htm. Countries pledging not to use the system as an importing country were: Australia, Canada, the European Communities, Iceland, Japan, New Zealand, Norway, Switzerland, and the United States. However, there is some question concerning whether the pledges are legally binding. *See* Andrew Mitchell & Tania Voon, *The TRIPS Waiver as a Recognition of Public Health Concerns*

The drugs that may be made under license for export pursuant to Article 31*bis* may be more limited than for compulsory licenses for domestic use. Article 31*bis* refers to products "needed to address the public health problems recognized in paragraph 1 of the Doha Public Health Declaration."[68] This paragraph, in turn, does not list any specific drugs. However, it states that members recognize "[t]he gravity of the public health problems afflicting many developing and least-developed countries, especially those resulting from HIV/AIDS, tuberculosis, malaria and other epidemics."[69] This paragraph does not strictly limit compulsory licenses of drugs to the listed diseases. However, some countries may nonetheless view the list as necessarily so restricted.

In addition to the limits on what countries can use the procedures, and the possible limits on the interpretation of applicable drugs, there are a number of specific procedures that must take place in the importing and exporting countries.

Countries can only import drugs from a country that has amended its laws to permit compulsory licenses for export—something that only a handful of countries have done.[70] Even with respect to countries that have laws that permit compulsory licenses for export, these laws may impose additional restrictions that minimize the utility of the license. For example, exporting countries may limit eligible drugs or the duration of the license.[71]

The country that issues a compulsory license for export also must comply with additional procedures beyond the basic TRIPS rules for compulsory licenses. For example, the compulsory license is limited to the amount necessary to meet the needs of the eligible importing member(s), and all drugs made under the compulsory license must be exported.[72] In addition, products made under the license must

---

in the WTO, *in* INCENTIVES FOR GLOBAL PUBLIC HEALTH: PATENT LAW AND ACCESS TO ESSENTIAL MEDICINES 66 (Thomas Pogge et al. eds., 2010) (suggesting that the statement has no interpretive force because a "corrigendum" was added to it that indicates that the statement is "for information purposes only and without prejudice to Members' legal rights and obligations").

68. 2003 WTO Council Decision, *supra* note 5, ¶ 1; Proposed TRIPS Annex, *supra* note 5, ¶ 1(a).

69. Doha Public Health Declaration, *supra* note 44, ¶ 1.

70. Countries that have amended their laws include Switzerland, EU member states, Canada, China, India, Korea and Norway. *See, e.g.*, CYNTHIA M. HO, ACCESS TO MEDICINE IN THE GLOBAL ECONOMY: INTERNATIONAL AGREEMENTS ON PATENTS AND RELATED RIGHTS 210 (2011).

71. For example, China currently limits the scope to infectious diseases. Order #37 (promulgated by the State Intellectual Prop. Office, Nov. 2005, effective Jan. 1, 2006) (2005) (China), *available at* http://www.cptech.org/ip/health/cl/china-order37.html (unofficial translation of the Chinese Patent Act with commentary). Canada's existing law imposes all of the other requirements. *See* Canada Patent Act, *supra* note 48, art. 21.03.

72. 2003 WTO Council Decision, *supra* note 5, ¶ 2; Proposed Article 31*bis*, *supra* note 5, ¶ 2(b).

be clearly identified to help ensure that products are not diverted to unintended markets. To further prevent diversion, the licensee must post to a website details concerning the distinguishing features and quantities supplied to relevant destinations before the licensed product is shipped.[73]

In addition, even for potential exporting countries that have appropriately amended laws, there must be a company within that country willing to seek a compulsory license for export. This may be challenging if domestic laws limit the scope of the compulsory license, such that it only applies to a single country, or for a limited shipment, as a limited number of generic drugs would not result in a profit, and may actually result in a net loss. Indeed, in the only ever shipment of drugs under this process, the manufacturing company, Apotex, ultimately provided drugs to Rwanda at a price that was below its original estimated cost.[74]

In addition to all of the above, there are reporting requirements for both the importing and exporting country. Before any drugs can be traded, the importing country must notify the TRIPS Council regarding the expected quantity of desired product and the exporting country must notify the TRIPS Council of the actual quantities of the drug to be exported, together with additional details, such as distinguishing features of the product, the quantity, the country to which it is supplied, and the duration of the license.[75] In addition, the exporting country also has reporting requirements. In particular, it must notify the TRIPS Council of the specific details concerning the compulsory license, including not only the product, but the licensed quantities, scope and duration of the license, and the country or countries to which the product will be supplied.

IV. Perspectives in Action

This section considers the impact of patent perspectives on both the basic compulsory license provision under TRIPS Article 31 and the proposed Article 31*bis* provision that is currently operating as a waiver of TRIPS Article 31(f). Due to space constraints, only a few issues are discussed. Nonetheless, these issues provide ample examples of the impact of perspectives.[76]

---

73. 2003 WTO Council Decision, *supra* note 5, ¶ 2(c); Proposed TRIPS Annex, *supra* note 5, ¶ 2(c).

74. Apotex supplied HIV drugs to Rwanda for only $0.195 per tablet, which is about half of its originally proposed "no profit" price. *See, e.g.*, Amir Attaran, *Why Canada's Access to Medicines Regime Can Never Succeed*, 60 U. New Brunswick L.J. 150 (2010).

75. 2003 WTO Council Decision, *supra* note 5, ¶ 2(c); Proposed TRIPS Annex, *supra* note 5, ¶ 2(c).

76. For additional examples, see Cynthia M. Ho, *Unveiling Competing Patent Perspectives*, 46 Hous. L. Rev. 1047 (2009) [hereinafter Ho, *Unveiling Competing Patent Perspectives*]

A. Article 31 Perspectives

1. Emergency Confusion or Conflation

A. COMPULSORY LICENSE ONLY IN CASE OF AN EMERGENCY

Those holding an uber-right view may suggest that compulsory licenses must be limited to emergency situations—even though nothing in TRIPS requires an emergency, as discussed above. Not surprisingly, those representing patent-owning drug companies are prone to such statements. For example, Jon Pender of GlaxoSmithKline stated that "although compulsory licensing is legal, TRIPS rules allow it only under limited circumstances, such as national health emergencies, and only after lengthy efforts to negotiate with firms."[77] As stated earlier, TRIPS does not generally limit compulsory licenses. Moreover, TRIPS explicitly waives any obligation to first negotiate with patent owners in the case of an emergency.[78]

An uber-right view is also expressed by some academics. For example, in a paper arguing against compulsory licenses of drugs patents, Professors Epstein and Kieff assert that TRIPS allows compulsory licenses "in times of national emergency" to ensure no delays to needed patented technology that might arise under voluntary negotiations.[79] In addition, they criticize the compulsory licenses issued by the Thai government in 2007, for example, for failing to meet the definition of an emergency because they were directed toward chronic and long-term conditions. Although the professors never explicitly state that compulsory licenses under TRIPS are limited to emergencies, that seems implicit in these statements. However, as discussed earlier, TRIPS is not so limited. The gap between what TRIPS permits versus what some scholars perceive as proper could

---

(providing more examples of the impact of patent perspectives regarding basic compulsory licenses under TRIPS Article 31).

77. *A Gathering Storm*, ECONOMIST, June 9, 2007, at 71. In addition, similar statements appear in the mainstream media, although those statements could reflect the views of those assumed to be knowledgeable—patent-owning drug companies. *See, e.g.*, Harish Mehta, *Cheap Life-Saving Drugs: Thailand Shows the Way*, BUS. TIMES SINGAPORE, Oct. 18, 2007, *available at* http://www.timesdirectories.com/biotech/news/cheap%20life%20saving%20drugs%20thailand%20shows%20the%20way/107882 (asserting a national emergency is the only ground for issuing a compulsory license without prior consultation with the patent owner); Nicholas Zamiska, *Thai Move to Trim Drug Costs Highlights Growing Patent Rift*, WALL ST. J., Jan. 30, 2007, at A8 ("WTO rules allow a government to unilaterally declare an emergency and make or sell patented drugs without the permission of the drug companies.").

78. TRIPS, *supra* note 1, art. 31(b).

79. Epstein & Kieff, *supra* note 30; *see also* Kristina M. Lybecker & Elisabeth Fowler, *Compulsory Licensing in Canada and Thailand: Comparing Regimes to Ensure Legitimate Use of the WTO Rules*, 37 J.L. MED. & ETHICS 222, 233 (2009) (criticizing Thai license of heart

be explained by the power of patent schemas. An alternative explanation is that Epstein and Kieff understand TRIPS requirements, but only selectively emphasize the portions they agree with. However, this would still be consistent with the theory of patent perspectives as schemas that shape our perceptions—whether conscious or unconscious.

B. NATIONAL EMERGENCY REQUIRED TO WAIVE PRIOR NEGOTIATION

Another incomplete characterization of TRIPS is the suggestion that a national emergency is the only condition that may waive the usual prior negotiation requirement.

The best example of this is an article coauthored by an employee of Eli Lilly and presented at a conference sponsored by the Intellectual Property Owners Association, which represents the interests of patent owners. The article correctly quotes the full text of TRIPS, but then concludes that it is much more limited. In particular, the article correctly quotes TRIPS as stating: "This requirement may be waived by a Member in the case of national emergency or other circumstances of extreme urgency *or in cases of public non-commercial use*."[80] However, the very next sentence states: "[t]his provision indicates that a WTO member can grant a compulsory license without first seeking authorization from the patent holder *only in situations of national emergency or other circumstances of extreme urgency*."[81] This clear omission of public noncommercial use may reflect the influence of an uber-right view.

There are also other examples of incomplete or inaccurate representations of when TRIPS permits waiver of the prior negotiation requirement. Professor Sykes states that TRIPS "limits compulsory licensing without prior negotiation to genuinely extreme circumstances," but mentions only a "national emergency."[82] In addition, Roger Bate, a fellow of the conservative think-tank American Enterprise International (AEI), stated that compulsory licenses are permissible if "efforts to obtain authorization from the right holder on reasonable commercial

---

medication as not a health emergency, on the assumption that such an emergency is required); *Bangkok's Drug War, Round Two*, WALL ST. J., Feb. 27, 2008, at 13 (mentioning only a national emergency as a possible exception to prior negotiation).

80. Christopher E. Chalsen & Lawrence T. Welch, *Growing Hostility Toward Intellectual Property?*, Paper Presented at First IPO–JIPA Asian Practice Int'l Congress 8 (Sept. 14, 2005), *available at* http://www.ipo.org/AM/Template.cfm?Section=Home&Template=/CM/HTMLDisplay.cfm& ContentID=14899 (emphasis added).

81. *Id.* (emphasis added).

82. Alan O. Sykes, *TRIPS, Pharmaceuticals, Developing Countries, and the Doha Solution*, 3 CHI. J. INT'L L. 47, 56 (2002).

terms and conditions have failed, or in cases of national emergency."[83] Although Bate recognizes that prior negotiation is not required in the case of a national emergency, he once again fails to state that a public noncommercial use is also a proper ground for a compulsory license without prior negotiation.

C. UNDUE NARROWING OF APPROPRIATE EMERGENCIES

The incorrect assumption that TRIPS requires an emergency—whether as a general matter, or to waive prior negotiation with the patent owner—is often closely tied to attempts to second-guess what constitutes an emergency. For example, when Thailand issued a series of compulsory licenses, many criticized the licenses as improper, based on scepticism concerning whether Thailand had an appropriate emergency. Some suggested that licenses on HIV drugs were inappropriate because Thailand had been too successful in controlling HIV infections, such that its infection rates were not the highest among WTO members.[84] In addition, compulsory licenses of drugs to treat other conditions such as heart disease and cancer were also criticized for not being emergencies[85] —even though Thailand never relied on this ground in issuing the licenses without prior negotiation; rather, Thailand did so on the ground that the licenses were for public noncommercial use.[86]

Although those with an uber-right view may predictably want to limit what constitutes an emergency, this is contrary to a proper interpretation of TRIPS and the Doha Public Health Declaration. As noted earlier, although the Doha Declaration expressly lists HIV/AIDS as an *example* of what constitutes an emergency, it clearly states that each nation has discretion to determine what constitutes an emergency.[87] In addition, TRIPS states that emergencies and public

---

83. Roger Bate, *Thailand's Drug Wars*, AMERICAN (Mar. 12, 2008), *available at* http://www.american.com/archive/2008/march-02-08/thailand2019s-drug-wars (internal quotation marks omitted).

84. *See, e.g.*, Horner, *supra* note 39, at 1 (suggesting that Thailand cannot qualify for the national emergency exception because of its "comparatively low rate of AIDS infection" with less than 1 percent of the population infected as opposed to other WTO members with rates as high as 20 percent).

85. *See, e.g.*, Roger Bate, *Thailand's Patent Attack*, N.Y. SUN (Feb. 13, 2007), *available at* http://www.nysun.com/opinion/thailands-patent-attack/48499 (asserting that heart disease and leukemia are not epidemics); Nirmal Ghosh, *Thailand Breaks Patents for Heart and AIDS Drugs*, STRAITS TIMES (SING.), Jan. 31, 2007 (asserting that heart disease afflicts the affluent, such that a compulsory license is not needed).

86. *See* THAI MINISTRY OF PUBLIC HEALTH AND THAI NAT'L HEALTH SEC. OFFICE, FACTS AND EVIDENCES ON THE TEN BURNING ISSUES RELATED TO THE GOVERNMENT USE OF PATENTS ON THREE PATENTED ESSENTIAL DRUGS IN THAILAND (Vichai Chokevivat ed., 2007), *available at* http://www.moph.go.th/hot/White%20Paper%20CL-EN.pdf.

87. Doha Public Health Declaration, *supra* note 44, ¶ 1.

noncommercial use are both valid grounds to issue a compulsory license without prior negotiation with the patent owner.[88]

2. Emergency Confusion Masking Public Noncommercial Use Issue

The confusion concerning national emergencies may reflect an uber-right view that rejects public noncommercial use as an appropriate consideration for issuance of a compulsory license without prior negotiation. For example, it may be more appealing to paint Thailand's actions as failing to meet the requirement of an emergency—even if this is not an issue—rather than admit that a public noncommercial use could qualify. In particular, the criticisms concerning licenses of drugs other than for HIV, such as the heart medication Plavix®, can be easily replaced by criticism against an overbroad interpretation of public noncommercial use. For example, the Plavix® license has been sharply criticized as the first step down a slippery slope toward licensing any and all patents.[89] However, it would seem equally apropos for the pharmaceutical industry to say that if public noncommercial use were a permissible ground for a compulsory license, it would be a slippery slope toward licensing all patents. Perhaps the public noncommercial use is the true slippery slope that terrifies patent owners such that they have displaced their concerns onto the emergency criterion. In addition, that might explain why criticism of Thailand's compulsory licenses often ignores that the licenses were stated to be for public noncommercial use,[90] or quickly dismisses this option as clearly inapplicable.[91]

---

88. TRIPS, *supra* note 1, art. 31(b).

89. Cass, *Thai Patent Turmoil*, *supra* note 40, at 13.

90. *See, e.g.*, Roger Bate & Kathryn Boateng, *Drug Pricing and Its Discontents: At Home and Abroad*, HEALTH POL'Y OUTLOOK 4 (Aug. 2007), *available at* http://www.aei.org/docLib/20070808_22039HPO09Bate_g.pdf; *Partnering for Better Health: An Industry Perspective*, PHARM. RESEARCH & MFRS. ASSN. 26 (2007) (suggesting by omission that the Doha Public Health Declaration only affirms the right to use flexibilities for health crises and emergencies).

91. *See, e.g.*, Cass, *Thai Patent Turmoil*, *supra* note 40, at 13 (suggesting that public noncommercial use is for public research programs and is inapplicable to Thailand's situation of permitting a for-profit government agency to manufacture drugs); *The Thai Flu*, WALL ST. J., Mar. 14, 2007, at A14 (asserting that for Thailand to assert the application of public noncommercial use would be simply "taking advantage of vague language"). Although there could be other alternative explanations, possible confusion is presented as the most likely. For greater discussion of other possible critiques of the patent perspectives theory, see Ho, *Unveiling Competing Patent Perspectives*, supra note 76, at 1095–110.

3. Alternative Explanation

An alternative explanation to patent perspectives may be that they simply represent confusion, at least with respect to compulsory licenses. After all, the assumption that an emergency is required to waive the usual prior negotiation with patent owners is made by both scholars with uber-right and privilege views of patents.[92]

Some confusion may stem from the first global controversy involving whether a nation's use of a compulsory license complied with TRIPS. Shortly after TRIPS was enacted, South Africa faced a major HIV epidemic and enacted a law that promoted broad use of compulsory licenses to help address the health crisis.[93] Although South Africa's law was controversial, there was no question it was facing an HIV epidemic. Thus, news reports justifying South Africa's actions as being proper in the face of an emergency may have left the impression that an emergency was required.[94]

Confusion regarding compulsory licenses may also be exacerbated by the fact that scholarly and public discussion on compulsory licenses has primarily focused on the 2003 waiver of the "predominantly domestic use" requirement (that is now proposed Article 31*bis*), which includes a number of limitations that could be conflated with the regular requirements for compulsory licenses. For example, the waiver is known to apply to any least-developed country (LDC) member,[95] perhaps giving credibility to the assumptions of some that all compulsory licenses should be limited to such countries.[96] In addition, a number of countries agreed to only use the procedure in case of national emergency or other circumstances of

---

92. For example, in an article discussing the history of TRIPS, as well as the discussion of access to medicines that is sympathetic to the interests of developing countries, Article 31 of TRIPS is stated as requiring prior negotiation except "in the event of a national emergency or other circumstances of extreme urgency," without any mention of public noncommercial use. *See* Duncan Matthews, *Is History Repeating Itself? The Outcome of Negotiations on Access to Medicines, the HIV/ADIS Pandemic and IPR in the WTO*, L. Soc. Justice & Global Dev. 5, 5 (2004).

93. Medicines and Related Substances Control Amendment Act 90 of 1997 (S. Afr.).

94. *See, e.g.*, Merrill Goozner, *Third World Battles for AIDS Drugs*, Chic. Trib., Apr. 28, 1999 ("[G]lobal rules of trade do allow for what is known as compulsory licensing if it is done to combat a national emergency."); Karl Vick, *African AIDS Victims Losers of a Drug War*, Wash. Post, Dec. 4, 1999 (stating TRIPS provided a "loophole" for countries facing a health emergency to make patented drugs).

95. Proposed TRIPS Annex, *supra* note 5, ¶ 1(b) (defining "eligible importing member").

96. For example, criticism of Thailand for using a compulsory license as a middle-income country appeared not only in editorials, but also was put forth by some academics. *See, e.g.*, Amy Kazmin & Andrew Jack, *Thailand Breaks Patent for AIDS Drug to Cut Costs*, Fin. Times, Nov. 30, 2006, at 9; Lybecker & Fowler, *supra* note 79, at 223.

extreme urgency, perhaps giving rise to confusion about whether an emergency is required.

Another possible source of confusion that promotes the uber-right view that compulsory licenses should be limited to national emergencies could be based on an assumption that TRIPS reflects what was originally proposed by patent-owning companies. After all, TRIPS was the brainchild of US companies seeking to protect their global interests in intellectual property, and most agree the final language in TRIPS often reflected the desires of developed countries, such as the United States.[97] In the context of compulsory licenses, the United States did suggest that licenses be limited to national emergencies.[98]

*B. 2003 Decision; Proposed Article 31*bis

Differing views of patent perspectives are also reflected in the development of the 2003 Decision, proposed Article 31*bis*, and national implementations of this process. As noted below, the different perspectives continue to be reflected in current discussions and developments concerning the process at both global and domestic levels.

1. Development

Recognition that there was a problem with the TRIPS framework for some countries was, not surprisingly, first raised by countries with a privilege view of patents. Although all countries at the 2001 WTO Ministerial Conference agreed that countries that lacked adequate manufacturing capacity could not adequately use a compulsory license, as permitted under TRIPS exceptions, the solutions posed to this problem tend to reflect different views of patents, as noted below.

A. PATENTS AS PRIVILEGE

The view of patents as a privilege was reflected in proposals to address the situation of countries without the ability to effectively use compulsory licenses. In contrast to the complexity of the current waiver of the TRIPS requirement that compulsory licenses be predominantly for domestic use, developing countries consistently proposed more streamlined solutions, but ones that reflected a privilege view.

For example, a group of developing countries proposed that a compulsory license issued by one member state could be given effect in another member state,

97. *See, e.g.*, Matthews, *supra* note 92, at 4 (stating the final version of TRIPS "was perceived to closely reflect the demands of developed countries and their industries").

98. Draft Agreement on the Trade-Related Aspects of Intellectual Property Rights—Communication from the United States, art. 27, MTN.GNG/NG11/W/70 (May 11, 1990).

to permit export of a licensed product without infringement.[99] There is no real explanation for how a compulsory license granted in one country can be automatically given effect in another country, let alone how this would be consistent with the TRIPS rules for compulsory licenses. It seems a stretch to not consider this a violation of the requirement that a compulsory license be predominantly for domestic use.[100] Interestingly, the proposal does not directly address this problem and instead simply states that it does not infringe the basic patent rights.

Another privilege-based solution is the suggestion that under TRIPS Article 30, members can make and export drugs without the consent of the patent owner to address the health needs of another country.[101] This proposal essentially uses an expansive interpretation of Article 30 to achieve what is not possible under TRIPS Article 31. This approach seems questionable based on traditional tools of treaty interpretation. Different provisions of TRIPS should be assumed to apply to different situations. It is widely recognized that Article 31 applies to compulsory licenses. Moreover, the text of Article 31 explicitly states that it applies to situations not covered by Article 30.[102] In addition, this interpretation seems inconsistent with a prior WTO panel decision that emphasized that Article 30 is a very limited exception to patent rights.[103]

B. UBER-RIGHT

An uber-right view of patents is reflected in the Waiver that was adopted. As discussed earlier, the Waiver is extraordinarily complex. In addition, some of the specific procedures, such as special marking of licensed products, favor the interests of the uber-right to segregate markets, such that they can maximize their profits in each separate market.

---

99. WTO Council for Trade-Related Aspects of Intellectual Property Rights, Proposal from a Group of Developing Countries, *Draft Ministerial Declaration on the TRIPS Agreement and Public Health*, ¶ 5, IP/C/W/312; WT/GC/W/450 (Oct. 4, 2001) [hereinafter Developing Country Draft Ministerial Declaration] ("A Compulsory license issued by a Member may be given effect by another Member. Such other Member may authorize a supplier within its territory to make and export the product covered by the license predominantly for the supply of the domestic market of the [importing] Member granting the license. Production and export under these conditions do not infringe the rights of the patent holder.").

100. Although the proposal uses the phrase "predominantly for the supply of the domestic market," it is actually referring to the domestic market of a country that is only importing the licensed drug.

101. *See, e.g.*, Developing Country Draft Ministerial Declaration, *supra* note 99, ¶ 9; *see also* WTO Council for Trade Related Aspects of Intellectual Property Rights, Submission by the African Group et al., ¶ 34, IP/C/W/296 (June 29, 2001).

102. TRIPS, *supra* note 1, art. 31.

103. Panel Report, *Canada—Patent Protection of Pharmaceutical Products*, WT/DS114/R (Mar. 17, 2000).

Negotiations subsequent to the 2003 Decision also reflect an uber-right view. For example, proposals to limit the scope to only a narrow range of diseases or eligible countries reflects the uber-right view that patent rights should only be restricted in very limited circumstances.[104] Along similar lines, suggestions that proposed Article 31*bis* should be limited to use in the case of emergencies by developing countries also reflects an uber-right view that exceptions to patent rights be narrow.[105]

2. Recent Developments

Current discussions continue to reflect both perspectives. For example, developing countries and those with a privilege view of patents largely consider the 2003 Decision unworkable because drugs have only been exported once since 2003 under the procedure.[106] On the other hand, many developed countries—or, at least those with an uber-right perspective—tend to suggest that the single use suggests that the procedure works fine and perhaps is not even needed. Moreover, the differing perspectives are also reflected by the fact that countries do not even agree on whether there is a problem to discuss.[107]

The competing perspectives were also reflected in a recent debate concerning a bill to amend Canada's existing patent laws concerning compulsory licenses for

104. *See, e.g.*, WTO Council for Trade-Related Aspects of Intellectual Property Rights, *Second Communication from the United States: Paragraph 6 of the Doha Declaration on the TRIPS Agreement and Public Health*, IP/C/W/358 (July 1992) (suggesting a moratorium on WTO disputes for LDCs only in the case of infectious epidemic diseases). A more moderate view that still tilts more to the uber-right perspective was offered in the European Union's suggestion that the Waiver not be limited to epidemics, but to a set of diseases that could be subject to expansion in specific cases, depending on WHO's determination of the severity of the problem. *See* International Trade Center, Negotiations on Paragraph 6 of the Doha Ministerial Declaration on the TRIPS Agreement and Public Health: Progress Report 14 (2003) (citing Communication from the European Communities and their Member States on Paragraph 6 of the Doha Declaration on the TRIPS Agreement and Public Health, JOB(03)/9 (Jan. 24, 2003), *available at* http://www.intracen.org/btp/issues/trips/publications/para6_august2003.pdf).

105. This was reflected in controversy over whether a statement read by the WTO Council should be part of the formal amendment to TRIPS. *See*, e.g., Ian Swanson, *WTO Members Re-open Fight over Substance of TRIPS Health Agreement,* INSIDE U.S. TRADE (Mar. 12, 2004), *available at* http://www.cptech.org/ip/wto/p6/insideustrade03122004.html.

106. *See, e.g.*, Brook Baker, *Arthritic Flexibilities for Accessing Medicines: Analysis of WTO Action Regarding Paragraph 6 of the Doha Declaration on the TRIPS Agreement and Public Health*, 14 IND. INT'L & COMP. L. REV. 613 (2004); Press Release, Oxfam, WTO Patent Rules Will Still Deny Medicines to the Poor (Aug. 27, 2003).

107. *See, e.g.*, Kaitlin Mara, *Efficacy of TRIPS Public Health Amendment in Question at WTO*, IP-WATCH (Mar. 1, 2010); Kaitlin Mara, *Medicines Access Again Captures Attention at WTO as Progress Urged in Round*, IP-WATCH (Oct. 30, 2009), *available at* http://www.ip-watch.org/2009/10/30/medicines-access-again-captures-attention-at-wto-as-progress-in-round-urged.

export.[108] Supporters of the bill, who likely have a privilege view of patents, suggested a "one-license" solution to streamline the process of issuing compulsory licenses for export to more countries. Canada's existing law requires that the application for a compulsory license for export must identify the importing country, as well as the maximum amount of drugs to be manufactured.[109] This can be problematic because in many developing countries, drugs are purchased through a competitive bidding process. A company that wanted to make drugs for export under a compulsory license likely would not be competitive, as its bid would be conditional on obtaining a compulsory license. Accordingly, the bill proposed that a license for export could simply identify the drug to be exported under a compulsory license first, with the idea that importing countries and drug quantities would later be specified. This would enable the manufacturer of drugs made under a license for export to potentially achieve economies of scale in making a low-cost product while simultaneously promoting access to more countries.

There is some question as to whether a one-license solution is consistent with the 2003 Decision—even among those who hold a privilege view of patents. For example, one group that presently argues that the one-license approach is consistent with the 2003 Decision had earlier admitted that the approach departed from the 2003 Decision, but was permissible based upon TRIPS Article 30.[110] In addition, Doctors Without Borders, a group that advocates access to medicines and generally has a privilege view of patents, criticized the 2003 Decision as requiring a "drug-by-drug, country-by country" process that thwarted economies of scale.[111] In other words, a group with a privilege view assumed that there could *not* be a one-license approach.

Holders of the strongest view of patents as a privilege could nonetheless be expected to interpret the 2003 Decision in a manner most compatible with ensuring that patents are consistent with current access to medicines. Accordingly, although the 2003 Decision requires that the compulsory license

---

108. Bill C-393, Fortieth Parliament, 3d Sess. (2011).

109. Canada Patent Act, *supra* note 48, § 21.04(2)(e) (requiring that applicant for compulsory license for export must state "the name of the country or WTO member to which the pharmaceutical product is to be exported").

110. Canadian HIV/AIDS Legal Network, *Getting the Regime Right: Compulsory Licensing of Pharmaceuticals for Export*, Brief to the House of Commons, Standing Committee on Industry, Science and Technology Regarding Canada's Access to Medicines Regime, 6–7 (Apr. 18, 2007), *available at* http://www.aidslaw.ca/publications/interfaces/downloadFile.php?ref=1058 (suggesting that the streamlined compulsory license mechanism is consistent with TRIPS Article 30, even though it departs from the 2003 WTO Decision).

111. Doctors Without Borders, *Neither Expeditious, Nor a Solution: The WTO August 30th Decision Is Unworkable* (Aug. 1, 2006), *available at* http://www.doctorswithoutborders.org/news/hiv-aids/WTO_chretien.pdf.

by the exporting member be issued for "only the amount necessary to meet the needs of the eligible importing Member(s)," this has been interpreted to be consistent with a single license solution that is issued *before* importing countries are identified and to also permit a license to apply to multiple countries.[112] These opinions tend to suggest that the "amount necessary to meet the needs of the eligible importing member[s]" does not refer to a set numerical amount and that compulsory licenses can mirror this language to satisfy it.

On the other hand, those with an uber-right view predictably suggest that the solution is inconsistent with TRIPS and also causes other problems. A number of commentators have criticized the proposal as inconsistent with TRIPS, although generally without explaining how it is inconsistent.[113] This omission possibly could be explained by an uber-right view of patents. For example, some who opposed the amendment suggested that patents are not a problem to accessing medicine—a view of the uber-right—and thus amendments to Canada's patent law were unnecessary.[114] In addition, some suggested that the bill would negatively impact development of new drugs in Canada, or result in companies fleeing Canada to countries that support patent protection.[115]

---

112. *See, e.g.*, Canadian HIV/AIDS Legal Network, *Reforming Canada's Access to Medicines Regime (CAMR): Bill C-393—Finding the Expeditious Solution*, Feb. 18, 2010 (suggesting that references to importing "member(s)" must indicate that a license can be issued to multiple countries); Frederick M. Abbott, *Prepared Testimony Before House of Commons Standing Comm. on Indus., Sci., and Tech., Parliament of Canada, Regarding Bill C-393, An Act to Amend the Patent Act* 4–5 (Oct. 26, 2010), *available at* http://www.aidslaw.ca/EN/camr/documents/Abbott_INDUCommitteeC-393withAnnex_26Oct2010__EN.pdf (suggesting that there is no explicit requirement to identify an importing country before a compulsory license is issued, although this conclusion is based on the separate notification requirement).

113. *See, e.g.*, Dale Smith, *Tony Smith Behind Senate Delay, Leaked Document Shows*, NATIONAL (Mar. 24, 2011), http://www.xtra.ca/public/National/Tony_Clement_behind_Senate_delay_leaked_document_shows-9929.aspx (providing leaked text of Tony Clement's memo to conservatives that suggests that the proposed bill would be "out of step with out international trade obligations" if passed, without explanation).

114. *See, e.g.*, Paul N. Lucase, *Medicines to Africa: How Do We Improve* (Feb. 28, 2011), http://www.canadapharma.org/CMFiles/Media%20Centre/CEO%20Spotlight/Feb282011_OP-ED_Paul_Lucas_Medicines_to_Africa_FINAL.pdf (stating GlaxoSmithKline's position that Canada's law does not need "fixing" and that the proposed bill would actually make it unworkable); *see also* Attaran, *supra* note 74 (criticizing the existing Canadian law in stating that "the fundamental reason for CAMR's failure today is pricing, not patents").

115. *See, e.g.*, Gloria Galloway, *Tony Clement Urges Senators to Block Generic-Drug Legislation*, GLOBAL & MAIL (Mar. 24, 2011), http://www.theglobeandmail.com/news/politics/ottawa-notebook/tony-clement-urges-senators-to-block-generic-drug-legislation/article612864/ (noting that Industry Minister Tony Clement suggested that the bill threatens

Conclusion

The examples here should help to illustrate that the compulsory license conundrum may actually be more complex than simply choosing a preferred interpretation and labeling the other one as "wrong." Although public debates tend to polarize discussions, this is consistent with social science evidence that people tend to adhere to the views of their group and assume that other groups are biased—without seeing their own bias.

In addition, although this chapter focused on how competing patent perspectives impact views of compulsory licenses under TRIPS Article 31 and proposed Article 31*bis*, the perspectives approach is certainly not limited to these topics.[116] Rather, the genesis of TRIPS, as well as some of the ambiguous language in TRIPS, reflects the push and pull of the two approaches. Similarly, "TRIPS-plus" free trade agreements that require more protection than does TRIPS likely reflect an uber-right view of patents. Although such agreements restrict flexibilities under TRIPS an uber-right view may nonetheless view these agreements as consistent with the Doha Public Health Declaration. Indeed, this is the position that the United States and the European Union have taken with respect to free trade agreements that allow fewer restrictions on patentability and limit exceptions to patent rights.[117] To those with a privilege view, the United States and European Union simply are being disingenuous because they believe that the Doha Declaration enshrines the right to promote "access to medicine for all." However, although the Doha Declaration contains such language, it also states that it affirms the obligations under TRIPS. One fundamental element of TRIPS that is often overlooked by those with a privilege view in criticizing TRIPS-plus agreements is that TRIPS is a minimum standards agreement that necessarily contemplates that stronger rights may be permissible. However, those with an uber-right perspective may also have an overly narrow view of the Doha Declaration as being promoted so long as the proposed Article

current patents, such that "patent holders will leave Canada seeking shelter in countries which value patent protection").

116. For additional discussion of patent perspectives in a number of areas, see Ho, *supra* note 70, at 325–72 (discussing the relevance of competing perspectives to recent global controversies concerning access to medicines, as well as proposals to improve access).

117. *See, e.g.*, United States–Peru Trade Promotion Agreement, art. 16.13, US–Peru (Apr. 12, 2006) *available at* http://www.ustr.gov/webfm_send/1031; CAFTA Understanding Regarding Certain Public Health Measures, US–Costa Rica–Dom. Rep.–El. Sal.–Guat.–Hond.–Nicar. (Aug. 5, 2004); Side Letter on Public Health Letter from Taib Fihhri, Minister of Morocco, to Ambassador Zoellick (June 15, 2004), *available at* http://www.ustr.gov/sites/default/files/uploads/agreements/fta/morocco/asset_upload_file258_3852.pdf.

31*bis*/waiver remains intact—even though that is only one issue addressed by the Declaration.

There are no easy answers on how to reconcile the existing and divergent patent perspectives. However, recognizing the existence and influential impact of these perspectives hopefully provides an important first step toward a better understanding that can lead to sustainable long-term solutions to balance patent rights and public health in the global arena.

16 BALANCING "INCENTIVE TO INNOVATE" AND "PROTECTION OF COMPETITION": AN AFRICAN PERSPECTIVE ON INTELLECTUAL PROPERTY RIGHTS AND COMPETITION LAW

Mor Bakhoum*

Introduction

This chapter deals with the interface(s) between intellectual property rights and competition law. The theory of complementarity, according to which intellectual property and competition share the common goal of dynamic competition—or competition for innovation—reconciles the two disciplines around the "innovation paradigm."[1] The very essence of the intellectual property system is to promote innovation by granting limited protection. The core goal of competition law, on the other hand, is to promote and protect dynamic competition,[2] in complement to static price competition, for the benefit of the consumer. Although they pursue the same objective of promoting innovation, intellectual property law and competition law use different—yet

* Senior Research Fellow, Max Planck Institute for Intellectual Property and Competition Law.

1. On the theory of complementarity between intellectual property law and competition law, see Josef Drexl, *IP in Bilateral Trade Agreements: Some Ideas on How They Promote Market Power and Distort International Competition*, *in* INTELLECTUAL PROPERTY AND MARKET POWER: ATRIP PAPERS 2006–2007, at 526–27 (Gustavo Ghidini & Luis Mariano Genovesi eds., 2008) [hereinafter Drexl, *IP in Bilateral Trade Agreements*]; *see also* Josef Drexl, *IMS Health and Trinko-Antitrust Placebo for Consumers Instead of Sound Economics in Refusal to Deal Cases*, 35(7) IIC 788 (2004) [hereinafter Drexl, *IMS*]. *See generally* Hanns Ullrich, *The Interaction Between Competition Law and Intellectual Property Law: An Overview*, EUR. U. INST. (2005), *available at* http://www.eui.eu/RSCAS/Research/Competition/2005/200612-CompUllrichOVERVIEW.pdf.

2. *See* J.G. Sidak & D.J. Teece, *Dynamic Competition in Antitrust Law*, 5 J. COMPET. L. & ECON. 587 (2009).

complementary—tools to achieve it. Intellectual property rights provide a limited exclusivity to their owner and thus, to a certain extent, diminish competition. For the duration of the patent, for instance, the patentee is sheltered from price competition from generic companies.[3] Such privilege only guarantees a certain kind of market power, namely to enter and commercialize one's products in a given market.[4] To the contrary, competition law promotes and protects dynamic competition as a tool to achieve innovation. In other words, innovation is likely to take place if markets are open and competitive forces released. At first sight, there is a tension between the two rationales: intellectual property excludes competition to promote innovation, and competition law promotes competition as a means to encourage innovation. The contradiction is, however, only superficial. Intellectual property protection in itself is only a *privilege* that has to be exercised in the marketplace. When entering the market, intellectual property owners have to abide by the rules of the game and respect competition norms.

Given that the exploitation of intellectual property rights could unduly limit competition, competition law stands as an "external" legal tool of correction that aims at keeping markets open and competitive. It is well established in EU case law that Article 86 (now Article 102) of the Treaty on the Functioning of the European Union (TFEU)[5] also applies to cases of abuse involving the exploitation of an intellectual property right. In *Magill*,[6] the European Court of Justice (ECJ) developed the conditions under which the exercise of an intellectual property right could amount to an abuse of a dominant position. The court concluded that an abuse of a dominant position is established when the intellectual property owner prevents the marketing of a new product for which there is consumer demand. The court's reasoning was based on an innovation rationale, namely that the final objective of intellectual property protection is to facilitate the introduction of new and innovative products in the marketplace.

European Union case law has likewise developed what is now known as the "essential facilities" doctrine. This doctrine has been introduced in the competition laws of some Sub-Saharan developing countries, most notably South Africa.

---

3. Unless the laws of a particular jurisdiction allow for compulsory licenses or other use without the authorization of the patent holder, or the patent right is terminated prior to the expiration of the notional term.

4. *See* Josef Drexl, *The Relationship Between the Legal Exclusivity and Economic Market Power: Links and Limits*, *in* INTELLECTUAL PROPERTY, MARKET POWER AND THE PUBLIC INTEREST 16–18 (Inge Govaere & Hanns Ullrich eds., 2008).

5. Treaty on the Functioning of the European Union, Mar. 30, 2010, 2010 O.J. (C 83) 171.

6. Case C-241/91 P, Radio Telefis Eireann (RTE) v. Comm'n of the Euro. Cmtys., 1995 E.C.R. I-00743 [hereinafter *Magill*], *available at* http://eur-lex.europa.eu/LexUriServ/LexUriServ.do?uri=CELEX:61991J0241:EN:HTML.

However, although the doctrine is expressly integrated in Section 8 of the South African Competition Act,[7] it has not yet been applied in practice. In a case that explicitly raised the essential facilities doctrine, *DW Integrators v. SAS Institute*,[8] the South African Competition Commission sidestepped the issue. In that case, the respondent was a large software firm that owned valuable intellectual property in certain software programs, while the claimant was a firm that provided consulting services to the licensees of the respondent's programs. The complainant argued that the software in question was an essential facility and requested an order directing the respondent to issue a license. The claimant argued that the respondent's refusal to grant a license amounted to a violation of Section 8(b) of the South African Competition Act. The Competition Tribunal, however, held that the complainant had failed to accurately define the relevant market and was thus not required to make a finding on the claimant's allegations that the use of the respondent's software was an essential facility.

This chapter "reopens" this inaugural case in South Africa and provides some insights on the legal challenges that the dispute would have raised if it had followed the *Magill* approach. Analyzed against the backdrop of the goals of competition law, with the essential facilities doctrine as a starting point, this chapter discusses the fledgling South African experience with the interface between competition and intellectual property law. In particular, it focuses on whether concerns about *access* to the fruits of innovation should triumph over *innovation* per se, at least in some sectors, such as pharmaceuticals. The issue is particularly relevant in the context of Sub-Saharan Africa, where access to public goods is arguably as critical for development as innovation. Technology-intensive products protected by intellectual property rights that are commercialized in Sub-Saharan African markets pose a particular challenge in terms of access to medicines,[9] food security,[10] and access

7. South Africa Competition Act, Nov. 30, 1998, as amended, *available at* http://www.compcom.co.za/assets/Files/pocket-book-2005-R.pdf.

8. DW Intergrators CC and SAS Institute (Pty) Ltd. (14/IR/Nov99), [2000] ZACT 16 (1 May 2000), *available at* http://www.saflii.org/za/cases/ZACT/2000/16.html.

9. *See, e.g.*, Hazel Tau v. GlaxoSmithKline, Statement of Complaint in Terms of Section 49B(2)(b) of the Competition Act 89 of 1998, *available at* http://www.section27.org.za/wp-content/uploads/2010/10/TauvGSKevidenceAndLegalSubmissions.pdf (last visited Sept. 4, 2013); *see also* Mor Bakhoum, *TRIPS, Patent Rights and Right to Health: "Price" or "Prize" for Better Access to Medicine, available at* http://papers.ssrn.com/sol3/papers.cfm?abstract_id=1619215 (last visited Sept. 4, 2013).

10. *See, e.g.*, Pioneer Hi-Bred Int'l Inc. v. Competition Comm'n, Case No. 113/CAC/NOV11, May 28, 2012, *available at* http://www.comptrib.co.za/assets/Uploads/113CACNov11-Pioneer-Pannar.pdf. On the general interface between intellectual property and food security, see Hans Morten Haugen, Manuel Ruiz Muller & Savita Mullapudi Narasimhan, *Food Security and Intellectual Property Rights: Finding the Linkages*, *in* INTELLECTUAL PROPERTY AND HUMAN DEVELOPMENT (Tzen Wong & Graham Dutfield

to technology,[11] all of which could potentially be limited by overly strong intellectual property protection. This chapter looks at the potential for African countries to use their competition laws as a balancing mechanism between access needs and the importance of appropriate innovation incentives through intellectual property rights. By defining the general economic framework under which intellectual property rights are commercialized, I argue that competition law constitutes a safeguard to the freedom to compete and makes even intellectual-property–intensive sectors contestable.[12]

The chapter is organized as follows: Section I discusses the essential facilities doctrine in the South African context in a comparative perspective with the EU case law; Section II discusses the issue of innovation and access against the backdrop of the essential facilities doctrine. Section III sheds some light on the challenges young African competition authorities will likely face when applying their competition laws and suggests ways for addressing intellectual-property–related restrictions on competition in Sub-Saharan African economies.

I. The Essential Facilities Doctrine as Dealt with in the Case Law in South Africa

In the *GSK/BI* case, a complaint was lodged before the South African Competition Commission against GlaxoSmithKline South Africa (GSK) and Boehringer Ingelheim (BI), alleging excessive pricing in contravention of Section 8(a) of the South African Competition Act.[13] Following its investigation, the Commission expanded the case to include possible violations of Sections 8(b) and (c) of the Competition Act, which deal respectively with the essential facilities doctrine and exclusionary conduct. The case was eventually settled. The following analysis sheds some light on the application of the concept of essential facilities in South Africa.

---

eds., 2011), *available at* http://www.piipa.org/files/Book_Content/Chapter%203%20-%20IP%20and%20Human%20Development.pdf.

11. *See infra* Section II.A.

12. *See* Andreas Heinemann, *The Contestability of IP-Protected Markets*, *in* RESEARCH HANDBOOK ON INTELLECTUAL PROPERTY AND COMPETITION LAW 54 (Josef Drexl ed., 2008).

13. For a discussion of the case, see Mfundo Ngobese & Liberty Mncube, *Competition Policy in South Africa's Pharmaceutical Sector: Balancing Competition and Innovation* (2011) (on file with author).

A. The Legal Challenges of Applying the Essential Facilities Doctrine

Section 8 of the South African Competition Act provides—

It is prohibited for a dominant firm to—

(a) charge an excessive price to the detriment of consumers;
(b) *refuse to give a competitor access to an essential facility when it is economically feasible to do so*;
(c) engage in an exclusionary act, other than an act listed in paragraph (d), if the anti-competitive effect of that act outweighs its technological, efficiency or pro-competitive gains."[14]

The South African Competition Appeal Court (CAC) in *Glaxo Wellcome v. National Ass'n of Pharmaceutical Wholesalers*[15] set forth the legal framework for the interpretation of Section 8(b). According to the Appeal Court, to allege a contravention of Section 8(b) a complainant will have to aver, in its complaint that—

1. the dominant firm concerned refuses to give the complainant access to an infrastructure or a resource;
2. the complainant and the dominant firm are competitors;
3. the infrastructure or resource concerned cannot reasonably be duplicated;
4. the complainant cannot reasonably provide goods or services to its competitors without access to the infrastructure or resources; and
5. it is economically feasible for the dominant firm to provide its competitors with access to the infrastructure or resource.

The following discussion focuses on the *GSK/BI* case against the backdrop of the Act and the requirements set forth in *Glaxo Wellcome*.

GSK held patents in South Africa on antiretroviral drugs (ARVs) AZT, Lamivudine, and AZT/Lamivudine. Boehringer Ingelheim (BI) held patents on Nevirapine. The patent holders did not provide the three-drug fixed dose

---

14. South Africa Competition Act, *supra* note 7, § 8.

15. Glaxo Wellcome v. Nat'l Ass'n of Pharm. Wholesalers, Case No. 15/CAC/Feb02, *available at* http://www.comptrib.co.za/assets/Uploads/Case-Documents/Glaxo%20Wellcome%20 15CACFeb02.pdf (last visited Sept. 4, 2013). For commentary on the case, see James Brand, *Intellectual Property and Abuse of Dominant Position in South African Competition Law*, 122 S. Afr. L.J. 910 (2005).

combinations[16] necessary for highly active antiretroviral therapy, or HAART. Among other things, GSK and BI were accused of the following anticompetitive conduct:

- Abusing a dominant position in the market for ARVs by charging excessive prices for the product;
- Making the product inaccessible to the general public;
- Refusing to supply a competitor access to an essential facility;
- Allowing a dramatic difference in price between ARVs sold in South Africa and generic alternatives sold outside South Africa.

The Competition Commission concluded its investigation with a finding that GSK and BI had abused their dominant position by charging excessive prices, *refusing to grant access to essential facilities to a competitor*, and engaging in exclusionary conduct. The matter did not come before the Competition Tribunal, because GSK and BI accepted a settlement opportunity, which resulted in a drastic reduction in the prices of the pharmaceuticals in South Africa.

As part of the settlement, GSK and BI agreed to: (1) grant licences to generics manufacturers; (2) permit the licensees to export the relevant ARV medicines to Sub-Saharan African countries; (3) where the licensee did not have manufacturing capability in South Africa, permit the importation of the ARV medicines for exclusive distribution in South Africa provided all the regulatory approvals were obtained; (4) permit licensees to combine the relevant ARVs with other ARV medicines; and (5) not require royalties in excess of 5 percent of the net sales of the relevant ARVs. [17]

The settlement of the case left open a number of questions as to how the Competition Tribunal would have dealt with the legal issues. The approach taken by the *Glaxo Wellcome* court in defining the guidelines for the application of the essential facilities doctrine seems to ignore the specificities of the intellectual-property–related facilities. Even when referring to the comparative case law in the European Union and the United States, the focus of the Court seems to have been only on the physical facilities. Hence, the Tribunal did not mention the leading EU case law on access to an essential facility involving an intellectual property right, in particular *Magill* and *IMS Health*. The new rule it established concerning products for which there is a consumer demand and which are not provided by the holder of the intellectual property rights is thus

16. Ngobese & Mncube, *supra* note 13, at 10.

17. *See id.* at 13.

seemingly not part of the requirement for the applicability of Section 8(b) of the Act.

Some commentators have argued in the context of the *GSK/BI* case that the conditions for the violation of Section 8(b) were met because:

1) GSK and BI's patents are non-duplicable resources, and generics wishing to enter the market have to obtain the necessary licenses from the originator;
2) Access to GSK and BI's patents is necessary to provide goods;
3) GSK and BI prevented access to their patented technology by refusal to license;
4) It was economically feasible for GSK and BI to grant access to their license through reasonable royalties.[18]

In South Africa, neither the Competition Act nor the case law appears to recognize a special treatment of intellectual-property–related cases when applying the essential facilities doctrine. As noted above, an intellectual property right provides an opportunity to enter a market and, in some cases, could lead to market dominance. In fact, the very nature of the exclusivity granted by an intellectual property right rationalizes a certain dominance in a market. The privilege of a patent holder to market the invention excluding all others, and to set the price, is part of the reward for her efforts and investment. In this regard, refusal by a right holder to grant a license, in itself, does not constitute an abuse of dominant position. Only an abusive refusal to license would qualify.

B. The GSK/BI Case and the Application of the Essential Facilities Doctrine to Intellectual-Property–Related Cases in South Africa

The exercise of the exclusive intellectual property right should not foreclose competition in the upstream or downstream market,[19] or hinder innovation. In the European Union, special treatment has been given to the essential facilities doctrine in intellectual-property–related cases. To find that a particular innovation is an essential facility, EU case law requires "exceptional circumstances" and that the refusal to license by the intellectual property owner is preventing the introduction of new products for which there is consumer demand.[20] This new product rule developed by the ECJ is the result of a balancing test weighing the freedom of the right holder to refuse to license against the interest in protecting

18. *Id.*

19. Heinemann, *supra* note 12, at 66–70.

20. *Magill*, *supra* note 6.

competition.[21] Let us now turn to the assessment of the *GSK/BI* case against the backdrop of the need to balance a right holder's *innovation incentive* and the necessity to protect *free competition.*

For the duration of a patent, competition in the upstream market is in principle limited given the exclusivity granted to the patent holder to market her invention. In the case of pharmaceuticals, generic companies usually are allowed to enter the market and to engage in price competition only after expiration of the patent term. In *GSK/BI,* this exclusivity was the very "privilege" that was challenged by generics manufacturers.

Considering the protected patent on the pharmaceuticals as an essential facility that should be shared or open to competitors without any safeguards would amount to denying the monopoly granted to the patent holder. If a patented product is considered an essential facility that must be licensed, it would allow competitors to engage in price competition by offering the same product for a cheaper price during the term of the patent. This would amount to denying the patent holder the very right to exclude that was granted by the legislature.

A refusal of a patent holder to license under reasonable commercial conditions to competitors who intend to create a new product for which there is a consumer demand could be considered a barrier to the access to the very invention for which the society has paid a price by the grant of the exclusive right. This behavior would diminish the innovation rationale of intellectual property law. However, the monopoly position created by an intellectual property right could lead to abuses in the form of excessive pricing, which undermines access to the innovation.

C. Rationale Underlying the GSK/BI *Case: "Innovation" or "Access"*

1. The *GSK/BI* Case and the "Innovation Rationale"

As noted above, in the European Union, the "innovation surplus"[22] realized by requiring the creation of a product for which there is consumer demand is a precondition for allowing competitors access to a patented technology. In the *GSK/BI* case, the South African Competition Commission's main concern was the high price of the ARVs for which the defendant had large market shares. [23] Had the Commission based its argument on the legal test put forward by the CAC in

---

21. Drexl, *supra* note 4, at 30.

22. *See id.*

23. Ngobese & Mncube, *supra* note 13, at 10 ("GSK had 100 percent market share in South Africa for the manufacture and sale of AZT, the manufacture and sale of 3TC and the manufacture and sale of Combir. BI had a 100 percent market share in South Africa for the manufacture and sale of Nevirapine.").

Glaxo Wellcome with regard to the essential facilities doctrine (which does not take into account the specificities of intellectual property rights and the innovation paradigm),[24] the Commission could have built a strong case. However, if the requirement of the creation of a new product was applied by the Competition Tribunal, it would have been hard for the Commission to find a violation of Section 8(b) of the Act. In requesting licenses on reasonable terms from the patent holders, the main goal of the generic companies was to engage in authorized price competition, not to create a new product.

Commenting on the case, a Commissioner stated that "the Competition Commission acknowledged that incentives can be reflected in a higher royalty rate and thus higher prices for generic equivalents, but price must be reasonable" and that it is "*not reasonable to put a premium on the incentive to innovate* which may result in restrictions on general access to treatment."[25] He added that the only concern of the Competition Commission was to either have GSK and BI reduce the prices or to grant licenses to generics manufacturers.[26] Given that these were the dominant concerns, the Commission could have succeeded simply by focusing its analysis on abuse of dominance by charging excessive prices, without wading into other issues.

Alternatively, a close look at the facts of the *GSK/BI* case could have supported a different reading of the case and may have advanced the innovation surplus argument of the Competition Tribunal. From the facts of the case, GSK held patents on three ARVs: AZT, Lamivudine (branded as 3TC), and AZT/Lamivudine. The patent holders did not provide the fixed combination of the three ARVs. If the creation of a new product for which there was consumer demand were to be required, the Commission could have argued that the patent holder, by refusing to grant licenses to generics competitors, prevented the emergence of a new product. Because the three drugs' fixed-dose combination had to be taken by the patient as a cocktail and not a single pill, the consumer demand easily existed.[27] It is therefore possible to argue that by refusing to license to generics companies that would provide a fixed-dose combination of ARVs in a single pill, the patent holders prevented the emergence of a new product. As thus presented, the facts of this case look very much like those in *Magill.* It is,

---

24. *Id.* at 11 (arguing that the conditions for the legal test are met).

25. Shan Ramburuth, Competition Law and Its Interface with Intellectual Property—A South African Perspective, Presentation to the WIPO Regional Seminar, Pretoria, South Africa (Apr. 14, 2011) (emphasis added).

26. *Id.*

27. Ngobese & Mncube, *supra* note 13, at 10.

however, not clear whether the generics manufacturers would have in fact created a single-dose pill for the benefit of the consumers.

A related question that arises is whether a single-dose pill combining three ARVs would be offered at a cheaper price than the three ARVs individually. If the cost of a single-dose pill is higher than the cost of a cocktail, one should balance the "innovative features" of the new product with the additional cost that the consumer will have to bear. This again raises the issue of innovation and price competition. Although the Commission could have argued that there was a barrier to innovation through refusals to license, its focus was more on access through affordable prices.

2. The *GSK/BI* Case in Light of the Goals of Competition Law: Price Competition or Innovation (Dynamic Efficiency)?

The *GSK/BI* case offers a very interesting example of the policy choices with which a legislature has to wrestle, and the enforcement approach of a competition authority, especially for intellectual-property–related cases. Should the emphasis be put on access with lower prices for the consumer (static or allocative efficiency), or should innovation (dynamic efficiency) take precedent? If static efficiency is given precedence, price competition, which would result in improved access to affordable medicine for consumers, would be privileged. Favoring static efficiency, the competition authority's concern would be to make pharmaceuticals accessible to a large number of consumers at a reasonable price. However, this static goal has to be balanced with the innovation rationale in cases involving intellectual property rights. As Josef Drexl has pointed out, "intellectual property cases are not only about static efficiency but about dynamic efficiency, in particular. Whereas it can relatively easily be assessed what would be lost by a decrease of price competition (allocative efficiency), nobody can predict what kind of innovation would be possible if the market were more dynamic (dynamic efficiency)."[28] If dynamic efficiency cannot be reached, or is very uncertain, a second-best solution should be to promote allocative efficiency,[29] especially withregard to the provision of public goods such as health and food security. In countries where the reality of diseased and impoverished populations often overwhelms the economic rationale for the protection of innovation as a first order policy preference, then allocative efficiency should be viewed as a preferred option. As one commentator has argued—

> ...innovation competition, just like price competition, is merely an intermediate goal in order to promote consumer welfare (one of the end goals,

28. Drexl, *supra* note 4, at 28.

29. Drexl, *IMS, supra* note 1, at 806.

> if not *the* end goal of competition policy). Even if compulsory licensing were found to reduce innovation incentives, and therefore dynamic efficiency, this negative effect may be outweighed by the "static" gains in allocative efficiency, so that the whole, consumer welfare is increased.[30]

Dynamic efficiency is of course important. However, in cases where access to the actual technology at an affordable price from the right holder is the focal point of dispute, or where the diffusion of innovation through licensing responds to consumer welfare interests, compulsory licenses should be readily granted. Diffusion and access is an integrated part of the current intellectual property system as proven by the licensing mechanism,[31] which aims at making technology accessible. TRIPS competition-related rules follow the same rationale.[32]

Distinguishing between technologies and invention in balancing static efficiency and dynamic efficiency is also an approach that could be useful for African countries. For sensitive patented inventions that have a direct impact on the living conditions of people, such as life-saving drugs or patented seeds, access through price competition should be the first concern. A patent for the results of inventive activity by the innovator is a result of a compromise between the incentive to innovate and the benefits that are provided to society. For this balance not to remain only theoretical, the consumer should have access to the invention in the first place. If the patent affords a monopoly that allows the owner to set prices at a level that makes access to the invention very limited or absent, one can raise the

---

30. Cyrill Ritter, *Refusal to Deal and "Essential Facilities": Does Intellectual Property Require Special Deference Compared to Tangible Property?*, 28 WORLD COMPET.: L. & ECON. REV. 1 (2005) (emphasis added).

31. As pointed out by Josef Drexl:

> Licensing agreements are generally considered to be efficient in economic terms. Licensing agreements allow most efficient allocation of resources (allocative efficiency), by enabling another undertaking (licensee) to produce at lower costs than the licensor. Simultaneously, licensing agreements spread technology and incite follow-on innovation by the licensee (dynamic efficiency). However, the right holder will only transfer technology, if effective protection of the IP right is available in the recipient country. Conversely, competition law is needed in the recipient country in order to protect the local licensee against abuses of the IP right by the rightholder.

Josef Drexl, *Intellectual Property and Competition: Sketching a Competition-Oriented Reform of TRIPS*, *in* FESTSKRIFT TILL MARIANNE LEVIN 265 (A. Bakardjieva Engelbrekt, U. Bernitz, B. Domej, A. Kur & P. J. Nordell eds., 2008).

32. *See* Agreement on Trade-Related Aspects of Intellectual Property Rights, arts. 8, 31(k), 40, Apr. 15, 1994, Marrakesh Agreement Establishing the World Trade Organization, Annex 1C, 33 I.L.M. 81 (1994) [hereinafter TRIPS Agreement or TRIPS].

question whether that balance, conceptualized in theory, finds any application in practice.

Internal legal tools within the intellectual property system, such as voluntary licensing and the regulation of potential abuses by owners are directed at making an invention accessible to the general public. If the intellectual property owner exercises rights in a manner that violates these internal balancing mechanisms, competition law provides a further legal tool that the legislature can use to maintain the necessary balance between innovation and access. Licensing mechanisms should be considered as a means of diffusion of the invention, not only a tool the right holder can use at his discretion.

In the European Union, when dealing with intellectual-property–related cases, emphasis is placed by courts on dynamic efficiency with the concern of protecting innovation, thereby complementing the innovation rationale of intellectual property. However for societies with chronic undersupply of public goods, even if some intellectual property rights are at stake, maybe access should take precedence over innovation. For example, with regard to the acute health crisis,[33] most African countries are facing, allocative efficiency in the sense of price competition that would facilitate access to essential medicines. Such access, should take precedence over dynamic efficiency. This approach gives deference to the welfare of those seeking access to the innovation. If we look again at the *GSK/BI* decision, the emphasis of the Commission on price competition and accessibility of the ARVs seems to be more in line with the actual needs of the consumers. Moreover, if a legislature chose to favor dynamic efficiency with regard to intellectual property rights, the expected future innovation should, to some extent, meet the needs of the consumers.[34]

---

33. *See* World Health Org. (WHO), The African Regional Health Report: The Health of the People (2006), *available at* http://whqlibdoc.who.int/afro/2006/9290231033_rev_eng.pdf.

34. This is not always the case, for the consumer's innovation needs do not always drive innovation efforts or focus. R & D efforts are not necessarily oriented toward finding, for instance, new drugs for neglected diseases. As pointed out in the EU Pharmaceutical Sector Inquiry Report of 2009, pharmaceutical companies compete by R & D activities in the therapeutic fields for which the demand is particularly high. *See* EU Comm'n, Pharmaceutical Sector Inquiry Report ¶¶ 1084–85 (July 2009), *available at* http://ec.europa.eu/competition/sectors/pharmaceuticals/inquiry/staff_working_paper_part1.pdf; *see also* Global Comm'n on HIV and the Law, Final Report on HIV and the Law 8 (July 2012), *available at* http://www.hivlawcommission.org/resources/report/FinalReport-Risks,Rights&Health-EN.pdf ("IP protection is supposed to provide an incentive for innovation but experience has shown that the current laws are failing to promote innovation that serves the medical need of the poor.").

D. *Considerations for Sub-Saharan African Countries*

The emphasis in the EU case law regarding the duty to license does not suggest that South Africa or other African countries by extension should follow the EU's approach. The European Union favors a clear distinction and differentiated treatment of tangible and intangible property, with the latter afforded stronger protection, as evidenced by the requirement of the new product doctrine. Such a requirement would in practice constitute a barrier to constructing a duty to license a patented product that constitutes an essential facility. In South Africa, the CAC appears not to follow the test set up in *Magill* with the requirement of new product as a precondition for a duty to license, but instead seems to apply the same treatment to tangible and intangible goods. Further, the EU line of case law generally emphasizes more the need to promote innovation, as evidenced by the "innovation surplus" requirement. However, for countries that are importers of technology where local innovation is limited or nonexistent, the "innovation surplus" requirement would simply hinder access to existing technology. Technology transfer through licensing would be limited in part for some of the same reasons access to the products that embody the technology would be limited, namely price. From a consumer-centered perspective of a developing or importing country, access to technology, in the short run, is more relevant than innovation. Following the EU "innovation approach" in the application of the essential facilities doctrine would hinder access to existing technology or to patented products.

II. Interfacing "Promoting Innovation" and "Protecting Competition" from an International Perspective

The theory of complementarity according to which intellectual property rights and competition law share the common goal of protecting dynamic competition is generally accepted in the scholarship.[35] This commands the intervention of competition law when the exercise of intellectual property rights threatens the very essence of, and the rationale for, granting exclusive rights—the promotion of innovation. By keeping the market open, competition law promotes competition by substitution. In that sense, the application of competition law, by imposing a duty to license, also aims at keeping the market open. However, from an international perspective, competition is not applied as an "innovation instrument" in complement to intellectual property rights. This is evidenced by the

35. See sources cited *supra* note 1.

competition-related provisions in the TRIPS Agreement, which only use competition law as a tool to correct abuses of intellectual property rights.[36]

A. The Approach of TRIPS' Competition-Related Provisions: Do They Promote Innovation and Protect Free Competition?

It is questionable whether the competition-related provisions embodied in the TRIPS Agreement are directed at promoting innovation. The relevant provisions in TRIPS Articles 8(2), 31(k), and 40(1) do not set an international or comprehensive framework organizing the commercialization and exercise of intellectual property rights in the marketplace. They merely safeguard and control the exercise of intellectual property rights and facilitate transfer of technology to developing countries.[37] With the conclusion of the TRIPS Agreement, it became relevant for developing countries to ensure that the protection of intellectual property rights at the international level is balanced by appropriate access mechanisms. By giving developing countries the possibility of applying their competition laws to intellectual-propery–related restrictions of competition, TRIPS struck a balance—whose fairness is questionable—between protection of intellectual property rights and transfer of technology. However, the flaws of this approach are apparent from the fact that there is no international mechanism guaranteeing that members make use of those technology transfer facilitating provisions. Hence, the capacity of members to use their competition laws as a tool to facilitate technology transfer depends largely on the institutional capacities of each country. Whereas in advanced jurisdictions, the competition laws are strong enough to address the abuse of intellectual property rights, developing and least-developed countries—for whom the competition-related provisions are designed—often lack the institutional capacities to apply soundly their competition laws and thereby facilitate technology transfer. This issue is especially relevant in the context of Sub-Saharan Africa, which still lacks strong competition authorities with, of course, the exception of South Africa.

Moreover, technology transfer itself is the prerequisite, but not the only condition, for fostering local innovation. Three levels of technological infrastructures have been considered by Josef Drexl with regard to a country being able to absorb a technology and make follow-up innovations: (1) *at the lowest level*, where countries do not have the capacity to manufacture technology-related goods; (2) *at the intermediary level*, where countries can easily build (their) own manufactured goods based on known technology, although these countries do not have the

36. *See* TRIPS Agreement, *supra* note 32, arts. 8, 31(k), 40.

37. Ullrich, *supra* note 1.

capacity of generating local innovation; and (3) *at the highest level*, where the countries actually possess capacity to absorb technology.[38] Drexl argues that it is only at the highest level of technological development that the potential for local innovation exists and, for those countries, technology transfer could foster such innovation.

The TRIPS Agreement did not take into account this differentiated technological capacity among recipient countries. The majority of the African countries belong to the lowest or intermediary-level countries when it comes to their technological capacity. The argument of a potential follow-up innovation resulting from technology transfer is therefore very questionable. Follow-up innovation is not necessarily the direct consequence of technology transfer. Those countries are more concerned with access to existing technology and technology-embodied goods than using technology for follow-up innovation. The level of technological development that would justify having strong intellectual property rights as a local incentive for innovation does not exist.

Certainly, from the perspective of the least technologically advanced countries in Sub-Saharan Africa, access to existing technology at affordable prices is more relevant, at least in the short run, than access to technology for follow-up innovation. To guarantee access to existing technology at affordable prices, robust competition needs to be protected. Analyzed form the standpoint of developing countries, one may contend that the TRIPS competition provisions do not promote innovation, especially for those countries that lack competition laws.

B. The Discrepancies between "IP Protection" and the "Protection of Freedom to Compete" from an International Perspective

Two aspects of the current development of intellectual property and competition law from an international perspective will be discussed in this section. On the one hand, there is a marked tendency toward more and more intellectual property protection, in particular in the European Union and the United States, which export high intellectual property protection standards into the legal framework of their trading partners through bilateral and regional trade agreements.[39] On the other hand, there is a lack of parallel checks on potential abuses of intellectual property rights. This is true not only with regard to the TRIPS competition provisions,[40] but also with regard to the approach taken by competition provisions

38. *See* Josef Drexl, *Competition Law and Transfer of Technology in the EC–CARIFORUM EPA* (forthcoming).

39. On this aspect, see Drexl, *IP in Bilateral Trade Agreements*, *supra* note 1.

40. *See* Hanns Ullrich, *Expansionist Intellectual Property Protection and Reductionist Competition Rules: A TRIPS Perspective*, *in* International Public Goods and

included in North–South trade agreements.[41] This discrepancy between strong intellectual property protection and limited competition control, noticeable at the international level, is especially harmful for developing jurisdictions, which are still at the beginning of adopting and enforcing their competition laws. The situation is more alarming in Sub-Saharan Africa, where the trend toward more intellectual property protection is occurring in a competition law vacuum. Where competition laws exist, they are still at an embryonic stage[42] or severely resource constrained and thus ineffective.

In the context of global free trade, countries have opened up their markets and committed themselves to respect minimum standards of intellectual property protection as required by the TRIPS Agreement. As Drexl points out, "WTO law at large has created the basis for competition in emerging international markets."[43] Although markets are open and competition is taking place on an international scale, free competition is also at risk by the very behavior of the transnational economic actors. Developed country markets are well protected by strong competition authorities, which are able to apply their law internationally with the effects doctrine. As to developing countries, seldom are there any who dare to challenge the anticompetitive practices of multinational firms. In its competition-related provisions, TRIPS gives WTO members only the possibility of applying their competition laws to intellectual-property–related restrictions of competition. There is no binding rule at the international level that obliges WTO members to act against outbound restrictions of competition initiated by intellectual property right holders. As a leading commentator notes, "TRIPS, by only authorizing WTO Members to apply their competition law, does not contribute anything to protecting competition against restrictive behavior."[44] The issue of the discrepancy between intellectual property protection and lax competition control at the international level is captured in the following passage:

Transfer of Technology Under A Globalized Intellectual Property Regime 727–56 (Keith E. Maskus & Jerome H. Reichman eds., 2005).

41. *See, e.g.*, Mor Bakhoum, *Commerce International, Politique de Concurrence et Accords de Partenariat Economique—Réflexions sur les Enjeux et Perspectives d'un Triptyque*, *in* Agence Africaine pour le Commerce et le Développement (2011); *see also* Josef Drexl, *Perspectives Européennes sur la Politique de Concurrence dans L'espace OHADA*, 3 RIDE 298 (2011) (providing a critique of the European policy approach in the EPAs).

42. For an overview of the state of competition legislation in Sub-Saharan Africa, see Africa Competition Forum, Full Needs Assessment Report (Feb. 24, 2011).

43. *See* Drexl, *supra* note 31, at 266.

44. *See* Drexl, *supra* note 38.

Still, by concluding TRIPS the Contracting Parties of the former GATT agreed only on one of the two necessary elements for a "pro-competitive" system promoting dynamic efficiency on the global level. Strong efforts have been made for raising the international standards of IP protection (first element), whereas there are no sufficient international safeguards against the anticompetitive effects of the international IP system (second element). TRIPS has entrusted the interest in fighting anticompetitive behavior of right holders to individual WTO members without any obligations to fight such restraints internationally.[45]

C. The Role of Competition Law in the Innovation Dynamic: A Sub-Saharan African Perspective

Intellectual property and competition law are not stand-alone disciplines with different goals and different legal and economic rationales. The theory of complementarity between intellectual property and competition law commands a legislative approach that takes into account the internal incentive structure of the intellectual property system, but also, and more important, the need to put safeguards in place that protect the competitive structure of the markets where intellectual property rights are commercialized. Keeping the intellectual-property–related markets competitive is a task vested in part to competition law. In this part of the text, I shed some light on whether and how the theory of complementarity is dealt with in the context of Sub-Saharan Africa. This will be done, on the one hand, by looking at how intellectual-property–related restrictions of competition are dealt with in some competition laws, and, on the other hand, by looking at relevant provisions in North–South trade agreements.

1. Intellectual Property and Competition Law from a Legislative and an Enforcement Perspective

Competition law and policy is relatively new in most African countries. According to a need assessment conducted by the African Competition Forum (ACF),[46] no fewer than twenty-four countries have competition laws. Although some countries have a long tradition of having a kind of competition law, which in fact was often related to price control regulations, an approach in which competition law guarantees the protection of free competition by a prohibition of cartels and abuses of dominance is relatively new. It was only in the 1990s,

---

45. *See* Drexl, *supra* note 31, at 266.

46. *See supra* note 42.

following a wave of liberalization and privatization, that most African countries started to draft and implement competition laws. Even Nigeria, the largest economy in West Africa, does not yet have a competition law or a competition authority. Moreover, when it comes to enforcement, limited results have been achieved. This is in part due to the lack of institutional, human, and financial capacities of the newly created competition authorities.

The treatment of intellectual-property–related restrictions of competition is rather rudimentary in many developing countries. Some laws do not even refer to the exercise of intellectual property rights as a potential form of anticompetitive behavior.[47] However, this does not mean that competition authorities do not have the legal tools to address intellectual-property–related restrictions on competition. Hence, anticompetitive licensing agreements or abuses of dominant positions involving intellectual property rights could be dealt with by a competition authority. This should be an area of important policy advocacy in Sub-Saharan Africa.

47. In Zambia for instance, Section (3)(a) of the Competition and Consumer Protection Act asserts that the Act shall not apply to "an agreement or conduct insofar as it relates to intellectual property rights including the protection, licensing or assignment of rights under, or existing by virtue of, a law relating to copyright, design rights, patents or trade marks." Zambia Competition and Consumer Protection Act, No. 24 of 2010, *available at* http://www.ccpc.org.zm. In the Seychelles, except in Section 7(4)(c) of the Fair Competition Act, which exempt abuses of dominant power resulting from the exercise of intellectual property rights, there is no comprehensive framework addressing the interface between intellectual property and competition law. *See* Seychelles Fair Competition Act, Supplement to the Official Gazette 219, Nov. 3, 2009, *available at* http://www.ftc.sc. In the Gambia, intellectual-property–related matters are expressly exempted from the application of the Competition Act. *See* Gambia Competition Act, Sch. 1 ¶5, No. 4 of 2007, Sept. 3, 2007, *available at* http://www.gcc.gm/downloads/COMPETITIONBILL2007.pdf. In contrast, in Kenya, the Competition Act does apply to intellectual-property–related restrictions unless an exemption is granted by the Competition Commission according to Section 28(1) of the Act. *See* Kenya Competition and Consumer Protection Act, No. 24 of 2010 315, *available at* http://www.ccpc.org.zm/index.php/legislation/competitiona-consumer-protection. Also, in Mauritius, the national competition law is applicable to intellectual-property–related restrictions of competition. Extended guidelines (CCM 7 Guidelines on General provisions deal with the issue of intellectual property rights) deal with the specific issue of the application of the Act to intellectual-property–related restrictions. *See* Mauritius Competition Act, Act No. 25/2007, Dec. 20, 2007, *available at* http://www.ccm.mu/English/Documents/Legislations/Competition%20Act%202007_Amended_050213.pdf. In Malawi, the Competition Act does not apply to licensing agreements involving an intellectual property right. *See* Malawi Competition and Fair Trading Act, § 3(d), Cap. 49:02, *available at* http://www.malawilii.org/mw/legislation/consolidated-act/4902. In Namibia, the Competition Act does apply to intellectual-property–related restrictions unless they are exempted by the Namibian Competition Commission. *See* Namibian Competition Act, § 30, Act No. 2 of 2003, *available at* http://www.nacc.com.na/downloads/competition-act-2003.pdf. In all of the above jurisdictions, there is no case law yet on the application of the competition law to intellectual-property–related restrictions.

The issue is more complex, and any possibility of enforcement lost, if a competition statute expressly exempts intellectual property from its application or gives the option for an exemption. In Gambia, for instance, the Competition Act expressly exempts intellectual-property–related issues from its application. To the contrary, Mauritius has comprehensive provisions in its Competition Act dealing with intellectual property. In South Africa, intellectual-property–IP-related restrictions of competition are expressly dealt with in the Competition Act. However, the Act provides the possibility of applying for an exemption from its application.

Exempting intellectual property rights from the application of competition law is a rather problematic approach, especially for developing countries. This is in contradiction with the very rationale of the international intellectual property system and the interests of developing countries to use competition law as a balancing tool and thereby to facilitate technology transfer. Fighting intellectual-property-rights–related abuses of competition aims at facilitating technology transfer. An absence or a weak enforcement of competition law in developing countries against abuses of intellectual property rights hinders technology transfer and subsequent potential follow-up innovation in countries with the requisite capacity. In this regard, one could argue that the legislative and enforcement approach to competition law in Sub-Saharan Africa does not yet fulfill its role of fostering innovation or access to protected goods. This is also true with regard to the treatment of competition law in North–South trade agreements, as exemplified by the EU–CARIFORUM Economic Partnership Agreement (EPA).[48]

2. Competition Provisions in the African EPAs: Approach and Implications

No fully fledged agreement encompassing rules on trade, intellectual property, and competition law has been concluded yet between the European Union and African countries. The interim agreements[49] signed by some states bilaterally with the European Union do not concern intellectual property or competition law. In the ongoing EPA negotiations between the European Union and African countries, intellectual property and competition law are part of the so-called

48. Economic Partnership Agreement Between the CARIFORUM States, of the one part, and the European Community and its Member States, of the other part, O.J. L289/13, Oct. 30, 2008, *available at* http://eur-lex.europa.eu/LexUriServ/LexUriServ.do?uri=OJ:L:2008:289:0003:1955:EN:PDF [hereinafter EU–CARIFORUM EPA].

49. *See, e.g.*, Interim Agreement with a View to an Economic Partnership Agreement Between the European Community and its Member States, of the one part, and the Central Africa Party, of the other part, OJ L57, Feb. 28, 2009, at 2, *available at* http://eur-lex.europa.eu/JOHtml.do?uri=OJ:L:2009:057:SOM:en:HTML.

"rendezvous" clause, which means the parties agree to discuss these clauses in the future. Despite the fact that there is no agreement yet signed between the European Union and the different regional economic communities, one can already discuss the potential approach the European Union will be taking when it comes to competition matters. The approach taken by the European Union in the agreement it has signed with the Caribbean Forum (CARIFORUM) provides some useful insights.[50]

The EU–CARIFORUM Agreement requires CARIFORUM states to introduce competition laws and create national competition authorities.[51] With regard to the substantive approach, the principle of non-discrimination, which does not take into account the need and competitive constraints of local companies, must be applied. Do the EPA's competition rules foster innovation and protect free competition? It is very doubtful. There is a differentiated treatment of intellectual property rights and competition law in the EPAs, as evidenced by the EU–CARIFORUM Agreement. Strong emphasis is put on intellectual property protection, whereas competition law remains marginal. Competition law is not approached as a complement of intellectual property law that would enhance innovation by protecting free competition. Instead, competition law provisions are viewed principally as an additional market-access tool.[52] In fact, in the EU–CARIFORUM Agreement, intellectual property is not expressly mentioned in the competition law provisions, as one would have expected given the complementarity of the two fields. This should not, however, in principle prevent countries from using their competition law against abuses of intellectual property rights. This possibility is only optional, as "the competition law rules of the EPA contain no express obligation to fight anticompetitive behavior relating to IPRs . . . and do not provide any guidance on how to apply competition law in IP-related cases."[53]

One may argue in the context of North–South trade agreements that raising the intellectual property protection standards provides an incentive to right holders to innovate by preventing misappropriation. However, there is also a risk that a high level of protection, or a monopoly power derived from an intellectual property right, will lead to market foreclosure. As Drexl observes, "these

50. *See* Bakhoum, *supra* note 41; *see also* South Centre, Fact Sheet No. 8: Competition Policy in Economic Partnership Agreements (CARIFORUM Text) (Apr. 2008), *available at* http://www.southcentre.org/index2.php?option=com_docman&task=doc_view&gid=707&Itemid=68 [hereinafter CARIFORUM Text].

51. *See* EU–CARIFORUM EPA, *supra* note 48, arts. 127–29. For a detailed analysis of the obligations, see CARIFORUM Text, *supra* note 50.

52. *See* Bakhoum, *supra* note 41; South Centre, *supra* note 50.

53. *See* Drexl, *supra* note 38.

standards are not necessarily pro-competitive in the sense that IP exclusivity will trigger dynamic competition; they may well foreclose markets for what may be more innovative competitors in favor of incumbent right-holders."[54]

III. Looking Forward: How to Better Address Intellectual-Property–Related Restrictions of Competition in Sub-Saharan Africa

Competition law and policy is still taking shape in Sub-Saharan Africa. Apart from South Africa, the enforcement experience is still very sparse. Despite the close link between intellectual property and competition law, and the crucial importance of competition law in keeping in check potential abuses of intellectual property rights, the legislative and enforcement approaches of Sub-Saharan countries often ignore intellectual-property–related considerations. As noted earlier, intellectual-property–related restrictions of competition are even exempted in some competition laws. Consistent with the competition provisions in the TRIPS Agreement that allow member states to use their competition policies as a tool to fight abuses of intellectual property rights, African states must integrate intellectual-property–related competition offenses in their competition law. Mauritius has done it[55] and others should follow. However, enforcing a competition law against strong intellectual property holders will certainly be a big challenge for many young competition authorities.

With regard to the ongoing EU/US approach of dealing with intellectual property issues through bilateral and regional trade agreements, it is of crucial importance that competition law and policy be approached differently by developing countries. As evidenced by the EU–CARIFORUM EPA, competition law is viewed primarily as a market-access tool by the developed countries. By requiring signatory states to only introduce competition law, to create competition authorities, and to apply the principle of non-discrimination, competition-related provisions in North–South trade agreements do not fully address the needs of developing countries. A few recommendations of importance in the context of negotiating EPAs include:

- Competition law and policy should not be considered as a mere market-access tool allowing the enterprises of the intellectual-property–exporting countries better access to the markets of the partner countries.

54. *See* Drexl, *supra* note 31, at 269.

55. *See supra* note 47.

- Competition provisions in the EPAs should take into account the political, institutional, economic, and political environment of the hosting country. The EU competition principles should not just be exported to the partner countries as are the intellectual property protection and enforcement standards.
- In line with the complementarity between intellectual property and competition, which both aim at fostering dynamic competition, competition law in the EPAs should create or "regulate" the environment within which intellectual property rights are marketed. Competition provisions should be crafted in a way that they will be applicable to exercises of intellectual property rights that threaten competitiveness of the markets.

Given the rudimentary character of competition laws of EU partner countries and the lack of sound enforcement institutions, EU partner countries should request that the European Union, in the first place, fight intellectual-property–related anticompetitive behavior originating from the European Union. In this vein, Article 142(2) of the EU–CARIFORUM Agreement has been interpreted as creating an obligation for the EU to fight anticompetitive agreements stemming from licensing agreements. Article 142(2) reads:

> The EC Party and the Signatory CARIFORUM States shall take measures, as appropriate, to prevent licensing practices or conditions pertaining to intellectual property rights which may adversely affect the international transfer of technology and that constitute and abuse of intellectual property rights by right holders or an abuse of obvious information asymmetries in the negotiation of licences.[56]

It has been argued with regard to this provision that "under Article 142(2) EPA, the EC is obliged to control licensing agreements between a licensor in the EU and a licensee in the Signatory CARIFORUM States which only restrict the use of intellectual property rights in these States. In other words, it seems that the EC would have to apply its own law with only extraterritorial effects."[57]

Practically, given the well-known effects doctrine in competition law, it would be rather difficult to implement such an obligation. This however should not prevent EU partner states from pushing for an obligation from the European Union to fight anticompetitive practices related to intellectual property rights.

---

56. EU–CARIFORUM EPA, *supra* note 48, art. 142(2).

57. *See* Drexl, *supra* note 38.

Conclusion

Competition law can and should play an important role in advancing the interest sof Sub-Sharan African countries with regards to accessing the benefits of innovation. The urgency of addressing the severe shortage of public goods, particularly where public health interests are at issue, requires policy initiatives that can effectively balance the freedom to innovate and the protection of competition. Neither the TRIPS Agreement nor existing national laws provide an appropriately robust framework for leveraging competition law in the interests of access or innovation. Even at the level of regional trade agreements, the role of competition law has been confined to a market opening tool. As countries in sub-Saharan Africa remain entrenched in a vicious cycle of trade and intellectual property contests, competition law could usefully be a less disruptive policy tool to ensure that the balance between trade and innovation remains safeguarded.

V TRIPS COMPLIANCE, PATENT ENFORCEMENT, AND PATENT REMEDIES

17 PATENTABILITY CRITERIA AS TRIPS FLEXIBILITIES: THE EXAMPLES OF CHINA AND INDIA

Daniel J. Gervais*

Introduction

Much has been said, and much more will be, about TRIPS "flexibilities," which one might loosely define as space available to policymakers making decisions about TRIPS implementation, including in particular exceptions and limitations on rights and/or remedies.[1] A list of such flexibilities typically includes not only patentability criteria and subject matter exclusions (e.g., morality-related exclusions or specific subject matter, such as methods of medical treatment), but also compulsory licenses, government use exceptions, exhaustion of rights, research exemptions, and a regulatory review ("Bolar") regime.[2]

Not surprisingly, the demanders who pushed for TRIPS—mostly pharmaceutical companies with significant support from entertainment and software companies—realized that there may be more there than meets the eye.[3] In response, they have been pushing in bilateral and other

---

* Professor of Law and Director, Vanderbilt Intellectual Property Program, Vanderbilt Law School.

1. The TRIPS Agreement is the Agreement on Trade-Related Aspects of Intellectual Property Rights, Apr. 15, 1994, Marrakesh Agreement Establishing the World Trade Organization, Annex 1C, 33 I.L.M. 81 (1994). On flexibilities generally, see Carolyn Deere, The Implementation Game (2009), and on pharmaceuticals in particular, see UNAIDS, *Using TRIPS Flexibilities to Improve Access to HIV Treatment* (2011), *available at* http://www.unaids.org/en/media/unaids/contentassets/documents/policy/2011/JC2049_PolicyBrief_TRIPS_en.pdf.

2. *See* World Intellectual Prop. Org. (WIPO), *Patent Related Flexibilities in the Multilateral Legal Framework and Their Legislative Implementation at the National and Regional Levels*, WIPO Doc. CDIP/5/4 (Mar. 1, 2010). *See* TRIPS Agreement, supra note 1, arts. 6, 27, 30, 31, 44.2.

3. *See* Robert Weissman, *A Long, Strange TRIPS: The Pharmaceutical Industry Drive to Harmonize Global Intellectual Property Rules, and the Remaining WTO*

trade agreements for "TRIPS-plus" standards that amount to efforts to cabin such flexibilities.[4]

TRIPS flexibilities stem from three major sources: (1) undefined terms in the Agreement, (2) specified limitations or exceptions (or tests therefor such as the three-step test), and (3) language that imposes a formal requirement only, such as the famous "authorities shall have the authority" phrase used sixteen times in the enforcement part of TRIPS.[5]

The last area of flexibility (adopting facially compliant legislation but with limited application in practice) is possible because the non-violation clause in the GATT does not apply to TRIPS disputes, reinforced by the recent interpretation of enforcement provisions by a WTO dispute-settlement panel in *United States v. China* and, *a contrario,* the Appellate Body's first TRIPS ruling in *United States v India.*[6] Essentially, most WTO members have been able to avoid WTO sanctions by complying with TRIPS as a formal matter, absent evidence that the application of the paper-compliant rule is clearly subpar.[7]

Legal Alternatives Available to Third World Countries, 17 U. PA. J. INT'L ECON. L. 1069, 1096 (1996).

4. On the European Commission, see Henning Grosse Ruse-Khan, Thomas Jaeger & Robert Kordic, *The Role of Atypical Acts in EU External Trade and Intellectual Property Policy*, 21 EUR. J. INT'L L. 901, 925 (2010). On the United States, see Sisule F. Musungu & Cecilia Oh, *The Use of Flexibilities in TRIPS by Developing Countries: Can They Promote Access to Medicines?* (2005), *available at* http://www.who.int/intellectualproperty/studies/TRIPSFLEXI.pdf, at 8; *see also* Carlos M. Correa, *Implications of Bilateral Free Trade Agreements on Access to Medicines* 399–404 (2006), *available at* http://www.who.int/bulletin/volumes/84/5/399.pdf; Thomas A. Faunce & Kathy Shats, *Bilateral Trade Agreements as Drivers of National and Transnational Benefit from Health Technology Policy: Implications of Recent US Deals for Australian Negotiations with China and India,* 62 AUSTL. J. INT'L AFF. 196–213 (2008). They are also using friendly governments to restrict the actual use of available flexibilities.

5. It is also used three more times in Section II. *See also infra* note 8.

6. China—Measures Affecting the Protection and Enforcement of Intellectual Property Rights, WTO Doc. WT/DS362/R (Jan. 26, 2009); India—Patent Protection for Pharmaceutical and Agricultural Chemical Products, WTO Doc. WT/DS50/AB/R (Dec. 19, 1997). In the US–China case, it can fairly be said that the panel required that, in the copyright portion of the case, the Chinese statute be facially TRIPS-compliant, but was more lenient in the application of the phrase "the authorities shall have the authority." This was interpreted as an obligation to have the authority, not to exercise it. In the India case, the Appellate Body did not consider de facto compliance via administrative measures sufficient and similarly required the presence of compliant statutory language. For comments, see DANIEL J. GERVAIS, THE TRIPS AGREEMENT: DRAFTING HISTORY AND ANALYSIS 168–72 (3d ed. 2008); Jayashree Watal, *US–China Intellectual Property Dispute: A Comment on the Interpretation of the TRIPS Enforcement Provision*, 13 J. WORLD INTELL. PROP. 605 (2010).

7. *See id.* In the US–China case, the panel noted that the United States had failed to provide evidence that behavior exempted from criminal sanctions due to thresholds in Chinese law was "on a commercial scale," as the expression is used in TRIPS Article 61.

Entire books have already been written on the second area of flexibility (limitations and exceptions), and two reports by WTO dispute-settlement panels have now interpreted the three-step test (which, in various incarnations, limits exceptions and limitations to copyright, design, and patent rights in TRIPS).[8] That is a relevant debate and it must continue. The extent to which it is informed by non-WTO norms is an area of particular interest, as the process to determine proper normative interfaces is still in development.[9]

However, exceptions and limitations in TRIPS are made possible by, not imposed on members.[10] Put differently, exceptions and limitations may be considered unregulated policy space at the international level.[11] To a large extent, the geometry of exceptions and limitations (hereinafter referred to as "E&Ls") in a WTO member depends not on TRIPS per se, which offers policy space but little normative guidance, but rather on domestic policymaking, academic and other studies of the role and optimal scope of exceptions, and of course bilateral or plurilateral negotiations aimed at cabining recourse to E&Ls. From that perspective, the TRIPS Agreement acts more as a filter for acceptable domestic measures than as a policy guide.

The first of the three areas mentioned above (namely the role of undefined terms) is, however, the main focus of this chapter. It is also perhaps the least explored of the three.

The matter is complex due to a web of semantic accommodations, sometimes masking unresolved issues. For example, TRIPS negotiators used the fiction of synonymy to declare that WTO members somehow saw eye-to-eye on terminology,

8. *See, e.g.*, Edson Beas Rodrigues Jr., The General Exception Clauses of the TRIPS Agreement: Promoting Sustainable Development (2012); Martin Senftleben, Copyright, Limitations and the Three-Step Test: An Analysis of the Three-Step Test in International and EC Copyright Law (2006); M. Ficsor, *How Much of What: The "Three-Step Test" and Its Application in Two Recent WTO Dispute-Settlement Cases*, 192 Revue Internationale du Droit d'Auteur [RIDA] 111 (2002); Daniel J. Gervais, *Towards a New Core International Copyright Norm: The Reverse Three-Step Test*, 9 Marq. Intell. Prop. L. Rev. 1 (2005).

9. *See* Graeme B. Dinwoodie & Rochelle C. Dreyfuss, A Neofederalist Vision of TRIPS: The Resilience of the International Intellectual Property Regime (2012).

10. Because the Berne Convention contains one mandatory exception (Article 10.1), and because the substantive provisions of that Convention were incorporated into TRIPS (Article 9.1), it is arguable that one exception is mandatory for WTO members, subject to how one interprets TRIPS Article 1.1.

11. *See* P. Bernt Hugenholtz & Ruth L. Okediji, Conceiving an International Instrument on Limitations and Exceptions to Copyright: Final Report (2008), *available at* http://www.ivir.nl/publications/hugenholtz/limitations_exceptions_copyright.pdf; Daniel J. Gervais, *Making Copyright Whole: A Principled Approach to Copyright Exceptions and Limitations*, 5 U. Ottawa L. & Tech. J. 1 (2008), *available at* http://ssrn.com/abstract=1825342.

such as when TRIPS states in a footnote that the North American and rest of the world patentability criteria "may be deemed to be synonymous" (non-obviousness and utility versus inventive step and industrial applicability).[12] This follows in the footsteps of other arranged marriages, such as the idea that copyright in common law jurisdictions going back to the 1710 Statute of Anne is somehow the same as authors' rights in the rest-of-the-world.[13] In other cases, dispute-settlement panels have turned to the Vienna Convention and the *Oxford Dictionary* to elucidate the "plain meaning" of undefined terms.[14] This has worked, to some degree, such as when everyday terms such as "commercial" or "normal" are used. It may not work as well when technical intellectual property terminology must be defined.

In this chapter, I tackle the lack of definitions of two of the three patentability criterion enunciated in TRIPS Article 27, namely non-obviousness and utility and their deemed synonyms, inventive step and industrial applicability.[15] For a variety of reasons, including space, I will leave aside novelty but it is clear that (1) discrepancies exist as to how it is defined, and (2) there are major policy impacts that result from adopting a particular definition.[16] To mention but one example, in the original patent legislation of India, the geographic scope of the

---

12. *See* TRIPS Agreement, *supra* note 1, art. 27.1 n.5.

13. The copyright system is a successor to, and helped transition from, the Stationers Company monopoly. Its economic focus is thus unsurprising. By contrast, the authors' rights system was born from Hegelian and other sources that tended to emphasize the author's importance in philosophical and cultural terms.

14. For example, in Section 110(5) of the US Copyright Act, WTO Doc. WT/DS160/R (June 15, 2000), the panel referred to the Oxford Dictionary in ¶¶ 6.108 and 6.109 and to the Vienna Convention in several instances (e.g., ¶¶ 6.37 and 6.43).

15. Readers in the United States might argue that there are four, not three, and add enablement. *See* 35 U.S.C. § 112 (2012). However, TRIPS Article 27.1 only provides that "patents shall be available for any inventions, whether products or processes, in all fields of technology, provided that they are new, involve an inventive step and are capable of industrial application." A footnote adds that, "For the purposes of this Article, the terms 'inventive step' and 'capable of industrial application' may be deemed by a Member to be synonymous with the terms 'non-obvious' and 'useful' respectively." Enablement is addressed in Article 29, which provides in part that "Members shall require that an applicant for a patent shall disclose the invention in a manner sufficiently clear and complete for the invention to be carried out by a person skilled in the art." *See* TRIPS Agreement, *supra* note 1, arts. 27.1, 29.

16. For example, the United States only considers sale or use in the United States as a bar to patentability. *See* 35 U.S.C. §§ 102(a)–(b). On policy impacts, see, e.g., Sean B. Seymore, *Rethinking Novelty in Patent Law*, 60 DUKE L.J. 919, 946–48 (2011):

> The invention landscape changed around the time of World War II, when key breakthroughs in antibiotic, vitamin, and hormone research spawned the "therapeutic revolution" and led to the discovery of many first-generation "wonder drugs." During this period, pharmaceutical companies quickly switched from a manufacturing to a research-based model and secured patents that allowed them to dominate sectors of

novelty inquiry was confined to the United Kingdom and India,[17] rather than to use or publication anywhere in the world.

The backdrop for my analysis is the realization, now also well documented, that TRIPS was initially implemented in many developing countries as a set of standard legislative implants—typically in the form of model laws.[18] The initial focus was *compliance* with TRIPS and avoidance of disputes. After the AIDS/malaria debacle, which led to the adoption of the only (still unratified) amendment to TRIPS (Article 31*bis*)[19], and owing in part to the emergence of more and better economic data and analyses on the role and impact of stronger intellectual property protection in developing countries, several developing countries are now re-implementing TRIPS with a view to fostering domestic innovation and development *and* to reducing the welfare costs of stronger standards.[20] I refer to this process as calibration.

The calibration process takes the form of a complex set of policy equations that include a focus on the country's comparative advantages (e.g., solid film or software industry basis; Ayurvedic medicine knowledge in India; or the availability of governmental resources and a large cadre of highly educated researchers as in China), but also on training for key officials, setting up nongovernmental or private–public organizations and partnerships, educational tools, venture capital (private or public), and intellectual- property–specific measures such as patent pooling and making resources available for local inventors.[21] In short, intellectual

---

> specific therapeutic markets. This, in turn, quickly forced the Patent Office and the courts to wrestle with fields key to drug research, like organic chemistry. [...] This shoehorning led to nonsensical outcomes and a disconnect between the judicial bench and the laboratory bench. [...] Perhaps the most important unresolved issue is whether and under what circumstances the appearance of a chemical name or structure in the prior art anticipates a subsequent inventor's claim for the compound.

I will also leave aside here the important issue of morality, which, in my mind goes to subject matter eligibility. For a discussion, see Margo A. Bagley, *Patent First, Ask Questions Later: Morality and Biotechnology in Patent Law*, 45 WM. & MARY L. REV. 469 (2003).

17. Peter Drahos, *The Jewel in the Crown: India's Patent Office and Patent-Based Innovation*, *in* INTELLECTUAL PROPERTY POLICY REFORM 80–81 (Christopher Arup & William van Caenegem eds., 2009).

18. *See* Daniel J. Gervais, *Of Clusters and Assumptions: Innovation as Part of a Full TRIPS Implementation*, 77 FORDHAM L. REV. 2353 (2009).

19. The purpose of which is to broaden the scope of patent compulsory licensing for pharmaceutical products needed by least-developed countries.

20. *See* Gervais, *supra* note 18.

21. *See* Ed Levy, Emily Marden, Ben Warren, David Hartell & Isaac Filaté, *Patent Pools and Genomics: Navigating a Course to Open Science?*, 16 B.U. J. SCI. & TECH. L. 75, 87–88 (2010).

property policy has become part of a much broader picture, possibly in the form of an enhanced industrial policy itself ensconced in developmental objectives.

From this perspective, re-implementing TRIPS means setting priorities and implementing those priorities within the flexibilities of TRIPS.[22] This, in turn, requires a determination of the exact scope of available flexibilities, each of which may be portrayed as a decision point for policymakers. It is an almost impossible task at a theoretical level to consider all flexibilities in one fell swoop. In this chapter, I thus limit my analysis to the way in which TRIPS patentability criteria were implemented and are applied in practice in two major emerging nations, China and India, with different but equipollent technological proficiency.[23] This partial look hopefully can serve as an illustration of the use of other similar TRIPS flexibilities.

Getting multilateral deals on substantive patent rules and definitions, especially at WIPO and WTO, may have become irredeemably difficult. Countries such as China and India, but also Argentina, Brazil, South Africa, Thailand, and many others, have become more assertive. It is obvious that their relative clout measured in trade terms has grown exponentially.[24] The work at WIPO on substantive patent harmonization is a testament to the level of difficulty.[25] Additionally, major developing countries now have the same level of sophisticated intellectual property knowledge as more industrialized nations, thus defeating the ignorance narrative to which the TRIPS "deal" has been partially attributed.[26]

Because multilateral efforts to agree on higher hard law standards (e.g., via a TRIPS amendment) seem bound to fail, some countries now resort to other methods. As already mentioned, bilateral trade deals are emerging, in some cases establishing incompatible rules, possibly as a strategy to influence multilateral outcomes.[27] In other cases, countries can produce and try to export soft law.

---

22. *See* Gervais, *supra* note 18.

23. *See* Shamnad Basheer & Annalisa Primi, *The WIPO Development Agenda: Factoring in the "Technologically Proficient" Developing Countries* 112–13 (2009), *available at* http://ssrn.com/abstract=1289288.

24. *See* Bruno Salama, *Pharmaceutical Patent Bargains: The Brazilian Experience*, 18 Cardozo J. Int'l & Comp. L. 633 (2010); P.K. Yu, *Access to Medicines, BRICS Alliances, and Collective Action*, 34 Am. J.L. & Med. 345, 355–58 (2008).

25. Indeed this WIPO agenda item has essentially been on hold since 2006. *See* WIPO, *Draft Substantive Patent Law Treaty*, *available at* http://www.wipo.int/patent-law/en/harmonization.htm (last visited Sept. 3, 2013).

26. *See* P.K. Yu, *TRIPS and Its Discontents*, 10 Marq. Intell. Prop. L. Rev. 369, 375–76 (2006).

27. *See* Gregory C. Shaffer & Mark A. Pollack, *Hard vs. Soft Law: Alternatives, Complements, and Antagonists in International Governance*, 94 Minn. L. Rev. 706 (2010).

A salient example is the use of patent examination guidelines issued by major patent offices trying to influence the meaning of patentability criteria.[28] One might argue that developing countries will increasingly look to other, more advanced developing nations when considering which set of rules to adopt or follow. As we will see below, China is an increasingly active player at that game.

Let us start, then, by considering, in Section I, the non-obviousness/inventive step requirement. In Section II, I turn to utility/industrial applicability. In Section III, I consider lessons that one might draw from the analysis of both topics. Wherever possible, existing reports by WTO dispute-settlement panels and the Appellate Body will be included in the analysis.

I. Non-Obviousness/Inventive Step

This patentability criterion matters a great deal. In fact, non-obviousness/inventive step has been referred to as the "ultimate condition of patentability"[29] and as "one of the most critical aspects of a patent regime, as it determines the level of technical contribution required to obtain a patent and the corresponding limitation on competition."[30] It is, or could be, the filter that separates inventions worthy of patent protection (which one might be tempted to describe as essential "consideration" if patents are looked at from a contractarian perspective) from the normal, incremental advances in technology that would likely happen anyway.[31] A glut of "unworthy"

---

28. A number of examples are given in the following sections of this chapter.

29. G.S. Rich, Nonobviousness: The Ultimate Condition of Patentability 1:201–213 (John F. Witherspoon ed., 1980).

30. Carlos Correa, *Guidelines for the Examination of Pharmaceutical Patents* 4 (Int'l Ctr. for Trade and Sustainable Dev. Working Paper, Jan. 2007), *available at* http://ictsd.org/downloads/2008/06/correa_patentability20guidelines.pdf.

31. *See* William Robinson, The Law of Patents for Useful Inventions § 22 (1890); Jennifer Nock & Sreekar Gadde, *Raising the Bar for Nonobviousness: An Empirical Study of Federal Circuit Case Law Following* KSR, 20 Fed. Cir. B.J. 369, 373–74 (2011):

> The requirements for patentability are designed to make patent protection available only to beneficial inventions that would not otherwise have been devised or disclosed.... The nonobviousness requirement ensures that a patented invention represents a substantial technical contribution to society.... A low standard for patentability harms the public because granting monopoly rights for obvious inventions "withdraws what already is known into the field of [the patent] monopoly and diminishes the resources available to skillful men."

It has also been suggested that:

> [the] nonobviousness requirement serves another very important purpose where, as is realistically nearly always the case, the social value of research projects substantially exceeds their private value. When this is the case, the socially preferable level of invention exceeds the privately optimal choice even when patents are available at both levels.

patents may, in fact, slow innovation, thereby reducing a nation's overall competiveness.[32] As a more practical matter, this criterion will directly affect whether patents are granted on genes (simply for discovering that a given gene may have a role to play in susceptibility to a particular disease) or on combinations or preexisting objects.[33]

Based on a desire to maintain incentives to discover genes, the US Court of Appeals for the Federal Circuit has "force[d] the square peg of the obviousness analysis of chemical compounds into the round hole of the obviousness analysis of claims to genetic sequences."[34] It may be, in fact, that non-obviousness is not one concept, but rather many different things depending on the type of invention.[35] One thing is

---

> The nonobviousness threshold may be used as a "stick" to induce researchers to pursue more difficult, socially preferred research projects.

Michael J. Meurer & Katherine J. Strandburg, *Patent Carrots and Sticks: A Model of Nonobviousness*, 112 Lewis & Clark L. Rev. 547, 549 (2008).

32. *See* Adam B. Jaffe & Josh Lerner, Innovation and Its Discontents: How Our Broken Patent System Is Endangering Innovation and Progress (2004); Nat'l Research Council of the Nat'l Acads., A Patent System for the 21st Century 1–2 (Stephen A. Merrill, Richard C. Levin & Mark B. Myers eds., 2004); Joshua D. Sarnoff, *Bilcare, KSR, Presumptions of Validity, Preliminary Relief, and Obviousness in Patent Law*, 25 Cardozo Arts & Ent. L.J. 995, 1051 (2008); Fed. Trade Comm'n (FTC), To Promote Innovation: The Proper Balance of Competition and Patent Law and Policy (2003), *available at* http://www.ftc.gov/os/2003/10/innovationrpt.pdf.

33. The former issue was recently addressed by the US Supreme Court in *Ass'n for Molecular Pathology v. Myriad Genetics, Inc.*,133 S. Ct. 2107 (2013), and the latter was dealt with in *KSR Int'l Co. v. Teleflex Inc.*, 550 U.S. 398 (2007).

34. Mark Polyakov & Eugene Goryunov, *(Non)Obviousness of Claims to Genetic Sequences: Finding the Middle Ground*, 26 Santa Clara Computer & High Tech. L.J. 1 (2009–10). The authors explain:

> By way of background, the CAFC devised a special test for chemical compounds, under which a prima facie case of obviousness requires "structural similarity between claimed and prior art subject matter ... where the prior art gives reason or motivation to make the claimed compositions." Once the prima facie case is established, an examiner must show some motivation to modify the prior art compound. Usually, this prong does not present a high burden since the motivation may be found in the knowledge that structurally similar compounds often display similar properties.... [T]his framework is ill-suited for genetic sequences because it does not consider the special nature of genes. A PHOSITA attempting to determine the DNA sequence of a protein does not generally start his search by looking at structurally similar DNA sequences. Instead, he starts by determining the amino acid sequence of the protein. Once the amino acid sequence has been determined, a PHOSITA can design and use nucleotide probes to uncover the protein's DNA sequence. It is irrelevant to the analysis whether the prior art discloses a structurally similar cDNA molecule.

Id. at 21–22.

35. *See* Gregory Mandel, *The Non-Obvious Problem: How the Indeterminate Nonobviousness Standard Produces Excessive Patent Grants*, 42 U.C. Davis L. Rev. 57, 111–12 (2008):

certain: there is no shortage of critiques of the Federal Circuit's approach, and other countries might want to take a fresh look at their own approach instead of copying the US one (whatever it might be as of this writing).[36]

By contrast, the European Patent Convention provides that[37] an invention "shall be considered as involving an inventive step if, having regard to the state of the art, it is *not obvious to a person skilled in the art*." Thus in the EPO context the synonymy established by TRIPS between inventive step and non-obviousness seems defensible. Still, the term "inventive" arguably points beyond nonobviousness, especially in the case of discoveries. Finding and isolating a DNA sequence requires much work but it is (arguably at least) not "inventive."[38] However, in an effort to harmonize with Japanese and US practice, a Directive was adopted by the European Union (and implemented by the EPO) to make isolated or artificially produced natural substances patentable.[39]

Inventive step has also been used in Europe as a reason to exclude computer programs "as such" from patentability.[40] In addition, it has also been used in a famous case to exclude from patentability a claimed invention where a large

---

> Some inventions may be non-obvious in their conception, though once conceived may be easy to achieve. Post-It notes provide an example... Other inventions are obvious to conceive, but identifying operative means for carrying them out is non-obvious. An HIV vaccine is an example.... A third nonobviousness category concerns inventions where potential operative means are obvious, but the field is uncertain enough that actually reducing the invention to practice is non-obvious. This could occur where certain operative means appear obvious, but do not actually work... For example, other inventors developed incandescent light bulbs before Edison, but their filaments burned out quickly, rendering the light bulbs impractical.

36. In a paper proposing substantial reforms, Professor Barton noted that non-obviousness has been "greatly weakened" in the United States in recent years and failed "the common-sense test," adding that "[i]f these patents are non-obvious to the person of reasonable skill in the art, that person simply isn't very bright." John Barton, *Non-Obviousness*, 43 IDEA 475, 477, 482 (2003). The patentability of unmodified but isolated gene sequences obviously raises subject matter questions as well.

37. Convention on the Grant of European Patents of 5 October 1973, as revised by the Act revising Article 63 EPC of 17 December 1991 and the Act revising the EPC of 29 November 2000, Article 56 [hereinafter European Patent Convention or EPC] (emphasis added).

38. This is reminiscent of the debate on copyright applying to sweat of the brow works, which, though non-creative, required a significant investment of time and/or effort. The EPC, *supra* note 37, specifically excludes discoveries (Article 52(2)). *See* Gerard Porter, *The Drafting History of the European Biotechnology Directive*, *in* EMBRYONIC STEM CELL PATENTS: EUROPEAN LAW & ETHICS 1, 8–9 (Aurora Plomer & Paul Torremans eds., 2009).

39. *See id.*; Directive 98/44/EC of the European Parliament and of the Council of 6 July 1998 on the Legal Protection of Biotechnological Inventions, [1998] OJ L213/13.

40. *See* Peter Stone, *Copyright Protection of Computer Programs in Europe*, *in* Law, Computer Science and Artificial Intelligence 146–47 (1998).

number of people were all working toward one goal and the means of achieving the goal. Getting there first was not necessarily a sign of an inventive step.[41]

Let us now consider how the criterion is applied in China and India.

A. China

China has tried to craft a patentability policy with particular attention to pharmaceuticals. Although desiring to provide access to products by consumers, it also wanted to develop both a "Western-style" biotechnology industry *and* an industry based on traditional medicinal knowledge.[42] That is not an easy task.

"Inventiveness," the Chinese application of the TRIPS "non-obviousness/inventive step" criterion,[43] is defined as meaning that "compared with the technology existing before the filing date of the application, the invention has prominent and substantive distinguishing features and represents a marked improvement, or the utility model possesses substantive distinguishing features and represents an improvement."[44] As with the non-obviousness requirement in the United States, the purpose of the inventiveness requirement is to "distinguish major innovations worthy of government-granted monopolies from ordinary and pedestrian advances that are not."[45] The *Chinese Patent Examination Guidelines* provide that the question arises only after a determination of novelty has been made.[46]

In examining the substantive features of a patent application, an examiner inquires "whether or not there exists such a ***technical motivation*** in the prior art as to apply the said distinguishing features to the closest prior art in solving the existing technical problem (that is, the technical problem actually solved by the invention), where such motivation would prompt a person skilled in the

41. *Genentech Inc.'s Patent*, [1998] RPC 147.

42. *See* C. Grace, *Update on China and India and Access to Medicines*, Dep't for Int'l Dev. (Nov. 2005); Xuan Li & Weiwei Li, *Inadequacy of Patent Regime on Traditional Medicinal Knowledge: A Diagnosis of 13-Year Traditional Medicinal Knowledge Patent Experience in China*, 10 J. World Int. Prop. 125 (2007).

43. Louis S. Sorell, *A Comparative Analysis of Selected Aspects of Patent Law in China and the United States*, 11 Pac. Rim L. & Pol'y J. 319, 326 (2002).

44. Patent Law of the People's Republic of China, *as revised by* Decision Regarding the Revision of Patent Law of the People's Republic of China, art. 22, promulgated by the Standing Comm. Nat'l People's Cong., Aug. 25, 2000 [hereinafter China Patent Act], *available at* http://www.sipo.gov.cn.

45. Some International Aspects of Patent Protection § 2A.06, 2004 WL 3804734.

46. *See* Li & Li, *supra* note 42, at 136; *see also* Cheng Wenting & Zhang Feifei, *Third Revision of China's Patent Law: Patent Examination Guidelines: A Comparative Analysis (2006 and 2010)*, IPR2 (Dec. 2010), *available at* http://www.ipr2.org/storage/Patent_Examination_Guidelines_comparative_2006_and_2010845.pdf.

art, when confronted with the technical problem, to improve the closest prior art and thus reach the claimed invention. If there exists such a technical motivation in the prior art, the invention is obvious and thus fails to have prominent substantive features (emphasis added)."[47] In assessing "notable progress," an examiner inquires "whether or not the invention produces ***advantageous technical effects*** (emphasis added)."[48] In addition to prior art, examiners may also take into account secondary indicia of inventiveness, including "providing a solution to a long-felt problem, overcoming technical prejudice, unexpected results, and commercial success."[49] Put differently, if the invention can be arrived at by a simple extrapolation in a straightforward way from the known art, it does not involve an inventive step.[50] Although "prominent substantive features" and "notable progress" carry equal weight in the examination process, the judiciary's assessment of inventiveness tends to place more emphasis on "prominent substantive features" than "notable progress."[51]

The *Examination Guidelines* provide three sets of factors beyond the general assessment criteria explained above. First, they differentiate among inventions opening up a whole new field, invention by combination, invention by selection, invention by diversion, invention of a new use of a known product, and invention by changing elements (changing relations between elements, replacing elements, or omitting elements).[52] On inventions by combination, one might wonder whether the *Guidelines* got it better than the US Supreme Court in *KSR*.[53] The *Guidelines* provide that if a claimed invention "is merely an aggregation or juxtaposition of certain known products or processes, each functioning in its routine way, and the overall technical effect is just the sum of the technical effects of each part without any functional interaction between the combined technical features, that is, the claimed invention is just a mere aggregation of features, then the invention by combination does not involve an inventive step" but if the "combined technical features functionally support each other and produce a new technical effect, or in other words, if the technical effect after combination is greater than

47. SIPO Guidelines for Examination ¶ 3.2 (2010), *available at* http://www.gechengip.com/information/content2010_en.htm. The Guidelines provide several illustrations of technical motivation.

48. *Id.*

49. Deming Liu, *Now the Wolf Has Indeed Come! Perspective on the Patent Protection of Biotechnology Inventions in China*, 53 Am. J. Comp. L. 207, 232–33 (2005).

50. SIPO Guidelines, *supra* note 47 ¶ 4.3.

51. Latest Developments in Trial of Patent Cases Before the Beijing Higher People's Court in 2009, China Patents & Trademarks (2009) [hereinafter Latest Developments in Patent Trial].

52. SIPO Guidelines, *supra* note 47 ¶¶ 4.1–4.6.3.

53. KSR Int'l Co. v. Teleflex Inc., 550 U.S. 398 (2007).

the sum of the technical effects of the individual features, then such combination has prominent substantive features and represents notable progress, and thus the invention involves an inventive step."[54]

The *Guidelines* list other factors to be considered in the examination of inventive step, namely: solving a long-felt but unsolved technical problem, overcoming a technical prejudice, producing unexpected technical effect, and achieving commercial success. When any of these secondary factors are present, the examiner is instructed not to make a "rash determination" of the absence of an inventive step.[55] Finally, the *Guidelines* instruct examiners—in a way that US experts may not find altogether surprising—to "note" the following points in the examination of the inventive-step criterion: generally avoid considering how an invention is accomplished even if by accident; avoid ex post facto analysis and the influence of subjectivity; do not overemphasize any unexpected technical effect; and finally, direct the analysis to the technical solutions as defined in the claims but considering the technical solution as a whole, rather than the individual technical features.[56]

SIPO assesses inventiveness differently for invention patents and utility model patents: whereas inventiveness for inventions requires "prominent substantive features" and "notable progress" over the state of the art at the filing date, utility model patents require only "substantive features" and "progress."[57] In fact, utility models are not examined as to substance, so "a large number of inventions-creations that are not inventive enough" are filed as utility model patent applications.[58] Judges have more discretion with utility model patents in determining the scope of prior art to reference.[59]

With respect to chemical inventions (including genes), if the claimed invention has a similar structure to any known compound, then in order to show inventiveness, it must be shown that the claimed invention has an "unexpected use or effect."[60] With respect to computer software, the Examination Guidelines require that

---

54. SIPO GUIDELINES, *supra* note 47, ¶ 4.2. The USPTO also uses this rationale based on *KSR*, 550 U.S. at 398.

55. SIPO GUIDELINES, *supra* note 47, ¶ 5.

56. *Id.* ¶¶ 6–6.4. This is also similar to the US approach.

57. Raymond M. Gabriel, *The Patent Revolution: Proposed Reforms in Chinese Intellectual Property Law, Policy, and Practice Are the Latest Step to Bolster Patent Protection in China*, 9 ASIAN-PAC. L. & POL'Y J. 323, 334 (2008).

58. Meng Fanxin, *Application of Equivalent Doctrine in Utility Model Patent Infringement Lawsuit*, 2006 CHINA PATENTS & TRADEMARKS 15.

59. Li Yonghong, *How to Define the Height of Inventiveness of Utility Model?*, 2008 CHINA PATENTS & TRADEMARKS 26–28.

60. *See id.*; *see also* Liu Xuqiang, *On the Inventiveness of Compounds Structurally Similar to Known Compounds*, 1997 CHINA PATENTS & TRADEMARKS 32–35 (discussing how SIPO assesses inventiveness of homologues, isomers, and analogues).

a computer-related invention "is intended to solve a technical problem, uses a technical means and is capable of producing a technical result" to meet the inventiveness requirement.[61] This is similar to the European approach to evaluating non-obviousness of software inventions, though "technical contribution" is better defined in the EPO jurisprudence.[62]

The case of *Schneider v. PRB and Chint* is an example of how the judiciary evaluates the inventiveness of a claimed invention. In that matter, the Beijing Higher People's Court concluded that in determining inventiveness of patents in the field of electricity, account should be taken not only of the connection of the circuit, but also the working state of the circuit. Differences in technical conception and technical solution resulting from a different working state generally produced an effect different from the prior art. The court determined that one skilled in the art would know that the value tested at one point in one of the references cited was the induction electric signal resulting from the change in the magnetic conductivity of the electromagnetic coil, after cutting off the supply of field current through the electromagnetic coil; and the tested value at another point was the change in the voltage at the artificial median point of the electric line output end at the disconnection of the contacts. By contrast, the patent in suit tested the field current going through the electromagnetic coil. This difference showed that the two technical solutions were in a different working state. In other words, the patent in suit, which was directed at resolving the technical problem of testing the dissipation loss of the switch, did so in a way quite different from the cited reference 1 and afforded a technical solution of different technical conception. It achieved a technical effect better than the prior art, and hence the Beijing Higher People's Court concluded it possessed inventiveness.[63]

China's inventive-step standard has been described as higher than the corresponding US and EPO standards.[64] For example, compared with the EPO approach, in China the problem, solution, *and effect* are interpreted in combination. [65] By contrast, the EPO, has adopted a problem-and-solution approach that focuses on the nearest prior art concerning the problem solved, identifies the solution disclosed in the application, considers whether the solution fits the

---

61. Li Yonghong, *Patent Protection for Software from the Perspective of EC Proposal for a Directive*, 2003 China Patents & Trademarks 26–30.

62. *Id.*

63. Latest Developments in Patent Trial, *supra* note 51; *see* Beijing Higher People's Court's Administrative Judgment No. Gaoxingzhongzi 225/2009; Beijing No.1 Intermediate People's Court's Administrative Judgment No. Yizhongxingchuzi 1156/2008.

64. *See* Li & Li, *supra* note 42, at 136.

65. *See id.* at 137; X.D. Zhang, Substantial Condition of Patentability (2002).

problem identified in the prior art, and then decides whether a person skilled in the art would find it obvious.[66]

B. India

The Indian Patents (Amendment) Act of 2005 made significant changes to the Indian Patent Act of 1970, including the introduction of product patents in the pharmaceutical field but also an exclusion in Section 3(d) of patents for "the mere discovery of a new form of a known substance which does not result in the enhancement of the known efficacy of that substance".[67] This followed a debate over what should be patentable in India, and took place against a backdrop of significant concerns about healthcare costs and other public health ramifications.[68] The government is trying to balance the emerging research-based industry in biotechnology and those using traditional (e.g., Ayurvedic) principles to develop new treatments, while providing access to antiretroviral and other patented drugs to the public[69] For example, India was not obligated to extend patents to pharmaceuticals until 2005, but the requirement that so-called "mailbox" applications should receive "exclusive marketing rights" was implemented by providing up to five years (or until the issuance or rejection of the patent application) a limited right to exclusively market the drug or medicine in question.[70]

The 2005 amendment to the Patents Act defines an "inventive" step to mean "a feature of an invention that involves technical advance as compared to the

---

66. *See* AMANDA WARREN-JONES, PATENTING RDNA: HUMAN AND ANIMAL BIOTECHNOLOGY IN THE UNITED KINGDOM AND EUROPE (2001). A factor not taken into consideration is the amount of effort expended to reach the solution if the work to achieve the solution was otherwise straightforward. *See* Unilever, EPO Technical Board of Appeal, T 939/92 (OJ EPO 1996, 309).

67. *See* India Patents Act, No. 39 of 1970, *as amended by* Patents (Amendment) Act, No. 15 of 2005, INDIA CODE (2005) [hereinafter India Patent Act]. In 1999, the Indian Parliament passed the Patents (Amendment) Act of 1999, made retroactive to January 1, 1995, to comply with the WTO decision that required a statutory provision for the mailbox/exclusive marketing rights provided for in TRIPS Article 70.9. For a more complete history, see Peter Drahos, *supra* note 17, at 81–84; Shamnad Basheer, *India's Tryst with TRIPS: The Patents (Amendment) Act 2005*, 1 INDIAN J. L. & TECH. 22 (2005).

68. *See* Grace, *supra* note 42.

69. *See* Padmashree Gehl Sampath, *India's Product Patent Protection Regime: Less or More of "Pills for the Poor"?*, 9 J. WORLD INT. PROP. 694, 703–10 (2006).

70. Patents Act, 1970, § 24B. The extra transitional period (until January 2005) is contained in TRIPS Article 65.4. The mailbox system is provided in Article 70.9. On those provisions, see GERVAIS, *supra* note 6, at 545–46. On the Indian application, see Brook K. Baker, *India's 2005 Patent Act: Death by Patent of Universal Access to Second and Future-Generation ARVS?*, GLOBAL AIDSLINK (Sept./Oct. 2005), *available at* http://lists.essential.org/pipermail/ip-health/2005-September/008301.html.

existing knowledge or having economic significance or both and that makes the invention not obvious to a person skilled in the art."[71] This "non-obviousness–plus" standard, a departure from India's former precedents, has been criticized as being a " 'vague and arbitrary' definition that fails to 'reflect the distilled stock of knowledge' on what nonobviousness means."[72] The "technical advance" criterion, for example, "invites qualitative comparisons with the prior art but is devoid of any requirement that such comparisons be made from the perspective of a person of ordinary skill in the art at the time of application filing."[73] As for the "economic significance" requirement, it is unclear whether this is analogous to the US "commercial success" secondary consideration or simply redundant with the utility requirement.[74]

Section 3(d) of the Patents Act as amended, which excludes "derivatives of a known substance" from patentability unless the claimed invention "differ[s] significantly in properties with regard to efficacy," seems to be aimed at "preventing frivolous patents that are only trivial modifications of existing inventions."[75] This provision was the subject of the recent Novartis litigation. The Patent Office refused to grant Novartis a patent for Gleevec (imatinib mesylate).[76] The Patent Office reasoned that "because imatinib mesylate was a salt form of the free base imatinib, and Novartis claimed all pharmaceutical salt forms of imatinib" in earlier patents in the United States and elsewhere, the Indian patent therefore lacked novelty and inventiveness.[77] Further, the Patent Office held that, although the Novartis invention "had a 30% increase in bioavailability (the percentage of the drug absorbed into the bloodstream) as compared with imatinib," this was insufficient to meet the "enhanced efficacy" requirement of Section 3(d).[78] Finally, the Patent Office noted that the discovery of imatinib mesylate was obvious because "once the free base was disclosed in the 1993 Patent, it would have been 'obvious

71. Gervais, *supra* note 6, at 545–46.

72. *Id.*

73. *Id; see also* Vijay Yalamanchili, *State of India's TRIPS-Compliant Patent Regime*, 26 Biotech. L. Rep. 211, 217 (2007) ("The Indian MPP suggests using the following question during examination: 'Would a non-inventive mind have thought of the alleged invention?'").

74. Mueller, *supra* note 22. Since many patent applications are filed well before the invention has achieved any commercial success, "strict imposition of the new 'economic significance' language is likely to prejudice independent inventors and small business entities." *Id.*

75. Linda L. Lee, *Trials and TRIPS-Ulations: Indian Patent Law and* Novartis Ag v. Union of India, 23 Berkeley Tech. L.J. 281, 294–95 (2008).

76. *Id.*

77. *Id.*

78. *Id.*

for a person skilled in the art to prepare corresponding pharmaceutically acceptable salts.' "[79]

Novartis, which had obtained from the Madras High Court an interim order for Exclusive Marketing Rights while its patent application was under examination, filed two writ petitions after the rejection: an appeal to the rejection order and a challenge to the validity of Section 3(d) on the grounds of incompatibility with India's obligation under TRIPS. [80] Novartis also contended that the patentability standards were arbitrary and in violation of the Equal Protection provisions of the Indian Constitution.[81] In response, the government argued that examiners had used scientific tools to make a determination on efficacy and, in any event, that there were higher forums to appeal an examiner's decision.[82]

The Madras Court ruled that Section 3(d) was constitutional, noting that "ambiguity was intended by the Parliament in order to avoid fixing a specific formula to be applied in all situations.... The Court noted that the legislative body, who were not technical experts, intended to provide the Patent Controller with a high degree of discretion to deal with both present and future technologies on a case-by-case basis."[83] On the compatibility with TRIPS, the Court declared that it had no jurisdiction to entertain a challenge based on TRIPS, a view with which two noted commentators took exception.[84]

On the substance of the patent dispute, the Court referred the matter to the Intellectual Property Appellate Board (IPAB). On June 6, 2009, the IPAB of Chennai rejected the lawsuit against the decision of the Patent Office. The IPAB ruled that Novartis's patent application covering Gleevec was not patentable under the Patents Act, noting that Section 3(d) is a heightened inventive-step standard.

---

79. Johanna Sheehe, *Indian Patent Law: Walking the Line?*, 29 Nw. J. Int'l L. & Bus. 577, 586 (2009).

80. *See* Shamnad Basheer & T. Prashant Reddy, *The "Efficacy" of Indian Patent Law: Ironing out the Creases in Section 3(d)*, 5 SCRIPTed 232, 237 (2008), *available at* http://www.law.ed.ac.uk/ahrc/script-ed/vol5-2/basheer.asp.

81. Constitution of India, art. 14, *available at* http://www.indiacode.nic.in/coiweb/welcome.html (last visited Sept. 3, 2013)

82. *See* Susan Fyan, *Pharmaceutical Patent Protection and Section 3(D): A Comparative Look at India and the U.S.* (2010), 15 Va. J.L. & Tech 198.

83. *Id.* The argument that patent offices should have more power to adapt the patentability criteria to evolving innovation practices and technological developments is supported by a number of commentators. *See* Dan L. Burk & Mark A. Lemley, The Patent Crisis and How the Courts Can Solve It 95–109 (2009); Michael J. Burstein, *Rules for Patents*, 52 Wm. & Mary L. Rev. 1747 (2011).

84. *See* Shamnad Basheer & Prashant Reddy, *"Ducking" TRIPS in India: A Saga Involving Novartis and the Legality of Section 3(d)*, 20 Nat'l Law Sch. of India Rev. (2008), *available at* http://ssrn.com/abstract=1329201.

The IPAB also held that the efficacy referred to in Section 3(d) was therapeutic efficacy. Novartis's product may possess improved bioavailability, thermodynamic stability, improved flow properties, and lower hygroscopicity, but according to IPAB this did not amount to an increase in "therapeutic efficacy."[85]

As a consequence, "Section 3(d) as applied to date appears to exclude from patentability new salt forms as well as new polymorphic forms of previously known substances. Exclusion of such new forms seems to be without regard to any improvement in properties such as bioavailability or physical or chemical stability of the drug."[86] However, the requirement of enhanced efficacy means that although some evergreening is not possible, patents for drug versions with even incremental efficacy should be available.[87]

Interestingly, recent Federal Circuit opinions in the United States do not seem to be radically different in result, if not reasoning. Pfizer's patent for the besylate salt of the drug amlodipine was invalidated after the Federal Circuit determined that a skilled artisan would have been motivated to combine prior art references to achieve the claimed invention, and that it would have been obvious to one skilled in the art to try substitution of the besylate salt for the earlier maleate salt form that had created issues of sticky tablets.[88]

Then, in *Schering Corp. v. Geneva Pharms, Inc.*, the Federal Circuit invalidated a patent for an active metabolite of the allergy drug loratadine (Claritin), on the grounds that "a limitation or the entire invention is inherent and in the public domain if it is the 'natural result flowing from' the explicit disclosure of the prior art."[89] According to one commentator, "it is conceivable that the concept of anticipation through inherency could be applied to new salt and polymorphic forms as well."[90]

With respect to genetic inventions, "there is unlikely to be an inventive step in identifying from within a sequenced genome any new gene, even those without known homologues," for it is "obvious to trawl the genome for previously unidentified genes, and any skilled worker would have some expectation of

---

85. *Id.* at 134–35.

86. Fyan, *supra* note 82.

87. *See* Shamnad Basheer, Limiting the Scope of Pharmaceutical Patents and Micro-Organisms: A TRIPS Compatibility Review (2005). On the notion of evergreening, see *infra* note 121. The Indian Supreme Court affirmed that Gleevec is not patentable in India. Novartis AG v. Union of India, Nos. 2706–16 (S.C. Apr. 1, 2013) (India), *available at* http://judis.nic.in/supremecourt/imgs1.aspx?filename=40212.

88. Pfizer, Inc. v. Apotex, Inc., 480 F. 3d 1348, 1361–68 (Fed. Cir. 2007).

89. Schering Corp. v. Geneva Pharms, Inc., 339 F.3d 1373, 1379–80 (Fed. Cir. 2003).

90. Fyan, *supra* note 82.

success."[91] In *Genentech,* the court noted that a genetic invention was obvious if "'the materials in question were lying in the road and ready for a research worker to use,'" even if the skilled man faced a number of obstacles in proceeding to his goal. However, if overcoming these obstacles required "'a spark of imagination . . . beyond the imagination properly attributable to him as a man skilled in the art' then there may be some element of inventive step."[92]

II. Utility/Industrial Applicability

Thomas Jefferson is said to have devised the utility requirement in the 1793 Patent Act of the United States.[93] A high threshold of utility might be defined, based on the plain meaning of the words, as requiring evidence that the claimed invention accomplishes a socially desirable goal. Industrial applicability seems a stricter notion. The evidence required here might be that the invention can in fact be used in at least one "industry." Due in part to a perception that the Patent Office is ill-equipped to gauge actual utility of an invention, the current US standard is extremely low: basically an invention is useful if it "does something."[94] Yet, as the US Supreme Court noted in the famous case of *Brenner v. Manson*, "[T]he basic quid pro quo contemplated by the Constitution and the Congress for granting a patent monopoly is the benefit derived by the public from an invention with substantial utility."[95]

In Europe, the industrial applicability criterion is applied liberally and "industry" in that context is basically any economic activity.[96] In keeping with TRIPS,

---

91. Comptroller Gen. of Patents Designs & Trademarks, Report of the Technical Expert Group on Patent Law Issues (2006), *available at* http://ipindia.nic.in/ipr/patent/mashelkar_committee_report.doc.

92. *Id.*

93. "Although the Patent Act was amended, revised or codified some 50 times between 1790 and 1950, Congress steered clear of a statutory set of requirements other than the bare novelty and utility tests reformulated in Jefferson's draft of the 1793 Patent Act." Graham v. John Deere Co., 383 U.S. 1, 10 (1966) (citing Thomas Jefferson, V Writings 47 (Paul Leicester Ford ed., 1895)); *see also* Jeffrey H. Matsuura, Jefferson vs. the Patent Trolls: A Populist Vision of Intellectual Property Rights 37–38, 41 (2008).

94. The US Manual of Patent Examining Procedure (MPEP) (last revised July 2010) states that "[i]f the applicant has asserted that the claimed invention is *useful for any particular practical purpose* (i.e., it has a 'specific and substantial utility') and the assertion would be considered credible by a person of ordinary skill in the art, do not impose a rejection based on lack of utility." USPTO, MPEP, § 2107(II)(B)(1) (8th ed., 2001, as amended) (emphasis added), *available at* http://www.uspto.gov/web/offices/pac/mpep/mpep-2100.pdf. It all seems to boil down to a somewhat credible assertion by the applicant.

95. 383 U.S. 519, 534–35 (1966).

96. In the words of Justice Hitchin of the English High Court: "All manufacturing, extracting and processing activities of enterprises that were carried out continuously, independently

all that is required is *potential* applicability.[97] However, there are limits. First, the exploitation of a patent on human biological materials might violate the morality provision.[98] Second, inventions destined to be used in private (e.g., a contraceptive method) would not be "industrially" applicable. In the case of DNA sequences, industrial applicability can have a significant impact. The English Court of Appeals recently noted that "just describing the existence of a protein and its structure is not enough. Nor is it enough to describe the function at a high level of generality—e.g., that the compound must have a significant function biologically and so it (or its antibodies) may be usable to treat some sort of disease. You have to say what it is for with more particularity."[99] Interestingly, the court signaled that it was in a better position to make a determination of industrial applicability, which in that case required a fact-intensive inquiry that a patent office was ill-equipped to perform, mostly relying instead on the applicant's assertions.[100] The UK Supreme Court reversed, noting that plausibility of a claimed use, or an "educated guess," could suffice, adding that otherwise there might be a "chilling effect on investment in bioscience."[101]

A. China

"Practical applicability," deemed the equivalent of "utility,"[102] means that the invention "can be made or used and can produce effective results."[103] This requires that the invention be repeatable and not be contrary to established natural law and, not unlike the US standard, not be negative or harmful.[104] An invention must

---

and for commercial gain." Eli Lilly and Company v Human Genome Sciences Inc. [2008] EWHC 1903, conf. by [2010] EWCA Civ 33 (Eng. C.A.), *rev'd*, [2011] UKSC 51. Article 57 of the EPC, *supra* note 37, extends industry to "any kind of industry, including agriculture." *See also* Aurora Plomer, *Towards Systematic Legal Conflict: Article 6(2)(c) of the EU Directive on Biotechnological Inventions*, *in* EMBRYONIC STEM CELL PATENTS: EUROPEAN LAW & ETHICS 189 (Aurora Plomer & Paul Torremans eds., 2009); Chiron Corp. v. Murex Diagnostics Ltd. [1996] R.P.C. 535, 607.

97. TRIPS Article 27.1 says "*capable* of industrial application." TRIPS Agreement, *supra* note 1, art. 27.1; *see also* Plomer, *supra* note 96, at 190.

98. EPC, *supra* note 37, art. 53(a); *see also* Directive 98/44/EC of the European Parliament and of the Council of 6 July 1998 on the Legal Protection of Biotechnological Inventions, 98/44/EC, art. 6 & pmbl. ¶ 37, 1998 O.J. (L 213), 13.

99. Human Genome Scis. v. Eli Lilly, [2008] EWHC 1903, ¶ 64.

100. *Id.*

101. *Id.* ¶¶ 122, 145, 171.

102. Louis S. Sorell, *Comparative Analysis of Selected Aspects of Patent Law in China and the United States*, 2002 PAC. RIM LAW & POL'Y J. 319, 326 (2002).

103. China Patent Act, *supra* note 44, art. 22.

104. *See* Gabriel, *supra* note 57; *see also* Yahong Li, *Imitation to Innovation in China: The Role of Patents in Biotechnology and Pharmaceutical Industry* 132 (2010); Zhang Xiaodu,

generally meet the needs of society, which inventions that pollute the environment, seriously waste energy or other resources, or injure human health would not.

Here we see that even a standard expressed in the statute as a fairly high threshold (evidence of positive effects or results) has apparently been watered down to a not negative standard by the Patent Re-examination Board (PREB).[105] In the pharmaceutical field, human clinical trials are not required, as expectation of applicability to humans can be presumed from animal or other efficacy tests.[106]

Still, China's practical applicability requirement seems broader than the "industrial applicability" adopted by some other countries, allowing more inventions to qualify for patent protection.[107] For instance, in Europe, where "an invention shall be considered as susceptible of industrial application if it can be made or used in any kind of industry, including agriculture," an invention that could be manufactured, but not industrialized, would not pass the "industrial application" test; however, such an invention would meet China's utility requirement.[108] This broader standard is reflective of the role of economic development policies in the formation of Chinese patent law.[109]

In the case of pharmaceuticals, the invention must "produce positive effects," which implies effects that are stable and repeatable.[110] Traditional medicines that can only be made manually are not industrially applicable, nor are methods of medical treatment.[111]

B. India

An invention has "industrial applicability" in India if the product can be (1) made, (2) used in at least one field of activity, and (3) reproduced with the same characteristics as many times as necessary.[112] Usefulness depends not on whether the

---

Practical Applicability and Full Disclosure of Technical Solutions, 2002 CHINA PATENTS & TRADEMARKS 23–28.

105. *See* Decision 2758, June 18, 2002, international category A01N57/14; Li, *supra* note 104, at 132–33.

106. Li, *supra* note 104, at 133.

107. Kong Qingjiang, *The Political Economy of the Intellectual Property Regime-Building in China: Evidence from the Evolution of the Chinese Patent Regime*, 21 PAC. MCGEORGE GLOBAL BUS. & DEV. L.J. 111, 121 (2008).

108. Xuqiang, *supra* note 60.

109. Qingjiang, *supra* note 107.

110. *See* Li & Li, *supra* note 42, at 142.

111. *See id.* at 143.

112. Yalamanchili, *supra* note 73.

invention will have commercial success, but whether the invention "had utility at the date of filing and will actually have the effects the patentee professed."[113]

According to the Patent Office Manual, "industry" should be "understood in its broad sense (as it is in the EPO), which includes any useful and practical, as distinct from intellectual or aesthetic, activity. As such, it does not necessarily imply the use of a machine or the manufacture of a product and covers such thing as a process for dispersing fog or a process of converting energy from one form to another."[114] However, "vague and speculative indication of possible objectives that might or might not be achievable by carrying out further research with the tool as described is not sufficient for fulfillment of the requirement of industrial applicability" because "the purpose of granting a patent *is not to reserve an unexplored field of research* for an applicant."[115] The reader of the patent should not need to look for ways "to exploit it in industry by carrying out work in search for some practical application."[116]

Although the exclusion applies to inventions that defy the laws of nature (a typical example being a perpetual motion machine) or machines that cannot perform the function described in the application, in India it also excludes from the realm of patentability methods of treatment of the human or animal body by surgery or therapy or of diagnosis practiced on the human or animal body and parts/pieces of the human or animal body to be used in transplants.[117]

India also excluded from patentability a number of potential inventions that might have fallen on the sword of utility. Article 3 of the Patent Act provides inter alia that the following shall not be patentable:

- An invention that is frivolous or that claims anything obvious, contrary to well-established natural laws. The notion of frivolity was interpreted as applying to an application for making in one piece, an article previously made in two or more pieces, noting that "mere usefulness" was insufficient.[118]

---

113. *Id.*

114. European Patent Office (EPO), Guidelines for Examination in the European Patent Office 5.1 (June 2012), *available at* http://www.epo.org/law-practice/legal-texts/guidelines.html.

115. Decision of the Boards of Appeal of the European Patent Office, Case T0870/04, ¶ 64 (May 11, 2005), *available at* http://www.epo.org/law-practice/case-law-appeals/pdf/t040870eu1.pdf.

116. *Id.* ¶ 19.

117. *Id.* ¶ 59. On inventions that cannot work as described and claimed, see Eastman Kodak Co. v Am. Photo Booths Inc., BLO/457/02. Such applications might also fail on the disclosure and enablement fronts.

118. Indian Vacuum Brake Co. Ltd. v. Laurd, AIR 1962, Cal 152.

A perpetual motion machine alleged to be giving output without any input is not patentable as it is contrary to natural law.

- An invention the primary or intended use or commercial exploitation of which could be contrary to public order or morality or which causes serious prejudice to human, animal or plant life or health or to the environment;[119]
- The mere discovery of a scientific principle or the formulation of an abstract theory or discovery of any living thing or nonliving substances occurring in nature; and, as already discussed, the mere discovery of a new form of a known substance that does not result in the enhancement of the known efficacy of that substance or the mere discovery of any new property or new use for a known substance or of the mere use of a known process, machine, or apparatus unless such known process results in a new product or employs at least one new reactant.[120] It has become common practice in the pharmaceutical industry to repackage a molecule (active ingredient) with known salts, metabolites etc., in some cases with little if any demonstrated advantage, to claim a new patent, one of the forms of "evergreening" (the prolongation of rent from otherwise out-of-patent branded medications) that has been criticized.[121] Pharmaceutical companies have argued that the cost of research requires protection beyond the twenty-year term on a new drug, much of which is not used to commercialize the drug because of regulatory approval processes.[122] This is an empirical matter that is beyond the scope of this chapter, but one on which different countries, at different stages of development and with different industrial setups and public health concerns, might examine from dissimilar perspectives.

---

119. India Patent Act, *supra* note 67, art. 3b.

120. *Id.* arts. 3c–d.

121. "Evergreening" techniques writ large include getting laws enacted requiring generic manufacturers to notify brand name competitors of their intention to enter the market and requiring government regulators not to give marketing approval for a generic medicine unless no contrary patent claims can be established (the Canadian system); getting a patent covering the capsule or gel of the drug, instead of its contents; threat of litigation; introducing once-a-day versions of a drug just before patent expiration to replace a three-times-a-day form or bringing a single isomer version of a drug that was previously marketed as a racemic isomer; asking doctors to attack generic products in academic journals; contractual agreements in which the generic manufacturer agrees not to enter the market in return for financial remuneration from the brand name manufacturer; and last but not least using data exclusivity forbidding generic companies from using the original safety and efficacy data for a period of time. *See* Thomas Faunce, Submission to the Department of Foreign Affairs and Trade Review of Export Policies and Programs 53–54 (May 2, 2008) (Austl.), *available at* www.dfat.gov.au/trade/export_review/submissions_received/TheAustralianNationalUniversityDrThomasFaunce.doc.

122. *See* Robert Chalmers, *Evergreen or Deciduous? Australian Trends in Relation to the "Evergreening" of Patents*, 30 MELB. U. L. REV. 29, 32 (2006).

- A substance obtained by a mere admixture resulting only in the aggregation of the properties of the components thereof or a process for producing such substance;[123]
- The mere arrangement or rearrangement or duplication of known devices each functioning independently of one another in a known way.[124]

It seems fair to say that India has decided not to rely only, or even mainly, on patentability criteria to exclude from patentability matter that it considers unpatentable. The best example is Article 3(d) of the Patent Act, which has been described as "crude, but constitutional."[125] It may be, however, that the definitional flexibility resulting from the undefined nature of the three main patentability criteria in TRIPS Article 27 would be the hook to justify the compatibility of the measure with TRIPS, in keeping with Article 1.1. Another option is to look at Article 30 as an exception, but Article 30 seems designed to provide exceptions to patent rights, not subject matter.

III. Lessons

As an early expert on US patent law, Willard Phillips noted that deciding what kinds of subjects should be covered by patents is a very difficult task.[126] Additionally, patentability rules—like most rules—can rapidly become obsolete because by definition patent law will encounter subject matter not before seen.[127]

Are general standards better than more specific ones? The debate about the flexibility of patentability standards enunciated in TRIPS, namely inventive step and industrial applicability, which may be "deemed to be synonymous" with non-obviousness and utility, respectively, is linked to the debate about rules versus standards. At the very least, echoes of that debate reverberate here, as well.[128]

Although the demands of global industries, especially pharmaceutical companies, to have as much uniformly high protection as possible, and to allow as many forms of evergreening as possible, are understandable, a world in which patent offices may be said to apply a strict patentability rule simply is not upon us.[129]

---

123. India Patent Act, *supra* note 67, art. 3e.

124. *Id.* art. 3f

125. Basheer & Reddy, *supra* note 80, at 265.

126. Willard Phillips, The Law of Patents for Invention 76 (1873).

127. *See* Thomas Webster, The Law and Practice of Letters Patent for Inventions 9 (1841).

128. TRIPS Agreement, *supra* note 1, art. 27.1; *see* Ruth Okediji, *Public Welfare and the Role of the WTO: Reconsidering the TRIPS Agreement*, 17 Emory Int'l L. Rev. 819, 902–03 (2003).

129. *See* Dianne Nicol, *Strong Patent Rights, Weak Standards and Innovation in Biomedicine*, *in* Intellectual Property Policy Reform 55–56 (Christopher Arup & William van Caenegem eds., 2009).

Patent offices, at least those that examine incoming applications, and the courts that review their decisions, must retain a degree of flexibility.

Echoes of the "rules versus standards" debate can be heard in the demands of TRIPS-plus advocates to cabin any flexibility in *refusing* patents for non-obviousness (for example patents on new forms of known substances) or lack of utility (for example because the application discloses no known treatment, as in many genetic sequence patents). One might say a rule is then superimposed on a standard to limit the field of application of that standard. This is problematic according to the traditional distinction because it is assumed to have a temporal component. A rule does not differ from a standard simply because of the specificity of the command, but also because it typically provides content ex ante, whereas a standard is used to consider conduct ex post.[130]

TRIPS demanders typically assert that patents are necessary to empower market forces to create incentives for innovation.[131] However, this does not make a case for a uniform standard across industries and technologies, because each may have its own market structure, thus requiring flexible standards.[132] Moreover, even if a good degree of predictability seems a priori desirable, one should distinguish the predictability of the notice function (e.g., the scope of the property rights claimed) from the application of the patentability criteria.[133] Greater ex ante predictability of the scope of a patent (essentially, rules concerning the construction of claims) allows easier and more efficient private ordering, legal advice, etc. However, strict patentability *rules* may not take account of changes and variations in a technological field and differences among technological fields.[134] Indeed, decisions about patentable subject matter are often informed by policy judgments about the desirability of patents in a given area.[135] Here, a standard often may seem fairer because it allows the decision-maker to factor in relevant circumstances, but it may also involve greater transaction and evidentiary costs.

---

130. *See* Louis Kaplow, *Rules versus Standards, An Economic Analysis*, 42 DUKE L.J. 557, 559 (1992).

131. *See* JOEL MOYKR, THE GIFTS OF ATHENA: HISTORICAL ORIGINS OF THE KNOWLEDGE ECONOMY 76 (2002).

132. *See* BURK & LEMLEY, *supra* note 83, at 95.

133. *See* Craig A. Nard, *Legal Forms and the Common Law of Patents*, 90 B.U. L. REV. 51, 80 (2010).

134. *See id.* at 99 ("[T]he judge, in the Hayekian sense, is more closely tethered to industry customs and norms, and this is more likely to develop doctrine that reflects the parties' and the relevant industry's expectations.")

135. *See* John F. Duffy, *Rules and Standards on the Forefront of Patentability*, 51 WM. & MARY L. REV. 609, 617 (2009–10) (discussing Justice Breyer's opinion in Lab. Corp. of Am. Holdings v. Metabolite Labs., Inc., 548 U.S. 124 (2006)).

Leaders in some industries believe that greater harmony in defining what is patentable will lead to more uniform global markets, reduced transaction costs, and more possibilities for rent extraction from more countries. They might refute the access to medicine concerns, for example, by stating that getting the patent is only a first step, and that measures to allow access by poorer patients can be taken as a second step, and that indeed taking that step (e.g., providing low-cost or free drugs) will be easier if patent rents are paid by more affluent patients in the developing world.

This cuts both ways however. First as to the method: India has shown that it can use rules to remove the flexibility to *grant* a patent on, for example, new forms of known substances.[136] We see this in the ruling by the Madras High Court, which noted the considerable degree of (necessary) discretion left to examiners and the Patent Appeal Board. Second, on the normative underpinning, one might counter that granting a patent and then having to go through a possible compulsory license exercise imposes transaction costs on the state and causes delays. Empirically, those who take this position might point to the sharp drop in new molecules in the approval pipeline in spite of their having the highest level of protection for the industry.[137] That said, countries who take a position viewed as unfavorable by the "TRIPS- plus" demanders should expect repeated visits by European and US officials and possible threats of retaliatory action, which in itself is a transaction cost worth pondering.

Indeed, the Indian decision concerning Novartis's Gleevec seems informed by the belief that fewer patents should result in a stronger indigenous industry, particularly in the area of pharmaceuticals and chemicals.[138] Article 3(d) of the Patents Act, presented as requiring a higher threshold of inventiveness, is a policy implementation of that belief, as are provisions on compulsory licensing, the fairly wide experimental use exception, and the parallel import provisions contained in Indian law.

The policy issue at the core of the debate on pharmaceutical patents is optimal, cumulative innovation, which one may situate between banal incremental innovation and rare substantive departures from the prior art ("breakthroughs").[139] Although there are occasional major breakthroughs in biopharmaceutical research as in other areas, most patents are granted to follow-on innovators. Economists have been actively debating for several years the best way to allocate efficiency between original and follow-on

---

136. India Patent Act, *supra* note 67, art. 3d.

137. At least for antibacterial research. *See* David J. Payne, Michael N. Gwynn, David J. Holmes & David L. Pompliano, *Drugs for Bad Bugs: Confronting the Challenges of Antibacterial Discovery*, 6 Nature Revs. Drug Discovery 39 (2007).

138. *See* Shamnad Basheer, *"Policy Style" Reasoning at the Indian Patent Office*, 3 Intell. Prop. Q. 309 (2005).

139. *See* Correa, *supra* note 30, at 4.

inventors.[140] The risks at both ends of the policy spectrum—that is, extreme under- or overprotection—include a lack of incentive for innovation and patent thickets with anticommons effects (holdups), respectively.[141] There are several options between those extremes, including the application of compulsory licensing for dependent patent use and the application of competition law measures in cases of patent abuse.[142]

The binary nature of the debate is itself a significant part of the problem. As with TRIPS, the first two phases of which (namely first, an attempt to get as high a level of protection as possible followed by an attempt to resist TRIPS and lower protection) were both ill-advised as a single source for policy development.[143] Calibration is required here as in so many other fields of regulatory endeavor. The debate is not about getting as many proprietary rights to support privately funded (closed) innovation with high proprietary barriers versus making innovation available at the lowest cost. China, India, and other countries, including many ranked as industrialized, should experiment with other forms of innovation and adapt patentability standards accordingly. Open innovation, sharing of information, possibly allocating public funding among countries instead of competing as in the closed model may indeed lead to much more promising developments while giving private biopharmaceutical companies the leeway to continue to invent as they see fit. This broader approach may work better for orphan, tropical, or low-market diseases, as recognized by the World Health Organization.[144]

Patent pools and licensing will not solve all issues that open source innovators might face. A private party may patent a basic innovative step and prevent the exploitation of (more advanced) results funded by public and/or open source innovators. Although excesses in this sphere may be addressed in some cases by (typically complex) competition law measures, the scope of potential abuses may also be reduced by applying patenting standards and possibly exceptions to exclusive rights informed by the need to allow basic scientific progress. [145]

---

140. Nicol, *supra* note 129, at 69–70.

141. The evidence of anticommons effects exists but is reportedly counterbalanced by "prolific licensing activity." *See id.* at 71; *see also* Rebecca S. Eisenbeg, *Genomic Patents and Product Incentives*, *in* HUMAN DNA: LAW AND POLICY 373–74 (Bartha M. Knoppers, C.M. LaBerge & Marie Hirtle eds., 1997).

142. *See* UNAIDS, *supra* note 1.

143. *See* Gervais, *supra* note 18.

144. *See* WORLD HEALTH ORG. (WHO), THE PUBLIC HEALTH, INNOVATION AND INTELLECTUAL PROPERTY: GLOBAL STRATEGY AND PLAN OF ACTION (2009). Paragraphs 17 through 20 focus especially on essential medical research and development (R & D) relevant to diseases that disproportionately affect developing countries.

145. *See* Margo A. Bagley, *Academic Discourse and Proprietary Rights: Putting Patents in Their Proper Place*, 47 B.C. L. REV. 217, 223–24 (2006); Rebecca S. Eisenberg, *Proprietary Rights and the Norms of Science in Biotechnology Research*, 97 YALE L.J. 177, 182–84 (1987).

Developing countries, in implementing patentability standards as part of a broader, systemic innovation strategy can and should avail themselves of opportunities that more industrialized countries, ensconced in strict constraints of private, market-based innovation, are unable to exploit. Public health is, one might suggest, about more than commerce but this is not my point.[146] Developing countries may have lower development costs for new drugs and can focus on orphan and tropical diseases that have a smaller market in industrialized countries. Current research focuses heavily on diseases not because they are serious or life-threatening but, rather, because they affect a significant number of potential buyers (patients). Some might see this as ideologically desirable or at least inevitable. Orphan-disease subsidies meant to correct this imbalance are not working well. [147] This notion points clearly to an advantage for lower-cost research environments, and the patent rents that those drugs could generate even in small markets in industrialized countries should be proportionally much more substantial if the drug was developed at a lower cost.

Beyond pharmaceuticals, the data concerning the benefits of patents are mixed in the aggregate. Bessen and Meurer report, for example, that "patents are neither the only nor even the most important means of encouraging innovation. On average, patents made a rather small contribution in this regard. Claims that the US patent system is responsible for current US technological leadership are exaggerated."[148] Still, a number of companies have benefited enormously from patents, and some would not exist but for the patent system. Proponents of strong patent protection can usually point to jobs created, etc. as a result of their patent-led growth, a good way to mollify policymakers, especially in hard economic times. However, it behooves those policymakers to consider the impact on their country's economy as a whole in making forward-looking policy, as opposed to anecdotal (but not necessarily insignificant, of course) examples in the rear-view mirror. Developing countries in particular might look at studies specific to

---

146. Though it is often made in this context. *See* Jeffrey D. Shelley, *Patent Pools for Orphan Diseases*, 18 Annals Health L. Advance Directive 141, 149 (2009) ("[T]he social and economic benefits stemming from orphan disease research and development, and global access to related drugs, outweigh for-profit interests and is well worth the endeavor.").

147. *See, e.g.*, the Orphan Drug Act, 21 U.S.C.A. §§ 360aa–ee (2007); European Union Regulation on Orphan Medicinal Products, Regulation (EC) No. 141/2000, OJ L18, 22/1/2000; Therapeutic Goods Regulations, 1997, Part 3B (Austl.). Most commentators agree that, although those measures were helpful, much more is needed. Suggestions include more public research, patent pools, and biobanking. *See* Dan Phair, *Orphan Drug Programs, Public-Private Partnerships and Current Efforts to Develop Treatments for Diseases Of Poverty*, 4 J. Health & Biomed. L. 193 (2008); Shelley, *supra* note 146; Brian Su, *Developing Biobanking Policy with an Oliver Twist: Addressing the Needs of Orphan and Neglected Diseases*, 66 La. L. Rev. 771 (2006).

148. James Bessen & Michael J. Meurer, Patent Failure 118 (2008).

their situation in crafting a patent policy most likely to develop their economy and local innovative potential.

India wants to leverage the fact that it has highly trained researchers but much lower costs than US and European laboratories.[149] Seen in this light, there is no reason not to support significant increases in the use of the patent system, "a rent extraction machine, which, used correctly, allows domestic firms to dominate foreign markets."[150] Yet winning the global innovation game by imposing lower patentability standards is a risky endeavor, at least temporarily: more patents likely will be granted to foreign applicants thus imposing additional rents and licensing constraints on Indian researchers and users.[151]

If a country decides not to invest in substantive patent examination, it has a number of options. One is to grant all patents applied for, and let courts debate their validity in case of infringement. Another is to have the applications examined by a foreign patent office. In that latter scenario, a country would have the option of going to a major industrialized Office (e.g., EPO, JPO, USPTO) or to the office of a developing countries with kindred views on development. In either case, however, the matter of whether a patent is valid and infringed will appear before courts sooner in later. Developing countries should be training judges and members of the local bar, providing them with interpretive tools and at least an overview of the underlying policy issues. Courts should adopt rules allowing themselves to appoint a neutral expert if necessary, as is the case in a number of European systems.[152]

A developing country might well decide to adopt fairly high standards, so as to exclude inventions that have no clear usefulness or that seem such a small incremental step that they could be considered "obvious." Yet, the patent system was not designed only for pioneer inventions, those major technological breakthroughs capable of transforming an industry or indeed establishing a new one; the obviousness and utility criteria are more typically filters that apply to "close calls." If the standard is too high, local inventors might be discouraged. Efforts in China and India to develop indigenous standards should be welcomed by policymakers in the developing world as they are likely to broaden the palette of options available to them in that regard.

Patent policy might be used to exclude subject matter, such as "pure" software or business methods for example. That may be made partially dependent on

---

149. *See* Drahos, *supra* note 17, at 90–91.

150. *Id.* at 91.

151. Drahos reports that the 1250 in-house R & D institutions in India had collectively filed only 743 applications, a much lower number than their counterparts in China. *Id.*

152. Strict rules are also required to ensure the neutrality of the expert.

industrial applicability for example, but may also be excluded at the subject matter level. Excluding software is a good example of a risk worth calculating: in many developing countries, access to computers and an abundance of bright programmers may lead to a rapidly emerging industry, one that is much less costly to establish than, say, a biotechnological research industry. [153] To make appropriate policy, one should consider the evidence of the impact of patents on the software industry. This should include consulting with, and providing information to, students in engineering and computer science schools and in small and medium businesses. Still, obviousness and utility will apply directly to genetic and many pharmaceutical reformulation patents, and those decisions may have considerable welfare and trade impacts and should thus be properly calibrated. More important perhaps, if a country wishes to develop its innovation potential, much more than calibrated patentability standards are required. Inventors must understand the patent system and have access to the necessary assistance to use it. In particular applicants who wish to obtain protection in foreign markets often need venture-type funding to finance a Patent Cooperation Treaty (PCT) process. At an even more fundamental level, education systems must reflect the value of inventiveness and creativity and provide the intellectual toolkits to enable citizens to innovate effectively. Patents can be barriers for many inventors because the threshold of inventiveness and especially novelty is usually judged in light of worldwide prior art. This means that local inventors must be global innovators.

Conclusion

In this chapter, I situated the "semantic" flexibilities following from the use of undefined terms in the TRIPS Agreement in the broader context of other forms of policy space available to WTO members in their effort to implement TRIPS and calibrate such implementation to maximize innovation outcomes while trying to minimize welfare and transition costs. I then considered two specific undefined terms, non-obviousness/inventive step and utility/industrial applicability, and how those definitions may be implemented in national law to embody domestic policies specific to a country's priorities. Such national design flexibility in applying these international standards may produce significantly different outcomes and better align patent law to the innovation goals with which it is often associated without compromising interests in making public goods readily accessible. Definitional flexibility, as seen in India and

---

153. *See* Robert E. Thomas & Larry A. DiMatteo, *Harmonizing the International Law of Business Method and Software Patents: Following Europe's Lead*, 16 Tex. Intell. Prop. L.J. 1, 37 (2007).

China, allows for policy experimentation in supporting research in new fields of scientific inquiry, providing adequate levels of incentives for those industries that depend more heavily on patents and, ultimately, in assuring that the patent system produces net gains that are meaningful in the specific context of each country.

18 PROOF OF PROGRESS: THE ROLE OF THE INVENTIVE STEP/NON-OBVIOUSNESS STANDARD IN THE INDIAN PATENT OFFICE

Feroz Ali Khader and Srividhya Ragavan*

Introduction

In patent law, the standard of non-obviousness has a dubious distinction. As a workable rule of law, this standard creates more questions than answers.[1] Justice Learned Hand perhaps has provided the most apt description of this standard yet: "[The inventiveness test is] as fugitive, impalpable, wayward, and vague a phantom as exists in the whole paraphernalia of legal concepts."[2] For those who take on the task of attempting a definition of this concept, Justice Jackson provides a cautious detail of the territory: "[T]he concept of invention is inherently elusive.... This, together with the imprecision of our language, have counseled courts and text writers to be cautious in affirmative definitions or rules on the subject."[3]

And yet, if there is one concept that serves to distinguish the *flash of genius* from *the skill of the calling*, it is the concept of non-obviousness.[4] The ability to so distinguish has made the inventive step or the concept of non-obviousness the centerpiece of every patent puzzle. In the

---

* Feroz Ali Khader is the Coordinator, MHRD IPR Chair at the Indian Institute of Technology (IIT), Madras, India. Srividhya Ragavan is Professor of Law, University of Oklahoma College of Law.

1. Martin J. Adelman et al., Cases and Materials on Patent Law 14 (1998).

2. Harries v. Air King Prods. Co., 183 F.2d 158, 162 (2d Cir. 1950).

3. Great Atl. & Pac. Tea Co. v. Supermarket Equip. Corp., 340 U.S. 147, 151 (1950).

4. Cuno Eng'g Corp. v. Automatic Devices Corp., 314 U.S. 84, 91 (1941).

United States with its fetish to ensure the patent system inures to its objectives of advancing progress, the development of non-obviousness from *Graham v. John Deere*[5] to *KSR v. Teleflex*[6] provides a keyhole view of not just the length of time but also how legislators, courts, and examiners have struggled to inject more objectivity and definiteness into the concept.[7]

This chapter examines the inventive step as a *new* protagonist to the Indian patent saga. In reality, while the world has been engrossed and entertained by the interpretation and application of the famed section 3(d), a provision of the Indian patent statute that restricts the grant of a patent for new forms and new uses of known substances, not much has been discussed or written about the standard of non-obviousness in India.[8] In an effort to avoid the pitfalls of embodying an ambiguous standard, the Indian patent statute has attempted a bolder definition, which, we assert, alters the standard of patentability by heightening it while being more definitive.

Is India's legislative effort a case of mere readjustment of the criterion to determine non-obviousness, or is it a meaningful step toward a more objective standard to measure the naturally subjective concept that is at the heart of patent law? In analyzing the non-obviousness provision of the Indian patent statute, the authors attempt to define the role of this provision per se as well as in comparison with the inventive step as established in the United States. An important distinction of our approach is that our analysis is based on the grant opinions of the patent office as opposed to judicial decisions.

The grant mechanisms—a system carefully engineered to enable innovation in a manner conducive to and balanced with its local realities—are unique to India. Any third party can oppose a patent application at the administrative level before a patent is granted. Even after grant, a patent can be challenged by an interested party. Thus, India has a pre-grant opposition mechanism that is complemented by post-grant opposition. The grant opinions are critical in India for practitioners and applicants to appreciate the predilections and predispositions of the first responders to innovation, namely the patent office. Grant opinions conserve administrative time otherwise spent on examining a patent application that could later be invalidated, in addition to preserving judicial time. Further, the grant structure represents a mechanism used uniquely to get around one of India's debilitating constraints, being the backlog in the judicial system. Consequently,

---

5. 383 U.S. 1 (1966).

6. 550 U.S. 398 (2007).

7. KSR Int'l Co. v. Teleflex Inc., 550 U.S. 398 (2007).

8. Indian Patents Act, No. 39 of 1970, *as amended by* Patents (Amendment) Act, No. 15 of 2005, India Code (2005), §3 (d) [hereinafter Patents Act, 2005].

the grant opinions in India may have more economic value, when compared to the USPTO's administrative opinions, for instance, not least because there are few judicial opinions on the question of inventiveness but also perhaps because of the influence of a combination of other factors such as the time taken to resolve disputes in India.[9] We also focus on grant opinions dealing with a pharmaceutical patent application because of its international character (the same product has the potential to be a subject of a patent application in several jurisdictions), the perceived sophistication of the parties involved, and the high value attached to these inventions.

Section I of this chapter examines the definition of non-obviousness (hereinafter, inventive step) and the classic ingredients required to prove the presence of inventive step under the Indian Patents Act, 1970. This section also examines how inventive step is determined, focusing on the guidelines given under the Patent Office Manual of Practice. Section II presents the "Proof of Progress" argument that has crept into the inventive step debate in Indian patent law. In doing so, this section discusses the future trajectory of the inventive step in India. Section III outlines a comparative study of how inventive step is determined in the United States to analyze whether and how the Indian standard is at variance with the American standard.

The authors conclude that the Indian patent system does embody a more definitive standard but without eliminating the subjectivity that is the hallmark of the inventive step requirement. Consequently, the authors conclude that the inventive step concept in India is efficient in performing the function of weeding out minor innovations. This is done, primarily, by requiring the patent applicant to demonstrate technical advancement—a concept that appears to represent technical superiority over the prior art. Thus, the inventive step requirement has become the new balancing mechanism to preserve private and public domains commensurate with national priorities. On the practice side though, the expectation is for applicants to be conversant with the enhanced standard and adduce relevant evidence for proving the additional requirements of "technical advancement" or "economic significance." However, the authors conclude that patent applicants have been apathetic in adducing material to prove these two requirements. Instead, applicants tend to provide inadequate proof to substantiate the requirements of Section 2(j)(a) and the *Manual of Patent Practice and Procedure* (MPPP) guidelines to prove the presence of the inventive step.[10]

---

9. *See, e.g.*, Jayanth K. Krishnan, *Social Policy Advocacy and the Role of the Courts in India*, 21 Am. Asian Rev. 91, 92–93 (2003).

10. *See* Indian Patents Act, No. 39 of 1970, *as amended by* Patents (Amendment) Act, No. 38 of 2002, § 2(1)(ja); *see also* Office of Controller General for Patents, Designs and Trademarks,

I. The Evolution of a New Standard of Inventive Step

In India, like in other WTO member jurisdictions, every invention—be it a product or a process—should be novel, embody an inventive step, and be capable of industrial application.[11] In lay terms, "inventive step" refers to the ingenuity in making the leap from the closest prior art to a non-obvious innovation. The terms "inventive step" and "not obvious" were generally interconnected and used synonymously in the sense that if an invention involves an inventive step, it should be non-obvious to a person skilled in the art.[12] The 1970 Indian Patent statute equated obviousness to the lack of inventive step—with inventive step meaning "a feature of the invention that makes the invention not obvious to a person skilled in the art."[13]

A. The New Statutory Standard Defined

The 2005 amendment defines an invention as a new product or process involving an inventive step and capable of industrial application.[14] Thus, inventive step remains a critical requirement to determine the patentability of an application. However, the definition originally codified in Section 2(1)(ja) was amended by the Patents (Amendment) Act, 2005 to introduce new elements.[15] Under the new standard, an *inventive step* means a feature of an invention that: (1) involves a technical advance as compared to the existing knowledge or, (2) has economic significance or, (3) both, *and that* makes the invention not obvious to a person skilled in the art. The section thus creates specific standards for determining inventive step: an application should embody a significant technical advancement and/or economic significance, and it should be of such an extent that it can be considered non-obvious over the existing prior art.[16] Notably, the definition leaves open the possibility that a technical advancement present in an invention can nevertheless be considered obvious under some circumstances. This

---

Manual for Patent Procedures and Practice (draft version 2011) (India) [hereinafter MPPP (2011)], *available at* http://www.ipindia.nic.in.

11. *See* Indian Patents Act, No. 39 of 1970, *as amended by* Patents (Amendment) Act, No. 38 of 2002, § 2(1)(ja).

12. *See id.*; Feroz Ali Khader, The Law of Patents—With a Special Focus on Pharmaceuticals in India 511 (2007) (for a detailed account on the inventive step standard in Indian law).

13. *See* Indian Patents Act, No. 39 of 1970, § 2(1)(j).

14. *Id.*

15. Patents Act, 2005, *supra* note 8, § 2.

16. F. Hoffmann-La Roche Ltd. v. Cipla Ltd., (2008) 37 P.T.C. 71 (Delhi H.C.) (India).

suggests that the standard of inventive step in India is both rigorous and fairly definitive.

The terms "technical advancement" and "economic significance" were not associated with the definition of inventive step in India until their introduction by the Patents (Amendment) Act of 2005. These two expressions were understood and used in the 1970 Patents statute in a different context. The phrase "technical advancement of considerable economic significance" was used in the context of a grant of compulsory license. Specifically, the expression was used in relation to another patent to compare two patents held by the same patentee, wherein one would be subject to compulsory licensing. For instance, the current Section 88(3) of the Patents Act as amended in 2002 reads as follows:

> 88. Powers of Controller in granting compulsory licences:
> (3) Where two or more patents are held by the same patentee and an applicant for a compulsory licence establishes that the reasonable requirements of the public have not been satisfied with respect to some only of the said patents, then, if the Controller is satisfied that the applicant cannot efficiently or satisfactorily work the licence granted to him under those patents without infringing the other patents held by the patentee and if those patents involve important *technical advancements of considerable economic significance* in relation to the other patents, he may, by order, direct the grant of a licence in respect of the other patents also to enable the licensee to work the patent or patents in regard to which a licence is granted under section 84. (emphasis added)

Interestingly, an identical reproduction of the expression is found in Section 48A of the 1977 U.K. Patents Act also in the context of compulsory licenses. The relevant provision states:

> (b) that by reason of the refusal of the proprietor of the patent concerned to grant a licence or licences on reasonable terms–
> (i) the exploitation in the United Kingdom of any other patented invention which involves an important *technical advance of considerable economic significance* in relation to the invention for which the patent concerned was granted is prevented or hindered... (emphasis added)

Arguably, Section 88(3) of the Indian patent statute deals with a situation of patent hoarding where one of the patents of the owner cannot be effectively worked unless a second patent, which embodies a considerable technical advancement of economic significance, can also be licensed. In granting compulsory licenses, the Controller would take into account whether there is a technical advancement of economic importance.[17] In other words, the statute indirectly acknowledges the higher value of patents embodying technical advancements and their economic significance. Perhaps it was this acknowledgment that inspired the inclusion of these phrases to create a new standard for obviousness.

The statute is complemented by the MPPP, versions of which are issued by the Patent Office from time to time.[18] The MPPP (2008)[19] demonstrates that the presence of inventive step can be determined by answering a simple question: "[W]ould a person with ordinary skills in the art have thought of the alleged invention?"[20] A negative answer to the question signifies the presence of an invention, even if the application material is very simple.[21] The MPPP also adds two other components: (1) to determine "whether the claim would inhibit the rights of a skilled workman to carry out routine modifications of what is already in the public domain?" and (2) to determine "whether the invention goes beyond the normal progress of technology?"[22] Although these 2008 MPPP guidelines on determining non-obviousness are missing from the 2011 version, no contradictory guidelines were issued in 2011, implying that there has been no derogation from the 2008 standards.

B. Finding an Inventive Step in India during Prosecution

During prosecution, the primary burden of finding a lack of an inventive step is on the examiner or the Controller.[23] Although this is similar to the practice in the United States (and in several other countries), in India that initial charge is not treated as a procedural tool of examination to allocate the burden going

17. *See* MPPP (2011), *supra* note 10; *see also* Office of Controller General for Patents, Designs and Trademarks, Manual for Patent Procedures and Practice 13.02.01.06 (draft version 2008) (India) *available at* http://www.ipindia.nic.in [hereinafter MPPP (2008)] (using the expression "important technical advancement or considerable economic significance"). The patent manual treats this as the same standard as "technical advancement or economic significance." Patents Act, 2005, *supra* note 8, § 2(j)(i).

18. *See* MPPP (2008), *supra* note 17. It contained certain guidelines on determining non-obviousness, which no longer exist in the 2011 version. MPPP (2011), *supra* note 10.

19. *See* MPPP (2008), *supra* note 17; *see also* MPPP (2011), *supra* note 10. Since its first issuance in 2005, the MPPP continues to remain in draft form. For our purposes, we have relied on the 2008 and 2011 versions of the MPPP.

20. *See* MPPP (2008), *supra* note 17, § 3.10.5–7.

21. *Id.* § 3.10.6.

22. *Id.* § 3.10.7.

23. *See* Patent Act, 2005, *supra* note 8, §§ 14–16.

forward.[24] Thus, in both India and the United States, the examiners are charged with the duty of determining non-obviousness; however in the United States, the examiner has the burden of showing a prima facie case of non-obviousness guided by two important dictates: first, to prevent a hindsight examination (that is, some inventions seem easier to assess in hindsight); and second, "rejections on obviousness cannot be sustained with mere conclusory statements; instead, there must be some articulated reasoning with some rational underpinning to support the legal conclusion of obviousness."[25] Termed as ex post facto analysis in relation to inventive step, the 2008 Indian MPPP also cautions against hindsight evaluations, but there are no guidelines or parameters to which the examiner must adhere.[26]

In India, all complete specifications filed should meet the requirements of Section 10, which, among other things, requires that the specification should end with a claim or claims defining the scope of the invention for which protection is sought; the claims should fully and particularly describe the invention and its operation or use, and disclose the best method of performing the invention known to the applicant and for which he or she is entitled to claim protection.[27] Generally, there is no obligation on the applicant to disclose the prior art. A request for examination made by the applicant causes the Patent Office to examine the application. [28] Section 12 of the Patent Act authorizes the examiner to determine if the application material qualifies for patent protection.[29] Where the application lacks an inventive step, the examiner can per se object that the application is obvious.[30] He then alerts the Controller to his findings.[31]

24. U.S. Dep't of Commerce, Manual of Patent Examining Procedure, Patent and Trademark Offices § 2142 (8th ed. 2010) [hereinafter MPEP (USA)].

25. *Id.*; *see also In re* Kahn, 441 F.3d 977, 988 (Fed. Cir. 2006).

26. *See* MPPP (2008), *supra* note 17, § 3.12.1 ("The examiner (or any other person) who is considering the question of whether or not an invention is obvious has to bear in mind that it is an ex-post facto analysis. He can be very easily misled by a line of reasoning involving taking the solution and working backwards to the problem by a succession of easy steps. In considering a prior publication, the examiner has to avoid looking at the document under the influence of the application he is examining, and should attempt to place himself in the shoes of the skilled person faced with the problem at hand."). The MPPP (2011) is surprisingly silent on the question of ex post facto analysis. *See* MPPP (2011), *supra* note 10.

27. *See* Patent Act, 2005, *supra* note 8, § 10.

28. *Id.* § 11B.

29. *Id.* § 12.

30. *Id.* Under Section 12(1)(b), the examiner is empowered to make a report on whether there are lawful grounds of objection to the grant of the patent under the Act in pursuance of the application. *Id.*

31. *See* Patent Act, 2005, *supra* note 8, § 14.

The applicant will be informed of the decision of the patent office (which is usually made by the Controller either in deference to the examiner's report or independently under Section 15 of the patent statute) and where necessary, a hearing will be scheduled.[32] When a question of obviousness is raised by the patent office, the applicant is required to produce proof of an inventive step.

C. Inventive Step Issues in Grant and Revocation Proceedings

Regardless of the prosecution process, the question of inventive step becomes significant during two other proceedings where third parties can raise objections involving the lack of inventive step—the opposition and revocation proceedings. This is of course different from litigation where lack of inventive step can be raised as a defense in an infringement suit.[33]

1. Inventive Step during Opposition Proceedings

As mentioned earlier, India's patent system provides two instances of administrative opposition to a patent application—a pre-grant and a post-grant opposition. An application that is subject to either a pre- or a post-grant proceeding can be safely termed a high-value application simply because unless the application material is of value, it is less likely third parties will oppose it. The convenience of administrative review coupled with the inconvenience of drawn out judicial proceedings has caused potential litigants to prefer the opposition mechanism as the first instance to determine the validity of the application.[34]

An objection of lack of inventive step can be asserted during opposition proceedings to request denial of the application. [35] In general, third parties can oppose a patent application after publication and before the grant of a patent (pre-grant) on several grounds, including for wrongfully obtaining the invention,[36] prior publication,[37] prior public use,[38] obviousness,[39] lack of adequate application disclosures

---

32. *Id.* §§ 14–15; *see also* Indian Patent Rules (Gazette of India, section II(3)(ii)) (Apr. 20, 1972), *as amended by* Indian Patent Rules (Gazette of India, section II(3)(ii)) (May 20, 2003), at Rules 28(3), 129 [hereinafter Indian Patent Rules (2003)], *available at* http://www.ipindia.nic.in/ipr/patent/patent_re_03.pdf.

33. *See* Patent Act, 2005, *supra* note 8, § 107.

34. *See generally* Feroz Ali Khader, The Touchstone Effect: The Impact of Pre-Grant Opposition on Patents (2008).

35. *See* Patent Act, 2005, *supra* note 8, § 25(1)(e), (2)(e).

36. *Id.* § 25(1)(a).

37. *Id.* § 25(1)(b).

38. *Id.* § 25(1)(d).

39. *Id.* § 25(1)(e).

including written description and specification requirements,[40] and false information.[41] A pre-grant opposition in India should be filed ideally within six months from the date of publication of the patent application,[42] although it is possible to file a pre-grant request any time before the patent is granted.[43] Once an opposition has been filed by a third party along with an evidentiary statement, the Controller shall give notice of the opposition to the applicant.[44] If the applicant submits his statement along with evidence within three months, a hearing is constituted.[45]

Notwithstanding the pre-grant opposition, any interested person may by way of a post-grant opposition oppose the patent within one year of the grant of patent.[46] The grounds for post-grant opposition are similar to the grounds for pre-grant opposition.[47] Both pre- and post-grant opposition can be made on the ground that the invention is anticipated by knowledge or practices of a local or indigenous community in India or improper use of biological material in the invention.[48] Arguably, the post-grant procedure in India is an improvised imitation of the much criticized re-examination proceedings in the United States under its 1952 Patent Act, but, without requiring the party to present a *substantial new question of patentability* or involving options of seeking an ex parte proceeding.[49]

2. Inventive Step during Revocation Proceedings

If lack of inventive step is proven during revocation proceedings, the patent will be revoked.[50] Interestingly, Section 64(1)(f), which details the ground of

---

40. *Id.* § 25(1)(g).

41. *Id.* § 25(1)(h).

42. *See* Indian Patents Rules (2003), *supra* note 32, Rule 55(1A).

43. *See* Patent Act, 2005, *supra* note 8, § 25(1)

44. *Id.*

45. Indian Patents Rules (2003), *supra* note 32, Rule 55(4), (6).

46. *See* Patent Act, 2005, *supra* note 8, § 2(1)(t). Person interested includes a person engaged in, or promoting, research in the same field as that to which the invention relates. Any person, including an organization, that has a manufacturing or trading interest in the goods connected with the patented article or who has a financial interest in manufacturing such goods, or who possesses patents relating to the same subject, is considered a person interested. *Id.*

47. Press Release, Ministry of Commerce and Industry, Press Information Bureau, Important Changes Incorporated in the Patents (Amendment) Bill, 2005 as Compared to the Patents (Amendment) Bill, 2003 (Mar. 23, 2005) [hereinafter Press Release, PIB], *available at* http://pib.nic.in/release/release.asp?relid=8096.

48. *See* Patent Act, 2005, *supra* note 8, § 25(k); *see also* Indian Patents Rules, (2003), *supra* note 32, Rule 55.

49. *See* 35 U.S.C. §§ 301–03 (2012).

50. *See* Patent Act, 2005, *supra* note 8, § 64(1)(f).

revocation, uses the expressions "obvious" as well as "does not involve an inventive step." It provides for revocation in cases where "the invention so far as claimed in any claim of the complete specification is obvious or does not involve any inventive step...."

The statute presumably uses the former expression to explain the latter.[51] That is, an invention will be regarded as involving an inventive step if it is "not obvious to a person skilled in the art." When read with Section 2(j), the statute suggests that inventive step can be proven from the existence of a non-obvious technical advancement or economic significance or both. Further, in stating that the lack of inventive step can be proved by "having regard to what is publicly known or publicly used in India or what was published in India or elsewhere before the priority date of the claim," this provision ties in the concept of utilizing the same prior-art material used for novelty determinations to determine inventive step.[52]

D. Judicial Decisions and Inventive Step

India, like the United States, has struggled with creating a standard for determining inventive step. The Indian system has tended to borrow extensively from British precedents, which establish that if an "alleged invention is or was at the material date obvious as that word is commonly used, then it follows that it did not involve any inventive step."[53] The policy behind the above provision is to ensure that a person "should not be prevented by a statutory monopoly from doing something which, at the date of the patent, was obvious."[54]

A 1945 Indian decision from the erstwhile Oudh High Court, Ganendra Nath Banerjee, dealing with a patent on the process of making pots with topall corks, cited Halsbury's Law of England[55] and observed that to determine what is obvious, the "general criterion seems to be whether that which is claimed lies within the limits of development of some existing trade, in the sense that if it is such a development as an ordinary person skilled in that trade could, if he wished so to do, naturally, make without any inventive step."[56]

Not much guidance other than the above can be culled from the Indian precedents except for *M/s. Bishwanath Prasad Radhey Shyam Appellant v. M/s.*

---

51. *Id.* § 2(1)(ja).

52. *See id.* § 64(1)(f).

53. *See* Benmax v. Austin Motor Co. Ltd., (1953) 70 R.P.C. 284, 288 (India).

54. *See* Brugger v. Medic-Aid Ltd., (1996) R.P.C. 635, 653 (India).

55. 24 Halsbury's Laws of England 586 (2d ed. 1937).

56. *See* Ganendro Nath Banerji v. Dhanpal Das Gupta, (1945) 32 A.I.R. 6 (Oudh) 9.

Hindustan Metal Industries, a 1982 decision that laid down the foundational rules of assessing inventive step.[57] Justice Sarkaria of the Supreme Court of India, in dealing with an invention relating to a new method of making utensils, held that inventiveness should be "...something more than a mere workshop improvement." The court went on to add that in order to be patentable the invention must "produce a new result, or a new article or a better or cheaper article than before...involving the exercise of any inventive faculty."[58] Thus, the emphasis remained on the *improvement* over the prior art. Other than highlighting that a simple innovation could also qualify for a patent,[59] very few cases in India have dealt directly with the question of inventive step.

One of the clearest expositions on non-obviousness after 1995 can be found in the case of *F. Hoffmann-La Roche Ltd. v. Cipla Ltd.*[60] where Justice Bhat acknowledged that there is a certain amount of elusiveness in what is obvious.[61] In considering the obviousness argument the court held that "the test of obviousness cannot be that the material or formula was published, but that having regard to the existing state of prior art or the published material, was it possible to a normal but unimaginative person skilled in the art to discern the step on the basis of the general common knowledge of the art at the priority date."[62] Another deciding factor, he added, was to determine "whether the differences between the prior art would, without knowledge of the alleged invention, constitute steps which could have been obvious to the skilled man or whether they required any degree of invention."[63] Importantly, the case heavily relied on the British Court of Appeal decision in *Windsurfing International Inc. v. Tabur Marine (GB) Ltd.*[64]

57. 1982 A.I.R. 1444 (S.C.) (India).

58. *Id.*

59. *See, e.g.*, Canadian Gen. Elec. Co. Ltd., v. Fada Radio Ltd., 1930 A.I.R. 1 (P.C.) (India), (holding that a small inventive step is nevertheless, an inventive step); *see also* Gillette Indus., Ltd. v. Yeshwant Bros., 1938 A.I.R. 347 (Bom.) (India) (holding that simplicity of an invention does not dilute the inventiveness).

60. (2008) 37 P.T.C. 71 (Delhi H.C.) (India).

61. *Id.*

62. *Id.*

63. *Id.*

64. (1985) R.P.C. 59 (stating that obviousness is proved by: (1) identifying the inventive concept embodied in the patent; (2) imputing to a normally skilled but unimaginative addressee what was common general knowledge in the art at the priority date; (3) identifying the differences if any between the matter cited and the alleged invention; and (4) deciding whether those differences, viewed without any knowledge of the alleged invention, constituted steps that would have been obvious to the skilled man or whether they required any degree of invention).

and further alluded to *Graham v. John Deere Co.*[65] to highlight the four cardinal determinations necessary regarding non-obviousness (reflected in both jurisdictions) as: (1) the scope and content of the prior art, (2) the level of ordinary skill in the prior art, (3) the differences between the claimed invention and the prior art, and (4) objective evidence of non-obviousness.[66]

Hoffmann-La Roche merely dealt with the question of non-obviousness in the interlocutory stage to determine whether an interim injunction could be granted. The case raised the question whether the single distinction of an ethynyl group substituted at the meta position of a phenyl ring would amount to fulfilling the inventive step requirement (the product itself was directed to a 4-(substituted phenylamino) quinazoline derivative with an alleged inventive step being the methyl substitute in the third position). At such a preliminary stage, Justice Bhat explained, the endeavor was to merely consider whether the defendant had raised an arguable case (a material issue) in response to the arguments asserted by the plaintiff. Once such a material issue has been raised, the plaintiff has to "answer a credible challenge to the patent, raised by the defendant." In so stating, the court, for the first time, disclosed a heightened burden on the plaintiff to prove inventive step when a *credible challenge* is raised against a patent.

II. Proof of Progress in Action

Sir Nicholas Pumfrey J., in determining the obviousness of Glaxo Group Ltd.'s patent over the drug Seretide or Advair, outlined the danger in too liberal an approach to obviousness, on the one hand, and too strict an approach on the other:[67]

> Both the Scylla of considering nothing obvious except that to which the skilled man is driven and the Charybdis of considering every invention obvious that can be decomposed into a sequence of obvious steps must be avoided. The former is unfair to industry because it stifles natural development. The latter is unfair to inventors and not countenanced by English patent law.[68]

---

65. 383 U.S. 1 (1966).

66. *Id.*

67. *See* Cipla Ltd. & Ors v. Glaxo Grp. Ltd., [2004] EWCH (Pat) 477 (Eng.).

68. *Id.*

In other words, obviousness is the tipping point at which the competing rights of inventors on the one hand, and industry (indeed the society at large) on the other must be balanced.[69]

A. The Proof of Progress Requirement in Indian Law

The statutory language of inventive step in the Patents (Amendment) Act, 2005 requires that a "technical and/or economic advancement" should also be non-obvious. Arguably this is a heightened standard, one that perhaps reduces the requirement of "nonobviousness" into a subset of an inquiry into whether an innovation embodies a technical or economic advancement. We call this new standard the *proof of progress* requirement. In other words, the applicant needs to positively demonstrate *proof of* [inventive] *progress* of his or her application over existing prior art. Technical or economic superiority can be significant to demonstrate progress that would ultimately lead to a finding that the invention is non-obvious or embodies an inventive step. When an applicant is called upon to prove inventiveness, the proof of progress requirement elevates the bar for patentability in India. Perhaps, it is the plethora of minor inventions for which patents have been granted and that has plagued jurisdictions such as the United States that has caused India to make the patentability bar higher and more definitive.

The proof of progress standard can be outlined in three specific requirements. The first requirement, technical advancement over existing knowledge, has been recognized by the courts as an inherent characteristic of inventive step.[70] The second requirement is economic significance, which courts in India have not dealt with directly as a consideration in determining its inventiveness. Nonetheless, the commercial success of an invention may hold some evidentiary value in determining inventiveness.[71] Courts have tried to keep the element of commercial success as a mere secondary consideration or rather, a fallback criterion used to demonstrate that the technical advancement is indeed more than merely trivial because of the market response to the invention.[72] Considering this, courts have gone as

---

69. Susan Crennan, Obviousness: Different Paths through Scylla and Charybdis, Intellectual Property Society of Australia and New Zealand (IPSANZ) Conference (Sept. 16, 2007), *available at* http://www.hcourt.gov.au/assets/publications/speeches/current-justices/crennanj/crennanj_16sep07.pdf

70. *See* Richardson-Vicks Inc's Patent, [1997] R.P.C. 888, 897; Windsurfing Int'l Inc. v. Tabur Marine (Great Britain) Ltd., [1985] R.P.C. 59, 72; Johns-Manville Corpn's Patent, [1967] R.P.C. 479.

71. *See, e.g.*, Hybritech, Inc. v. Monoclonal Antibodies, Inc., 802 F.2d 1367 (Fed. Cir. 1986).

72. *Id.*

far as holding that obviousness refers to technical obviousness and not commercial obviousness.[73] Generally, where courts are confronted with a commercially successful invention with arguably minor technical advancement, in order to prevent hindsight rejections, the evidence of commercial success of the invention usually leads to an investigation of the reasons for the success. Courts are aware of the possibility that the commercial success of an invention can be unrelated to the non-obviousness determination.[74] Similarly, success in the marketplace does not automatically prove technical superiority or non-obviousness of the material.[75] Thus, like in the United States, economic significance or commercial success in India can, at best, be regarded as a secondary consideration in determining obviousness. As for the third and last requirement, whether or not the invention is "nonobvious," the analysis is directed at whether a demonstrated advancement proves that the invention is technically or practically non-obvious as opposed to a normal advancement of skill.[76]

B. Proof of Progress at the Patent and Trademarks Office

1. The *Valcyte* Post-Grant Story

The *Valcyte* story began when Hoffmann La Roche, a Swiss-based pharmaceutical company, submitted its patent application in July 1995 for a mono-L-valine ester of ganciclovir and its pharmaceutically acceptable salt.[77] The *Valcyte* (brand name for valganciclovir) application was first subject to a pre-grant opposition in July 2006, which the applicant successfully overcame, resulting in a pre-grant order and issuance of the Valcyte patent.[78] No hearing was conducted before the pre-grant order was issued. When the opponents raised the lack of a hearing, the patent office replied that the grounds for objections raised in the statements were considered and dismissed.[79] Dissatisfied with having been denied a hearing, the opponents filed a writ petition in October 2008 before the Madras High Court[80]

73. *See* Hallen Co v. Brabantia (U.K.) Ltd., [1989] R.P.C. 307, 327.

74. *See* Mölnlycke AB v. Procter & Gamble Ltd., [1994] R.P.C. 49, 113.

75. *See* Haberman v. Jackel Int'l Ltd., [1999] F.S.R. 683, 699.

76. *See* Hallen Co. v. Brabantia (U.K.) Ltd., [1989] R.P.C. 307, 327.

77. Indian Patent Application No. 959/MAS/1995 (filed July 27, 1995).

78. Roche's Application (Opposition by Tamil Nadu Networking People with AIDS), [2009] P.F.R. 4; *see* Indian Patent No. 207232.

79. *See* Sai Deepak, *Pre-Grant Objections Against Valganciclovir Application: Were They Frivolous?*, SPICY IP (Mar. 1, 2009), http://www.spicyipindia.blogspot.com/2009/03/pre-grant-objections-against.html.

80. Indian Network of People Living with AIDS v. Union of India, Writ Petition No. 24904 of 2008 (C.H.C.) filed under INDIAN CONSTITUTION, art. 226 (1950).

seeking a writ of mandamus to quash the patent and direct the patent office to conduct a grant hearing.[81] The Madras High Court held that the pre-grant order and the patent were legally unsustainable on grounds that there was a "blatant violation of statutory procedure by the statutory authority, which is acting in a quasi-judicial capacity."[82] Meanwhile, the patentee appealed the Madras High Court Order to the Supreme Court on a Special Leave Petition.[83] Forced by both courts to conduct a hearing, the Controller of the Chennai patent office heard the opponents in the pre-grant procedure but ruled against them.[84] Patent number 207232 was again granted in January 2009.

Immediately thereafter, a group of generic drug companies along with non-governmental organizations, as interested parties, filed a post-grant opposition against the grant of the patent.[85] The first ground of opposition, which failed, was for lack of novelty, asserting that the patent was for the ester form of a known drug—ganciclovir, which was first disclosed in a 1980's US patent.[86] Opponents also alleged that the US '339 patent[87] disclosed the L-valine ester of ganciclovir hydrochloride, while the specifications and the examples in European Patent '329, whose owner enjoyed the benefit of a Supplementary Protection Certificate claiming valganciclovir hydrochloride, disclosed the method of making the application material.[88] The UK Patent Office in that case determined that the invention was novel in the light of the cited prior art.[89]

---

81. *Id.* at 2.

82. *Id.* at 51.

83. *See Roche's Plea: SC Asks Patent Office to Decide by January 31*, Bus. Standard (Dec. 19, 2008), *available at* http://www.business-standard.com/india/news/roches-plea-sc-asks-patent-office-to-decide-by-january-31/343715/.

84. *See* Roche v. Indian Network for People Living with HIV/AIDS and Tamil Nadu Networking People with HIV/AIDS, [2009] P.F.R. 4 (India).

85. *See* Hoffman Roche (Post-grant opinion) (Apr. 30, 2010) (on file with authors) [hereinafter Hoffman Roche (Post-grant opinion)]. Opponents include Ranbaxy, Cipla, Matrix, and Bakul Laboratories along with Delhi Network of Positive People, Indian Network of People Living with HIV/AIDS and the Tamil Nadu Network of People with HIV/AIDS. *Id.*

86. U.S. Patent No. 4,355,032 (filed May 21, 1981).

87. U.S. Patent No. 5,043,339 (filed Dec. 18, 1989).

88. European Patent No. EP0375329; *see also* Hoffman Roche (Post-grant opinion), *supra* note 85. It was asserted that the specification and the examples in EP '329 disclosed that by reducing the amount of amino acid added to the reaction relative to the ganciclovir in this case, mono-ester compounds as in the case of the application would ensue. In this regard, the opponents further submitted a Supplementary Protection Certificate, which had been granted to the owner of the EP '329 patent claiming valganciclovir hydrochloride. *Id.* ¶¶ 8–26.

89. *See* Intellectual Prop. Office, U.K., Supplementary Protection Certificates: Guide for Applicants, *available at* http://www.ipo.gov.uk/spctext.pdf. A supplementary protection certificate (SPC), obtainable individually for different countries

Next, the opponents alleged that the patent lacked an inventive step. They asserted that ganciclovir was present in the prior art (albeit for intravenous administration) as an antiviral to treat HIV infections. The L-valine ester of "ganciclovir" is the ester form that is used for oral administration on account of its sustained bioavailability. In order to prove that the L-valine form of ganciclovir is an obvious variation, the opponents used the example of acyclovir—a molecule comparable in structure and function (with anti viral properties) to ganciclovir. A more bioavailable form of acyclovir was obtained by esterizing select amino acids to result in the L-valine ester of acyclovir called valacyclovir. It was alleged that a person skilled in the art would follow the route of L-valine ester of acyclovir (i.e., valacyclovir) and apply the same to ganciclovir to get valganciclovir. Valacyclovir and valganciclovir are nucleoside analogs having similar structure and used for the same treatment—their hydrochloride salt forms were used successfully to treat ailments.[90]

2. Proof of Progress in *Valcyte*

The patent office endorsed the opponent's arguments that several nucleosides were converted to ester (preferably L-valine) forms to improve bioavailability. Most interestingly though, the patent office highlighted that the comparison between the ester and salt of ester (when hydrochloric acid is added) used in the specification to demonstrate inventive step was inadequate, because it is obvious that hydrochloride salt forms are more soluble in water, especially when compared with esters.[91] The office ruled that all salt forms would be expected to demonstrate increased bioavailability and this application was no exception. It added that there is no difference between the hydrochloride salts of other forms of esters and the hydrochloride salt of of ganciclovir. In other words, the improved bioavailability was lent from the salt form (of the L-valine).[92] The L-valine form itself was obvious from the prior art, and nothing in the specification showed improved bioavailability obtained from anything other than the salt form.[93] The

---

of the European Union, is an extension of a patent given to compensate for regulatory delays that the product suffered. An SPC comes into effect after the patent expires. Although an SPC can extend up to a maximum of 5.5 years, the total market exclusivity of the product cannot extend beyond a total of 15.5 years. *Id.*

90. *See* Hoffman Roche (Post-grant opinion), *supra* note 85. Valacyclovir was then combined with hydrochloric acid to result in Valacyclvoir Hydrochloride (sold as Valtrex, an anti herpes drug by GSK). Valgancyclovir when combined with hydrochloric acid to result in Valgancyclovir Hydrochloride (sold as Valcyte by Roche). *Id.* ¶ 34.

91. *Id.* ¶ 41

92. *Id.* ¶ 42.

93. *Id.* ¶ 45.

technical advancement demonstrated from the pro-drug form (less active form of the drug) was considered inadequate to demonstrate inventive step.

C. Inventive Step in the Light of Proof of Progress

The statutory language and the *Valcyte* grant opinion/decision give rise to at least two questions. First, will a satisfactory showing of technical advancement or economic significance automatically lead to a conclusion of non-obviousness? Second, is a showing of technical advancement or economic significance an absolute necessity? The statutory language and the grant opinion suggests that there is a mandatory requirement to show technical advancement or perhaps, economic significance in a manner that outweighs the inadequacies of technical advancement. Even the *Valcyte* decision strongly suggests an obligation to delineate and disclose a technical advancement. As for the first query of whether a demonstration of technical advancement per se would lead to a conclusion of non-obviousness, the grant opinions discussed below indicate otherwise.

1. Role of Technical Advancement—The *IPCA* Application

The role of technical advancement to prove inventive step can best be appreciated by examining the application filed in the Mumbai patent office in 2005 by IPCA Laboratories Limited, an Indian company, for a "Delayed Release Anti-Malaria Composition."[94] The Controller rejected the application on the grounds that it was not novel and that it lacked an inventive step.[95] During prosecution, the claims were amended and limited as follows: "A pharmaceutical tablet comprising β-arteether core coated by an enteric coat to delay the release; wherein, the enteric coat comprises hypromellose phthalate 12 to 14%; dibutyl phthalate 1 to 2%; titanium dioxide 0.3 to 0.7% and talc 0.3 to 0.7% by weight of the formulation."[96]

In sum, the invention is a medication that consists of β-arteether in the core and embodies an enteric coat over the core comprising hypromellose phthalate, dibutyl phthalate, titanium dioxide, and talc.[97] The coat is responsible for the delayed release of the drug into the intestine, instead of at the stomach.[98] However, the Controller cited six items of prior art, bearing numbers D1 to D6.

---

94. Indian Patent Application No. 1225/MUM/2005 (filed Sept. 30, 2005).

95. Mumbai Patent Office, *In the Matter of an Application for Patent by IPCA Labs Limited, Patent Application No. 1225/MUM/2005* (Feb. 24, 2011) (on file with author) [hereinafter Anti-Malaria Patent Rejection].

96. *Id.*

97. *Id.*

98. *Id.*

The Controller asserted that the prior art disclosed the antimalarial agent including beta arteether, the direct compression process for the preparation of the tablet in claim 1, and a pharmaceutical composition comprising microcrystalline cellulose and/or cross carmellose sodium.

The Controller outlined that the subject matter of D1—the first item of prior art—differed from the application material in respect only of the specific polymer used.[99] Further, the Controller added that use of an enteric coating over arteether was covered by D2 and D3.[100] And, although D4 addressed the same problem as the application, the examiner felt that D5 and D6 dealt with enteric coating to deliver the antimalarial compound in combination with other agents such as artemsinin, artesunate, artemether, etc., to the intestine.[101] Reading them together, the Controller opined that for a person skilled in the art, combining the teachings of D1 to D6 would reasonably result in an expectation of success. The Controller added that the "applicant has failed to demonstrate any surprising effect or technical advancement of the claimed invention over the products of the prior art."[102] The Controller concluded that the invention disclosed and claimed by the applicant is an obvious combination of the prior art.[103] The opinion added that the applicant failed to demonstrate "any surprising effect or technical advancement of the claimed invention over the products of the prior art."[104]

---

99. *Id.*; World Intellectual Prop. Org. (WIPO), Methods and Compositions for the Treatment of Helicobacter Pylori-Associated Diseases Using Endoperoxide Bridge-Containing Compounds, WIPO Doc. WO 2005/048912 (June 2, 2005) (tiled D1, this invention relates to methods and compositions for the treatment of pathological conditions associated with ferrous-dependent bacteria, such as, Helicobacter pylori).

100. Anti-Malaria Patent Rejection, *supra* note 95; WIPO Doc. WO 02/078714 (titled D2, this invention relates to gastric-juice–resistant formulations consisting of anti-infective compounds that inhibit the 2-C-methylerythrose-4 metabolic pathway and its salts and esters); U.S. Patent No. 5,578,637 (filed May 3, 1995) (titled D3, this application discloses methods for inhibiting cancer cells from compounds having an endoperoxide moiety by enhancing intracellular iron concentrations).

101. Anti-Malaria Patent Rejection, *supra* note 95; WIPO Doc. WO 92/02217 (representing D4 disclosed a synergistic antimalarial combination of benflumetol with a antimalarial agent from the artemisinine group such as artemether); U.S. Patent No. 6,284,772 (filed Sept. 28, 1999) (representing D5 describes Indolo [2, 1-b]quinazoline-6,12-dione, and substituted derivatives, which exhibit potent in-vitro antimalarial activity against Plasmodium falciparum); U.S. Patent No. 6,486,199 (filed June 21, 2001) (representing D6 describing a method for treatment of malaria using a Spiro or dispiro 1, 2, 4-trioxolane).

102. Anti-Malaria Patent Rejection, *supra* note 95, at 9.

103. *Id.*

104. *Id.*

The Controller observed that in the absence of "any experimental proof" to demonstrate otherwise, the invention remained obvious.

Although the standard for proof of technical advancement is not always a subject of rigorous discussion, grant opinions such as those of *Valcyte* and *IPCA* clarify the need for clear proof of technical advancement.

2. Role of Technical Advancement—The *NATCO* Application

The Controller's Report issued in relation to an application filed by the Indian generic drug company, Natco Pharma, to patent a controlled release gastric floating matrix containing imatinib and its polymorph demonstrates, albeit without discussion, the constituent elements of the technical advancement requirement.[105] Generally, the solubility of imatinib mesylate is low especially in the intestine due to the neutral/alkaline medium of the region. However, bioavailability of imatinib would be high if it is continuously released in the stomach. The new formulation was a controlled release gastric floating matrix containing a gel forming hydrophilic polymers to achieve the controlled release floating matrix. The inventive feature was the dissolution profile or the controlled release of the drug, proof of which was submitted from in vitro (animal) studies.

The applicant asserted that such a controlled release reduced the risks of the side effects associated with uncontrolled releases that invariably necessitated a higher dose of the medication in patients. The inventive step of a formulation, the applicant asserted, is established from an in-vitro dissolution profile that demonstrated the ability of the invention to provide controlled release. But, the dissolution results of the in-vitro study did not speak about the side effects simply because it was an in-vitro as opposed to an in-vivo study. The Controller opined that "the drug is known, the polymer for control release is also known, the method of making the controlled release tablet and dissolution method to check the control release is also known in the art. Therefore, the dissolution result for the drug release profile is an obvious and expected result." What was needed in the specification was proof that the invention resulted in eliminating or reducing the side effects—and that was not present in the application.

In response, the applicant submitted the USFDA Guidance Document that highlighted the relevance of in-vitro studies to predict the in-vivo response.[106] Rejecting them as regulatory guidelines focused on the safety and toxicity of the drug substance, the Controller held that the report is qualitative in nature highlighting the release profile of the tablet and does not establish the *activity of the drug.* The opinion differentiates between release profile and pharmacological

---

105. Indian Patent Application No. 1049/CHE/2004 (filed Oct. 11, 2004).

106. *Id.* ¶ 10.

activity of a drug to state that merely because there is a predicted release profile of a drug whose constituents are known, it does not follow that pharmacological activity will be presumed. Thus, dissolution tests (which provide the solubility of the substance) are not considered as indicative of the desired pharmacological action, which has to be proved using a cell culture test detailing the pharmacological activity of the drug. The Controller added that although there is no stringent requirement to submit an in-vivo animal or human study, it is mandatory to show proof of the claimed activity of the substance.[107] The application material, the Controller added, did not meet the required inventive step threshold. Notably, the Controller's decision was rendered under Section 15 of the Patents Act, 1970. The application was not subjected to any third party opposition.

3. Does Technical Advancement Automatically Result in the Presence of Obviousness?

Arguably, a showing of technical advancement or economic significance will greatly enhance the chances of an applicant successfully traversing an inventive step objection. However, the mere presence of a technical advancement does not automatically guarantee that the application will be considered non-obvious. Thus, the applicant has to co-relate the technical advancement with proof that, *the feature* of the invention embodying the technical advancement indeed makes the invention non-obvious. At least one decision of the newly constituted Intellectual Property Appellate Board (IPAB) has confronted this issue.

In the case relating to the famed but ill-fated patent application of Novartis for a cancer drug named Gleevec (or "Glivec" in markets except the United States)[108] discussed later in this chapter, Mr. Anand Grover who appeared on behalf of the Cancer Patients Association of India asserted that "a mere showing of an added advantage or avoidance of a disadvantage" is not sufficient to meet the requirement of inventive step in India. Assuming that such advantage or avoidance of disadvantage constituted technical advance, an applicant still has to confirm that the alleged invention is not obvious to a person skilled in the art."[109] However, the IPAB did not specifically address this query—instead, it held that the application material shows "a technical advance as compared to the existing knowledge by way of demonstration of polymorphism, isolation, characterization of beta (and alpha) crystal forms of imatinib mesylate, identifying suitable properties in the

---

107. *Id.* ¶ 13.

108. *See generally* Ashling O'Connor, *Activists Protest at Novartis's Patent Law Challenge*, TIMES (UK), Mar. 7, 2007, at 59.

109. *See* Novartis v. Union of India & Others, IPAB Order No. 100/2009 (June 26, 2009) (India), *available at* http://www.ipab.tn.nic.in/Orders/100-2009.htm.

beta crystal form usable in the making of an oral solid drug formulation for curing cancer."[110] Inventive step notwithstanding, the IPAB held that the application material still had to be evaluated to determine whether the claimed innovation is "not an invention under §3(d) of the Act," thus creating a new standard.[111] Gleevec was deemed to satisfy the inventive step threshold because of the several elements of technical advancement(s) that the Board identified. The interaction of Section 3 with the inventive step requirement is discussed further below, giving rise to the query whether the inventive step requirement is subject to Section 3 of the patents statute in India. Alternately, should fulfillment of Section 3 of the patents statute be considered a threshold question?

4. Can a Patent That Clears the Inventive Step Threshold Still Be "No Invention"?

Sections 3 and 4 of the Indian Patent statute set forth items that are "not inventions"—that is subject matter excluded from patentability notwithstanding the presence of the other elements of patentability. The *Valcyte* opinion, for instance, links the finding of inventive step with a finding of increased efficacy, typically a requirement for granting patents for new forms of known substances. This was a standard first enunciated in *Novartis* under Section 3(d) of the Indian Patent Act.[112] That is, after the opponents raised the issue that the application material lacked an inventive step, they further asserted that as a new form of a known substance, the application did not disclose an invention under Section 3(d). After exhaustively discussing the issue of inventive step, the grant opinion considered whether the application material also has to fulfill the Section 3(d) threshold—which in turn, requires proof of enhanced efficacy—to be eligible for a patent.[113]

The recent highly publicized case of Novartis AG v. Union of India[114] illustrates the nexus between Sections 2(j) and 3(d)—that is, the role of patentability standards in light of the exclusions outlined in Sections 3 and 4. Novartis's ill-fated patent application for a cancer drug named Gleevec,[115] although non-obvious,

---

110. *Id.*

111. *Id.*

112. *See* Novartis v. Union of India, [2009] P.F.R. 86 (I.P.A.B.) (India); *see also* Patent Act, 2005, *supra* note 8, §§ 14–16.

113. *See generally* F. Hoffmann-La Roche Ltd. v. Cipla Ltd., (2008) 37 P.T.C. 71 (Delhi H.C.) ¶ 47 (India) (stating that "[D]ata provided in the complete specification... is not sufficient to meet the requirement of significant enhancement of therapeutic efficacy") *see also id.* ¶ 49 ("Ester of known substance *prima facie* will get a patent only if it shows significant enhancement of efficacy.")

114. Novartis AG v. Union of India, (2009) 2 M.I.P.R. 345, 445 (India).

115. O'Connor, *supra* note 108, at 59.

was rejected under Section 3(d) of the amended Indian patent statute as a new form of a known substance.[116] Interestingly, Novartis's first application for the free base of Gleevec—Imatinib—issued in 1996.[117] Gleevec is the beta crystalline form of imatinib mesylate, which is a salt form of the free base.[118] An opposition was filed by Cipla, Ranbaxy, Natco Pharmaceuticals, and others, claiming that imatinib mesylate was anticipated by US Patent No: 5,521,184 and that the imatinib mesylate salt form existed inherently in the crystelline form.[119] Rejecting the argument of Novartis that the free base was chemically changed into a salt form by human intervention, the Patent Controller concluded that the application was obvious in light of the prior art.[120] Eventually, the patent application landed before the newly constituted IPAB to which all patent appeals are transferred.[121] The IPAB ruled that the application embodied an inventive step. However, the IPAB went on to add that:

> [T]he inventive step thus evaluated for a pharmaceutically active substance in particular, as the case herein, would finally have to be re-evaluated as to whether or not the claimed products viz. substances, compositions fall within the category of "not an invention under Section 3(d) of the Act which says a new form/derivative of a known substance will be considered same substance unless it displays significant enhancement in properties with regard to efficacy.[122]

In 2009, the Appellate Board dismissed the Novartis appeal again under Section 3(d) of the Indian patent statute. On the final appeal, the Supreme Court

---

116. *See* Patent Act, 2005, *supra* note 8, § 3(d).

117. Pyrimidine Derivatives and Processes for the Preparation Thereof, U.S. Patent No. 5,521,184 (filed Apr. 28, 1994 as a continuation–in-part of an abandoned application filed on Apr. 2, 1993) (issued May 28, 1996)

118. *See* Novartis AG v. Natco Pharma and Others, Controller of Patents and Designs, Indian Patent Office, Application No. 1602/MAS/1998 (Jan. 25, 2005) *available at* http://indiankanoon.org/doc/1352538 [hereinafter Novartis v. Natco].

119. Novartis v. Natco, *supra* note 118, ¶ 5.

120. *Id.* ¶¶ 6, 7.

121. *See* Notification No. 12/15/2006-IPR-III (2/4/2007), Ministry of Commerce & Industry, Government of India, *available at* http://ipindia.nic.in/ipr/patent/gazetteofindia_apr2007.pdf.

122. *See* Novartis v. Union of India & Others, IPAB Order No. 100/2009 (June 26, 2009), *available at* http://www.ipab.tn.nic.in/Orders/100-2009.htm. The IPAB also specifically stated that all patent applications should be "screened through the test of patentability under Sections 3 and 4 of the Act," which provides a list of materials not susceptible to patent protection in India. *Id. See also* Novartis AG v. Union of India, (2009) 2 M.I.P.R. 345, 445 (India).

clarified that the term "efficacy" in Section 3(d) relates to "therapeutic efficacy," proof of which would be weighed by strict and narrow standards.[123] In the case of pharmaceutical compounds, the Supreme Court noted that external and internal factors should clearly demonstrate therapeutic efficacy established by research data. Thus, "the mere change of form with properties inherent to that form would not qualify as 'enhancement of efficacy' of a known substance."[124] Though the court did not directly deliberate on whether the Section 3(d) issue is a threshold question, it did refer to it as "a second tier of qualifying standards" for pharmaceutical products, and noted that Section 3(d) in combination with the definition of invention in Section 2(1)(j) and (ja) raised the invention threshold higher.[125]

5. Obvious to Try, Proof of Progress, and TissueGene

The *Valcyte* opinion discusses whether *obvious to try* would be considered obvious in India. The discussion indicates that when the prior art makes the application obvious to try, there is no technical advancement or *progress* to demonstrate inventive step. Another interesting opinion that treats "obvious to try" as obvious relates to an application filed by M/s. Tissuegene, Inc., of the United States for "[a] method for production of a novel mammalian recombinant cell."[126] In considering the application, the examiner objected on the grounds that the subject matter lacked an inventive step in view of U.S. Patent No. 636605.[127] Consequently, the applicant amended the claims and restricted it to a method for preparation of a pharmaceutical composition for generation or regeneration of bone at a bone defective site.[128] The application method envisaged inserting a gene encoding a protein belonging to the Transforming Growth Factor-B (TGF-B) family having a bone regenerating function to encode Bone Marrow Protein (BMP-2) into a retroviral plasmid vector operatively linked to a promoter, and transfecting or transducing a population of mammalian fibroblasts, chondrocytes, or bone progenitor cells in vitro with the said recombinant vector.[129]

---

123. Novartis AG v. Union of India, Nos. 2706–16, slip op. at 90–91 (S.C. Apr. 1, 2013) (India), *available at* http://judis.nic.in/supremecourt/imgs1.aspx?filename=40212.

124. *Id.; see also* Srividhya Ragavan & Aju John, *Nothing Wrong with Setting High Standards of Patentability*, MYLAW.NET (May 10, 2013), *available at* http://www.mylaw.net/Article/Nothing_wrong_with_setting_high_standards_of_patentability/#.UljoIFCTiSo.

125. *Novartis*, slip op. at 56–57.

126. Indian Patent Application No. 2860/Delnp/2004 (filed Sept. 23, 2004).

127. Tissuegene Inc's Application, [2010] P.F.R. 4 (Del.) (India) (Jan. 19, 2010) [hereinafter TissueGene].

128. *Id.* at 1.

129. *Id.* at 4.

In determining whether there was an inventive step, the Controller looked at three references. Of these, the first reference taught (using a rabbit spine fracture model) a method of inserting a gene encoding BMP-2 into a retroviral vector operatively linked to a promoter and transfecting a population of connective cells in vitro with said recombinant vector.[130] The next reference disclosed that the administering cell population comprising the fibroblasts, which are transduced with BMP-2 to hind limb of the SCID mouse, results in bone generation.[131] The last reference taught a method for generating cartilage in a mammal comprising the steps of inserting a gene encoding TGF-B-1 into a recombinant vector, transfecting said vector into a fibroblast cell line, then transferring the population of fibroblast to the cartilage defect site in vivo, wherein the cartilage generated is disclosed.[132]

When the prior art was read together, the Controller determined that the application was obvious. In denying the presence of an inventive step, the grant opinion states that "[i]n view of [the prior art] it would be obvious for a person skilled in the art to try to use fibroblast cell population to deliver BMP-2 to the bone defect site because optimal delivery is one of the key aspects in gene therapy."[133] Elsewhere the opinion states that "[i]t would be an obvious choice to try because Musgrave et al. [the second reference] teaches that the fibroblast population has optimal delivery capacity and Noh et al. demonstrated the successful delivery of TGF-B-1 in vivo to generate cartilage."[134] Thus, it would not be unpredictable—in fact, it would be an *obvious choice to try*—to get results in bone formation by increased delivery of BMP-2 through the claimed composition.

TissueGene is supported by several opinions in India that equate obvious to try as lacking inventive step. For instance, a pre-grant opposition that Cipla,[135] the famed Indian generic drug company filed against Novartis's patent application for *Dispersible Tablets Comprising Deferacirox*,[136] notes that a "[m]ere selection or choice of the excipient which is available in the art and manipulating the range of the excipient [to suit the API] for preparing [a] dispersible tablet is routine experiment and it is not inventive," again suggesting that obvious to try would fail the inventive step test in India.[137]

---

130. *See* 68 Calcified Tissue Int'l 87–94 (2001).

131. TissueGene, *supra* note 127, ¶ 5.

132. U.S. Patent No. 6,315,992 (filed June 30, 1999).

133. TissueGene, *supra* note 127, ¶ 5.

134. *Id.* ¶ 8

135. Novartis v. Cipla, Pre-grant opposition (on file with author).

136. Indian Patent Application No. 593/CHENP/2005 (filed Apr. 11, 2005).

137. Novartis v. Cipla, Pre-grant opposition, *supra* note 135, ¶ 20.

III. A Comparative Study of the Threshold of Inventiveness

The inventive step requirement in India is seemingly more definitive and robust in application than in the United States. Indeed, at the time India started to implement process patents, several of the so-called generic drugs in India arguably would have been able to secure a patent in the United States because the generic drugs—derogatorily referred to as *copy cats*—more often than not, presented a change that would qualify as being non-obvious in the United States.[138] In short, India has traditionally been a country with a higher standard of inventiveness than in the United States where different modern claiming techniques have allowed applications distinguished by a *single* claim limitation from an existing patent to be awarded different and independent patents.[139]

To provide a background of the current standard of non-obviousness in the United States, it is important to appreciate how that country moved from the *Graham* test to the TSM (teaching, suggestion, and motivation) test. The watershed moment, according to Professor John Duffy, perhaps occurred in 1984 when the Federal Circuit, in *ACS Hospital System, Inc. v. Montefiore Hospital*, enunciated the test that laid the groundwork for the teaching, suggestion, or motivation test.[140] The court held that teachings of prior art references could be combined to prove obviousness only if there is some teaching, suggestion, or motivation to do so in the prior art.[141] By 1985, the rule been elevated to a standardized prescription from which examiners were unable to derogate on account of a Federal Circuit decision that prevented district court(s) from holding a patent invalid for non-obviousness unless it "elucidate[d]" "factual teachings, suggestions or incentives from th[e] prior art that show[] the propriety of [the patented] combination."[142]

That standard would be celebrated as *the* TSM (teaching, suggestion, and motivation) test, which became an important procedural tool to allocate the burden to prove non-obviousness going forward.[143] In sum, an examiner discharges

---

138. *See* Srividhya Ragavan, *A "Patent" Restriction on Research and Development: Infringers or Innovators?*, 1 Ill. J.L. Tech. & Pol'y 73 (2004), *available at* http://www.soonerlawyeronline.org/faculty/facfiles/patent_restriction.pdf.

139. *Id.* at 89–94.

140. 732 F.2d 1572, 1577 (Fed. Cir. 1984).

141. *Id.* at 1577.

142. John F. Duffy & Robert P. Merges, Patent Law and Policy: Cases and Materials 664 (4th ed. 2007); *see also* Ashland Oil v. Delta Resins & Refractories, 776 F.2d 281, 297 (Fed. Cir. 1985).

143. *See* U.S. Patent & Trademark Office, Manual of Patent Examining Procedure, Patent and Trademark Offices § 2142 (7th ed. 1998).

the initial burden of proving that a patent application is prima facie obvious only if there is a teaching, suggestion, or motivation of or for the claimed combination in the prior art. However, this test ostensibly eliminated the need for the most critical element—common sense—that an examiner would use to make the obviousness determination. Although the test takes credit for eliminating hindsight rejections, it clearly permits blind allowance of patent applications. Basically, as long as a combination of prior art does not overtly show a teaching, suggestion, or motivation of all of the claim limitations, the examiner can clear the patent application as being prima facie non-obvious; the application would clear the non-obvious threshold even when there is a strong indication that it may in fact be obvious to a person of ordinary skill in the art. The standard resulted in creating a distinction between legal obviousness and common-sense–based obviousness. That is, what would be obvious to a person of ordinary skill in the art or even to a person of no skill in the art became non-obvious in law until the Supreme Court intervened in *KSR v. Teleflex*[144] to impose on the patent community the use of common sense in determining non-obviousness. *In Re Dembiczak*, which held a Halloween designed trash bag to be an invention on the grounds that there was no prior art showing a "clear and particular" teaching to use all of the claim limitations (namely, the use of a plastic bag, in pre-manufactured orange with facial indicia) is an outstanding example to demonstrate the point.[145] *Dembiczak* was by no means an aberration but instead, was part of a steady stream of cases where the line between obvious and non-obvious was determined by what was typecast in the prior art as opposed to the perspective of a person of ordinary skill in the art. Consequently, the United States became and unfortunately, remains a jurisdiction where it is relatively easy to traverse the Section 103 requirement.

In any event, the quality of this threshold requirement for patent protection has been a subject of concern in the United States. In 2006, Justice Kennedy lamented specifically with reference to the vagueness and the suspect validity of business method patents, noting that a "[s]ignificant problem that plagues business method patents is that they tend to be of poor overall quality."[146] Judge Mayer of the Federal Circuit, in 2008, in a dissent referenced the low threshold for patent eligibility by referring to patents granted in the wake of *State Street,* as ranging from "the somewhat ridiculous to the truly absurd."[147] Although the insights were in reference to the quality of patents flowing from patent subject matter eligibility

---

144. 175 F.3d 994 (1999).

145. *Id.*

146. *See* eBay Inc. v. MercExchange, LLC, 547 U.S. 388, 397 (2006) (Kennedy, J., joined by Stevens, Souter, and Breyer, JJ., concurring).

147. *In re* Bilski, 545 F.3d 943, 1004 (Fed. Cir. 2008).

standards, it is interesting to note that Roche's Valcyte—rejected for patent protection in India—was able to secure a patent in the United States.[148] Indeed, in the United States, "new forms of a known substance" such as metabolites, salts, polymorphs, or esters of existing compounds (specifically not patentable under Section 3(d) of the Indian statute as embodying an inadequate inventive step to qualify for a separate patent) remain protectable. For instance, in *Schering Corp. v. Geneva Pharmaceuticals*, the court noted that "[t]his court's conclusion... does not preclude patent protection for metabolites of known drugs. With proper claiming, patent protection is available for metabolites of known drugs."[149]

Notwithstanding the above, it is common knowledge that each jurisdiction tailors its individual requirement to determine the constituents of an inventive step. For instance, a problem–solution approach is advocated under the European Patent Convention to determine the presence of an inventive step. The system consists of three steps:

(i) determining the "closest prior art";
(ii) establishing the "objective technical problem" to be solved and;
(iii) considering whether or not the claimed invention, starting from the closest prior art and the objective technical problem, *would* (as opposed to *could)* have been obvious to the skilled person.[150]

Notably, the question is whether the skilled person would have arrived at the application material because the prior art prompted him to do so.[151] In Europe, the standard is that an *implicit prompting or an implicitly recognisable incentive* is adequate to justify the showing that a skilled person would have combined the elements from the prior art.[152] Unlike the United States, the evidentiary burden on the first responders (examiners) in Europe is lesser, thereby allowing examiners the ability to use more of their subjective faculties in determining the presence of an inventive step.

Like Europe, India's seeming imposition of a higher burden on the applicant is well within the country's rights and is very much in alignment with

---

148. U.S. Patent No. 6,083,953 (filed Mar. 4, 1997).

149. Schering Corp. v. Geneva Pharms., 339 F.3d 1373, 1381 (Fed. Cir. 2003); *see also In re* Kratz, 592 F.2d 1169, 1174 (C.C.P.A. 1979) (noting that a naturally occurring strawberry constituent compound does not anticipate claims to the substantially pure compound).

150. European Patent Office, Guidelines for Examination in the European Patent Office pt. G, ch. VII, § 5 (2013), *available at* http://www.epo.org/law-practice/legal-texts/html/guidelines/e/g_vii_5.htm.

151. *See* Case T 2/83, Simethicone Tablet, [1984] O.J. (EPO) 265 (Mar. 15, 1984).

152. *See* Case T 0257/98, Lipase-containing Detergent Composition/UNILEVER (Sept. 9, 2002); *see also* Case T 0035/04, Paper Surface Sizing/AVEBE, (Jan. 18, 2006).

the requirements under the TRIPS Agreement.[153] Notably, the standard of non-obviousness is one of the filters nations use to delineate the private and the public domains. The United States has traditionally used novelty as one such filter; Section 102 (b) for instance, filtered out materials that have been on sale or in public use anywhere in the world, save for the United States.[154] For any country, the use of such filters is important to ensure a strong balance between the private and public realms—India is no exception. Given the financial and organizational constraints faced by the patent offices of the developing countries, having a higher patentability standard may in fact be an appropriate approach for at least two reasons: first, the higher patentability standard combined with the requirement that the applicant demonstrate *proof of progress* shifts part of the cost of search from the patent office to the applicant. Second, higher patentability standards in a patent system with multilevel challenge mechanisms, such as the pre- and post-grant oppositions, increases the possibility of validity challenges, thereby addressing the information asymmetry problems characteristic of patent offices. In fact, in countries such as India, the quest for innovation and development has to be carefully balanced with existing social issues;[155] creating such a carefully planned balance is not just a constitutional requirement, but is necessitated by several factors described below:

1. Given the social and economic issues that are unique to India, it makes sense to install a higher invention bar to justify a private domain by way of patent protection. Further, as India transitions into a full-fledged patent regime, it is well worth remembering Justice Breyer's cautionary note in *Laboratory Corporation v. Metabolite*:[156] "[S]ometimes [the] presence [of patents] can discourage research..., by requiring complex licensing arrangements, and by raising the costs of using the patented information, sometimes prohibitively so."[157] He advocates that patent law should carefully seek to avoid the dangers of overprotection just as surely as it should avoid the diminished incentive

---

153. Agreement on Trade-Related Aspects of Intellectual Property Rights, art. 27, Apr. 15, 1994, Marrakesh Agreement Establishing the World Trade Organization, Annex 1C, 33 L.L.M. 1197 (1994) ("Subject to the provisions of paragraphs 2 and 3, patents shall be available for any inventions, whether products or processes, in all fields of technology, provided that they are new, involve an inventive step and are capable of industrial application.").

154. *See* 35 U.S.C. § 102(b) (2012).

155. *See generally* INDIA CONSTITUTION, *supra* note 80, pmbl.

156. Lab. Corp. of Am. Holdings v. Metabolite Labs., Inc., 548 U.S. 124 (2006) (Breyer, J., joined by Stevens & Souter, JJ., dissenting).

157. *Id.*

from under protection.[158] As India moves from an under-protection regime to a sophisticated patent system, the end of the exercise is more innovation (as opposed to more granted patents), and this should dictate the trajectory of patent policies.

2. Instead of allowing the market to act as the gatekeeper, thereby depending on infringement and patent validity suits to weed out patents whose inventiveness has not been subject to rigorous scrutiny at the patent office, there is arguably a much higher benefit to the public if the first responders—the patent office—weeds out applications that cannot cross the inventive step threshold. In India, litigation remains expensive, and tends to be time-consuming and beyond the reach of many local companies and the average citizen. *Pfizer's* experience with its famed US patent on amlodipine besylate (sold under the trademark Norvasc®) demonstrates the point.[159] The question of obviousness arose in the light of Pfizer's own earlier patent on amlodipine,[160] and two other patents that disclosed the besylate acid and its use in a pharmaceutical compound.[161] During prosecution, Pfizer successfully traversed a prima facie obviousness rejection at the patent office by filing a Rule 1.132[162] declaration, affirming that the besylate salt of amlodipine is a unique compound because each salt form exhibits unique properties. When Apotex sought approval to market generic amlodipine besylate tablets, Pfizer sued Apotex, which prompted Apotex to counterclaim arguing patent invalidity. In concluding that Pfizer's patent was obvious, the Federal Circuit determined that the suggestion, teaching, or motivation to combine the relevant prior art "does not have to be found explicitly in the prior art references sought to be combined," but rather "may be found in any number of sources, including common knowledge, the prior art as a whole, or the nature of the problem itself."[163]

The later invalidation of the patent on the salt form, after weighing the prior art in light of common knowledge, was a precursor to the common sense standard that that Supreme Court would later enunciate in *KSR v. Teleflex*.[164] For a country such as India, it would serve the patent regime well if such

158. *Id.*

159. Pfizer, Inc. v. Apotex Inc., 488 F.3d 1377 (Fed. Cir. 2007); Pfizer, Inc. v. Apotex, Inc., 488 F.3d 1377 (Fed. Cir. 2007).

160. U.S. Patent No. 4,572,909 (filed Feb. 3, 1984).

161. U.S. Patent No. 3,816,612 (filed Feb. 10, 1970); U.S. Patent No. 4,032,637 (filed Oct. 31, 1974).

162. 37 C.F.R. § 1.132.

163. *Pfizer*, 480 F.3d at 1362.

164. 127 S. Ct. 1727 (2007).

a patent is rejected by the first responders rather than after several years in litigation as in the case of Pfizer. Mechanisms such as the pre-grant opposition not only provide the patent office with an opportunity to scrutinize inventiveness, but they also promote innovation by providing competitors an opportunity to challenge a patent application before the grant and thus, serve as a check on the grant of unjustified exclusive rights. Given that litigation is often protracted, allowing opponents to cite the relevant prior art and disclose industry standards during opposition uniquely transforms competitors and opponents in India into amici providing useful and valuable information and assisting in the examination process, albeit for their own benefit. This again raises the threshold for patents granted in India, thereby creating a unique set of checks and balances to the scrutiny of patent applications.

3. The standard for determining inventive step, which has arguably increased the burden on the applicant, compliments India's current position on Section 3,[165] which requires proof that an invention is deserving of monopoly privileges, or it simply will be deemed as "no invention." Further, the standard complements the Indian judicial sentiment that has repeatedly favored higher standards of invention as a balancing mechanism.

Conclusion

The higher burden of requiring a showing of technical advancement and/or economic significance as a demonstration of the proof of progress is distinctive to India. Notably, the proof of progress requirement has contributed to enhancing the disclosure requirement for patent applicants. Not only does a prospective applicant have to mention the closest prior art, but he or she also has to show that there is a non-obvious technical advancement over the cited prior art. This heightened threshold is a pervasive aspect of the Indian patent system. As previously noted, under Section 3(d), application materials involving a new form of a known substance should demonstrate enhanced efficacy, that is, efficacy that is *significantly* different from what is known in the prior art, in order to clear the patentability threshold. Arguably, the higher standard of inventiveness in India makes it harder to get patents. This fact has to be viewed in light of another interesting development—enforcing a patent is also subject to stricter standards in India. For instance, in *F. Hoffmann-La Roche Ltd. v. Cipla Ltd.*,[166] involving a patent over the lung cancer drug Erlotinib, the Delhi High Court refused to grant an interlocutory injunction restraining the defendant from manufacturing

---

165. *See* Patent Act, 2005, *supra* note 8, § 3.

166. (2008) 37 P.T.C. 71 (Delhi H.C.) (India).

and marketing the drug in India, on the grounds that the defendant raised a *credible challenge to the patent.* The decision was upheld on appeal by the Division Bench of the Delhi High Court.[167] Similarly, the Supreme Court in *Bajaj Auto Ltd. v. TVS Motor Co. Ltd.*[168] held that infringement suits relating to patents and other intellectual property rights should be decided expeditiously by the trial court instead of concentrating on the interlocutory reliefs. Thus, where there is a credible challenge to the validity of the patent, the balance in India skews toward minimizing the unfairness to the public domain should the patent be invalidated in the future.[169] Hence, there is a preference for a final decision rather than providing interlocutory relief. At the end of the day, as a jurisdiction where it is hard to get a patent and even harder to sustain it, India seems to be sending a strong message: an invention in India is something to be earned.

167. F. Hoffmann-La Roche Ltd. v. Cipla Ltd., (2009) 40 P.T.C. 125 (Delhi H.C.) (India).

168. (2009) 41 P.T.C. 398 (S.C.) (India).

169. *See* Bilcare Ltd. v. Supreme Indus., [2007] P.F.R. 2 (Delhi H.C.) ("No presumption of validity attaches to a patent granted by the Controller under the Act notwithstanding examination and investigation made under Sections 12 and 13."). The Patent Act, 2005, *supra* note 8, § 113, discusses obtaining a certificate of validity of the specification to determine costs in a suit—which indicates that not all granted patents are automatically presumed valid. *Id.*

19 PHARMACEUTICAL PATENT ENFORCEMENT: A DEVELOPMENTAL PERSPECTIVE

Shamnad Basheer, Jay Sanklecha and Prakruthi Gowda*

Introduction

The advent of the Agreement on Trade-Related Aspects of Intellectual Property Rights[1] (TRIPS Agreement) triggered a number of scholarly papers on a theme popularly referred to as "intellectual property (IP) and development."[2] This "developmental" clarion call peaked to such an extent that the World Intellectual Property Organization (WIPO) was forced to recalibrate its approach and adopt a Developmental Agenda premised, inter alia, on the notion that the standards of intellectual property protection cannot be uniform, but ought to correspond with an individual nation's socioeconomic status and technological capabilities.[3]

Although standards for the grant of intellectual property rights often take center stage in the literature on intellectual property and

---

* Shamnad Basheer is the Ministry of HRD Professor of IP Law at WB NUJS. Jay Sanklecha and Prakruthi Gowda are recent graduates of WB NUJS. We wish to thank Prashant Reddy, Jaideep Reddy, Mihir Naniwadekar, and Abira Gupta for extensive help with this piece.

1. Agreement on Trade-Related Aspects of Intellectual Property Rights, Apr. 15, 1994, Annex 1C, 33 I.L.M. 1197 (1994) [hereinafter TRIPS Agreement].

2. *See generally* THE DEVELOPMENT AGENDA: GLOBAL INTELLECTUAL PROPERTY AND DEVELOPING COUNTRIES (Neil Weinstock Netanel ed., 2009); Jeremy De Beer & Sara Bannerman, *Foresight into the Future of WIPO's Development Agenda*, 1 WIPO J. (2010), *available at* http://papers.ssrn.com/sol3/papers.cfm?abstract_id=1726153.

3. At the 2007 General Assembly, the member states of the World Intellectual Property Organization (WIPO) adopted 45 recommendations (of a total of 111 proposals) made by the Provisional Committee on Proposals Related to a WIPO Development Agenda. A Committee on Development and Intellectual Property has been established for the implementation of the 45 adopted recommendations.

development,[4] intellectual property enforcement is largely ignored. This chapter seeks to fill this gap, albeit to a limited extent, by focusing on the standards for the grant of injunctions in patent infringement suits.

This is particularly relevant, as a number of developing countries, such as India, are faced with burgeoning patent disputes:[5] disputes that have enormous implications for the future of innovation and the issue of access to patented goods, notably pharmaceuticals. Given that interim injunctions are largely dispositive of intellectual property disputes in many cases, this chapter focuses largely on such injunctions.

The standards for the grant of injunctions ought to be calibrated in a manner that appropriately balances the interests of the patentee in securing timely and effective enforcement of her rights with the public interest in guarding against erroneous injunctions (i.e., where the patent turns out to be invalid or not infringed after trial). Such wrongly granted injunctions harm not only competitors against whom they are granted, but also consumers who are forced to pay a monopoly price during the subsistence of the injunction/restraining order.

It would appear that much in line with US and UK jurisprudence, the grant of a temporary restraining order (or injunction) in India flows from "equity"[6] and is conditioned upon the satisfaction of the following prerequisites:

1. there is a prima facie case in favor of the plaintiff;
2. the plaintiff is likely to suffer an irreparable injury if the defendant is not restrained;

See Proposal by Argentina and Brazil for the Establishment of a Development Agenda for WIPO, WIPO Doc. WO/GA/31/11 (Aug. 27, 2004), *available at* http://www.wipo.int/edocs/mdocs/govbody/en/wo_ga_31/wo_ga_31_11.doc; WIPO, Assemblies of the Member States of WIPO, Forty-Third Series of Meetings, General Report Adopted by the Assemblies, Geneva, September 24 to October 3, 2007, WIPO Doc. A/43/16 (Nov. 17, 2007).

4. *See, e.g.*, Amy Kapczynski, *Harmonization and Its Discontents: A Case Study of the TRIPS Implementation in India's Pharmaceutical Sector*, 97 Calif. L. Rev. 1571 (2009).

5. A study conducted by us reveals that a total of eighty-five patent infringement litigations were decided by courts up until July 2012. It must be noted that most of these were interim decisions, with only two or three having gone through trial. Of these, the number of decisions from 1914 till 1995 was only thirty. The number jumps dramatically in the post-TRIPS era. From 1995 till 2012, Indian courts decided fifty-six cases. Between 2005 and 2012 (post–pharmaceutical product patent era), they decided forty-three cases. The data set for the purposes of the above-mentioned study included only cases reported in the Manupatra online database (www.manupatra.com).

6. *See* N.R. Dongre v. Whirlpool Corp., (1996) 5 S.C.C 714, 716 ("Injunction is a relief in equity and is based on equitable principles."). On the "equity" origins of preliminary injunctions in English Chancery Courts, see Morton Denlow, *The Motion for a Preliminary Injunction: Time for a Uniform Federal Standard*, 22 Rev. Litig. 495, 500 (2003) ("A preliminary injunction is an equitable remedy derived from the principles of the system of judicial remedies devised and administered by the English Court of Chancery").

3. the balance of convenience lies in favor of the plaintiff; and
4. public interest would not be disserviced by the grant of the injunction.

None of these factors has been explicated in a conceptually sound manner by Indian courts. Illustratively, it is not clear what the appropriate standard for determining the establishment of a prima facie case is.[7] This is particularly problematic, because an interim injunction has a potentially "draconian" effect[8] on competitors, who are forced to desist from manufacturing and selling the allegedly infringing good, as well as consumers, who are denied access to cheaper goods. This latter concern is most starkly felt in the pharmaceutical context, where the relevant intellectual property good in question could be a life-saving drug.

This chapter attempts to delineate the prima facie standard, keeping in mind the level of deference that ought to be accorded to decisions of the patent-granting authority, namely the Indian Patent Office. More important, it advocates dispensing with the interim phase insofar as complex patent disputes (where a credible challenge is lodged against the validity of the patent or the factum of infringement) are concerned, as it is very difficult in such cases to arrive at a robust "prima facie" determination—one of the cardinal prerequisites for the grant of a temporary injunction.

We recommend that, when faced with a complex patent dispute where it is difficult to legitimately assess the strength of each party's case at the interim stage and effectively predict who is more likely to win at trial, courts move directly to the trial stage—a suggestion that is coming to be increasingly adopted by the Indian Supreme Court.[9] We argue that this is a TRIPS flexibility that developing countries, such as India, can legitimately exploit. We also highlight issues of institutional capacity and ask: Should developing countries such as India institute specialist intellectual property courts to decide patent infringement suits? Would this make for more optimal intellectual property adjudication? Though specific to India, most of the suggestions in this chapter could prove useful for a number

---

7. *See* Power Control Appliances v. Sumeet Machs. Pvt. Ltd., (1994) 2 S.C.C 448. In this trademark and copyright infringement case, the Indian Supreme Court discussed the prima facie standard: "This does not mean, however, that the Court should examine in detail the facts of the case and anticipate or prejudice the verdict which might be pronounced after the hearing of the suit or that the plaintiff should make out a case which would entitled him at all events to relief at the hearing." *Id.* at 463. The Supreme Court subsequently spoke of "a strong prima facie case" in a 2001 trademark decision in *Uniply Indus. Ltd. v. Unicorn Plywood Pvt. Ltd.*, (2001) 5 S.C.C 95.

8. Michael Burdon & Kristie Sloper, *Granting Emergency Interim Injunction Pending Trial,* 2004 J. Generic Meds. 167.

9. *See* Bajaj Auto Ltd. v. TVS Motor Co. Ltd., (2009) 41 P.T.C. 398 (S.C.) (India).

of other developing countries, particularly those that follow common law and are yet to experience a significant number of patent infringement cases.

This chapter is structured as follows. In Section I, we introduce the concept of injunctions (both temporary and permanent) and discuss the standards for establishing a prima facie case. In Section II, we discuss *Roche v. Cipla*, an Indian patent infringement case that supports our thesis that given the complexity of patent cases and the relative inexperience of Indian judges in adjudicating them, it is far more optimal to dispense with the interim injunction phase.[10] In Section III, we recommend dispensing with the interim phase and moving directly to the trial stage. We also address the various issues stemming from such a dispensation and propose broad policy reforms to ensure a speedy and effective enforcement of patent rights. Finally, in Section IV, we explore the various flexibilities under the TRIPS regime that enable member states to fashion their injunction jurisprudence to serve important policy goals.

I. Patent Injunctions

"*Ubi jus ibi remedium*" is a Latin phrase that roughly translates to "*where there is a right, there must be a remedy.*"[11] In keeping with this cardinal jurisprudential tenet, most legal regimes provide for remedies, both civil and criminal, to prevent patent infringement. Civil remedies could be further divided into injunctions and damages.[12] Of these remedies, an injunction to restrain the defendant from engaging in the allegedly infringing act is the more potent one. In the words of the US Court of Appeals for the Federal Circuit, "without the right to obtain an injunction, the right to exclude granted to the patentee would have only a fraction of the value it was intended to have, and would no longer be as great an incentive to engage in the toils of scientific and technological research."[13]

---

10. Hoffman-La Roche Ltd. v. Cipla Ltd., (2008) 37 P.T.C. 71 (Delhi H.C.) (India) (hereinafter referred to as "Single Judge" or "Trial Court" order). This order was then appealed to a Division Bench of the Delhi High Court. Hoffman La-Roche Ltd. v. Cipla Ltd., (2009) 40 P.T.C. 125 (Delhi H.C.) (India) (hereinafter referred to as "Division Bench" or "Appellate Court" order).

11. Black's Law Dictionary 1520 (6th ed. 1990).

12. For a useful comparison of the patent remedies available in G7 countries, see Larry Coury, Note, *C'est What? Saisie! A Comparison of Patent Infringement Remedies Among the G7 Nations*, 13 Fordham Intell. Prop. Media & Ent. L.J. 1101 (2003); *see also* Thomas F. Cotter, Comparative Patent Remedies: A Legal and Economic Analysis (2013). It is pertinent to note that many countries provide for criminal penalties as well. However, this chapter focuses only on civil remedies.

13. Smith Int'l, Inc. v. Huges Tool Co., 718 F.2d 1573, 1578 (Fed. Cir. 1983); *see also* B. Webster & S. Walmley, *Unclean Hands and Preliminary Injunctions: The Effects of Delay in Bringing Patent Infringement Cases*, 84 J. Pat. Trademark Off. Soc'y 291 (2002).

It is important to note that there are two kinds of injunctions: a temporary or preliminary injunction issued soon after the institution of a suit,[14] and a permanent or final injunction issued after trial. A temporary injunction is an extraordinary relief devised to maintain the status quo and protect the interests of the patentee during the course of the trial.[15] It is customarily granted on the basis of abridged procedures and evidentiary requirements that are less formal than those associated with a full-fledged trial.[16]

A temporary injunction in a patent suit is predicated on the assumption that if the patentee were to ultimately win at trial, the non-grant of a temporary injunction would detrimentally impact the patentee and her rights in the interim.[17] The need for such injunctions is particularly acute in developing economies, where patent infringement suits often take years to litigate, and the lack of an injunction at the interim phase all but divests the patentee of a validly granted statutory monopoly. A permanent injunction, on the other hand, is awarded only after trial concludes and the patent is found to be valid and infringed.[18]

---

14. The law relating to injunctions in India is contained in the Specific Relief Act, 1963, and the Code of Civil Procedure, 1908. In accordance with Section 37(1) of the Specific Relief Act, temporary injunctions are regulated by the Code of Civil Procedure. Section 94(c) of the Code of Civil Procedure provides that in order to prevent the ends of justice from being defeated, the Court may grant temporary injunctions. Order 39, Rule 1, of the Code of Civil Procedure lays down the cases in which a temporary injunction may be granted. These are: (1) any property in dispute in a suit is in danger of being wasted, damaged, or alienated by any party to the suit, or wrongfully sold in execution of a decree; (2) defendant threatens, or intends to remove or dispose of his property with a view to defrauding his creditors; or (3) defendant threatens to dispossess the plaintiff or otherwise cause injury to the plaintiff in relation to any property in dispute in the suit.

15. *See* Wander Ltd. v. Antox India Pvt. Ltd., 1990 (Supp.) S.C.C. 727, para. 5:

> [U]sually, the prayer for grant of an interlocutory injunction is at a stage when the existence of the legal right asserted by the plaintiff and its alleged violation are both contested and uncertain and remain uncertain till they are established at the trial on evidence. The Court, at this stage, acts on certain well settled principles of administration of this form of interlocutory remedy which is both temporary and discretionary. The interlocutory remedy is intended to preserve in status quo, the rights of parties which may appear on a prima facie case....

See also Gujarat Bottling Co. Ltd. v. Coca Cola Co., 1995 A.I.R. 2372 (S.C.) (India). For a US perspective, see *Cordis Corp. v. Medtronic, Inc.* 835 F.2d 859, 863 (Fed. Cir. 1987), which held that the purpose of a temporary injunction is "to preserve the status quo and protect the respective rights of the parties pending a determination on merits."

16. *See Cordis*, 835 F.2d at 859; *see also* Univ. of Tex. v. Camenisch, 451 U.S. 390, 395 (1981).

17. *See* Webster & Walmley, *supra* note 13, at 292.

18. *See* Cotton Corp. of India v. United Indus. Bank, (1983) 3 S.C.R. 962, 969 (India):

> Temporary injunctions are such as are to continue until a specified time, or until the further order of the Court, and they may be granted at any stage of a suit, and are regulated

A number of common law jurisdictions, such as the United States and the United Kingdom, provide that a temporary injunction ought to be granted only if the plaintiff proves the following to the satisfaction of the court:[19]

1. A prima facie case in favor of the plaintiff;
2. That in the absence of an injunction, the plaintiff would suffer irreparable injury; and
3. That the balance of convenience lies in favor of the plaintiff.[20]

In addition to the above, some countries (e.g., the United States) also provide for a fourth requirement—that an injunction ought not to issue if it contravenes the public interest.[21]

At the stage of grant of a permanent injunction, the laws of the United States (and some other countries[22]) stipulate that the above factors remain the same, barring the requirement of a prima facie case. The reason is obvious: given that

by the Code of Civil Procedure, 1908. Permanent injunctions can only be granted by the decree made at the hearing and upon merits of the suit and thereby defendant in the suit is perpetually enjoined from assertion of a right or from commission of an act, which would be contrary to the rights of the plaintiffs.

19. This is generally the typical case across several jurisdictions. *See* Hybritech Inc. v. Abbott Labs., 849 F.2d 1446, 1451 (Fed. Cir. 1988); *see also* Am. Cyanamid v. Ethicon Ltd., [1975] A.C. 396 (H.L.) 407–08 (U.K.).

20. The inconvenience occasioned to the plaintiff as a result of the non-grant of an injunction would be greater than the inconvenience occasioned to the defendant if the injunction were to be granted.

21. *See* eBay, Inc. v. MercExchange, LLC, 547 US 388, 392 (2006). Indian courts appear to recognize this fourth factor. *See* Mahadeo Savalvam Shelke v. Pune Mun. Corp., (1995) 3 S.C.C. 33; Shiv Kumar Chadha v. Mun. Corp. of Delhi (1993) 3 S.C.C. 161; Tata Sons Ltd. v. Greenpeace Int'l, (2011) 178 D.L.T. 705; *see also* Novartis AG v. Mehar Pharma, (2005) 3 Bom. C.R. 191, 201 ("If the defendants are also restrained from manufacturing and marketing their anti-cancer drug in the market, it would cause great prejudice to public health and public interest and create a grave public health crisis with disastrous consequences. . . ."). However, some cases appear to treat "public interest" not as a separate, stand-alone factor, but as an important ingredient of the "balance of convenience" factor. *See* Hoffman-La Roche Ltd. v. Cipla Ltd., (2008) 37 P.T.C. 71 (Delhi H.C.) (India) (citing Franz Xaver Huemer v. New Yash Eng'rs, 1997 A.I.R. 79, 81 (Delhi H.C.) (India)) ("Balance of convenience also has an important role. Stultification of the defendant's investment, loss of employment, public interest in the product (such as a life saving drug), product quality coupled with price, or the defendant being smaller in size may go against the plaintiff."). It bears noting that the United Kingdom does not recognize this as a separate ground. *See generally* McDermott, Will & Emery, EUR. IP BULL. July 2006, *available at* http://www.mwe.com/info/news/euroip0706.pdf.

22. *See, e.g.*, Ireland (where though public interest is not a criterion, the equitable principles in the determination of temporary and permanent injunctions remain the same), https://www.aippi.org/download/commitees/219/GR219ireland.pdf; Malaysia, https://www.aippi.org/

the plaintiff has proved her case in "full," there is no need for a prima facie assessment of her case.

The United Kingdom takes a different approach in that the plaintiff does not need to establish any of these prerequisites when requesting a permanent injunction. Barring exceptional circumstances where the effects of the injunction are likely to be oppressive—that is, the effect of the injunction would be "grossly disproportionate"[23] to the intellectual property right protected—UK courts grant an injunction upon being satisfied that the patent is valid and infringed.[24]

According to India's patent legislation, "the reliefs which a court may grant in any suit for infringement include an injunction (subject to such terms, if any, as the court thinks fit) and, at the option of the plaintiff either damages or an account of profits."[25] This provision leans toward the UK approach, wherein proof of infringement is sufficient to obtain an injunction. [26]

In the next part, we focus on the "prima facie" prerequisite and argue that developing countries, such as India, ought to adopt a more rigorous threshold for a determination on this count than do developed countries such as the United Kingdom.

A. Pegging the Prima Facie Standard

The prima facie requirement is one of the essential prerequisites for the grant of a temporary injunction. For a large part of the twentieth century, English courts routinely held that a prima facie case was established only when the court undertook a relative assessment of the strength of each party's case and found that the plaintiff was more likely to win at trial (hereafter referred to as the "relative

---

download/commitees/219/GR219mexico.pdf; Mexico, https://www.aippi.org/download/commitees/219/GR219mexico.pdf; New Zealand, https://www.aippi.org/download/commitees/219/GR219ireland.pdf. All of the above resources are responses from National Groups to the Working Committee of the "Injunctions in cases of infringement of IPRs."

23. UK courts have pointed out that the use of the word "grossly" "avoids any suggestion that all that has to be done is to strike a balance of convenience." Virgin Atl. v. Premium Aircraft, [2009] EWCA (Civ) 1513, ¶¶ 24–25.

24. *See generally* Nicola Dagg, *Injunctions in English Patent Litigation—A Big Thing or Small, a Winner Takes It All*, WHO'SWHOLEGAL (May 2010), http://www.whoswholegal.com/news/features/article/28371/injunctions-english-patent-litigation-big-thing-small-winner-takes-all; Lorna Brazzel at al., *Injunctions in Cases of Infringement of IPRs*, AIPPI (Apr. 25, 2011), https://www.aippi.org/download/commitees/219/GR219united_kingdom.pdf.

25. Patent Act, § 108(1), No. 39 of 1970, INDIA CODE (2005) [hereinafter India Patent Act].

26. *See* Farbewerke Hoechst v. Unichem Labs., 1969 A.I.R. 255 (Bom.) (India); Mahesh Gupta v. Tej Singh Yadav, (2009) 41 P.T.C. 109 (Del.) (India).

assessment" standard). However, in 1975, Lord Diplock effectively watered down the threshold by suggesting that the plaintiff simply had to demonstrate a "serious issue to be tried" and that his case was not frivolous or vexatious:

> Your Lordships should in my view take this opportunity of declaring that there is no such rule. The use of such expressions as "a probability," "a prima facie case," or "a strong prima facie case" in the context of the exercise of a discretionary power to grant an interlocutory injunction leads to confusion as to the object sought to be achieved by this form of temporary relief. The court no doubt must be satisfied that the claim is not frivolous or vexatious, in other words, that there is a serious question to be tried. . . .[27]

As evident, merely proving the existence of a "serious issue to be tried" (hereafter "triable issue"[28] standard) is a lower threshold than proving that, on a relative assessment of the strength of each party's case, the plaintiff is more likely to win.

Lord Diplock's support for a lower threshold emanated from his concern that a temporary injunction phase is not inherently suited to a rigorous assessment of arguments and evidence and that any such effort could risk converting the interim proceeding into a mini-trial.[29] Lord Diplock also relied on the fact that patents were now subject to a more elaborate patent examination procedure, necessitating a presumption of validity in favor of granted patents:

> Historically there was undoubtedly a time when in an action for infringement of a patent that was not already "well established," whatever that may have meant, an interlocutory injunction to restrain infringement would

---

27. Am. Cyanamid Co. v. Ethicon Ltd., [1975] A.C. 396 (H.L.) 399, 407 (U.K.). The rule in question, as articulated by Lord Diplock, is that the court ought to be satisfied that "if the case went to trial upon no other evidence than is before the court at the hearing of the application, the plaintiff would be entitled to judgment for a permanent injunction in the same terms as the interlocutory injunction sought." *Id.* at 407. Put another way, the court necessarily had to consider all the submissions on record and determine as to who was more likely to win at trial. Although this prima facie rule was most directly articulated in *J.T. Stratford & Son Ltd. v. Lindley*, [1965] A.C. 269 (H.L.) 331 (U.K.), Lord Diplock does not refer to it even once in his judgment, leading Lord Denning to remark that "I find it impossible to reconcile this statement with the statements in the House 10 years ago in J.T. Stratford & Son Ltd. v. Lindley [1965] A.C. 269. Yet the House did not even mention J.T. Stratford & Son Ltd. v. Lindley. I do not like to suggest that this was per incuriam." *See* Fellowes & Son v. Fisher [1976] Q.B. 122, 132 (Eng.).

28. The term "triable issue" was used in *Johnwalker & Sons v. Rothmans Int'l*, [1978] F.S.R. 357, 361–62 (U.K.), and *Management Publications v. Blenheim Exhibitions*, [1991] F.S.R. 348, 352 (U.K.), referring to Lord Diplock's new standard.

29. In particular, affidavit evidence submitted during the interim phase was not tested on the basis of oral examinations, as a result of which the veracity of such evidence could

> not be granted if counsel for the defendant stated that it was intended to attack the validity of the patent. Relics of this reluctance to enforce a monopoly that was challenged, even though the alleged grounds of invalidity were weak, are to be found in the judgment of Scrutton LJ as late as 1924 in Smith v Grigg Ltd ([1924] 1 KB 655 at 658), but the elaborate procedure for the examination of patent specifications by expert examiners before a patent is granted, the opportunity for opposition at that stage and the provisions for appeal to the Patent Appeal Tribunal in the person of a patent judge of the High Court, make the grant of a patent nowadays a good prima facie reason, in the true sense of that term, for supposing the patent to be valid, and have rendered obsolete the former rule of practice as respects interlocutory injunctions in infringement actions....[30]

Thus, the "triable issue" standard was in some way premised on the notion that granted patents ought to be presumed to be valid. Although later decisions professed allegiance to the "triable issue" standard,[31] judges often found clever ways of circumventing it. Illustratively, Lord Denning sought refuge in the "special factors" exception made in the *American Cyanamid*[32] case, to revert to the earlier prima facie standard.[33] He outlined several reasons for this, including the fact that in the vast majority of cases, the interim stage was dispositive and settled the matter once and for all.[34] Therefore, it was imperative for the court to scrutinize

---

not be fully ascertained in accordance with established norms of procedural justice. As a result, interim injunctions were being rendered on the basis of incomplete evidence and mini-trials, which were consuming so much time that they virtually destroyed the very rationale of interlocutory injunctions. Am. Cyanamid v. Ethicon Ltd., [1975] A.C. 396 (H.L.) 406 (U.K.).

30. *Id.* at 405-06.

31. *See, e.g.*, Foseco Int'l v. Fordath Ltd., [1975] F.S.R. 507 (U.K.); Potters-Ballotini v. Weston Baker, [1977] R.P.C 202 (U.K.).

32. *Am. Cyanamid,* [1975] A.C. at 409 ("[T]here may be many other special factors to be taken into consideration in the particular circumstances of individual cases.").

33. Fellowes & Son v. Fisher, [1976] Q.B. 122, 132 (Eng.) ("In considering whether to grant an interlocutory injunction, the right course for a judge is to look at the whole case. He must have regard not only to the strength of the claim but also to the strength of the defence, and then decide what is best to be done.").

34. *Id.* at 133:

> [T]hat sentence points the way. These individual cases are numerous and important. They are all cases where it is urgent and imperative to come to a decision. The affidavits may be conflicting. The questions of law may be difficult and call for detailed consideration. Nevertheless, the need for immediate decision is such that the court has to make an estimate of the relative strength of each party's case. If the plaintiff makes out a prima facie case, the court may grant an injunction. If it is a weak case, or is met by a strong

the issues more carefully in order to determine which party had a stronger probability of winning at trial.[35] Similarly, Justice Laddie also bypassed the lower "triable issue" standard in the *Series 5* case,[36] holding that nothing prevented the court from engaging in a "relative assessment" in appropriate circumstances:

> In my view Lord Diplock did not intend by the last quoted passage to exclude consideration of the strength of the cases in most applications for interlocutory relief. It appears to me that what is intended is that the court should not attempt to resolve difficult issues of fact or law on an application for interlocutory relief. If, on the other hand, the court is able to come to a view as to the strength of the parties' cases on the credible evidence then it can do so.[37]

Although the rulings above demonstrate the willingness of UK courts to deploy the "relative assessment" standard more frequently than Lord Diplock might have expected, it stands to reason that one could do so only when the case was amenable to such assessment during the short interim window.[38] Where such an assessment was not possible, a judge would necessarily have to fall back on Lord Diplock's lower "triable issue" standard.

Although the standard is one that is easily workable by courts and provides some measure of certainty to a patentee, it carries with it the significant risk of error where the patent is finally found to be invalid or not infringed. Such mistakes impact not just the defendant who is prevented from entering the market till the trial concludes, but also patients who are unable to access cheaper versions of generic drugs. As such, the harm is not one that is readily amenable to monetary evaluation,[39] making it even more imperative that the court avoid costly errors at the interim stage.

> defence, the court may refuse an injunction. Sometimes it means that the court virtually decides the case at that stage. At other times it gives the parties such good guidance that the case is settled. At any rate, in 99 cases out of 100, the matter goes no further.

35. He also noted that Lord Diplock left open the scope to engage in a relative assessment while assessing the balance of convenience limb, when the extent of uncompensable damage on either side would not differ widely. *See id.* at 134 ("There are many cases in which either party could suffer great disadvantages which could not be adequately compensated in damages. In all these it is permissible to consider the relative strength of each party's case.").

36. Series 5 Software Ltd. v. Clarke, [1996] C.L.C. 631 (Ch.) 641 (Eng.).

37. *Id.* at 641.

38. *Id.* at 640.

39. *See* Ofer Grosskopf & Barak Medina, *Remedies for Wrongfully Issued Preliminary Injunctions: The Case for Disgorgement of Profits*, 32 SEATTLE UNIV. L. REV. 903, 920 (2009)

B. The Appropriate Standard for Developing Countries?

Our main thesis is that developing countries such as India should embrace the more rigorous "relative assessment" threshold, as opposed to the lower triable issue standard. The primary reason for this suggestion is that various institutional infirmities in the patent grant process in developing countries militate against the conferral of a presumption of validity in favor of granted patents. Notably, the Indian Patent Office suffers a chronic shortage of adequately trained and specialized personnel and often lacks access to leading prior art databases to undertake a thorough and comprehensive examination of complex patent applications.[40] Several scholars have expressed concern over the fact that developing country patent offices, including India's, rubber-stamp decisions by international authorities such as the various Patent Cooperation Treaty (PCT) offices,[41] rather than subjecting them to national patent standards.[42]

Bhaven Sampat argues that, for the most part, India's higher patent standards (in the form of Section 3(d)) may not have actually been translated to practice: "Despite much discussion about the novelty and international uniqueness of India's patent laws, in practice I find the 3(d) provision has little effect on patent prosecution. This may not be surprising to legal scholars who emphasize the differences between laws on the book and laws and practice...."[43]

(arguing that a wrongfully issued preliminary injunction causes deadweight losses that are socially irreparable).

40. *See* SHAMNAD BASHEER ET AL., EU-INDIA TRADE AND INVESTMENT DEVELOPMENT PROGRAMME: CURRENT STATUS REPORT 89-90 (2006) (on file with author).

41. The Patent Cooperation Treaty (PCT) is an international treaty under which patent applicants may designate all or any states party to the PCT as countries where they wish to apply for a patent while availing of the earliest filed priority date. Under the PCT, certain major national patent offices act as International Search Authorities (ISAs), which identify prior art and make an opinion on patentability. It is this opinion that is allegedly rubber-stamped by developed countries' patent offices. For details of the procedure under the PCT, see WIPO, Protecting Your Inventions Abroad: Frequently Asked Questions About the Patent Cooperation Treaty (PCT), *available at* http://www.wipo.int/pct/en/basic_facts/faqs_about_the_pct.pdf (last visited Sept. 3, 2013).

42. *See* Peter Drahos, Trust Me: Patent Offices in Developing Countries, Centre for Governance of Knowledge and Development Working Paper (Nov. 2008), *available at* http://papers.ssrn.com/sol3/papers.cfm?abstract_id=1028676; Kapczynski, *supra* note 4; Janice M. Mueller, *The Tiger Awakens: The Tumultuous Transformation of India's Patent System and the Rise of Indian Pharmaceutical Innovation*, 68 UNIV. PITT. L. REV. 491 (2007).

43. *See* Bhaven Sampat, *Institutional Innovation or Institutional limitation? The Impact of TRIPS on India's Patent Law & Practice*, WIPO 29 (July 2010) http://www.wipo.int/edocs/mdocs/mdocs/en/wipo_ip_econ_ge_6_10/wipo_ip_econ_ge_6_10_ref_sampat.pdf.

Second, the Indian Patent Office has often faced charges of corruption and "cosy" linkages between patent agents and patent examiners/controllers.[44] A Thomson-Reuters report noted that there were "significant questions about the quality of what is coming out of the Indian IP Office."[45] The chances of error in decision-making are, therefore, high.

One might, therefore, suspect a bias in favor of patent grants in such countries. As a patent examiner tellingly remarked to one of the authors of this piece:

> Examiners are often in short supply and confronted with applications in areas of technology completely unfamiliar to them. In such cases, it is far easier to grant the patent, since a rejection would mean appreciating the technology in issue, and then spelling out clear reasons for the said rejection, which will likely be contested before appellate fora.[46]

Although this insight may suggest that the Indian Patent Office ought to begin demanding reasoned decisions even when patents are granted, it is an issue beyond the scope of this chapter.

The short point is that the various institutional infirmities offer us reason enough to desist from fully trusting the patent office to always make the right decision.[47] This is more so in the case of developing countries that suffer more severe resource constraints than their developed-country counterparts. However, it bears noting the Indian Patent Act of 1970 provides for a fairly robust opposition procedure, where any person is entitled to oppose a patent at the pre-grant stage or at the post-grant stage on a variety of grounds, including lack of novelty, inventive step, or utility.[48] One might suggest that the decision-making process tends to improve if a patent has been subjected to a rigorous opposition proceeding.

---

44. *See* Austria Wirtschaftsservice, *India IPR Inquiry 02/2008*, *in* BHAVEN SAMPAT, INSTITUTIONAL INNOVATION OR INSTITUTIONAL LIMITATION? THE IMPACT OF TRIPS ON INDIA'S PATENT LAW & PRACTICE (2010); *see also* C.H. Unnikrishnan & Khushboo Narayan, *Cozy Deals and Conflict of Interest Mark Indian Patent Grant Process*, LIVEMINT (Oct. 7, 2008), http://www.livemint.com/Industry/tdY0ALeaV5v3ln0X7LFFPK/Cozy-deals-and-conflict-of-interest-mark-Indian-patent-grant.html.

45. *See* THOMSON REUTERS, PATENT FOCUS REPORT (2009), *available at* http://science.thomsonreuters.com/news/2009-02/8502356/.

46. Authors' interview with senior patent examiner (Sept. 20, 2011).

47. *See* Drahos, *supra* note 42.

48. India Patent Act, *supra* note 25, §§ 25–28; *see also* Shamnad Basheer, *India's Tryst with TRIPS: The Patents (Amendments) Act, 2005*, 1 INDIAN J.L. & TECH. 15, 26 (2005).

We therefore argue that patents that have been granted after being opposed ought to be accorded some level of deference. The underlying rationale for this suggestion is best articulated by Professors Lemley and Lichtman:

> If some initial decision maker has made a decision about an issue, and if there is reason to believe that the decision is probably right, a presumption works to avoid wasteful reconsideration. This might in the aggregate reduce accuracy; but the point is that the first decision is sufficiently good that the odds of improving it are small and thus the costs of a second look are unwarranted. . . .[49]

Such deference is in conformity with the basic tenets or principles of administrative law under which administrative decisions are accorded at least some level of deference.[50] Also relevant is the fact that Indian courts have accepted the applicability of administrative law to patent law and other intellectual property decisions.[51]

The precise standards of deference turn significantly upon whether the issue being appealed or challenged qualifies as an issue of law, fact, or a mixed question of law and fact.[52] A discussion of what these terms mean and how they might apply to various patent office determinations is beyond the scope of this chapter. Suffice it to note that some broad level of deference ought to be accorded to patent office decisions that have undergone rigorous scrutiny in the form of oppositions, particularly when the invalidity defense mounted by the defendant encompasses the same set of issues that had already been considered once and disposed of by the patent office in a well-reasoned decision.[53]

---

49. Doug Lichtman & Mark A. Lemley, *Rethinking Patent Law's Presumption of Validity*, 60 Stanford L. Rev. 45 (2007).

50. *See* Nagendra Nath Bora v. The Commissioner of Hills Division and Appeals, Assam, 1958 A.I.R. 398 (S.C.) (India); *see also* Stuart Benjamin & Arti Rai, *Who's Afraid of the APA? What the Patent System Can Learn from Administrative Law*, 95 Geo. L.J. 269 (2007).

51. *See* Indian Network for People Living with HIV/AIDS v. Union of India, (2009) 1 C.T.C. 32 (Madras H.C.) (India); Whirlpool Corp. v. Registrar of Trademarks, MANU/SC/0664/1998; Bhikhubhai Vithalbhai Patel v. State of Gujarat, MANU/SC/7399/2008.

52. For a general overview of the distinction between questions of law and questions of fact, particularly in patent cases involving complicated claims, see Michael Muerer & Katherine Strandburg, *Patent Carrots and Sticks: A Model of Nonobviousness*, 12 Lewis & Clark L. Rev. 549, 565–69 (2008).

53. There is, however, no possibility of res judicata operating here, as the patent office is not a court of concurrent or exclusive jurisdiction.

Although the total number of oppositions filed in India (as a percentage of the total number of applications filed) leaves much to be desired,[54] it would appear that an opposed patent is subjected to a higher standard of scrutiny. Our study focusing on a data set of thirty-four patent oppositions filed from 2005 to 2008[55] found that twenty-five of the said oppositions resulted in rejections—that is, a significantly high percentage (73.5 percent) of the opposed applications were rejected![56]

The statistical percentages remained broadly similar for a later study on this same theme, where it was found that of the total number of fifty-eight opposition decisions till September 2010, 72.5 percent (forty-two oppositions) resulted in rejections.[57]

These numbers indicate that any fear of bias in favor of a grant is significantly reduced in cases where the patent has been opposed. We therefore argue that in all such cases (where the defendant raises an invalidity defense along the same lines as that raised before the patent office and disposed of in an adequately well-reasoned decision), the court should lean in favor of finding a prima facie case, unless the defendant raises a credible challenge to the factum of infringement and it is difficult to assess who might win at trial on this count.

It bears reiteration that patent office decisions ought to merit deference only when they have been adequately reasoned and all conclusions find some support in the text of the decision itself. The level of deference ought to be higher for a post-grant proceeding, as this involves a more intensive process of scrutiny than the pre-grant process. In a post-grant proceeding, the controller forms an opposition board consisting of three members, and nominates one of them as chairman. The opposition board conducts an examination of the notice of opposition along with the replies, and submits a report to the controller with a recommendation to

---

54. Our analysis revealed that the total number of pharmaceutical applications filed between 2005 and 2008 was around 9719. The number of pharmaceutical patents granted during the above time frame was around 2734. Of the 9719 pharma applications filed, only 34 were challenged—an abysmal. 3 percent! *See* Shamnad Basheer, *Patent Oppositions in India: The Efficacy of Section 3(d)*, SpicyIP (Sept. 16, 2009), http://spicyipindia.blogspot.in/2009/09/patent-oppositions-in-india-efficacy-of.html.

55. Out of the thirty-four challenges, one was a post-grant opposition (in favor of Roche covering Pegasus) and the remaining thirty-three were pre-grant challenges. *Id.*

56. The patent applicant won in only nine cases, that is, 9/34 or 26.5 percent of the time. It bears noting that twenty out of the twenty-five rejections above were based on Section 3(d). In other words, 80 percent of the rejections were based on Section 3(d), indicating the "efficacy" of this controversial section. *Id.*

57. *See* Tahir Amin, Challenging TRIPS: The Role of Patent Oppositions, Presentation Delivered at the TRIPS@10 Conference, Columbia Univ., Nov. 16–18, 2010, *available at* http://www.earth.columbia.edu/tripsat10/sitefiles/file/Panel5_Amin_Tahir_PowerPoint.pdf.

maintain, amend, or revoke the patent. The controller thereafter schedules a hearing between the parties, and makes a final decision after taking into account the report of the board. In contrast, in a pre–grant proceeding, the controller decides the entire matter by himself without any assistance from an opposition board.

It must be noted that we do not support a blanket statutory presumption of validity for all granted patents. Rather, our suggestion is that some level of deference ought to be accorded, depending on whether the patent has been opposed and the opposition disposed of on adequately reasoned grounds. A close reading of the Indian Patent Act suggests the possibility of such deference/presumption.

Section 13(4) stipulates as follows:

> The examination and investigations required under section 12 and this section *shall not be deemed in any way to warrant the validity of any patent*, and no liability shall be incurred by the Central Government or any officer thereof by reason of, or in connection with, any such examination or investigation or any report or other proceedings consequent thereon.[58]

This section merely suggests that there is no warranty that the patent is a valid one; it does not in any way prohibit a presumption of validity in appropriate cases. The statutory negation of a specific "warranty" is to protect government officials responsible for the grant. The latter part of Section 13(4) makes this clear: "no liability shall be incurred by the Central Government or any officer thereof by reason of, or in connection with, any such examination or investigation or any report or other proceedings consequent thereon."[59]

Although Section 13(4) is clear that a patentee cannot claim a presumption of patent validity as a matter of statutory entitlement, a court is at liberty to presume validity in appropriate cases. A presumption is nothing more than a rule of evidence, to help avoid wastage of resources in readjudicating the same set of issues. However, where a strong invalidity attack is mounted against the patent, and the challenger raises issues that have never been raised before the patent office, the presumption lifts and the plaintiff has to now defend the patent afresh.[60]

To summarize the above discussion, the prima facie threshold must be a high one and a judge must, where the materials so permit, make a determination as to who is more likely to win at trial. A temporary injunction should be granted

---

58. India Patent Act, *supra* note 25, § 13(4) (emphasis added).

59. *Id.*

60. This is true even in the United States, where a strong presumption exists. *See* Lemley & Lichtman, *supra* note 49; *see also* Circle Film Enters. Inc. v. Canadian Broad. Corp., [1959] S.C.R. 602 (Can.) (holding that the presumption of validity disappears when a credible challenge is mounted). This aspect is discussed in greater detail later in this chapter.

only where it is clear that the plaintiff has a higher probability of winning at trial. In complex patent cases, where such a determination is not realistically possible at the interim stage, an injunction ought not to be granted and the matter must move to trial at the earliest. If a judge is forced to assess the relative merits of a complex patent case at the interim stage, one of two things could happen:

i) the proceedings run the risk of getting unduly delayed, as the judge takes a longer time to decide, and may even call for more evidence and documents, effectively converting the proceeding to a trial.
ii) the judge arrives at a wrong assessment (an assessment that is finally overturned at trial). Given the relative inexperience of Indian judges with complex patent disputes, the error rate is likely to be higher if judges are forced to decide within a short interim window.

This latter concern is not merely a theoretical one, but has already played out in *Roche v. Cipla,*[61] the first pharmaceutical patent infringement case in post-TRIPS India.

II. *Roche v. Cipla*: Patent Proficiency?

Roche, OSI, and Pfizer, three leading global pharmaceutical firms, sued Cipla, an Indian generic company, before the Delhi High Court, alleging infringement of a patented anti-cancer drug. More specifically, the suit claimed that "Erlocip," a generic anti-cancer drug manufactured by Cipla corresponding to Roche's originator drug, "Tarceva," violated OSI's Indian patent (IN196774) (hereafter the '774 patent).[62]

Both drugs are based on "Erlotinib Hydrochloride," a compound now known to exist in several polymorphic forms. The trial judge[63] declined on the ground of "public interest" to grant a temporary injunction, holding that the patented drug was very expensive and that the lower-priced generic version manufactured

---

61. F. Hoffman-La Roche Ltd. v. Cipla Ltd., (2008) 37 P.T.C. 71 (Delhi H.C.) (India).

62. OSI jointly owned a patent with Pfizer Products Inc. in respect of an HER/EGFR Inhibitor, popularly known as "Erlotinib." OSI and Roche subsequently entered into a development collaboration and licensing agreement, whereby Roche was granted a license to use and sell and offer for sale the licensed products of the former, including Erlotinib, marketed under the brand name "Tarceva." Roche was further licensed and authorized to cause enforcement of any infringement of property rights of any of the licensed products.

63. In the Delhi High Court as well in several High Courts in India, cases are first heard by a Single Judge. An appeal may be preferred from this decision to a Division Bench of two Judges. Decisions may further be appealed to the Supreme Court of India, if special leave to appeal is granted by the Supreme Court.

by Cipla fostered greater access to this life-saving drug in India.[64] Although Cipla challenged the validity of OSI's patent in this suit (a claim that the judge found to be a "credible" one) and denied that it infringed the specific claims, the trial judge found that the plaintiff had established a prima facie case of infringement.[65]

The appellate court agreed with the Single Judge and refused to enjoin Cipla.[66] However, it noted that in view of the credible challenge to the validity of the patent and the fact that Cipla's product does not appear to infringe the suit patent, a prima facie case could not be said to be established. It also noted that even if a prima facie case were established, the injunction would still be denied on the ground of "public interest."

The Single Judge may have endorsed the low "triable issue" threshold enunciated in the *American Cyanamid*[67] decision to support his finding that a prima facie case had been made out. However, this is a mere surmise, as the judge does not elaborate on this aspect or provide reasons for his conclusion. Further, it bears noting that the Indian Supreme Court has, much like Justice Laddie in *Series 5*, effectively bypassed the "triable issue" standard. In *Colgate Palmolive v. Hindustan Lever*;[68] it adopted Justice Laddie's understanding of the *Cyanamid* principles, noting that "lately the 'triable issue concept' as introduced by Lord Diplock in *Cyanamid's* case has been thought to be much too rigid and wide even conceptually."[69]

However other cases have not been this straightforward and it is not clear whether the courts, while expressing allegiance to the precedential value of *Cyanamid*, were following *Cyanamid* strictly, or Justice Laddie's opinion distinguishing the

---

64. *See* Single Judge order, *supra* note 10, ¶¶ 84–85.

65. *See id.* ¶ 78 ("All that can be therefore said is that the plaintiff's case though arguable and though disclosing prima facie merit, has to answer a credible challenge to the patent raised by the defendant.").

66. Hoffman-La Roche Ltd. v. Cipla Ltd., (2009) 40 P.T.C. 125 (Delhi H.C.) (India).

67. Am. Cyanamid v. Ethicon Ltd., (1975) 1 All E.R. 504 (U.K.).

68. Colgate Palmolive (India) Ltd. v. Hindustan Lever Ltd., 1999 A.I.R. 3105 (S.C.) (India).

69. *Id.* at 3110:

> [W]e record our concurrence with the views expressed by Laddie J. in Series 5 Software Vs. Clarke and Others in 1996 (1) ALL ER 853 wherein the learned Judge has explained the judgment of *American Cyanamid* with extreme competency and in our view also correctly Laddie, J. observed: "In many cases before *American Cyanamid* the prospect of success was one of the important factors taken into account in assessing the balance of convenience. The courts would be less willing to subject the plaintiff to the risk of irrecoverable loss which would befall him if an interlocutory injunction was refused in those cases where it thought he was likely to win at the trial than in those cases where it thought he was likely to lose. The assessment of the prospects of success therefore was an important factor in deciding whether the court should exercise its discretion to grant interlocutory relief. It is this consideration which *American Cyanamid* is said to have prohibited in all but the most exceptional case.

Cyanamid principles.[70] The *Roche v. Cipla* decision also suffers from this flaw, in that it does not articulate the precise standard for prima facie assessment.

In the following section, we deal with the appellate court's finding on non-infringement to highlight its relative lack of familiarity with patent concepts.

The appellate court denied interim relief to Roche on the main ground that they had not demonstrated an infringement of their '774 patent, and therefore a prima facie case could not be said to have been established in their favor.[71] An exploration of the patents at issue reveals the court's flawed understanding of basic jurisprudential tenets underlying patent law.

The '774 patent, which claims "Erlotinib Hydrochloride," was granted on July 6, 2007.[72] A later patent application claiming a specific polymorph B of Erlotinib Hydrochloride was filed in India (hereinafter the '507 application) in 2002.[73] This application was opposed by Cipla and was ultimately rejected in December 2008 on the ground that the invention was not patentable under Section 3(d) of the Indian Patent Act.

Whereas the '774 patent covers Erlotinib Hydrochloride, the '507 application claimed a specific polymorph of Erlotinib Hydrochloride, namely Polymorph B. Cipla argued that the suit patent envisaged Erlotinib Hydrochloride existing as a combination of Polymorphs A and B. It stressed that the Tarceva tablet corresponded to Polymorph B alone, ass this polymorph was more stable and capable of being made into tablet form (solid form oral dosage). Based on this, Cipla argued that there was no infringement of Roche's registered '774 patent.

The Division Bench agreed with Cipla, holding that the drug Tarceva corresponded with Polymorph B, which was the subject matter of the '507 application, an application that was ultimately rejected. According to the Division Bench:

> [T]he plaintiffs are admittedly not commercially exploiting the patent granted in their favour for a compound which is a mixture of

---

70. *See* S.M. Dyechem Ltd. v. Cadbury (India) Ltd., 2000 A.I.R. 2114 (S.C.) 2121 (India) (holding that "in trade mark matters, it is now necessary to go into the question of 'comparable strength' of the cases of either party"). This was affirmed in *Cadila Healthcare Ltd. v. Cadila Pharmaceuticals Ltd.*, 2001 A.I.R. 1952 (S.C.) (India). However, other cases appear to have relied on a much lower "triable issue" standard. For a more detailed survey of the Indian case law in this regard, see Ashutosh Kumar, *Patent Protection Through Interim Injunctions*, 20 Nat'l L. Sch. India Rev. 202 (2008).

71. It is pertinent to note that the court also ruled that the patent at issue faced a credible challenge of invalidity and was therefore vulnerable. Additionally, it held that "public interest" dictated that no injunction ought to be granted on the facts of the case. However, because the public interest aspect is more or less a reiteration of the trial court order, it is not dealt with in this chapter.

72. This was listed as Application no. 537/DEL/1996. The corresponding U.S. patent is No. 5,747,498, granted in 1998.

73. IN/PCT/2002/00507/DEL. This corresponds with US Patent No. 6,900,221.

> Polymorphs A and B, since the tablet form corresponds to Polymorph B of the said compound Erlotinib Hydrochloride.... [T]his Court holds that to the extent that the defendant has raised a serious doubt whether the plaintiffs in fact hold a patent for the product sold in the tablet form as Tarceva, the plaintiffs must be held not to have been able to cross the first hurdle of showing that they have a prima facie case in their favour for grant of an order restraining the defendant from marketing Erlocip....[74]

Accordingly there was no infringement of the '774 patent.

A. Whither Claim Construction?

That the reasoning of the appellate court is infirm is evident from the fact that the judge does not refer to the claim in issue even once. As a leading patent scholar once noted, "the name of the game is the claim."[75] It is expected that when faced with a patent infringement suit, a court would first turn to the claim and construe it.[76]

The first claim of the '774 patent reads: "A novel [6,7-bis(2-methoxyethoxy) quinazolin-4-yl]-(3-ethynylphenyl) amine hydrochloride compound of the formula A."[77] A literal reading of this claim suggests that it covers the compound "Erlotinib Hydrochloride." This compound may exist in several polymorphic forms, but all such forms are likely to be subsumed within the patented parent compound.[78] Therefore, even assuming that Cipla's drug relies on only one particular polymorphic form of the Erlotinib Hydrochloride compound (Form B), a correct appreciation of the principles of claim construction would lead to a

---

74. *See* Division Bench order, *supra* note 10, ¶¶ 46–47.

75. *See* Giles S. Rich, *The Extent of the Protection and the Interpretation of Claims—American Perspectives*, 21 INT'L REV. INDUS. PROP. & COPYR. L. 497, 499 (1990) ("To coin a phrase, the name of the game is the claim.").

76. *See* Netword, LLC v. Centraal Corp., 242 F.3d 1347, 1350 (Fed. Cir. 2001) ("An infringement analysis entails two steps. In the first step the court construes the patent claims, by establishing the scope and boundaries of the subject matter that is patented, as a matter of law, and in the second step the trier of fact applies the construed claims to the accused device.").

77. This Formula A is the chemical formula of the molecule in question and separate from "Form A," the polymorph mentioned earlier.

78. *See generally* Linear Tech. Corp. v. Int'l Trade Comm'n, 566 F.3d 1049 (Fed. Cir. 2009) ("[E]ven in situations when only one embodiment is disclosed, the claims generally should not be narrowed to cover only the disclosed embodiments or examples in the specification."). On a parity of reasoning, the first claim of the '774 patent specification would have been sufficient to cover any polymorphic forms of Erlotinib Hydrochloride.

finding that Cipla's drug necessarily infringes the main compound patent,[79] that is, Cipla could not have created the said polymorphic form without infringing the main patent covering Erlotinib Hydrochloride. Therefore, it ought not to matter that the drug "Tarceva" (and Cipla's generic version) was primarily based on the Polymorph B version, specifically claimed in the later '507 application. Making Tarceva, or a generic equivalent thereof, would necessarily infringe the parent '774 patent.

Broadly, the court committed two fundamental errors that militate against basic patent jurisprudence:

1. Rather than relying on the patent claim and specification to determine the scope of the patent (and then assessing whether the defendant's product falls within the said scope), the appellate court relies on the composition of the patented drug.
2. The court misses the point that a polymorph of a parent compound could still infringe the patent claiming the parent compound, if one cannot make the said derivative without first making the parent compound.

This jurisprudential flaw may have stemmed from the relative inexperience of the court with patent cases and an inability to appropriately assess the merits or otherwise of the dispute during the limited window that is normally available during the interim stage.[80]

---

79. Both the US and UK courts suggest that when a claim is expressed in clear language, that language should not be cut down by reference to the rest of the specification (but there are also Court of Appeals for the Federal Circuit cases holding the opposite). *See* THORLEY ET AL., TERRELL ON THE LAW OF PATENTS 130 (2000). Further, courts have held that "if the claims have a plain meaning in themselves, then advantage cannot be taken of the language used in the body of the specification to make them mean something different." Elec. & Musical Indus. v. Lissen, [1939] 56 R.P.C. 23 (H.L.) 57 (U.K.) (Lord Porter); *see also* Interactive Gift Express, Inc. v. CompuServe Inc., 256 F.3d 1323, 1331 (Fed. Cir. 2001) ("If the claim language is clear on its face then the consideration of the rest of the intrinsic evidence is restricted to determining if a deviation from the clear language is specified.").

80. Unfortunately, this same erroneous reasoning permeated the court's ruling even after a full-fledged trial. In this case, the court committed yet another error. After upholding the challenged claim as valid, it went on to narrow its scope of the patent during the infringement analysis. *See* F. Hoffmann-La Roche Ltd. v. Cipla Ltd., (2012) 52 P.T.C. 1 (Delhi H.C.) (India). For a critique of this decision, see Shamnad Basheer, Roche vs. Cipla: *Pondering the Puzzle*, SPICYIP (Sept. 12, 2012), http://spicyipindia.blogspot.in/2012/09/roche-vs-cipla-pondering-puzzle.html.

III. Dispensing with the Interim Phase

Our thesis is that in complex patent disputes where a defendant mounts a credible argument in terms of non-infringement or invalidity (or both), the interim phase ought to be dispensed with, and the matter should proceed directly to trial. A logical corollary of this proposition is that mere assertion of invalidity or of non-infringement will not suffice—rather, the assertion must be cogently reasoned and substantiated such that a judge is convinced that there is some merit in the defendant's contention.

Similarly, in cases where the defendant agitates the very same grounds that had been taken up during an earlier opposition proceeding and decided upon in a well-reasoned ruling from the patent office, we propose that the court accord some deference to the patent office decision. In all such cases, the court ought to proceed on the basis that the prima facie standard has been satisfied, unless the defendant raises a serious doubt about the factum of infringement.

Even in cases where a court is able to arrive at a quick prima facie view of the merits in favor of the plaintiff, there cannot be an automatic grant of an injunction. Rather, courts are to additionally determine whether the other prerequisites for the grant of a temporary injunction have been met—namely, that a prospect exists of the plaintiff suffering irreparable injury without an injunction, that the balance of convenience tilts in favor of the plaintiff, and that the public interest will not be disserviced by the grant of an injunction.

In the event that a court finds the defendant's challenge to be credible and directs the matter to trial in accordance with our proposal above, we argue that it should compel the defendant to keep an account of sales of the allegedly infringing product. This would ensure that in the event of the plaintiff ultimately prevailing at trial, he or she is compensated from these amounts for the damage suffered in the interim.[81]

Similarly, if the court decides to issue a temporary injunction or restraining order, it should ask the patentee to post bond, which can be forfeited if the defendant ultimately prevails. Given that an erroneously issued preliminary injunction

---

81. In *Roche v. Cipla*, the trial judge, while rejecting the request for an ad interim injunction, directed the defendant to maintain accounts of the sales of its product "Erlocip." *See* Single Judge order, *supra* note 10, ¶ 87:

> The result of the above discussion is that the plaintiff is not entitled to claim an ad interim injunction, in the terms sought. However this court is not unmindful of the fact that if no equitable balancing order protecting its interest is made at this stage, there is a likelihood of the plaintiff being prejudiced at the final stage. Therefore the defendant is hereby directed to . . . maintain faithful accounts of its sale on the product Erlocip and file quarterly accounts in this court supported by the affidavit of one of its directors, affirming the veracity of the same.

could inflict considerable injury on the defendant, this is an important requirement that must be insisted upon.[82]

It is interesting to note that the above proposal for a dispensation with the interim phase is one that has been endorsed by the Supreme Court of India in a few intellectual property cases, as discussed below.

A. Bajaj v. TVS

Bajaj Auto, Ltd. (hereafter Bajaj), a leading automobile manufacturer in India, sued its competitor, TVS Motor Company (hereafter TVS), before the Madras High Court, alleging infringement of its patent covering DTS-i (dual twin spark ignition) technology.[83] Pending the suit, Bajaj prayed for a temporary restraining order against TVS. TVS claimed it did not infringe, counterclaimed invalidity, and also separately petitioned for revocation of Bajaj's patent (No. 195904) before the Intellectual Property Appellate Board (IPAB) under Section 64 of the Patent Act.[84] Although the Single Judge granted the injunction, the Division Bench of the Madras High Court, sitting as the appellate court, reversed. Bajaj then appealed to the Supreme Court, which declined to intervene and instead directed that the matter proceed to trial.[85] The Court was particularly influenced by the fact that the interim proceedings lasted for almost two years:[86]

> It is evident that the suit is still pending before the learned Single Judge of the Madras High Court. We are unhappy that the matter has been pending in the High Court at the interlocutory stage for such a long time as

82. *See* Grosskopf & Medina, *supra* note 39, at 905; *see also* Fed. R. Civ. P. 65(c) ("The court may issue a preliminary injunction or a temporary restraining order only if the movant gives security in an amount that the court considers proper to pay the costs and damages sustained by any party found to have been wrongfully enjoined or restrained."). Indian cases have also insisted on this requirement. *See* Bilcare Ltd. v. M/s Supreme Indus., (2007) 34 P.T.C. 444 (Delhi H.C.) (India).

83. Indian Patent No. 195904. The suit was filed under Section 108 of the Patent Act which provides that courts may grant "an injunction... and, at the option of the plaintiff either damages or an account of profits." India Patent Act, *supra* note 25, § 108.

84. Section 64 of the Indian Patent Act provides for revocation of patents by the Appellate Board or on a counterclaim in a suit for infringement of the patent by the High Court on grounds mentioned in the section. *See* India Patent Act, *supra* note 25, § 64.

85. Bajaj Auto Ltd. v. TVS Motor Co. Ltd., (2009) 41 P.T.C. 398 (S.C.) (India).

86. The suit and the application for interim injunction were filed on December 17, 2007, and the injunction was granted by the Single Judge on February 16, 2008. The Appellate Court reversed the Single Judge order on May 18, 2009. The matter was then appealed to the Supreme Court, which gave its decision (directing that the matter to trial) on September 16, 2009, a good twenty-one months after the suit had first been filed.

> the suit was filed in December, 2007 and yet even written statement has not been filed.
>
> Recently, we have held in the case of *M/s. Shree Vardhman Rice Gen Mills v. M/s Amar Singh Chawalwala* as follows: "Without going into the merits of the controversy, we are of the opinion that the matters relating to trademarks, copyrights and patents should be finally decided very expeditiously by the Trial Court instead of merely granting or refusing to grant injunction. Experience shows that in the matters of trademarks, copyrights and patents, litigation is mainly fought between the parties about the temporary injunction and that goes on for years and years and the result is that the suit is hardly decided finally. This is not proper. Proviso (a) to Order XVII Rule 1(2) C.P.C. states that when the hearing of the suit has commenced, it shall be continued from day-to-day until all the witnesses in attendance have been examined, unless the Court finds that, for exceptional reasons to be recorded by it the adjournment of the hearing beyond the following day is necessary.
>
> The Court should also observe clauses (b) to (e) of the said proviso. In our opinion, in matters relating to trademarks, copyright and patents the proviso to Order XVII Rule 1(2) C.P.C. should be strictly complied with by all the Courts, and the hearing of the suit in such matters should proceed on day to day basis and the final judgment should be given normally within four months from the date of the filing of the suit....
>
> As has been observed by us in the aforesaid case, experience has shown that in our country, suits relating to the matters of patents, trademarks and copyrights are pending for years and years and litigation is mainly fought between the parties about the temporary injunction. This is a very unsatisfactory state of affairs, and hence we had passed the above quoted order in the above-mentioned case to serve the ends of justice. We direct that the directions in the aforesaid order be carried out by all courts and tribunals in this country punctually and faithfully....

This philosophy of dispensing with the interim phase and moving directly to trial was endorsed again by the Supreme Court in *Cipla,* where it declined to interfere with the interim ruling of the lower court and requested "the learned single judge dealing with the civil suit to conclude the trial as expeditiously as possible...."[87] Other courts have also leaned toward this approach,[88] with the

---

87. *See* Hoffman-La Roche Ltd. v. Cipla Ltd., Petition for Special Leave to Appeal (Civil) No. 20111/2009 (S.C. Aug. 28, 2009) (India), *available at* http://courtnic.nic.in/supremecourt/temp/sc%202011109p.txt.

88. *See Generally* Allergan Inc. v. Ajanta Pharma Ltd., CS(OS) No. 1302/2011 (Delhi H.C. Mar. 6, 2011) (India), *available at* http://delhihighcourt.nic.in/dhcqrydisp_o.

Madras High Court noting in a patent infringement suit that "where the main issue to be decided in the stage of the interim injunction application is the same as the one to be tried in the suit the best course will be to expedite the trial of the suit, so that neither party would be put to any irreparable damage or loss of prejudice."[89] It thereafter proceeded to direct that trial court to dispose of the matter "as expeditiously as possible" and preferably before the expiry of two months from the date of the order.[90]

Although we agree with the overall conclusion of the Supreme Court in *M/s. Shree Vardhman Rice Gen Mills v. M/s Amar Singh Chawalwala* that a direct recourse to trial may be more optimal in certain intellectual property infringement cases, we disagree with the broad scope and ambit of its order: "[W]e are of the opinion that the matters relating to trademarks, copyrights and patents should be finally decided very expeditiously by the Trial Court instead of merely granting or refusing to grant injunction."[91] Whereas the Court suggests that the interim phase be dispensed with in all cases relating to intellectual property, we advocate that dispensation take place only for complex patent cases, where it is difficult to arrive at a robust prima facie assessment within the limited interim window available.

B. Expediting Trials

Any dispensation with the interim phase comes at a cost, particularly for patent owners who desire a speedy and effective enforcement of their rights. We outline broad policy suggestions to help mitigate delays and provide effective relief to patent owners.

The most obvious disadvantage of eliminating the interim phase is that an intellectual property owner is now subject to the vagaries of protracted trials. The purpose of an interim injunction is to mete out timely justice to an intellectual property owner, given that a full-fledged trial is likely to take several years, during which time the intellectual property owner's exclusive rights are gradually

---

asp?pn=114897&yr=2011; Hoffman-La Roche Ltd. v. Reddy's Lab., IA No. 628/2010 (Delhi H.C. Oct. 2, 2011) (India), *available at* http://delhihighcourt.nic.in/dhcqrydisp_o.asp?pn=31443&yr=2011; Bayer Corp. v. Cipla Ltd., CS(OS) No. 523/2010 (Delhi H.C. Sept. 17, 2013) (India); Prashant Reddy, *Update on Bayer–Cipla Patent Infringement: Nexavar Patent Infringement Suit Before the Delhi High Court Goes Directly to Trial*, SpicyIP (Sept. 14, 2010), *available at* http://spicyipindia.blogspot.in/2010/09/update-on-bayer-cipla-patent.html. In *Cipla*, CS(OS) No. 523/2010, Justice Bhat persuaded the parties to agree to a direct move to trial.

89. V. Manioka Thevar v. Star Plough Works, 1965 A.I.R. 327 (Madras H.C.) ¶ 8 (India).

90. *Id.*

91. Bajaj Auto Ltd. v. TVS Motor Co. Ltd., (2009) 41 P.T.C. 398 (S.C.) (India).

eroded. The problem is particularly acute in technology sectors where products may have short shelf lives, and loss of exclusivity for even a short period could cause serious injury to the patentee in terms of loss of market share and/or price erosion.[92]

One possible way of redressing this harm to the patentee is by expediting trials. This approach has been endorsed by the Supreme Court in *Bajaj*, where it directed that the trial court conclude proceedings within four months.[93] Unfortunately, despite more than two years having gone by, the court of first instance is yet to complete the trial.[94] This suggests that the Supreme Court cannot rest easy with merely laying down guidelines, but must ensure that they are complied with.

In order to ensure that trials are expedited and completed in a timely manner, we propose the following:

1. An abridged trial;
2. Specialist courts; and
3. Specialist advisers.

We deal with each in turn.

1. Abridged Trials

We propose that trials be abridged in the following way:

(1) Parties must reduce all their submissions to writing. In particular, courts must insist on evidence by way of affidavit;[95]
(2) Only a limited number of experts (on both sides) must be permitted to testify;
(3) As far as possible, evidence ought to be recorded by court-appointed commissioners. This enables a speedier adjudication of the disputes. However, care

92. Price erosion refers to instances where the patentee would be forced to bring down her prices owing to the presence of the allegedly infringing competitor. *See* Webster & Walmley, *supra* note 13, at 292. However, the loss of market share and price erosion does not always follow from the competitors' entry into the market. Rather, these are case-specific and have to be independently established by the patentee.

93. *Bajaj*, (2009) 41 P.T.C. at 398.

94. The suit (Civil Suit No. 1111/2007) is listed as "pending." *See Case Status of High Court of Chennai*, Indian Courts Case Status, http://courtnic.nic.in/courtnic_chennai/default.aspx (last visited Sept. 3, 2013).

95. It must, however, be noted that a court order is necessary to prove any fact as an affidavit. *See* M.P. Jain, The Code of Civil Procedure 131 (2005).

must be taken to select commissioners who have extensive experience with recording trial evidence. Illustratively, in *Garware Wall-Ropes ltd. v. Anant Kanoi*,[96] the parties to a patent infringement suit agreed to appoint a retired judge to record the oral evidence. The court directed the Commissioner to record the evidence and complete the same within a period of two months from the date of his appointment;[97]

(4) The hearings ought to proceed on a daily basis, as recommended by the Supreme Court in *Bajaj*.[98] The Court rightly noted that under the proviso (a) to Order XVII, Rule 1(2) of the Code of Civil Procedure, 1908, barring exceptional circumstances, a suit should continue on a day-to-day basis until all the witnesses in attendance have been examined; and

(5) Adjournments ought to be granted only in exceptional cases. One of the primary causes of litigation delay is the frequency with which adjournments are requested and granted, even those based on utterly inadequate grounds.[99] Rule 1, Order XVII of the Code of Civil Procedure, empowers the court to adjourn proceedings only if sufficient cause is shown. In any case, not more than three adjournments can be granted to a party during the hearing of a suit.[100] To guard against frequent requests for adjournments at the trial stage, the court can record the statements of witnesses and pass such orders as it thinks fit, dispensing with examination/cross-examination where the pleader or party is either not present or not ready.[101] Care must be taken to strictly

---

96. *See* Garware Wall Ropes Ltd. v. Anant Kanoi, No. CS/4/2005 (Gujarat H.C. Apr. 29, 2011) (India), *available at* http://indiankanoon.org/doc/1974764. In the *Cipla* trial, the Delhi High Court appointed a retired judge to record evidence in the matter. *See* Shamnad Basheer, *Breaking News: Supreme Court Dismisses Roche "Tarceva" Petition*, SpicyIP (Aug. 28, 2009), http://spicyipindia.blogspot.in/2009/08/breaking-news-supreme-court-dismisses.html.

97. *See* F. Hoffman-La Roche Ltd. v. Cipla Ltd., (2009) 40 P.T.C. 125 (Delhi H.C.) (India) (appointing commissioners to record evidence).

98. *See* Shree Vardhman Rice v. Amar Singh Chawalwala, (2009) 10 S.C.C. 257 (India); Bajaj Auto Ltd. v. TVS Motor Co. Ltd., (2009) 9 S.C.C. 797 (India).

99. *See* Committee on Reforms of the Criminal Justice System, Government of India, Ministry of Home Affairs, ¶ 9.14.1 ("A notorious problem in the functioning of courts, particularly the trial courts, is the granting of adjournments, mostly on flimsy grounds.").

100. Code of Civil Procedure, 1908, Order XVII, Rule 1 ("Court may grant time and adjourn hearing.-(1) The Court may, if sufficient cause is shown, at any stage of the suit grant time to the parties or any of them, and may from time to time adjourn the hearing of the suit for reasons recorded in writing: Provided that no such adjournment shall be granted more than three times to a party during hearing of the suit.").

101. *Id.* Proviso (e) to Order XVII, Rule 1(2).

comply with these measures and ensure timely disposal of suits, despite the difficulty of executing them in an adjournment entrenched ecosystem.[102]

Our interviews with leading patent counsels suggest that if the above measures are adopted, trials could realistically conclude in about a year's time in India.[103]

2. Specialist Courts

Patent infringement suits often involve complex technical issues, requiring some level of specialist knowledge on the part of an adjudicator.[104] Specialized courts/tribunals have been heralded as offering numerous advantages over generalist courts,[105] including expedited hearings and disposals, reduced costs for litigants, improved decision-making, and greater predictability.[106]

However, critics have cautioned against the creation of a "collegial" atmosphere between adjudicators and the specialist litigators, drawing into question the objectivity of the specialist court.[107] Second, lobbying by interested parties for adjudicatory positions may occur, as the payoff is likely to be more direct in a

---

102. *See* Kishan Lal Gupta v. Dujodwala Indus., 1977 A.I.R. 49 (Delhi H.C.) 51 (India) ("Adjournment is an exception and continuous hearing the rule. It is true that the courts, in practice have buried the rule fathoms deep and have been granting adjournments on the flimsiest of grounds. It is rare indeed when a court holds a trial continuously in terms of this rule. But then this is no reason why this rule should not be followed and enforced in practice.").

103. Interview with Pravin Anand, Managing Partner, Anand and Anand (Feb. 20, 2012); Interview with Lakshmikumaran, Managing Partner, Lakshmikumaran and Sridhar (Mar. 5, 2012).

104. *See* Robert M. Sherwood, *Specialised Judicial Arrangements for Intellectual Property*, 36 Rev. ABPI 38 (1998).

105. *See* Zuallcobley et al., Study on Specialized Intellectual Property Courts (2012), *available at* http://iipi.org/wp-content/uploads/2012/05/Study-on-Specialized-IPR-Courts.pdf ("The results of these case studies suggest a positive correlation between specialized IPR courts and the efficient and effective resolution of IP cases.").

106. Central European and Eurasian Law Initiative (CEELI), *Specialized Courts: A Concept Paper*, A.B.A. 11 (June 25, 1996), *available at* http://apps.americanbar.org/rol/publications/concept_paper_specialized_courts_1996.pdf; Martin J. Kroeze, *Ensuring Judicial Infrastructure: The Theory and Practice of Specialised Courts*, *in* Enforcement of Corporate Governance in Asia: The Unfinished Agenda 63 (2007).

107. *See* Markus B. Zimmer, *Overview of Specialized Courts*, 2 Int'l J. Court Admin. 46 (Aug. 2009); David Rottman, *Does Effective Therapeutic Jurisprudence Require Specialized Courts (and Do Specialized Courts Imply Specialist Judges)?*, 37 C. Rev. 22, 24 (2000) (stating that "[s]uch potential costs have not been taken into account in evaluations of special courts").

narrower setting.[108] Last, specialist courts run the risk of specialist jurisprudence developing in isolation from the general body of law, leading to stagnation.[109]

Notwithstanding these concerns, a recent survey indicates that there is an increasing trend in favor of specialized adjudication globally.[110] Even in those countries without specialist intellectual property courts, practitioners are overwhelmingly in favor of their creation.[111]

We are of the view that the advantages associated with a specialist adjudicator (particularly in terms of speed of adjudication) outweigh the disadvantages, particularly when it comes to developing countries and the need to dispense with interim relief and have expedited trials. However, the precise institutionalized form of such specialized adjudication is debatable. One possibility is to formalize existing informal arrangements in Indian courts where intellectual property specific benches are formed.[112] As it stands now, the highest volume of patent litigation is spread across courts in five cities: New Delhi, Chennai, Mumbai, Kolkata, and Gujarat, which accounted for seventy-four of the total eighty-five patent infringement cases.[113] Therefore, specialist benches could be instituted at each of these courts.[114]

---

108. *See* Deepak Somaya & Christine A. McDaniel, *Tribunal Specialization and Institutional Targeting in Patent Enforcement*, ORG. SCI., June 17, 2011, at 12, 16–18; *see also* AUSTRALIAN LAW REFORM COMM'N, MANAGING JUSTICE: A REVIEW OF THE FEDERAL CIVIL JUSTICE SYSTEM, 406, 493 (2000), *available at* http://www.alrc.gov.au/sites/default/files/pdfs/publications/ALRC89.pdf.

109. *See* Stefan Kirchner, *Legal Unity Through Specialized Courts on a European Level?*, 6 GERMAN L.J. 217 (2005), *available at* http://www.germanlawjournal.com/pdfs/Vol06No01/PDF_Vol_06_No_01_217-226_Developments_Kirchner.pdf; Lawrence Baum, *Specialization and Authority Acceptance: The Supreme Court and Lower Federal Courts*, 47 POL. RES. Q. 693 (1994).

110. *See* INT'L BAR ASS'N, INTERNATIONAL SURVEY OF SPECIALISED INTELLECTUAL PROPERTY COURTS AND TRIBUNALS 3 (2007), *available at* http://www.ibanet.org/Document/Default.aspx?DocumentUid=7F5A1221-6C07-4CE7-A628-1F457A2433A5.

111. *Id.* A poll conducted on a popular Indian blog, SpicyIP, also showed that the majority of those who took the poll preferred a specialist body. Only 3 percent voted in favor of steering clear of specialist courts. *See* Shamnad Basheer, *A Specialist IP Court: Knowing More about Less?*, SPICYIP (Apr. 2, 2012), http://spicyipindia.blogspot.in/2012/04/specialist-ip-court-knowing-more-about.html.

112. Basheer, *supra* note 111.

113. *See supra* note 5 (highlighting our study showing that only eighty-five patent infringement litigations have been decided by courts up until July 2012).

114. Pravin Anand, a leading intellectual property counsel in India, opines that such a specialized bench operating within the jurisdictional superintendence of the High Courts is far more preferable to a stand-alone intellectual property tribunal. Justice Ruma Pal, a retired Supreme Court judge, endorses this view. A majority of voters who took the poll on the SpicyIP blog preferred a specialized intellectual property bench within existing high courts to a specialized tribunal such as the IPAB. *See* Basheer, *supra* note 111.

The advantage of this approach is that the specialist benches fall within the overall framework of regular courts, necessitating no amendments to Indian law.[115] However, one must devise appropriate eligibility criteria for judges who are to sit on this specialist bench. Second, the judges must be given intensive intellectual property training. Last, it must be ensured that these judges sit on the intellectual property bench for a significant period of time, without being made to rotate every two to three years.[116]

A recent Commercial Division of High Courts Bill, 2009, proposed the setting up of specialist "commercial" benches in every High Court to hear commercial disputes, including intellectual property disputes valued at Rs 5 Crores or higher.[117] However, the Bill attracted sharp criticism in the upper house of Parliament, owing to poor drafting.[118]

Alternatively, a distinct intellectual property court or tribunal could be created, such as the Court of Appeals for the Federal Circuit in the United States,[119] or the Patents Court of the High Court and the Patents County Court in the United Kingdom.[120]

It bears noting that India has a specialist intellectual property tribunal called the Intellectual Property Appellate Tribunal (IPAB), with the jurisdictional authority to hear appeals in certain cases from decisions of the patent and trademark office. However, it does not have the competence to hear disputes pertaining to infringement.[121] Unfortunately, the IPAB suffers from serious design flaws

---

115. Such matters could be effectively regulated through court rules. *See, e.g., Bombay High Court (Original Side) Rules*, BOMBAY HIGH CT., http://bombayhighcourt.nic.in/libweb/rules/OSrules/bhcosrules.html (last visited Sept. 3, 2013).

116. Unfortunately, under the current framework, specialist intellectual property judges are made to rotate to other benches within a short span of being on the intellectual property bench. The assignments are usually fixed by the Chief Justice of the concerned high court. *See, e.g., id.*

117. *See* PRS Legislative Brief, The Commercial Division of High Courts Bill (2009), *available at* http://www.prsindia.org/administrator/uploads/media/Division%20High%20Courts/Brief%20on%20Commercial%20division%20of%20High%20Courts%20Bill%202009.pdf.

118. *See* Sumathi Chandrashekara, *Parliament Returns "Badly Drafted" Commercial Division*, SPICYIP (Dec. 14, 2011), http://spicyipindia.blogspot.in/2011/12/parliament-returns-badly-drafted.html.

119. The US Court of Appeals for the Federal Circuit has nationwide jurisdiction of all patent appeals from trial courts, and also appeals from the Patent Trial and Appeal Board, and has been characterized as the "most powerful and influential force in the US Patent system." *See* R. Polk Wagner & Lee Petherbridge, *Is the Federal Circuit Suceeding? An Empirical Assessment of Judicial Performance*, 152 U. PENN. L. REV. 1105 (2004), *available at* https://www.law.upenn.edu/fac/pwagner/FedCir/claimconstruction/wagner_federal_circuit.pdf.

120. *See* INT'L BAR ASS'N, *supra* note 110, at 31.

121. Under current law, it can only decide appeals from the intellectual property office and revocation petitions. *See generally Jurisdiction*, INTELL. PROP. APP. BOARD (IPAB), http://

and is currently the subject of a constitutional challenge before the courts.[122] The challenge alleges that the IPAB is not sufficiently independent of the executive; that the qualifications for the appointment of judicial members on the board are contrary to the law of the land; and that irregular appointments have been the norm rather than the exception.[123] If such infirmities are remedied in the future, the IPAB may offer scope for the vesting of specialist intellectual property adjudicatory functions.[124]

3. Specialist Advisers

We recommend that courts adjudicating a patent dispute appoint an independent expert to help them appreciate the "technical" aspects of the dispute, such as the technology underling the invention and the distance between it and the closest prior art. The Patent Act provides for the appointment of such specialist advisers by the court, either *suo moto* or on the application of parties.[125]

However, the assistance of the expert is limited to questions of fact or opinion expressly formulated by the court and does not extend to an interpretation of the law.[126] This provision has already been invoked by courts in some cases, including *Cipla*, a heavily contested patent litigation where an expert was appointed to help the court appreciate the scientific and technical nuances of the case before it.[127]

---

www.ipab.tn.nic.in/Jurisdiction.htm (last visited Sept. 3, 2013); Writ Petition Challenging the Constitutionality of the IPAB in *Shamnad Basheer v. Union of India, available at* www.spicyip.com/docs/IPABwrit.doc [hereinafter Writ Petition] (last visited Sept. 3, 2013).

122. *See* Prashant Reddy, *Madras High Court Admits the Petition Challenging the Constitutionality of the IPAB and the Copyright Board*, SpicyIP (Jan. 27, 2011), http://spicyipindia.blogspot.in/2011/01/breaking-news-madras-high-court-admits.html.

123. *See generally* Writ Petition, *supra* note 121.

124. Lakshmikumaran, a leading intellectual property counsel in India, is of the view that a properly constituted IPAB would make for a good specialist intellectual property tribunal within which to vest all intellectual property adjudicatory functions.

125. *See* India Patent Act, *supra* note 25, § 115. More than a year ago, the Indian Patent Office released a list of such advisors after calling for applications from interested parties. *See List of Scientific Advisers*, http://ipindia.nic.in/List_ScientificAdvisors_30August2010.pdf (last visited Sept. 3, 2013).

126. *See* India Patent Act, *supra* note 25, § 115 ("Scientific Advisers (1) In any suit for infringement or in any proceedings . . . the court may . . . appoint an independent scientific adviser to assist the court or to enquire and report upon any such *question of fact or of opinion (not involving a question of interpretation of law)* as it may formulate for the purpose . . . .") (emphasis added).

127. Bayer Corp. v. Cipla Ltd., CS(OS) No. 523/2010 (Delhi H.C. Sept. 17, 2013) (India).

IV. TRIPS Compliance

One might argue that dispensing with an interim phase in complex patent disputes, as proposed in this chapter, contravenes the minimum mandatory obligations under the TRIPS Agreement. On a close reading however, it is clear that TRIPS provisions admit some flexibility in this regard.

Article 50 of the TRIPS Agreement deals with provisional measures, and provides, inter alia, that "[j]udicial authorities shall have the authority to order prompt and effective provisional measures."[128] The provision contemplates interlocutory or interim injunctions.[129] It bears noting that we do not advocate a dispensation with interim injunctions in all cases—rather, we propose that it be dispensed with in only those complex cases where it is difficult to arrive at a prima facie assessment of the strength of each party's case. The judicial authority continues "having the authority" to order provisional measures.[130]

Article 50(3) of the TRIPS Agreement is also relevant in this context, providing that:

> The judicial authorities shall have the authority to require the applicant to provide any *reasonably available evidence* in order to satisfy themselves with *a sufficient degree of certainty* that the *applicant is the right holder* and that the *applicant's right is being infringed or that such infringement is imminent*, and to order the applicant to provide a security or equivalent assurance sufficient to protect the defendant and to prevent abuse."[131]

This provision envisages a situation where a member state court refuses to grant a temporary injunction on the ground that it is not satisfied with a sufficient degree of certainty that the applicant is the right holder and that the applicant's rights are being infringed. In other words, a court could refuse to grant an interim injunction on the ground that it is incapable of correctly assessing the relative merits of a complex patent case at the interim stage.

Last, Article 44(2) of TRIPS reads as below:

> Notwithstanding the other provisions of this Part and provided that the provisions of Part II specifically addressing use by governments, or by

128. TRIPS Agreement, *supra* note 1, art. 50(1).

129. *See* Daniel Gervais, The TRIPS Agreement: Drafting History and Analysis 466, ¶ 2.401 (2008).

130. *Id.* at 447, ¶ 2.366.

131. TRIPS Agreement, *supra* note 1, art. 50(3) (emphasis added).

> third parties authorized by a government, without the authorization of the right holder are complied with, Members may limit the remedies available against such use to payment of remuneration in accordance with subparagraph (h) of Article 31. In other cases, the remedies under this Part shall apply or, where these remedies are inconsistent with a Member's law, declaratory judgments and adequate compensation shall be available.[132]

As seen from the above, Article 44(2) specifies that, in cases where the grant of an injunction is inconsistent with the laws of the member state, declaratory judgments and adequate compensation shall be the available remedies. In other words, member states are free to craft laws denying injunctions in specific patent cases and providing for adequate monetary compensation.

Conclusion

Interim injunctions are potent weapons in the enforcement arsenal of a pharmaceutical patent owner. However, where wrongly granted, they cause significant losses to generic competitors and consumers. Therefore, courts ought to be extremely cautious when granting this extraordinary remedy.

The prevalent UK "triable issue" threshold for determining the existence of a prima facie case for the grant of an interim injunction is an abysmally low one; adopting it carries the significant risk of reversal at the final stage, where the patent is found to be either invalid or not infringed. Given this risk, this chapter advocates for a more rigorous prima facie standard, where the court assesses the arguments of each side to determine who is more likely to win at trial. However, because a short window may not be conducive to such a robust relative assessment in many complex patent cases, we advocate that countries—particularly developing countries that are relatively new to complex patent matters—dispense with the interim injunction phase in such cases and move directly to trial.

We are mindful of the fact that such dispensation may be disadvantageous to an intellectual property owner who is interested in expeditious enforcement. We therefore offer suggestions for expediting trials by, inter alia, ensuring that all evidence is reduced to writing and that only a limited number of experts be permitted to testify in court. We also propose specialist intellectual property courts/adjudicators to improve the speed and accuracy of intellectual property adjudication and to enhance the technical capacity of courts through the appointment of technical experts.

---

132. *Id.* art. 44(2).

Although the Indian Supreme Court leans toward a dispensation with the interim phase, we take issue with the broad manner in which the Court advocates the proposition, that is, that it apply to all intellectual property cases, including those where it is relatively easy to assess the existence of a prima facie case in favor of the plaintiff.

Many of the policy reforms advocated in this chapter could be suitably and contextually adapted to suit the needs of other developing countries that face similar institutional constraints to India.

20 A RESEARCH AGENDA FOR THE COMPARATIVE LAW AND ECONOMICS OF PATENT REMEDIES

Thomas F. Cotter*

Introduction

Over the past two decades, a small but growing number of law-and-economics scholars have begun to apply the standard tools of economic analysis to a field that long had evaded scrutiny by the law-and-economics community, namely that of comparative law.[1] Research within this still-emerging discipline of comparative law and economics thus far has addressed a wide variety of foreign and domestic legal rules, standards, and institutions, from both a "positive" perspective (using the tools of economic analysis to explain, or predict the consequences of, observed similarities and differences among relevant

---

* Briggs and Morgan Professor of Law, University of Minnesota Law School. Since completing an earlier version of this chapter, I also have benefited from additional research conducted at the Max Planck Institute in Munich, Germany, in connection with my forthcoming book project on the law and economics of comparative patent remedies, and from presentations I have made on this chapter at the University of San Diego and University of Minnesota Law Schools. I also thank Suzanne Thorpe, Piper Walters, and David Zopfi-Jordan for excellent assistance in tracking down sources, and Theresa Stadheim and Sanjiv Laud for research assistance. Any errors that remain are mine.

1. For overviews of the field of comparative law and economics, see, e.g., Raffaela Caterina, *Comparative Law and Economics*, *in* ELGAR ENCYCLOPEDIA OF COMPARATIVE LAW (Jan M. Smits ed., 2006); Gerrit De Geest & Roger Van den Bergh, *Introduction*, *in* 1 COMPARATIVE LAW AND ECONOMICS, at ix (Gerrit de Geest & Roger van den Bergh eds., 2004); Florian Faust, *Comparative Law and Economic Analysis of Law*, *in* OXFORD HANDBOOK OF COMPARATIVE LAW (Mathias Reimann & Reinhard Zimmermann eds., 2006); UGO MATTEI, COMPARATIVE LAW AND ECONOMICS (1997); Ugo Mattei et al., *Comparative Law and Economics*, *in* ELGAR ENCYCLOPEDIA OF LAW AND ECONOMICS (2000); Gillian K. Hadfield, *The Strategy of Methodology: The Virtues of Being Reductionist for Comparative Law*, 59 U. TORONTO L.J. 223 (2009); Ralf

rules, standards, and institutions) as well as a "normative" one (using economic tools to evaluate or critique those similarities and differences).[2] And although some observers have critiqued certain applications of comparative law and economics for their alleged oversimplifications,[3] influential scholars within the field have recognized the need for serious comparative work to account not only for "law on the books" but also for the "law in action,"[4] have stressed the importance of institutional design and path dependency on the types of solutions legal systems develop to solve similar problems,[5] and have cautioned against making facile distinctions between civil and common law approaches in evaluating possible grounds for divergence among different systems.[6]

To date, however, comparative law and economics scholars have devoted relatively little attention to the law of intellectual property—a lacuna that is in some ways surprising, given the professedly instrumental[7] character of (much of) intellectual property law and policy. This chapter presents a modest effort to begin remedying this state of affairs, by presenting a proposed research agenda for a comparative law and economics analysis of one specific set of issues within the law of intellectual property: the law of patent remedies. (As of this writing, I am

---

Michaels, *The Second Wave of Comparative Law and Economics?*, 59 U. Toronto L.J. 197 (2009).

2. For some recent examples, see, e.g., Nuno Garoupa & Tom Ginsburg, *The Comparative Law and Economics of Judicial Councils*, 27 Berkeley J. Int'l L. 52 (2009) (presenting an economic explanation of judicial councils as a response to a principal–agent problem, and predicting the effects of different versions of the judicial council model on judicial independence and accountability); Alberto Monti, *Environmental Risk: A Comparative Law and Economics Approach to Liability and Insurance*, 9 Eur. Rev. Priv. L. 51 (2001) (evaluating and critiquing different approaches to environmental liability under US, German, and French law); Francesco Parisi et al., *The Comparative Law and Economics of Pure Economic Loss*, 27 Int'l Rev. L. & Econ. 29 (2007) (explaining the underlying economic logic of facially inconsistent European rules relating to tort recoveries for economic loss).

3. *See, e.g.*, Gillian K. Hadfield, *The Levers of Legal Design: Institutional Determinants of the Quality of Law*, 36 J. Comp. Econ. 43 (2008) (critiquing research purporting to attribute differences in economic performance among nations to their adherence to a common law or civil law system); Michaels, *supra* note 1, at 200–02 (similar).

4. *See* Mattei et al., *supra* note 1, at 518–19 (distinguishing "the working rule, that is the rule applied in a given case (traditionally known as law in action)" from "the legal justifications that are needed in the system to ground the application of this rule").

5. *See* Caterina, *supra* note 1, at 164–65; Mattei et al., *supra* note 1, at 521–27.

6. *See* Mattei et al., *supra* note 1, at 519–21; Hadfield, *supra* note 1, at 226–27.

7. That is, the view of intellectual property law as primarily a means to an end. To be sure, natural rights theories also pervade many aspects of intellectual property law, most notably copyright law as it exists outside the United States. In most countries, however, patents and trademarks are viewed largely as a means to an end—in the case of patents, as a means for encouraging investment in invention, disclosure, innovation, and the like.

in the process of implementing this agenda myself in connection with a recently published book on the comparative law and economics of patent remedies.) A comparative analysis of the law of patent remedies is particularly apt at this time in history, given the increasing importance of, and changes to, this body of law both within and outside the United States over the last decade or so. Within the United States in particular, in recent years critics have argued that US patent remedies law has sometimes exacerbated the phenomenon of "patent holdup," whereby some patent owners have been able to exploit their market power ex post by threatening to enjoin the marketing of widely distributed products or services that (often inadvertently) incorporate one or more allegedly infringing components.[8] By the mid-2000s, concerns over patent holdup were among the reasons cited by critics advocating various substantive and procedural reforms of the US patent system; as discussed below, US courts have embraced some of these proposed reforms on their own in lieu of legislative action, including certain modifications of the law of remedies. Debates over the reform of patent remedies law nevertheless are likely to continue within the United States, as courts struggle to apply standards that remain, in many important respects, works in progress. If nothing else, then, the experience of other countries in confronting (or avoiding) the types of problems that have given rise to calls for patent reform in the United States should be of interest to policymakers considering further changes to a still-evolving US system. More important, perhaps, as patent protection continues to become an increasingly globalized phenomenon, both US inventors seeking patent protection and enforcement abroad, and their foreign counterparts seeking protection and enforcement in the United States, would do well to better understand the similarities and differences between different nations' approaches to patent remedies. A systematic comparative analysis of patent remedies among the world's major patent systems therefore is long overdue. To the extent economic analysis can be of use both in explaining the relevant differences and similarities, as well as, where appropriate, in suggesting reasons to favor one approach over another, it may perform a valuable service in rendering patent laws better able to achieve the right balance between the provision of incentives to invent and disclose, on the one hand, and ensuring access to innovation, on the other.

Section I lays out a proposed protocol for the use of economic analysis to better understand, evaluate, and critique the law of patent remedies as it exists both in the United States and abroad. Section II focuses on the two principal remedies for patent infringement: permanent injunctions and monetary damages.[9]

---

8. For a discussion of patent holdup and its implications for the law of remedies, see Thomas F. Cotter, *Patent Holdup, Patent Remedies, and Antitrust Responses*, 34 J. Corp. L. 1151 (2009).

9. My recently published work explores several other remedies as well, including provisional injunctions and declaratory judgments, as well as the topics mentioned above in greater depth.

In particular, Section II outlines the similarities and differences among various nations' approaches to injunctions and damages for patent infringement; suggests some possible explanations for, and consequences of, the perceived differences; and proposes some areas for further research.

I. A Four-Step Protocol

In this section, I set out a proposed protocol for engaging in the comparative analysis of patent remedies. The four steps consist of (1) identifying differences and similarities with respect to the competing rules, standards, or institutions at issue; (2) positing possible reasons for the differences, if any; (3) evaluating the plausibility of the posited reasons; and (4) where possible, reaching a normative judgment as to the merits of the competing rules, standards, or institutions. The following paragraphs elaborate upon these points.

A. Identifying Differences and Similarities

In theory, the first step of identifying the differences and similarities with respect to the competing rules, standards, or institutions at issue should be the simplest: one simply observes how two or more countries' patent systems calculate reasonable royalties, for example (or whether they award them at all), and tally up all the points of similarity and difference. In practice, of course, matters are rarely quite so simple. Most obviously, there can be practical obstacles, such as language barriers and the accessibility of relevant materials, that can render the process of comparison difficult and expensive. Additionally, the researcher must keep in mind that even if the articulated legal rules followed by two or more countries are similar, their application could be very different in practice. Two countries may employ similar criteria in awarding reasonable royalties, for example, and yet weigh those factors differently; average awards in two countries therefore could be quite different, even in cases that appear facially similar. Moreover, if one takes a broader view of the law as consisting not only of formal rules and their application but also of the institutions that generate and apply these rules, and of other empirical realities (e.g., how often are patents litigated?), the points of potential similarity and of difference will be magnified. Of course, depending on the problem at issue one may choose to focus on a subset of similarities and differences and consider some of the other datapoints (e.g., frequency of litigation, number of patents issued, rules for claim construction) fixed—though in reality no datum is likely to be entirely independent of the others. A country that awards a relatively large number of patents per capita, for example, may attract more lawsuits, giving rise to more high-profile damages awards, which in turn may attract more patent filings, etc.

B. Positing Possible Reasons for the Differences

The second step is to posit possible reasons for the observed differences in rules, standards, or institutions. Suppose, for example, that courts in two countries employ different techniques for calculating reasonable royalties. If each methodology is based firmly and explicitly on a statutory text or binding judicial opinion, the question of why each country employs a different method is, in one sense, easily resolved: judges are simply following the law. In another sense, however, this answer merely pushes the inquiry back one level—why did the legislature or the earlier court choose the methodology at issue, and why hasn't it changed?—though there may not always be a clearly discernible answer to this question. Where the difference at issue is not clearly traceable to any binding legislative, judicial, or constitutional constraint but is more a matter of judicial, legislative, or administrative discretion, then one needs to ask why the relevant policymaker selected the rule in question over some alternative. Again, there may turn out to be no definitive answer, but at this stage the task is merely to generate hypotheses. Is the difference a matter of political judgment? Different understandings of the efficiency consequences of the rule (or different weights accorded to different types of efficiency, e.g., static and dynamic)? Might administrative or error costs play a role in favoring one rule over another? Does fairness play a role, and if so what is it likely to mean in the context at issue? Could it be a matter of training or expertise on the part of the relevant participants in the decision-making process? Do path dependence or other institutional constraints make it difficult to change course once a particular rule has gained ground? Or perhaps it is all a matter of different legal or judicial cultures—though one should exercise caution before indulging in facile assumptions about cultural differences, and should be cognizant that explaining the otherwise unexplainable by reference to culture may simply be another way of saying "I don't know."

Alternatively, some differences may be more a matter of outcomes than of legal doctrine as such. Imagine, for example, that two countries purport to follow the same factors in deciding whether to grant permanent injunctions to prevailing patentees, but that in one country prevailing patentees almost always obtain injunctions, whereas in the other patentees succeed in obtaining injunctions only about half the time. What might account for this discrepancy? Could it be a matter of the types of patents granted or litigated in each country (e.g., does one country have more or less reason to be concerned about potential patent holdup)? Could the practical availability of injunctive relief be dependent, in part, on courts' confidence in their ability to award appropriate prospective damages (a type of compulsory license)? Are there other plausible explanations?

C. Positive Analysis

Having assembled the possible reasons for the observed differences, the researcher then may attempt to evaluate each for its plausibility. Going back to the reasonable royalty example, if two countries employ different methodologies, how likely is it that the difference arises merely from a difference of opinion as to which rule better promotes innovation? Might the routine availability (or unavailability) of discovery in a given system result in differences in the ways that damages are calculated? Are administrative costs likely to be higher in one system or another, and if so is this a possible causal factor? Alternatively, given the rules that exist, what consequences (if any) are likely to follow? Might one expect one nation or the other to have a marginally higher rate of innovation as a result of following its favored rule, or are any such differences unlikely to be traceable to any single feature of the patent system? Can any of the hypotheses generated under step 2 be subjected to rigorous empirical testing?

D. Normative Analysis

Finally—depending on how far the analysis has proceeded, and in part perhaps on the researcher's taste for this sort of thing—one may venture to offer some tentative conclusions as to whether one nation's solution to a given problem is, on balance, better than another's. To be sure, one needs to exercise caution in this regard. The efficacy of any one rule may depend in large part on a host of other supporting rules and practices that lie beyond the scope of the researcher's understanding. Moreover, the underlying rationale for a rule, or the consequences of its abolition, may not be apparent even to participants within the system. As Burke, Hayek, and many others have periodically reminded us throughout history, sometimes tradition really might embody the wisdom of the many;[10] and as contemporary comparative law and economics also teaches, local conditions may be more or less favorable at any given time for the success of certain "legal transplants."[11] To return to the reasonable royalty example, suppose that one concludes that the difference in approach between Country A and Country B is most likely a result of administrative costs—in particular, that the limited availability of discovery in Country B affects the way in which royalties are calculated because one cannot routinely employ the litigants' own internal financial data in crafting the award. As a result, Country B resorts more to self-reported industry

10. *See* Thomas F. Cotter, *A Burkean Perspective on Patent Eligibility*, 22 BERKELEY TECH. L.J. 855, 856–57 (2007); Thomas F. Cotter, *A Burkean Perspective on Patent Eligibility Part II: Reflections on the (Counter)Revolution in Patent Law*, 11 MINN. J.L. SCI. & TECH. 365, 369–70 (2011).

11. *See, e.g.*, Mattei et al., *supra* note 1, at 521–25.

averages or arbitrary "rules of thumb" than does Country A. In such a case—and depending on the content of Country A's rule, of course—each country's rule may be adequately suited to its own unique legal environment. Given the constraints under which Country B's system operates, changing Country B's rule to resemble Country A's might result in awards that depart even farther from some hypothetical ideal—or at least, one may be unable to draw any more definitive conclusion given the limitations of the data.

On the other hand, sometimes traditions are dysfunctional, too, and sometimes countries *ought* to consider a transplant to keep the patient healthy, as it were.[12] A certain modesty in assessing the efficacy of one's own, or another nation's, laws and institutions hardly means that one should offer no opinions whatsoever. Particularly if two nations both espouse the same rationale for a given rule (e.g., encouraging innovation) and yet the bulk of evidence strongly suggests that one rule is much better suited to this end than is another, there is no obvious reason to be reticent in recommending reform. Even in less stark cases, one may respectfully offer revisable, tentative opinions on whether one nation's rule is likely to better serve the goals of the patent system than is another's—though as noted above, it may be the case that both rules are adequately suited for their unique environments. And even in the example presented above, it might not be inappropriate for the researcher to suggest that Country B's limits on discovery hamper the efficacy of Country B's damages rules, and that more liberal discovery rules (though hardly costless) might improve conditions ceteris paribus. (As always, though, one may question how likely it is that, in the real world, those *ceteres* would be *pares*.)

Finally, any researcher engaged in normative analysis probably should make clear his or her normative premises at the outset—particularly if those premises themselves are likely to exert some influence on the data to be considered or the weighting of plausibility in the positive analysis. In this sense, the division among the four steps can never be sharply maintained; one's judgment on the positive side is likely to be affected by one's normative preferences, and vice versa. In the following section, I will try to make clear my own normative views, as expressed in previous work,[13] while recognizing that some of the normative recommendations

---

12. *See* Ugo Mattei, *Efficiency in Legal Transplants: An Essay in Comparative Law and Economics*, 14 INT'L REV. L. & ECON. 3, 16 (1994).

13. *See* Federal Trade Commission (FTC), Hearings on the Evolving IP Marketplace (Dec. 5, 2008) (testimony of Thomas F. Cotter), *available at* http://www.ftc.gov/bc/workshops/ipmarketplace/dec5/081205transcript.pdf (discussing relevant material at pages 131–40); ROGER D. BLAIR & THOMAS F. COTTER, INTELLECTUAL PROPERTY: ECONOMIC AND LEGAL DIMENSIONS OF RIGHTS AND REMEDIES (2005); Cotter, *supra* note 8; Thomas F. Cotter, *An Economic Analysis of Enhanced Damages and Attorneys' Fees for Willful Patent*

I have made against the background of the US patent system may not always translate well into other patent systems having substantially different background features and assumptions. Indeed, even within the US context alone there are serious researchers whose conclusions on any given issue will differ from mine,[14] and some of my policy recommendations depart from the rules that US courts routinely continue to follow. In the pages that follow I will acknowledge these competing views.

II. A Preliminary Research Agenda

This section outlines the two principal remedies for patent infringement, namely permanent injunctions and damages. With respect to each, I will describe the ways in which some of the world's major patent systems approach certain common problems, propose some possible explanatory hypotheses and some possible consequences of the different approaches, and suggest some ways in which further research may help to test the hypotheses or lead to new ones.

A. Injunctions

From the patentee's perspective, the most important remedy is often an injunction, that is, a court order demanding that the infringer cease infringing. Indeed, a patentee who prevails at trial almost always demands what is referred to in the United States as a permanent injunction; on occasion, patentees also seek various types of interim injunctions pending trial.[15] As I have discussed in other work, one fairly mainstream view within the US law-and-economics literature has

---

Infringement, 14 FED. CIR. B.J. 291 (2004); Thomas F. Cotter, Symposium Review, *Four Principles for Calculating Reasonable Royalties in Patent Infringement Cases*, 27 SANTA CLARA COMPUTER & HIGH TECH. L.J. 725 (2011), *available at* http://ssrn.com/abstract=1736147 [hereinafter Cotter, *Four Principles*]; Thomas F. Cotter, *Patent Remedies and Practice Reason*, 88 TEX. L. REV. 125 (2010) [hereinafter Cotter, *Patent Remedies*].

14. *See generally* BLAIR & COTTER, *supra* note 13, at 61–66 (responding to an argument set forth in Ian Ayres & Paul Klemperer, *Limiting Patentees' Market Power Without Reducing Innovation Incentives: The Perverse Benefits of Uncertainty and Non-injunctive Remedies*, 97 MICH. L. REV. 985, 1028–31 (1999)); Cotter, *supra* note 8, at 1177–78 n.137 (responding to other models of patent damages).

15. In the United States, plaintiffs sometimes seek temporary restraining orders and/or preliminary injunctions pending trial. The U.S. Court of Appeals for the Federal Circuit has recently clarified the framework for awarding preliminary injunctions in patent litigation. *See* Titan Tire Corp. v. Case New Holland, Inc., 566 F.3d 1372 (Fed. Cir. 2009) (stating that a plaintiff must show "[1] that he is likely to succeed on the merits, [2] that he is likely to suffer irreparable harm in the absence of preliminary relief, [3] that the balance of equities tips in his favor, and [4] that an injunction is in the public interest," and setting forth a framework

been that protecting patent and other intellectual property rights by means of "property" rules (that is, conferring upon the patentee a right to obtain injunctive relief) is preferable to protecting them by means of "liability" rules (entitling would-be users to infringe at will and pay damages).[16] In large part, the rationale for preferring a property-like entitlement comes down to a matter of comparative information costs. Because intellectual property rights are unique and can be difficult to value, the argument goes, and because the owner and would-be user are likely to have superior information (in comparison with a court or other governmental agency) as to what that value is, the intellectual property system is better served by encouraging the parties to negotiate a license than by setting the terms of such a license for them. A right to injunctive relief, in turn, encourages licensing insofar as it provides the would-be user with an incentive to seek out and negotiate a license for a desired technology—to bargain in the shadow of the law—rather than merely to appropriate the technology and wait for a court to assess damages.[17]

---

for determining whether the defendant has presented a "substantial question" on patent validity sufficient to call into question the plaintiff's likelihood of success on the merits). Other countries generally permit preliminary injunctions as well, subject to conditions that vary somewhat from one place to another. For general discussion, see, e.g., Larry Coury, Note, *C'est What? Saisie! A Comparison of Patent Infringement Remedies Among the G7 Nations*, 13 FORDHAM INTELL. PROP. MEDIA & ENT. L.J. 1101 (2003); *see also* Directive 2004/48 of the European Parliament and of the Council of 29 April 2004 on the Enforcement of Intellectual Property Rights, art. 9.1.a, 2004 O.J. (L 195) 16, 22 (EC) [hereinafter EC Enforcement Directive] (requiring member states to "ensure that judicial authorities may... issue against the alleged infringer an interlocutory injunction"); *Analysis of the Application of Directive 2004/48/EC of the European Parliament and the Council of 29 April 2004 on the Enforcement of Intellectual Property Rights in the Member States*, at 13–14, SEC (2010) 1589 final (Dec. 22, 2010) [hereinafter Analysis of Enforcement Directive]. Other types of interim relief, including injunctions requiring the provisional discovery and/or seizure of allegedly infringing merchandise (*Anton Piller* orders or *saisies contrefaçon*), and the preservation of assets pending trial (*Mareva* orders) also can be important in European patent litigation. *See* EC Enforcement Directive, *supra*, arts. 7, 9.2; Analysis of Enforcement Directive, *supra*, at 7–8; Coury, *supra*, at 1120–21, 1125, 1152–54. Also important is the question of whether or when courts in one state may issue cross-border injunctions. *See, e.g.*, Analysis of Enforcement Directive, *supra*, at 18; Jochen Bühling, *Cross-Border Injunctions in Patent Infringement Cases: Paradise Lost?*, *in* BUILDING AND ENFORCING IP VALUE (2007), *available at* http://www.buildingipvalue.com/07EU/p.172-175%20Krieger.pdf; Marketa Trimble, *Cross Border Injunctions in U.S. Patent Cases and Their Enforcement Abroad*, 13 MARQUETTE INTELL. PROP. L. REV. 331, 354–67 (2009). As noted above, in my forthcoming book, I hope to focus on these issues in addition to the matters discussed in the text above.

16. *See* BLAIR & COTTER, *supra* note 13, at 38–41. The property rule/liability rule framework comes from Guido Calabresi and A. Douglas Melamed's seminal work, *Property Rules, Liability Rules and Inalienability: One View of the Cathedral*, 85 HARV. L. REV. 1089 (1972).

17. *See* BLAIR & COTTER, *supra* note 13, at 40–41. In addition, in a case in which the would-be user is a less efficient user of the intellectual property at issue (e.g., a higher-cost manufacturer of products covered by the patent in suit) than is the right holder, the better solution is for

In recent years, however, this rationale has come under scrutiny in two principal contexts. First, in some cases the public interest in accessing a product covered by an intellectual property right may be sufficiently high to justify the imposition of a compulsory license on terms set by a governmental body. A common example, even in the United States, is the US government's assertion of the power to "take" a license to use patents and copyrights (for example, for national defense) in exchange for fair market value.[18] Another example, traditionally more common in other countries and the subject of intense debate among members of the World Trade Organization since the beginning of this century, is the national interest in ensuring widespread access to certain types of innovations, such as patented essential medicines.[19] Second, in recent years many observers have expressed concern over the phenomenon of "patent holdup," whereby a patent owner (often a nonpracticing entity) asserts patent rights in a relatively small-value part of a complex end product. When such rights are asserted after the end product has achieved a degree of market success (and the patented component at issue may be difficult to immediately design around), the patent owner who can obtain injunctive relief potentially stands to extract a settlement that is far out of proportion to the ex ante value of the patent in suit.[20] Although in theory, producers could protect themselves against patent holdup simply by identifying and, where necessary, negotiating licenses to relevant patents in advance of launching a new product, in reality such preclearance is often infeasible, particularly in many high technology markets.[21] Thus in some instances the social interest in fostering innovation may be better served by awarding an ongoing reasonable royalty rather than by enjoining a defendant's use.[22]

Concerns of this nature in large part explain the course of US law, as it relates to permanent injunctions, over the past two decades. Traditionally, US courts awarded prevailing patentees permanent injunctions in the vast majority of cases, including

---

the patentee to exclude the user altogether rather than to license her use. *See id.* at 52–56. A property-rule entitlement enforces the patentee's interest in exclusion in such a case. *See also* Edmund S. Kitch, *The Nature and Function of the Patent System*, 20 J.L. & ECON. 265, 286–87 (1977) (arguing that compulsory licensing destroys the patentee's ability to coordinate investment in follow-up improvements). Finally, some observers have expressed concerns that courts or other governmental bodies are apt to systematically undervalue intellectual property rights. If such concerns are valid, the risk arises that the patent incentive will be suboptimal.

18. *See* 28 U.S.C. § 1498 (1998); Thomas F. Cotter, *Do Federal Uses of Intellectual Property Implicate the Fifth Amendment?*, 50 FLA. L. REV. 529 (1998).

19. *See* Thomas F. Cotter, *Market Fundamentalism and the TRIPS Agreement*, 22 CARDOZO ARTS & ENT. L.J. 307 (2004).

20. *See* Cotter, *supra* note 8, at 1181–82.

21. *See id.* at 1181 n.148.

22. *See id.* at 1182; Cotter, *Four Principles*, *supra* note 13, at 12.

cases in which the patentee marketed no products.[23] Indeed, in recent years, the US Court of Appeals for the Federal Circuit (the court that hears virtually all appeals in patent matters) had announced that prevailing patentees were automatically entitled to permanent injunctions, absent "exceptional circumstances."[24] By the middle part of the preceding decade, however, concerns over the risk of patent holdup had led some observers to challenge the rigidity of the Federal Circuit's rule. In 2006, the US Supreme Court sided with the critics in *eBay Inc. v. MercExchange, LLC*,[25] and held that before a court may award a permanent injunction:

> [a] plaintiff must demonstrate: (1) that it has suffered an irreparable injury; (2) that remedies available at law, such as monetary damages, are inadequate to compensate for that injury; (3) that, considering the balance of hardships between the plaintiff and defendant, a remedy in equity is warranted; and (4) that the public interest would not be disserved by a permanent injunction.[26]

In the wake of *eBay*, courts within the United States have begun to scrutinize requests for permanent injunctive relief much more carefully. At present, US district courts are awarding permanent injunctions in only about three-quarters of cases in which patentees prevail.[27] Nonpracticing entities in particular are much

---

23. *See* Cont'l Paper Bag Co. v. E. Paper Bag Co., 210 U.S. 405 (1908). Cases in which courts concluded that the public interest weighed against the award of a permanent injunction existed, but appear to have been few and far between. *See, e.g.*, Foster v. Am. Mach. & Foundry Co., 492 F.2d 1317, 1324 (2d Cir. 1974) (affirming decision not to award nonpracticing patentee an injunction); Vitamin Technologists, Inc. v. Wis. Alumni Research Found., 146 F.2d 941, 945 (9th Cir. 1944) (suggesting, in dicta, that court would have declined to enjoin infringement of patent relating to vitamin fortification had it found patent to be valid); City of Milwaukee v. Activated Sludge, Inc., 69 F.2d 577, 593 (7th Cir. 1934) (declining to award injunctive relief, where consequence would have been the emission of raw sewage into Lake Michigan).

24. *See* MercExchange, LLC v. eBay, Inc., 401 F.3d 1323, 1339 (Fed. Cir. 2005), *rev'd*, 547 U.S. 388 (2006).

25. 547 U.S. 388 (2006).

26. *Id.* at 391. In support of this holding, the Court stated that the relevant text of the Patent Act confers equitable discretion upon the courts to award injunctions; it does not create a mandatory rule. *See id.* at 392; *see also* 35 U.S.C. § 283 (2012) (stating that "[t]he several courts having jurisdiction of cases under this title *may* grant injunctions in accordance with the principles of equity to prevent the violation of any right secured by patent, on such terms as the court deems reasonable") (emphasis added). In a concurring opinion, Justice Kennedy expressed concern over both patent holdup and the vagueness with which many business method patents in particular are drafted as reasons for courts to consider denying injunctions in some cases. *eBay*, 547 U.S. at 396–97 (Kennedy, J., concurring).

27. Patstats.org lists 144 cases granting permanent injunctions post-*eBay*, as of August 11, 2011, and 48 denying them. Note, however, that in proceedings before the United States

less likely to be awarded injunctive relief.[28] In cases in which courts decline to issue permanent injunctions, they now have to grapple with the question of how to calculate reasonable royalties for the infringer's ongoing use of the patented invention.[29]

If we shift to the comparative realm, the TRIPS Agreement appears to provide considerable leeway for nations to decide whether, or under what conditions, permanent injunctions should be awarded.[30] The secondary literature nevertheless strongly suggests that permanent injunctions remain the norm in many countries.[31] To the extent most other nations remain more firmly rooted

---

International Trade Commission (USITC) pursuant to 19 U.S.C. § 1337 (2004), the entry of an order excluding infringing merchandise from entering the United States is automatic, absent a finding that the effects of one or more of the statutorily enumerated public interest factors (effect on public health and welfare, competitive conditions in the US economy, the production of like or directly competitive articles in the United States, or US consumers) counsel otherwise. No consideration is given to the four-factor *eBay* test, *see* Spansion, Inc. v. ITC, 629 F.3d 1331, 1357–60 (Fed. Cir. 2010), though in one recent case the USITC delayed entry of an exclusion for four months to afford HTC some time to design around one of Apple's patents. Conceivably, the provision of a form of automatic injunctive relief in USITC proceedings may lead some patent owners to choose the USITC, where possible, over litigation in federal district court.

28. *See* Steve Malin & Ari Rafilson, *Empirical Analysis of Permanent Injunction Decisions Following* eBay, FTC.GOV (2011), http://www.ftc.gov/bc/workshops/ipmarketplace/feb11/docs/smalin.pdf (reviewing a wide variety of factors that are relevant to awards of injunctive relief post-*eBay*).

29. *See, e.g.*, Paice LLC v. Toyota Motor Corp., 504 F.3d 1293 (Fed. Cir. 2007). Several courts have expressed the view that the royalty rate should be different from the rate used to calculate the reasonable royalty awardable for past infringement. *See, e.g.*, Amado v. Microsoft Corp., 517 F. 3d 1353, 1362 (Fed. Cir. 2008) (stating that post-trial damages "should take into account the change in the parties' bargaining positions, and the resulting change in economic circumstances, resulting from the determination of liability"); Paice LLC v. Toyota Motor Corp., 609 F. Supp. 2d 620, 630–31 (E.D. Tex. 2009). I have been critical of this development. *See* Cotter, *Four Principles*, *supra* note 13, at 12.

30. *See* Agreement on Trade-Related Aspects of Intellectual Property Rights, art. 44.1, Apr. 15, 1994, Marrakesh Agreement Establishing the World Trade Organization, Annex 1C, 33 I.L.M. 81 (1994) [hereinafter TRIPS Agreement] (stating that "[t]he judicial authorities shall have the authority to order a party to desist from an infringement," and that "[m]embers are not obliged to accord such authority in respect of protected subject matter acquired or ordered by a person prior to knowing or having reasonable grounds to know that dealing in such subject matter would entail the infringement of an intellectual property right"); CARLOS M. CORREA, TRADE RELATED ASPECTS OF INTELLECTUAL PROPERTY RIGHTS: A COMMENTARY ON THE TRIPS AGREEMENT 424 (2007) ("Given ... that Article 44.1 only mandates members to give judicial authorities the power to grant injunctions, such authorities can exercise their discretion in deciding particular cases."). Article 50 of TRIPS requires members to authorize provisional remedies (e.g., preliminary injunctions), and Article 51 measures to prevent the importation of infringing merchandise. TRIPS Agreement, *supra*, arts. 50–51.

31. *See, e.g.*, Joachim Feldges & Birgit Kramer, *Patent Law*, *in* BUSINESS LAWS OF GERMANY § 13.74 (2010) ("If the court finds the patent to be infringed and there is a risk that infringing

than the United States post-*eBay* in the "property rule" camp, however, there could be a number of plausible explanations that merit further analysis. One that immediately leaps to mind is the somewhat greater receptivity of some patent systems to compulsory licensing. Even post-TRIPS, many industrialized nations have retained the option of compulsory licensing as a solution to the blocking patents phenomenon.[32] Many of them also have in place more expansive exceptions for experimental use than are available in the United States.[33] Conceivably, these exceptions could provide a needed safety valve in a substantial number of cases in which the public interest (in the language of *eBay*) would be "disserved" by injunctive relief, and in this manner relieve some of the pressure to deviate from the injunctive relief norm in cases that do not fall within the scope of these exceptions; in other words, some cases may self-select out and wind up being negotiated rather than litigated.[34] It also could be that the phenomenon of patent

---

acts may recur (as it is the rule), the court will grant an injunction upon a request by the plaintiff without further consideration. There is no further requirement to obtain an injunction. In particular, the court will (and, indeed, must) not take any equitable factors or weigh up the involved interests of the parties."); Mistrale Goudreau, *Canada*, *in* 3 INTERNATIONAL ENCYCLOPEDIA OF LAWS: INTELLECTUAL PROPERTY § 269 (Hendrik Vanhees ed., 2004) (stating that "[a] successful plaintiff can normally obtain a permanent injunction" (citing IMMANUEL GOLDSMITH, PATENTS OF INVENTION § 532 (1981))); Junichi Kitahara & Hirokazu Honda, *Substantive Issues in a Patent Infringement Case*, *in* JAPANESE PATENT LITIGATION § 3.49 (2010) (stating that "Japanese courts automatically order permanent injunctions in most cases when they find an infringement," and that although courts may deny injunctive relief on the basis of the doctrine of "abuse of right," as yet no such cases exist; however, by posting security, defendants may delay enforcement of an injunction pending appeal); Joanna Schmidt-Szalewski & Nicolas Bouche, *France*, *in* 3 INTERNATIONAL ENCYCLOPEDIA OF LAWS: INTELLECTUAL PROPERTY 156, *supra* (stating that "an injunction to discontinue the working of the infringing technique is ordered by the court, at the plaintiff's request"). *Compare* Michael C. Elmer, *U.S. and Global Patent Litigation Forum Shopping*, COLO. BAR. ASS'N IP SEC. 6 (Aug. 27, 2009), http://www.ipsectioncolorado.org/content/PowerPoint/08-27-2009_-_Global_Patent_Litigation_Strategy.ppt (reporting that permanent injunctions are automatic in Germany and the United Kingdom, but that prevailing patentees are awarded permanent injunctions in only 80 percent of cases in France and 65 percent in Japan), *with* EC Enforcement Directive, *supra* note 15, art. 12 (stating that members "may provide that ... judicial authorities may order pecuniary compensation to be paid to the injured party instead of applying [corrective measures and injunctions] if [the defendant] acted unintentionally and without negligence, if execution of the measures in question would cause him/her disproportionate harm and if pecuniary compensation to the injured party appears reasonably satisfactory"), *and* Analysis of EC Enforcement Directive, *supra* note 15, at 21 (listing member states that have implemented this article).

32. *See* Robert P. Merges, *Intellectual Property Rights and Bargaining Breakdown: The Case of Blocking Patents*, 62 TENN. L. REV. 75, 103–05 (1994).

33. *See* Henrik Holzapfel & Joshua D. Sarnoff, *A Cross-Atlantic Dialog on Experimental Use and Research Tools*, 48 IDEA 123, 149–99 (2008).

34. *See* Merges, *supra* note 32, at 104 (reporting that disputes involving compulsory licensing for blocking patents are rarely litigated).

holdup is either not (yet?) perceived to be a problem in many countries other than the United States, in which case the need to confer discretion upon judges to deny requests for permanent injunctions may not appear pressing. Further investigation that lent support to this explanation would, in turn, demand its own explanation. Are there other mechanisms in place—such as more restrictive standards relating to patentable subject matter or non-obviousness (inventive step), alternative methods of claim construction, or an absence of continuation procedures and thus of their arguable abuse—that reduce the perceived risks of holdup in other countries?[35] Alternatively, might the asserted preference for injunctive relief be an artifact of past times, when patentees in some nations were required to "work" their patents (thus eliminating the risk of patent trolling, though not without some costs)? As Mattei has argued, legal systems sometimes evolve in ways that are evocative of evolution in the natural world, where path dependency plays a major role in determining which incremental mutations are feasible at any one time.[36] Comparative law and economics could play a useful role in predicting whether a given feature of a nation's patent laws, such as a near-automatic entitlement to injunctive relief, might become dysfunctional when it fails to evolve in synch with other changes in the patent "ecosystem."

Equally important—and this point cuts across the entire landscape of patent remedies, not just injunctions—are questions concerning matters such as the annual number of patents issued by a given patent office, the number of lawsuits filed in a given country, the percentage of cases that settle, the percentage of cases that patentees win, and so on. Some of this information (e.g., number of patents issued by a given patent office) is a matter of public record and easy to access. Other relevant information, such as the number of lawsuits filed and the settlement and win rates, can be much more difficult to obtain.[37] Recent studies by Michael Elmer and Stacy Lewis, however, present statistics on patent filings and win rates from a

---

35. *See* Dietmar Harhoff, *Challenges Affecting the Use and Enforcement of Intellectual Property Rights*, U.K. Intell. Prop. Office (IPO) 8 (May 31, 2009), www.ipo.gov.uk/ipresearch-challenges-200905.pdf (asserting that "[t]he weaker presence of 'trolls' is presumably due to the fact that the patent system in Europe deviates from the US system in several crucial points. In Europe, (i) court proceedings are much less costly, (ii) cost allocation favours the winning party, (iii) damage awards are not excessive, (iv) most courts have sought a careful balance between the rights of the parties and have not followed a blind pro-patent posture, (v) injunctions are not issued automatically, and (vi) the quality of patent examination has been considerably better than in the US, despite some weakening").

36. *See* Mattei et al., *supra* note 1, at 521–25.

37. Joseph Straus, *Patent Litigation in Europe—A Glimmer of Hope? Present Status and Future Perspectives*, 2 Wash. U. J.L. & Pol'y 403, 407–08 (2000) (reporting on the difficulty, as of 2000, of obtaining reliable information on the number and nature of cases litigated involving European patents).

variety of countries.[38] Not surprisingly, perhaps, the United States outpaced all other countries in the number of patent cases filed (China was a distant second)[39] and "offers the best opportunity for obtaining high patent damages."[40] The overall US win rate (taking into consideration cases disposed of on pretrial motions) appears comparable to the global average of 35 percent.[41] The authors report considerable variation among countries, however, with Germany appearing to be a relatively favorable venue for patentees,[42] and England and Japan much less favorable (though correspondingly more favorable for declaratory judgment plaintiffs).[43]

Assuming that the authors' methodology and data-gathering are sound, these studies and the databases they reflect may provide very useful information for testing various hypotheses to explain different countries' approaches to patent remedies. At present, however, no single hypothesis clearly outshines the others as a possible explanation for some countries' apparent adherence to an automatic, or near-automatic, injunction rule.

---

38. *See* Elmer, *supra* note 31; Michael Elmer & Stacy Lewis, *Where to Win: Patent-Friendly Courts Revealed*, Managing Intell. Prop. (Sept. 1, 2010), http://www.managingip.com/Article/2660946/Where-to-win-patent-friendly-courts-revealed.html. These studies are based on Finnegan, Henderson's Global IP Project, which appears to be a proprietary database compiled from "summaries of first-instance final decisions of *inter partes* patent infringement cases by calendar year from participating countries using a standardized case summary sheet." Elmer, *supra* note 31, at 24; *see also* Elmer & Lewis, *supra*, at 6 (explaining the authors' methodology).

39. *See* Elmer & Lewis, *supra* note 38, at 2.

40. *See id.* at 9; *see also* Elmer, *supra* note 31, at 23 (reporting average damages awards in various countries over an unspecified time period).

41. *See* Elmer, *supra* note 31, at 6 (reporting an overall US win rate of 35 percent for the years 1997–2007); Elmer & Lewis, *supra* note 38, at 9 (reporting a global win rate, not including the United States, of 35 percent based on 2006–2008 data from sixteen countries). Elmer also reports win rates of 39 percent for France and of 34 percent for China, for the years 1997–2007, though based on partial data. *See* Elmer, *supra* note 31, at 6.

42. *See* Elmer, *supra* note 31, at 6 (reporting an overall win rate of 29 percent in Germany for the years 1997–2007, though based on incomplete data); Elmer & Lewis, *supra* note 38, at 4 (reporting a win rate of 63 percent on the issue of infringement from a sample of cases filed in the Düsseldorf court of first instance between 2006 and 2009, and of 57 percent on validity challenges in the German Federal patent Court for the years 2003–2007); *see also id.* at 9 (noting some limitations of the German data).

43. *See* Elmer, *supra* note 31, at 6 (reporting overall win rates in England of 22 percent and in Japan of 20 percent for the years 1997–2007); Elmer & Lewis, *supra* note 38, at 2 (reporting win rates in Tokyo of 26 percent and in Osaka of 16 percent for the years 2006–2008); *id.* at 9 (reporting that the Patent Courts in London are "the least patentee-friendly forum in Europe" and that alleged infringers prevailed in twenty-five out of twenty-nine cases from 2006 to 2009). Elmer and Lewis also report some limited data on an industry-specific basis. *See id.* at 7 (noting, inter alia, that pharmaceutical patents fare better in the Netherlands than do electrical and chemical/materials patents, whereas in France the win rate was lower for medical devices for 2006–2008 than the overall average); *see also* Elmer, *supra* note 31, at 30–31 (reporting medical device and mechanical patent win rates for selected countries).

B. Damages

The comparative law of patent damages reveals a number of similarities, but also some important differences, among the world's major patent systems. On the one hand, it appears that most patent systems offer the prevailing patentee the opportunity to recover compensatory damages in the form of lost profits or reasonable royalties.[44] As discussed in brief below, however, there are some significant differences in the ways in which countries calculate lost profits and reasonable royalties. Some countries also condition an award of damages upon the infringer's having

---

44. The TRIPS Agreement requires member states to confer upon "judicial authorities... the authority to order the infringer to pay the right holder damages adequate to compensate for the injury the right holder has suffered because of an infringement of that person's intellectual property right by an infringer who knowingly, or with reasonable grounds to know, engaged in infringing activity." *See* TRIPS Agreement, *supra* note 30, art. 45. In many countries, the relevant statutory text (like TRIPS Article 45) does not specify what forms compensatory damages must take. *See, e.g.*, 35 U.S.C. § 284 (2012) ("Upon finding for the claimant the court shall award the claimant damages adequate for the infringement, but in no event less than a reasonable royalty for the use made of the invention by the infringer, together with interest and costs as fixed by the court."); Patents Act 1990 (Cth) § 122(1) (Austl.) [hereinafter Australian Patent Act] (authorizing awards of "damages"); Patents Act, 1977, c. 37, § 61(1)(c) (U.K.) [hereinafter UK Patents Act] ("[A] claim may be made for... damages in respect of the infringement."); Patent Act, R.S.C. 1985, c. P-4, § 55(1) (Can.) [hereinafter Canadian Patent Act] ("A person who infringes a patent is liable to the patentee and to all persons claiming under the patentee for all damage sustained by the patentee or by any such person, after the grant of the patent, by reason of the infringement."). Case law in these countries nevertheless generally recognizes awards of lost profits or reasonable royalties. *See, e.g.*, Lucent Techs., Inc. v. Gateway, Inc., 580 F.3d 1301, 1324 (Fed. Cir. 2009); Consolboard Inc. v. MacMillan Bloedel (Saskatchewan) Ltd. (1982), 63 C.P.R. 2d 1 (Can. Fed. Ct.), *aff'd in part, rev'd in part on other grounds*, (1983), 74 C.P.R.2d 179 (Can. Fed. Ct. App. Div.); Ultraframe (UK) Ltd. v. Eurocell Bldg. Plastics. Ltd., [2006] EWHC (Pat) 1344 (Eng.); Schmidt-Szalewski & Bouche, *supra* note 31, at 156 (noting that, under French law, "ordinary rules of civil liability" as set forth in the Civil Code apply to the assessment of damages in patent cases, and that in the patent context such damages may include lost profits or royalties); *see also* EC Enforcement Directive, *supra* note 15, art. 13(1) (stating that, in setting damages "appropriate to the actual prejudice suffered... as a result of the infringement," judicial authorities in member states shall take into account, inter alia, lost profits or, alternatively, award the royalties that "would have been due if the infringer had requested authorization"); Analysis of EC Enforcement Directive, *supra* note 15, at 22, 23 (noting that all member states already "take rightholders' lost profits into account" and "generally provide for lump sum damages calculated on the basis of the amount of royalties that would have been due if the infringer had sought an authorization (licence) from the rightholder"); Gregory J. Urbanchuk & James R. Tumbridge, *Patent Damages: The European Landscape*, J. INTELL. PROP. L. & PRAC. 1 (July 10, 2008) (surveying the rules in the United Kingdom, France, Germany, the Netherlands, and Italy). In some other countries, the statutory text explicitly alludes to lost profits or reasonable royalties, though typically without extended discussion of how to calculate these damages. *See, e.g.*, Patentgesetz [PatG] [Patent Act], as amended July 31, 2009, BGBL I at 2521, art. 139(2) (Ger.) [hereinafter German Patent Act]; Tokkyo jōrei [Patent Act], Law No. 121 of 1959, as amended, art. 102 (Japan) [hereinafter Japanese Patent Act], *translated version available at* http://www.cas.go.jp/jp/seisaku/hourei/data/PA.pdf (last visited Sept. 3, 2013); *see also* EC Enforcement Directive, *supra* note 15, art. 13.1.

acted intentionally or at least negligently, though what this sort of requirement means in practice can vary considerably from one place to another.[45] In addition, certain types of damages are recoverable in some systems but not others. In 1946, for example, the United States eliminated the remedy of an accounting of defendant's profits as a recoverable category of damages for the infringement of utility patents.[46] Many other countries, by contrast, permit this form of recovery,[47] though the frequency with which it is awarded, and the rules surrounding its proper

45. *See, e.g.*, Australian Patent Act, *supra* note 44, § 123 ("A court may refuse to award damages, or to make an order for an account of profits, in respect of an infringement of a patent if the defendant satisfies the court that, at the date of the infringement, the defendant was not aware, and had no reason to believe, that a patent for the invention existed."); U.K. Patents Act, *supra* note 44, § 62(1) (stating that neither damages nor profits shall be awarded "against a defendant or defender who proves that at the date of the infringement he was not aware, and had no reasonable grounds for supposing, that the patent existed"); Japanese Patent Act, *supra* note 44, § 102 (appearing to permit damages awards only for intentional or negligent infringement); *id.* § 103 ("An infringer of a patent right or exclusive license of another person is presumed negligent in the commission of the said act of infringement."); Peter Meier-Beck, *Damages for Patent Infringement According to German Law—Basic Principles, Assessment, and Enforcement*, 35 Int'l Rev. of Intell. Prop. & Competition L. 113, 115 (2004) (noting that under § 139(2) of the German Patent Act "the obligation to pay damages to the prejudiced party only applies to persons whose infringing actions were carried out willfuly or by negligence," but asserting that in practice this rule does not affect damages awards because courts assume that the defendant could have avoided infringement by making a more through pre-infringement investigation); *see also* TRIPS Agreement, *supra* note 30, art. 45 (requiring member states to award damages at least in cases in which the infringer "knowingly, or with reasonable grounds to know, engaged in infringing activity"); EC Enforcement Directive, *supra* note 15, art. 13(1) (similar). There also can be significant variations in procedure, with some countries (such as Germany and the United Kingdom) requiring plaintiffs to institute separate proceedings on damages. *See* Meier-Beck, *supra*, at 114 (Germany); Chelton D. Tanger et al., *Accounting for Damages in Intellectual Property Litigation*, IP Value 4 (2004) (U.K.), *available at* http://www.buildingipvalue.com/05_SF/354_359.htm.

46. *See* Roger D. Blair & Thomas F. Cotter, *An Economic Analysis of Damages Rules in Intellectual Property Law*, 39 Wm. & Mary L. Rev. 1585, 1647–48 (1998). United States law still permits awards of defendant's profits for the infringement of design patents, however. *See* 35 U.S.C. § 289 (2012). United States law also permits awards of defendant's profits in copyright and in some trademark and false advertising cases. *See* 17 U.S.C. § 504(b) (2012) (copyright); 15 U.S.C. § 1117(a) (2012) (trademark); Thomas F. Cotter, *Optimal Fines for False Patent Marking*, 17 Mich. Telecom. & Tech. L.J. 181, 192–97 (2010).

47. *See, e.g.*, Australian Patent Act, *supra* note 44, § 122(1) (authorizing an account of profits); U.K. Patents Act, *supra* note 44, § 61(1)(d) (stating that the patentee may claim "an account of the profits derived by [the defendant] from the infringement"); Analysis of EC Enforcement Directive, *supra* note 15, at 22–23 (discussing the availability of awards of defendants' profits among EC member states); Meier-Beck, *supra* note 45, at 118 (citing German case law permitting awards of defendants' profits); *cf.* Japanese Patent Act, *supra* note 44, § 102(2) ("Where a patentee or an exclusive licensee claims against an infringer compensation for damage sustained as a result of the intentional or negligent infringement of the patent right or exclusive license, and the infringer earned profits from the act of infringement, the amount of profits earned by the infringer shall be presumed to be the amount of damage sustained by the patentee or exclusive licensee."); EC Enforcement Directive, *supra* note 15, art. 13(2) ("Where the

calculation, vary.[48] On the other hand, US law permits awards of enhanced damages in some cases;[49] other countries generally do not,[50] though there are some exceptions.[51] Other countries nevertheless generally award the prevailing party its attorney's fees;[52] the United States does so only in "exceptional" cases.[53]

---

infringer did not knowingly, or with reasonable grounds to know, engage in infringing activity, Member States may lay down that the judicial authorities may order the recovery of profits or the payment of damages, which may be pre-established.").

48. *See infra* Section II.B.3.

49. *See* 35 U.S.C. § 284 (2012) ("[T]he court may increase the damages up to three times the amount found or assessed."); *In re* Seagate Tech., LLC, 497 F.3d 1360, 1371 (Fed. Cir. 2007) (en banc) (holding that, for an infringement to be deemed "willful," and therefore potentially deserving of enhanced damages, the infringement must have been both objectively and subjectively reckless); *see also infra* Section II.B.3.

50. *See, e.g.*, Catnic Components Ltd. v. Hill & Smith [1983] F.S.R. 512 (Ch. D.) 537–41 (Eng.); Toshiko Takenaka, *Adequate Compensation for Patent Infringement Damages: A Comparative Study of Damage Measurements in Japan and the United States*, *in* PATENT LAW AND THEORY: A HANDBOOK OF CONTEMPORARY RESEARCH 584, 602 (Toshiko Takenaka ed., 2008) (discussing Japanese law); Christian W. Appelt & Heinz Goddar, *Enforcement of Patents in Europe—Germany as an Example*, PRO INNO EUROPE 17 (Nov. 27, 2006), *available at* http://www.ip4inno.eu/index.php?id=187&jumpurl=uploads%2Ftx_cmsip4inno%2FAppelt_Patent-Enforcement-EP_Presence.pdf&juSecure=1&mimeType=application%2Fpdf&locationData=187%3Atx_cmsip4inno_document%3A56&juHash=42944254de22f4f224ac07f6539761d12388d794 (discussing German law); Schmidt-Szalewski & Bouche, *supra* note 31, at 156 (discussing French law).

51. *See* Australian Patent Act, *supra* note 44, § 122(1A) (authorizing aggravated damages on the basis of factors such as flagrancy of the infringement, deterrence, and the defendant's conduct); Eurocopter (Société par Actions Simplifiées) v. Bell Helicopter Textron Canada Limitée, ¶¶ 417–56 (Can. Fed. Ct. Jan. 30, 2012) (holding that patentee was entitled to punitive damages, in an amount yet to be determined, under Canadian law); *cf.* Analysis of EC Enforcement Directive, *supra* note 15, at 24 (stating that some member states authorize multiple (mostly double) royalties, principally for copyright and related rights infringement, and that at least one member state (Slovenia) reports providing for punitive damages). I thank Professor Norman Siebrasse for calling the recent Canadian case to my attention.

52. *See, e.g.*, Analysis of EC Enforcement Directive, *supra* note 15, at 24 (discussing the (often only partial) reimbursement of legal expenses in member states); EC Enforcement Directive, *supra* note 15, art. 14 (requiring members to "ensure that reasonable and proportionate legal costs and other expenses incurred by the successful party shall, as a general rule, be borne by the unsuccessful party, unless equity does not allow this"); NOBUHIRO NAKAYAMA, INDUSTRIAL PROPERTY LAW: VOLUME 1, PATENT LAW 342 (1993) (Intellectual Property Institute tr. 2000) (stating that, under Japanese law, "compensation can be claimed for attorney's fees and infringement investigation fees in some cases"); David Vaver, *Canada's Intellectual Property Framework: A Comparative Overview*, 17 INTELL. PROP. J. 125, 183 (2004) ("In Canada, the successful party also usually recovers its reasonable costs (i.e., attorney's fees) and disbursements from the losing party. In practice, costs awards cover only part (perhaps only a third) of the actual expenditure"); Ted Wright, *Australia: A Need for Clarity*, 20 JUST. SYS. J. 131, 146 (1999) (noting that attorney's fees generally are recoverable in Australia).

53. *See* 35 U.S.C. § 285 (2012); Wedgetail, Ltd. v. Huddleston Deluxe, Inc., 576 F.3d 1302, 1304 (Fed. Cir. 2009) (stating that only a limited number of circumstances justify fee awards

In the pages that follow, I will focus briefly on some of the more important similarities and differences between US and other countries' rules relating to the calculation of patent damages. Presumably these rules play some role (but how much of a role, in comparison with other features of the patent system?) in the fact, as reported by several commentators, that average damages awards are much higher in the United States than elsewhere.[54] I will suggest some possible explanatory hypotheses for the differences, as well as some avenues for further research.

1. Lost Profits

As noted above, most countries authorize courts to award the prevailing patentee its lost profit, that is, the profit the plaintiff can prove it lost as a result of the infringement. As I have argued in previous work, conceptually, an award of lost profits is appropriate in a case in which the patentee and the infringer compete in the same market and, but for the infringement, the patentee would have excluded the infringer from using the patented invention (rather than licensing its use).[55] Logically, if one of the relevant goals of the patent damages system is to ensure that the patentee is rendered no worse off than it would have been, but for the infringement, an award of lost profits ideally should be identical to the profit the patentee would have earned on sales of products that it did not make as a result of the defendant's infringing competition.[56] In reality, of course, even with the aid of sophisticated economic models, attempts to restore the patentee to the position it would have occupied but for the infringement are likely to involve an

---

in patent cases, including "inequitable conduct before the PTO; litigation misconduct; vexatious, unjustified, and otherwise bad faith litigation; a frivolous suit or willful infringement"). Other possible points of similarity and difference, which I intend to discuss in forthcoming works, include such topics as the recovery of prejudgment interest, *see* General Motors Corp. v. Devex Corp., 461 U.S. 648, 651–53 (1981) (authorizing the routine awarding of interest to the prevailing patentee); the recovery of damages for the unauthorized use of an invention in between the dates on which the patent application is published and the date on which it is granted, *see* 35 U.S.C. § 154(d) (2012); patent marking requirements as a precondition to the recovery of damages sustained prior to the defendant's receipt of actual notice of infringement, *see* 35 U.S.C. § 287(a), (b) (2012); and penalties for false patent marking, *see* 35 U.S.C. § 292 (2012).

54. *See* Elmer & Lewis, *supra* note 38, at 9; *see also* Appelt & Goddar, *supra* note 50, at 16 (stating that from the standpoint of a US observer, "the amount of damages that can be collected in a country like Germany sometimes appears as ridiculous"); Takenaka, *supra* note 50, at 607 (stating that average damages awards in Japan are much lower than average awards in the United States).

55. *See* BLAIR & COTTER, *supra* note 13, at 49–59, 212; Cotter, *Four Principles*, *supra* note 13, at 2–3 & n.10; Cotter, *Patent Remedies*, *supra* note 13, at 130–31, 134 & n.58.

56. As a matter of economic theory, one might argue that the patentee's damages should be either higher or lower than this baseline amount, for various reasons. In my previous work, however, I have argued that as a general matter courts should attempt simply to restore to the

element of guesswork. Moreover, as is demonstrated in many other areas of the law, efforts to improve accuracy can be costly; sometimes bright-line rules may be a preferable second-best solution than a more finely grained, administratively costly system of case-by-case determinations. One way of approaching comparative damages rules, then, may be to consider whether a given system appears to tilt more in the direction of accuracy or administrative cost reduction, and whether other features of the patent system influence, reinforce, or counteract this implicit choice.

As a general matter, US courts today adhere to the general principle that lost profits awards should restore the patentee to the position it would have occupied but for the infringement, and they have, by and large, abandoned more simplistic methods of calculating exactly what that restoration should entail.[57] In *Rite-Hite Corp. v. Kelley Co.*,[58] for example, the en banc Federal Circuit held that the law of patent damages should reflect principles of general tort law by awarding the patentee the lost profit that was both caused-in-fact and proximately caused by the infringement.[59] Put another way, the patentee is entitled to recover the profits it would have earned but for the infringement (cause-in-fact), subject to the qualification that such damages must not be too attenuated or unforeseeable (proximate cause).[60] Although the general principle seems, at first blush, unremarkable, its application can lead to results that may strike some observers as counterintuitive. Principal among these results is the possibility that the patentee may recover lost profits not only on forgone sales of products embodying the patented invention, which sales it can prove it would have made but for competition from defendant's infringing goods, but also on forgone sales of products that did *not* embody the patented invention but that also competed with the defendant's infringing goods. In *Rite-Hite* itself, for example, the defendant Kelley's

---

patentee the provable profits it would have earned but for the infringement. Awarding more (absent a need for enhanced damages to counter a probability that the infringement would not have been detected) threatens to over-deter would-be users from lawfully designing around; awarding less undermines the substantive patent law's implicit trade-offs between access and incentives. Even if the trade-offs implicit in a given patent system are suboptimal (which is surely likely to be the case in the real world), I have argued that, for rule-of-law and institutional competency reasons, in awarding damages courts should take the contours of substantive patent law as a "given." *See* Cotter, *Four Principles*, *supra* note 13, at 9–10.

57. At one time, for example, courts were willing to compare the plaintiff's profits before and after the infringement over some relevant time period, and award the plaintiff the difference; or to assume that the plaintiff would have made a sale for every sale the defendant made. Neither approach is favored today. *See* Blair & Cotter, *supra* note 13, at 213.

58. 56 F.3d 1538 (Fed. Cir. 1995) (en banc).

59. *See id.* at 1544–46.

60. *See id.*

infringing product competed not only with a Rite-Hite product (the "MDL-55") that incorporated the patent in suit but also with another Rite-Hite product (the "ADL-100") that did not.[61] The court held that Rite-Hite could recover lost profits on sales of both MDL-55s and ADL-100s, insofar as Rite-Hite was able to show that competition from the infringing product proximately caused Rite-Hite to lose sales on both product lines.[62] In other work, I have praised this result in *Rite-Hite* as economically sound, if the goal is to restore the patentee to the position it would have occupied in the absence of infringement.[63] The result does not sit well with everyone, however; according to the dissenting judges in *Rite-Hite*, awarding damages on sales of unpatented products expands the scope of the patent grant.[64]

A second implication of *Rite-Hite* is that an accurate calculation of lost profits requires that the court also take into account what the defendant's likely course of conduct would have been, but for the infringement. In the context of *Rite-Hite*, for example, what if the defendant Kelley had chosen to compete with Rite-Hite by selling products that did not incorporate the patent in suit, but which simply copied Rite-Hite's ADL-100? Unfortunately for Kelley, that option was not on the table either, because the ADL-100 incorporated *another* patent or patents owned by Rite-Hite.[65] Rite-Hite was therefore able to show that, but for the infringement, it could have excluded Kelley altogether from the market for products that would have cut into sales of both ADL-100s and MDL-55s. More generally, though, the presence (or absence) of non-infringing alternatives can be an important factor in the accurate calculation of patent damages. Thus, in a case in which the defendant could have avoided infringement altogether by turning to a non-infringing alternative, the patentee may be unable to prove *any* lost

---

61. *See id.* at 1542–43.

62. *See id.* at 1543–49.

63. *See* Blair & Cotter, *supra* note 13, at 244.

64. *See Rite-Hite*, 56 F.3d at 1556–75 (Nies, J., dissenting). A variation on this theme, which in my view presents a genuinely difficult problem, is whether the patentee should be allowed to recover lost profits on sales of unpatented goods that compete with infringing goods, in a case (unlike *Rite-Hite*) in which the patentee does not sell *any* products covered by the patent in suit. Allowing the patentee to recover arguably follows from the logic of *Rite-Hite*, and in other work Roger Blair and I have tentatively endorsed this approach; other scholars have argued, however, with some force that allowing the recovery of lost profits in such a case may encourage technology suppression. For discussion, see Blair & Cotter, *supra* note 13, at 246–54; *see also* Kurt M. Saunders, *Patent Nonuse and the Role of Public Interest as a Deterrent to Technology Suppression*, 15 Harv. J. L. & Tech. 389, 426–27 (2002) (critiquing *Rite-Hite*).

65. *See Rite-Hite*, 56 F.3d at 1543. *But see id.* at 1572–73 (Nies, J., dissenting) (noting that no court had passed judgment on the validity of the patents allegedly covering the ADL-100s, and arguing that there were other unpatented alternatives available to Kelley).

profits resulting from the infringement; in such a case, the infringement was not a but-for cause of any harm to the patentee, because in a world in which the defendant did not infringe it would have competed just as effectively against the patentee, using the non-infringing alternative. To be sure, in some cases it can be quite difficult to determine whether a non-infringing alternative was available during the time period at issue, as well as whether the defendant and consumers would have considered it to be an adequate substitute for the patented invention; differences in cost or quality may have led consumers to prefer the patented invention after all. Nevertheless, US courts have long required the plaintiff to demonstrate the non-existence of adequate non-infringing alternatives if the plaintiff hopes to obtain a lost profits award.[66] Once again, my previous work has approved of this policy as economically rational.[67]

A third implication, which follows from the preceding consideration of non-infringing alternatives, is that sometimes a non-infringing alternative may be adequate from the standpoint of some but not all consumers. To illustrate, suppose that a manufacturer has a patent on a particular feature or component that it incorporates into a complex end product (for example, a cell phone), which it makes and sells to consumers. For some consumers, the presence of the feature or component may be determinative of whether they buy the phone at all; for others, the purchasing decision may depend on how much the added feature contributes to the phone's price; yet other consumers may not care much about the feature at all. Now suppose that another cell phone manufacturer incorporates the patented feature into its phones, without authorization from the patentee, and thus is able to sell a product that is comparable to the patentee's product at a comparable (or lower) price. In such a case, it stands to reason that some of the consumers who care about the feature at issue might buy from the infringer instead of from the patentee; for these consumers, a phone without the feature in question would not have been an adequate substitute, and thus in the absence of infringement these consumers would have bought the patentee's phone. The patentee has clearly lost profits on these forgone sales. For those consumers who are more price-sensitive, however, or who do not care about the feature at all, the infringement may not have caused the patentee to lose any sales. Absent the infringement, in other words, the defendant could have sold a non-infringing alternative (a phone without the patented feature) that would have been adequate to some, but not all, consumers. Accurate lost profit damages therefore ought to take into account the possibility that the infringement may have caused the plaintiff to lose

---

66. *See, e.g.*, Grain Processing Corp. v. Am. Maize-Prods. Co., 185 F.3d 1341, 1350–56 (Fed. Cir. 1999); Panduit Corp. v. Stahlin Bros. Fibre Works Inc., 575 F.2d 1152, 1156 (6th Cir. 1978).

67. *See* BLAIR & COTTER, *supra* note 13, at 214.

sales, but that the defendant (and other firms within the relevant product market) might have retained some degree of market share even in the absence of infringement. United States' case law in recent years has been receptive to this sort of analysis as well, allowing plaintiffs to recover lost profits proportionate to the amount of market share they lost as a consequence of the infringement, and only a reasonable royalty on the remainder of the defendants' infringing sales.[68] My previous work has argued that this framework as well is consistent with a rational system of patent damages.[69]

The fourth implication of *Rite-Hite* relates to lost profits on lost sales of unpatented "collateral" goods, that is, complementary products that the plaintiff likely would have sold, either at the same time it sold the products against which the infringing goods competed, or as follow-up sales (e.g., spare or repair parts).[70] On one view, the logic of *Rite-Hite* suggests that the patentee should be able to recover lost profits on lost sales of these products too, as long as the lost sales were attributable to and proximately (e.g., foreseeably) caused by the infringement. The *Rite-Hite* majority nevertheless was unwilling to extend the law of patent damages quite this far, adopting instead a rule that patentees may recover lost profits on lost sales of unpatented collateral goods only if such goods "function with the patented invention."[71] Collateral goods that do not satisfy this condition, by contrast, but that are sold "only as a matter of convenience or business advantage," cannot give rise to compensable lost profits.[72] Although the rule is somewhat difficult to square with the overall thrust of the *Rite-Hite* opinion, with its emphasis on restoring the patentee to the position it would have occupied but for the infringement, the rule could be optimal if, for example, the amount by which it reduces what might otherwise be substantial administrative costs exceeds the

---

68. *See* BIC Leisure Prods. v. Windsurfing Int'l Inc., 1 F.3d 1214, 1217–19 (Fed. Cir. 1993); Atlantic Thermoplastics, Co. v. Faytex Corp., 970 F.2d 834, 847 (Fed. Cir. 1992).

69. *See* Blair & Cotter, *supra* note 13, at 221–22. The more traditional approach to a case in which the plaintiff sought lost profit damages resulting from the infringing of a patent that covered only part of an end product was to engage in some form of damages apportionment—unless the patented feature drove the demand for the end product, in which instances courts would apply the "entire market value rule" and award the patentee damages equal to the patentee's entire lost profit. *See id.* (arguing that the logic of cases awarding damages based on market share should lead to the demise of the entire market value rule, in favor of a simple but-for approach); *see also* Cotter, *Four Principles*, *supra* note 13, at 19–20 (similar).

70. In the case law, the products the patentee would have sold at the same time as it sold the principal goods are referred to as "convoyed" goods; the products it would have sold later are referred to as "derivative" goods. *See* Carborundum Co. v. Molten Metal Equip. Innovations, Inc., 72 F.3d 872, 881 n.8 (Fed. Cir. 1995).

71. Rite-Hite Corp. v. Kelley Co., 56 F.3d 1538, 1549 (Fed. Cir. 1995) (en banc).

72. *See id.*

marginal benefit to the patent incentive system that full compensation would promote.[73]

This brief overview of the US law of lost profits demonstrates that, over the last two decades in particular, the courts have come to adopt a lost profits framework that is largely consistent with the principle that patentees should be restored to the position they would have enjoyed in the absence of the infringement; and that the courts' analysis of exactly what it means to carry out this principle, although surely imperfect in some respects, has become increasingly sophisticated. All of this effort at precision, however, does come with a cost, and one possible negative consequence is that courts may apply the lost profits framework so strictly that patentees who likely have lost profits as a result of the infringement may opt nevertheless to recover only the more easily provable measure of damages, namely reasonable royalties.[74] Secondary literature on other nations' patent systems suggests that patentees in other countries often make precisely the same choice—though sometimes for the additional reason that, in countries in which extensive discovery is much less routine than in the United States, patentees may prefer not to disclose the sort of information relating to sales volume and costs that would be necessary for a court to calculate lost profits with any degree of confidence.[75] Reasonable royalties therefore appear to be a much more common measure of damages in many countries, and the case law on lost profits remains somewhat underdeveloped.

One exception, however, is the law of lost profits as it exists in England and other countries of the British Commonwealth, whose courts have generated a small but nontrivial number of reported decisions on lost profits for patent infringement. In these countries, the law on lost profits is in many respects quite similar to US law. Courts have stated that, as a general matter, the law should restore the patentee to the position it would have occupied in the absence of infringement, and have stressed the applicability of general torts principles of cause-in-fact and proximate cause in attaining this goal.[76] English courts also have allowed patentees to recover lost profits on sales of convoyed goods, where such losses are proximately attributable to the infringement, without the limitation imposed by U.S courts that the convoyed merchandise be functionally related to the patented products.[77]

---

73. *See* Blair & Cotter, *supra* note 13, at 258–59.

74. *See* Mark A. Lemley, *Distinguishing Lost Profits from Reasonable Royalties*, 51 Wm. & Mary L. Rev. 655, 656, 661 (2009).

75. *See, e.g.*, Meier-Beck, *supra* note 45, at 117; Tanger et al., *supra* note 45, at 6.

76. *See* Ultraframe (UK) Ltd. v. Eurocell Bldg. Plastics Ltd., [2006] EWHC (Pat) 1344, [47] (Eng.); Gerber Garment Tech. Inc. v. Lectra Sys. Ltd., [1997] R.P.C. 443 (A.C.) 452–56 (Eng.).

77. *See Gerber*, [1997] R.P.C. at 453–56.

A major area in which the US and Commonwealth courts diverge, however, is with respect to the significance of non-infringing alternatives. In 1888, the Law Lords in *United Horse Shoe & Nail Co. v. John Stewart & Co.*[78] concluded that the defendant's purported ability to have avoided infringement by resort to a non-infringing alternative was a matter of no consequence; the plaintiff was entitled to recover lost profits on the assumption that it would have made all or most of the sales the defendant made during the infringement period.[79] Courts in the United Kingdom,[80] Canada,[81] and Australia[82] have continued to apply this rule in the modern day, apparently without recognizing that in doing so they are making the patentee *better off* than it would have been in the absence of infringement. The courts' adherence to the rule, other than as a matter of stare decisis, is not easy to understand; in other respects these same courts demonstrate considerable economic sophistication. A few possible explanations nevertheless come to mind. One is that the question simply has not appeared important enough to generate much pressure, from potentially interested parties or from legal scholars, for a legislative or judicial modification. The number of cases in which the issue has come up still appears to be quite small, and thus may not be distorting the patent incentive scheme to any appreciable degree.

Another possible explanation is that, as noted above, the Patent Acts in England and Australia do not authorize awards of damages unless the defendant is at fault.[83] Perhaps courts in those countries are simply less concerned about potentially overrewarding patentees in cases in which the defendant clearly *is* at fault. This explanation does not appear overly persuasive, however. For one thing, in the typical lost profits case, the defendant is a competitor of the plaintiff and thus (arguably) unlikely to qualify as an innocent infringer. The defendant's innocence, or lack thereof, therefore may not be a salient point of distinction between lost profits cases as litigated in

---

78. [1888] App. Cas. 401 (H.L.) (appeal taken from Scot.).

79. *See id.* at 408–09 (opinion of Lord Halsbury, L.C.); *id.* at 416 (opinion of Lord Macnaghten). The third Law Lord, Lord Watson, was of the view that it had not been proven, and ought not to be presumed, that the infringer would have been able to make as many sales by using a non-infringing alternative. *See id.* at 414 (opinion of Lord Watson).

80. *See Ultraframe*, [2006] EWHC (Pat) at [93]; Coflexip SA Coflexip Stena Offshore Ltd. v. Stolt Offshore MS Ltd., [2003] EWCA (Civ) 296 (Eng.); Catnic Components Ltd. v. Hill & Smith Ltd., [1983] F.S.R. 512 (Ch. D.) 521, 524–25 (Eng.).

81. *See* Domco Indus. Ltd. v. Armstrong Cork Can. Ltd. (1986) 10 C.P.R. 3d 53, 61–62 (Can. Fed. Ct.) (dicta). The Canadian Supreme Court has now rejected this rule in the context of awards of defendant's profits, however, as discussed in note 123 *infra*, logic would suggest that the rule should no longer be applicable in the context of lost profits either.

82. *See* Advanced Bldg. Sys. Pty. Ltd. v. Ramset Fasteners (Aust.) Pty. Ltd., [2001] F.C.A. 1098, [126]–[28] (Austl.).

83. *See* Australian Patent Act, *supra* note 44, § 123(1); U.K. Patents Act, *supra* note 44, § 62(1).

the United States or the British Commonwealth. Moreover, Canadian courts at times have expressed agreement with the English rule concerning the irrelevance of non-infringing alternatives and yet the Canadian Patent Act, like the US Act, does not provide a damages exemption for innocent infringement.[84] On the other hand, English courts do not award enhanced damages in patent cases,[85] so perhaps adding a "kicker" to the plaintiff's true lost profits is a way of compensating the plaintiff for otherwise noncompensable losses or for attaining an adequate deterrent. Enhanced damages are potentially recoverable in Canada and Australia, however.[86] And courts in all three countries typically do award the prevailing party its attorney's fees,[87] an option that is usually not available in the United States.

Yet another explanation might be that, in each of the cases in which Commonwealth courts have expressed the above rule, there may have been reason to doubt that the defendant really could have made the same number of sales by resorting to a non-infringing alternative.[88] This explanation still leaves open the question why, in each instance, the courts felt the need to express a rule that not only went beyond the facts of the case but that also undermines the principle that patentees should be restored to the position they would have enjoyed but for the infringement. Indeed, the principle that defendants cannot avoid paying lost profits even if they *are* able to show that they would have resorted to a non-infringing alternative, and thus would have remained on the market, could compound problems arising from patent holdup in a case in which the patented invention is but a small component of the final product. In some such instances, however, Commonwealth courts have appeared open to applying some sort of damages apportionment,[89] in a manner analogous to the approach US courts have taken in some cases.[90] As a result, the practical differences between the US and Commonwealth approaches may be somewhat less stark than at first they appear.

---

84. *See* Canadian Patent Act, *supra* note 44, § 55(1); *see also* Coury, *supra* note 15, at 1131. Moreover, even in England and Australia a defendant cannot assert innocent infringement in mitigation of damages if the plaintiff properly "marked" its products. *See* Australian Patent Act, *supra* note 44, § 123(1); U.K. Patents Act, *supra* note 44, § 62(1).

85. *See supra* note 50.

86. *See supra* note 51.

87. *See supra* note 52.

88. *See, e.g.*, *United Horse Shoe*, [1888] App. Cas. at 414 (opinion of Lord Watson); *Advanced Bldg. Sys.*, [2001] F.C.A. at [128]–[129] (Austl.) (describing the English precedent as, in large part, dicta, "albeit very strong dicta").

89. *See, e.g.*, Jay-Lor Int'l Inc. v. Penta Farm Sys. Ltd. (2007), 59 C.P.R. 4th 228, 289–90 (Can. Fed. Ct.); *Coflexip*, [2003] EWCA (Civ) at [37] (Eng.).

90. *See supra* note 69.

2. Reasonable Royalties

The principal alternative to lost profits is an award of reasonable royalties. Conceptually, this measure of damages is appropriate when the patentee either does not or cannot prove that the defendant's unauthorized use of the patented invention caused the patentee to lose profits as a result of fewer sales or lower prices.[91] In theory, there are a variety of ways to go about calculating a reasonable royalty. In the United States[92] and many other countries,[93] a common approach is to employ the so-called "willing licensor–willing licensee" framework to estimate the royalties the patentee and the defendant would have agreed to, as of the date the infringement began. Critics occasionally have taken issue with this approach, arguing for example that it fails adequately to deter infringement and is unrealistic.[94] In other work, however, I have defended the framework as economically rational, on the assumption that (as with lost profits) the appropriate goal of a system of patent damages is to render the patentee neither better nor worse off than it would have been, but for the infringement.[95] In addition, the economic logic that leads courts to limit lost profits based upon the existence of adequate non-infringing alternatives should apply as well to the calculation of reasonable royalties.[96] A rational would-be user would never agree to pay more to license

---

91. *See* Cotter, *Four Principles*, *supra* note 13, at 2–3.

92. *See* Lucent Techs., Inc. v. Gateway, Inc., 580 F.3d 1301, 1324 (Fed. Cir. 2009).

93. *See, e.g.*, *Advanced Bldg. Sys.*, [2001] F.C.A. at, [191]–[203] (Austl.); *Consolboard*, 63 C.P.R. 2d at 8 (Can.); *Ultraframe*, [2006] EWHC (Pat) at [47] (Eng.); EC Enforcement Directive, *supra* note 15, art. 13(1)(b); Meier-Beck, *supra* note 45, at 118 (Germany); Schmidt-Szalewski & Bouche, *supra* note 31, at 156 (France). Both German and French authorities note, however, that judges retain some discretion to depart from this framework in certain respects. *See* Appelt & Goddar, *supra* note 50, at 19; Meier-Beck, *supra* note 45, at 119; Schmidt-Szalewski & Bouche, *supra* note 31, at 156–57 (France) (stating that in some cases French courts have decided "that the 'compensatory royalty' to be paid by the infringer may be higher than an agreed royalty").

94. *See, e.g.*, Maxwell v. J. Baker, Inc., 86 F.3d 1098, 1109 (Fed. Cir. 1996) (referring to the willing licensor–willing licensee framework as "absurd," "when the parties were previously unable to come to an agreement, *i.e.*, were not 'willing,'" and that use of the framework "risks creation of the perception that blatant, blind appropriation of inventions patented by individual, nonmanufacturing inventors is the profitable, can't-lose course") (citations omitted); Jerry Hausman et al., *Patent Damages and Real Options: How Judicial Characterization of Noninfringing Alternatives Reduces Incentives to Innovate*, 22 Berkeley Tech. L.J. 825, 839–46 (2007) (arguing that the approach encourages would-be users to adopt a wait-and-see attitude); Meier-Beck, *supra* note 45, at 118.

95. *See* Cotter, *Four Principles*, *supra* note 13, at 11–13 (arguing, inter alia, that the approach probably does not encourage much infringement once the risk to the use of paying attorney's fees and of incurring enhanced damages are taken into account; and that estimating royalties as of the date infringement began is preferable to using some other date as the baseline).

96. *See* Grain Processing Corp. v. Am. Maize-Prods. Co., 893 F. Supp. 1386, 1392–93 (N.D. Ind. 1995) (Easterbrook, J., sitting by designation) (using the defendant's cost savings from

a patented technology than the value it expected to derive from the use of that technology (in terms of enhanced profits or lowered costs) in comparison with the next-best alternative.[97]

Determining that value with precision nevertheless can be extremely difficult. As a result, courts do (and perhaps must) resort to a variety of indirect measures of the terms the parties likely would have struck. Courts in the United States frequently consider the *Georgia-Pacific* factors, a long list of considerations that may be relevant to estimating the amount on which the willing licensor and licensee would have agreed.[98] Some other countries consider

---

the use of the patented invention, in comparison with the next-best available alternative, for purposes of estimating a reasonable royalty), *rev'd on other grounds*, 108 F.3d 1392 (Fed. Cir. 1997) (mem.); Cotter, *Four Principles*, *supra* note 13, at 14–16.

97. *See* Cotter, *Four Principles*, *supra* note 13, at 14.

98. The factors are:

1. The royalties received by the patentee for the licensing of the patent in suit, proving or tending to prove an established royalty.
2. The rates paid by the licensee for the use of other patents comparable to the patent in suit.
3. The nature and scope of the license, as exclusive or non-exclusive; or as restricted or non-restricted in terms of territory or with respect to whom the manufactured product may be sold.
4. The licensor's established policy and marketing program to maintain his patent monopoly by not licensing others to use the invention or by granting licenses under special conditions designed to preserve that monopoly.
5. The commercial relationship between the licensor and licensee, such as, whether they are competitors in the same territory in the same line of business; or whether they are inventor and promoter.
6. The effect of selling the patented specialty in promoting sales of other products of the licensee; the existing value of the invention to the licensor as a generator or sales of his non-patented items; and the extent of such derivative or convoyed sales.
7. The duration of the patent and the term of the license.
8. The established profitability of the product made under the patent; its commercial success; and its current popularity.
9. The utility and advantages of the patent property over the old modes or devices, if any, that had been used for working out similar results.
10. The nature of the patented invention; the character of the commercial embodiment of it as owned and produced by the licensor; and the benefits to those who have used the invention.
11. The extent to which the infringer has made use of the invention; and any evidence probative of the value of that use.
12. The portion of the profit or of the selling price that may be customary in the particular business or in comparable businesses to allow for the use of the invention or analogous inventions.
13. The portion of the realizable profit that should be credited to the invention as distinguished from non-patented elements, the manufacturing process, business risks, or significant features or improvements added by the infringer.

analogous factors.[99] Unless courts structure the analysis in some manner, however, these multifactor approaches can be so malleable as to provide little if any specific guidance or predictability in their application.[100] Moreover, although factors such as "comparable" licenses (*Georgia-Pacific* factor number two) can be quite probative of the terms the patentee and the defendant would have reached,

---

14. The opinion testimony of qualified experts.
15. The amount that a licensor (such as the patentee) and a licensee (such as the infringer) would have agreed upon (at the time the infringement began) if both had been reasonably and voluntarily trying to reach an agreement; that is, the amount which a prudent licensee—who desired, as a business proposition, to obtain a license to manufacture and sell a particular article embodying the patented invention—would have been willing to pay as a royalty and yet be able to make a reasonable profit and which amount would have been acceptable by a prudent patentee who was willing to grant a license.

Georgia-Pacific Co. v. U.S. Plywood Co., 318 F. Supp. 1116, 1120 (S.D.N.Y. 1970), *modified*, 446 F.2d 295 (2d Cir. 1971). The fifteenth factor, the "amount that a licensor…and a licensee…would have agreed upon (at the time the infringement began) if both had been reasonably and voluntarily trying to reach an agreement," is probably best viewed as the ultimate issue to which the other factors are relevant. *See* Lucent Techs., Inc. v. Gateway, Inc., 580 F.3d 1301, 1324–26 (Fed. Cir. 2009) (considering, arguably, the *Georgia-Pacific* factors in light of this overarching framework); Daralyn J. Durie & Mark A. Lemley, *A Structured Approach to Calculating Reasonable Royalties*, 14 Lewis & Clark L. Rev. 627, 643 (2010) (arguing that *Georgia-Pacific* factor fifteen "represents the ultimate question all of the other factors are trying to establish").

99. *See, e.g.*, AlliedSignal Inc. v. DuPont Can. Inc. (1998), 78 C.P.R. 3d 129, 176–81 (Can. Fed. Ct.) (listing, and approving, a set of thirteen factors the defendant's expert had used to calculate a royalty); General Tire & Rubber Co. v. Firestone Tyre & Rubber Co., [1975] F.S.R. 273 (H.L.) 280–81 (appeal taken from Eng.) (opinion of Lord Wilberforce) (stating that judges should "take into account any licences actually granted and the rates of royalty fixed by them, to estimate their relevance and comparability, to apply them so far as he can to the bargain hypothetically to be made between the patentee and the infringer, and to the extent to which they do not provide a figure on which the damage can be measured, to consider any other evidence, according to its relevance and weight, upon which he can fix a rate of royalty which would have been agreed"); *id.* at 292–93 (opinion of Lord Salmon) (stating that, "when there is no established market rate, damages for infringement can be assessed only on the basis of what royalty a willing licensee would have been prepared to pay and a willing licensor to accept"); Michael Gross, *Actual Royalty Rates in Patent-, Know-How- and Computer Program-License Agreements*, 4 CASRIP Newsletter 2 (Univ. of Wash. Sch. of Law, Ctr. for Advanced Study and Research on Intellectual Prop.) (stating that "German Courts make use of factors which are quite similar to the *Georgia-Pacific* factors"); Takenaka, *supra* note 50, at 605 (listing, as relevant factors under Japanese law, "a legally negotiated and agreed upon royalty, an average royalty in the industry of the invention, the significance of the patented invention, the act of infringement, profits made by the infringer from the infringement, the relationship between the patentee and infringer in the relevant market and the patentee's market strategies").

100. *See* Cotter, *Four Principles*, *supra* note 13, at 4; Durie & Lemley, *supra* note 98, at 632.

finding licenses that are truly "comparable" can be challenging. To their credit, US courts recently have begun to subject this factor to careful scrutiny, noting for example that different types of licenses (lump-sum, running royalty, portfolio, licenses entered into in settlement of other litigation) may not be economically comparable inter se.[101]

In other instances, the evidence that courts have considered has had little if any relationship to the type of bargain the parties realistically would have struck. In some cases, for example, US courts have applied the entire market value rule (originally a lost profits doctrine, as discussed above)[102] for use in calculating reasonable royalties—the idea being that if the patented feature is what drove the demand for the end product, it is appropriate to use the value of the end product as the royalty base.[103] Sometimes courts have even taken the value of collateral products into account in setting the royalty base.[104] Critics, however (including me), have argued that this application of the entire market value rule artificially inflates the value of royalty awards, and have argued that courts should instead simply focus on reconstructing the royalty base (and rate) the parties themselves would have chosen, had they opted for a running royalty over a lump sum.[105] Until recently, U.S. courts also permitted experts to testify on the basis of a so-called "rule of thumb" that, as a starting point in negotiations, parties generally proceed on the assumption that the user will pay a royalty equal to 25 percent of the profits to be earned from the use of the invention.[106] In January 2011, however, the Federal Circuit ruled that the 25 percent rule of thumb did not rest upon an adequate scientific basis to render testimony based upon the rule admissible in federal court.[107] In these and other respects, then, the US law

---

101. *See, e.g.*, Lucent, 580 F.3d at 1325–3232 (reversing a damages judgment based on a jury's award of a lump-sum royalty in the amount of $350 million; other lump-sum licenses offered in evidence were not comparable, because one was for a large portfolio and others were for PC-related technology, and running royalty licenses were not comparable because they were not lump-sum); *see also* Wordtech Sys., Inc. v. Integrated Networks Solutions, Inc., 609 F.3d 1308, 1320 (Fed. Cir. 2010).

102. *See supra* note 69.

103. *See, e.g.*, Rite-Hite Corp. v. Kelley Co., 56 F.3d 1538 (Fed. Cir. 1995) (en banc) (suggesting that, when the elements of the entire market value rule are in place, a court may use the entire market value of end products incorporating the patented invention as the royalty base).

104. *See* Fresenius USA, Inc. v. Baxter Int'l, Inc., 582 F.3d 1288, 1303 (Fed. Cir. 2009).

105. *See* Cotter, *Four Principles*, *supra* note 13, at 5, 19–23.

106. *See* i4i Ltd. P'ship v. Microsoft Corp., 598 F.3d 831, 853–55 (Fed. Cir. 2010), *aff'd*, 131 S. Ct. 2238 (2011).

107. *See* Uniloc USA, Inc. v. Microsoft Corp., 632 F.3d 1292, 2011 WL 9738, at *17–19 (Fed. Cir. 2011).

on reasonable royalties has become more economically rational than in the past, though it still admits of certain anomalies that are difficult to justify from an economic perspective.[108]

Other countries follow a variety of practices for calculating reasonable royalties, many of which are analogous to US practice. For example, many countries apply multifactor tests reminiscent of the *Georgia-Pacific* factors.[109] Some countries, including Germany and Japan, sometimes also have made use of published, industry-wide royalty standards.[110] In addition, courts in some countries on occasion have relied on rules of thumb like the one rejected in *Uniloc*.[111] And, as in the lost profits arena, English courts appear disinclined to consider non-infringing alternatives as a constraint upon what the defendant would have agreed to, despite the obvious relevance of this factor in real-world negotiations.[112]

Further research could deepen our understanding of the similarities and differences among various nations' methods for assessing royalties. Do courts in certain countries weigh some factors more heavily than do courts in others, for example? Do countries that employ rules of thumb employ the *same* rules of thumb? Do those rules ever vary from one industry to another? For that matter, are royalties consistently higher or lower for comparable inventions in different countries? Do procedural differences among countries, which may play a role in settlement rates,[113] play any role in the magnitude of any such differences? Do countries that encourage settlement of damages disputes generate too little

---

108. In some cases, for example, courts have permitted royalties to exceed the amount that the defendant would have expected to leave it with any profit. *See* Cotter, *Four Principles*, *supra* note 13, at 5, 12, 14. Another contestable feature of US patent damages is the recent trend toward awarding prospective royalties, in lieu of a permanent injunction, based on a higher royalty rate than was used to calculate prejudgment royalties. *See id.* at 12. One unrealistic assumption that courts use but which makes sense, however, is that the parties would have negotiated ex ante on the assumption that the patent was valid and infringed. *See id.* at 23–24 (explaining why this assumption avoids a double counting problem).

109. *See supra* note 98.

110. *See* Appelt & Goddar, *supra* note 50, at 18–19 (discussing the use in patent litigation of published royalty rates used for compensating employees for inventions made on the job); Takenaka, *supra* note 50, at 598, 600 (discussing Japanese practice prior to the 1998 revision to the Japanese Patent Act).

111. *See* Appelt & Goddar, *supra* note 50, at 19 (stating that typical German royalty rates "rather closely follow the well-known 25% rule"); *see also* Jay-Lor Int'l Inc. v. Penta Farm Sys. Ltd. (2007), 59 C.P.R. 4th 228, 272–77, 284–85 (Can. Fed. Ct.) (applying a 30 percent royalty rate); AlliedSignal Inc. v. DuPont Canada Inc. (1998), 78 C.P.R. 3d 129, 178, 180 (Can. Fed. Ct.) (crediting expert testimony that "a reasonable royalty for patented technology would be approximately 25 to 33 1/3 % of profit before tax").

112. *See Catnic*, [1983] F.S.R. at 532.

113. *See* Meier-Beck, *supra* note 45, at 124 (noting that, in Germany, most patent cases settle after trial on infringement, rather than proceed to a separate trial on damages).

precedent to guide future courts? Does the lack of jury trials outside the United States result in the use of smaller royalty rates and royalty bases? Answers to these and other questions would help to illuminate the workings of the global patent system; more facts in turn could help to generate more hypotheses and the evidence by which to evaluate them.

3. Defendant's Profits and Enhanced Damages

In addition to offering the prevailing patentee the option of recovering lost profits or a reasonable royalty, many countries also offer as a third option an award of the defendant's profits attributable to the infringement.[114] Patent case law involving awards of defendant's profits nevertheless appears to be relatively sparse. The Patents Court's 1999 decision in *Celanese Int'l Corp. v. BP Chemicals Ltd.*,[115] for example, is said to have been the first English patent case in over 150 years to have awarded this type of remedy,[116] though more such cases can be found in some of the other countries of the Commonwealth.[117] Awards of the defendant's profits also could become more common in Germany, where a 2000 decision interpreting the German Industrial Designs Act held that, for purposes of calculating defendants' profits attributable to infringement, defendants may deduct from their gross revenues only the variable costs of producing infringing products, and not allocable fixed costs such as overhead.[118]

United States' patent law at one time permitted awards of defendant's profits too, but Congress put an end to the practice in 1946 on the rationale that proceedings

---

114. *See supra* note 47 and accompanying text.

115. [1999] R.P.C. 203 (Ch.) (Eng.).

116. *See* Urbanchuk & Tumbridge, *supra* note 44, at 3.

117. *See* Dart Indus. Inc. v. Decor Corp. Pty. Ltd. (1993), 179 C.L.R. 101 (Austl.); Teledyne Indus. Inc. v. Lido Prods. Ltd. (1982), 68 C.P.R. 2d 56 (Can. Fed. Ct.); *see also* Neil R. Belmore et al., *Canadian Patent Infringement Remedies*, BELMORE LAW 8–9 (Oct. 13, 1994), *available at* http://www.belmorelaw.com/pdf/05.pdf (discussing the increasing use of the remedy in Canada); Coury, *supra* note 15, at 1131–33 (discussing Canadian law); Goudreau, *supra* note 31, at 119 (same).

118. *See Gemeinkostenanteil [Share of Overheads]*, 145 BGHZ 366 (2000), *translated in* 33 IIC 900, 903 (2002). More precisely, the court stated that defendants may deduct "merely the variable costs...for the production and distribution of the infringing objects but not the fixed costs." The latter presumably would have been incurred in any event, though "[i]f and to the extent that fixed costs and variable overheads can in exceptional cases be directly ascribed to the infringing objects, these are...to be deducted from the proceeds when calculating the profit from the infringement...." *Id.* at 902–04. For discussion, see Appelt & Goddar, *supra* note 50, at 20–21; Meier-Beck, *supra* note 45, at 120–22. Both commentators note the difficulty of second-guessing the infringer's calculations, however, traditionally due to the general lack of discovery in Germany. *See* Appelt & Goddar, *supra* note 50, at 21; Meier-Beck, *supra* note 45, at 120. The extent to which this may change under new provisions of the German Patent Act, added in response to the Enforcement Directive, remains to be seen.

were becoming too drawn out and expensive.[119] In theory, however, the option of awarding defendant's profits might seem desirable in some cases. In particular, where the user's expected profits exceed the patentee's own lost profits (such that licensing is the desired outcome), an award of defendant's profits would work to channel the parties into voluntary licensing ex ante; awarding the defendant's profits also would obviate the need to engage in the complexity of estimating reasonable royalties.[120] Moreover, as noted above, Congress has continued to authorize awards of defendant's profits in the analogous fields of copyright and trademark law, as well as design patents.[121] Nevertheless, in recent work I have expressed doubt that US utility patent law should resurrect this remedy, for two reasons. One is that these types of awards may pose some risk of over-deterrence, particularly if the profit the defendant earned from the use of the patented invention is not calculated as the surplus over and above what it would have earned from the next-best alternative, or is difficult to separate from the defendant's overall profit from the sale of a product incorporating many patented inventions.[122] In the English *Celanese* decision, for example, the court refused to calculate the defendant's profit as the surplus over the profit the defendant would have earned from the use of the next-best alternative[123] —in parallel with the Commonwealth courts' refusal to consider next-best alternatives in calculating lost profits or reasonable royalties—with the result that, from an economic perspective, the amount awarded may have exceeded the economic benefit the defendant actually derived from the use of the invention.[124] Analogously, the German court's[125]

119. *See* BLAIR & COTTER, *supra* note 13, at 71–72. More precisely, Congress put an end to restitutionary awards in cases involving utility patents. Such awards are still permissible in cases involving design patents. *See* 35 U.S.C. § 289 (2012).

120. *See* BLAIR & COTTER, *supra* note 13, at 73–74.

121. *See supra* note 46.

122. *See* Cotter, *Four Principles*, *supra* note 13, at 13 n.68.

123. *See*, Celanese Int'l Corp. v. BP Chemicals Ltd., [1999] R.P.C. 203 (Ch.) (Eng.). The court in *Celanese* cited with approval a Canadian case, *Baker Energy Resources Corp. v. Reading & Bates Construction.* (1994), 58 C.P.R. 3d 359, 368 (Can. Ct. App.), that followed the same principle. A recent decision of the Canadian Supreme Court, however, appears to overrule *Baker Energy* on this point. *See* Monsanto Can. Inc. v. Schmeiser, [2004] 1 S.C.R. 902, 938–39 (Can.) (reversing an award of profits, on the ground that defendant would have earned the same profit had he planted non-infringing canola, and stating that "[a] comparison is to be made between the defendant's profit attributable to the invention and his profit had he used the best non-infringing option").

124. However, after deducting allocable costs and engaging in an apportionment analysis, the court ultimately awarded only £567,840—only a fraction of the £180 million the plaintiff had sought. So perhaps at the end of the day, concerns about overcompensation can be exaggerated.

125. *See supra* note 118 and accompanying text. In copyright and trademark cases, U.S. courts often do permit a deduction for allocable overhead. *See, e.g.*, Hamil Am., Inc. v. GFI, 193 F.3d

refusal to deduct allocable overhead from the defendant's gross revenue potentially could generate awards that are excessive.[126] Second, if the correct amount of an award of profits is the amount the defendant earned from the use of the patent in comparison with the amount if would have earned by using the next-best available alternative, the calculation may not be any less complex than the calculation of the hypothetical license.[127] At the very least, countries that offer the patentee the option of recovering the defendant's profit must confront calculation issues that defy simplistic analysis and resolution. In my forthcoming work, I hope to provide a more thorough analysis of the comparative efficiencies of the various options relating to awards of defendants' profits.

If foreign law is marginally more plaintiff-friendly than US law in offering prevailing patentees the option of an award of defendants' profits, US law arguably favors patentees by offering the prospect of enhanced damages (up to three times the amount of actual damages) for willful infringement.[128] In theory, awards of enhanced damages make sense in cases in which an award of actual damages is likely to underdeter infringement, such as when the act of infringement was of a type that was likely to avoid detection.[129] In practice, however, US law has not conditioned awards of enhanced damages on this factor but rather on the supposed egregiousness of the defendant's conduct.[130] Nevertheless, in recent years the Federal Circuit has tightened the standards for awarding enhanced damages; patentees now are entitled to damages enhancements only in cases in which the defendant's conduct was both objectively and subjectively reckless.[131] Fewer cases therefore qualify for enhanced damages under the new standard, in which a substantial defense to infringement or validity often demonstrates that the defendant's conduct was not reckless.[132]

92, 103–07 (2d Cir. 1999) (copyright); Manhattan Indus., Inc. v. Sweater Bee by Banff, Ltd., 885 F.2d 1, 7–8 (2d Cir. 1989). However, there are some cases to the contrary. English and Australian case law appears to be consistent with the Second Circuit's approach in this regard. *See Celanese*, [1999] R.P.C. at 235 (Eng.); Dart Indus. Inc. v. Decor Corp. Pty. Ltd. (1993), 179 C.L.R. 101, 123–33 (Austl.). Some Canadian case law appears to be consistent with the German approach. *See* Teledyne Indus. Inc. v. Lido Prods. Ltd. (1982), 68 C.P.R. 2d 56, 83 (Can. Fed. Ct.).

126. For a discussion of the economic arguments for and against the type of rule applied in *Gemeinkostenanteil*, see Stephen E. Margolis, *The Profits of Infringement: Richard Posner v. Learned Hand*, 22 Berkeley Tech. L.J. 1521 (2007) (arguing that the Second Circuit's approach properly takes into account the defendant's opportunity costs).

127. *See* Cotter, *Four Principles*, *supra* note 13, at 13 n.68.

128. *See supra* note 49 and accompanying text.

129. *See* Cotter, *Four Principles*, *supra* note 13, at 13–14.

130. *See* Read Corp. v. Portec, Inc., 970 F.2d 816, 826–27 (Fed. Cir. 1992).

131. *See In re* Seagate Tech., LLC, 497 F.3d 1360, 1371 (Fed. Cir. 2007) (en banc).

132. Spine Solutions, Inc. v. Medtronic Safamor Danek USA, Inc., 620 F.3d 1305, 1319 (Fed. Cir. 2010) (stating that the "'objective' prong of *Seagate* tends not to be met where an accused infringer relies on a reasonable defense to a charge of infringement").

Foreign courts, in keeping with their general disinclination to award punitive damages, for the most part appear equally disinclined to award enhanced damages in patent cases.[133] Also in keeping with foreign practice, however, foreign courts generally *do* award prevailing parties at least some portion of their attorney's fees.[134] Their US counterparts, by contrast, generally do not award prevailing patentees their attorney's fees except in cases of willful infringement.[135] Whether US or foreign courts systematically over- or undercompensate patentees is consequently difficult to determine in the abstract. To the extent US courts are less likely to award enhanced damages today than they were just a few years ago, however, this particular point of distinction may become increasingly less relevant as the years go by. The US practice with respect to the recoverability of attorney's fees, on the other hand, not only is consistent with the general US policy that each party bear its own fees (the so-called "American rule") but is embodied in the Patent Act itself.[136] As a result, this particular distinguishing feature, the welfare effects of which are ambiguous, seems unlikely to disappear anytime soon.[137]

Conclusion

A comparative law and economics approach to the field of patent remedies would assist in developing a deeper understanding of the ways in which different countries and regions enforce patent rights, both as a matter of law on the books and as a matter of law in action. As discussed above, different countries' laws share many commonalities but also are characterized by many distinguishing features, particularly with respect to the recoverability and calculation of certain categories of monetary damages. Further research may assist in better understanding both the reasons for the differences and their practical consequences. Understanding the consequences, in turn, may be of benefit to policymakers—both in the developed world and in such rapidly developing countries as China and India—as they work to reform and adapt their patent laws and policies to the challenges posed by new technologies, new business models, and a new and ever-evolving

133. *See supra* notes 50–51 and accompanying text.

134. *See supra* note 52 and accompanying text.

135. *See supra* note 53 and accompanying text.

136. *See* 35 U.S.C. § 285 (2012) (authorizing awards only in "exceptional cases").

137. This could change if Congress enacts HR 6245, the Saving High-tech Innovators from Egregious Legal Disputes (SHIELD) Act of 2013, which, as proposed, would include fee-shifting provisions.

set of legal issues. I hope that the approach sketched out above will provide at least a modest contribution to this effort, and that other scholars, policymakers, and practitioners in both the United States and abroad will devote attention to what promises to be an intellectually rich, and practically important, new field of inquiry. For one thing is surely true: in a world in which knowledge really *is* power, insularity will not be a viable option.

21 THE RULE OF PATENT LAW (RPL) AS ESTABLISHED BY THE TRIPS AGREEMENT AND ITS ROLE IN PROMOTING TRADE RATHER THAN INVENTION

Nuno Pires de Carvalho*

Introduction

A few decades ago, in a seminal and often-quoted paper commissioned by the US Senate, which enquired about the usefulness and efficiency of the patent system in promoting invention, Fritz Machlup, an Austrian economist, wrote:

> Patents, by giving their owners exclusive rights to the commercial exploitation of inventions, secure to these owners profits (the so-called "quasi-rents") which are ultimately collected from consumers as part of the price paid for goods and services. The consumers pay; the patent owners receive. Are the consumers—the non–patent-owning people—worse off for it? "No, they are not," says one group of economists. Patents are granted on inventions which would not have been made in the absence of a patent system; the inventions make it possible to produce more or better products than could have been produced without them; hence, whatever the consumers pay to the patent owners is only a part of the increase in real income that is engendered by the patent-induced inventions. "Wrong," says another group of economists. Many of the inventions for which patents are granted would also be made and put to use without any patent system.[1]

---

* The author serves in the International Bureau of the World Intellectual Property Organization (WIPO) in Geneva, Switzerland. All views expressed are the author's and not necessarily those of the Secretariat or member states of WIPO.

1. Fritz Machlup, An Economic Review of the Patent System, Study No. 15 of S. Comm. on the Judiciary, Subcomm. on Patents, Trademarks, and Copyrights, 85th Cong. 44 (Comm. Print 1958).

Both views are, actually, based on very simplistic approaches, particularly if we look at the patent system from an international perspective. It definitely is not true that corporations based in the United States, Germany, or Japan count on least-developed-country consumers to pay for whatever increased rent such corporations could get from the use of their inventions. Nor is the patent system, as a proprietary regime, confined to acting as bait for companies that, in its absence, would be reluctant to sink resources into inventing. The patent system performs a relevant role even as regards inventions that would be made in its absence—that of reducing transaction costs. The value of patents cannot be assessed in isolation, but rather by comparison with two other regimes that are available to induce invention: trade secrets and government sponsorship (either by means of pre-invention grants or post-invention prizes).

Patents reduce transaction costs as far as they permit market competitive forces to operate and provide for a better evaluation of inventions. Patents increase certainty—as compared to alternative methods of promoting invention—by *quantifying* the technology (through specifications and claims) and by *qualifying* the rights.

Quantification is provided by the description of the invention "in a manner sufficiently clear and complete out by a person skilled in the art."[2] This requirement may be supplemented by an indication of "the best mode for carrying out the invention known to the inventor at the filing date."[3] Claims, which point out the specific aspects that the inventor regards as his/her invention, contribute to the precise identification of the subject matter. To some extent, patents describe inventions very much as deeds describe the geographical limits of a piece of real estate.

Qualification of rights is provided by two factors. First, rights created *erga omnes* are property rights, which generate a negative duty to refrain from trespassing. Second, patent terms predetermine the extent of any injunction that courts may issue against infringers; this stands in sharp contrast to the indefinite duration of secrecy available for trade secrets.[4]

This means that the patent system may perform a significant role even in those countries where there is low inventive activity, but that engage in importing technology. Patents facilitate legal security with regards to imported technology

2. Agreement on Trade-Related Aspects of Intellectual Property Rights, art. 29.1, Apr. 15, 1994, Marrakesh Agreement Establishing the World Trade Organization, Annex 1C, 33 I.L.M. 1143 (1994) [hereinafter TRIPS Agreement].

3. *Id.*

4. A more extensive discussion on this topic can be found in Nuno Pires de Carvalho, *The Primary Function of Patents*, 2001 U. ILL. J.L. TECH. & POL'Y 25, 44–51 (2001).

and submit it to the pressure of market forces (provided that competing technology is available from other sources).

Machlup's quote above reflects what mainstream economists thought of the patent system at that time. But Machlup's own views were different. Answering the query whether the patent system "as a whole" is effective in promoting invention, he wrote:

> [T]he safest "policy conclusion" is to "muddle through"—either with it, if one has long lived with it, or without it, if one has lived without it. If we did not have a patent system, it would be irresponsible, on the basis of our present knowledge of its economic consequences, to recommend instituting one. But since we have had a patent system for a long time, it would be irresponsible, on the basis of our present knowledge, to recommend abolishing it. This last statement refers to a country such as the United States of America—not to a small country and not a predominantly non-industrial country, where a different weight of argument might well suggest another conclusion.[5]

Today the level of knowledge about the effects of the patent system remains the same as in Machlup's time, with a difference: other countries have joined the United States in using the patent system intensively, and that fact has not harmed the competitiveness of their economies—one might even suggest that the adoption of the system has increased their competitiveness. However, the patent system is just a small part of the whole picture and its economic performance depends on many factors. The motivation for invention is also grounded on different reasons—patents operate differently across industrial sectors and across national markets. One could say, for example, that the existence of a patent system in Switzerland is almost irrelevant as an incentive for its large pharmaceutical research-based companies to engage in inventive activity. Perhaps the patent system in Switzerland would be very important as bait for a start-up, a small company whose market does not go beyond Swiss borders. But for Novartis or Roche, arguably what matters are the patent systems in the European Union, the United States, Japan, China, and eventually in Brazil and India where, on the one hand, they can extract profits from a very broad base of well-off customers and health insurance companies and, on the other hand, patents have the ability to temporarily thwart local efforts to reverse engineer the companies' inventions, thus pushing the entry of generics back.

5. Machlup, *supra* note 1, at 80.

The purpose of this chapter is not to elaborate on the efficiency of the patent system in promoting invention. Serious doubts can be raised regarding whether it will ever be possible to clearly assess and measure such efficiency. After all, we are talking about the operation of law, and as happens with all branches of law, we only can make more or less educated guesses about how it operates. The social, political, cultural, and economic elements that interfere with the operation of law are so diverse and unpredictable that assessing the law's economic efficiency is as difficult as establishing an accurate model for weather forecasting. In both cases, one can identify general tendencies by resorting to past experiences, and thus deduce an overall view that may permit one to identify some trends—but that is all. The enquiry in this chapter is much more modest; it focuses instead on the possible effects of implementing a patent system that conforms to the harmonized standards of the Agreement on Trade-Related Aspects of Intellectual Property Rights (TRIPS Agreement).

I. The TRIPS Agreement and the Rule of Patent Law

Machlup's skeptical approach varies greatly from the more upbeat language of Article 7 of the TRIPS Agreement, which provides:

> The protection and enforcement of intellectual property should contribute to the promotion of technological innovation and to the transfer and dissemination of technology, to the mutual advantage of producers and users of technological knowledge, and in a manner conducive to social and economic welfare, and to a balance of rights and obligations.

Article 7 of the TRIPS Agreement is not about facts, but rather about a legitimate aspiration of those developing countries that subscribed to the Agreement that, in 1994, established the World Trade Organization (WTO). Article 7 was extracted from the text of a draft agreement submitted by a number of developing countries in 1990 (in reaction to the proposals then being submitted by the European Communities and the United States).[6] Because the draft was introduced after the initial Uruguay Round period, during which developing countries adamantly refused to accept higher standards of protection for intellectual property, Article 7 must be read not literally, but between the lines.

---

6. *See* GATT Secretariat, *Communication from Argentina, Brazil, Chile, China, Colombia, Cuba, Egypt, India, Nigeria, Peru, Tanzania, and Uruguay*, MTN.GNG/NG11/W/71 (May 14, 1990).

First, even though Article 7 mentions the objectives of protection (and enforcement) of intellectual property,[7] it is actually about the objectives of patents. Indeed, one cannot conceive how trademark protection in general can promote technological innovation, let alone technology dissemination. Moreover, geographical indications, which are a component of intellectual property in their own right, do not promote invention—on the contrary, they stimulate the permanence of practices and technical characteristics that result from the association of the goods with the name of the place of production or extraction. Any change in those characteristics and, in some cases, in the methods of production or extraction of those goods may lead to the loss of protection of geographical indications.

Second, where Article 7 refers to "the protection and enforcement of intellectual property," it does not refer to just any intellectual property regime, but rather, and specifically, to intellectual property as set in accordance with the standards adopted in the TRIPS Agreement. It would make no sense to use a TRIPS provision to praise the virtues of a regime of intellectual property different from the one that is established in the Agreement. This means, in other words, that developing countries have accepted the higher standards set in Section 5, Part II of the TRIPS Agreement with the expectation that those standards would lead to the promotion of technological innovation and to the transfer and dissemination of technology.

It is generally understood that the rule of law is integral to promote economic progress and development. As William J. Bernstein has put it, "[i]t is Institutions—property rights, individual liberties, the rule of law, the intellectual tolerance implicit in scientific rationalism, and capital-market structure—that matter."[8] This idea has come a long way, since at least Cicero,[9] but it was the economics school of neo-institutionalism initiated by Ronald Coase,[10] Douglass North,[11] and others that made a clear link between the rule of law and efficiency

---

7. The heading "Objectives" of Article 7 does not concern the TRIPS Agreement as a whole, but intellectual property. The objectives of the TRIPS Agreement are set in the very first paragraph of the preamble: to reduce distortions and impediments to international trade. TRIPS Agreement, *supra* note 2, pmbl. ¶ 1 ("*Desiring* to reduce distortions and impediments to international trade, and taking into account the need to promote effective and adequate protection of intellectual property rights, and to ensure that measures and procedures to enforce intellectual property rights do not themselves become barriers to legitimate trade; ...").

8. William J. Bernstein, The Birth of Plenty—How the Prosperity of the Modern World Was Created 189 (2004).

9. "It is clearly better that property should be private, but the use of it common; and the special business of the law maker is to create in men this benevolent disposition." Aristotle, Politics bk. 2, pt. V (Benjamin Jowett trans.) (c. 350 BCE), *available at* classics.mit.edu/Aristotle/politics.html (last visited Sept. 3, 2013).

10. *See* Ronald H. Coase, *The Problem of Social Cost*, 3 J.L. & Econ. 1 (1960); *see also* R.H. Coase, The Firm, The Market and the Law (1988).

11. *See* Douglass C. North, Structure and Change in Economic History (1981).

in reducing social costs. The patent system, remarkably, speaks to all the various elements noted by Bernstein:

- the patent system establishes property rights;
- the patent system is about individual freedom; the major shift in the patent legal regime from the late Middle Ages to the Industrial Age (from the seventeenth century on) was that the rights granted by patents changed from the right to use to the right to exclude due to the dismantling of the guild regime in the seventeenth and eighteenth centuries, and the advent of economic liberalism;
- patents have a connection with scientific rationalism to the extent that technology has been associated with rationalism since the scientific revolution introduced by Galileo, Bacon, and Descartes;[12] and
- patents do not make any sense outside of capitalist contexts; they are about the private appropriation of knowledge and are meaningless if they are not given a capital-market dimension.

Invention and investments in invention occur where the rule of law exists and is efficient. Where patent rules are efficient, predictable, and enforceable, we may call it the rule of patent law (RPL). The idea is that in the same way economic progress depends on the rule of law in general, technological progress depends on the adoption and enforcement of the RPL. Article 7 thus should be understood to mean the fostering of invention as well as the promotion of technological transfer, which does not take place only by means of technology licensing but also through foreign direct investment (FDI), which depends on the adoption and the implementation of the RPL.

But what patent system are we talking about? What kind of patent system is efficient so as to promote inventions and bring them to the market? These are two extremely difficult questions because the patent system operates differently depending on many factors. Moreover, industries respond differently to the availability of patents, depending on the level of rivalry as well as on the existence of other barriers that can replace legal fences at a lower cost (e.g., trade secrets, head start, high fixed costs, or government regulation). In this light, we can speculate that there must be benchmarks above which marginal increases in the standards of protection of patents make no difference in making the system more useful—or more used.

---

12. *See* Robert Friedel, A Culture of Improvement—Technology and the Western Millennium 149–69 (2007).

The RPL is established in the TRIPS Agreement as a set of minimum standards, but as implied by Article 1.1, higher standards could be likewise effective in promoting invention. However, the RPL is a benchmark, and for this purpose it is the minimum standards that count.

II. The RPL as Established in the TRIPS Agreement

Not all patent standards contained in the TRIPS Agreement may be deemed essential as elements of persuasion for prospective inventors. For example, the provisions of Article 32 (on the availability of judicial review of decisions to revoke or forfeit patents) and Article 34 (on the reversal of the burden of proof in the case of infringement of process patents) are relevant as instruments of a good, predictable patent system, but alone they are not sufficient to attract the attention of a prospective inventor. Moreover, the rule of Article 32 was already available in countries, including many developing countries, and the rule of Article 34 was well-known and used in countries with an advanced enforcement regime.

What elements of RPL would a prospective inventor—that is, a person with a technical solution for a certain problem in his/her mind and who has access to the capital necessary to fund the inventive activity—deem sufficiently persuasive?

The elements in the TRIPS Agreement that may encourage an inventor to engage in research and development (R & D), in the absence of or in combination with other reasons he/she may have, are of a double nature, substantive and procedural. I will overlook procedural guarantees of protection, for the simple reason that their effectiveness depends very much on national systems of law enforcement. The TRIPS Agreement itself (in Articles 1.1 and 41) acknowledges that this is an area where harmonization is elusive.

There are four standards in Part II, Section 5 of the TRIPS Agreement that may be significantly persuasive bait from an inventor/investor's point of view. They are those that concern:

(a) the term of protection;
(b) the identification of patentable subject matter, with special relevance to the mandatory absence of discrimination as to the field of technology (in a positive or negative sense);
(c) the restrictions regarding internal exceptions and limitations (protection against confiscation); and
(d) the restrictions regarding external exceptions and limitations (absence of encumbrances to license and assign as well as to use, sell, and set price; absence of the local exploitation requirement, etc.).

A. *Term of Protection*

Inventions take a certain time to mature before they reach the market and start producing sufficient revenues to justify R & D investment. This may not be true in the field of consumer goods (except where consumers take time to accept the new products or technologies),[13] but it is certainly so in areas where the implementation of the invention depends on high fixed costs (e.g., basic industries, such as metallurgy, energy, mining) or where government regulation may delay the entry of products (pharmaceuticals, agrochemicals). Before the TRIPS Agreement entered into force, certain areas were particularly discriminated against in this regard. Where those areas were not simply excluded from patentability, patents for certain inventions had an extremely short duration, to the point they were irrelevant, such as five years for processes of making pharmaceutical products (in India) or one year for pharmaceutical products (in Costa Rica).

The RPL, as established by the TRIPS Agreement, is based on a twenty-year term of protection from the date of filing. Because many inventions take some time to reach the market, including those the commercialization of which is subject to regulatory approval, and because in most countries patent rights can be enforced only after the grant, the twenty-year term is actually just a term of reference. It does not correspond to a fixed term of protection. In some cases, even after the twenty-year term has expired, the patent may still produce some lasting effects in the event commercial acts by third parties are based on acts of infringement practiced before the expiry date. In sum, the duration of a patent corresponds to the average period of life that society attributes to the subject technology. This period may vary according to each specific technical field. In the steel industry, for instance, technology changes more slowly than in communications. In the pharmaceutical sector, inventions keep their technical and economic utility for much longer. However, for practical purposes, patent laws

---

13. As Norman Pounds noted, society tends to resist the introduction of new technologies because these disturb social order and interfere with a series of intertwined activities. *See* NORMAN J.G. POUNDS, LA VIDA COTIDIANA—HISTORIA DE LA CULTURA MATERIAL 281 (Jordi Ainaud trans., 1999). This may be particularly true as regards completely new technology, which may change social life. In general, it is society that commands invention—from its inception to its acceptance—and this makes society the constructor of invention. As Fernand Braudel noted, "[n]ine times out of ten the patents of inventions, serious or not, recorded on the pages of registers and dossiers in the Venetian senate corresponded to the particular problems of the city: they sought to make the waterways converging on the lagoon navigable; dig canals; raise water; drain swampy land; turn mills without resorting to hydraulic power (unobtainable in this world of stagnant); work saws, millstones, hammers to powder the tannin or raw materials used to produce glass. Social considerations were uppermost." FERNAND BRAUDEL, CIVILIZATION & CAPITALISM—15TH–18TH CENTURY, THE STRUCTURES OF EVERYDAY LIFE 433–34 (Siân Reynolds trans., 1992). The original version (in French) of the last sentence is more emphatic: *La société commande.* Society commands.

adopt an average life-span criterion rather than create a specific duration for each industrial branch.

The result is that the twenty-year term is arbitrary as regards its role in inducing innovation across all fields. Nevertheless, it is an integral part of the RPL.

B. Identification of Patentable Subject Matter

The substantive standards of the TRIPS Agreement were included in the Uruguay Round because of the pressure of the US and European pharmaceutical sectors. In general, in other fields of technology there was no significant discrimination. However, certain large developing countries continued to impose restrictions on the patentability of inventions in the fields of chemistry and foodstuffs. The pharmaceutical sector viewed such discrimination against their inventions as a disincentive to the enormous expenditures spent on innovation. Moreover, they viewed such discrimination as a barrier against their exports to some large markets, in particular (but not exclusively) in developing countries. Article 27.1 and its prohibition against discrimination as to the fields of technology is thus a core provision of the Agreement. All other provisions on substantive standards emerged as addenda to that major concern. The RPL, in this sense, emerged with an industry-specific focus.

Very importantly, the RPL contains a provision directed at the arbitrary use of major social concerns as a pretext for denying patents. Article 27.2, in somewhat complex and oblique language, says that only inventions whose commercialization is prohibited (on grounds of public policy) can be excluded from patentability. Moreover, only exclusions necessary to enforce the prohibition of commercialization can be adopted. This is a two-step test that ensures inventors that they will not be subject to random and erratic views of patent examiners as regards the nature of their inventions. Instead, commercial considerations should buttress moral or ethics-related criteria on the scope of patentability.

C. Restrictions as Regards Internal[14] Exceptions and Limitations

The third element on which the RPL is based is the guarantee that there will be no taking of inventions without due compensation—in other words, that there will be no confiscation of rights. This is a basic guarantee of private property.

---

14. *Internal* exceptions and limitations are those that concern the rights granted. They are different from *external* exceptions and limitations, which concern the economic use or exploitation of the rights. In the field of patents, external exceptions and limitations are mentioned in Articles 8.1 and 40.2. TRIPS Agreement, *supra* note 2, arts. 8.1, 40.2.

Property is not absolute, and no GATT negotiator would challenge the rights of governments to impose rules that put private property at the service of public interests. Even so, however, governments cannot simply diminish those rights to the extent that they become economically insignificant. The rules that limit and/or impose conditions on the granting of compulsory licenses, such as in Article 31, have the purpose of prohibiting confiscation, and constitute part of a "takings clause" framework. Other pieces of the framework include Article 28 (on the rights guaranteed), Article 30 (on the three-step test that internal exceptions must pass), and Article 45 (on damages).

The protection against confiscation is perhaps one of the most important effective elements of the RPL. Firms choosing to invest in costly R & D efforts are strongly influenced by this factor.

D. Restrictions as Regards External Exceptions and Limitations

National patent laws can sometimes express certain idiosyncrasies that run blatantly against the spirit of the RPL. Two of those expressions are the obligation to locally exploit patented inventions and the imposition of sanctions against patent owners who exercise their fundamental right to say "no" to licensing or other commercial opportunities.

Let us start by briefly looking at the second aspect. To obtain and to use patent rights are matters that generally belong to private businesses and are governed by business considerations. However, one provision in the TRIPS Agreement is commonly understood as allowing WTO members to take measures that diminish patent rights, for the sake of the protection of public health and nutrition, and to promote the public interest in sectors of vital importance as provided in Article 8.1. However, this provision adds a crucially important expression: "provided that such measures are consistent with the provisions of this Agreement." If the measures in question are consistent with the other provisions of the Agreement, why would there be a need to expressly permit WTO members to adopt them? The answer is that Article 8.1 is not about permitting limitations to rights, but rather setting the ground for non-violation complaints for the purposes of Article XXIII of the GATT 1947.[15] In other words, Article 8.1 says that

---

15. Article XXIII.1 of GATT 1947 reads:

> If any contracting party should consider that any benefit accruing to it directly or indirectly under this Agreement is being nullified or impaired or that the attainment of any objective of the Agreement is being impeded as the result of
>
> (*a*) the failure of another contracting party to carry out its obligations under this Agreement, or
>
> (*b*) the application by another contracting party of any measure, whether or not it conflicts with the provisions of this Agreement, or

in taking measures to protect public health and nutrition, and to promote the public interest in sectors of vital importance, WTO members should not do so in a manner that nullifies or impairs the rights ensured by the Agreement. The fact that non-violation complaints in the context of the TRIPS Agreement are under a moratorium does not change this. When Article 8.1 was negotiated, its proponents did not expect that Article 64.2 would be suspended almost one decade later.

The obligation to locally exploit an invention means that the patent owner is required to manufacture the patented product or to use the patented process to manufacture a product within the geographical borders of the country that grants the patent. In the latter case, the patentee would not be obliged to perform all industrial phases of manufacturing the end product within the country; he/she would be obliged only to perform that phase which requires the use of the patented process. Likewise, the obligation to manufacture a patented product would not demand that all production phases occur within the country—countries might authorize the importation of certain components and intermediary inputs, provided that those imports would not replace a large part of the production process.

The local exploitation requirement has been present in the patent system for centuries. It has a strong medieval flavor given that it was ancillary to the first patents granted in Europe. These were mainly granted to lure foreign artisans to introduce new technologies so as to expand the industrial base and, hence, the tax collection base. Only artisans enrolled in local guilds or corporations could perform industrial activities. But foreign artisans were welcome (by the authorities, not necessarily by the guilds) if they could bring and teach local artisans new knowledge. The obligation of local exploitation was linked to the requirement of foreign artisans to teach the new technology to local artisans; mere importation would deny the possibility of local artisans learning by doing. Two conditions were attached to those early patents: the new technology could not interfere with activities of local guilds (it is in this requirement that one can discern the origins of the novelty requirement), and introducers of the new techniques were given a period for commencing the exploitation of the invention; otherwise the city or

---

(*c*) the existence of any other situation,

the contracting party may, with a view to the satisfactory adjustment of the matter, make written representations or proposals to the other contracting party or parties which it considers to be concerned. Any contracting party thus approached shall give sympathetic consideration to the representations or proposals made to it.

General Agreement on Tariffs and Trade, art. XXIII.1, Oct. 30, 1947, 61 Stat. A-11, T.I.A.S. 1700, 55 U.N.T.S. 194 [hereinafter GATT 1947].

the country could give a patent for another foreign artisan willing to introduce the same technology.

The local exploitation requirement has survived in national patent laws, with a rationale that is not very different from the original one. It is true that, on the one hand, modern patents disclose the invention, which dispenses with the need of teaching the invention by personal contact between the inventor and local artisans; but, on the other hand, some countries understand that patents granted to foreigners should not serve merely as tools to guarantee a monopoly over importing goods manufactured abroad.[16]

However, in a global economy, the local exploitation requirement on its own will rarely encourage a foreign inventor to export his/her knowledge and his/her capital to engage in local manufacture. If the domestic market offers conditions for a profitable manufacture of the invention, there is no need for the patent system to have that requirement. Indeed, the obligation to exploit the invention in all countries where he/she applies for a patent may discourage the inventor to invent altogether if the invention has to be applied on a global scale. Alternatively, the local exploitation requirement simply discourages the inventor from filing for a patent in the countries that adopt it.

Accordingly, the prohibition against adoption of a local working requirement under Article 27.1[17] fits in the RPL to the extent that it contributes to encouraging global companies "to invent globally," that is, to find new technologies of interest to a wide variety of countries and economies, all linked by inflows of foreign goods.

The freedom of patent holders to engage in licensing agreements is also a significant component of the RPL. Before the TRIPS Agreement, a number of countries adopted laws, regulations, and guidelines imposing on intellectual property holders stringent obligations as regards licensing, under the justification (or pretext, one could say) of preserving competition. That trend receded considerably with the wave of privatization that has swept through some of the major

---

16. Section 83 of the Indian Patents Act, No. 39 of 1970, INDIA CODE (2003), *as amended by* The Patents (Amendment) Act 2005, No. 15 of 2005, INDIA CODE (2005), says so explicitly:

> Without prejudice to the other provisions in this Act, in exercising the powers conferred by this Chapter, regard shall be had to the following general considerations, namely—
>
> (a) that patents are granted to encourage inventions and to secure that the inventions are worked in India on a commercial scale and to the fullest extent that is reasonably practicable without undue delay;
> (b) that they are not granted merely to enable patentees to enjoy a monopoly for the importation of the patented article....

17. *See* TRIPS Agreement, *supra* note 2, art. 27.1 ("[P]atents shall be available and patent rights enjoyable without discrimination as to... whether products are imported or locally produced.").

economies but it was a major concern when the TRIPS Agreement was negotiated. In this regard, the Agreement has two provisions of relevance: Article 28.2 and Article 40.2. Like Article 7, the apparently innocuous and straightforward language of those provisions has a significant meaning that calls for an educated reading.

Indeed, Article 28.2 seems to state that patent owners shall have the right to assign and to license their patents. If it said only that, it would be redundant, having in view the language of Article 28.1—if patent owners are granted the right to say "no," it follows that they are automatically bestowed with the right to say "yes." But what Article 28.2 says really is that the right of patent owners to assign and to license their inventions should not be unreasonably curtailed or impaired by national legislative activity. Article 40.2 goes exactly in the same direction. It seems to say that members are authorized to adopt measures providing for the ex officio and ex ante monitoring of potentially anticompetitive clauses in intellectual property licensing agreements. However, this provision demands a more sophisticated reading to the extent it indicates measures "consistent with the other provisions of this Agreement." The same question asked earlier can be repeated here: If the measures in question are consistent with the other provisions of the Agreement, why would they need to be authorized? The answer is that, like Article 8.1, Article 40.2 is not about an authorization, but rather about non-violation complaints for the purposes of Article XXIII of the GATT 1947. In other words, Article 40.2 could be understood as proscribing measures to prevent or control anticompetitive practices in licensing agreements that nullify or impair the rights ensured by the Agreement.

For these reasons, the protection of patent holders from arbitrary intervention of governments in a way that reduces their assets to economic insignificance is an important component of the RPL. The lack of such protection may operate as a disincentive to engage in costly inventive activities, in licensing opportunities, and/or in investing capital abroad.

III. The Efficacy of the RPL in Ensuring the Objectives Set Forth in Article 7 of the TRIPS Agreement

Having outlined the key elements of the RPL, let us assess how it contributes to achieving the objectives set forth in Article 7. As noted earlier, Article 7 links the protection of patents to the capacity of countries to enable an inventive climate, to encourage technology transfer, and to attract foreign direct investment. The enquiry is to see whether the adoption of the RPL is effective in that regard.

It was also earlier noted that the operation of law cannot be accurately measured by means of numbers and statistics. However, economic data can provide some guidance as to how certain effects of the law can be detected. Before the data is analyzed, some preliminary points on how the RPL can be set should be made. The national laws of twenty WTO members have been randomly selected for comparison. Those twenty members comprise some of the largest economies in both developed and developing countries along with certain small economies. In alphabetical order, these economies are: Argentina, Australia, Brazil, Canada, China, Denmark, France, Germany, Hong Kong, India, Indonesia, Israel, Japan, Oman, Singapore, South Africa, Spain, Switzerland, Thailand, and the United States.

At the outset, it should be emphasized that the values attributed to the RPL for each WTO member should not be seen as an evaluation of how they have transposed into their national laws the minimum standards established by Section 5 of Part II of the TRIPS Agreement. In other words, what follows is not an examination of the TRIPS-consistency of the patent statutes of these economies. Such an examination, for the sake of fairness and methodological rigor, would require a much more exhaustive consultation of national laws, including regulations, guidelines, and their practices by the executive and judicial branches of governments. Given that such an exercise would be extremely long, I have settled for a general assessment of the RPL, which, although it is not a thorough evaluation of TRIPS compliance, still gives a good idea of how the four elements operate to achieve the objectives set in Article 7.

A. The First Element of the RPL: A General Assessment

In general, the twenty economies selected respect the minimum term of protection as set by Article 33 of the TRIPS Agreement. Companies that are more RPL-sensitive (pharmaceutical, electronic, and consumer goods manufacturers) might be concerned, however, with the provisions contained in a number of domestic laws that provide for a twenty-year term from the filing date that is non-extendable, regardless of the time it takes to obtain a patent. Such provisions may indicate a lack of concern with Article 62.2, which provides that members shall provide that procedures for the grant of patents should not generate an "unwarranted curtailment of the period of protection."[18] A number of countries provide for the limited (up to five years) extension of patent terms in order to compensate for delays in the patent application procedures. In general, that extension is limited to pharmaceutical products, which diminishes the RPL because it discriminates against other fields of technology. A few countries, however, provide

18. *See id.* art. 62.2.

for an extension regardless of the field of technology. One WTO member provides for extensions of patent terms (without any restrictions as to the field of technology) to compensate for procedural delays that go beyond ten years from the date of filing—which is an unacceptable admission that the effective term of patents may be curtailed by a long period, provided it is less than ten years. Some countries provide for the possibility of provisional protection, which means that patent applicants may initiate proceedings before the patent is granted. This may include the determination that courts should take into consideration in the allocation of damages infringements that occurred before the grant of the patent. This is a sort of alleviation of the problems that may result from undue delays, but it does not address the condition of Article 62.2, which calls for measures that avoid the curtailment of periods, not for measures that reduce their negative effects. Another country establishes that disputes regarding the right to patent shall not affect the term. In practice this means that the counting of the patent term should be stopped during those disputes. That is an interesting (and uncommon) measure that could strongly count for the RPL, provided that the disputes do not take too long, otherwise when the decision to grant is awarded the invention in question may have become technically and/or commercially obsolete.

B. The Second Element of the RPL: A General Assessment

The second element of the RPL is the identification of patentable subject matter, or the scope of patentability. In this regard, the national statutes of WTO members follow a pattern: they provide for specific technological-bound exclusions, they provide for invention-nature bound exclusions, and they provide for exclusions imposed by ethical and public-policy grounded exclusions. In general, there are no significant RPL inconsistencies in connection with the two first categories of exclusions. WTO members tend to exclude from patentability inventions in fields authorized by the TRIPS Agreement, namely biotechnology and medical methods (diagnostic, therapeutic, and surgical methods). This may be a discouragement for biotechnology companies, but it is still a matter left outside the RPL. Certain WTO members also use the leeway that stems from the loose language of Article 27.1 (on patentability requirements) and exclude incremental inventions from patentability, such as second uses of known substances. One WTO member, for example, extends patentability to modified genes but only if the modifications were introduced directly by human intervention. These are examples of TRIPS-consistent exclusions, and therefore they may not be characterized as contrary to the RPL.

The third category of exclusions concerns those that are ethical and public-policy based. In this area it is worth emphasizing that only four WTO members out of the twenty that are considered in this analysis have provisions that are RPL-compliant, and only one of them has explicitly adopted language that squarely translates the two-step test established by Article 27.2.[19] As mentioned earlier, under the two-step test, in order for a WTO member to exclude inventions from patentability on grounds of protection of morality or public policy (*ordre public*), it is necessary to establish that the exploitation of those inventions is prohibited and that the exclusion from patentability is necessary to enforce the prohibition. The first step, therefore, is evidence of necessity. The second step is one of proportionality: exclusion from patentability, even if proven necessary, is only acceptable where less trade-restrictive measures are not available.

It is striking to note that so many WTO members have overlooked these basic components of the RPL. However, even if it is expected that authorities in those WTO members will show restraint in resorting to ethical values and public policy considerations—such as security requirements, health, and the environment—to deny patents, the fact that their statutes are inconsistent with Article 27.2 may raise concerns that certain inventions may be subject to arbitrary discrimination.

Other WTO members provide for exclusions that are equally RPL-diminishing, such as those grounded on lack of compliance with certain national regulations—for example, national statutes dealing with access to genetic resources.

C. The Third Element of the RPL: A General Assessment

This is an important component of the RPL and is subject to a wide variety of challenges by the statutes of WTO members. Out of the twenty statutes considered, only one seems to transpose this RPL component without any shadow of a flaw. And flaws are indeed abundant. For example, one WTO member has established criteria (with seemingly charitable motives) for determining royalties for compulsory licenses of pharmaceutical product patents under paragraph 6 of the Doha Declaration on the TRIPS Agreement and Public Health.[20] That same WTO member does not ensure that compulsory licenses to remedy anticompetitive practices shall be preceded by adjudication. Several WTO members impose

---

19. Article 27.2 reads:

> members may exclude from patentability inventions, the prevention within their territory of the commercial exploitation of which is necessary to protect *ordre public* or morality, including to protect human, animal or plant life or health or to avoid serious prejudice to the environment, provided that such exclusion is not made merely because the exploitation is prohibited by their law.

Id. art. 27.2.

20. *See* Declaration on the TRIPS Agreement and Public Health, WT/MIN(01)/DEC/2 (Nov. 20, 2001); Implementation of Paragraph 6 of the Doha Declaration on the TRIPS Agreement and Public Health, WT/L/540 (Sept. 2, 2003).

exceptions and limitations to rights that may discourage a would-be inventor, such as the use of patented pharmaceutical products for preparations in drugstores. Sometimes, that exception seems to be conditioned by the fact that such preparation must be based on a medical prescription and must be individual—that is, the production must not be serial. However, many individual preparations are still too many preparations. What is an exception may eventually become a rule—and a disregard of the RPL.

A significant number of member states adopt the so-called prior user exception without any safeguard for patent owners. Under this exception, a person or company that has started exploiting an invention, or has taken serious preparations to exploit it, shall be immune from liability on grounds of infringement if the same invention is independently created and patented by another person or company. The prior user exception is therefore a departure from the traditional bargain of "a patent for a disclosed invention." The TRIPS Agreement does not provide for a clear prohibition of this exception—which, as said, is widespread among WTO members—but it is arguably contrary to Article 70.4.[21] As an element of a regime that seeks predictability and security—the key words for would-be investors in invention and innovation—the possibility that a patent owner is not entitled to collect damages from a third party, who without the owner's permission uses a patented invention, is troublesome. And when patent owners are small companies, the risk of ambush may be indeed a discouraging threat.

Certain WTO members fail to provide a robust RPL because they adopt measures such as the possibility of compulsory licenses in the case where the patent owner uses the invention in a way that harms the local economy, local production, or exports. The vagueness of these grounds can be used as an instrument of repression of particularly competitive patent owners. A small number of the twenty WTO members sampled adopt provisions in disregard of the RPL that are more focused: for example, some of them speak directly against biotechnological inventions by diminishing the rights of inventions that are deemed research tools (one might wonder why that same country does not provide for any measure against microscopes), or, more generally, that consist of DNA-recombinant technologies.

---

21. Article 70.4 provides:

> In respect of any acts in respect of specific objects embodying protected subject matter which become infringing under the terms of legislation in conformity with this Agreement, and which were commenced, or in respect of which a significant investment was made, before the date of acceptance of the WTO Agreement by that member, any member may provide for a limitation of the remedies available to the right holder as to the continued performance of such acts after the date of application of this Agreement for that member. In such cases the member shall, however, at least provide for the payment of equitable remuneration.

TRIPS Agreement, *supra* note 2, art. 70.4.

D. The Fourth Element of the RPL: A General Assessment

A notorious factor that diminishes the RPL and threatens to discourage global inventors is the local exploitation requirement. However, it is found in the laws of a number of WTO members in spite of its being notoriously outdated (and prohibited).[22] A number of WTO members deny patent owners the right to refuse to license—which, if stated in the reverse, means that patent owners are obliged to license. In other words, in those WTO members the exclusivity of the patent right—which is translated into the right to say "no"—is denied. That is a powerful blow to the RPL. It should be noted that in this particular regard the patent statutes of one WTO member has a provision that explicitly acknowledges the fact that owning a patent does not give rise in itself to market power and that there is nothing wrong (in principle) in using and asserting the right to exclude. This is a powerful component of the RPL that can only be seen positively by would-be inventors.

Certain countries also threaten compulsory licenses against patent owners who charge excessive prices for patented products (in certain cases, limited to the pharmaceutical sector), and such countries make no reference to prior regulatory caps imposed on those prices—an obvious case of disregard to the core principle *nullum crimen sine lege*.

Another diminution to the RPL is the adoption by WTO members of mechanisms for monitoring patent licensing agreements, including schemes of pre-registration that are conditions of enforceability of contracts. When those schemes include per se approaches, or situations that are deemed anticompetitive, hence unlawful and unenforceable, the RPL is seriously impaired to the extent the patent owner loses the freedom to manage his/her assets according to subjective business considerations.

The RPL is also diminished by those few WTO members who make disclosure of the origin of genetic resources used in the making of the patented invention a condition of the grant or the validity of the rights. This is a condition that goes beyond Article 27.1 of the TRIPS Agreement and aims at putting in practice the provisions of a treaty—the Convention on Biological Diversity (CBD)—that is not trade-oriented and whose thrust is far away from legal security and predictability.

Having the considerations above in view, it is possible to draw a table that classifies WTO members in accordance with their level of adherence to the RPL. The same table shows the corresponding levels of achievement of targets that are mentioned by Article 7 as objectives of patent protection. Those targets

---

22. One of those WTO members imposes the local exploitation requirement upon government-funded inventions only. That is still TRIPS-inconsistent but does not diminish RPL because it does not discourage private inventors.

were identified in the table that follows as the capacity of innovation of countries (as set by the World Economic Forum as a compound of a number of elements, which are sophisticated and complex enough to make believe that there is a beginning of accuracy and economic reality in the figures provided) and the index of patent applications filed by residents. After all, if patent protection is promised as a tool to foster domestic invention, its efficacy should be translated into patent applications filed by residents, rather than by foreign inventors. The table also shows the correspondence of the RPL to the volume of FDI. As noted before, one of the ways in which technology transfer occurs is FDI. In general, internationally harmonized patent protection is antagonistic to FDI to the extent that it ensures a foreign patent owner gains from exporting the products that embody the patented inventions. But, on the other hand, there are factors in international harmonization that operate the other way around, in the sense that foreign inventors obtain protection against free riding by local competitors where they decide to benefit from local comparative advantages. FDI, therefore, may count for indirect transfers of technology: foreign direct investors not only need to train local workers and engineers but also have to teach local suppliers. Additionally, increased technology transfer would be translated into more patent licenses and increased levels of royalty payments.

Taking into account the elements above, this is what the impact of the RPL on the achievement of Article 7 objectives in twenty WTO members looks like:

Table 1[23] The Relation of the RPL with Article 7–Related Objectives in Twenty WTO members

Country	RPL Compliance Index	Capacity for innovation (2010–2011)	FDI per million inhabitants (2008) (USD)	Payments of Royalties/ capita (2008) (USD)	Patent Applications by Residents/ million inhabitants (2008)
Oman	9.8	3.0	966	n.a.	n.a.
Singapore	9.6	4.3	5'675	3'084.2	161.8
Japan	9.4	5.8	190	143	2'603

(*continued*)

23. Data was extracted from THE ECONOMIST, POCKET WORLD IN FIGURES (2008) (population data); U.N. Conference on Trade & Dev. (UNCTAD), UNCTAD Handbook of Statistics—2009, UN Doc. TD/STAT.34, UN Sales No. E/F.09.II.D.10, *available at* www.unctad.org/en/docs/tdstat34_enfr.pdf (data on FDI); WORLD ECON. FORUM, THE GLOBAL COMPETITIVENESS REPORT, 2010–2011 (2011), *available at* http://www3.weforum.org/docs/WEF_GlobalCompetitivenessReport_2010-11.pdf (data on capacity for innovation);

Table 1. *(continued)*

Country	RPL Compliance Index	Capacity for innovation (2010–2011)	FDI per million inhabit-ants (2008) (USD)	Payments of Royalties/ capita (2008) (USD)	Patent Applications by Residents/ million inhabitants (2008)
Israel	9.3	5.3	1'600	184.5	241
South Africa	9.3	3.4	191	35.6	n.a.
Switzerland	9.3	5.7	2'485	n.a.	231.7
Germany	9.2	5.9	303	153.4	578.6
Spain	9.2	3.4	1'523	77.5	75.8
United States	9.2	5.3	1'060	86.5	809.3
Denmark	9.0	4.9	2'100	n.a.	307.4
India	9.0	3.6	32	1.3	4.8
Thailand	9.0	3.1	156	39.9	14.7
Argentina	8.9	3.0	232	38	n.a.
China	8.8	4.2	83	7.9	116.3
Indonesia	8.8	3.7	35	5.9	n.a.
France	8.7	5.1	1'958	90.8	243.3
Australia	8.6	4.1	2'338	151.3	134.5
Canada	8.5	4.2	1'397	270	154.7
Brazil	8.2	3.8	242	14.5	21.5
Hong Kong	8.2	3.3	9'000	230	22.8

The obvious conclusion that one can extract from the table above is that there is no correspondence whatsoever between the levels of RPL and the achievement of the objectives set by Article 7 of the TRIPS Agreement. The RPL may even be deemed as superfluous for WTO members such as Hong Kong, China, and Oman, as far as the objectives of Article 7 are concerned. In Hong Kong and

---

Royalty and Licensing Fees, Payments, THE WORLD BANK, data.worldbank.org/indicator/BM.GSR.ROYL.CD (data on payment of royalties); *Patent Applications by Patent Office, Broken Down by Resident and Non-resident (1883-2009)*, WIPO http://www.wipo.int/ipstats/en (last visited Oct. 12, 2012).

China, the RPL is low and yet FDI is very high.[24] Oman has a very high RPL—actually, the highest among those twenty WTO members and perhaps one of the highest among all WTO members—and yet all the indicators of achievement of Article 7 objectives are very low.[25] By contrast, Singapore, Israel, and Japan show a very high RPL as well as a very high level of success in pursuing Article 7–related objectives. A strong RPL could have a role in this, but the lack of correlation of these factors where other similarly situated WTO members are concerned may indicate that is not much more than a coincidence. A similar balanced proportion appears as regards Switzerland—a relatively high level of RPL and high figures in relation to technology transfer and innovation capacity. But what can one say of the importance of the RPL in the United States, for example? That country is in the middle of the ranking of the selected countries as far as the RPL and technology transfer (including FDI) are concerned. The US RPL is high but it is not one of the highest, and yet its capacity for innovation is one of the highest. Which means that for a country such as the United States, compliance with the standards, as harmonized by Section 5 of Part II of the TRIPS Agreement, is not important. In the United States, locally adapted levels of protection are combined with factors that, together, are able to generate a high capacity of innovation.

IV. What Purpose Does the RPL Serve?

If the RPL does not have a specific or significant role in promoting the achievement of Article 7 objectives, then what is its role? The answer to this question, actually, is very much obvious in view of what was said earlier about the main motivation for the inclusion of substantive intellectual property standards to be included in the Uruguay Round, namely the need to eliminate discriminatory treatment in many patent laws against pharmaceutical inventions and reducing impediments to international trade of pharmaceutical products.

The lack of patent protection may, in certain circumstances, be seen as a non-tariff barrier to trade. Suppose a company spends 100 million dollars in inventing and testing a new pharmaceutical product. That company will file and obtain patents around

24. It can be submitted that a high RPL can be found at 9.0 and above. Of course, the 1.0 value missing could arise from just one or two serious failures in transposing Section 5, Part II standards into national law. For example, a country might comply rigorously with all the provisions of the RPL except one—say, it might establish patent terms for five years. However, such country would not have been able to maintain this position before being obliged by its trading partners to change its law. This means that values deducted from RPL correspond to a myriad of small and sometimes subtle inconsistencies—in the case of certain WTO members, they are many; in others, they are few.

25. The fact that there is no data available as regards patent applications in 2008 does not mean that those figures are unknown—it simply means that the few patent applications filed until then with the Omani Ministry of Economy and Trade were simply not processed for lack of bureaucratic structure. Inventors wishing to file patent applications in Oman could seek the

Table 2[26] Imports of Pharmaceuticals (US dollar at current prices (millions))

Country	RPL Compliance Index	1998	2002	2009
Oman	9.8	2	1	20
South Africa	9.3	n.a.	96	1687
India	9.0	384	542	2005
Thailand	9.0	341	543	1566
Argentina	8.9	782	478	1400
China	8.8	534	1434	6699
Indonesia	8.8	174	196	591
Brazil	8.2	1606	1886	5124

the world so that patients in as many countries as possible pay for the money sunk into R & D. Now suppose that in a certain country, where a local manufacturer has know-how to reverse engineer the latest technology, pharmaceutical inventions are not patentable subject matter. The foreign company will reverse engineer that product and will commercialize it within the borders of the country (and will export it to countries where patent protection is likewise unavailable). Later the original inventor will try to export its own pharmaceutical products to that other country but the fact that there is a local competitor selling an identical or bioequivalent product at a much lower price (because he/she does not need to recoup investment in R & D) will prevent it. This means, in trade-specific jargon, that the unavailability of patent protection in that other country and the resulting possibility of local manufacturers to free ride on foreign technology arguably constitutes a non-tariff barrier to trade.

And indeed, it is not difficult to find a relation of cause and effect between the adoption of the RPL and the increase in dollar-value of imports of pharmaceutical products by developing countries. Table 21.2 below is illustrative of this:

All the eight countries above had an enormous increase of imports of pharmaceutical products if we compare the years of 1998, when they were still enjoying the

GCC (Gulf Cooperation Council) avenue, whose patent office is based in Riyadh, Saudi Arabia. Under the GCC Patent Regulations, it is possible to obtain patents enforceable across the six member states (Bahrain, Kuwait, Oman, Qatar, Saudi Arabia, and the United Arab Emirates).

26. Data extracted from the WTO website, www.wto.org (last visited Oct. 12, 2013).

transitional period, and 2009 when all of them had already fully implemented their obligations under TRIPS.[27]

Of course, the data above concern eight developing countries, and thus they do not offer a sufficiently wide base that permits one to extract final observations across the entire developing country membership of the WTO. Moreover, a normal increase in the value of imported goods should be expected, as a consequence of the overall increase of global trade in that period. Additionally, some of the newly introduced pharmaceuticals are technically very complex, which has a necessary impact on prices. Furthermore, the exchange rate of the US dollar has fallen consistently in recent years, and this factor alone could justify a determined increase in the dollar value of imports. On the other hand, for the sake of consistency, it should be noted that trade in pharmaceuticals does not mean trade of patented pharmaceuticals. Actually, a large number of pharmaceuticals are off-patent.[28]

And yet, the RPL certainly has a role in a twenty-fold increase in imports of pharmaceutical products by a country such as Oman, and seventeen-fold in imports by South Africa—the two developing countries that rank higher in the classification of RPL.[29] These increases are so significant that the impossibility of access to generic products (which would be naturally imported from countries such as India or Jordan, were patents absent), as a consequence of the introduction of patent protection, should have a role. Strikingly, India, which holds a relatively high position in the RPL ranking, ranks third (out of those eight developing countries) in connection with increase in the value of imports—3.6 times

27. Brazil actually anticipated the protection of pharmaceutical products in 1996. Pakistan and India used their extended transitional period until 2005. And China only joined the WTO in 2001. South Africa started implementing TRIPS earlier than most other developing countries because it is considered, in the WTO, to be a developed country. Therefore, it was not able to benefit from transitional periods beyond 1996.

28. Off-patent pharmaceuticals are not the same as off-intellectual property pharmaceuticals. Even when they are not covered by patents, pharmaceuticals are still the subject matter of intellectual property rights, such as trademarks and repression of unfair competition (both in general and in particular, that is, as far as protection of test data submitted in the course of regulatory marketing approval procedures is concerned). The linkage of test data (the standards of whose protection developing country WTO members have introduced for the first time as a consequence of the implementation of Article 39.3 of the TRIPS Agreement) with imports of pharmaceutical products is much more subtle than patents. Actually, countries that show significant increases in imports have not yet modified their laws to incorporate Article 39.3 provisions, such as India and Brazil (the Brazilian statute that applies to test data has left pharmaceutical products for human consumption outside its scope).

29. Singapore and Israel can be politically identified as developing countries, but the indicators they reveal of GDP per capita and Human Development Index associate them with developed economies.

between 2002 and 2009. This is less than the increase in imports of pharmaceuticals by China—4.6 times in the same period—but there is a difference: China adopted the RPL in 2001 whereas India only did so in 2005.

On the other hand, tariffs can help, to some extent, explain the increase in imports of pharmaceuticals by WTO members. Data concerning average tariffs in WTO members imposed on chemical products (including pharmaceuticals) show there is no correlation between the RPL and the increase of imports and tariffs. Oman egregiously ranks, again, first in all three categories (lower tariffs representing a better mark than heavy ones): highest RPL, highest increase in imports since it has introduced the RPL, and lowest tariffs (average, 5.4 percent).[30] South Africa ranks second as regards the RPL; it ranks second in increase of imports of pharmaceuticals, and it ranks third in tariffs (12.3 percent). Brazil levies high tariffs (21.1 percent) and has the lowest level of consistency with the RPL among the twenty countries analyzed. One might submit that it was not a coincidence that Brazil had the smallest increase (as compared to the other seven developing countries) in the value of imports of pharmaceuticals between 2002 and 2009.[31]

Before concluding, however, it should be made clear that the RPL (as set by the TRIPS Agreement) may not be a decisive factor for enabling and/or triggering the capacity to invent in WTO members. But we should not jump to the opposite conclusion: the RPL may not be decisive, but it does not appear to hinder the capacity to invent either. Singapore, Japan, Israel, Switzerland, Germany, Denmark, and the United States are examples of countries that observe a very high RPL (9 and above) and yet show a very high rate of innovation capacity (5 and above).

This discussion is beyond the narrow scope of this chapter, but a second conclusion seems evident: the capacity to invent and the capacity to put the RPL at its service depend on factors that are external to the RPL. The RPL is certainly important, because of the role that predictability and security of legal institutions play in reducing transaction costs. However, that is not the RPL as established by Section 5 of Part II of the TRIPS Agreement. Indeed, the acceptable levels of security are not even identical across the world. Eventually, an efficient RPL would need to be adjusted to the legal culture of different countries, yet without losing sight of the need to provide security and predictability to global transactions. The

30. In 2006, Oman signed a Free Trade Agreement with the United States, which contains a number of TRIPS-plus provisions in the field of patents. However, those higher standards have not had an impact on the increase of imports of pharmaceuticals by 2009 because the statute that enacted them was passed in 2008 only.

31. The data on tariffs of imported articles are available on the WTO website, www.wto.org (last visited Oct. 12, 2013).

TRIPS Agreement itself acknowledges this, in the last sentence of Article 1.1.[32] However, this apparent tolerance only applies to "the method of implementing the provisions" of the TRIPS Agreement, not to the choice of standards.

Conclusion

In conclusion, developing countries may have really believed that their proposal to include in the TRIPS Agreement the language that is now in Article 7 could be persuasive in transforming higher standards of patent protection into an operative and efficient tool for the promotion of technology creation and transfer. But they were flatly wrong in their belief. Instead, those standards, collectively referred in this chapter as the RPL, served the purpose set in the first paragraph of the preamble of eliminating a barrier to trade of pharmaceutical products, generated by lack of or insufficient patent protection.

32. TRIPS Agreement, *supra* note 2, art. 1.1 ("Members shall be free to determine the appropriate method of implementing the provisions of this Agreement within their own legal system and practice.").

INDEX

A2K mobilization, 79
Abbreviated New Drug Application (ANDA), 348–51, 353, 359
Aboriginal people *See* indigenous developmental networks and the non-developmental state
ACA (Patient Protection and Affordable Care Act), 354
Acemoglu, Daron, 293
ACLU (American Civil Liberties Union), 154, 157, 402–3
ACS Hospital System v. Montefiore Hospital (1984), 595
ACTA *See* Anti-Counterfeiting Trade Agreement
ACT UP, 124
Acycrobil (2002), 433, 459
Adams v. Burke (1873), 433–34
Administrative Council (Europe), 328, 330
Administrative Procedure Act (APA), 167
AEI (American Enterprise International), 503
Africa *See* innovation and protection of competition (Africa); *Specific countries*
African Competition Forum (ACF), 531
Agreement on Trade Related Aspects of Intellectual Property Rights *See* TRIPS Agreement
Agreement relating to Community Patents (CP Agreement), 430–34
AIA *See* Leahy–Smith America Invents Act
AIDS *See* HIV/AIDS
Alaska Natives, 288
Aldous, Justice, 262, 266
Alter, Karen, 117
America Invents Act *See* Leahy–Smith America Invents Act (AIA)
American Civil Liberties Union (ACLU), 154, 157, 402–3
American Cyanamid v. Ethicon (1975), 611, 619–20
American Enterprise International (AEI), 503
American Indians, 288
American Society for the Prevention of Cruelty to Animals, 158
Amgen, 365
AMP case *See Ass'n for Molecular Pathology v. Myriad Genetics* (2013)
ANDA (Abbreviated New Drug Application), 348–51, 353, 359

Andean Pact, 101, 124
Andean states, 68–69
Andean Tribunal, 12*n*61, 117
Andrews, Lori B., 159
Animal Legal Defense Fund, 157
animal patents, 152, 157–58, 327, 335–36, 435
animal rights organizations, 335
AntiCancer, 407
Anti-Counterfeiting Trade Agreement (ACTA), 86, 113–15, 114*n*69, 145
antitrust law, 88–90
APA (Administrative Procedure Act), 167
APEC (Asia-Pacific Economic Cooperation), 76
Apotex, Pfizer v. (2007), 557, 599
Apple, 46
Argentina
  access to medicine rights in, 119
  ACTA and, 115
  Development Agenda and, 78
  Friends of Development membership, 143*n*65
  patent law in, 109*n*40
  pharmaceutical imports, 694*t*
  plant patents in, 230, 235
  RoundUp Ready soybeans, 181, 182, 194
  trademarks in, 77
  WIPO and, 79, 143
  WIPO Development Agenda and, 117
  See also emerging economies and international IP lawmaking
Ariad Pharmaceuticals v. Eli Lilly ("Ariad I") (1973), 278–79
Ariad Pharmaceuticals v. Eli Lilly ("Ariad II") (2010), 279–80, 281, 282
Arkwright, R v. (1785), 282
Arnold, Justice, 264, 270
Aro Mfg. v. Convertible Top Replacement (Aro 1, 1961 & Aro 2, 1964), 452–53
Asia, indigenous people in, 288
Asian Tigers and technology, 131
Asia-Pacific Economic Cooperation (APEC), 76
Ass'n for Molecular Pathology v. Myriad Genetics (2013), 154, 156, 158, 159, 160, 165, 166, 184, 207, 208, 209, 210
Ass'n for Molecular Pathology v. USPTO (2012), 206
Association Internationale pour le Protection de la Propriété Industrielle (AIPPI), 133
AstraZeneca, 300
Auction Method/Hitachi (2004), 204, 205, 209
Australia
  damages in, 662
  exhaustion in, 424
  free trade agreements, 110, 111
  lost profits in, 661
  plant patents in, 215, 217, 218–19, 241
  research exceptions in, 382*n*11
  TPP and, 86
  treaties with the US, 111, 112
  See also indigenous developmental networks and the non-developmental state
Australian Manual of Patent Examination and Procedure, 219
Australian Patent Act, 653*n*45, 654*n*51
Austria
  chemical substance patents in, 325
  EFTA and, 329
  Strasbourg Convention and, 327
Aztecs, 291

Badische Anilin und Soda Fabrik v. Isler (1906), 457
Bagley, Margo A., xvi, xxii–xxiii, 149–85
Bahrain, treaties with the US, 86, 111, 112
Bajaj Auto v. TVS Motor (2009), 601, 624–26, 624*n*86, 627
Baker, Brook K., 491
Bakhoum, Mor, xvii, xxviii–xxix, 515–37
Bangladesh, TRIPS and, 30*n*153, 106*n*31
Barnett, Michael, 103
Basel Committee on Banking Supervision, 56
Basheer, Shamnad, xviii, xxxi, 603–35
Bate, Roger, 503
Bayer AG, 372
Bayer CropScience, 176
Bayh Dole Act, 63
BBC Car Wheels (1997), 431–32, 460, 466
Beauregard, In re (1995), 200
Belgian Patent Act (BPA), 404–5
Belgium
  CoE and, 323
  the EEC and, 328
  exhaustion in, 445
  research exceptions in, 404–5, 409–10
  Strasbourg Convention and, 327
Bell, Alexander Graham, 244*n*4
Berne Convention
  copyright, 57, 60, 61
  ICJ and the, 69
  original signatories to the, 104, 104*n*19
Bernstein, William J., 677, 678
Bessen, James, 567
best mode, 168, 214*n*6, 245, 258, 271–72, 274–77, 283, 674
Betts v. Willmott (1871), 425–26, 463–64
Bhar, Justice, 582
Bilateral Investment Treaties (BITs)
  developed countries and, 58
  EU treaties, 110, 535
  meta-norms and, 81
  proliferation of, 80, 109–15, 110*nn*44, 47–48, 112*n*58, 113*n*61, 114*n*69
  US treaties, 86, 110, 112–13, 535
Bilski, In re (2008), 153–54, 184, 206
Biodefense and Pandemic Vaccine and Drug Development Act, 372
biodiversity protection laws, 176
  See also indigenous developmental networks and the non-developmental state
Biogen v. Medeva (1997), 266, 268, 282
biologics, 345–46, 353–55, 353*n*54, 354*n*57, 360, 364–69
Biologics Price Competition and Innovation Act (BPCIA), 354–55, 357, 362, 363
Biomedical Patent Mgmt. v. California Department of Health Services (2007), 399
biotechnology
  Canadian inventions, 189–90, 191, 192, 192–93
  China and, 550
  ethics and, 337–41
  European patents, 156–57, 333–34, 337–41
  human patents, 151, 152, 163, 327
  See also DNA/RNA patents; European Union Biotechnology Directive; subject matter eligibility in the United States; subject matter eligibility in transatlantic context
Biotech Patent Directive (Europe), 333–34
BITs *See* Bilateral Investment Treaties (BITs)
black-letter patent law, 197

Blair, Tony, 146
Blanco-White, T.A., 260, 261
Board of Patent Appeals and Interferences of the USPTO, 216
Boesch v. Graff (1890), 466–67, 468, 479
Bolar exception, 107–9, 140, 387, 401, 401*nn*95–96, 406, 409, 411
Bolivia, Friends of Development membership, 143*n*65
Bonn Guidelines, 297
Bowman, Vernon, 156
Bowman v. Monsanto (2013), 156, 172, 239–40, 435–36
BPCIA *See* Biologics Price Competition and Innovation Act
Braithwaite, John, 120
Brave New World (Huxley), 436
Brazil
  access to medicine rights and, 123
  ACTA and, 115
  biodiversity protection law in, 176
  copyright law in, 32
  Friends of Development membership, 143*n*65
  generic drugs in, 119
  harmonization and, 135
  import markets of, 28
  indigenous people in, 313
  Paris Convention and, 104*n*19
  pharmaceutical imports, 694*t*, 696
  pharmaceutical industry and, 675
  SPLT and, 143
  TRIPS and, 78
  US and, 9*n*44
  WIPO and, 77, 79
  WIPO Development Agenda and, 117
  See also emerging economies and international IP lawmaking
breast cancer susceptibility genes (BRCA1 and BRCA2), 154, 157, 207, 335
Brennan, William J., Jr., 453
Brenner v. Manson (1966), 558
Breyer, Stephen, 153, 205, 598
broad claims in English Law, 265–71, 266*nn*106, 112, 268*n*118
Brunei, TPP and, 86
Bryson, Judge, 159
Burk, Dan L., xxiii, 187–211
Burke, Edmond, 642
Bush (George W.) administration, 125

CAFC *See* Court of Appeals for the Federal Circuit
Calabresi, G., 387–88, 389, 390, 409, 412, 414
Caldwell v. Vanvlissengen (1851), 463
California, patent actions against, 399
Cambodia, treaties with the US, 86
Canada
  biotech inventions in, 189–90, 191, 192, 192–93
  compulsory licenses in, 509–10, 511
  damages in, 662
  indigenous people in, 287, 288, 305, 311–14, 312*n*106, 313*n*109
  lost profits in, 661
  pharmaceutical patents in, 35–36, 36*n*182
  plant patents in, 181, 215, 220, 225–28, 225*n*69, 227*n*85, 231, 236, 236*n*128, 237, 237*n*134
  See also subject matter eligibility in transatlantic context
Canada Patent Act, 225
Canada–Patent Protection of Pharmaceutical Products, 71, 71*n*92, 76, 108
Canadian Handicrafts Guild, 312
Cancer Patients Association of India, 590
Canon Ink Tank (2007), 442, 443–45, 454, 455, 455–56
Carpmael, William, 254

CBD *See* Convention on Biological Diversity
CCPA *See* Court of Customs and Patent Appeals
CDIP (Committee on Development and Intellectual Property), 78, 81, 83
Cecil, William (Lord Burghley), 248
Cefetra, Monsanto v. (2010), 181–83, 194, 209, 210, 235–36
Celanese Int'l v. BP Chemicals (1999), 668, 669, 669*nn*123–24
Center for Open Innovation at the University of California, Berkeley, 393
Central American FTA with the Dominican Republic, 111, 112
chemical substance patents, 324–25
Chesbrough, Henry, 393–94
Chile
- TPP and, 86
- treaties with the US, 86, 111, 112
- TRIPS and, 30*n*153, 76

China
- ACTA and, 115
- biodiversity protection law in, 176
- exhaustion in, 458
- free riding on innovations, 100
- innovation and patent policies, xxi, 91–96
- novelty in, 550
- patent filings, 138–39, 651
- pharmaceutical imports, 694*t*
- pharmaceutical industry and, 675
- plant patents in, 215, 228–29, 228*n*94
- practical applicability requirement, 559–60
- Southeast Asia and, 79
- TRIPS flexibilities and, xvii, xxx, 542, 542*nn*6–7, 545–47, 550–54, 554*n*66, 559–60, 566, 568–70

China, United States v. (2009), 542, 542*nn*6–7
China
- US and, 85, 91–96
- *See also* emerging economies and international IP lawmaking

China-Enforcement (2009), 69*n*85, 72–73
Chinese Ministry of Finance, 93
Chinese Patent Examination Guidelines, 550–52
Chinese State Intellectual Property Office (SIPO), 92, 93, 228–29, 552
Chiron v. Organon (No. 3) (1994), 265–66
Chon, Maggie, 60, 81
Cicero, 677
Cipla, 618–19, 620–22
Cipla, Hoffmann-La Roche v. (2008), 581, 600, 618, 620–22, 625–26, 632
Cipro, 372
CJEU *See* Court of Justice of the European Union
claim construction, 181–84
Classen Immunotherapies v. Biogen (2011), 161, 173
Climate, Community, and Biodiversity Standards, 313
clinical exemption, 401–3, 402*nn*97, 100, 407
CLS Bank International v. Alice Pty (2013), 162, 173
Coase, Ronald, 677
CoE (Council of Europe), 323–28, 329
Coffee Pads (Senseo) (2005), 449–51, 452, 456
Cohen, Julie E., 199
Colchicine, 374
Colgate Palmolive v. Hindustan Lever (1999), 619
Colombia
- access to medicine rights in, 119
- treaties with the US, 86

colonization, 291–95
Columbia University, 369
Comalco mining, 293
Commercial Division of High Courts Bill, 631
commercial/noncommercial use, research exceptions
  Bolar type exception, 409
  experimental use, 385
  goal/purpose of research, 386–87
  overview, 414–17
  statutory research exception, 400, 404*n*111, 405, 407–8
  university research defense, 397
  zero-price liability rule, 410–12
Committee on Development and Intellectual Property (CDIP), 78, 81, 83
competition
  Africa and, xvii, xxviii, 12
  enforcement, 531–33, 532*n*47
  in Europe, 2
  Europe vs. US, 23
  IP system and, 515–18
  legislation and, 531–33, 532*n*47
  patents and, 2
  TRIPS and, 525, 528–31, 537
  See also innovation; innovation and protection of competition (Africa)
compulsory licenses, generally, 392, 392*n*52, 512, 575
compulsory licenses on drug patents
  Doha Declaration and, 107–8, 107*n*33, 496–98, 497*nn*45–46, 499*n*65, 500, 512–13, 688–89
  emergencies and, 502–7, 502*n*79, 504*n*84, 505*n*91, 506*nn*92, 94
  in Europe, 325
  existing patent theories, 489–90, 489*nn*16–17, 490*nn*18–19
  FTAs and, 111
  overview, xvii, xxviii, 485–87, 489*n*14, 512–13
  patent perspectives impact, 501–11, 502*nn*77, 79, 504*nn*84–85, 505*nn*90–91, 506*nn*92, 94, 96, 507–8, 508*nn*99–100, 104, 510*n*109, 511*nn*113–15
  patents as an uber-right, 494–96, 494*nn*31–32, 495*nn*34–38, 496*n*43, 508–9, 509*nn*104–5
  patents as privilege, 491–93, 491*n*23, 493*nn*27–28
  schemas (unconscious biases) regarding, 487–89, 487*n*9, 488*n*13, 489*nn*14–15
  TRIPS on emergencies, 502–7, 506*n*94
  TRIPS requirements, 485, 486–87, 490, 491, 491*n*23, 493, 496–501, 496*n*44, 497*nn*45–46, 50, 498*nn*51, 54–56, 499*nn*65, 67, 500*n*71, 507–12
computer-implemented inventions and business methods patents, 152–54, 163–64
computer software *See* software
conditional sales in civil law countries, 459–61
consent based exhaustion, 469–72, 472*nn*100–101
contract and exhaustion, 456–62
Controlling Pension Benefit System PBS Partnership (2001), 204, 209
Convention for the European Patent (CPC) (Luxembourg Community Patent Convention), 403–4, 406
Convention on Biological Diversity (CBD), 39, 55, 75, 294, 297–98, 299, 690
Cooperative Research and Technology Enhancement (CREATE) Act (2004), 12

copyright
Berne Convention and, 57, 60
in common law vs. non-common law jurisdictions, 544, 544*n*13
creativity, new types of, 63–64
Internet and, 65
vs. patent law, 31–32, 198
WIPO and, 76
Costa Rica
access to medicine rights in, 119
treaties with the US, 86, 111
Cotter, Thomas F., xviii, xxxi–xxxii, 637–72
Coulter, M., 254
Council of Europe (CoE), 323–28, 329
Court of Appeals for the Federal Circuit (CAFC)
on damages, 670
on exhaustion, 435–36
on injunctions, 606, 647
on inventiveness, 595
jurisdiction of the, 631, 631*n*119
on obviousness, 548–49
patent eligibility cases, 15–63, 152, 153, 155, 167, 174–75, 200, 201
percolation of patent law issues and the, 172–74
on research exceptions, 397–98, 401, 411
on sufficiency and written description, 259, 271, 272–74, 279–81
Court of Customs and Patent Appeals (CCPA), 271, 274, 279
Court of Justice of the European Union (CJEU), 181–83, 188, 195, 235–36
C-Path (Critical Path Institute), 380, 414
CPC *See* Convention for the European Patent (CPC) (Luxembourg Community Patent Convention)
CREATE Act (2004), 12
Creative Commons, 64, 80
creative industries, developing countries and, 59–61
Critical Path Institute (C-Path), 380, 414
Crouch, Dennis D., 161, 169
CSTP (Japanese Council for Science and Technology Policy), 407
Cuba, Friends of Development membership, 143*n*65
Cyprus, Strasbourg Convention and, 327

damages
defendant's profits and enhanced damages, 668–71, 668*nn*117–18, 669*nn*119, 123–25, 670*n*132, 671*n*137
lost profits, 655–62, 655*n*54, 56, 656*n*57, 657*nn*64–65, 659*nn*69–70, 661*nn*79, 81, 662*n*84
overview, 652–55, 652*n*44, 653*nn*45–47, 654*nn*49, 51–53
reasonable royalties, 663–68, 663*nn*93–96, 664*n*98, 665*n*99, 666*nn*101–3, 667*nn*108–10
Darcy v. Allein (1602), 253
Darwin, Charles, 314
de Carvalho, Nuno Pires, xviii, xxxii–xxxiii, 673–97
declaratory judgment proceedings, 168–70, 169*nn*79, 82, 170*n*84
Deere, Carolyn, 124
de facto research exception, 398–400
defendant's profits and enhanced damages, 668–71, 668*nn*117–18, 669*nn*119, 123–25, 670*n*132, 671*n*137
Dembiczak, In re (1999), 596
Denmark
CoE and, 323
EFTA and, 329
Strasbourg Convention and, 327
Denning, Lord, 611

Department of Health and Human Services (HHS), 89, 412
Department of Justice (DOJ) Antitrust Division, 89–90
developed countries
  Berne Convention original signatories, 104, 104*n*19
  defense of the patent system by, 7*n*28, 8
  domestic patent policy in, 11–12
  gene-rich developing countries and, 176
  IP concept in, 104–5
  least-developed countries and inventions of, 674
  patent policy in, 6
  pharmaceutical patents, 105
  trade agreements, 58
  TRIPS and, 45, 47–48, 53–54, 58–62, 65*nn*59–60, 63
  welfare interests and patent harmonization, 16
  WIPO and the, 16
  See also Specific countries
developing countries
  benefits of patents in, 567–69
  economic and social rights TLO, 116–26
  foreign direct investment and, 7*n*28
  free riding on innovations, 99–100
  generic and counterfeit drugs and, 13–14, 13*n*70
  gene-rich, 176
  harmonization and, xvi, 134–39, 141–44
  HIV/AIDS in, 40*n*203
  infringement as crime in, 134
  Paris Convention original signatories, 104*n*19
  pharmaceutical patents and, 38*n*193, 567
  skepticism regarding patents in, 2
  Transnational Legal Ordering and access to medicines, 120–26, 123*nn*108, 110, 125*n*121
  treaties with the US, 110–15
  TRIPS and, 17, 29–33, 30*nn*153–54, 39, 45, 47–48, 70, 106–9, 106*nn*27, 31, 108*n*36, 109*n*40, 110*n*44, 676
  vaccines and, 37
  WIPO and, 17, 43
  WTO agreements and, 493
  See also emerging economies and international IP lawmaking; innovation and protection of competition (Africa); least-developed countries; *Specific countries*
development, levels of development argument, 134–36
Diamond v. Chakrabarty (1980), 149–50, 151, 152, 157–58, 160, 175, 180, 190–92, 193, 196, 197, 199, 209, 215, 219, 223, 364
Diamond v. Diehr (1981), 152, 196, 199, 202, 203, 206, 209
Dinwoodie, Graeme, 65
Diplock, Lord, 610, 612, 619
disclosure, 233–34, 245
Dispute Settlement Body (DSB), 54, 62, 65, 69–70, 76–77
DNA/RNA patents
  breast cancer susceptibility genes (BRCA1 and BRCA2), 154, 157, 207, 335
  broad claims, 267–68
  inventive step and, 549
  isolated genomic, 159–60, 165, 183, 207–9
  limiting, 211
  obviousness and, 593–94
  process inventions and, 204
  reproduction and exhaustion, 435
  in the US, 364
  USPTO and, 152

written description and, 280–82
See also biotechnology
Doctors Without Borders, 124, 510
Doha Declaration on the TRIPS Agreement and Public Health
compulsory licenses and, 107–8, 107*n*33, 496–98, 497*nn*45–46, 499*n*65, 500, 512–13, 688–89
on epidemics/emergencies, 498
exhaustion under, 481, 484
on HIV/AIDS, 498, 504
India and the, 74
public health and the, 35, 36
social welfare concerns and the, 117
WIPO's response to the, 17
Doha Ministerial Declaration on TRIPS, 119
DOJ Antitrust Division, 89–90
domestic actors and institutions
innovation and, 46–48
public welfare role of, 11–17, 12*nn*61, 63, 13*nn*65, 67, 70, 14*n*72, 16*n*79
TRIPS and, 2
Dominican Republic, Friends of Development membership, 143*n*65
Dominican Republic–Central American Free Trade Agreement (DR–CAFTA), 111, 112
Dongaria Kondh, 303
Drahos, Peter, xvi, xxiv–xxv, 120, 123, 145, 287–320
DR–CAFTA, 111, 112
Dreamtime, 295
Drexl, Josef, 524, 525*n*31, 528–29, 534–35
Dreyfuss, Rochelle C., xvi, xx–xxi, 53–84, 171, 172, 173
Drug Price Competition and Patent Term Restoration Act *See* Hatch–Waxman Act (1984)
DSB *See* Dispute Settlement Body
Duffy, John F., 170
Dunlop Pneumatic Tyre v. Neal (1999), 453–54
Dutfield, Graham, xvi, xxii, 127–46
DW Integrators v. SAS Institute (2000), 517
Dyk, Judge, 272, 274

eBay v. MercExchange (2006), 121, 647, 649
ECJ (European Court of Justice), 151, 328, 469–71, 472*nn*100–101
economic development *See* developing countries; indigenous developmental networks and the non-developmental state; innovation and protection of competition (Africa); public welfare and the international patent system; remedies, comparative law and economic analysis
Economic Partnership Agreements (EPAs), 80, 533–35
economics of patent remedies *See* remedies, comparative law and economic analysis
Ecuador
access to medicine rights in, 119
Friends of Development membership, 143*n*65
Edison, Thomas, 139
education policies, 32, 32*n*165, 60
EEC (European Economic Community), 328–29, 403
EFTA (European Free Trade Association), 328, 329
Egypt
Friends of Development membership, 143*n*65
TRIPS and, 30*n*153
WIPO Development Agenda and, 117
Eisenberg, Rebecca, 345, 360

Eleventh Amendment, 398
Eli Lilly, 503
Elizabeth I, 248–49, 253
Elmer, Michael, 650
El Salvador
  access to medicine rights in, 119
  Paris Convention and, 104*n*19
  treaties with the US, 111
Embrex v. Service Engineering (2000), 397
emergencies (national) *See* epidemics/emergencies
emerging economies and international IP lawmaking
  developed countries and, 62–65, 65*nn*59–60, 63
  developing administrative law norms, 79–83
  developing countries and, 58–62, 61*n*42
  developing IP norms, 66–67, 66*nn*65, 67
  developing IP norms, domestic legislation, 67–69, 67*nn*71, 73
  developing IP norms, international adjudication, 69–73, 70*nn*85, 89, 71*nn*90, 92, 72*n*101, 73*nn*105, 110
  developing IP norms, international lawmaking, 74–79, 75*n*121
  overview, xvi, xx–xxi, 53–58, 84
enablement, sufficiency and written description
  English early development, 248–55, 249*nn*18, 20, 250*nn*23, 28, 251*n*31, 252*n*40
  invalidation of patent for failure to satisfy the written description, 367–68
  overview, xvi, xxiv, 243–47, 244*n*4, 245*nn*5–6, 246*nn*9–10, 247*nn*11–13, 282–83, 282*n*188
  in the UK, 245, 259–71, 260*n*77, 261*n*80, 266*n*106, 268*n*118, 276
  in the US, 245, 271–81, 271*n*132, 274*n*146, 275*n*149, 279*nn*168–69, 281*n*178
  US early development, 255–59, 256*nn*58, 60, 257*nn*61–62, 64–65, 258*nn*68–69, 259*n*72
enforcement
  in China, 69*n*85, 72–73
  competition law, 531–33, 532*n*47
  inventor's entitlement, 388–93, 388*n*29, 389*nn*35–36, 390*n*42, 392*n*52, 393*f*
  plant patents, 234–41, 237*nn*131–34, 238*n*137, 240*nn*146–47
  TPP and, 145
  under TRIPS, 44–46, 57
  See also pharmaceutical patent enforcement in India, injunctions
England
  colonization by, 292–93
  damages in, 668
  exhaustion in, 421, 424–26, 437–40, 478
  implied license in, 453–54
  injunctions in, 609–12
  lost profits in, 660–69
  novelty in, 249
  patent case win rate in, 651, 651*n*43
  patent law development in, 248–55, 249*nn*18, 20, 250*nn*23, 28, 251*n*31, 252*n*40
  royalties in, 667
  technological growth of, 20
enhanced damages and defendant's profits, 668–71, 668*nn*117–18, 669*nn*119, 123–25, 670*n*132, 671*n*137
EPAs (Economic Partnership Agreements), 80, 533–35
EPC *See* European Patent Convention

epidemics/emergencies
compulsory licenses on drug patents, 502–7, 502*n*79, 504*n*84, 505*n*91, 506*nn*92, 94, 94
Doha Declaration on, 498
malaria, 35, 303, 319, 498, 500, 545
as national emergency, 35–36
TRIPS on emergencies, 502–7, 506*n*94
tuberculosis, 35, 303, 353, 498, 500
See also HIV/AIDS
Epogen, 365–66, 369
Epstein, Richard A., 494, 502–3
essential element, 445–53, 446*n*44, 455–56
essential facilities doctrine, 516–17, 518–27, 522*n*23, 525*n*31, 526*n*34
ethics and biotech, 337–41
EU–CARIFORUM Economic Partnership Agreement (EPA), 533–36
Europe
biotech subject matter eligibility in, 156–57
competition and patents in, 2
competition with the US, 23
compulsory licenses in, 325
computer-implemented inventions and business methods patents in, 152
computer/software patents in, 201–4, 334–35
developing countries and, 124–25
emerging economies and, 71
first patents granted in, 683
harmonization and, 130, 131, 132, 134
India and, 79
industrial application requirement in, 558–59
inventive step in, 549–50, 597–98
Japan and, 138
nineteenth century, 24
non-discrimination as a basis of patent harmonization and, 21
novelty in, 323, 324–25, 326, 478
plant patents in, 215, 217, 220–25, 220*n*46, 221*nn*48, 52, 223*nn*58, 60, 224*nn*63–64, 225*nn*67–68, 237, 237*n*133
property rights system of, 294
research exceptions in, 381–83, 384, 395–96, 400, 403–7, 404*nn*111–12, 405*n*119, 406*n*120, 408, 409–17, 413*n*140
trade promotion in, 69
TRIPS and, 28
See also social and political context of European patent system; *Specific countries*; subject matter eligibility in transatlantic context
European Commission for a unitary EU patent, 405–6
European Court of Human Rights, 69
European Court of Justice *See* ECJ
European Economic Community (EEC), 328–29, 403
European Free Trade Association (EFTA), 328, 329
European Industrial Research Management Association, 327
European Parliament, 330, 333–36, 336, 341
European Patent Convention (EPC)
inventive step in the, 597
origins of the, 329–30
patentability requirements under the, 246–47
plant patents and the, 220, 223
revision to the, 12
revocation of patent under the, 265
Strasbourg Convention and the, 327
subject matter exclusions, 201–4
sufficiency under the, 267, 269
UK Patent Act and the, 260

European Patent Office (EPO)
developing countries and the, 123
EPC and the, 201–4
on human embryonic stem cells patents, 151
inventive step standard, 553–54
Japan and, 145
national patents and the, 130
patent law development and the, 188
plant patents and the, 176, 221–25, 229
subject matter eligibility and the, 157, 165
sufficiency and the, 265, 266, 270
time allowed for patent challenges, 167
US and, 145
utility patent system and, 221
See also social and political context of European patent system
European Patent Organization, 330
European Union
ACTA in, 115
developing countries and the, 105
essential facilities doctrine in the, 516–17, 521–22
free trade agreements and the, 512
generic drugs and, 119
harmonization and the, 141, 143, 145
innovation and IP in the, 526, 527
international IP rules shaped by the, 104
IP exploitation in, 516
microorganism patents in the, 150
pharmaceutical industry and the, 675
plant patents in the, 181–83, 194–95
PPPs in the, 380
precedents used by developing countries, 121
protectionism of, 80
royalties in, 663
TRIPS and the, 106, 676
unitary EU patent, 405–6
US patent system vs. EU system, 126–27, 133
European Union Biotechnology Directive
plant and animal patents and the, 235, 237*n*133, 340, 435–36
subject matter eligibility and the, 151, 156, 182, 194–95, 404
Evans v. Eaton (1818), 256–57, 256*n*58, 278
evergreening, 562*n*121
exceptions and limits to patent protection
benefits of, 6–7
TRIPS and, 76, 681–85, 681*n*14, 684*n*17
WIPO and, 76
See also compulsory licenses on drug patents; exhaustion; exhaustion, international; research exceptions
exclusive rights, *See also* regulatory exclusivities
Exela PharmaSciences, LLC v. Kappos (2012), 167
exhaustion
contract and, 456–62
definition and use of the term, 421
doctrinal concepts regarding, 421–24
domestic exhaustion principles, 424–34
implied permission and, 456
overview, xvii, xxvii–xxviii, 419–20
reconstruction and essential element, 455–56
repair and, 436–45
repair, exhaustion and essential element exchange, 445–53, 446*n*44
repair, exhaustion and social perception, 453–56
reproduction and, 434–36, 435*n*25

exhaustion, international
consent based, 469–72, 472*nn*100–101
GATT Agreement and, 482–83
implied license doctrine, 463–66
overview, 462–63, 482–84, 484*n*127
Paris Convention and, 478–79, 479*n*119
remuneration based, 472–77, 477*n*113
TRIPS Agreement and, 420, 480–81
under US law, 466–69
experimental use doctrine, 396–98
extractive institutions and practices *See* indigenous developmental networks and the non-developmental state
EXXON/Fuel Oils (1994), 266, 267

Farey, John, 255
FCN treaties, 24, 24*nn*123–24
FCTC (Framework Convention for Tobacco Control), 37
FDA *See* Food and Drug Administration
Federal Circuit Court *See* Court of Appeals for the Federal Circuit (CAFC)
"Federal Participation in the Development and Use of Voluntary Consensus Standards and in Conformity Assessment Activities" (OMB), 95
Federal Trade Commission (FTC), 89–90, 122
Finger, J. Michael, 15–16
first sale doctrine, 423, 433–34
Fisher, Matthew, xvi, xxiv, 243–83
Flook, Parker v. (1978), 152, 196, 199, 202, 203, 206
Floyd, Justice, 264
Flügelradzhler/Impeller Flow Meter (2006), 446–47, 449
Food, Drug, and Cosmetic Act, 374
Food and Drug Administration (FDA)
Bolar exception and the, 401, 411
generic approval by the, 88
Orphan Drug Act and the, 373–74
regulatory exclusivities and the, 347–54, 356–57, 358–60, 365, 368, 375
terrorism and the, 372
food patents, in Europe, 324–25
Forest Stewardship Council and the Fairtrade Labelling Organizations International, 312, 313*n*109
Framework Convention for Tobacco Control (FCTC), 37
France
CoE and, 323
the EEC and, 328, 329
exhaustion in, 421, 474, 475, 478
IPI and, 323
Patent Act, 437
pharmaceutical patents in, 127*n*3
royalties in, 663
Strasbourg Convention and, 327
TRIPS and, 30*n*154, 686, 692*t*
Frankel, Susy, 62
free trade agreements (FTAs)
compulsory licenses under, 111, 512
developed countries and, 58
proliferation of, 80
regulatory exclusivities and, 347
TRIPS-minus and TRIPS-plus provisions in, 110–15
US and, 28*n*146
Friendship, Commerce, and Navigation (FCN) treaties, 24, 24*nn*123–24
Friends of Development, 143–44, 143*n*65
FTAs *See* free trade agreements
FTC (Federal Trade Commission), 89–90, 122
Fuji Camera v. Jazz (2000), 440–42, 467–68

GAIN Act, 352–53
Gambia, Competition Act of, 533
Garware Wall-Ropes v. Anant Kanoi (2005), 628
GATT (General Agreement on Tariffs and Trade)
exhaustion and the, 462, 480, 482–83, 484
harmonization and, 132
Japan and US dispute under the, 23
property law and, 682
public health and, 682–83
US and, 23, 131
Gay, In re (1962), 274, 276–77
The Gazette (EPO), 337
Genentech, MedImmune v. (2006), 157, 158, 558
Genentech's Patent (1987), 265
General Agreement on Tariffs and Trade *See* GATT
Generating Antibiotic Incentives Now (GAIN) Act, 352–53
generic drugs
counterfeit drugs and, 13–14, 13*n*70, 86
emerging economies and, 75, 675
India and, 13, 67–68
US and, 88–90
See also compulsory licenses on drug patents; pharmaceutical patents; regulatory exclusivities
Generics (UK) v. Lundbeck (2009), 268, 268*n*118
genetic resources, 58
Genetic Resources Action International (GRAIN), 140
genetic testing, 402–3
Genomic Research and Diagnostic Accessibility Act (proposed), 400, 402, 411
Georgia-Pacific factors, 664–67, 664*n*98, 665*n*99
Germany
damages in, 668, 668*n*118, 669–70
the EEC and, 328
exhaustion in, 420, 421, 422, 440, 445, 446–51, 448*n*48, 450*n*55, 451*n*56, 455, 456, 474, 475–76, 477
first patent law, 140
harmonization and, 137, 138
implied permission in, 456
inventions of, 674
Paris Convention and, 18–19
patent case win rate in, 651, 651*n*41
patents filed at WIPO by, 93
pharmaceutical and chemical patents in, 127, 127*n*
repair, exhaustion and essential element exchange in, 446–51, 448*n*48, 449*n*49, 450*n*55, 451*n*56
royalties in, 663, 667, 667*nn*111, 113
TRIPS and, 30*n*154
Gervais, Daniel J., xvii, xxix–xxx, 541–70
Glaxo Group Ltd. Seretide patent, 582
GlaxoSmithKline South Africa (GSK), xvii, xxviii, 518–24, 526
Gleevac, 68, 119, 119*n*94
Global Business Regulation (Braithwaite and Drahos), 120
Gore, Al, 125
Gottschalk v. Benson (1972), 152, 196, 197, 198, 199, 200, 201, 202, 209
Gowda, Prakruthi, xviii, xxxi, 603–35
Graham v. John Deere (1966), 572, 582, 595
GRAIN (Genetic Resources Action International), 140
Grain Pool v. Commonwealth (2000), 219
Gray, Elisha, 244*n*4
Greece, Strasbourg Convention and, 327
Greenpeace, 333, 335, 343, 358
Griffith University, 300
Grover, Anand, 590

Guatamala
Paris Convention and, 104*n*19
treaties with the US, 111

Hamamatsu Medical School, 407
Hand, Learned, 571
harmonization, limits of substantive patent law
counter-counter harmonization, 143–46, 144*n*71
counter-expertise and the postponement of the SPLT process, 140–43, 141*n*56, 142*n*60
differentiation argument, 136–39, 138*n*46
levels-of-development argument, 134–36, 134*n*29, 136*n*35
nobody-can-be-sure-what's-best-fo r-any-country argument, 139–40
overview, xvi, xxii, 127–34, 127*n*3, 128*n*7, 130*n*14, 131*n*16, 146
harmonization
US executive branch patent policy, xxi, 85–96
See also emerging economies and international IP lawmaking; public welfare and the international patent system
Harvard College v. Canada (2002), 189–90, 191, 192, 193, 194, 226, 227
Harvard Oncomouse case, 189–90, 335–36
Hatch–Waxman Act (1984)
FTAs and the, 111
regulatory exclusivities and the, 347–53, 347*n*12, 350*n*24, 353, 359, 368, 368*n*149, 374–75, 376
Roche v. Bolar and the, 11
validity challenges by generics and the, 88–89
Hayek, Friedrich, 642
Health Global Access Project, 491
Heath, Christopher, xvii, xxvii–xxviii, 419–84
Helfer, Laurence, 102, 117, 119
HHS (Department of Health and Human Services), 89, 412
Hibberd, Ex parte, 215–16
Hindmarch, W.M., 254
Hitachi (2004), 204, 205, 209
HIV/AIDS
activism and, 125
Doha Declaration on, 498, 504
Health Global Access Project, 491
as national emergency, 35, 500
PEPFAR and, 90
pharmaceutical patents/access to medicine and, 100, 107, 119
price of medicines for, 40*nn*199, 203
in South Africa, 12, 506
in Thailand, 504
TRIPS and, 545
Ho, Cynthia M., xvii, xxviii, 485–513
Hoffmann, Lord, 267
Hoffmann La Roche, 584
Hoffmann–La Roche v. Cipla (2008), 581, 600, 618, 620–22, 625–26, 632
Holland *See* Netherlands
Honduras, treaties with the US, 111
Hong Kong, TRIPS RPL and, 686, 692–93, 692*t*
Hopi, 310
Houston, James, 311–12
Hricik, David, 169
Huawei Technologies (China), 139
Humane Farming Association, 158
human organism patents, 163
Huxley, Aldous, 436

ICESR (International Covenant on Economic Social and Cultural Rights), 116, 494
ICJ (International Court of Justice), 69

Imaflora, 313
IMF (International Monetary Fund), 56
IMI (Innovative Medicines Initiative), 380, 414
implied license
in England, 420, 421–22, 437–38, 453–54
exhaustion, international and, 463–66
first-sale-doctrine and, 423, 433–34
implied permission, 456
Incas, 291
India
ACTA and, 115
biodiversity protection law in, 176, 177
compulsory licenses in, 575
domestic legislation of, 67–69
Europe and, 79
harmonization and, 135
import markets of, 28
industrial applicability requirement in, 560–63
novelty in, 123*n*110, 585, 613, 614
pharmaceutical imports, 694*t*
pharmaceutical industry and, 675
pharmaceutical patents and, 12–13, 73, 105*n*21, 119, 119*n*94, 492
plant patents in, 215, 228, 229–30, 230
preliminary injunctions in, 121–22
process patents in, 595
public welfare costs of foreign patent protection in, 5
Traditional Knowledge Digital Library (TKDL) initiative, 298, 302–4, 319
TRIPS and, xvii, xxx, 45–46, 78, 542, 544–47, 554–58, 554*n*67, 555*nn*73–74, 560–63, 562*n*121, 565–66, 568–70
See also emerging economies and international IP lawmaking; non-obviousness/inventive step in India; pharmaceutical patent enforcement in India, injunctions
India, United States v. (1997), 542, 542*n*6
Indian Biological Diversity Act, 177
Indian Patent Act
domestic needs and the, 67–68
inventive step/non-obviousness under the, 573–74, 574, 577, 590, 591–92
patent opposition/challenge under the, 614, 617
pharmaceutical patents under the, 12
specialist advisers under the, 632
TPP and the, 87
Indian Patent Office, 605, 613–14
Indian Patents (Amendment) Act, 554–55, 574–75, 583
Indian Patents Act, 229–30
Indian Supreme Court
injunctions and the, 605
on the interim phase, 624, 635
triable issue standard and the, 619
indigenous developmental networks and the non-developmental state
adaptive governance and developmental networks, 305–14, 308*n*88, 310*n*96, 312*n*106, 313*n*109
extractive property rights, 195*n*34, 290–304, 292*n*18, 293*n*26, 296*n*40, 297*n*44, 46, 299*n*56, 304*n*78
overview, xvi, xxiv–xxv, 287–90, 289*nn*8–9, 319–20
scientists and indigenous people, 314–19
Indonesia
pharmaceutical imports, 694*t*
industrial applicability *See* utility/industrial applicability requirement
information and communications technology (ICT) firms, 93

information technology industry, 63–64
infringement
  business method patents, 153, 163–64
  claims to protect domestic markets, 23, 26
  competitor alleged infringers, 158–59, 167
  as crime, 134
  description and, 257
  Internet and copyright, 65
  literary and artistic works, 20
  non-competitor alleged and civil society, 156–58, 156*n*33
  See also remedies, comparative law and economic analysis
injunctions
  permanent, 644–51, 644*n*15, 645*n*17, 647*nn*23, 26–27, 648*nn*29–31, 650*nn*35–36, 651*nn*38, 41–43
  property rules and, 389
  in the UK, xxxi
innovation
  addressing restrictions of competition, 535–36
  in developing countries, 123, 123*n*110
  free riding on, 99–100
  harmonization and, 41–44, 42*nn*206–7
  in India, 123, 123*n*110
  IP system and, 515–18
  pharmaceutical, 347–55, 347*nn*11–12, 350*n*24, 353*n*54, 354*n*57
  in the UK, 1*nn*1–2
  See also competition; indigenous developmental networks and the non-developmental state; non-obviousness/inventive step; plant utility patent protection; regulatory exclusivities; research exceptions; social and political context of European patent system; technology;
innovation and protection of competition (Africa)
  essential facilities doctrine and, 516–17, 518–27, 522*n*23, 525*n*31, 526*n*34
  international perspective regarding, 527–35, 532*n*47
  overview, xvii, xxviii–xxix, 515–18, 516*n*3, 537
Innovative Medicines Initiative (IMI), 380, 414
intellectual property (IP)
  competition and, 515–18, 529–37
  exploitation in the EU, 516
  innovation and, 515–18
  international IP rules shaped by the EU and US, 104
  technology transfer and, 134
  TLO for, 103–15, 105*nn*21, 25, 106*nn*27, 31, 107*nn*33, 35, 108*n*36, 109*n*40, 110*nn*44, 47–48, 112*n*58, 113*n*61, 114*n*69
  See also emerging economies and international IP lawmaking
Intellectual Property Appellate Board (IPAB), 556–57, 590–91, 624, 631
Intellectual Property Owners Association, 503
International Association for the Protection of Intellectual Property, 327
International Centre for Trade and Sustainable Development, 339
International Chamber of Commerce, 327
International Convention for the Protection of New Varieties of Plants *See* UPOV Convention
International Court of Justice (ICJ), 69
International Covenant on Economic Social and Cultural Rights (ICESCR), 116, 494

international exhaustion *See* exhaustion, international
International Federation of Patent Agents, 327
International Monetary Fund (IMF), 56
International Panel on Climate Change, 314
International Patent Classification (IPC) system, 302
International Patent Institute (IPI), 323, 329
International Trade Commission (ITC), 23, 46
Internet service providers, 64
Inuit art market, 311–12, 319
invention, tangible, 199–201
inventive step *See* non-obviousness/ inventive step
IPAB *See* Intellectual Property Appellate Board
IPC (International Patent Classification) system, 302
IPCA Laboratories Limited, 587–89
IPI (International Patent Institute), 323, 329
Iran, Friends of Development membership, 143*n*65
Ireland
  CoE and, 323
  injunctions in, 608*n*22
  pharmaceutical patents in, 127*n*3
  Strasbourg Convention and, 327
Israel, TRIPS and, 686, 692*t*, 693, 696
Issue Management Groups (IMGs), 341
Italy
  CoE and, 323
  the EEC and, 328
  patents detrimental to health in, 325
  pharmaceutical patents in, 127*n*3
ITC (International Trade Commission), 23, 46
J.E.M. Ag Supply v. Pioneer Hi-Bred (2001), 152, 156, 216
Jackson, Justice, 571
Jacob, Lord Justice, 260, 268, 269, 270
Janis, Mark D., xvi, xxiii–xxiv, 213–42, 278
Japan
  competition with the US, 23
  conditional sales in, 460–61
  economy of, 91
  essential element in, 455
  EU and, 145
  exhaustion in, 420, 421, 431–33, 440–42, 443–45, 454, 455–56, 460–61, 473, 474, 475, 476
  free riding on innovations, 100
  harmonization and, 130, 131, 132–33, 137, 138, 141, 143
  inventions of, 674
  multifactor test in, 454
  patent filing case win rate in, 651
  patents filed at WIPO by, 93
  pharmaceutical industry and, 675
  pharmaceutical patents in, 127*n*3
  plant patents in, 215, 217–18, 220, 232, 233, 241, 241*n*152
  product identity in, 440–42, 443–45
  research exceptions in, 381, 383, 384, 395–96, 400, 407–9, 409–17, 413*n*140
  royalties in, 667, 667*n*110
  technology and, 131
  TPP and, 86
  US and, 30*n*155, 145
Japanese Council for Science and Technology Policy (CSTP), 407
Japanese Patent Act (JPA), 407, 408, 431–32
Japan Patent Office (JPO), Examination Guidelines, 218
Jarlmadangah Burru Aboriginal Community, 305

Jefferson, Thomas, 558
Jordan, treaties with the US, 86, 111, 112
journalism, 64–65

Kapczynski, Amy, 79
Keen, Ian, 296
Kennedy, Anthony, 596
Kenya
  access to medicine rights in, 119
  Friends of Development membership, 143*n*65
  human right to health in, 14, 14*n*72, 119*n*98
  public health and patent law in, 13–14
  public welfare costs of foreign patent protection in, 5
Khader, Feroz Ali, xviii, xxx, 571–601
Kieff, F. Scott, 494, 502–3
Kim, Linsu, 134
King Pharmaceuticals v. Eon Labs (2010), 206
knowledge
  product access, 56
  in public domain, 99, 99*n*4
  of technology, 30
  TKDL initiative, 298, 302–4, 319
  traditional knowledge and folklore, 53, 58
  Yolgnu system of, 296
  See also indigenous developmental networks and the non-developmental state
Kohler, Josef (Doctrine of Exhaustion), 420, 422–23, 426–33, 456, 472–73
Kondhs, 303
Korea, free riding on innovations, 100
KSR v. Teleflex (2007), 88, 551, 572, 596, 599
Kuhn, Thomas, 315
Kur, Annette, 82
Kuuku I'yu people, 307
Laboratory Corporation v. Metabolite (2006), 153, 205–6, 598
Laddie, Justice, 612, 619
Lall, Sanjaya, 136
Lardil People, 296
Latin America, innovations in, 123
Laufkranz (2006), 447, 449
Leahy–Smith America Invents Act (AIA)
  best mode and the, 275–76
  debate and enactment of the, 367, 370–71, 403
  subject matter eligibility and the, 163, 166, 167, 170, 175
least-developed countries
  compulsory licenses and, 506
  emerging economies and, 56
  HIV/AIDS in, 500
  skepticism regarding patents in, 2
  TRIPS and, 39, 45, 47–48, 106, 106*nn*27, 31
  WIPO and, 17, 81
  See also Specific countries
legislation, competition law and, 531–33, 532*n*47
Lemley, Mark A., 199, 356, 391, 410, 412–13, 615
Lenzing AG's European Patent (UK) (1997), 12
Levin, Marianne, 82
Lewis, Stacy, 650
Liardet v. Johnson (1778), 251–52
Lichtman, Doug, 615
Lincoln, Abraham, 171
Linn, Judge, 278
Linux, 63
Lloyd, Lord Justice, 262–64
Longchambon, Henri, 324
lost profits *See* profits, lost
Lourie, Judge, 159, 279–80
Luxembourg
  CoE and, 323

Luxembourg *(Cont'd.)*
the EEC and, 328
Strasbourg Convention and, 327
Luxembourg Community Patent Convention, 403–4, 406

Machlup, Fritz, 673, 675, 676
Madey v. Duke (2002), 397–98, 400, 401, 410, 415
Magill (1995), xxix, 516, 517, 523, 527
malaria, 35, 303, 319, 498, 500, 545, 587–88, 588*n*101
Malaysia, TPP and, 86
Mallinckrodt v. Medipart (1972), 458–59
Mansfield, Lord, 252
Manual of Patent Practice and Procedure (MPPP), 573, 576, 577
Mars Bar, 308–10
Marx, Karl, 315
Maskus, Keith, 30, 136
Masur, Jonathan, 174
Mattei, Ugo, 650
Matthews, Duncan, 79, 80
Mauritius, Competition Act of, 533
Mayer, Judge, 596
Mayo Collaborative Services v. Prometheus (2012), 154, 155, 159, 165, 168, 173, 184, 205, 206
McKnight, David, 296
Medicaid and Medicare, regulatory exclusivities and, 355, 365, 374
MedImmune v. Genentech (2006), 157, 158, 558
MedImmune v. Novartis (2011), 270
Melamed, A.G., 387–88, 389, 390, 409, 412, 414
Menck, National Phonograph v. (1911), 424, 457, 464–66
Mentor v. Hollister (1993), 261–63, 264
Merck v. Integra (2005), 401
Merck v. Primecrown (1995), 470–71
Merck v. Stephar (1981), 470–71
Merges, Robert P., 161, 388, 390
Merkel, Angela, 339
Meurer, Michael J., 567
MFN (most-favored-nation), 4, 21, 22
Millennium Development Goals (United Nations), 77
Millennium Ecosystem Assessment (MA), 314–15
mining, 293, 303
Modernizing Our Drug & Diagnostics Evaluation and Regulatory Network (MODDERN) Cures Act, 375
monopoly, patent
enforcement of the, 611
exhaustion and, 470
patents as, 282*n*188
public health and, 325
utility and, 558
written disclosure as the price of, 250–52, 250*nn*23, 28, 251*nn*31–32, 252*nn*35, 38, 40
Monsanto, 156, 157, 172, 177, 179–80, 184, 217, 217*n*20, 236–40
Monsanto, Organic Seed Growers and Trade Association v. (2013), 156, 157, 158, 172, 179–80, 184, 233–34
Monsanto Canada v. Schmeiser (2004), 156, 181, 192–93, 209, 226–28
Monsanto v. Cefetra (2010), 181–83, 194, 209, 210, 235–36
Monsanto v. McFarling (2002), 238, 239, 240
Monsanto v. Scruggs (2006), 238, 238*n*139
Moore, Judge, 165, 166, 208
morality exceptions, 325–26, 327
Morocco, treaties with the US, 110, 111, 112
Morphy, Howard, 296
most-favored-nation (MFN), 4, 21, 22

MPPP (*Manual of Patent Practice and Procedure*), 573, 576, 577
M/s. Bishwanath Prasad Radhey Shyam Appellant v. M/s Hindustan Metal Industries (1982), 580–81
M/s. Shree Vardhman Rice Gen Mills v. M/s Amar Singh Chawalwala (2009), 625–26
M/s. Tissuegene, Inc., 593
Mughals, 291
multifactor test, 454
Myriad Genetics, Inc., 154, 156, 158, 159, 160, 165, 166, 184

NAFTA (North American Free Trade Agreement), 101
Nagoya Protocol on Access to Genetic Resources, 297
Nasmith, John, 250–51
Natco Pharma, 589–90
National Phonograph v. Menck (1911), 424, 457, 464–66
National Research Development Corp. v. Commissioner of Patents (1959), 219
national treatment (NT), patent harmonization framework and, 4
natural law, 197
NCE *See* New Chemical Entity
NDA *See* New Drug Application
Nestec S.A. v. Dualit (2012), 451–52, 451*n*56
Netherlands
  CoE and, 323
  the EEC and, 328
  exhaustion in, 445–46
  generic drugs and, 119
  IPI and, 323
  Paris Convention and, 18–19
  patent law in, 127, 139
  plant patents in, 235
  repair, exhaustion and essential element exchange in the, 445–46, 446*n*44
  Strasbourg Convention and, 327
new chemical entity (NCE), 349–51, 355–56, 357, 362, 374–75
New Drug Application (NDA), 347–51, 374
Newman, Judge, 272
New Zealand
  indigenous people in, 287, 288, 305
  TPP and, 86
NGOs *See* nongovernmental organizations
Nicaragua, treaties with the US, 86, 111
Nicholls, Donald, 283
Nigeria, competition authority in, 532
No-Fume v. Pitchford (1935), 261
non-discrimination as a basis of patent harmonization, 18, 21–27, 22*nn*111–12, 24*nn*122, 124–25
  See also most-favored-nation (MFN)
nongovernmental organizations (NGOs)
  developing and emerging economies and, 79, 124–26
  emerging economies and, 56
  on harmonization, 140–41, 142
  on seed patents, 176–77
  standards developed by, 313
non-obviousness/inventive step
  in China, 550–54
  external restrictions and, 247
  in India, 614
  local inventors and, 569
  pharmaceutical patents and, 87, 584–94
  plant patents, 176, 218, 231–32, 231*nn*106–7, 232*n*108
  TRIPS and, 106–7, 151, 544, 547–58, 547*n*31, 548*nn*34, 35, 549*n*36, 555*nn*73–74

non-obviousness/inventive step *(Cont'd.)*
US law regarding, xxx, 88, 169*n*82, 571–73, 576–77, 579, 580, 583–84, 585, 595–600
See also enablement, sufficiency and written description; plant utility patent protection
non-obviousness/inventive step in India
inventive step new standard evolution, 574–82, 576*n*17, 577*nn*26, 30, 581*n*64
overview, xviii, xxx, 554–58, 555*nn*73–74, 571–73, 600–601, 601*n*169
proof of progress, 582–94, 585*nn*85, 88–89, 586*n*90, 588*nn*99–101, 591*n*113, 592*n*122
threshold of inventiveness India vs. US, 595–600, 598*n*153
non-working and patent forfeiture, 493
North, Douglass, 677
North America
Japan and, 138
TLOs, 101
See also subject matter eligibility in transatlantic context
North American Free Trade Agreement (NAFTA), 101
Norway
CoE and, 323
EFTA and, 329
Strasbourg Convention and, 327
Novartis (2000), 221, 225, 226
Novartis AG v. Union of India (2013), 12–13, 68, 119*n*94, 555–57, 590, 591–93
Novartis v. Johnson & Johnson (2010), 268–69, 270
novelty requirement
absolute, 478–79
in China, 550
definition of, 544–45
in England, 249
in Europe, 323, 324–25, 326, 478
external restrictions and, 247
in India, 123*n*110, 585, 613, 614
local inventors and, 569
Novartis patent, 555
origins of the, 683
pioneer patents and, xii
plant patents and, 176, 223
printed matter doctrine, 198
TRIPs and, 106
in the US, 161, 167, 169*n*82, 204, 257, 598
WIPO and, 132
Nunhems, 176
Nussbaum, Martha, 81

OAPI (Organisation Africaine de la Propriété Intellectuelle), 124
Obama administration
on antitrust law, 88–90
on biologic exclusivity, 354–55, 360, 363, 374
ITC import ban and the, 46
subject matter eligibility and the, 204
Oddzon v. Just Toys (1997), 12
OECD (Organization for Economic Co-operation and Development) countries, 9, 32, 82, 110
Office of Management and Budget (OMB), 90, 95
Office of the United States Trade Representative (USTR), 114
Okediji, Ruth L., xix–xx, 1–50
Oman
pharmaceutical imports, 694*t*, 695, 696
treaties with the US, 86, 692
OMB (Office of Management and Budget), 90, 95
OMB Circular A-119, 95
Oncomouse case, 189–90, 335–36
open innovation *See* research exceptions

O'Reilly v. Morse (1854), 282
Organic Seed Growers and Trade Association v. Monsanto (2013), 156, 157, 158, 172, 179–80, 184, 233–34
Organization for Economic Co-operation and Development *See* OECD
Orphan Drug Act, 351–52, 363, 373–74
Oxfam, 124

Padel, Felix, 303
Paiter Suruí, 313
Panama, treaties with the US, 86
Paris Convention
  on compulsory licenses, 493
  developed countries and the, 16
  enforcement under, 44
  exhaustion and the, 462, 467, 474, 478–79, 479*n*119
  ICJ and the, 69
  justification for patent protection, 18–19
  justifications for harmonization under, 4
  most-favored-nation provisions and the, 21, 22
  national patent laws under the, 22, 41
  national policies under, 25–26
  non-patent norms and the, 10
  original signatories to the, 104, 104*n*19
  patents and trademarks under the, 57, 77
  public policy and, 325–26
  structure and norms imposed by the, 27–28
Parker v. Flook (1978), 152, 196, 199, 202, 203, 206
Parthasarathy, Shobita, xvii, xxv–xxvi, 321–43
patentability requirements
  in Europe, 324–28
  foundations of the patent system, 246
  morality exceptions, 325–26, 327
  TRIPS and, 544–63
  See also exceptions and limits to patent protection; non-obviousness/inventive step; novelty requirement; subject matter eligibility; TRIPS Agreement flexibilities; utility/industrial applicability requirement
patent acts *See Specific countries*
Patent Cooperation Treaty (PCT), 93, 130, 139, 143, 569, 613
patent theories
  generally, 489–90, 489*nn*16–17, 490*nn*18–19
  patents as an uber-right, 494–96, 494*nn*31–32, 495*nn*34–38, 496*n*43, 508–9, 509*nn*104–5
  patents as privilege, 491–93, 491*n*23, 493*nn*27–28
  subject matter eligibility, 170–80, 170*n*87, 174*n*102, 175*nn*104, 107, 176*n*109, 177*n*112, 178*nn*113, 115, 179*n*118
Patient Protection and Affordable Care Act (ACA), 354
PCDA (Provisional Committee on Proposals Related to a WIPO Development Agenda), 77–78, 81
PCT *See* Patent Cooperation Treaty
Pender, Jon, 502
Pension Benefit System PBS Partnership, Controlling (2001), 204, 209
PEPFAR (President's Emergency Plan for AIDS Relief), 90
percolation of patent law issues, 172–74
Peru
  access to medicine rights in, 119
  Friends of Development membership, 143*n*65
  TPP and, 86
  treaties with the US, 86

Pfizer v. Apotex (2007), 557, 599
pharmaceutical industry
developed countries patent systems and the, 675
developing countries and inventions of the, 675
in India, 587–90
patent-to-product ratio, 65
PPPs and the, 380
pursuit of patent related regulatory interests by, 145
trade associations, 104
pharmaceutical patent enforcement in India, injunctions
appropriate standard for developing countries, 613–18, 616*nn*54–56
interim phase, dispensing with the, 623–32, 623*n*81, 624*nn*82–84, 86, 627*n*92, 628*nn*99–100, 629*n*102, 630*nn*113–14, 631*nn*115–16, 119, 632*nn*124, 126, 635
overview, xviii, xxxi, 603–9, 603*n*3, 604*nn*5–6, 605*n*7, 607*nn*14–15, 608*nn*20–22, 634–35
patent proficiency, 618–22, 618*nn*62–63, 619*nn*65, 69, 620*nn*70–71, 621*nn*76–78, 622*nn*79–80
prima facie standard, 606–14, 610*nn*27–29, 611*nn*32–34, 612*nn*35, 39, 613*n*41, 615*n*53, 634
TRIPS compliance, 605, 633–34
pharmaceutical patents
access to medicines, 12, 12*n*63, 13, 34–37, 49, 59, 61, 67–68, 106–9, 106*nn*27, 31, 107*nn*33, 35, 108*n*36, 109*n*40, 120–26, 123*nn*108, 110, 125*n*122, 646
developing countries and, 38*n*193, 88, 567
in Europe, 324–25, 327, 470
in India, 67–68, 562, 587–94
innovation and, 347–55, 347*nn*11–12, 350*n*24, 353*n*54, 354*n*57
international vs. domestic positions on, 88–91
proof of progress and, 584–87
resistance to patent protection, 49
in South Africa, 519–25, 522*n*23
TPP on, 86–87
TRIPS and, xxxiii, 693–97, 694*t*
See also compulsory licenses on drug patents; generic drugs; indigenous developmental networks and the non-developmental state; regulatory exclusivities; transnational legal ordering (TLO) and access to medicines
pharmaceutical trade associations, 104
Philips Company, 139
Phillips, Willard, 563
Pioneer Hi-Bred v. Canada (1989), 225–26
Pipettensystem (2007), 447–49, 448*n*48
Plant Breeders' Rights Act (Canada), 226
plant patents
in Europe, 324, 327
in India, 544–45
reproduction and exhaustion, 435
in the UK, 545
in the US, 152
See also indigenous developmental networks and the non-developmental state
plant utility patent protection
eligibility, intermediate approaches, 220–28, 220*n*46, 221*nn*48, 52, 223*nn*58, 60, 224*nn*63–64, 225*n*69, 225*nn*67–68, 227*n*85
eligibility, permissive approaches, 215–20, 216*n*20, 217*n*24
eligibility, restrictive approaches, 228–30, 228*nn*91, 94, 229*n*96

non-obviousness/inventive step, 231–32, 231*nn*106–7, 232*n*108
overview, xvi, xxiii–xxiv, 213–15, 214*nn*7–9, 241–42
scope of, 234–41, 236*n*128, 237*nn*131–34, 238*n*139, 240*nn*146–48, 241*n*152
utility and disclosure, 233–34, 233*nn*115–18, 234*n*122
Plant Variety Protection Act (United States), 238
Plavix®, 505
plurilateral treaties
ACTA, 113–14
proliferation of, 109–15, 110*nn*44, 47–48, 112*n*58, 113*n*61, 114*n*69
US treaties, 86
politics, *See also* social and political context of European patent system
Pompidou, Alain, 337–38
Portugal, exhaustion in, 421
Post Grant Review (PGR) proceeding, 166, 167, 175
PPPs *See* Public Private Partnerships
Pregis v. Kappos (2012), 167
preliminary injunctions, 121–22
President's Emergency Plan for AIDS Relief (PEPFAR), 90
process patents
for biologics, 366
burden of proof for infringement of, 679
business method, 153–54
exhaustion and, 430
in India, 595
living organisms/biotechnology, 192, 207, 364*n*109, 366, 366*nn*129–30
medical, 402*n*97
product identity and repair, 437–45
profits
defendant's profits and enhanced damages, 668–71, 668*nn*117–18, 669*nn*119, 123–25, 670*n*132, 671*n*137
lost, 655–62, 655*n*54, 56, 656*n*57, 657*nn*64–65, 659*nn*69–70, 661*nn*79, 81, 662*n*84
Project Bioshield II Act, 372
Prometheus, Mayo Collaborative Services v. (2012), 154, 155, 159, 165, 168, 173, 184, 205, 206
Prometheus Labs, 154, 155, 159, 168, 205
proof of progress, 587–94, 588*nn*99–101, 591*n*113
property–liability dichotomy, 387–93, 388*n*29, 389*nn*35–37, 390*n*42, 393*f*
Provisional Committee on Proposals Related to a WIPO Development Agenda (PCDA), 77–78, 81
public domain, extractive, 295–301, 296*n*40, 297*n*44, 46, 299*n*56
public health
monopolies detrimental to, 325
TRIPS and, 682–83
See also compulsory licenses on drug patents; Doha Declaration on the TRIPS Agreement and Public Health; regulatory exclusivities
Public Patent Foundation (PubPat), 154, 157
Public Private Partnerships (PPPs), 36, 379–80, 395, 414, 416
public welfare and the international patent system
consolidation of patent harmonization and trade policy (TRIPS), 27–29
domestic actors and institutions role, 11–17, 12*nn*61, 63, 13*nn*65, 67, 70, 14*n*72, 16*n*79
harmonization process weaknesses, 5–10, 7*n*28, 8*n*37, 9*nn*40–42, 44, 10*n*49

public welfare and the international patent system *(Cont'd.)*
non-discrimination as a basis of patent harmonization, 18, 21–27, 22*nn*111–12, 24*nn*122, 124–25
overview, xvi, xix–xx, 1–5, 1*nn*1–2, 2*nn*4–5, 7, 4*n*14, 49–50
policy proposals regarding, 44–48
public welfare gaps on a global stage, 33–44, 34*n*172, 36*nn*182–83, 37*n*188, 38*nn*189, 191, 193, 39*nn*194, 197–98, 40*nn*199, 203, 42*nn*206–7
rationalizations and justifications for patent rights, 17–21, 18*n*88, 19*n*91, 20*nn*96, 100–101
shared view of importance of patent protection, 29–33, 29*n*148, 30*nn*153–55, 31*n*159, 32*nn*162, 165
Pumfrey, Nicholas, 582
qualified infectious disease product (QIDP), 352–53

Quanta Computer v. LG Elecs (2008), 238
Queensland Department of Environment and Resource Management, 300
Queensland Herbarium, 300
Queensland Museum, 300
Queensland's Biodiscovery Act, 298–301, 319

Rader, Judge, 161, 173
Ragavan, Srividhya, xviii, xxx, 571–601
Rai, Arti K., xxi, 85–96
Rainforest Alliance, 313
Regents of the University of California v. Eli Lilly (1997), 280–82
regional trade agreements, 535
regulatory exclusivities
overview, xvi–xvii, xxvi, 345–46, 345*n*1, 375–76
patents versus, 355–57, 355*n*60, 356*n*64, 357*n*67
pharmaceutical innovation and, 347–55, 347*nn*11–12, 350*n*24, 353*n*54, 354*n*57
policy implications of, 370
proposed reforms regarding, 370–75, 373*n*173
role of, 358–62, 358*nn*74–75, 359*nn*81–82, 361*n*91
theory versus practice, 362–69, 362*n*94, 363*n*99, 364*nn*103, 109–10, 365*n*117, 366*nn*129–30, 367*nn*133–34, 368*nn*151–52
remedies, comparative law and economic analysis
damages, 652–55, 652*n*44, 653*nn*45–47, 654*nn*49, 51–53
damages, defendant's profits and enhanced damages, 668–71, 668*nn*117–18, 669*nn*119, 123–25, 670*n*132, 671*n*137
damages, lost profits, 655–62, 655*n*54, 56, 656*n*57, 657*nn*64–65, 659*nn*69–70, 661*nn*79, 81, 662*n*84
damages, reasonable royalties, 663–68, 663*nn*93–96, 664*n*98, 665*n*99, 666*nn*101–3, 667*nn*108–10
injunctions, 644–51, 644*n*15, 645*n*17, 647*nn*23, 26–27, 648*nn*29–31, 650*nn*35–36, 651*nn*38, 41–43
overview, xviii, xxxi–xxxii, 637–40, 638*n*7, 671–72
protocol for, 640–44
remuneration based exhaustion, 472–77, 477*n*113
Rentzsch, Hermann, 243
repair and exhaustion, 420, 436–53, 446*n*44, 448*n*48, 449*n*49, 450*n*55, 451*n*56
reproduction and exhaustion, 434–36, 435*n*25

research exceptions
definition of, 384–85
in Europe, 381–83, 384, 395–96, 400, 403–7, 404*nn*111–12, 405*n*119, 406*n*120, 408, 409–17, 413*n*140
in Japan, 381, 383, 384, 395–96, 400, 407–9, 409–17, 413*n*140
in an open innovation context, 393–95, 395*n*64
overview, xvii, xxvii, 379–84, 380*n*6, 381*n*8, 382*nn*10–11, 414–17, 415*nn*141–43, 416*nn*144–45
parameters of, 386–87
rationale for, 385–86
statutory, 400, 403–6, 404*n*111–12, 405*n*119, 407–8
in the US, 381–83, 384, 395–403, 395*n*65, 397*n*74, 401*nn*95–96, 402*nn*97, 100, 103, 403*n*105, 409–17, 413*n*140
zero-price liability rule, 387–93, 388*n*29, 389*nn*35–37, 390*n*42, 392*n*52, 393*f*
Rich, Judge, 278, 279
Rite-Hite v. Kelley (1995), 656–59
RMail v. Amazon.com (2012), 168
RNA patents *See* DNA/RNA patents
Roche v. Bolar (1984), 11, 108, 121, 397, 401
Roche v. Cipla (2008), 581, 600, 618, 620–22, 625–26, 632
Round-Up *See Bowman v. Monsanto* (2013); *Monsanto Canada v. Schmeiser* (2004); *Monsanto v. Cefetra* (2010)
Roussel Uclaf S.A. v. Hockley (1996), 458
royalties, reasonable, 663–68, 663*nn*93–96, 664*n*98, 665*n*99, 666*nn*101–3, 667*nn*108–10
Ruschig, In re, 279
R v. Arkwright (1785), 282
SACGHS (*Secretary's Advisory Committee on Genetics, Health, and Society*), 400, 402, 411
Safrin, Sabrina, 175–76
Sampat, Bhaven, 613
Samsung, 46
Sanklecha, Jay, xviii, xxxi, 603–35
SAP Am. v. Versata Dev. Group (2013), 164, 167
SAPTO (South African Patent Office), 12
Sarkaria, Justice, 581
Sarnoff, Joshua D., 171
Sawin v. Guild (1813), 396
Scenarios project, 338–39
Schatz, Ullrich, 460
Schering v. Geneva Pharms (2003), 557, 597
Schmeiser, Monsanto Canada v. (2004), 156, 181, 192–93, 209, 226–28
Schmeiser, Percy, 156, 197, 226–27
Schneider v. PRB and Chint (2008), 553
Schumacher, E.F., 318
Schumpeterian creative destruction, 313–14
Schütz (UK) v. Werit. (UK) (2011), 438–39, 438*f*, 448, 452
scientists and indigenous people, 314–19
SCP (Standing Committee on Patents), 2
Secretary's Advisory Committee on Genetics, Health, and Society (SACGHS), 400, 402, 411
Sell, Susan K., xvi, xxi–xxii, 97–126
Sen, Amartya, 81
Senseo Coffee Pads (2005), 449–51, 452, 456
Sensitive Cases (SeCa) system, 339–41, 342
Shaffer, Gregory, xvi, xxi–xxii, 79, 97–126
Shapiro, Carl, 356

Shaw of Dunfermline, Lord, 464
Sierra Leone, Friends of Development membership, 143*n*65
Singapore
  TPP and, 86
  treaties with the US, 110, 111, 112
SIPO *See* Chinese State Intellectual Property Office
small and medium enterprises (SMEs), 93
Smith v. Grigg (1924), 611
social and political context of European patent system
  challenging the patent system, 332–37
  history of the patent system, 323–32
  overview, xvii, xxv–xxvi, 321–22, 342–43
  systemic responses to challenges, 337–41
social perception and exhaustion, 453–54
Society of Arts and Patents, 1*nn*1–2
software
  computer-implemented inventions and business method patents, 152–54, 163–64
  as essential facility, 517
  European patents, 201–4, 334–35, 549–50
  inventive step and, 549–50
  tangible inventions, 199–201
  See also biotechnology; subject matter eligibility in transatlantic context
South Africa
  access to medicines in, 12, 12*n*63, 13, 119, 123
  Friends of Development membership, 143*n*65
  HIV/AIDS in, 12, 506
  pharmaceutical imports, 694*t*, 695, 696
  See also innovation and protection of competition
South African Competition Act, 517, 518, 519, 521
South African Competition Appeal Court (CAC), 519
South African Competition Commission, xvii, xxviii, 12, 517, 520, 522–24
South African Patent Office (SAPTO), 12
South America
  plant patents in, 228, 230
  TLOs and, 101
Southeast Asia, China and, 79
South Korea
  harmonization and, 137, 138
  plant patents in, 215, 217, 219–20, 220*n*42, 241
  strong IPR protection and, 134
Soviet Union, 131
Spain
  pharmaceutical patents in, 127*n*3
specification as a mark of evolution, 253–55
Spence, W., 254
SPLT *See* Substantive Patent Law Treaty
Standing Committee on Patents (SCP), 2
state sovereign immunity, 398, 399
State Street Bank v. Signature Financial Group (1998), 152–53, 200, 596
Statute of Monopolies, 253, 424
statutory research exception, 400, 403–6, 404*n*111–12, 405*n*119, 407–8
  See also research exceptions
Story, Joseph, 256, 277, 326, 396
Strandburg, Kathy, 63
Strasbourg Convention on the Unification of Certain Points of Substantive Law on Patents for Invention, 327

subject matter eligibility
TRIPS and, 496, 681
See also plant utility patent protection
subject matter eligibility in the United States
biotech players, 155–64, 156*n*33, 158*n*42, 160*n*47, 161*n*51, 162*n*55, 163*n*61, 164*n*66
exclusions, 195–204
mechanisms of patent eligibility, 180–84, 183*n*133
overview, xvi, xxii–xxiii, 149–51, 184–85
theories regarding, 170–80, 170*n*87, 174*n*102, 175*nn*104, 107, 176*n*109, 177*n*112, 178*nn*113, 115, 179*n*118
timing of subject matter disputes, 164–66, 166*nn*69–70
types of biotechnology and computer-implemented inventions eligible, 151–55, 153*n*18, 154*n*20, 155*n*27, 190–92
venues for subject matter disputes, 164, 166–70, 169*nn*79, 82, 170*n*84
subject matter eligibility in transatlantic context
categorizing subject matter, 189–95
exclusions, 195–204
overview, xvi, xxiii, 187–89, 209–11
process and apparatus, 204–9
Substantive Patent Law Treaty (SPLT), 113, 113*n*61, 333
postponement and opposition of, 140–44, 146
sufficiency *See* enablement, sufficiency and written description
Swan, Joseph, 139
Sweden
CoE and, 323
EFTA and, 329
pharmaceutical patents in, 127*n*3
Strasbourg Convention and, 327
Switzerland
EFTA and, 329
exhaustion in, 475, 476, 477, 482–83
harmonization and, 138
Paris Convention and, 18–19
patent law in, 127
pharmaceutical and chemical patents in, 127, 127*n*3
pharmaceutical industry and, 675
Strasbourg Convention and, 327

Taiwan, harmonization and, 138
Tanzania, Friends of Development membership, 143*n*65
Technical Board of Appeal (TBA), 266–67
technology
advances and patent incentive, 170–71
Asian Tigers, 131
growth of, 19–21, 20*nn*96, 100–101
Japan emergence regarding, 131
knowledge regarding, 30
transfer, 134, 528–29
US supremacy regarding, 131
See also biotechnology; innovation; software
term of protection, in TRIPS, 680–81
terrorism countermeasures, 372–74, 373*n*173
Thailand
compulsory licenses in, 502, 504, 505
pharmaceutical imports, 694*t*
Thirteenth Amendment, 163
Thomas, John R., xvi, xxvi, 345–76
'318 Patent Infringement Litigation, In re (2009), 273
TKDL (Traditional Knowledge Digital Library) initiative, 298, 302–4, 319
TLO *See* transnational legal ordering (TLO) and access to medicines

TPP *See* Trans-Pacific Partnership Agreement (TPP)
trade
ACTA and, 115
Eastern Bloc and, 131
EFTA, 328, 329
free, 4, 46
FTAs and BITs, 58
limits to patent law and, 422
local, 69
patent harmonization and trade policy consolidation, 27–29
protectionist practices, 2, 18
weak vs. strong IPRs and, 137–39
See also listings at TRIPS; transnational legal ordering (TLO) and access to medicines
trade agreements *See* treaties
trademarks
Inuit art market, 311–12
Paris Convention and, 57, 77
patents vs. trademarks for plant compounds, 310
well-known, 77
trade secrets, 308–10
traditional knowledge and folklore
TRIPS and, 53, 58
See also indigenous developmental networks and the non-developmental state
Traditional Knowledge Digital Library (TKDL) initiative, 298, 302–4, 319
transatlantic *See* subject matter eligibility in transatlantic context
transnational legal ordering (TLO) and access to medicines
construction of the TLO for IP, 103–15, 105*nn*21, 25, 106*nn*27, 31, 107*nn*33, 35, 108*n*36, 109*n*40, 110*nn*44, 47–48, 112*n*58, 113*n*61, 114*n*69
developing countries and strategies regarding, 120–26, 123*nn*108, 110, 125*n*122
economic and social rights TLOs, 116–20, 118*n*90, 119*nn*94, 98
overview, xvi, xxi–xxii, 97–98, 126
rise of the TLO, 101–3, 101*n*12, 103*n*15
Trans-Pacific Partnership Agreement (TPP)
emerging economies and the, 91, 94
enforcement and the, 145
negotiations over, xxi
pharmaceutical patent provisions in draft, 86–87, 89
pre-grant opposition in draft, 112
TLOs and the, 101
TRIPS-plus provisions in the draft, 115
treaties
EPAs, 80, 533–35
EU–CARIFORUM EPA, 533–36
Friendship, Commerce and Navigation, 24, 24*nn*123–24
harmonization and, 189
north-south, 533–36
PCT, 93
regional trade agreements, 535
Treaty Presentment Clause, 115
See also bilateral investment treaties (BITs); free trade agreements (FTAs); plurilateral treaties; *Specific agreements and treaties*; transnational legal ordering (TLO) and access to medicines
Treaty Presentment Clause of the US Constitution, 115
trials, expediting, 626–32, 628*n*100, 629*n*102, 630*n*114, 631*n*116, 119
TRIPS Agreement
competition under the, 525, 528–31, 537

compulsory licenses and the, 485, 486–87, 490, 491, 491*n*23, 493, 496–501, 496*n*44, 497*nn*45–46, 50, 498*nn*51, 54–56, 499*nn*65, 67, 500*n*71, 502–12, 506*n*95
creation of the, 85–86
damages under, 652*n*44
developed countries and the, 62, 65
developing and least-developed countries and the, 17, 29–33, 30*nn*153–54, 58–62, 70, 105, 125
disclosure requirements under, 245
domestic public interests under, 2, 2*n*7
economic and social rights and the, 116–20
emerging economies and, 66, 67, 74–77
exceptions to exclusive rights under, 381*n*6
exhaustion and the, 420, 462, 476–77, 480–81, 484
flexibilities of, 106–9, 106*nn*27, 31, 107*nn*33, 35, 108*n*36, 109*n*40
harmonization achieved with the, 131
injunctions under, 648–49
innovation and, 41–42
inventive step under, 598
IP development and the, 603
least-developed countries and, 106, 106*nn*27, 31
minimum standards, 57
non-discrimination principles and, 18, 23, 25
overlapping and rival TLOs and the, 98
patent provisions under, 28–29
pharmaceutical patents under, 99, 177
plant patents under, 213–14, 213*n*3, 241–42
public goods under, 34–37, 34*n*172, 47–48
public policy exception under, 326
signatory's patent protection requirements under, 187
strengthening of enforceable patent norms in the, 7
suggestions for amending the, 82
types of inventions patentable under, 151
WTO and, 371
See also compulsory licenses on drug patents; Doha Declaration on the TRIPS Agreement and Public Health

TRIPS Agreement flexibilities
access to medicines, 106–9, 106*nn*27, 31, 107*nn*33, 35, 108*n*36, 109*n*40, 118–19, 119*nn*94, 98
China and, xvii, xxx, 542, 542*nn*6–7, 545–47, 550–54, 554*n*66, 559–60, 566, 568–70
India and, xvii, xxx, 45–46, 78, 542, 544–47, 554–58, 554*n*67, 555*nn*73–74, 560–63, 562*n*121, 565–66, 568–70, 633–34
lessons learned regarding, 563–69
non-obviousness/inventive step, 547–58, 547*n*31, 548*nn*34, 35, 549*n*36, 555*nn*73–74
overview, xvii–xviii, xxix–xxx, 541–47, 542*nn*6–7, 543*n*10, 544*nn*13, 15–16, 569–70
TRIPS-plus and TRIPS-minus provisions eliminating flexibilities, 110–11
utility/industrial applicability, 558–63, 558*nn*93–94, 96, 562*n*121

TRIPS-minus provisions, 110–11, 138

TRIPS-plus provisions/agreements
Andean Tribunal and, 117
defined, 103*n*15
developing and emerging economies and, 58, 62, 73*n*110, 86
minimum standard of TRIPS, 512

TRIPS-plus provisions/agreements *(Cont'd.)*
rules vs. standards debate and, 564–65
TLOs and, 116–17
TRIPS flexibilities and, 110, 542
UN MDG on, 116
in US treaties, 86
tuberculosis, 35, 303, 353, 498, 500
Turkey, Strasbourg Convention and, 327

UN *See* United Nations
Uniloc USA v. Microsoft (2011), 667
United International Bureaux for the Protection of Intellectual Property, 329
United Kingdom
CoE and the, 323
compulsory licenses in the, 575
EFTA and, 329
exhaustion in the, 424, 425, 438–39, 438*f*, 440, 448, 449, 451–52, 456
injunctions in the, xxxi, 604, 608, 609, 609*n*23, 634
innovation in the, 1*n*1
lost profits in the, 661
patent sufficiency and support in the, 245, 259–71, 260*n*77, 261*n*80, 266*n*106, 268*n*118
Strasbourg Convention and, 327
See also Specific countries
United Kingdom Patent and Designs Act, 261
United Kingdom Patents Act, 259–64, 265, 267, 270, 276, 575–76
United Nations
Millennium Ecosystem Assessment and the, 314
TLOs and the, 102
United Nations Commissioner of Human Rights, 491–92
United Nations Committee on Economic, Social and Cultural Rights, 116
United Nations Conference on Trade and Development (UNCTAD), 39, 39*n*198, 82, 83, 124
United Nations Educational, Scientific and Cultural Organization (UNESCO), 83
United Nations Environment Programme (UNEP), 82, 339
United Nations Framework Convention on Climate Change (UNFCCC), 38
United Nations High Commissioner on Human Rights (UNHCHR), 116
United Nations Human Rights Council, 116
United Nations Industrial Development Organization (UNIDO), 82
United Nations International Covenant on Economic, Social and Cultural Rights (ICESCR), 102–3, 119
United Nations Millennium Development Goals (MDG), 77, 116, 144
United Nations Special Rapporteur on the Right to Health, 116
United States
China and the, xxi, 91–92, 91–96
claim scope in the, 183–84
competition with Europe, 23
competition with Japan, 23
copyright law in the, 31
damages, 652–55, 653*n*46, 654*nn*49, 53
damages, defendant's profits and enhanced damages, 668–71, 669*nn*119, 125, 670*n*132, 671*n*137
damages, lost profits, 656–62, 657*nn*64–65, 659*nn*69–70

damages, reasonable royalties, 663–68, 663*nn*94–96, 664*n*98, 666*nn*101, 103, 667*n*108
developing countries and the, 105, 105*n*25, 123, 124–25
domestic actors and patent policy in the, 11–12
emerging economies and the, 71
enablement and written description in the, 245, 271–81, 271*n*132, 274*n*146, 275*n*149, 279*nn*168–69, 281*n*178
essential element in the, 452–53, 455
EU and the, 145
EU patent system vs. US system, 126–27, 133, 321, 322, 326, 328, 331–32, 336
executive branch patent policy, xvi, xxi, 85–96
exhaustion, international and the, 466–69
exhaustion in the, 423–24, 442, 452–53, 455, 458–59, 461–62
First Sale Doctrine in the, 433–34
free speech in the, 121
free trade agreements and the, 28, 110–15, 112*n*58, 512
GATT and the, 132
harmonization and the, 130–31, 134, 135, 141, 143, 145
indigenous people in the, 287, 288, 305, 310
injunctions in the, 604, 608–9, 644–51, 644*n*15, 645*n*17, 647*nn*23, 26–27, 648*n*29, 650*n*35, 651*n*41
innovation policy of, 30–31, 30*n*155
international IP rules shaped by the, 104
inventions of the, 674
Japan and the, 145
non-obviousness/inventive step in the, 88, 553, 571–73, 576–77, 579, 580, 583–84, 585, 595–600
novelty in the, 161, 167, 169*n*82, 204, 257, 598
Paris Convention and the, 26
patentability requirements in the, 247, 247*n*11
patent case win rate in the, 651, 651*n*41
patent filings, 93, 139, 651
patent law development in the, 140, 188, 255–59, 256*nn*58, 60, 257*nn*61–62, 64–65, 258*nn*68–69, 259*n*72
patent system use in the, 675
pharmaceutical and biotech patents and, 105, 177–80, 177*n*112, 178*nn*113, 115
pharmaceutical industry and the, 675
plant patents in the, 215–17, 216*n*20, 217*n*24, 218, 219–20, 225, 231–41, 237*n*133, 240*nn*147–48
PPPs in the, 380
precedents used by developing countries, 121–22
product identity in the, 442
protectionism of the, 46–47, 80
reform of patent policy in the, 9, 9*nn*40–41
remedies in the, 639
research exceptions in the, 381–83, 384, 395–403, 395*n*65, 397*n*74, 401*nn*95–96, 402*nn*97, 100, 103, 403*n*105, 409–17, 413*n*140
technological growth of the, 19–20, 20*n*96, 21*n*102
TRIPS and the, 25, 28, 30*n*154, 42, 42*n*206, 45–46, 106, 109–10, 676
utility requirement in the, 558
validity of patent presumption in the, 617*n*60
WIPO and the, 132
See also regulatory exclusivities; subject matter eligibility in the United States; subject matter eligibility in transatlantic context

United States Congress
ACTA and, 115
bilateral treaties and the, 112
Bolar exception and the, 401
clinical exception and the, 402–3
defendant's profits and the, 668–69
harmonization and the, 132
patent system creation by the, 255–56
PEPFAR and, 90
percolation of patent law issues and the, 172
pharmaceuticals and the, 345, 348–49, 351–52
on plant patents, 152
regulatory exclusivities and the, 345, 349, 362, 370–72, 375
subject matter eligibility and the, 163–64, 164*n*65, 185
terrorism countermeasures and the, 372–74
USPTO and, 330, 336–37
United States Federal Circuit Court *See* Court of Appeals for the Federal Circuit (CAFC)
United States International Trade Commission (USITC), 93
United States Patent Act, 255–58, 277, 358, 366–67, 400, 402, 579, 671
United States Patent and Trademark Office (USPTO)
AIA and the, 403
biotech inventions and the, 151–52
business method patents and the, 153
Congress and the, 330, 336–37
controversial patents and the, 157–58, 178–80
examination practices, 275
exclusivity for biologics and the, 367
patent inflation and the, 174–75
pharmaceutical patents, 88
plant patents and the, 216
SIPO vs., 92, 93
subject matter challenges and, 165, 169–70, 185
on the Thirteenth Amendment, 163
time allowed for patent challenges, 167
United States Supreme Court
on the Bolar exception, 401, 411
on exhaustion, 436
on the first-sale-doctrine, 433–34
on genetically modified organisms, 364–65
on injunctions, 647
on inventiveness, 596–97, 598, 599
patent eligibility cases, 149–50, 152–55, 153, 156–58, 157, 159–63, 160*n*47, 161*n*51, 165–66, 169, 170, 173, 175, 184
on patents stifling useful arts, 361
on state sovereign immunity, 399
written description and the, 258
United States Trade Representative (USTR), 90–91
United States v. China (2009), 542, 542*nn*6–7
United States v. India (1997), 542, 542*n*6
United Wire v. Screen Repair Servs. (Scot.) (2000), 437–40, 439–40, 456
Universal Declaration on Human Rights, 116, 494
UPOV Convention, 213–14, 214*n*9, 223, 230, 237, 237*nn*131–34, 240, 240*n*148, 241
Uruguay Round Implementation Act, 23
USPTO *See* United States Patent and Trademark Office
USTR (United States Trade Representative), 114
utilitarian theory of patents, 489–90
utility/industrial applicability requirement
in China, 559–60

failure to satisfy the, 367
in India, 560–63, 614
plant patents, 233–34
TRIPS and, 544–63, 547*n*31, 548*nn*34, 35, 549*n*36, 555*nn*73–74, 558*nn*93–94, 96, 562*n*121
in the US, 193, 204
See also plant utility patent protection

Valcyte patent, 584–87, 593, 597
Valensi v. British Radio Corporation (1973), 263–64
van Overwalle, Geertrui, xvii, xxvii, 379–417
van Zimmeren, Esther, xvii, xxvii, 379–417
Vas-Cath v. Mahurkur (1991), 259
Venezuela
access to medicine rights in, 119
Friends of Development membership, 143*n*65
Versata Development Group, 164, 167
Vienna Convention, 70
Vietnam
TPP and, 86
treaties with the US, 86, 111, 112
von Hippel, Eric, 63
Voyage of the Beagle (Darwin), 314

Walker, Alexis, xvii, xxv–xxvi, 321–43
Wands, In re (1988), 272–73
Weiser, Phil, 391–92, 410, 412–13
Weldon Amendment, 163
White House Office of Science and Technology Policy (OSTP), 95
White House Office of the Intellectual Property Enforcement Coordinator, 94
Whither Claim Construction, 621–22, 621*nn*76–78, 622*n*79–80
Whittemore v. Cutter (1813), 396
WHO (World Health Organization)
breast-milk substitutes code, 37
Commission on Intellectual Property Rights, 83
FCTC, 37
generic and counterfeit drugs and, 13*n*70
human rights agreements, 54
right to health and the, 116
WHO International Health Regulations, 37
Williamson, E.M., 318
Windsurfing International v. Tabur Marine (GB) (1985), 581
WIPO (World Intellectual Property Organization)
developing countries and, 43
emerging economies and, 74–78, 79, 80, 81, 82, 83
harmonization and, 2–3, 127, 132–33, 142–44, 146, 546
human rights agreements, 54–55
institutional innovation in, 8
norm-setting activities of, 47
patents filed at, 93
pharmaceutical trade associations and, 104
predecessor of the, 329
public welfare and, 38, 38*n*191, 40
TLOs and, 102
TRIPS minimum standards and, 57
welfare costs of the patent system and, 16–17
WIPO Development Agenda, 17, 603, 603*n*3
Women's Art Association of Montreal, 312
World Bank, 56
World Economic Forum, 691
World Health Organization *See* WHO (World Health Organization)
World Intellectual Property Organization *See* WIPO

written description *See* enablement, sufficiency and written description
WTO (World Trade Organization)
adjudication by the DSB, 54
developing countries and the, 62, 80
dispute settlement procedure, 69–70
on orphan, tropical, or low-market diseases, 566
PEPFAR and, 90
power shift toward the, 56
public welfare and, 40, 40*n*203
regulatory exclusivities and the, 347
TLOs and the, 102
transnational legal ordering and the, 97
TRIPS and, 371
See also listings at TRIPS

Yolgnu system of knowledge, 296

Zambia, competition law in, 532*n*47
zero-price liability rule, 387–93, 388*n*29, 389*nn*35–37, 390*n*42, 392*n*52, 393*f*, 413*n*140, 417
Zipher v. Markem (2009), 264
ZTE Corporation (China), 139